Atmospheric Radar

Richly illustrated, and including both an extensive bibliography and index, this indispensable guide brings together the theory, design, and application of atmospheric radar. It explains the basic thermodynamics and dynamics of the troposphere, stratosphere, and mesosphere, and discusses the physical and engineering principles behind one of the key tools used to study these regions – MST radars. Key topics covered include antennas, signal propagation, and signal processing techniques. A wide range of practical applications is discussed, including the use of atmospheric radar to study wind profiles, tropospheric temperature, and gravity waves. A detailed overview of radar designs provides a wealth of knowledge and tools, providing readers with a strong basis for building their own instruments. This is an essential resource for graduate students and researchers working in the areas of radar engineering, remote sensing, meteorology, and atmospheric physics, as well as for practitioners in the radar industry.

Wayne K. Hocking is a Professor of Physics at the University of Western Ontario and a Fellow of the Royal Society of Canada and of the Australian Institute of Physics. He has built over 40 radars world-wide and edited multiple special issues of journals. He is the recipient of the Medal for Outstanding Achievement in Industrial/Applied Physics from the Canadian Association of Physicists and the Pawsey Medal from the Australian Academy of Science. He has also received a citation from NASA for his work on the Space Shuttle re-entry environment.

Jürgen Röttger is a Fellow of the Royal Astronomical Society and holds the Minerva Medal of the Max Planck Society. He has also held the position of Chair Professor at National Central University. In the 1970s he was a leading developer of the SOUSY radar. In 1985 he headed atmospheric sciences at the Arecibo Observatory, and from 1986–1997 was the Director of EISCAT, where he was awarded the EISCAT Beynon Medal for his role in the development of the EISCAT Svalbard radar. He also led the design of the Chung-Li MST radar in Taiwan.

Robert D. Palmer is the Executive Director of the Advanced Radar Research Center and the Craighead Chair in the School of Meteorology at the University of Oklahoma. He also serves as the University's Associate Vice President for Research. He has published widely in the area of radar sensing of the atmosphere, with an emphasis on imaging problems, waveform design, clutter mitigation, and the application of array/signal processing techniques to observations of both the clear-air environment and severe weather. Professor Palmer is a Fellow of the American Meteorological Society.

Toru Sato is a Professor in the Graduate School of Informatics at Kyoto University. He has been engaged in data analysis of Jicamarca and Arecibo radars, and has contributed to the design and operation of Japanese MST/IS radars, notably the MU radar, Equatorial Atmosphere radar, and PANSY radar. He has published more than 160 journal papers, and in 2015 received the Commendation for Contributors to Promotion of an Oceanic State from the Prime Minister of Japan.

Phillip B. Chilson is a Professor in the School of Meteorology at the University of Oklahoma and a member of the University's Advanced Radar Research Center. He has been involved in atmospheric radar research and development for over 25 years and has helped to develop many advanced radar signal processing tools. Professor Chilson has previously held positions at the Max Planck Institute for Astronomy, the Swedish Institute of Space Physics, and the University of Colorado in Boulder.

Atmospheric Radar

Application and Science of MST Radars in the Earth's Mesosphere, Stratosphere, Troposphere, and Weakly Ionized Regions

WAYNE K. HOCKING
University of Western Ontario

JÜRGEN RÖTTGER
Max Planck Institute for Solar System Research

ROBERT D. PALMER
University of Oklahoma

TORU SATO
Kyoto University

PHILLIP B. CHILSON
University of Oklahoma

CAMBRIDGE
UNIVERSITY PRESS

CAMBRIDGE
UNIVERSITY PRESS

University Printing House, Cambridge CB2 8BS, United Kingdom

One Liberty Plaza, 20th Floor, New York, NY 10006, USA

477 Williamstown Road, Port Melbourne, VIC 3207, Australia

4843/24, 2nd Floor, Ansari Road, Daryaganj, Delhi – 110002, India

79 Anson Road, #06–04/06, Singapore 079906

Cambridge University Press is part of the University of Cambridge.

It furthers the University's mission by disseminating knowledge in the pursuit of education, learning and research at the highest international levels of excellence.

www.cambridge.org
Information on this title: www.cambridge.org/9781107147461

First published 2016

Printed in the United Kingdom by Bell and Bain Ltd, Glasgow

A catalog record for this publication is available from the British Library

Library of Congress Cataloging in Publication data
Names: Hocking, W. K., author. | Röttger, J. (Jürgen), author. | Palmer, Robert D., 1962- author. | Sato, Toru (Professor), author. | Chilson, Phillip B., 1963- author.
Title: Atmospheric radar : application and science of MST radars in the Earth's mesosphere, stratosphere, troposphere, and weakly ionized regions / Wayne K. Hocking (University of Western Ontario), Jürgen Röttger (Max Planck Institute for Solar System Research), Robert D. Palmer (University of Oklahoma), Toru Sato (Kyoto University), Phillip B. Chilson (University of Oklahoma).
Description: Cambridge, United Kingdom ; New York, NY : Cambridge University Press, 2016. | Includes bibliographical references and index.
Identifiers: LCCN 2016013389 | ISBN 9781107147461 | ISBN 1107147468
Subjects: LCSH: Atmosphere–Measurement. | Radar meteorology. | Atmospheric physics.
Classification: LCC QC973.5 .H63 2016 | DDC 621.3848–dc23
LC record available at https://lccn.loc.gov/2016013389

ISBN 978-1-107-14746-1 Hardback

Contents

We then focus in on the middle atmosphere, giving a moderately detailed discussion of the large-scale dynamical circulation of that region which could be of value even to a student of meteorology and atmospheric physics with little interest in radar. Chapter 2 then changes to the quite separate topic of the history of atmospheric radar, which has a fascinating chronology all of its own. Chapter 3 pertains to the refractive index of the atmosphere, and therefore to the fundamental mechanism that causes the radar backscattered signal. The chapter is mathematically complex but of use to a general student of optics and electromagnetic theory, and could be used as the basis of a small independent course on essential radio-optics and plasma processes.

At this point, we had two choices. One option was to further develop Chapter 1, and discuss the basic physics and fluid dynamics of the atmosphere, so that the reader could have a good background of the topics that can be studied with an atmospheric radar. With this strategy, discussions about radar techniques would be left till later. The alternative was to now launch into discussions of radar techniques, even though the applicable atmospheric physics was a little under-developed. Since the book is directed at radar users, we adopted the latter approach, leaving further details about atmospheric processes to Chapters 11 and 12. So the decision to split the discussions of basic fluid dynamics to the start and end of the book was a deliberate one. Sufficient detail is given in Chapter 1 to permit the reader to usefully apply the more engineering-based aspects of Chapters 4 to 6, but the focus of these three chapters is definitely on radar engineering and design.

Following detailed discussions of radar design and principles in Chapters 4 and 5, Chapter 6 gives several examples of design details of early and more recent radars. We present a mixture of large, powerful and expensive systems and low-cost units that can be built even by a modestly-funded research group. In Chapter 7, we start to unify the areas of atmospheric physics and radar engineering, discussing the important atmospheric parameters that can be measured using a radar. Signal processing is an important aspect of radar studies, not only at the native level of data acquisition, but also in the post-acquisition phase, so Chapter 8 focuses on this area.

One of the areas of greatest recent application has been that of spaced antenna and interferometric studies. This goes considerably beyond the simpler concepts of fixed-beam-pointing and Doppler studies, and allows studies at a more detailed level, including sub-pulse resolution and resolutions smaller than the radar beamwidth (subject to certain assumptions), so Chapter 9 is dedicated to this topic.

Of course the desire of any serious researcher is to produce publications and advance the state of human knowledge. This can be done with standard applications of the techniques developed in the foregoing chapters, but one common agenda of many researcher in the field, and the basis of many of the more significant papers, is the desire to "push the limits" of the radar studies into uncharted territory. Chapter 10 is all about such adventures into such extraneous activities, many of which in time have become mainstream areas of study.

Finally Chapters 11 and 12 bring us back to more complex extensions of Chapter 1, allowing us to delve more thoroughly into the atmospheric processes from waves and turbulence to general atmospheric flows, storms, and even severe weather. It is in these areas that the potential for research into the physical sciences is greatest, and these topics can form the basis of many theses and projects.

We hope there is something in this book for everyone, but at the same time that it can be a valuable learning tool for those new to the field and an important resource to the more experienced members of the research community.

Acknowledgments

The authors of this book would like to thank the following people for assistance with its development, including diagram preparation, proof reading, and general support: Jeffrey Hocking, Anna Hocking, Suzanne Hocking, Patrick Hatch, Ildiko Beres, Stephanie May, Sian Evans, Marcial Garbanzo-Salas, and Boonleng Cheong. JR would like to express his deep appreciation for the support provided by his late wife Rosi during preparation of this book and indeed for many years prior as well.

We are grateful to the following for advice of a scientific nature: Rolando Garcia, John Mathews, and Werner Singer.

We would also like to acknowledge our friend and colleague, Dr. Shoichiro Fukao, who was an early inspiration of this book before his untimely death. His passing was a tremendous loss to our field.

1 An overview of the atmosphere

1.1 Introduction

Many instruments have been used to study the atmosphere, both by in-situ and remote methods. From anemometers to satellites, chemical sensors to balloons and rockets, the array of tools is broad. Since the early 1900s, a key instrument for such studies has been radar. RADAR stands for Radio Detection And Ranging. Radars operating in a variety of frequency bands, from wavelengths of kilometers to wavelengths of millimeters, have all found application. They have been used to study the upper ionosphere and the neutral atmosphere, right down to ground level.

In this book, we will concentrate on a class of radar generally referred to as MST radar. In this description, M stands for Mesosphere, S for Stratosphere, and T for Troposphere, where these three "spheres" refer to different height-regimes of the atmosphere which collectively cover the region from ground level up to about 90 km altitude. More exact definitions will be given shortly. For now, consider the troposphere as the region from the ground to 12 km altitude, the stratosphere as the region from 12 to 50 km altitude, and the mesosphere the region from 50 to 90 km altitude. Under the narrowest definition, the term MST radar was originally used primarily to refer to radars operating in the VHF (very high frequency) band, with special emphasis on frequencies around 50 MHz, which could probe (at least in part) all three regions. More generally it has come to refer to any radars that can be used for studies of any of these three regions of the atmosphere. These radars include MF (medium frequency), HF (high frequency), VHF, and UHF (ultra-high frequency). They also include so-called meteor radars. Generally, precipitation radars (referred to as "Doppler radars" by the meteorological community) are not considered to be MST radars, although we will discuss them a little in this book. (As an aside, we will generally refer to these radars as precipitation radars in this book. The phrase "Doppler radar" is not a good one to describe these radars, since they are most certainly not the only Doppler radars! The term "Doppler radar" arises from the fact that these radars can measure the Doppler frequency-shift of reflected signals. As we will see, almost all MST radars are also Doppler radars.) As a rule we will consider MST radars to cover the frequency range from about 1 MHz to 1 GHz, with radars operating at frequencies beyond 500 MHz being discussed less completely than the others.

The middle atmosphere is generally considered to be the height region from 10 km altitude to 100 km altitude, and therefore includes the upper troposphere, the stratosphere, the mesosphere, and the lowest few kilometers (90–100 km) of the thermosphere. (The thermosphere, in a more general context, refers to the region above the mesopause, or in other words, above about 90 km altitude.)

In this book we will concentrate on the radars that cover the frequency range between 1 MHz to about 500 MHz in the main, which are used to probe the region between 0 and 100 km altitude. Many of the radars to be discussed can probe even higher in altitude, up into the ionosphere, but we will not consider these applications here. Nevertheless, since many MST-type radars had their origins in ionospheric studies, we will discuss some ionospheric aspects of the atmosphere at times.

The focus of this book is twofold. First, there is an engineering aspect, in which we will describe hardware aspects of these radars, and these descriptions will have wider-reaching relevance than applications specific to these radars. A variety of hardware aspects will be considered that are general to all types of radars. This focus also includes signal-processing techniques. The second focus will be on the types of studies that can be performed with these radars, including the physics and dynamics of the atmosphere. Within the book, the two focuses will be somewhat interleaved.

In this chapter, we will give a very brief introduction to radar as it relates to atmospheric studies, but will largely concentrate on giving an overview of the troposphere and middle-atmosphere, our region of particular interest. Chapter 2 will then focus on the ways in which radar came to be used for atmospheric studies. In the next 8 chapters, our focus will be more on various aspects of hardware, radar application, and signal processing. Chapters 11 and 12 will then return to observational aspects, discussing in some detail various dynamical and meteorological aspects of these radars.

We will begin the next section with a brief history of the origins of these types of radar, but will leave more detailed discussions to Chapter 2. The rest of this chapter will be devoted to an overview of the basic dynamics and thermodynamics associated with the troposphere and middle atmosphere.

1.2 The origins of radar

Over the past century, radiowaves have developed a major presence in industrialized society. We use them to communicate on a day-to-day basis (both locally and globally), to transmit and receive key information, to monitor space and our environment, to transmit television and voice information, to detect remote objects, and so on and so forth. Radiowaves are a part of the electromagnetic spectrum and can generally propagate freely through the atmosphere. The "discovery" of radiowaves can be attributed to the efforts of multiple people, including Faraday, Maxwell, Hertz, Bose, Marconi, Popov, and Tesla, among others. Hertz appears to have been the first to generate and detect radiowaves, but many others were involved in improving detection and transmission devices. From the point of view of global communication, however, there is no doubt that a key experiment was the transmission of a radio signal by Marconi across the Atlantic Ocean on 12 December, 1901. Marconi's devices utilized no less than 17 patents developed by Tesla, including the Tesla disruptive coil. Much debate exists about who should rightfully be considered the "inventor" of radio transmission. The awarding of the Nobel Prize to Marconi in 1911 was not without controversy, with Tesla laying strong claim. We will not dwell on these points here, except to comment that it is probable that the development of radio communication at the time was likely an idea whose "time had come," and it was inevitable that multiple scientists would develop similar ideas and hardware at the same or similar times.

Subsequent developments led to a variety of radio-transmission and radio-reception devices, including the famous discovery of extra-terrestrial signals by Jansky in 1931, which led, through the enthusiasm of Grote Reber, an amateur astronomer, to the very busy field of radio astronomy – still a very active field even today.

Figure 1.1 shows the relation between the electric and magnetic fields in a radiowave, and Figure 1.2 shows the approximate location of the so-called "radio band" relative to other types of electromagnetic radiation (EM) as a function of wavelength and frequency.

Radio work received a rapid jolt in pace during World War II, with all parties recognizing the value of detecting enemy aircraft tens and even hundreds of kilometers away. This led ultimately to substantial development of radar for the real-time detection of remote targets by transmission and reflection of radiowaves. Although radar has been

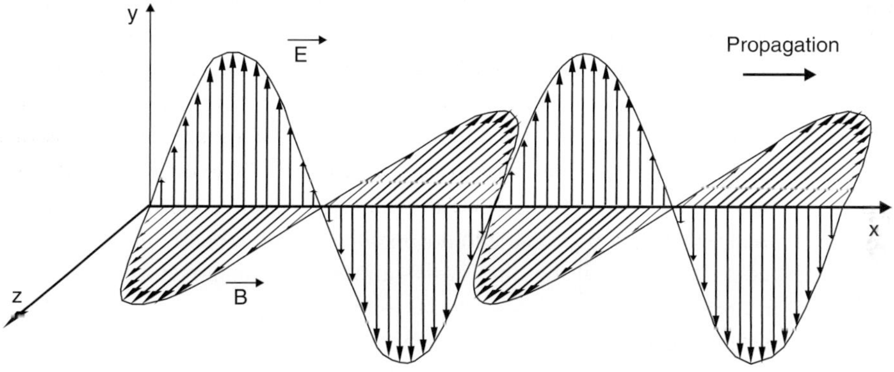

Figure 1.1 Electric and magnetic fields in an electromagnetic wave.

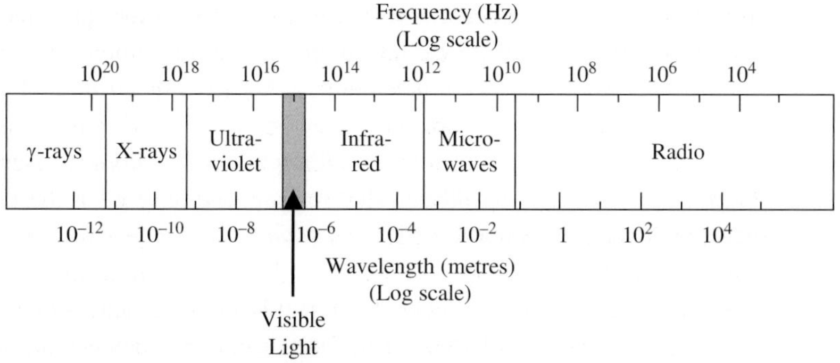

Figure 1.2 Location of radiowaves in the electromagnetic spectrum.

considered to be "invented" in 1936 by Watson-Watt (*Watson-Watt et al.*, 1936), this belief stems from the fact that he worked closely with developments of it during the 1930s and 1940s. In truth, simple radars had been demonstrated to have existed much earlier. In 1904 Christian Huelsmeyer gave public demonstrations in Germany and the Netherlands of a primitive "radar" using a simple spark-gap system as a transmitter. It is true, however, that the 1940s were when radar's real potential was realized. A key aspect of radar is the ability not only to receive signal reflected by targets, but also to determine the range to the target by using time-delay measurements, and to determine target direction. Of particular importance was the development of radars that operated at 3 000 and 10 000 MHZ (S-band and X-band), since only at these frequencies could radars with good directional capabilities be made small enough to be fitted onto aircraft and moving vehicles. Previous radars had worked at frequencies of 200 and 400 MHz. The development of the "cavity magnetron" was a key factor in allowing these high frequency S- and X-band radars to become viable, and the development of these radars in Britain and the USA well before the German development in 1943 gave the Allies a key advantage in World War II.

After World War II, radio work received another boost. Radio astronomy began in earnest and scientists started to use radar to track balloons released into the air, allowing upper level wind speeds to be determined. It was used for telemetry with rocket experiments, and of course for communication.

Radio work for human applications developed along several fronts, with two standing out – first, radar for aircraft and vehicle tracking, and second, studies involving communication. The second field of study actually arose as a result of Marconi's original transmission of radiowaves across the Atlantic Ocean, which should not have been possible since theoretical calculations suggested (after some erroneous early miscalculations!) that radiowaves should not be able to bend around the Earth. Subsequent studies led to the proposal that the Earth's atmosphere contained a layer of reflecting plasma at upper altitudes, which could be used to facilitate global communication. Heaviside and Kennelly proposed the existence of such a layer, but Appleton and Barnett were the first to prove its existence as early as 1925 (*Appleton and Barnett*, 1925; *Appleton*, 1930). At the same time as proving its existence, they were actually able to determine the height of radio reflection, using frequency adjustments of the British Broadcasting Corporation (BBC) radio transmission systems. The reflections were from what is now known as the "ionospheric E-region" (at one time called the Heaviside layer, with another layer higher up being labelled the "Appleton layer"). After a while, the plasma region was simply referred to as the "ionosphere," and that is the most common term used to describe it today. Appleton and Barnett used a swept-frequency method; an alternative method, using pulsed radar that was stepped through a variety of frequencies on successive pulses, was developed by *Breit and Tuve* (1926). Both methods are still used today.

After World War II, world-wide communication by radio signals reflected from the so-called "ionosphere" became a primary means for near-instantaneous world-wide communication. Sports matches in England were broadcast to Australia, for example. However, the ionosphere was not a stable reflector, and signals varied enormously

Table 1.1 Designations of radio bands.

Name	Frequency	Wavelength
Low frequency (LF)	30–300 kHz	10–1 km
Medium frequency (MF)	0.3–3 MHz	1000–100 m
High frequency (HF)	3–30 MHz	100–10 m
Very high frequency (VHF)	30–300 MHz	10–1 m
Ultra-high frequency	300–3000 MHz	1 m–10 cm
Microwave region	$\geq$ 3000 MHz	$\leq$ 10 cm
Further-subdivision		
L-band	1–2 GHz (1000–2000 MHz)	30–15 cm
S-band	2–4 GHz	15–7.5 cm
C-band	4–8 GHz	7.5–3.75 cm
X-band	9–12 GHz	3.75–2.5 cm
K-I (or K_u) band	12–18 GHz	2.5–1.7 cm
K-II (or K_a) band	27–40 GHz	1.2–0.75 cm

in strength, often showing fading on the order of seconds. Large scientific networks were set up to investigate and better understand this valuable transmission medium and were well funded until at least the more recent advent of under-sea cables and satellite radio-relay systems.

A variety of different radio frequencies were used for these studies, with lower frequencies generally being used for ionospheric work, and high frequencies (shorter wavelengths) being used for meteorology. Table 1.1 summarizes the main frequency bands.

The primary bands are the LF to UHF bands, but within the UHF band and into the microwave region, there are also some special frequencies which have extra designations. This nomenclature developed during World War II, and is listed under "further subdivision" in the table. Note that the "K-band" is split because there is strong water-vapor absorption between the two bands, meaning that this particular section is not useful for transmission within the Earth's atmosphere. There is also a class of amateurs who use radio communication for fun, and they have various bands designated for their purposes. Two such bands are the 30–50 MHz band (which they call the "low band" or the "six-meter band"), and a band at 148–174 MHz (which they call the "high band" or the "two-meter band").

A related effect of both radar applications and ionospheric studies was scientific investigation into the basic nature of the transmission and scattering processes, and of the nature of the media that transmitted and scattered the radio signals. In the lower atmosphere, centimeter-band radars were developed for studies of storms and precipitation, and eventually were proposed for studies of neutral turbulence (e.g., *Buehler and Lunden*, 1964; *Friend*, 1949). Studies of the ionosphere, and also other plasmas generated by processes like meteor intrusions into the atmosphere, were also vigorously pursued.

Interestingly, the prime topic of this book brings together aspects of both of these fields. MST (Mesosphere-Troposphere-Stratosphere) radars had their origins in the ionosphere, yet found strong application in the neutral lower atmosphere.

In order to understand these radar and radio techniques and to see how they have turned out to be so important in atmospheric studies, we first need to understand a little about the nature of the atmosphere itself. Although much of this book will be about studies of the neutral atmosphere, the ionospheric origins of these instruments cannot be denied. In addition, understanding radars often involves understanding plasmas. This is especially true with regard to the radar refractive index, a topic of great importance both for transmission and reflection.

Consequently, in the rest of this chapter we will introduce some fundamentals about the basic structure of the Earth's atmosphere and include not only the neutral composition but also the ionized portions. A more detailed discussion of the history of radar in atmospheric studies will follow in the next chapter.

1.3 The atmosphere – an overview

1.3.1 The Earth's neutral atmosphere and ionosphere

The atmosphere is a large system and the interactions involved are complex and intricate. From the (relatively) dense boundary layer to the tenuous remnants thousands of miles into space, there are a wealth of physical processes both fascinating and important to all inhabitants of the Earth. This complexity is indeed one of the reasons that radar studies are so useful – radar is one of the few tools that can remotely monitor many of these complex motions.

At large distances from the Earth's surface, free electrons and ions spiral freely along magnetic lines of force and interact strongly with a wind of similar particles flowing from the sun. This "plasma-sphere" extends far into space – up to around 25 to 30 Earth radii in places – and is a type of "outer region" for the atmosphere. It is is considerably beyond the scope of our topics for this book, but its presence should be recognized. Various reviews of the region exist, (e.g., *Bahnsen*, 1978). A diagram showing some of its main features can be found in Figure 1.3, and a more detailed three-dimensional version can be found in *Kelley* (1989), Figure 1.3.

In order to put things into perspective, we also show Figure 1.4, which shows the scale of the lowest 100 km of the atmosphere relative to the size of the Earth. Bearing in mind that the region of the air in which we live, the "troposphere," typically lies below 10–15 km in height (about one tenth to one sixth of the thickness of the region shown in the figure) and contains over 85 percent of the entire atmospheric mass, this figure reinforces just how thin the "practical atmosphere" is that we depend on for our existence. The huge extent of atmosphere shown in Figure 1.3 is somewhat misleading – while this region is, to some extent, under control of the Earth's gravity, it contains but a small fraction of the total atmospheric mass. Although this region can have significant impact on our lives, through auroras and magnetic storms which

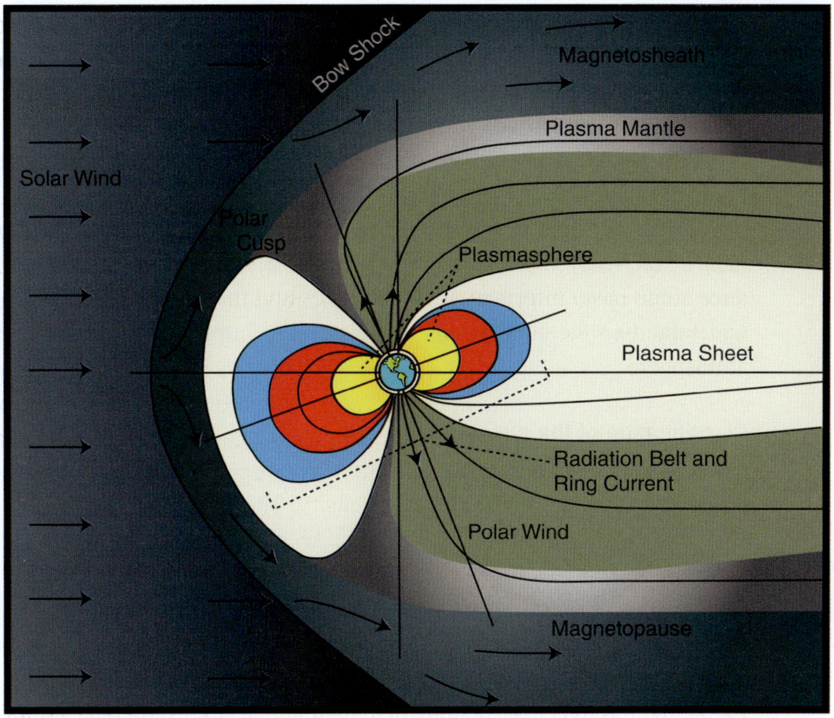

Figure 1.3 The Magnetosphere, adapted from *Bahnsen* (1978). The inner yellow region and outer blue cyan region show approximate locations of the inner and outer Van Allen radiation belts.

Figure 1.4 The depth of the atmosphere relative to the Earth. The thin shaded shell represents the atmosphere drawn to scale to a depth of 100 km.

may damage spacecraft and even, at times, Earth-bound power grids, the most important part is the thin shell shown in Figure 1.4. The radars discussed in this book will concentrate on that shell. The lowest regions of the atmosphere are the most dense, and the pressure and density decrease roughly exponentially as a function of height.

The reasons for this will be discussed later in this chapter – for now we accept it as a fact.

In the following sections, ways of classifying the atmosphere will be examined and some of its regimes will be briefly discussed. We will begin in the outer reaches and proceed down to the so-called middle atmosphere, which will eventually be the primary focus of this text.

We will concentrate especially on atmospheric flows, since this is something that radars can study especially well. But at the end of the chapter, we will also introduce some basic information about statics and thermodynamics which is important for understanding the atmosphere, and especially important for radar investigations of the atmosphere.

Classification of the atmosphere

Two of the most common classification schemes used in the atmosphere are based upon temperature structure and electron density. The former is more common, but for radio work it is important to acknowledge the second classification scheme.

Figure 1.5 shows these classifications. Temperature classifications give rise to the troposphere (lowest region of decreasing temperature), the stratosphere (region of increasing temperature above the troposphere), the mesosphere, (region of decreasing temperature above the stratosphere), and the thermosphere. The symbols D, E, F1, and F2 denote the ionospheric nomenclature. The E and F2 regions are local peaks in electron density, and F1 is a local peak at times (particularly during sunspot maximum summers). It is important to note that these sample profiles can only be approximate, since the temperatures and electron densities can vary substantially with time and location, and with sunspot number, especially in the ionosphere. Day-to-night variation can also be substantial. At night, T_n falls to about 600 K at sunspot minimum and about 900 K at sunspot maximum (*King-Hele*, 1978). Temperature maximum occurs at about 1600 hours local time, and minimum at about 0400 hours in the thermosphere (*King-Hele*, 1978). References for these data include *Houghton* (1977) (Appendix 5), *Roble and Schmidtke* (1979), and *Garrett and Forbes* (1978) (Figure 1). Even the heights of the various regimes can change – for instance, the tropopause height varies both latitudinally and with time of day. In the exosphere (about 500–1000 km), kinetic temperature is not a meaningful term since neutral atoms rarely collide (*King-Hele*, 1978). The hatching in the figure gives some idea of the variations in electron density which can occur. Data are taken from *Craig* (1965), Figures 9.11 and 9.15, and *Ratcliffe* (1972), Figure 3.3. Also shown are some typical E-region night-time electron densities.

One important region not presented on this diagram is the turbopause region. Up to about 100–115 km altitude, turbulence can play a major role in the dynamics of the atmosphere, but above this region, turbulence is a relatively rare phenomenon. The reason is that the mean free path of particles and the increase in temperature result in large molecular diffusion rates, and most small scale transfers of heat and particles occur by such molecular transport. The transition region between the turbulent dominated and non-turbulent regimes is quite narrow, and is called the turbopause. This is discussed in more detail elsewhere. Very rarely, patches of turbulence can be found above the

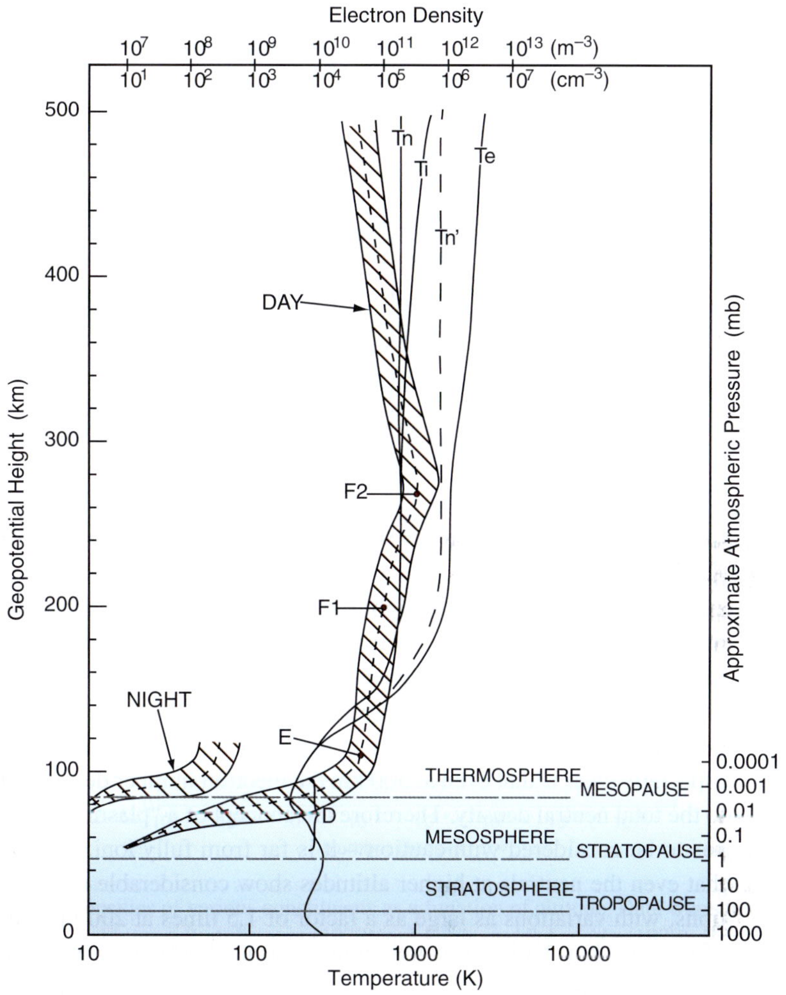

Figure 1.5 Electron-density-based and temperature classification schemes for the atmosphere. Typical daytime temperature profiles for neutral kinetic temperature T_n, the ion temperature (T_i), and T_e (the electron temperature) are also shown for sunspot minimum. T_n shows a typical daytime neutral temperature profile during sunspot maximum. Atmospheric pressure decreases approximately exponentially with increasing height, and typical values are shown on the right-hand side ordinate. There exists considerable latitudinal, seasonal, and day-to-night variability in all parameters. See text for more details. Shaded regions represent electron density profiles, with D, E, F1, and F2 layers emphasized.

turbopause. (Recently, reports of high-altitude turbulence measured by radar have been presented by *Fujiwara et al.* (2004), but these are due to erroneous interpretation of the data, not real effects, as will be demonstrated in later chapters.)

Figure 1.6 shows an expanded view of the temperature profile at below 100 km altitude, showing the temperature classifications in that region more clearly. Some brief annotation describing the reasons for the different regimes is given, and these will be elaborated upon in greater detail shortly.

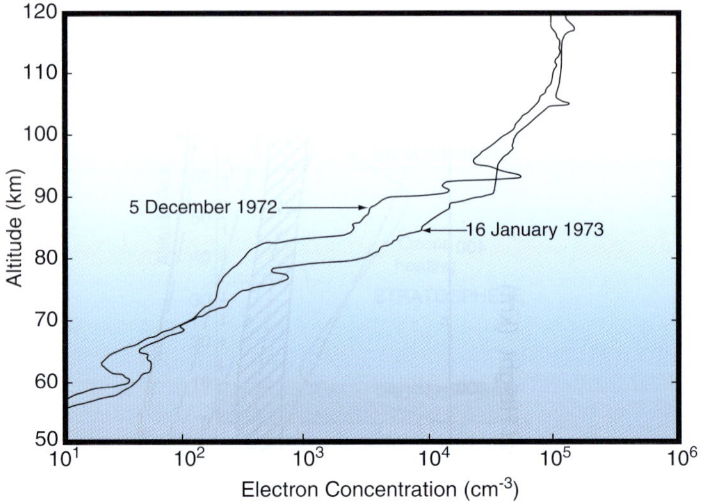

Figure 1.8 Sample electron density profiles at Wallops Is (38N, 75W) from *Smith et al.* (1978). The profile for 5 Dec 1972 is a "normal" profile, and the one for 16 January 1973 shows unusually large electron densities.

At ground level, the total number density is $2.5 \times 10^{25}\ m^{-3}$. The distribution (by number density) is 78.03% N_2, 20.95% O_2, 0.93% CO_2 (*Weast*, 1970).

The D-region is the lowest prominent quasi-ionized region. (Labelling of the ionospheric layers was chosen in such a way that room was left for the discovery of other layers C, B and A below the D-region, but no such layers were found. Occasionally there is talk of a C-layer, but it is very rare and not normally acknowledged.) The D-region has relatively low electron densities compared to the E- and F-regions, but is nevertheless important. Free electrons in the D-region are produced by extreme ultraviolet radiation, gamma rays, and cosmic rays. One reason why the D-region is important is that it is where these dangerous radiations are absorbed. This arises mainly due to the higher neutral densities here than higher up. Sample electron densities measured by rocket are shown in Figure 1.8. Figure 1.9 shows the main ions in the D-region. The D-region has quite complex chemistry, which involves both positive and negative ions. Some idea of the complexity of this chemistry can be found in *Chakrabarty et al.* (1978a, b).

Figure 1.8 shows in part another reason why the D-region is important. The figure shows a "normal" day and a day of higher electron density in the height region 80–95 km. Electron densities at these heights are 3–4 times higher in the case for 16 January 1973. Although electron densities are much lower here than in the E- or F-region, the neutral densities are much higher, and so the frequency of collision of electrons with molecules is much higher than in the upper regions. The result is that when radiowaves pass through this region, they are significantly absorbed, especially when the electron density is high. The D-region is in fact the main absorbing region for radiowaves. When absorption is strong, world-wide radio transmission is diminished and can even be blocked. At a time when world-wide communications depended strongly on ionospheric transmissions, knowledge of the absorptive character of the

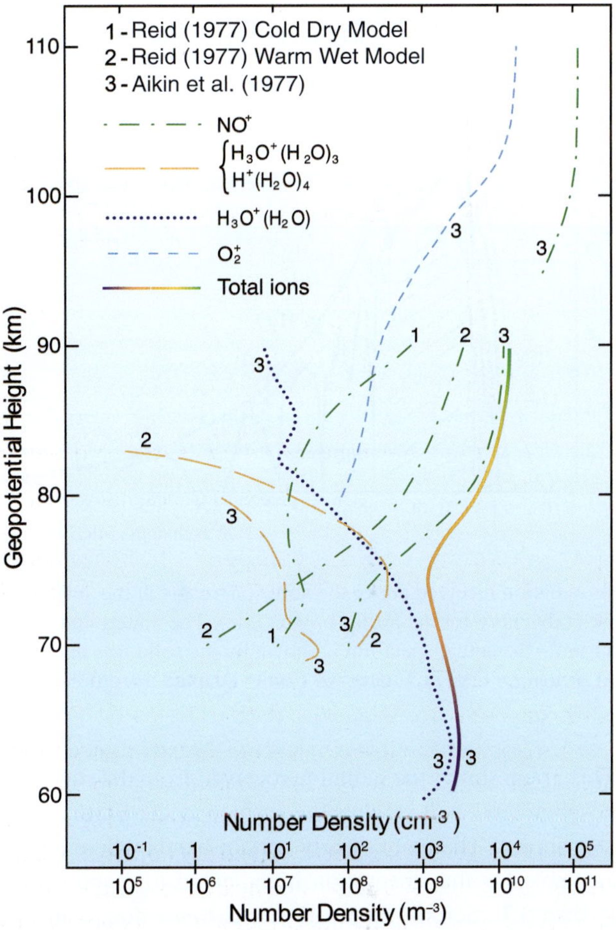

Figure 1.9 A representative sample of D-region ions. Briefly, there is an NO^+/O_2^+ dominance above 80 km, and a water-cluster ion dominance below 75 km. Data were taken from *Reid* (1977) and *Aikin et al.* (1977). Note that only a small selection of profiles are shown – and only a few representative water-cluster ions are presented – just enough to demonstrate the wide variablility of possible constituents and densities. For more detail, the reader is referred to the references given.

D-region was especially important. Variations with sunspot cycle are also important. *Mechtly et al.* (1972) shows a variety of generally representative D-region electron density profiles, and *McNamara* (1979) provides something of a larger scale atlas of electron density profiles for the D-region, though with relatively coarse resolution.

1.3.2 Causes of the temperature and density structures

The basic structure just described is complex, but at least some aspects of it can be explained with relatively simple physical arguments. Much of the structure relates to chemistry and solar radiation. First examine Figure 1.10.

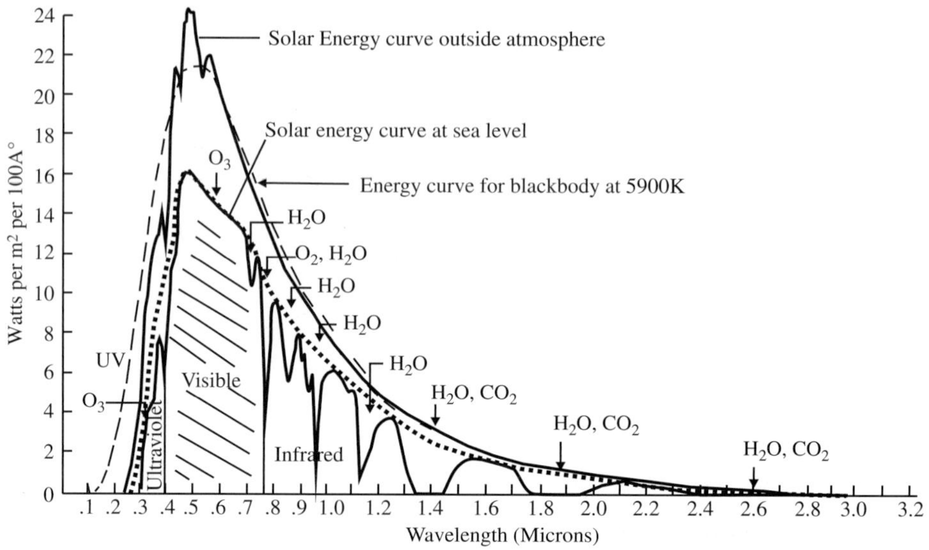

Figure 1.10 Solar radiation received above the atmosphere and at the surface (solid lines), and the theoretical black-body curve for the Earth (broken line). The dotted line shows the envelope of the sea-level spectrum – the actual spectrum is shown by the solid line underlying the dotted one. Adapted from *Houghton* (1977), Figure A8.1, who adapted it from *AFC-Laboratories* (1965).

This graph shows the radiation received from the sun, both before entering the Earth's atmosphere and as it reaches the surface of the Earth. The solar ideal blackbody curve is also shown. There are clearly certain bands where absorption is strong. The species responsible for this absorption is indicated for each band. At very short wavelengths (less than 0.3 microns) attenuation is extreme due to absorption of ultraviolet radiation (UV) by a variety of processes including ozone chemistry. Gamma rays and X-rays are removed by simple scattering processes due to the atmosphere, but longer wavelengths involve significant photochemistry.

As solar radiation enters the atmosphere, different wavelength bands are affected by different gaseous constituents. Absorption processes work at all heights down to the troposphere, in ways that we will discuss shortly. Within the troposphere, however, absorption and transmission processes become more complex, as re-emission and reflection from the surface cause some special radiation transfer physics to come into play.

For now, we will concentrate on processes from the top of the atmosphere down to the tropopause. The processes below the tropopause will be treated separately.

Although radiation is absorbed and scattered at all wavelengths as it enters the atmosphere, certain wavelengths are more affected than others. For example, the Lyman-Alpha (121.6 nm) and Lyman-Beta bands, which are associated with orbit-level changes in the hydrogen atom, are especially susceptible to oxygen photochemistry. At heights above 150 km, the oxygen number density is low, so the importance is low. By the time the Lyman-band radiation reaches 120 km, however, oxygen densities are high enough that absorption becomes relatively large. The UV radiation in these bands is absorbed

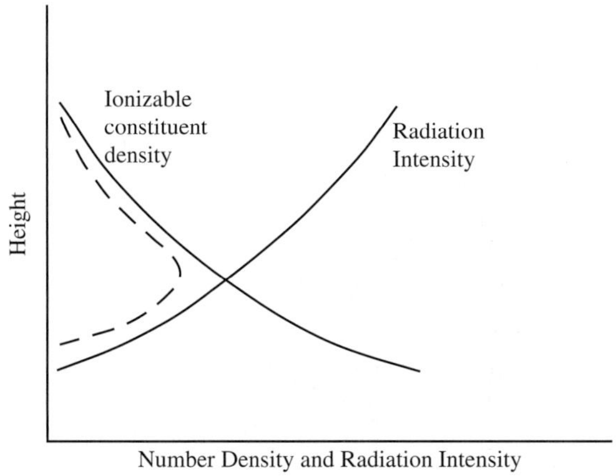

Figure 1.11 Production of Chapman layers.

out by reactions involving oxygen. In the process, electrons are produced, increasing the electron density at this altitude. This strong absorption means that the UV radiation is almost totally diminished at lower heights, so that reaction ceases lower down due to insufficient radiation. The result is a layer of enhanced electron density at 100–120 km, as hinted at in Figure 1.5. (In that case, the "layer" does not seem to be a true maximum, but rather a kink in the curve. This is because the diagram is only schematic – often the layer does show a real local peak.) A layer produced in this way is called a "Chapman layer." In both Figures 1.5 and 1.7, layers of enhanced density can be seen. In the first case, layers of locally increased electron density occur, and in the second, a layer of enhanced ozone density can be seen at about 20–40 km altitude in the stratosphere. The ozone is produced by absorption of UV radiation at wavelengths less than 246 nm, and, in the process, produces heating, thereby causing the stratosphere to be warmer then the tropopause air immediately under it.

Figure 1.11 demonstrates the process just described in a schematic manner, showing the decrease of number density of the basic constituent (oxygen, in the case of the E-region), the absorption of the ionizing radiation as one decreases in height, and the resultant production rate shown by the broken line. Chapman was the first to derive a theoretical expression for the shape of this profile, which takes the form

$$q = q_m \exp\{1 - \xi - e^{-\xi}\}, \tag{1.1}$$

where q is the production rate, q_m is the peak value, and ξ is the height normalized by division by the scale height H, where H is the vertical distance over which the pressure drops by a factor of e. This formula deals only with production rates – loss processes alter the profile again.

With regard to the ozone layer, the Chapman process somewhat oversimplifies the description. In truth, minor constituents such as NO, NO_2, and OH also play important roles, and UV radiation of wavelengths less than 1140 nm, particularly less than 310 nm, is also involved, being absorbed in the photo-dissociation of ozone. This production of O_3 absorbs out the UV, and thus at heights below 15–20 km, little of the radiation penetrates. These O_3 reactions produce molecular kinetic energy and hence atmospheric heating. Some of the heat is re-radiated by CO_2 at infrared wavelengths. The balance between O_3 heating and CO_2 losses produces a temperature peak at the stratopause (around 50 km altitude) and a maximum in the O_3 profile at around 25–30 km (*Houghton*, 1977, Sections 4.7 and 5.5).

The F-region is produced principally by radiation at 20–80 nm, ionizing N_2, O_2 and, particularly, O. The maximum of ion production is at heights of about 150 to 170 km (Fl height). Although the F-region still has production processes that resemble a Chapman layer, the region does not show a real peak – rather, the electron density continues to rise above this height. The reason for this is that attachment and recombination (i.e. the re-attachment of free electrons to surrounding ions) happens fast enough to distort the appearance of the layer. Different electron loss processes happen at different heights, complicating the simple "Chapman Layer" description once again.

The Chapman process is of course not the only process defining the atmospheric profile. Just as the Sun is a black body radiator, so is the Earth. It radiates at a typical temperature of 250–300 K. Its peak radiation intensity is in the infrared band. Carbon dioxide is an especially important radiator. At heights above the ozone layer, cooling occurs due to radiation to space by CO_2 of wavelengths around 15 microns. This cooling contributes significantly to a temperature decrease above the stratopause (e.g. see *Allen et al.*, 1979).

Some of these processes are noted succinctly in Figure 1.6.

At the very highest regions of the atmosphere, above 100 km, the temperature simply rises steadily, up to thousands of Kelvin (see Figure 1.5), due to direct absorption of radiation from the Sun. However, because the mean-free paths of the particles are so long, a spaceship passing through this region will not suffer excessive heating effects. The impact of a temperature of over 1000 K is something different to that which it would have had in the lower atmosphere.

1.3.3 Radiative transfer in the troposphere and greenhouse warming

At lower heights, below about 10 km altitude, the absorption process becomes even more complex. Some further radiation is absorbed directly by the air, as at higher altitudes, and some is reflected by clouds. Meanwhile, a large amount of radiation reaches the ground, especially in the visible regions of the spectrum (where the Sun produces its highest intensities), as indicated in Figure 1.10. Shortly, we will discuss this process in more detail, but before doing so, an examination of the expected tropospheric temperatures is warranted.

Incoming radiation from the sun encounters the earth and covers a disk of area πR_E^2, where R_E is the radius of the Earth (see Figure 1.12). Approximately 30% of this incident light is reflected by clouds, scattering from the atmosphere, and by reflection from the ground (snow, ice, water, sand, etc.). This percentage of reflected radiation is called the albedo. Assuming that the absorbed energy is the remaining 70%, this means the absorbed energy intensity is

$$I_A = 0.7 \times \pi R_E^2 F, \tag{1.2}$$

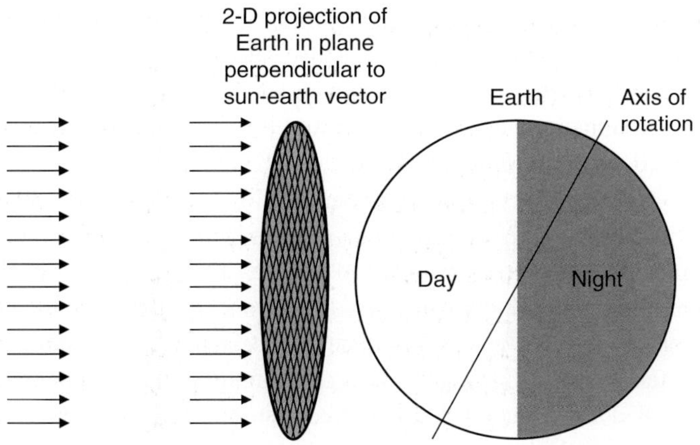

Figure 1.12 Determination of the mean temperature of the Earth's atmosphere.

where F is the incident flux from the Sun (called the solar constant) and equals 1367 Wm^{-2}.

Since the Earth's average temperature has been stable for thousands of years, we now assume that the outgoing radiation equals the absorbed radiation. It is important to recognize that this is only an approximation; on smaller time scales, exact balance may not occur. "Global warming" is currently a focus of some concern, and under this scenario, equilibrium may not be a valid assumption, although the differences between outgoing and incoming radiation should still be slight.

Treating the Earth as a black-body radiator, and assuming it radiates equally from all of its surface (including the night sector), outgoing radiation intensity equals

$$I_O = 4\pi R_E^2 \sigma T_e^4, \tag{1.3}$$

where T_e is the Earth's effective temperature, σ is the Stefan–Boltzmann constant ($\sigma = 5.67 \times 10^{-8}\ Wm^{-2}K^{-4}$), and $4\pi R_E^2$ is the surface area of the Earth.

Then equating the incoming and outgoing radiation gives

$$0.7\pi R_E^2 F = 4\pi R_E^2 \sigma T_e^4. \tag{1.4}$$

Solving for T_e gives

$$T_e = \sqrt[4]{\frac{0.7 \times 1367}{5.67 \times 10^{-8} \times 4}} \simeq 255\,K. \tag{1.5}$$

This number seems lower than our current typical temperature of 290–300 K (17 to 27 °C), but is in fact the temperature of the Earth as "seen" by satellites from space, which tend to see the top of the stratosphere. Temperatures in the lower troposphere seem warmer, and we now turn to understanding why.

To do this, return to the issue of the heating of the troposphere. As noted, some of this radiation is reflected by clouds, but a significant amount reaches the ground, especially in the visible regions of the spectrum. Some of this radiation is reflected by the ground, but over 50% of the total initial incident radiation is absorbed by the ground.

It then reradiates as infrared radiation. On this return path upward, absorption by the air becomes substantial, and it is this absorption of infrared radiation that leads to the tropospheric heating. For this reason, the temperature decreases with increasing height, as less radiation penetrates to greater heights. Convection and so forth also act, resulting in a temperature profile approximating, but not exacty equal to, the adiabatic lapse rate; the difference is typically 4 °C per km.

The term *adiabatic lapse rate* needs a little explanation. When a parcel of air moves up in the atmosphere, it moves to regions of reduced pressure relative to that below. The parcel expands, and in so doing, does work. By the first law of thermodynamics, doing work requires energy, so this energy is extracted from the internal heat storage of the system, thereby cooling the parcel. The higher the parcel goes, the cooler it gets. This is called an "adiabatic process" because total energy within the parcel is assumed to be conserved. With this concept in mind, it is clear that as the parcel rises, it gets colder and colder, and the gradient of the temperature of the parcel as a function of height has a particular value of close to 10 °C per km. This will be derived later. It is this temperature gradient that is referred to as the adiabatic lapse rate (or more precisely, the adiabatic lapse rate for dry air, since the existence of moisture changes its value (see Chapter 12)). If the air parcel descends, it heats, as it is externally compressed by the surrounding air, which increases in pressure as the parcel moves downward. Hence work is done on the parcel and this is stored as heat within the parcel. Any type of heating or cooling due to downward or upward parcel movement is referred to as "adiabatic" heating or cooling. Heating or cooling due to other processes, like chemistry or solar effects, is referred to as "diabatic" heating or cooling.

We now return to our discussion of typical temperature profiles. Of course, many local processes also act to produce local deviation from this picture. This region will not be discussed greatly here. The basic picture of the interplay between the different types of radiation is shown in Figure 1.13. Note that although the incident radiation is set at 100 units, some of the numbers involved in the right-hand picture exceed 100, because some units of radiation are absorbed and re-emitted at one level, then further absorbed and partially re-emitted at another level, thereby being counted more than once.

This complex process of absorption and re-emission is termed radiative transfer. It results in heating of the atmosphere above and beyond its normally expected value. This heating process is termed greenhouse warming. The name arises because in early days it was believed that the glass of a greenhouse kept the greenhouse warm by allowing visible light in but not allowing infrared radiation out, in a similar way to that described above. It turned out that this was untrue for a greenhouse, but the name persisted when describing atmospheric heating. Carbon dioxide, water vapor and various other "greenhouse gases" are responsible for the absorption and re-radiation of infrared radiation. Ultimately, infrared radiation from above the stratosphere to space keeps the Earth's temperature in approximate balance. Using principles of radiative transfer, it has been seen to be possible to determine reasonable first-order estimates of global temperatures and even height profiles of temperature, and these discussions will be expanded even further below.

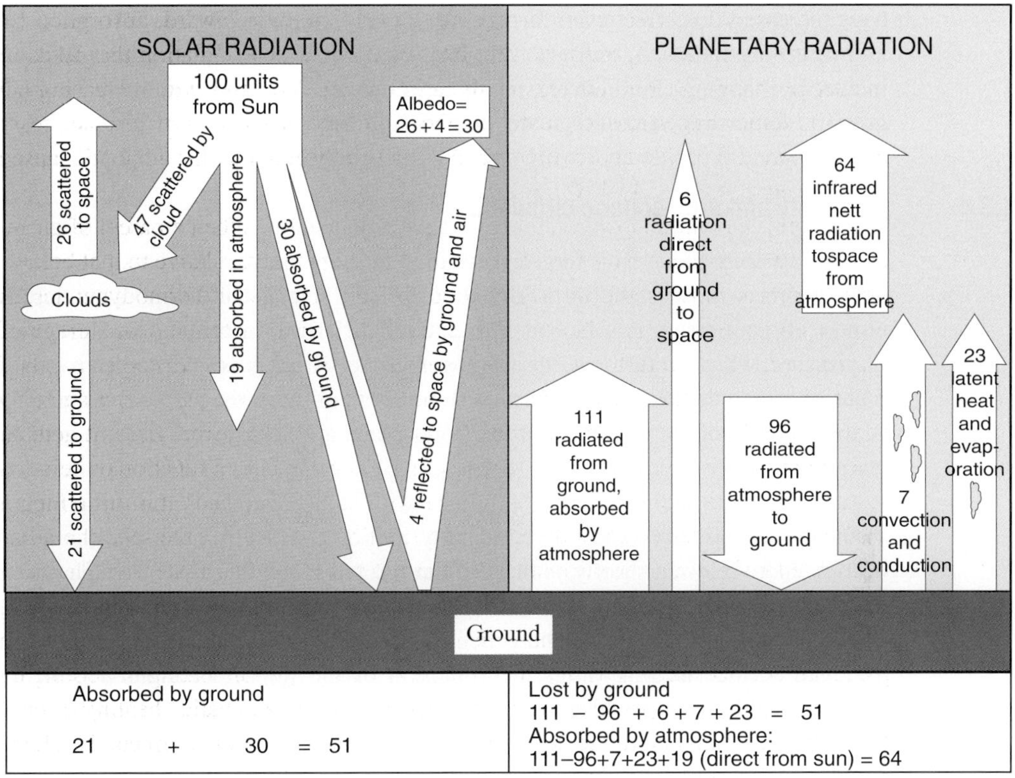

Figure 1.13 The left-hand figure shows the possible processes that may happen to visible light ("short-wave radiation") entering the top of the troposphere. The right-hand figure shows the distribution of infrared (long-wave) radiation as it is radiated by the ground. Note that the insolation absorbed by the ground on the left (51) matches the insolation lost by the ground (51) on the right.

Variations in this simple picture occur due to latitudinal, longitudinal, and temporal variations in albedo, variations in surface temperature, and breakdown of the assumption of equilibrium. Satellite measurements of outgoing longwave radiation do indeed show significant variability in space and time.

Since increased atmospheric concentrations of CO_2 lead to higher infrared absorption in the troposphere, there should also be an expectation that this should lead to increased *radiation* of infrared by CO_2 in the stratosphere and mesosphere, leading to cooling in these upper regions. Computer models (e.g., *Rind et al.*, 1990) seem to show evidence for this, though as yet no definitive experimental proof has been presented. Tentative evidence related to increased rates of occurrence of noctilucent clouds at 80–85 km altitudes may be suggestive of this, however. The issue of determining whether anthropogenic increases in CO_2 emissions are leading to excess global heating is an area of great importance for life on Earth.

Heating of the troposphere in this way elevates the temperature by typically 35 Kelvin above expected – without this process, temperatures would be too low to sustain life as we know it. The modern issue with so-called "global warming" is that increases in CO_2

have increased this effect even further, producing a non-equilibrium between outgoing and incoming radiation, and also altering the Earth's albedo (through various effects including changing cloud cover), resulting in enhanced heating which leaves the Earth growing somewhat warmer than we are currently used to.

1.3.4 Variability and atmospheric circulation

The discussions above have largely pertained to global mean situations. Of course, the atmosphere is never static on a global scale, and spatial and temporal variability in almost all parameters is a fact of life. Spatial variations in heating lead to variations in pressure, which in turn lead to air movement, or wind. It is with regard to measurement of winds that the radars discussed in this book have their greatest applicability. A short review of the dynamics of the troposphere and middle atmosphere is therefore warranted.

All motions in the atmosphere originate with solar heating. Without the Sun, there would be no gaseous atmosphere, nor any motions. A second key point, which will become more relevant shortly, is that the Earth rotates, and this also has huge impact on the production of "weather." We will begin by examining solar effects.

The motions of the troposphere begin essentially at the equator, where annually averaged surface heating (after incorporation of the impact of clouds) is strongest. Atmospheric heating produces vertical convective motions, which lift air up from the ground, carrying heat with it. The tops of the large tropical clouds formed in this way typically reach 14 km altitude or so. As the air rises, it eventually flows away from the equator, toward the north and south. It spreads out to non-equatorial latitudes and eventually begins to descend (*Held and Hou*, 1980). The falling motions begin at around 25–30° latitude north and south. At these latitudes, the descending air heats by adiabatic compression, thereby establishing an internal temperature profile immediately above the ground and extending up to a few hundred meters altitude, which warms with increasing altitude (see *Ahrens*, 1999, Figure 7.5). This heating produces strong stability within the air. As a consequence, the major deserts of the world occur at these latitudes – places of strong atmospheric stability and hence little precipitation.

As the air moves down towards the ground, it again spreads to the north and south. The equator-bound air moves back to its source and closes a vertical loop, as shown in Figure 1.14. Since the Earth is a non-inertial reference frame, due to its rotation, we need to include non-inertial pseudo-forces in our discussions. Chief amongst these is the Coriolis force. As the air moves back towards the equator, it experiences a Coriolis force to the right in the northern hemisphere, and to the left in the southern hemisphere. The net result is a wind flow out of the east-north-east in the northern hemisphere, and out of the east-south-east in the southern hemisphere. These winds are known as "trade winds," because of their predictable consistency, which enabled ancient mariners to rely on them for trade. The vertical cell and associated easterly (westward) winds are referred to as a Hadley cell. At or near the equator (or more specifically, along a slightly meandering line called the inter-tropical convergence zone, or ITCZ), the winds are quite light and irregular, leading to the term the "doldrums" to describe the weather there.

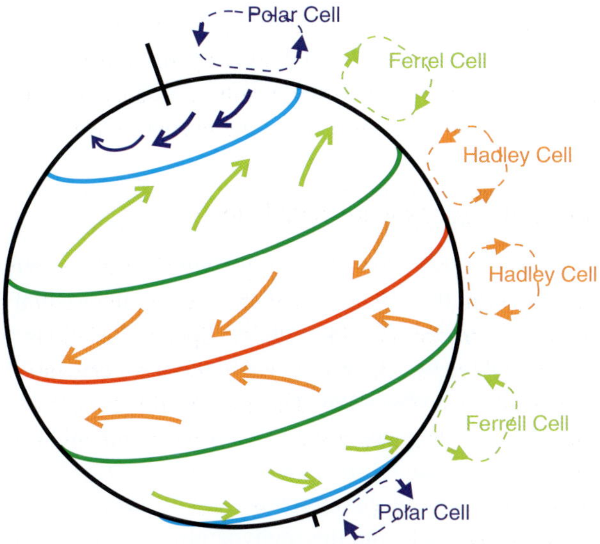

Figure 1.14 Air flow in the troposphere, showing the Hadley and Ferrel cells.

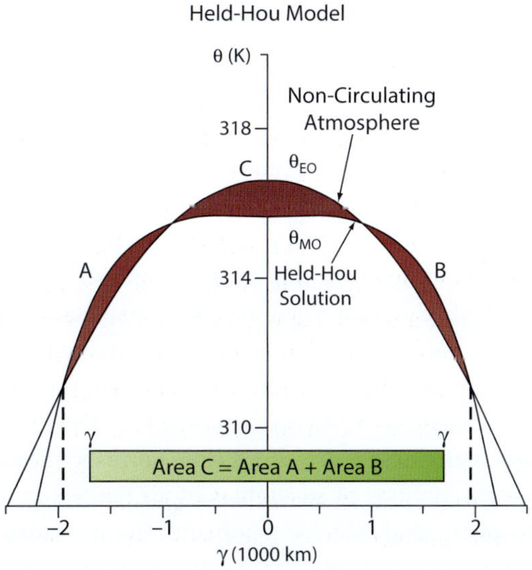

Figure 1.15 The reasons for the outer limits of the Hadley cell, adapted from *Held and Hou* (1980). See the text for details.

The fact that this Hadley cell has a circulation (driven by the Earth's rotation) also explains why it has the latitudinal limits that it does. The reasons why the air descends at around 30° latitude are essentially related to the latitudinal temperature distribution of the air relative to the distribution for a non-circulating atmosphere. At the equator, the air at upper altitudes in a realistic atmosphere is cooler than it would be for a simple non-rotating Earth (non-circulating flow). As it spreads northward and southward, it reaches latitudes where the reverse is true. The idea is illustrated conceptually in Figure 1.15, where it is seen that the air in the more realistic (circulating) situation has a flatter temperature distribution than for the non-circulating case. As the air spreads, it eventually reaches a point where the temperature of the circulating atmosphere is greater than that of the non-circulating case. Adjustment of the cross-over point to ensure that area "A" equals that of the sum of "B" and "C" defines the edge of the poleward flow of air. This places a thermodynamic limit on the poleward expansion, but of course as the air flows out from the equator, the air has to go somewhere, and it is the fact that the Coriolis force simultaneously twists the north-south flow into a zonal flow that allows the poleward meridional flow to be stifled. These combined effects also lead to the downward descent of the air, whereupon it circles back to the equatorial regions at lower altitudes and so completes the Hadley cell. Although Figure 1.15 somewhat oversimplifies the process, it gives a reasonable physical way to determine the latitudinal extent of the Hadley cell. The detailed structure needs to be solved using numerical models, as illustrated in greater detail by *Held and Hou* (1980), *Held* (2000), and *Frierson et al.* (2007), among others. Later in this chapter, we will further elaborate briefly on the dynamical processes (as distinct from thermodynamical processes) involved with constraining the Hadley cell, but that needs to wait till we have a little more familiarity with waves in the atmosphere.

Poleward of the Hadley cell exists a second zone of circulation, referred to as a Ferrel cell, which appears to show a generally eastward flow with some poleward contribution. This is also shown in Figure 1.14. It is tempting to believe that this might be due to the descending air at the outer edges of the Hadley cell flowing partly poleward as it reaches the ground, resulting in some poleward flow which then sets up the cell. However, this is not true at all. The reason is a mixture of effects. First, the region is one of strong wave activity, especially planetary waves. Rossby waves are particularly important. When maps of the wind flow like that seen in Figure 1.14 are developed, they are determined using so-called "Eulerian zonal means." This means that the average speed as seen from the ground is averaged around an entire great circle of the Earth at fixed latitude. However, waves are intrisically Lagrangian in form (i.e., they involve particle motions that involve large-scale movement across the Earth's surface). It is generally now accepted by workers in the field that there is little correspondence between Eulerian zonal means and the actual motions of ensembles of air parcels in the mid/high-latitude troposphere and stratosphere, and even for that matter in the oceans (e.g., *Plumb and Ferrari*, 2005). The apparent "reverse" cell at mid-latitudes in the atmosphere is a mathematical artifact due to the dominance of Rossby waves in these regions. Two papers which discuss this subject are *Matsuno* (1980) and *Edmon et al.* (1980). These papers also contain useful additional references.

The whole issue of Lagrangian vs. Eulerian description of flows is important for this region, but also important for all circulation on the Earth's surface. Many numerical models now use a formalism called "transformed Eulerian mean" (TEM). In this process, the wave- and small-scale motions are described within an Eulerian framework, but this reference frame itself moves with the mean flow. So the method uses two frames of reference at different scales, the smaller one embedded in the larger one. The idea was introduced by *Andrews and McIntyre* (1976, 1978), and is discussed in a more descriptive sense by *Andrews et al.* (1987) (see Chapter 3 of that book for a nice discussion) and *Brasseur et al.* (1990). We will not pursue this discussion further here. But the main consequence is that TEM is a much simpler (and more physically meaningful) formalism for describing the mean meridional circulation of the middle atmosphere.

While discussing such models, it is also useful to point out here one extra point of potential confusion. When describing wind flow directions, there are two conventions, depending on the user. Meteorologists, on one hand, use a convention describing where the winds have come from, so a wind from the west is called a "westerly" wind by meteorologists. However, another common usage, particularly used in radar circles, is to use the term "eastward" to describe a wind moving *toward* the east. We will use both conventions herein, and at times even use both. Usage will depend to some extent on the context of the application, and the field most relevant (meteorology vs. radar) to the discussion. We will *never* use the term "easterly" to describe a wind blowing *toward* the east; such a wind will be referred to as either eastward or westerly.

Momentum forcing and waves

Over much of the Earth, it is not possible to properly describe the large-scale motions without involving waves. We have seen this already in the preceding paragraphs.

Waves take a variety of forms, from gravity (or buoyancy) waves with periods of a few minutes and horizontal wavelengths of a few kilometers, to planetary scale waves with periods of many hours and even days, and wavelengths of thousands of kilometers. Waves can be either free or forced. Atmospheric free oscillations can take the form of gravitational modes, Rossby modes, mixed-Rossby and Kelvin modes (e.g. *Forbes et al.*, 1999). Forced waves can include atmospheric tides and Rossby modes. The causes of waves are also extremely varied. Gravity waves can be generated by convection, frontal systems, eclipses, flow over orography, geostrophic adjustment (a process by which geostrophic balance is maintained by the shedding of the geostrophically unbalanced part of the motion), and nonlinear interactions, among others. The main generators of planetary-scale Rossby waves (which are especially important in the discussions above) in the atmosphere are flow over topography, diabatic (i.e. non-adiabatic) variations, nonlinear interactions between synoptic scale motions, and long-wave baroclinic instability.

In view of the importance of these waves, we now take a few moments to discuss the effects of momentum transport and divergence on mean flows.

Any type of wave carries with it momentum flux. Eddy motions can also carry momentum flux. We first need an expression for this flux, and then need to see how it can force flows at different scales.

Terms like $\rho\overline{u'w'}$, $\rho\overline{v'w'}$, and $\rho\overline{u'v'}$ occur frequently in atmospheric studies. In turbulence studies, we often equate these terms to turbulent diffusion of momentum. In gravity wave and planetary wave studies, they are given different interpretations. For example, in this case the vector

$$\vec{\mathcal{F}} = -(0, \rho_0\overline{u'v'}, \rho_0\overline{u'w'}) \tag{1.6}$$

is given the very special name of the "Eliassen–Palm flux" (*Eliassen and Palm*, 1960). In dealing with the atmospheric flow, we often consider the Navier–Stokes equation.

In this chapter we have tried to keep the discussion descriptive, so have largely avoided too much mathematics, but here it is appropriate to introduce a little. More mathematical detail will be given in Chapter 11. The Eulerian Navier–Stokes equation, with the effects of molecular viscosity removed, and the impact of the Eliassen–Palm flux included, is written as

$$\frac{\mathrm{D}\vec{u}}{\mathrm{D}t} = -\frac{1}{\rho}\vec{\nabla}p - 2\vec{\Omega} \times \vec{u} + \vec{g} + \frac{1}{\rho_0}\vec{\nabla}_{\mathcal{F}}, \tag{1.7}$$

where $\vec{u}$ is the velocity vector, t is time, ρ is density, p is pressure, $\vec{\Omega}$ is the angular velocity of the Earth, $\vec{g}$ is the acceleration due to gravity. The term $\frac{\mathrm{D}}{\mathrm{D}t}$ is $\frac{\partial}{\partial t} + \vec{u} \cdot \vec{\nabla}$ and represents differentiation following the motion. The last term is considered as a "forcing term" which arises due to forcing due to the E–P flux $\mathcal{F}$. In turbulence studies, the same types of terms appear (e.g., $\rho\overline{u'w'}$), but then they are called *Reynolds' stresses*. In transformed Eulerian mean theory, the equations are slightly different but the forcing due to momentum flux divergence remains.

Quantities like $\rho\overline{u'w'}$ have dimensions of mass divided by length and divided by time squared, which is the same as the dimensions of momentum per unit area per second (i.e. momentum flux). In fact $\rho\overline{u'w'}$ represents the vertical flux of horizontal momentum (and also the horizontal flux of vertical momentum), and we will now show why this is true.

The term $\vec{\Omega} \times \vec{u}$ represents the Coriolis force. An astute student of physics will recognize that the Coriolis force is not a true force at all, and may wonder why this term is used here. A more correct approach would be to view the entire motion from an inertial reference frame outside the Earth, in which case no Coriolis force should arise. But mathematically, this is inconvenient, and the normal approach is to view the situation from the frame of reference of a non-inertial observer rotating with the Earth. Therefore the pseudo-forces need to be included – even a centrifugal force should be included, although due to its relatively weak effect we have not done so. So the Coriolis pseudo-force will be included in all of our treatments of fluid flow on the Earth.

With regard to flux, consider Figure 1.16(a). The flow has a vertical component w and a component in the x direction which we will denote by u. We assume that the box is

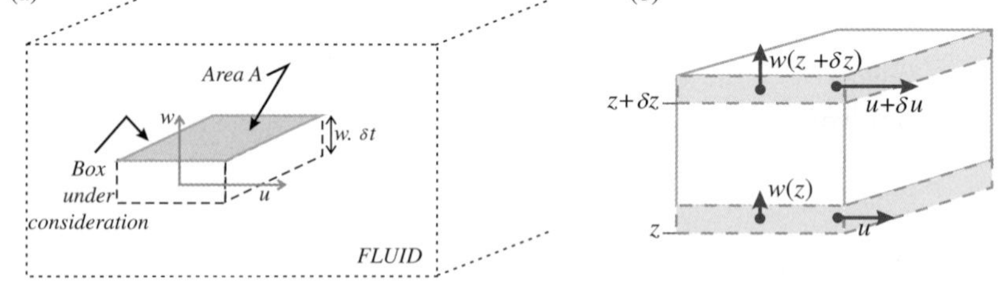

Figure 1.16 (a) A fluid with a sub-region showing the vertical extent covered by movement of a horizontal plane of the fluid in time δt. The figure is used to determine the rate of flow of momentum across the lower base of the box. (b) A small element of the fluid, showing the height variations of the horizontal and vertical components of velocity.

small enough that the velocity can be considered as constant within the box. We ask the following question: "how much horizontal momentum passes through the lower plane A in time δt?" In order to find this, we find the total momentum contained in a box of area of cross-section A and depth $w\delta t$. This will be exactly the amount that has flowed into the box in time δt, and equals the mass of fluid contained in the box multiplied by its horizontal speed, or

$$\mathcal{P}_x = \rho.(w.\delta t.A) \times u, \tag{1.8}$$

where $w.\delta t.A$ is the volume of the box.

In order to obtain the momentum flux, we divide by both A and δt, giving a flux of

$$\mathcal{F} = \rho uw. \tag{1.9}$$

If the velocities are fluctuating in time, we can write $u = \overline{u} + u'$ and $w = \overline{w} + w'$, and take a time average. Then the mean momentum flux is

$$\overline{\mathcal{F}} = \overline{\rho.(\overline{u} + u').(\overline{w} + w')} = \rho.\overline{u}\overline{w} + \rho\overline{u'w'}, \tag{1.10}$$

since terms like $\overline{\overline{u}w'}$ are zero by definition.

(In reality we should also recognize that we could write $\rho = \overline{\rho} + \rho'$, but this is usually a minor correction and is often ignored.)

Hence we see that terms like $\rho\overline{u'w'}$ do truly represent a vertical flux of horizontal momentum due to perturbation quantities.

We can do an identical analysis using a vertically orientated plane sheet to show that $\rho\overline{u'w'}$ is also the horizontal flux of vertical momentum.

We now consider again a box of fluid, as in Figure 1.16(a), but this time we allow the box to be large enough that the flow velocity may vary within the box, as seen in Figure 1.16(b). If the box has depth δz, and cross-sectional area A, then the momentum that flows out of the top of the box (at $z+\delta z$) is given by $\overline{\mathcal{F}(z + \delta z)}.A.\delta t$, and the momentum flowing into the box at the bottom is $\overline{\mathcal{F}(z)}.A.\delta t$. The net momentum flowing in is the difference, and equals $[\overline{\mathcal{F}(z)} - \overline{\mathcal{F}(z + \delta z)}].A.\delta t$. The net force on the box is given by the rate of change of momentum, so we divide by δt. Then we may divide and multiply by δz to give that the net force is

$$F_{\text{net}} = [\overline{\mathcal{F}(z)} - \overline{\mathcal{F}(z + \delta z)}]/(\delta z) \times \delta z.A = [\overline{\mathcal{F}(z)} - \overline{\mathcal{F}(z + \delta z)}]/(\delta z) \times \delta V, \tag{1.11}$$

where δV is the volume of the element of fluid. Hence we can write that the force per unit volume is

$$F_{\text{vol}} = -\frac{\partial}{\partial z}\mathcal{F}, \tag{1.12}$$

or

$$F_{\text{vol}} = -\frac{\partial}{\partial z}(\rho\overline{u'w'}), \tag{1.13}$$

the latter term being considered as a forcing associated with the Reynolds' stress term $\rho\overline{u'w'}$.

More commonly we seek the force per unit mass, so we divide by ρ to give

$$F_m = -\frac{1}{\rho}\frac{\partial}{\partial z}(\rho\overline{u'w'}). \tag{1.14}$$

Thus the height-derivative of Reynolds' stress terms (or Eliassen–Palm terms if dealing with stratospheric forcings) tells us the body force per unit mass on the fluid due to fluctuating x-motions. Of course other forcings exist – we could look at the y-derivative of the x-motion, or the z-derivative of the y-motions, or any other combinations. As a rule, things change most rapidly with height, so the height-derivative terms often dominate. We have in fact already seen these forcing terms in operation in regard to Equation (1.7).

Non-tropical circulation cells

We now return to our discussion of the different cells of the Earth's circulation. The latitude band around 30–35° marks a transition between trade winds and the predominant eastward winds (westerlies) of the mid and high latitudes. Here, winds can be quite variable, as they are influenced by frontal passages and different types of high and low pressure cells. At times, ancient sailors could be marooned there for long periods of time, when caught in regions of high pressure and weak winds. The region is known as the "Horse Latitudes." The reasons for this nomenclature are uncertain, but one theory is related to becalmed sailors having insufficient fresh water to sustain the horses being carried on board, and subsequently throwing dead and dying horses overboard, leading to an abundance of horses found floating in the oceans there. Other explanations exist as well, but the weather-related aspects of this explanation are of note. The interested reader is referred to Wikipedia on the world-wide-web for other competing explanations for the nomenclature.

This Ferrel cell extends to perhaps 50° of latitude, beyond which there is, in principle, a third cell established with associated polar easterly (westward) flow. Of the three cells, only the Hadley cell is a reasonably accurate description of the mean motions of ensembles of air parcels in the meridional plane. The other cells are complicated by a great deal of other flows associated with waves, orography, instabilities, vortical flows, and pressure cells, which even complicates the meaning of the "zonal mean."

It is now of interest to look at Figure 1.17. It may be seen from this figure that the height of the tropopause varies significantly with latitude. Near the equator it is around 16–18 km, and then shows a transition to a dual tropopause at around 25–30°latitude. Between 25° and 45°, tropopauses may exist at 16–20 km and also at 9–12 km. The lower one shows a distinct average decrease in height from 30° to 60°latitude. At higher latitudes and into the arctic regions, the tropopause height can be as low as 8 km and less.

The boundary between the Ferrel and polar cells represents a special region of the atmosphere. There is a considerable pressure difference across this boundary (at around 50°–55° latitude), and an associated jump in temperature, as illustrated in Figure 1.17. The associated north-south pressure gradient across this region, through the action of the Coriolis force, leads in turn to a strong eastward flow. This produces a river of air that flows along the boundary in a generally eastward direction. This river is called the "jet stream." In truth, it is far from fixed in latitude, but meanders extensively. The northern

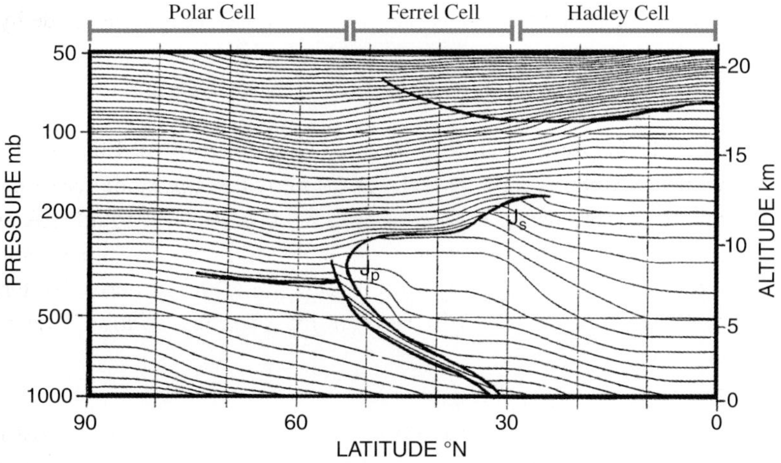

Figure 1.17 Equator-to-pole distribution of the temperature in the troposphere and lower stratosphere, showing the tropopause as a function of latitude for January 1956. Contours indicate adiabats at 5 °C steps, and thick curves show the tropopause(s). J_s and J_p represent the subtropical and polar jet streams respectively. The diagram has been adapted from *Yamanaka* (1992), who developed it using data from *Defant and Taba* (1957).

hemisphere jet stream is illustrated in Figure 1.18. A similar stream exists for the southern hemisphere. Yet another similar jet stream exists in the upper subtropics, at around 30° latitude, called the subtropical jet stream, but it does not have the same strength as the polar jet stream. Its maximum is at higher altitude than the polar jet stream, and it marks the boundary between tropics and mid-latitudes. It coincides with the tropopause break discussed in regard to Figure 1.17.

In the last few paragraphs we have seen that the dynamics outside the tropics has many differences compared to the equatorial regions. The importance of wave forcing has been especially noted. We also discussed the Held–Hou model for describing constraints on the latitudinal extent of the Hadley cell. With our new knowledge, we can also add some extra dynamical physics to this description.

Not only is wave forcing a part of the atmospheric circulation – it is a crucial part. The mean meridional circulation outside the tropics *must* be wave driven because there is no other way to overcome the angular momentum constraints on the flow. The ensuing discussion does not explain *why* the waves exist, but shows that they must be present if we are to have the flow regime that we do.

Although the full equations for fluid flow of the atmosphere are complex, there are simplifications that apply in particular circumstances. One such simplified equation is the zonal-mean momentum equation, which can be written as:

$$-\left(f - \frac{\partial u}{\partial y}\right) v^* = \vec{\nabla}_{\mathcal{F}}, \tag{1.15}$$

where $f = 2\Omega \sin\theta$ is the Coriolis parameter, Ω being the rotation rate of the Earth in radians per second, θ being the latitude, u being the zonal mean wind, v^* being the

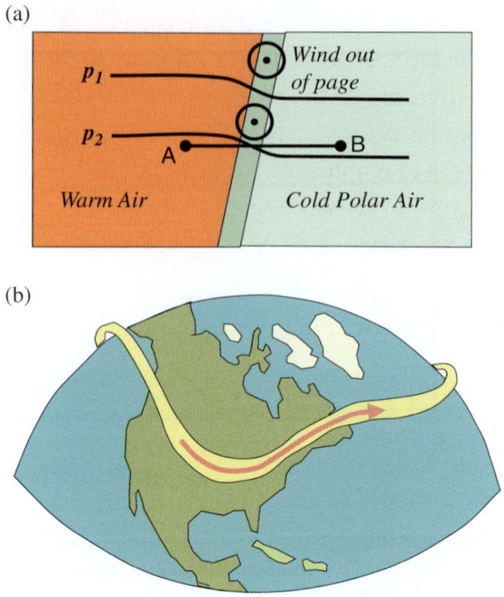

Figure 1.18 (a) Generation of the polar jet stream. The colder polar temperatures result in more rapid decrease in pressure with height (smaller scale height), so by the height of the tropopause, the atmosphere has lower pressures in the polar region than at mid-latitudes. This leads to strong north-south pressure differences, as can be seen along the line AB. (Lines p_1 and p_2 are isobars.) This meridional pressure drives a strong zonal wind-flow via the Coriolis force (winds drawn out of the page), which is the polar jet stream. (b) The jet stream is not fixed in latitude, but wanders around significantly.

residual meridional wind, and $\vec{\nabla}_{\mathcal{F}}$ being the Eliassen–Palm forcing. The term y represents distance along a meridian of longitude. The equation can be determined from (1.7) by ignoring pressure gradients and gravitational forces, and assume $\frac{\partial}{\partial t}$ is zero in $\frac{\mathrm{D}}{\mathrm{Dt}}$.

This equation says that meridional circulation results as a balance between the gradient of zonal wind as a function of latitude $\frac{\partial u}{\partial y}$, the Coriolis force (through f), and the wave forcing term.

The Earth can be divided into two key areas in regard to this equation. Firstly, outside the deep tropics $f \gg \frac{\partial u}{\partial y}$, so we can make the following good approximation:

$$-fv^* = \vec{\nabla}_{\mathcal{F}}. \tag{1.16}$$

This means that v^* can only exist if $|\vec{\nabla}_{\mathcal{F}}| \neq 0$. In other words, the only way for a non-zero meridional mean flow to exist (as we know occurs) is for the term $\vec{\nabla}_{\mathcal{F}}$ to be non-zero, which can only happen with the co-existence of waves.

Secondly, in the deep tropics, where f now becomes comparable to $\frac{\partial u}{\partial y}$, it is possible to have angular-momentum-conserving circulations even if $\vec{\nabla}_{\mathcal{F}} = 0$ because it is possible to have $v^* \neq 0$ if (and only if)

$$-(f - \frac{\partial u}{\partial y})v^* = 0 \qquad \text{or} \qquad f = \frac{\partial u}{\partial y}. \tag{1.17}$$

This is essentially what happens in the thermally driven Hadley cell. The Hadley cell cannot extend farther than 30° latitude or so because the u-distribution necessary to sustain the requirement $f = \frac{\partial u}{\partial y}$ is unstable beyond this latitude.

More elaborate details can be found in *Held* (2000) and *Frierson et al.* (2007). The key point here is to emphasize the fact that the Earth's circulation is critically dependent on wave motions, even with regard to determing the limits of the Hadley cell. At upper levels, which we will now discuss, waves are even more important.

1.3.5 Atmospheric circulation in the upper stratosphere and mesosphere

The atmospheric circulation above the tropopause comprises two different regimes – an essentially pole to pole flow above 45–50 km, and an equator to pole flow below about 30 km, with some intermediate types of flow in the adjoining altitudes (see Figure 1.19). The pole to pole flow above 45–50 km is actually a little easier to summarize and explain, so we will begin with that. The reasons for this flow were only finally understood in the 1980s. In the following pages, we will refer on several occasions to Figure 1.20, so the reader should keep that accessible.

Circulation above 45 km altitude

The region above 45 km altitude comprises the upper stratosphere and the mesosphere. In the stratosphere, the flow is considered as part of the so-called "Brewer–Dobson circulation" (BDC), which is the name given to the total stratospheric flow. The BDC is

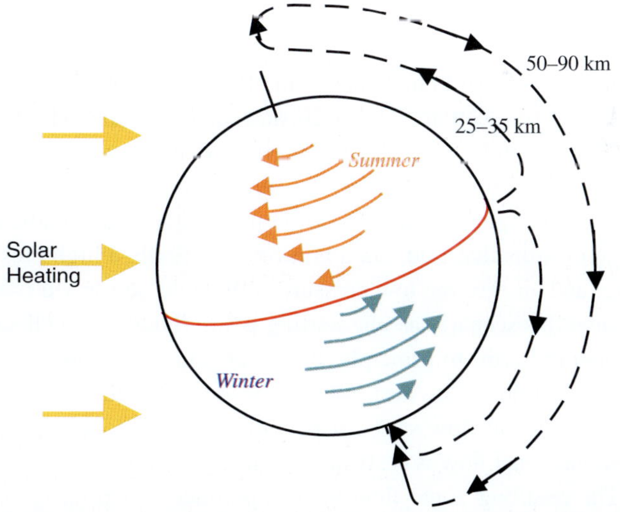

Figure 1.19 Showing wind flows in the upper stratosphere and mesosphere. Flow is predominantly to the east in the winter and to the west in the summer at above 50 km at midlatitudes. Zonal speeds can reach 60–70 ms^{-1}. North-south (meridional) flows are much weaker, being generally less than a few meters per second. The lower altitude flow is more complex, with outflow from the equator supplying the summer vertical motions, and the winter polar flow showing downwelling at all heights, where it merges with the tropospheric flow – see Figure 1.20.

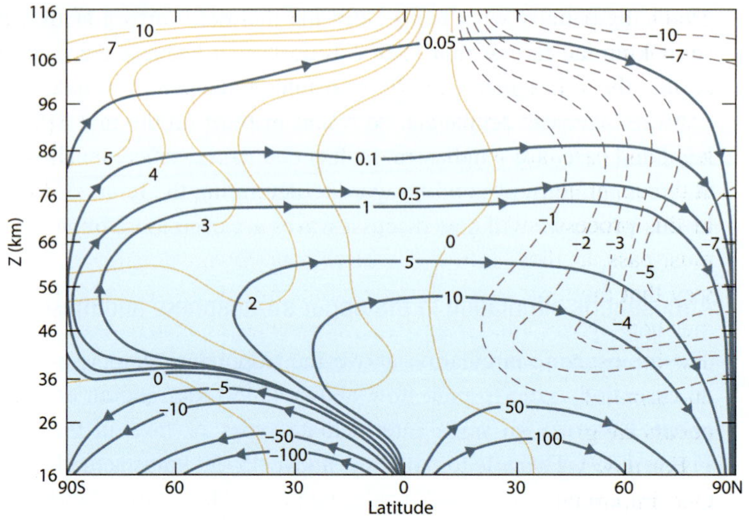

Figure 1.20 The solid blue lines show meridional atmospheric flows produced from a computer model (*Garcia and Solomon*, 1983). Specifically, the lines are model outputs of the "mass meridional stream function," though they may qualitatively be thought of as density-weighted meridional flow vectors. The lighter lines (orange and broken lines) show net diabatic (i.e. non-adiabatic) heating contours. The graph is for northern hemisphere winter solstice (southern hemisphere summer) conditions. Notice the merging of the upper atmosphere flow and the lower branch of the Brewer–Dobson circulation at about 30–35 km altitude – see the text for details. (Reproduced with permission form John Wiley and Sons.)

considered to have a lower and an upper branch. The portion under discussion here is actually the upper branch, and is in turn well coupled to the mesospheric flow.

The flow is seen in Figure 1.20, and can be explained by recognizing that heating over the summer pole occurs 24 hours per day. Without clouds to reflect the incoming radiation at these upper altitudes, the summer pole for a non-rotating Earth would become in principle quite warm. This of course requires a suitable absorber, but ozone in the region fulfills that role. On a non-rotating Earth, a high pressure cell would be formed here, and an extreme low pressure cell would be formed over the winter pole. We will assume this scenario as our starting point, though we will see that it will be altered by a rotating Earth. In principle, this arrangement should drive a summer-to-winter polar meridional flow. However, we now need to recognize that the Earth *is* rotating, so once this meridional flow starts, it is converted to a zonal flow by the Coriolis force. The resultant zonal flow is eastward in winter and westward in summer.

The resulting zonal flow in turn produces a Coriolis force back towards the summer pole, which cancels the summer-to-winter meridional flow, leaving behind a purely zonal flow.

Another process now comes into play. Wave activity, especially waves called gravity (or buoyancy) waves, produce extra forcing on the zonal flow, effectively reducing the winter-to-summer Coriolis force, destroying the north-south balance, and so allowing some limited meridional flow. Another way to view the effect is to consider that the waves allow air parcels to move across isolines of constant angular momentum, producing a divergence of angular momentum which allows meridional forcing and so making mean or steady-state extratropical meridional circulations possible (*Haynes et al.*, 1991). In the mid/upper mesosphere, there is always a meridional flow because gravity waves can propagate from their sources in the troposphere through both westward and eastward winds (unless they encounter critical levels) and they begin to break at the upper altitudes, imparting momentum flux to the mean flow. An important aspect of this process is the prevailing *stratospheric* winds through which the gravity-waves must pass, as these filter the spectrum of upward propagating gravity waves in such a way that the force they exert upon breaking in the mesosphere is eastward in the summer hemisphere and westward in the winter hemisphere, which is just what is needed to drive a pole-to-pole circulation by $-fv^* = \vec{\nabla}_{\mathcal{F}}$, where here $\mathcal{F}$ is the momentum flux due primarily to gravity waves. More specific details about the mechanism by which this occurs are given in Chapter 11.

For now we simply present the results, which are shown in Figures 1.19 and 1.20. One important result of this flow is the existence of rising air in the summer polar mesosphere, causing large amounts of adiabatic cooling, and making the air high in the mesosphere over the summer pole extremely cold (Figure 1.23).

The above description has been necessarily simplistic. We have ignored effects of some major waves, including various types of forced waves. On any one day, the actual winds and temperatures may deviate markedly from the mean, especially at the higher altitudes. Atmospheric tides can reach velocity amplitudes as high as 50 to 70 ms^{-1}, and at times even more. These are waves which are forced by solar heating effects. Planetary waves can also exist, and at times drive the daily mean stratospheric state substantially away from the "normal" values, during events called "sudden stratospheric warmings." These are often driven by strong planetary-wave activity, especially Rossby waves.

We cannot discuss all of these types of waves in this book. We will concentrate on areas in which radar studies are of greatest special value, so we will focus especially on gravity waves, since radars can resolve the temporal and spatial variability of these events with perhaps the best resolution of any instruments. Radars can also be of great value in resolving tides and planetary waves as well, of course, but since these events can also be studied well with satellites and rockets, we will choose the demarkation of wave-like events to be discussed in detail in this book at the gravity wave/tidal interface, and give greatest precedence to gravity waves. The interested reader is referred to *Andrews et al.* (1987), especially Chapter 4, as a useful starting point for further investigations of larger-scale types of wave events.

The lower branch of the Brewer–Dobson circulation

We now turn to the region below about 45 km altitude. In fact we will mainly concentrate on the region below about 25–30 km, since it is evident from Figure 1.20 that this has a very different circulation to that above 45 km altitude. This lower region is the lower branch of the BDC. The region in between 30 to 45 km can be considered as a transition between the upper and lower branches, although with some extra subtleties of its own.

The form of the lower branch differs from the upper branch. Both branches occur due to waves. As seen, the mesospheric circulation is dominated by gravity waves, while in the lower branch of the BDC it is especially planetary waves that matter the most.

This is not to say the relation is exclusive – all types of waves impact all areas of the atmosphere, but the stated ones are dominant in the regions indicated. However, gravity waves really only become dominant above about 70 km altitude, while the upper branch of the BDC is also still significantly affected by planetary waves.

The general importance of waves is crucial. While the Hadley cell is a thermally driven circulation and exists only in the tropics (in fact, it defines the tropics), meridional flows beyond the tropics and even above the Hadley cell must be largely wave driven.

In regard to the lower branch of the BDC, the primary initial flow tendency is for poleward flow out of the equator (a little analogous to the Hadley cell formation), but in this case the flow attempts to complete the circuit to the poles. However, as discussed already for the other levels, this motion quickly becomes a zonal flow, due to the Coriolis forces. This in turn produces (again) a north-south Coriolis pressure-gradient which cancels the initial tendency for poleward flow, resulting in a purely zonal flow to the east in both hemispheres at 15 to 25 km altitude (e.g. see Figures 1.21 and 1.22). However as before, this simple model is corrupted by the existence of wave forcing.

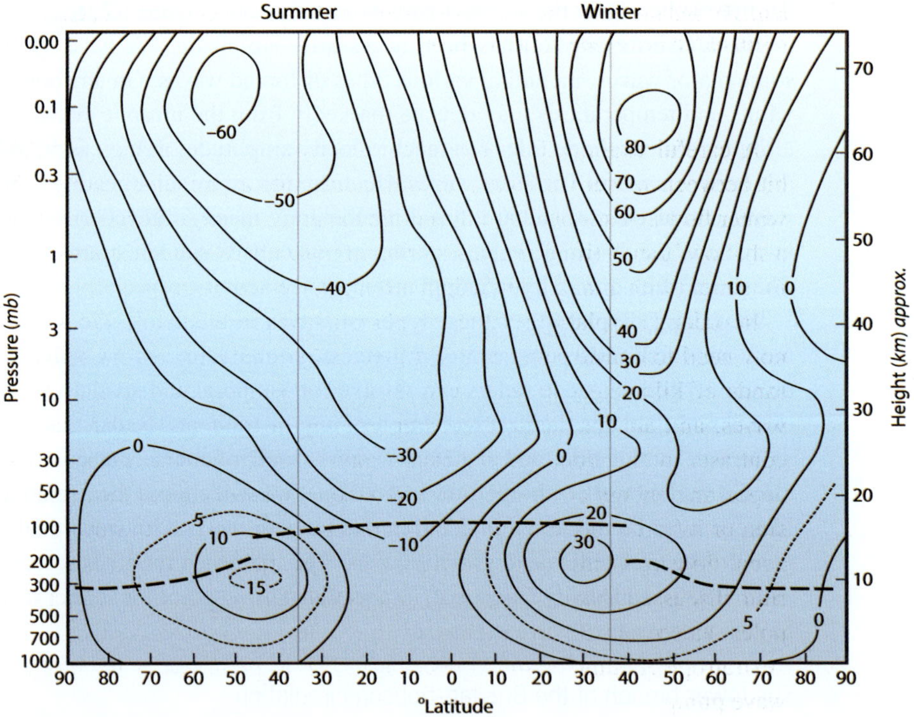

Figure 1.21 Contour plots of wind flows up to about 75 km altitude are shown for winter and summer. It has been assumed that symmetry exists between the hemispheres, and the situation simply flips between the two hemispheres every 6 months. (This is not entirely a valid assumption, and significant hemispheric differences do exist, but the assumption is acceptable for this simple overview.) Positive winds are eastward. The graph is adapted from *Murgatroyd* (1969). The broken lines show the height of the tropopause. A representative cross-section through 35° latitude for summer and winter is shown in Figure 1.22.

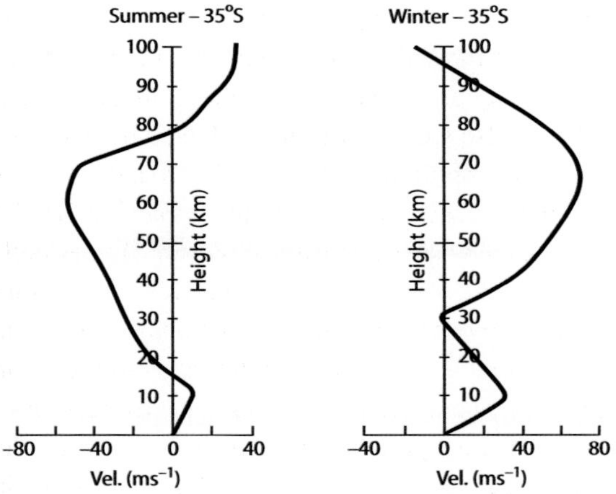

Figure 1.22 Representative wind profiles for a latitude of 35° (in this case 35° S), taken from Figure 1.21. Heights above 75 km have been supplemented from data presented by *Groves* (1969), Figures 1 and 2. Note that the wind direction reverses between summer and winter above 20–30 km altitude, and the flow is largely eastward (westerly) below 15–20 km in both seasons, which is a result of the Ferrel cell at this latitude.

A careful examination of Figure 1.20 shows that the lower branch differs quite a bit between winter and summer, with noticeable asymmetry between the summer and winter flows. A stronger and deeper flow exists in winter. These are often referred to as a shallow branch (in summer, with maximum altitude around 25 km), and a deep branch in winter (with a maximum height around 35 km or more).

In order to explain the final circulation structure which develops in this region, we now need to consider the role of waves. Large scale waves with wavelengths of thousands of kilometers propagate to these heights from below. They are called Rossby waves, and are forced by flow over large-scale topography and thermal (land-ocean) contrasts in the troposphere. These waves can be highly nonlinear, often breaking, depositing energy and momentum into the atmosphere. They also lose energy by emission of infrared radiation to space (radiative damping). They impart drag forces to the zonal flow, reducing the mean flow, and thus unbalance the north-south force equilibrium discussed above. The result is a residual north-south flow from the equator to both poles, as seen in Figure 1.20 at 15 to 25–30 km altitude. This is variously called an "extratropical pump" (*Holton et al.*, 1995) or, as preferred by *Plumb* (2002), a "Rossby wave pump."

This is somwhat oversimplified, however. It is true in an annual-mean sense, but there are clearly seasonal (summer/winter) differences, as discussed above. These arise due to different types of wave-forcing. The deep branch of the BDC is driven by planetary Rossby waves, so it can only exist in the winter hemisphere, since the waves cannot propagate into the summer easterly (westward) mean winds due to critical level absorption (see Chapter 11). The shallow, lower branch is driven by dissipation of tropical Rossby waves and synoptic-scale Rossby waves. These propagate to and dissipate in

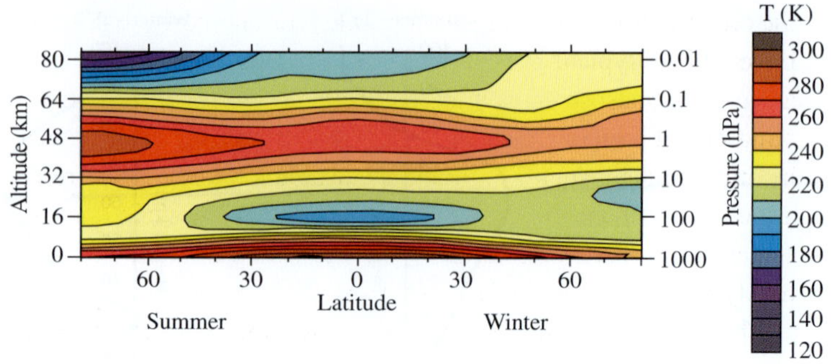

Figure 1.23 Temperature distribution as a function of height and time in the middle atmosphere.

the lowermost stratosphere (below 25 km) in both summer and winter, and so the lower shallow branch of the BDC exists year round (although its intensity varies seasonally). In winter, the shallow branch is enhanced to become a deep branch. For more about the details of this complex stratospheric interplay, the reader is referred to *Birner and Bönish* (2011); *Holton et al.* (1995); *Plumb* (2002).

Contour plots of the resultant wind fields in summer and winter are shown in Figure 1.21 (adapted from *Murgatroyd* (1969), and also presented by *Geller* (1979)), and a representative mid-latitude profile for 35° S is shown in Figure 1.22. The graphs were deliberately prepared using older data, to illustrate that even in the 1960s and 1970s, the mean circulation of the atmosphere was well documented. However, the reasons for the structures were not so well known, and it was in the 1980s and onward that much was learned about the causes of these particular circulation scenarios. Hemispheric differences also exist, which were not so well understood at the time and have not been presented in this figure. These differences especially arise due to the larger amounts of land mass in the northern hemisphere, giving rise to greater orographic forcing of the various waves.

The resulting temperature distribution is shown in Figure 1.23. Note the odd feature that the temperatures over the summer pole at 80 km altitude are in fact colder than those over the winter pole. In fact, the summer polar mesopause temperatures are so low that this is the coldest place anywhere in the Earth's atmosphere, with temperatures reaching as low as 120 K. This is also due to the gravity wave drag effects discussed in the previous paragraph, and will be explained more completely in Chapter 11. For now, we simply note that the reason is associated with gravity waves, and these are events that are ideally suited to studies by MST radar.

1.3.6 Synoptic and mesoscale flows

The dynamics and temperature/pressure structures discussed in the previous section occur on scales of thousands of kilometers, and on time scales of weeks and months. They are generally referred to as "global scale" motions. Smaller scale motions also exist. Phenomena that occur with spatial scales of hundreds of kilometers out to a few thousand kilometers or so are called "synoptic flows." They generally have lifetimes of the order of hours to days. Smaller scale motions, of the order of tens to hundreds of kilometers (sizes typical of the dimensions of a large city) and time scales of minutes to days are called "mesoscale motions." Smaller scale motions of the order of centimeters to hundreds of meters are called "microscale motions" (at least in a meteorological sense: in other studies, such as studies of turbulence, microscale refers to motions less than a meter and down to centimeters or so). Meteorological microscale motions have time scales of the order of minutes, seconds and less. These different regimes are summarized nicely in Table 10.1 of *Ahrens* (1999).

These synoptic, mesoscale, and microscale motions are well suited for studies by radar, and will arise frequently in this text. Examples include gravity waves (to be defined later), high and low pressure systems, frontal systems, hurricanes, storm-systems, lake and sea breezes, mountain and valley breezes, tornadoes, and turbulence, among others.

The origins of these motions are varied. One common source is the interaction of global-scale motions with local variability of features associated with the land and sea. For example, flow of the polar jet stream over the Rocky Mountains of North America often leads to the production of low-pressure cells and associated severe weather such as the Colorado Lows and the Alberta Clipper. Generation of large wind variation as a function of latitude can result in instabilities that lead to generation of wave events in the atmosphere, like Rossby waves. Differential heating over the land and sea leads to sea and land breezes. Water vapor can have a large impact on many flows as well. Water has a high latent heat capacity and when it transforms from liquid to gas can absorb large quantities of energy. The vapor may then travel large distances and re-condense as a liquid, releasing large quantities of heat locally. Such processes can be important in the generation of thunderstorms and hurricanes. Heat carried in this way is called "latent heat energy," while heat content carried as internal energy of air, which is proportional to its temperature, is called "sensible heat." Breakdown of synoptic and mesoscale flows can in turn lead to smaller scale waves and turbulence.

In general, there are many sources of these smaller scale motions, and radars are particularly useful for studying this type of activity. We will see a great deal about these types of dynamical events when we discuss application of radar later in this book.

1.4 Some important thermodynamics and statics

1.4.1 Introduction

In this section, we address several important characteristics of atmospheric stability and thermodynamics.

In particular, we show:

- that the atmospheric pressure decreases exponentially with a scale height H,
- that a parcel of air displaced adiabatically vertically cools as it rises with a rate called the "adiabatic lapse rate" Γ_a,

  and

- that a parcel of air in a stable atmosphere will oscillate with a frequency called the "Brunt–Väisälä frequency," denoted either as ω_B or sometimes N. We will mainly use ω_B. Sometimes (rarely) the Brunt–Väisälä frequency is also called the Väisälä–Brunt frequency, but the first usage is by far more common and we will persist with that.

We give explicit expressions for each of H, Γ_a and ω_B. We also define the "potential temperature," denoted by Θ.

1.4.2 Pressure as a function of height

Consider a slab of atmosphere, as shown in Figure 1.24. Then for a static parcel of cross-sectional area A, and mass Δm, Newton's second law states that the forces of gravity (down) and pressure (up) balance, so that

$$A\,dp = -\Delta m\, g. \tag{1.18}$$

Writing $\Delta m = \rho A \Delta z$, where ρ is the mass density, produces the differential equation

$$\frac{\mathrm{d}p}{\mathrm{d}z} = -\rho\, g. \tag{1.19}$$

Then the ideal gas equation, $pV = nRT$, where n is the number of moles of gas in volume V, is rewritten as $p = (nM/V)(R/M)T$, where M is the mass of one mole of gas, and where we recognize the term nM/V as the mass density ρ. Substituting for ρ in Equation (1.19), and now considering the terms like dp as differentials, gives

$$\frac{1}{p}\frac{dp}{dz} = -\frac{Mg}{RT} \tag{1.20}$$

or

$$\frac{\mathrm{d}(\ln p)}{\mathrm{d}z} = -\frac{Mg}{RT}. \tag{1.21}$$

Integrating gives

$$p = p_0 \mathrm{e}^{-\int \frac{Mg}{RT}\mathrm{d}z}, \tag{1.22}$$

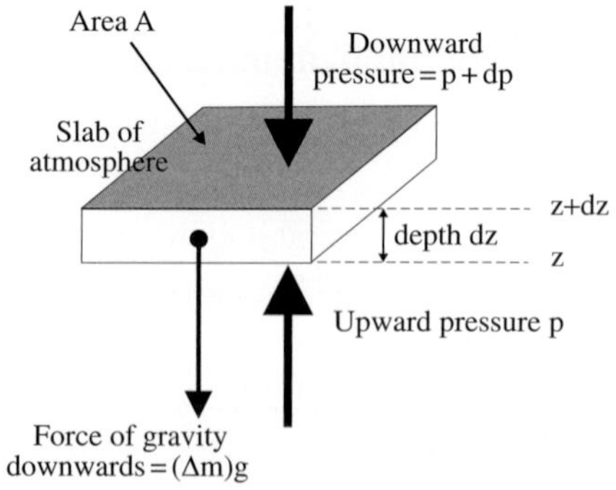

Figure 1.24 Pressure forces and gravitational forces on a slab of atmosphere.

where p_0 is the pressure at height $z = 0$, which shows that the pressure decays exponentially with height. This is often written as

$$p = p_0 \mathrm{e}^{-\int \frac{\mathrm{d}z}{H}}, \tag{1.23}$$

where $H = \frac{RT}{Mg}$ is called the scale height. This applies for an atmosphere comprising a single molecular species – if there are multiple species, R must be replaced by

$$R_* = \sum_{i=1}^{i=\nu} \frac{\rho_i}{\rho} \frac{R}{M_i}, \tag{1.24}$$

where the subscript "i" refers to the different species. Equation (1.24) is stated without proof, since in this chapter we simply wish to give an overview. A proof of this expression will be given in the later chapter on Meteorology (Chapter 12). Equation (1.23) is a quite general relation, and can be used if the temperature varies with height. For the special case of an isothermal atmosphere, H is constant and so

$$p = p_0 \mathrm{e}^{-\frac{z}{H}}, \tag{1.25}$$

and H is the vertical distance over which the pressure decreases by a factor of e. For an isothermal atmosphere, the density also decays exponentially with increasing height, with a scale height also equal to H.

1.4.3 Adiabatic expansion

If a volume of gas expands in such a way that no heat flows in or out of the region, the process is termed "adiabatic." This forces a particular relation between the pressure and volume of the gas. Using the first law of thermodynamics,

$$dQ = dU + p\,dV, \tag{1.26}$$

where dQ is the heat supplied to the parcel, which in this case is zero, and using the standard definition of the molar specific heat at constant volume, we write that for η_m moles,

$$0 = C_v\,\eta_m\,dT + p\,dV. \tag{1.27}$$

Recognizing that $T = pV/(\eta_m R)$, so that $dT = \frac{1}{\eta_m R}(Vdp + pdV)$ for fixed η_m gives

$$C_v\,\eta_m \frac{1}{\eta_m R}(p\,dV + V\,dp) = -p\,dV \tag{1.28}$$

or

$$\frac{C_v}{R} V\,dp = -\left(1 + \frac{C_v}{R}\right) p\,dV. \tag{1.29}$$

Mutiplying through by $\frac{1}{pV}$, and producing a common denominator for the bracketed term on the right-hand side, gives

$$\frac{C_v}{R}\frac{dp}{p} = -\left(\frac{R + C_v}{R}\right)\frac{dV}{V}. \tag{1.30}$$

Converting small elements like dp to differentials, and recognizing the term C_v+R as the molar specific heat at constant pressure, C_p, and that $\frac{dV}{V} = \mathrm{d}(\ln V)$ while $\frac{dp}{p} = \mathrm{d}(\ln p)$, and multiplying through by R gives

$$\mathrm{d}\ \ln p = -\frac{C_p}{C_v}\mathrm{d}\ \ln V. \tag{1.31}$$

Integrating both sides gives

$$\ln p = -\gamma \ln V + \kappa, \tag{1.32}$$

where $\gamma = \frac{C_p}{C_v}$ and κ is an arbitrary constant. Taking exponentials gives

$$p = \mathrm{e}^{\kappa} V^{-\gamma}. \tag{1.33}$$

Multiplying both sides by V^{γ} gives

$$pV^{\gamma} = \mathrm{e}^{\kappa}, \tag{1.34}$$

which tells us that pV^{γ} is a conserved quantity. This is an important property of adiabatic processes which will be used shortly.

1.4.4 Adiabatic lapse rate

If a parcel of air is displaced vertically in the surrounding air, it moves into a region in which the pressure of the surrounding air is less. The parcel therefore expands so that its internal pressure decreases until it equilibrates with that of the surrounding air. In the process of expanding, the parcel cools. The expansion is generally regarded to be an adiabatic process. Physically, the temperature decrease can be seen to be due to the fact that the walls of the parcel are expanding and therefore moving outward from the center, so that a molecule in the gas moving outwards, which reflects off this imaginary wall, will return with a reduced speed relative to its incident speed, thereby achieving a reduced kinetic energy and so a lowered temperature.

Mathematically, the degree of cooling can be determined from the the first law of thermodynamics, the adiabatic expansion expression (1.34), and the expression for pressure as a function of height (1.25).

The expansion process is shown in Figure 1.25.

To see this, consider a parcel of air, with cross-sectional area A, as shown in Figure 1.26, and consider that it is displaced vertically by a distance Δz. The parcel has an initial depth δz, and a final depth $\delta z + \delta\xi$. For simplicity we assume that the parcel expands only vertically; the only important thing is the overall volume change, so this is not too restrictive. For simplicity, consider the starting point as $z = 0$, with pressure p_0, and the final point to be at height Δz, with pressure p_1.

Then the adiabatic law (1.34) tells us that

$$p_0(A\delta z)^{\gamma} = p_1(A[\delta z + \delta\xi])^{\gamma} = p_1(A\delta z)^{\gamma}\left(1 + \frac{\delta\xi}{\delta z}\right)^{\gamma}, \tag{1.35}$$

where the last term arises from the second with δz taken outside of the brackets. But $p_1/p_0 = \mathrm{e}^{-\Delta z/H}$, and we may cancel the terms $(A\delta z)^{\gamma}$ on each side of the first and last expressions of (1.35), plus apply a binomial expansion to $(1 + \frac{\delta\xi}{\delta z})^{\gamma}$ to give

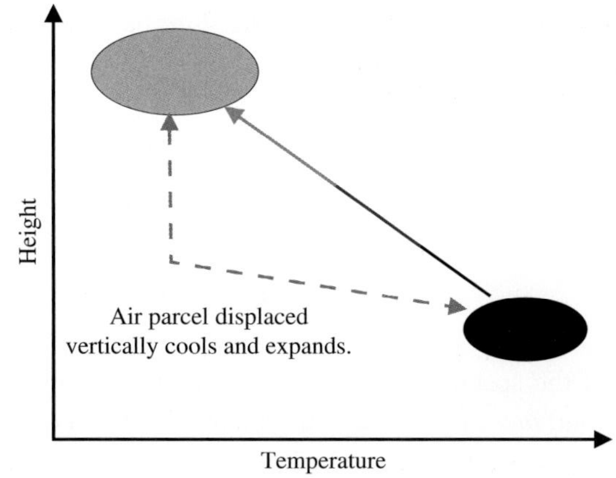

Figure 1.25 The effects of vertical movement on a parcel of air. The parcel is shown as a dark color initially, then moves vertically to the lighter grey parcel, where it has expanded and cooled.

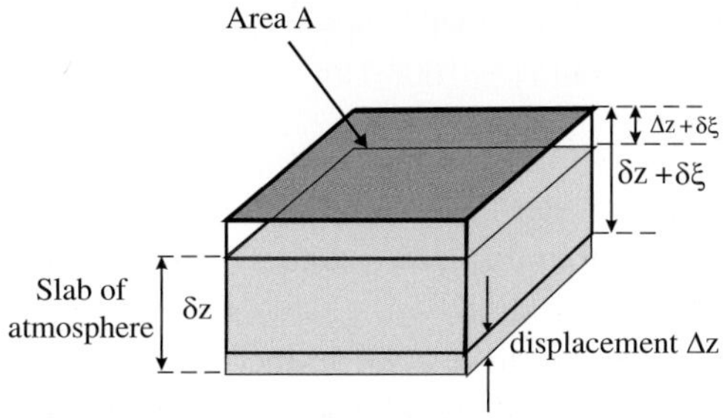

Figure 1.26 The effects of vertical movement on a rectangular slab of air, which is easiest to deal with mathematically. The parcel moves vertically and expands.

$$e^{\Delta z/H} = 1 + \gamma \frac{\delta \xi}{\delta z}. \tag{1.36}$$

Using a Taylor expansion $e^{\Delta z/H} = 1 + \frac{\Delta z}{H}$ leads to

$$\Delta z/H = \gamma \frac{\delta \xi}{\delta z}. \tag{1.37}$$

Since the change in depth of the parcel is $\delta \xi$, then the change in volume of the parcel is

$$\delta V = \delta \xi A = \frac{1}{\gamma} \frac{\Delta z \delta z}{H} A. \tag{1.38}$$

Assume that the temperature changes by ΔT. Now apply the first law of thermodynamics for an adiabatic expansion, viz. $C_v\, \eta_m\, \Delta T + p\, \delta V = 0$, and use $\eta_m = \frac{pV}{RT}$ to give

$$C_v \frac{pV}{RT} \Delta T = -p.\delta \xi . A = -p \frac{1}{\gamma} \frac{\Delta z \delta z}{H} A, \tag{1.39}$$

where we have replaced $\delta\xi.A$ with the right-hand-side of equation (1.38). Cancel the p terms, and write $H = \frac{RT}{Mg}$, to give

$$C_v \frac{A\delta z}{RT} \Delta T = -\frac{1}{\gamma} \Delta z A \delta z \frac{Mg}{RT}. \tag{1.40}$$

Cancelling obvious terms, and using $\gamma = \frac{C_p}{C_v}$ gives

$$\frac{\Delta T}{\Delta z} = -\frac{1}{C_v} \frac{C_v}{C_p} Mg. \tag{1.41}$$

Finally, if the specific heat per unit mole at constant pressure is C_p, then the specific heat per unit *mass* is $c_p = C_p/M$, where M is the molecular mass. (Note the change to a lower-case "c"). If we take the limit of small displacements, so that the left-hand side can be written as $\frac{\mathrm{d}T}{\mathrm{d}z}$, then

$$\frac{\mathrm{d}T}{\mathrm{d}z} = -\frac{g}{c_p}, \tag{1.42}$$

which gives us the adiabatic lapse rate, usually denoted Γ_a. The value for c_p in the atmosphere is about $c_p = 1006\,\mathrm{J\,kg^{-1}\,K^{-1}}$, so that

$$\Gamma_a = -\frac{g}{c_p} = 0.0098\,\mathrm{Km^{-1}} = 9.8\,\mathrm{K\,km^{-1}}. \tag{1.43}$$

Hence a parcel of air displaced adiabatically upward cools at a rate of 9.8 °C per km of ascent.

This has all assumed dry air. In the chapter on meteorology later in this book, we will examine the situation when the air contains water vapor, which leads to the so-called "moist (or wet) adiabatic lapse rate." This recognizes that water droplets in the air-volume may release or absorb latent heat as they condense or evaporate, and this can have profound effects on the amount of heat available to the air, thereby altering the adiabatic lapse rate compared to the dry-air case.

1.4.5 Brunt–Väisälä frequency

In Section (1.4.2), we assumed a static situation. However, it is possible that the gravity force and the pressure forces in that section might not balance, in which case the parcel will accelerate up or down. In this section, we determine the acceleration on such a parcel which has been displaced from its equilibrium, and consider its ensuing motion.

We therefore return to Figure 1.24, and once again examine the balance of forces. The upward force is again $[p - (p + dp)]A = -A\ dp$, and the downward force is again $mg = \rho_{\mathrm{parcel}}.A.dz.g$, so the net upward force is $-A\ dp - \rho_{\mathrm{parcel}}.A.dz.g$. However, we no longer assume that this is zero, but assume that the parcel is accelerating, so that Newton's second law applies, viz. $F = ma$ or

$$-A\ dp - \rho_{\mathrm{parcel}}.A.dz.g = m\frac{\mathrm{d}^2\zeta}{\mathrm{d}t^2}, \tag{1.44}$$

where ζ is the displacement of the parcel from equilibrium. In a stable region, we expect the parcel to oscillate, as first it rises into a region where it is more dense than its surrounds, then falls back through zero, and then becomes less dense than its surrounds, whereupon it is forced back up. We therefore expect a sinusoidal solution. We shall now verify this.

Recognizing that the mass is $m = \rho_{\text{parcel}}\Delta V = \rho_{\text{parcel}}A\ dz$, and making $\frac{dp}{dz}$ into a differential operator, we rewrite Equation (1.44) as

$$-A\ dz\left[\frac{\mathrm{d}p}{\mathrm{d}z} + \rho_{\text{parcel}}\cdot g\right] = (\rho_{\text{parcel}}A\ dz)\frac{\mathrm{d}^2\zeta}{\mathrm{d}t^2}. \tag{1.45}$$

The terms $A\ \ dz$ may be cancelled on either side. Recall that the hydrostatic equation gives $\frac{\mathrm{d}p}{\mathrm{d}z} = -\rho_{\text{air}}g$. We will use this, but change the subscript $_{\text{air}}$ to $_{\text{env}}$, where "env" means "environmental" – i.e. the value for the air surrounding the parcel, or its environment. So we write

$$\frac{\mathrm{d}p}{\mathrm{d}z} = -\rho_{\text{env}}g, \tag{1.46}$$

and produce

$$-\frac{g}{\rho_{\text{parcel}}}\left[\rho_{\text{parcel}} - \rho_{\text{env}}\right] = \frac{\mathrm{d}^2\zeta}{\mathrm{d}t^2}. \tag{1.47}$$

Now examine Figure 1.27, and in particular look at the upper graph. Let the point where the adiabatic lapse rate and the environmental lapse rate cross be $z = 0$. At this point, the parcel and the background air (environment) have the same density, $\rho(0)$. Then at any displacement ζ from this point, we may write

$$\begin{aligned}\rho_{\text{parcel}} &= \rho(0) + \frac{\mathrm{d}\rho_{\text{parcel}}}{\mathrm{d}z}\zeta \\ \rho_{\text{env}} &= \rho(0) + \frac{\mathrm{d}\rho_{\text{env}}}{\mathrm{d}z}\zeta,\end{aligned} \tag{1.48}$$

where the subscript "env" refers to the environmental (background) situation.

Now substitute these expressions into Equation (1.47) to give

$$\frac{\mathrm{d}^2\zeta}{\mathrm{d}t^2} = -\frac{g}{\rho_{\text{parcel}}}\left[\left(\frac{\mathrm{d}\rho}{\mathrm{d}z}\right)_{\text{parcel}} - \left(\frac{\mathrm{d}\rho}{\mathrm{d}z}\right)_{\text{env}}\right]\zeta. \tag{1.49}$$

This is an equation of the form

$$\frac{\mathrm{d}^2\zeta}{\mathrm{d}t^2} = -K\zeta, \tag{1.50}$$

where K is an *approximate* constant. If K is taken as an exact constant, this equation has a solution of the form $A\mathrm{e}^{\mathrm{i}\omega t} + B\mathrm{e}^{-\mathrm{i}\omega t}$, where $-\omega^2\zeta = -K\zeta$, or

$$\omega = \sqrt{K}. \tag{1.51}$$

Hence our resultant solution is an oscillatory (sinusoidal) motion with angular frequency

$$\omega_B = \sqrt{\frac{g}{\rho_{\text{parcel}}}\left[\left(\frac{\mathrm{d}\rho}{\mathrm{d}z}\right)_{\text{parcel}} - \left(\frac{\mathrm{d}\rho}{\mathrm{d}z}\right)_{\text{env}}\right]}. \tag{1.52}$$

In some texts this is denoted as N.

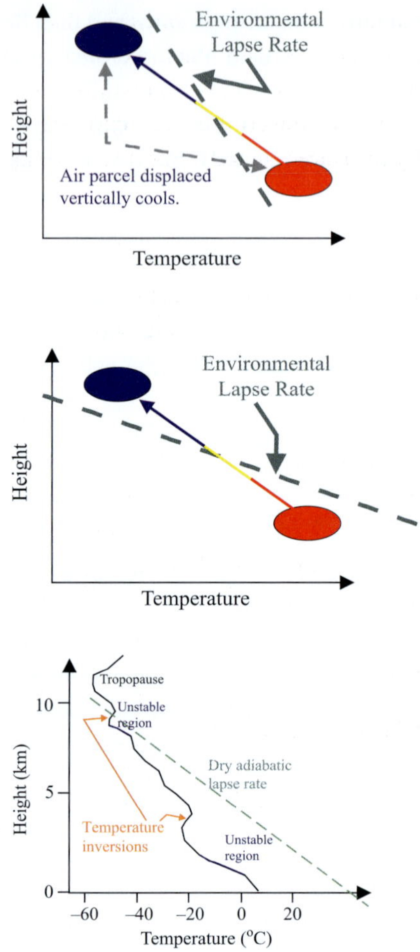

Figure 1.27 Environmental lapse rates (i.e. the rate of *decrease* of temperature with height for the background atmosphere), and the adiabatic lapse rate for a displaced particle. The upper figure shows a *stable* situation – a displaced parcel is cooler than its surrounds, and so denser. It therefore slows its vertical speed, stops, and returns toward equilbrium. It then overshoots the equilibrium, moves below the equilibrium, and then has a warmer temperature and lower density than its surrounds, so it is slowed and eventually forced back up. The result is an oscillation. In the second figure, a vertically displaced parcel is less dense than its surrounds, and continues to rise. This is an unstable situation. The upper figure is the most important for calculations of the Brunt–Väisälä frequency. The lowest figure shows a realistic atmosphere, with a mixture of stable and unstable layers.

It will be noted that the point was made that the term K is only approximately a constant. In truth, the density of the parcel is itself a function of vertical displacement, so K is not a true constant. Hence the derived frequency of oscillation has similar issues to those encountered in the derivation of the frequency of oscillation of a simple pendulum, where it is assumed that force on the pendulum is proportional to the displacement angle θ. In reality it is proportional to $\sin\theta$. As a result the true oscillation (both for our parcel and the pendulum) is not a true sinusoid, and the "frequency" is in fact amplitude-dependent. In the case of the parcel, we take the density to be the density when in its equilibrium position. This assumption will become important later.

Finally, we need to convert this to an expression involving temperature, since density is not normally measured directly. For this, we employ the ideal gas equation, written in the form

$$p = \rho K_{B*} T, \tag{1.53}$$

where K_{B*} is an equivalent Boltzmann constant for the case that the density is used in the ideal gas equation. Diffferentiating, we produce

$$\frac{\mathrm{d}p}{p} = \frac{\mathrm{d}\rho}{\rho} + \frac{\mathrm{d}T}{T}. \tag{1.54}$$

Although not completely valid (violation occurs at small scales in turbulence, and in sound waves), we can assume to first order that dp is zero (i.e. the pressure of the parcel adjusts almost instantaneously to that of the background – an assumption we have already made when we discussed the adiabatic expansion of the parcel). This is true for both the environmental changes with height and the changes for the parcel, so that

$$\frac{\mathrm{d}\rho}{\rho} = -\frac{\mathrm{d}T}{T} \tag{1.55}$$

in both cases. Hence (1.52) becomes

$$\omega_B = \sqrt{-\frac{g}{T_{\text{parcel}}}\left(\frac{\mathrm{d}T}{\mathrm{d}z}\right)_{\text{parcel}} - \frac{\rho_{\text{env}}}{\rho_{\text{parcel}}}\frac{g}{\rho_{\text{env}}}\left(\frac{\mathrm{d}\rho}{\mathrm{d}z}\right)_{\text{env}}}, \tag{1.56}$$

where we have employed (1.55) in the first portion, and multipied by $\frac{\rho_{\text{env}}}{\rho_{\text{env}}}$ in the second.

Now recall that following Equation (1.52), the validity of the assumption that ρ_{parcel} could be assumed to be a constant was discussed. It was concluded that this was not strictly a constant, but that we needed to assume it to be so in order to produce an analytical response to the differential Equation (1.49). The value chosen was the value at equilibrium. But at equilibrium, the densities of the parcel and the surrounding env are equal, so in Equation (1.56) we may, within the context of the approximations used to date, take $\frac{\rho_{\text{env}}}{\rho_{\text{parcel}}}$ as unity. (Of course the equilibrium point may not be at the height of interest, but for small oscillations it will not be too far removed from the current point and so is still a reasonably valid substitute.) Likewise the *temperature* of the parcel, when used as a constant, must be the value at equilibrium, so when T_{parcel} is used in a multiplicative or divisional way, we replace it with its mean value, and so may use T_{env}. We may then apply (1.55) to the term $\frac{g}{\rho_{\text{env}}}\left(\frac{\mathrm{d}\rho}{\mathrm{d}z}\right)_{\text{env}}$ to give

$$\omega_B = \sqrt{\frac{g}{T}\left[\left(\frac{\mathrm{d}T}{\mathrm{d}z}\right)_{\text{parcel}} - \left(\frac{\mathrm{d}T}{\mathrm{d}z}\right)_{\text{env}}\right]}, \tag{1.57}$$

with T in the term $\frac{g}{T}$ being the environmental temperature. This is a suitable expression given the approximations we have made in order to produce an analytical solution.

If we define $\Gamma_{\text{env}} = \Gamma_{\text{e}} = \left(-\frac{\mathrm{d}T}{\mathrm{d}z}\right)_{\text{env}}$ and $\Gamma_{\text{adiabatic}} = \Gamma_{\text{a}} = \left(-\frac{\mathrm{d}T}{\mathrm{d}z}\right)_{\text{parcel}} = \frac{g}{c_p}$ (Equation 1.42), then we may write

$$\omega_B \text{ (or } N) = \sqrt{\frac{g}{T}\left(\Gamma_{\text{a}} - \Gamma_{\text{e}}\right)}. \tag{1.58}$$

1.4.6 Potential temperature

For an adiabatic process,

$$pV^{\gamma} = \text{constant}, \tag{1.59}$$

where γ is the ratio of specific heats, $\gamma = \frac{c_p}{c_v}$ (see Equation (1.34)).

But if we assume the atmosphere is an ideal gas, then

$$pV = \eta_m RT, \tag{1.60}$$

where η_m is the number of moles in volume V, so we may write

$$(\eta_m RT)^{\gamma}\, p^{(1-\gamma)} = \text{constant}. \tag{1.61}$$

Hence

$$T^{\gamma} p^{1-\gamma} = \text{constant}, \tag{1.62}$$

or

$$\left[T^{\gamma} p^{1-\gamma}\right]^{\frac{1}{\gamma}} = \text{constant}^{\frac{1}{\gamma}}, \tag{1.63}$$

which is just another constant. Hence we may write

$$Tp^{\left(\frac{1}{1-\gamma}\right)} = \text{constant}. \tag{1.64}$$

If we consider a parcel of air of temperature T_0 at a pressure p_0 = 1000 hPa, and move it vertically to a pressure p, then

$$Tp^{\frac{1}{1-\gamma}} = T_0 p_0^{\frac{1}{1-\gamma}}. \tag{1.65}$$

Rearranging gives

$$T_0 = T\left(\frac{p}{p_0}\right)^{\frac{1-\gamma}{\gamma}} = T\left(\frac{p_0}{p}\right)^{\frac{\gamma-1}{\gamma}}. \tag{1.66}$$

Hence T_0 is the temperature that a parcel of air of at pressure p and temperature T *would have* if it were moved adiabatically to the ground (or more specifically, to a point where the pressure was p_0 = 1000 hPa.)

T_0 is called the *potential temperature*, and is usually denoted Θ. Surfaces of constant potential temperature are called *isentropes* since as a parcel moves along the surface, the potential temperature is unchanged, which means no heat is lost or gained, which means $dQ = 0$, so $\frac{\mathrm{d}Q}{T}$ is zero, or the entropy change is zero. Hence they are surfaces of constant entropy.

Therefore, for a parcel moving along the isentrope, the rate of change of potential temperature is zero, so that

$$\frac{D\Theta}{Dt} = 0, \tag{1.67}$$

where $\frac{D}{Dt}$ refers to differentiation following the motion.

This assumes no diffusion of the heat from the parcel. If diffusion is allowed,

$$\frac{D\Theta}{Dt} = \frac{\kappa}{\rho}\nabla^2\Theta, \tag{1.68}$$

where κ is a constant related to the thermal diffusion coefficient and ρ is the density.

Sometimes the Brunt–Väisälä frequency is also written in terms of the potential temperature, and (1.58) becomes

$$\omega_B = \sqrt{\frac{g}{\Theta}\frac{\partial\Theta}{\partial z}} = \sqrt{g\frac{\partial \ln\Theta}{\partial z}}. \tag{1.69}$$

1.4.7 Atmospheric stability and the Richardson number

Another parameter of considerable importance in atmospheric studies is the Richardson number, R_i. This is the ratio of "potential energy storage capability" to "available kinetic energy" in any region of the atmosphere. The atmosphere is unstable, and may break into turbulence if either: (i) the temperature gradient is unstable, producing a negative Richardson number; or (ii) the available kinetic energy exceeds the available potential energy storage capability. We will not derive the expression, although it is relatively straightforward. Mathematically, in one dimension,

$$R_i = \frac{\frac{g}{T}(\Gamma_a - \Gamma_e)}{\left(\frac{\partial\overline{u}}{\partial z}\right)^2}. \tag{1.70}$$

where T is the environmental temperature at the height in question and Γ_a and Γ_e are the adiabatic and environmental lapse rates defined following Equation (1.57).

For the case of the two-dimensional wind $\vec{\overline{u}}$ (assumed to be zero in the vertical) we can write this as

$$R_i = \frac{\frac{g}{T}(\Gamma_a - \Gamma_e)}{\mid \nabla_z\vec{\overline{u}} \mid^2}, \tag{1.71}$$

and also

$$R_i = \frac{\frac{g}{\Theta}\frac{\partial\Theta}{\partial z}}{\mid \nabla_z\vec{\overline{u}} \mid^2} = \frac{g\frac{\partial \ell n\Theta}{\partial z}}{\mid \nabla_z\vec{\overline{u}} \mid^2}. \tag{1.72}$$

The last statement follows because $\frac{1}{\Theta}\frac{\partial\Theta}{\partial z} = \frac{1}{T}(\Gamma_a - \Gamma_e)$. This can be seen as follows:

The potential temperature is defined as $\Theta = T\left(\frac{p}{p_0}\right)^{\frac{1-\gamma}{\gamma}}$. Differentiating with respect to z gives

$$\frac{d\Theta}{dz} = p^{\frac{1-\gamma}{\gamma}} p_0^{\frac{\gamma-1}{\gamma}} \frac{dT}{dz} + T p_0^{\frac{\gamma-1}{\gamma}} \frac{1-\gamma}{\gamma} p^{\frac{1-2\gamma}{\gamma}} \frac{dp}{dz}. \tag{1.73}$$

Now divide by Θ on the left-hand side, and by $T\left(\frac{p}{p_0}\right)^{\frac{1-\gamma}{\gamma}}$ (which is also Θ) on the right-hand side. This produces

$$\frac{1}{\Theta}\frac{d\Theta}{dz} = \frac{1}{T}\frac{dT}{dz} + \frac{1-\gamma}{\gamma}\frac{1}{p}\frac{dp}{dz}. \tag{1.74}$$

The first term on the right-hand side is simply $-\frac{1}{T}\Gamma_{\mathrm{e}}$.

To deal with the second term on the right, first recognize that $\frac{1-\gamma}{\gamma} = \frac{1-\frac{C_p}{C_v}}{\frac{C_p}{C_v}} = \frac{C_v - C_p}{C_p} = -\frac{R}{C_p}$. Also recognize that $\frac{1}{p}\frac{dp}{dz} = -\frac{g}{k_* T}$ from hydrostatic balance. Then (1.74) becomes

$$\frac{1}{\Theta}\frac{d\Theta}{dz} = -\frac{1}{T}\Gamma_{\mathrm{e}} + \frac{R}{C_p}\frac{g}{k_* T}. \tag{1.75}$$

But $\frac{R}{k_* C_p} = \frac{R}{(R/\mathcal{M}C_p)} = \frac{1}{c_p}$ where c_p is the specific heat at constant pressure per unit mass. Recognizing that $\frac{g}{c_p}$ is just Γ_a produces

$$\frac{1}{\Theta}\frac{d\Theta}{dz} = \frac{1}{T}(\Gamma_a - \Gamma_e). \tag{1.76}$$

This completes the proof. □

The parameter R_i that we have defined has the full title of the "gradient Richardson number," since it is based on wind-shear and temperature gradients. There is an alternative called the "flux Richardson number," often denoted as R_f, which relates to kinetic and potential energy fluxes. This is often used in modeling studies. We will not refer to this form much, and when we refer to the "Richardson number" we will generally mean R_i.

Conditions for instability are: (i) $R_i < 0$, which is convective instability; and (ii) $0 \leq R_i < 0.25$, which is dynamic instability. The first results in turbulence driven by convective motions, the second results in turbulence due to wind-shears overpowering the static stability. There seems to be some evidence that once turbulence starts, it can be maintained even up to Richardson numbers as high as 1.0.

This will be sufficient information for our current purposes. As noted, more extensive discussions about things like virtual temperature, moist adiabatic lapse rates, and atmospheric stability/instability will be considered in the chapter on meteorology.

2 The history of radar in atmospheric investigations

2.1 Introduction

The groundwork describing the atmospheric environment and the types of flows that radar can study in the Earth's atmosphere has been laid in the previous chapter. We now turn to a brief history of how radars came to be involved with studies of this type.

While most of this book is about MST radar, it is important that MST radar be seen in a broader context. We therefore begin this section on the history of the development of MST radar by looking not at MST radar itself, but rather at the development of meteorological radar. As indicated earlier, the period following World War II saw various developments of radar. Two primary streams were (i) ionospheric studies for world-wide communication, and (ii) studies of contaminants in radar detection for military and civil applications. The first stream of development led to extensive studies of the upper atmosphere and ionosphere, and the second led to more intensive investigations of the troposphere. Only with the development of MST radar did the two streams once again really merge.

Initially, there were two main aspects to radar detection – determination of range and, if possible, direction. Directional determination was achieved by using large antennas which concentrated the radar directionally, and range was generally found using time-of-flight delays.

The atmospheric radar principle for range-detection is basically fairly straightforward. A short pulse of an electromagnetic wave of typically several microseconds duration is transmitted from the radar antenna, whereupon it eventually may strike a target. It is then scattered back from the target to the radar antenna. The receive signal is called an echo, by analogy with the sound heard when your voice echoes from a distant object. Multiple radar echoes can be detected if there are multiple targets.

Echo samples are examined at consecutive time steps. Using early radars, this was done visually, whereas with more recent ones, digital sampling is used. Since the radar signal propagates with the speed of light c, the time t elapsed from the transmission of the pulse to the reception corresponds to a given range $r = ct/2$. We find, as an example, that echoes from backscattering targets at a range of, say 15 km, are received 100 microseconds after the pulse was transmitted. Echo samples are taken at a series of successive delays, called range gates.

A single such "snapshot" gives the range to the target and the strength of the backscattered power. But the target often changes with time, so to study this effect, the pulse is transmitted repeatedly. After the first pulse is transmitted, a second, then a third, is transmitted. This sequence is repeated every so-called interpulse period, which is (for VHF MST work at least) usually less than 1 millisecond, where 1ms corresponds to an unambiguous range of 150 km. The strength of the received signal fades up and down as the target changes.

In more sophisticated analyses, this time variation is recorded at a fixed range, and in modern radars it is stored as a series of complex data samples for each range gate. These data series are then transformed into so-called Doppler frequency spectra or covariance functions. These contain information about the scatter cross-section (i.e., a measure of the scattering area of an equivalent radar target), the Doppler frequency shift, and spread as a function of range or height. These processes were developed much later in the history of radar. These advanced techniques, which can be quite elaborate, are described in much more detail in Chapter 4.

We will begin by looking briefly at the meteorological developments and then discuss the ionospheric developments. Our discussion of the history of meteorological radar will be brief, since it has been very well covered by *Atlas* (1990). The reader is especially referred to the "Radar tree" on pages 95–96 in that book, depicting the history and evolution of radar meteorology. Developments pertaining exclusively to MST radars will represent a third subsection of this historical overview.

2.2 Meteorological radar

Early studies of lower atmosphere by radar seem to have been primarily for the purpose of improving radar detection of aircraft, ships, and man-made targets. It was found that spurious echoes often occurred on the radar screens, which generally seemed related to atmospheric phenomena. Storms could contaminate the signal, and hail seemed especially problematic. In addition, echoes often appeared that seemed to have no real physical source. These became known by various names such as "angels" and "pixies." Whilst these could still be due to "targets," the targets were not "hard" targets like aircraft, but were eventually shown to be variations of the atmospheric refractive index due to phenomena like turbulence and waves.

Tropospheric radiowave propagation also showed odd anomalies, with radiowave transmissions being ducted or refracted during their horizontal paths of hundreds of kilometers, resulting in signal fading and growing as a function of time, demonstrating the existence of large-scale refractive index variability (*Katz and Harney*, 1990). A great deal of debate ensued as to whether the observed radar echoes were due to small insects and/or birds, or whether they were truly atmospheric in origin (*Gossard and Yeh*, 1980; *James*, 1980). The truth turned out to be a little of both. One of the main consequences of these interfering echoes was the emergence of science out of radar studies. Investigations began into the sources of these echoes, and so the science of meteorological radar was born. Many studies of layering, ducting, clear-air echoes, turbulence breakdown, and so forth were carried out, improving understanding of the atmosphere (*Atlas*, 1990; *Probert-Jones*, 1990). *Gossard* (1990) especially shows some excellent examples of high-resolution studies of atmospheric scattering in the boundary layer by both insects and clear air, including some examples of Kelvin–Helmholtz instabilities, turbulent layers, angels, pixies, and wave events measured both by FMCW (particularly from Chadwick and Richter (e.g., *Chadwick and Gossard*, 1983; *Richter*, 1969)), and other types of meteorological radars. Examples from these references can be seen in Figures 2.1 and 2.2. Studies of turbulence in the clear air, termed "Clear Air Turbulence (CAT)," are of particular importance, since they are visually invisible and so a danger to aircraft. Such studies are one area where VHF/UHF radars can be valuable.

In addition, *Ryde and Ryde* (1945) and *Ryde* (1946) presented theoretical investigations of radiowave attenuation and scattering. In particular they addressed the difficult problem of scattering by rain and hail. It was generally considered that water should be a stronger scatterer than ice, since the dielectric constant (relative permittivity) of water is 81, whilst that of ice is only 1.3, but these authors were able to show, using Mie scatter (large particle) theory, that hail would produce very strong scatter if the particles were of the right size. Generally, small ice particles scatter

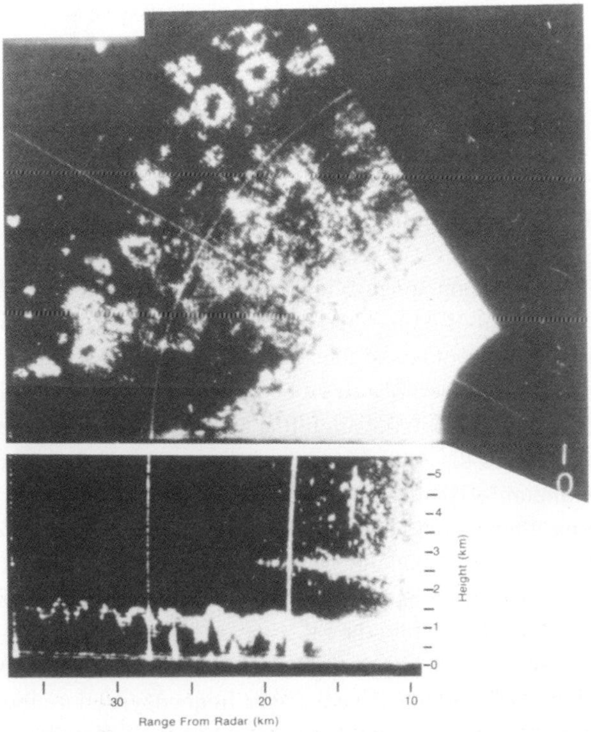

Figure 2.1 Some examples of early amplitude-scans using meteorological side-looking radars. Clear-air echoes are seen to the left. The upper graph shows an azimuthal sweep at fixed elevation, and the lower one shows a plot in x-z (horizontal-vertical) space. Capped convection is seen. From *Gossard* (1990). (Reprinted with permission from the American Meteorological Society.)

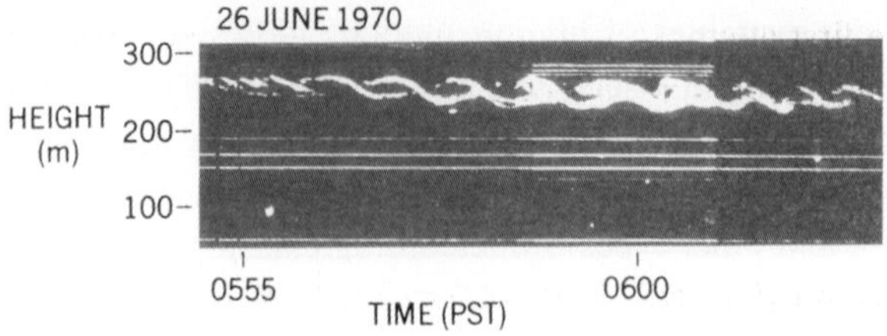

Figure 2.2 Example of the braided structure of Kelvin–Helmholtz instabilities, seen with a clear-air FMCW radar. From *Gossard* (1990). (Reprinted with permission from the American Meteorological Society.)

only one fifth as well as small water droplets, but spheres of ice become better scatterers than those made of water once the diameter exceeds 0.6 radar wavelengths (*Atlas*, 1990). The theoretical framework developed by the Rydes was an important aspect in allowing these radars to be used as quantitative scientific instruments, although it took over 15 years before the work was properly appreciated (*Atlas*, 1990).

All of these earlier studies used so-called "amplitude-only" radar. In other words, only the signal of the returned echoes was displayed. Nevertheless, some spectacular progress was made and some beautiful images of scattering processes were obtained (e.g., see *Gossard*, 1990). A major step forward ensued in the 1960s and 1970s – the gradual introduction of Doppler methods.

2.3 Doppler methods in radar meteorology

While recognizing, quantifying, and cataloguing atmospheric structures and developing statistics about different types of echoes were useful exercises, the ability to determine the dynamical motions of the scatterers, and the associated wind patterns, gave a whole new class of capabilities to radar techniques. Speed could of course be calculated by measuring the position of aircraft and tracking the position as a function of time, but this was limited to discrete objects, and did not work well when signal was received from a continuum of scattering objects in which no one scatterer stood out. However, any given scatterer should produce a scattered signal with a different frequency to the signal that arrives at the scatterer from the radar. For example, standard Doppler theory tells us that if a scatterer is moving away from the radar, and the component of the velocity along a line drawn from the radar to the scatterer equals v_{rad} (where the subscript refers to the "radial" velocity component), then the scattered (reflected radiowave, moving target) radiowave will have a lower frequency by an amount $\Delta f = (2/\lambda)v_{\mathrm{rad}}$. As an example, if the radar wavelength is 3 cm (radar frequency = 10 GHz), and the radial velocity is 15 ms^{-1}, then the change in frequency will be 1 kHz. If the frequency of the signal returned to the radar can be measured relative to the transmitted signal, the radial velocity of the scatterer can be determined.

The first attempts to do this were by using continuous-wave systems (no pulses and no range-resolution) and beating the received signal with the transmitted signal. To see how this works, consider adding two signals of the type $a_1 cos(\omega_1 t)$ and $a_2 cos(\omega_2 t)$. In the simplest case that $a_1 = a_2$, the resultant is

$$y(t) = a_1(\cos\omega_1 t + \cos\omega_2 t) = 2a_1\left[\cos\frac{(\omega_1+\omega_2)}{2}t\right] \times \cos\left[\frac{(\omega_1-\omega_2)}{2}t\right], \quad (2.1)$$

which is a rapidly oscillating sinusoidal oscillation of frequency $\frac{\omega_1+\omega_2}{2}$ with a more slowly varying amplitude of frequency $(\frac{(\omega_1-\omega_2)}{2})$. (This may readily be proved by using the substitutions $\omega_1 = \frac{\omega_1+\omega_2}{2} + \frac{\omega_1-\omega_2}{2}$ and $\omega_2 = \frac{\omega_1+\omega_2}{2} - \frac{\omega_1-\omega_2}{2}$ into the original summation.) In reality, the two initial sinusoidal oscillations may have different phases, but this will simply shift the position of the nulls, and not change the beat frequency.

If one were to watch this signal on a cathode ray oscilloscope, a rapid oscillation would be seen which varies in amplitude with time. For example, if the frequency difference were 2 Hz, the amplitude would pass through a null every 0.5 seconds (twice during each period $2/(\omega_1 - \omega_2)$). The closer the frequencies ω_1 and ω_2 are to each other, the longer the beat period.

Of course it is not entirely practical to measure beat oscillations like this visually, so better methods had to be developed. The first methods were all analogue, and a typical example is shown in Figure 2.3. The key ingredients of this circuit are: (i) the mixer; (ii) the low-pass filter; and (iii) the VFO (variable frequency oscillator). The mixer is represented by the circle with the cross in the middle, and the filter is represented with the unit called "LPF" (for low-pass filter). The circuit works as follows.

The RF (radio frequency) signal is fed jointly with a reference sinusoidal frequency into a mixer. A mixer is a non-linear device. If a signal $\zeta(t)$ is fed into the mixer, it produces an output of the form $\xi(t) = a_1\zeta(t) + a_2\zeta^2(t) + a_3\zeta^3(t)$ plus additional terms involving powers of 4, 5, 6, etc. Normally the coefficients a_1 and a_2 are large enough that the first two terms are the dominant ones. Consider what happens if the input signal is $\zeta(t) = b_1 cos(\omega_1 t) + b_2 cos(\omega_2 t)$. (We could allow these terms to have non-zero phases, but we choose not to for simplicity. The essential features remain as we will describe.) In our case, ω_1 might be the RF signal, and ω_2 might be the signal from the VFO. Then the output is

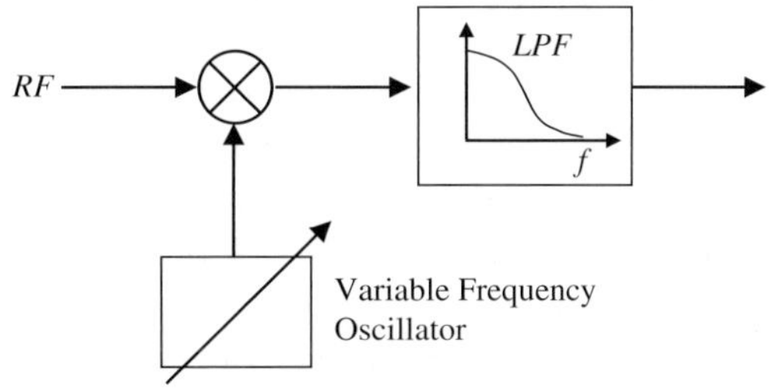

Figure 2.3 Circuit for a simple spectrum analyzer.

which has no range resolution. For some time it was believed that if Doppler measurements were required, then CW radar was necessary, and that pulsed radars (capable of range resolution) could not be used for Doppler measurements. The unification of the two (pulsed plus Doppler) was a major achievement.

Our focus here will be on the history of meteorological applications. The earliest attempt to measure the Doppler shift of the signal returned for meteorological applications appears to have been by *Barratt and Browne* (1953), who implemented a primitive (but effective) version of the spectrum analyzer discussed above. They used a CW radar and reflected a small portion from a nearby chimney, while directing most of the radiation upward into the air. They were in essence able to beat the signal reflected from the chimney with the signal returned from the air, with the signal from the chimney representing the reference. By determining the beat frequency, they were able to deduce the existence of downdrafts of air with speeds of 2 ms^{-1}. Although they attempted to publish their results, initial publications were rejected, and they were only able to publish a brief note in *Barratt and Browne* (1953). This history has been reported briefly by *Probert-Jones* (1990) and *Rogers* (1990). It is worth noting, however, that some of the first long-term applications of Doppler and phase measurements in atmospheric sciences belong not to any area of meteorology at all, but rather to upper atmospheric studies of meteors (*Manning et al.*, 1950; *Robertson et al.*, 1953). These authors used separate transmitter and receiver antennas (multi-static system) and beat the signals received from meteor trails with the ground-wave from the transmitter (in much the same way as Barrat and Browne did (discussed above)), to determine the phase and Doppler offset produced by the meteor drifts. This allowed them to determine both the meteor trail location and its radial velocity of motion, thereby allowing upper-level winds to be determined at 80–100 km altitude. Meteor radars will be discussed later in this chapter.

Returning to our discussion of meteorological systems, we recognize of course that a CW system has no range-resolution, and without the ability to determine range, the value of the radar is largely lost. The first reported proposal to use a pulsed Doppler system for meteorological applications appears to have been the report by Barratt and Browne (*Barratt and Browne*, 1953; *Probert-Jones*, 1990) followed closely by *Lhermitte and Atlas* (1961). One of the first measurements of winds by pulse-Doppler methods is shown in *Metcalf and Glover* (1990), Figure 7, as taken from *Lhermitte and Atlas* (1961). The measurements were made on May 27, 1961. Doppler shifts were determined and displayed by analogue techniques, using special display modifications on a modified standard radar screen (*Metcalf and Glover*, 1990).

Further discussions of the early development of Doppler radar have been given by *Rogers* (1990), and we will not elaborate on these developments here.

While these developments were no doubt important, the real power of Doppler processing of radar signals was revealed when it could be implemented on computers.

Independently, *Rummler* (1968) and *Woodman and Hagfors* (1969) developed methods for determination of Doppler shifts using computer techniques. Rummler was involved with weather radar, while Woodman and Hagfors were studying ionospheric phenomena. Woodman and Hagfors appear to have been the first to actually use the method. The procedure involved application of correlative techniques, and multi-pulse

(pulse-pair) methods, the technicalities of which will be discussed in later chapters of this book. The key importance of the method, however, lay in the fact that Doppler shifts could be determined using computer-digitized data, and that this could be done with sufficient speed that near-real-time estimates of radial scatterer velocities could be determined. This in turn allowed such determinations to become a routine aspect of radar measurements, rather than requiring special extra effort, and so revolutionized the world of atmospheric radar. An excellent summary of the history of such developments can be found in *Keeler and Passarelli* (1990). Further developments included application of spectral processing and the Fast Fourier Transform (FFT) (e.g., *Sirmans and Bumgarner*, 1975; *Gage and Green*, 1978; *James*, 1980) in place of autocorrelative techniques, but there can be no doubt that the general availability of computers in the early 1970s had a major impact on atmospheric radar in all fields. Detailed discussion of spectral and autocorrelative techniques will be given in Chapter 4.

Another important capability introduced in association with Doppler techniques was the ability to estimate atmospheric turbulence strengths. *Hitschfeld and Dennis* (1956) showed how the spectral width was related to the root-mean-square motion of the scatterers, and thence to the strength of atmospheric turbulence, but implementation was not really possible until Doppler methods were properly developed. *Frisch and Clifford* (1974) and *Labitt* (1979) developed the relationship between spectral width and turbulence strength more fully, but did not fully incorporate the limitations of the buoyancy (or outer) scale of turbulence. None of these papers considered the role of the broadening of the spectrum due to the mean wind motion tangential to the radar beam. *Atlas* (1964), *Sloss and Atlas* (1968) and *Atlas et al.* (1969), also discussed other contributors to the spectral width, which included mean motion of the scatterers across the beam (irrespective of turbulence). However, despite these various developments, they were not fully unified for routine turbulence estimates until the 1980s and later. This unification will be discussed in regard to MF D-region studies shortly.

From this point forth, Doppler methods were incorporated into meteorological radars almost as a standard, including (eventually) in the large networks of precipitation radars. We will not discuss the history of meteorological radars further, since it is at this point in time that the meteorological and ionospheric aspects of radar research began to meld, leading in part to the MST technique. We therefore now turn to the history of ionospheric radar, again leading up to the early and mid-1970s.

2.4 Ionospheric history pertaining to MST radar

As early as 1920, the potential for radiowave communication and radio detection methods was recognized. In the United Kingdom, the Department of Scientific and Industrial Research formed the Radio Research Board, which included some significant members such as Lord Rutherford. Its purpose was to encourage research of a fundamental nature in the area of radio, both for civilian and military applications. This led to active research in this area on many fronts, including, as mentioned earlier, Appleton and Barnett's measurement of the height of the ionosphere in 1925. They detected the so-called E-region, and the F-regions were discovered soon after.

The importance of monitoring these upper layers was quickly recognized, since they afforded a means to allow world-wide, near-instantaneous communication. A frequency-swept radar was developed to make these routine measurements, with initial measurements being made at Slough in the UK around 1930. On 11 January, 1931, a 90-day program of measurements was begun. Initially the technique used was based on the swept-frequency principle first used by Appleton and Barnett. Later, it was replaced with a pulsed method, initially developed by *Breit and Tuve* (1926).

The instrument, called an ionosonde, became the main workhorse for ionospheric studies, and networks were developed all over the world. Even today, networks of ionosondes are major tools for ionospheric research.

The pulsed ionosonde works on the principle of sending successive pulses of radiowaves up into the atmosphere, with each pulse being transmitted at a different frequency to the last, until the range from typically 2 to 20 MHz is covered. Each time a pulse is transmitted, it rises up through the atmosphere, and encounters a reflecting layer in the ionosphere. Lower frequencies are reflected from the E-region. Higher frequencies can pass through the E-region, where they then encounter the F-region, and reflect off that. The reflected pulse is received at the ground and recorded, with the time delay between transmission and reception being recorded.

Figure 2.6 shows the time-frequency structure that might be expected for such a situation. Radiowaves may reflect from a plasma, or pass through it, depending on the electron density. When the electron density exceeds some critical value, which depends on the radio frequency, the plasma represents a barrier to propagation. At lower electron densities, the radiowave is permitted to pass.

The left-hand diagram in Figure 2.6 shows a representative profile of electron density as a function of height, with ledges seen at approximately 100, 200, and 300 km. These ledges represent the E, F1, and F2 layers. When a radiowave of frequency 1.5 MHz propagates up into the atmosphere, it encounters a region just below 100 km altitude,

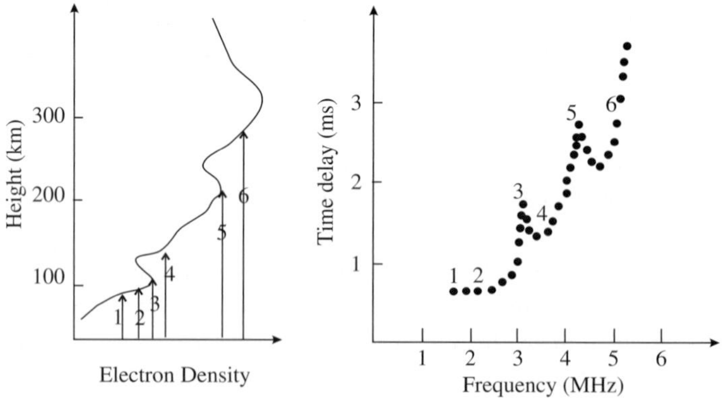

Figure 2.6 Representative profile of electron density in the ionosphere, and a sample ionogram.

where it reaches its critical plasma density. It is therefore reflected. This is shown by the vertical arrow labelled 1 in the left hand figure. At a frequency of say 2 MHz, a similar result happens, but the reflecting region is just marginally higher than for case 1 – this is shown as arrow 2 in the figure. The time delay for pulses of these frequencies is typically about 670 μsecs, so these register as small dots at a delay of 670 μsecs in the right-hand figure (the ionogram). In each case, the position of the arrow is chosen to lie immediately under the region where the associated frequency encounters its critical plasma density. However, at a frequency of just above 3 MHz (in this case), the critical plasma density happens to be exactly at the peak of the E-region. It cannot quite penetrate, and is reflected, but because it is close to the critical frequency, the speed of the pulse is substantially reduced. Hence the pulse takes a relatively long time to slow, turn around, and reflect back to the ground. This happens whenever the electron density changes only slowly with height. Hence the pulse arrives back at the ground with a delay of almost 2 ms. This effect is called group retardation. The situation is shown by the arrow labelled 3 in the left-hand figure. At a slightly higher frequency again, the critical electron density exceeds that in the E-region, so the radiowave passes through effortlessly, and finds its critical plasma density much higher in the ionosphere. This case is illustrated by the arrow number 4. Although the pulse actually travels further than pulse number 3, it does not suffer any substantial group retardation and so returns to the ground more quickly than pulse number 3. The description continues in a similar manner for higher frequency pulses, with number 5 encountering a critical level at the F1 layer, and number 6 passing through it.

The enhanced delays associated with group retardation in the electron density layers, where the electron density changes only slowly with increasing height, gives rise to the odd structure of the ionogram. From day to day and hour to hour, this structure continually changes, with layer heights and critical frequencies associated with the layers varying. These variations may be interpreted to infer information about the structure of the upper ionosphere. This diagram is somewhat oversimplified, and does not consider many aspects such as polarization of the radiation, but it is sufficient for our purposes.

For our applications, the structure of the ionogram is of less interest. Ionosondes were important because they were the basic tool used for probing the ionosphere, and it was through their use that the D-region of the ionosphere was discovered, thereby giving rise to the next important part of our history of MST radars.

When recording ionograms, it was occasionally noticed that extra, weak layers of reflected signal appeared to come from heights of 70 to 95 km, below the E-layer. These reflections were not due to encounters with critical electron densities, but simply due to reflections from sharp gradients in the electron density profile below the E-region. This region is the D-region, and although its electron densities are much lower than the E-region, it is capable of producing reflections. The difference between reflection from the E-region and from the D-region can be compared to reflections produced from a mirror (total reflection) and reflection produced by a piece of transparent glass (in which much of the light passes through, but a small portion is partially reflected). Reflections from the D-region were therefore called *partial reflections*.

2.5 D-region studies with MF and HF radar

In the following paragraphs, the early developments in studying the D-region – a major part of the MST region – will be briefly outlined. For a more extensive review, see *Hocking* (1981), pages 36–56.

Ellyett and Watts (1959) provided a review of the early discoveries of these partial reflecting layers and the weakly ionized regions in which they existed. As early as 1930, *Appleton* (1930) had found indications of their associated radio reflections. Early hints of such scattering heights also occurred in the period 1930 to 1950, when experiments using VLF, LF, and MF radiowaves suggested such structures.

Dieminger (1952) presented evidence of these echoes using conventional ionosondes, and appears to have been one of the earliest workers to have studied the echoes in any real degree of detail. Dieminger found that the echoes could be seen over heights of 75 to 90 km, and over a frequency range 1.6 to 4.0 MHz (this upper limit is most likely a result of the sensitivity of the radar system). The heights of the echoes varied diurnally, with a minimum at local noon; and the heights were frequency independent. The low echoes were most frequent in winter, and occurred in groups of days. Other studies were provided by *Gnanalingam and Weekes* (1952).

It was *Gardner and Pawsey* (1953), however, who provided perhaps the first detailed studies of these echoes, with an experiment specifically designed for investigation of these weak reflections. By using a single frequency, rather than sweeping across a range of frequencies, they were able to build antennas specifically tuned to their frequency, resulting in a system 30 to 40 dB more sensitive than Dieminger's ionosondes. Gardner and Pawsey used a frequency of 2.28 MHz, and a pulse length of 30 μsecs, giving a resolution of about 4.5 km. Half wavelength dipoles were used for transmission and reception. Different polarizations of radiation could be transmitted – either plane polarized radiation, or circularly polarized [Ordinary (O) or Extraordinary (X)] radiation. The transmitter had a peak power around 1 kW.

The echo strengths observed by Gardner and Pawsey (near Sydney, Australia) corresponded to reflection coefficients of typically 10^{-5} at heights near 70 km and 10^{-3} around 90 km. Echoes at 80–100 km generally appeared to come down in height until noon, then rise again afterwards. The "60–70 km" echoes appeared to rise in height until noon, then decrease again. An echo might appear at one height and remain visible (during the day-time) for periods ranging from hours to days. Then it would disappear. Echoes continually came and went. The echoes faded in amplitude with fading times of the order of seconds. However, an echo at any one height would still fluctuate in height around its mean. Echoes below about 80 km appeared only during the daylight hours, but above this height, night-time echoes did occur. These results discussed above provide an accurate assessment of many of the major characteristics of these D-region echoes.

Gardner and Pawsey also developed a special new experiment. By transmitting radiation with opposite senses of circular polarization, (known as O- and X-modes), and comparing the backscattered strengths as a function of altitude, it was possible to deduce electron density profiles in the D-region. Below about 70 km, the X-mode echo is often the stronger, but above this height, the X-mode radiation is strongly absorbed due to the large electron densities, and the corresponding echoes are quite weak. O mode absorption is not so strong, so the O echoes are still quite strong even when coming from heights around 90 km. Gardner and Pawsey used these different absorption characteristics to determine the D-region ionospheric electron densities, and thus devised the so-called "differential absorption experiment" (DAE), which was a major technique used in D-region studies for many years.

The work of Gardner and Pawsey led to something of a new field of research, allowing the ionospheric D-region to be studied in new detail.

Gardner and Pawsey did not continue the work reported in their paper. *Gregory* (1956, 1961), however, did carry on similar observations at Christchurch, New Zealand, using a similar experimental arrangement to that of Gardner and Pawsey, but on a frequency of 1.75 MHz. Gregory's work provided the first detailed analysis of seasonal variations in D-region echo occurrence, and he also began to look carefully at the problem of preferred heights.

Studies in this area continued into the 1980s. Several key areas were investigated. First, there seemed a preference for the layers to appear at certain key heights, at least at low and middle magnetic latitudes. Scatterers seemed most prevalent at heights of around 66, 70–74, 76–80 (at times), around 85 and at 90–95 km, although the preferred heights do vary somewhat seasonally. Key investigations in this area were undertaken by *Belrose and Burke* (1964); *Gregory and Vincent* (1970); *Fraser and Vincent* (1970); *Austin and Manson* (1969); *Lindner* (1972), and *Hocking* (1981), among others, and have been summarized by *Fraser* (1984). A related key investigation was determination of the depths of these scattering layers. Results suggested layer depths varying from 1 to 10 km, with thicker layers occurring at higher altitudes. It was not clear whether the layers were continuous within such depths, or comprised sub-layers which could not be resolved by the pulses used (which were typically 3–4 km wide). *Gregory and Vincent* (1970) were involved in such studies, and *Austin et al.* (1969), followed by *Chandra and Vincent* (1979), tried deconvolution procedures to determine the sub-pulse layer distribution. The general feeling was that the thicker layers often did comprise discrete sub-layers, especially at heights below 80 km.

The DAE, introduced by Gardner and Pawsey, received considerable attention. It was especially important because absorption of radiowaves is especially prevalent in the region 60 to 90 km altitude. This was because the region had much higher electron collision frequencies than higher up, leading to greater damping of the waves. Absorption was especially a problem during high sun-spot numbers, when electron densities generally increased, and knowledge of the absorption was important with regard to world-wide communication via radiowaves. Publications by *Titheridge* (1962); *Belrose and Burke* (1964); *Gregory* (1965) and *Gregory and Manson* (1967) are representative papers of the types of experiments performed with the DAE method, with *Belrose* (1970) giving a useful review. *Manson and Meek* (1984) give an overview of important results, and a discussion of advantages and pitfalls, of the DAE method.

Another key experiment, introduced in the mid-1960s, was the measurement of drifts in the D-region (e.g., *Fraser*, 1965, 1968). This technique allowed the motions of the

reliable (though there are times, when the E-Region echo is weak, that spaced antenna data can be considered reliable even up to 95 km and higher).

The previous paragraphs discussed the usefulness of the D-region scatter for measurement of wind speeds. However, in the 1960s and 1970s, the need to understand the physical nature of the scatterers, purely from a perspective of fundamental principles, was also considered of paramount importance. This was for several reasons. One was to allow better understanding of the spaced antenna drifts method. The second related to the differential absorption experiment. Despite its apparent usefulness, it too had problems with accuracy. The principle of the method relied on a one-dimensional theory, so effectively assumed that the scatterers were mirror-like reflectors. If the scatterers were three-dimensional, such as turbulent eddies, it may change the interpretation. Scatter from oblique off-vertical angles would occur, and it could not be assumed that all scatter was from directly overhead. Different results might be expected for radars with wide and narrow beams.

For the various reasons listed above, studies of the nature of the scatterers became an area of great importance. Were the scatterers due to turbulence, and so quasi-isotropic? If so, the DAE experiment would have to be re-examined. Or were the scatterers actually mirror-like reflectors, as required for DAE theory? If so, that fact may call into question some of the assumptions about the spaced antenna drifts method. What caused the scatterers? Was it related to gravity waves, and could this affect what was being measured?

The vertical distribution of the layers was one important parameter, and that has already been discussed. Layer thickness was also important, and we have also briefly considered that. Isotropy of the scatterers was a third important issue.

Multiple methods have been used to determine the degree of isotropy/anisotropy of the scatterers. The most common have revolved around determination of a parameter called θ_s, which parameterizes the ratio of the width to the depth of the scatterers. Essentially, if θ_s is small (less than say 3–5°), the scatterers are very wide compared to their depth, and, in the limit of the smallest values, can be considered to be stratified, flat reflectors. Large values of θ_s can be considered to be associated with more isotropic scatterers, with width to depth ratios of 2:1 or less. This class of scatterer is usually (but not always) considered to be due to turbulence. Physically, the most direct way to measure θ_s is to compare powers measured when the scatterers are observed with an off-vertical beam to those observed with a vertical beam – the powers recorded on the oblique beam are much weaker if θ_s is small. The initial theory associated with these concepts was presented by *Briggs and Vincent* (1973) and *Vincent* (1973), and was later extended by *Hocking* (1987a) to relate the value of θ_s to the length-to-depth ratio of the scatterers. *Doviak and Zrnić* (1984) did similar studies, but instead of using θ_s, they looked directly at the horizontal correlation scales of the scatterers.

While the most direct method to determine θ_s is to point the radar beam to different regions of the sky and compare powers, this requires radar beam-steering, which was not available on many MF radars prior to 1970, and only a few, such as the large Buckland Park array in Australia (*Briggs et al.*, 1969) were large enough to make beam-steering useful. Therefore, initial measurements of θ_s were made using indirect methods such as studies of the spatial autocorrelation function using spaced receivers (often produced as a by-product of the spaced antenna drifts analysis discussed earlier – in that case, the spatial correlation is referred to as the "pattern scale"). First results were presented by *Lindner* (1975a, b), and suggested that θ_s was small below about 80 km altitude, and larger above. Direct measurements using beam-pointing procedures were provided by *Hocking* (1979), and provided confirmation of Lindner's results. An adaptation of the main figure from *Hocking* (1979) is shown in Figure 2.7.

More extensive discussions of the procedures used and the results produced in studies of θ_s will be provided in Chapter 7. The main results world-wide are summarized in Figure 2.8, from *Reid* (1990), showing generally small θ_s at the lower heights with a rapid transition to larger values at heights above 80 km. Similar studies were also performed at VHF frequencies – these will be discussed shortly.

Interpretation of these results is not straightforward, as will be discussed in Chapter 7. A simplistic summary is to conclude that large values of θ_s can be associated with turbulence, and small values can be associated with flat, mirror-like reflectors (so-called "specular reflectors"), which correspond to horizontally stratified steps in electron density with vertical extents less than about 40 meters, but horizontally flat over distances of at least a kilometer. Interestingly, many of these same techniques were later adopted by researchers using VHF-MST radars, and they were able to show that the steps are often in fact less than 1 or 2 meters in vertical size – a profound result, with significant implications for the small-scale dynamics and thermodynamics of the atmosphere. VHF aspects will be discussed shortly. The reasons for the existence of these sharp steps in electron density are still hotly debated today, and will be further considered in Chapter 7.

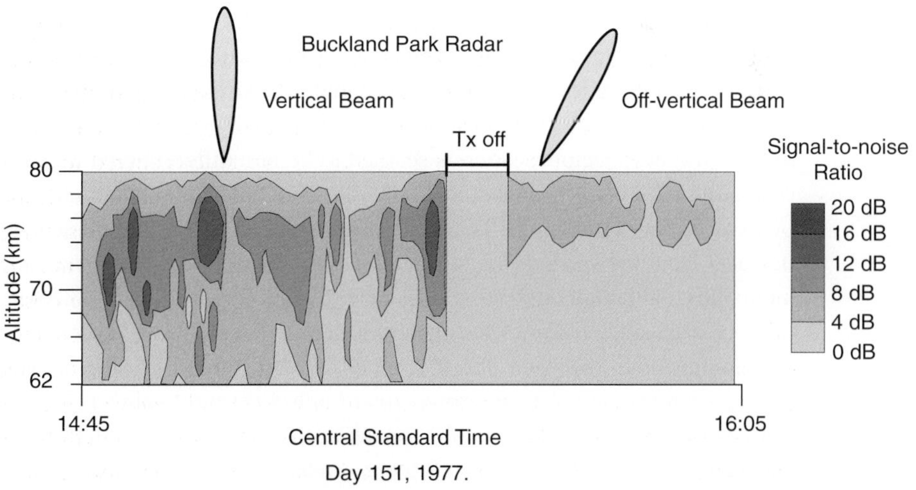

Figure 2.7 This graph shows mean powers as a function of height and time measured with a 2 MHz radar. The beam used was moderately narrow (±4.5° half-power half-width), and was pointed first vertically, and then off-vertically. It was possible to prove that the decrease in power was not simply a temporal change by monitoring the signal strength vertically on a second, broader beam throughout the whole period. From *Hocking* (1979). (Reproduced with permission from John Wiley and Sons.)

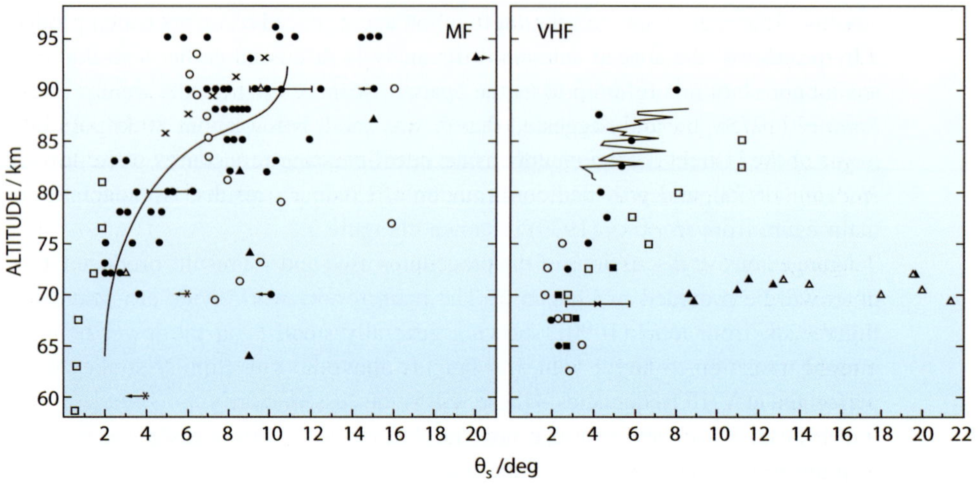

Figure 2.8 Height profile of mean θ_S values, taken from various references (from *Reid*, 1990). Most sites are plotted as points, but the continuous line is for Andoya. MF data are plotted on the left, and VHF data (to be discussed later) are plotted on the right.

One additional consideration needs to be recognized. Because computers had limited capability and small storage space in the 1970s, long-term data-runs were limited. For studies of scatterer characteristics, it was common to use data-lengths of only a few hours, up to a day or two in some cases, but generally not much longer. Because computer space was sparse, and extracting data from magnetic media could be tedious, it was not uncommon to watch the signals received on a cathode-ray oscilloscope, and wait for suitably strong echoes to appear. Then the observer would begin recording. This worked well, but had an unfortunate side-effect – only the strongest echoes were recorded and analyzed. As *Hocking* (1988) was able to resolve many years later, the characteristics of these stronger reflectors/scatterers seemed to be of a different type to the weaker, more common scatterers. It seems quite likely that even below 80 km altitude, a large number of weak but important quasi-isotropic scatterers exist, due to turbulence, with large values of θ_S, and the spectacular, strong quasi-specular echoes with small θ_S may well be anomalies – although interesting ones nevertheless. Figure 7.21 in Chapter 7 shows this quite clearly, since it shows the spectrum due to a very strong "specular" reflector and the simultaneous spectrum due to a much weaker, more isotropic background signal. Figure 2.8 must therefore be examined with this warning in mind.

Because of these curious results, which at the time were uncertain and could not be explained, studies of the nature of these scatterers was a topic of great interest in the 1970s. Other methods of analysis included rocket studies, studies of fading times, and examination of phase variations.

The need to be selective in data-acquisition intervals, discussed above, brings up another interesting point. Earlier data were recorded directly to large magnetic tapes, which were somewhat cumbersome to use, and had limited data storage capability (at least by modern standards). Raw data were recorded on tape, and subsequently analyzed off-line. Often the tapes needed to be taken to a central server, where they could wait for periods of hours and even days before they were played and analyzed. With the development of small computers that could be purchased by an individual laboratory (and ultimately the development of personal computers), it was even possible to do some of the initial processing at the radar site. Several laboratories worked towards this end, and of course nowadays it is commonplace to do a great deal of processing on-site. At the time, however, it was difficult. Much of the coding was done at the machine-language level, and various tricks were employed to speed up the processing. One such interesting technique was used to speed up calculation of the correlation functions used for the spaced antenna drifts analysis. Rather than record the actual digital levels, the numbers were converted to either a 0 or a 1, depending on whether they exceeded or fell below the mean. The correlations were performed on these single-bit numbers, allowing much faster on-line calculations. The method worked surprisingly well. Details about these type of issues have been discussed by *Fraser* (1984), for example.

Development of on-line analysis like this finally allowed data acquisition to be continuous, rather than on a campaign basis, and thereby helped remove some of the biases which could arise due to the selection strategies employed.

We now return to the issue of rocket studies. One of the key questions regarded whether high-profile rocket electron density measurements revealed structures that might be associated with the scatterers. It was possible that the scatter might be due to thousands of small scatterers, which would have electron density fluctuations too small to be resolved by in-situ rocket detection. Alternatively, it was possible that the scatter/reflections were due to a smaller number of structures which individually had larger electron density fluctuations which could be seen in rocket profiles.

Manson et al. (1969) presented statistical comparisons of echoes and high-resolution electron density profiles obtained by rockets. Small scale fluctuations in electron density, of the order of 1% to 20%, were observed in the rocket data, with a mode at about 3%. The fluctuations seemed largest at 60–70 km, and 80 km, which crudely agreed with some known preferred heights of radio scatter. Winter and autumn rocket firings showed more marked small-scale electron density irregularities, agreeing with the observation that partial reflections appear to be greatest in those months. On two days of anomalously high radiowave absorption, these density irregularities were found to be enhanced; this agreed with observations suggesting HF scatter is strongest on such days (*Gregory*, 1956). Calculations using the Sen-Wyller equations (*Sen and Wyller*, 1960), with corrections by *Manchester* (1965) suggested that each electron density fluctuation was capable of giving rise to scattered radio pulses with powers of the order of one third to one quarter of those observed by radio techniques. The combined effects of several of these perturbations were quite capable of causing the observed echo strengths. The paper thus concluded that the observed echoes, at least up to around 80 km, were due to fine-scale (less than 100 m) irregular variations of the electron density with height.

Hocking and Vincent (1982b) took this one step further. Using a detailed dispersive radiowave propagation model, they were able to determine the expected radiowave returns from electron-density profiles determined by rocket-based Langmuir-probe measurements (including the effects of radiowave absorption). The model was limited to one-dimensional simulations, however. Results again showed good correlation between electron density structures observed by the probe and radiowave echoes. The results would have been especially valid below 80 km, though may be considered perhaps to be only qualitative above that height, where the scatter was most likely to have been quasi-isotropic.

Another approach for examining the nature of the scatterers was the study of amplitude distributions. If radiowaves are scattered from a random distribution of moving scatterers, such as occurs in turbulence, the amplitude distribution follows a characteristic form called a Rayleigh distribution (*Rayleigh*, 1894). If, however, one scatterer dominates, the situation is akin to finding the probability distribution of a constant vector with many other small random vectors added on. The result is the so-called Rice distribution (*Norton et al.*, 1955; *Rice*, 1944, 1954; *Van der Ziel*, 1954). Then plotting amplitude distributions can indicate the type of scatterer being examined. Specular scatter, from a simple step in the electron density profile, with a certain degree of horizontal "roughness" superimposed, should be Rice-distributed with a large Rice parameter α (α being a measure of the specular component amplitude to the standard deviation of the random contribution). Turbulence would be expected to produce Rayleigh-distributed ($\alpha = 0$) cases.

Unfortunately, things are not so simple in practice. The case of two or three specular components of differing amplitudes is not covered by either case. Greater than five roughly equal specular scatterers would also give a close approximation of a Rayleigh-type distribution (*Goldstein et al.*, 1951; *Vincent and Belrose*, 1978). Further, if the components do not exhibit all possible phase differences with equal probability, the theory is not valid. Nevertheless, amplitude distributions did offer a possible tool for D-region investigations.

Von Biel (1971) at Christchurch, New Zealand, appears to have been the first author to attempt this procedure. *Mathews et al.* (1973) at Ottawa, Canada, followed this attempt, and *Newman and Ferraro* (1976) produced further results. All three groups of authors concluded that scatter appeared to be Rayleigh-like below about 80 km, but may have some specular contribution above this height. On the other hand, *Chandra and Vincent* (1979) produced Rice parameters which did indeed show strong specularity at the lower heights (below 80 km), and *Vincent and Belrose* (1978) compared their amplitude distributions to those expected for two or three specular components, and found that for typically 20% of the data the distributions appeared to indicate two or three principal specular scatterers. *Rastogi and Holt* (1981) also made Rice-distribution studies in Norway.

Results turned out, therefore, to be conflicting, and the observation of Rayleigh character below 80 km seemed at odds with the concept of specular reflectors at these heights. The resolution of the conflict in fact turned out to be associated with the data-lengths used to form the distributions. *Chandra and Vincent* (1979) had used 3 min data sets, while *Von Biel* (1971, 1981) and others used data-sets of 10 min or more. *Hocking* (1987b) showed that during a period as long as 10 or 20 min, the assumptions of statistical stationarity could not be supported, whereas use of short data lengths (less then a few minutes) provided insufficient independent points to produce reliable results (at least, insufficient points for comparisons with traditional Rice and Rayleigh distributions). A technique was developed to avoid both the issue of non-stationarity for longer data sets *and* the issue of insufficient points for short data-sets, which involved studies of the distributions of the Rice parameter itself for short data sets. This approach has somewhat unified the above apparently discrepant results, and again suggests that the dominant scatterers below 80 km do seem to be specular (*Hocking*, 1987b). More specific details are given in Chapter 7.

Another parameter of significance in examination of D-region HF echoes (and indeed echoes generally) is the fading time. This is generally taken as the time for the temporal autocorrelation of the data to fall to 0.5. Small values mean that the signal changes rapidly with time, large values mean the signal changes relatively slowly with time.

Echoes from turbulent scatterers may be expected to give signals at the ground which become uncorrelated after a time delay of 1 or 2 seconds (at around 2 MHz probing frequency). On the other hand specular echoes, perhaps due to steps in electron density of the order of kilometers in horizontal extent, may well give rise to much better spatially correlated signals, and thus longer fading times as they drift overhead. In addition, isotropic scatterers may produce scatter from a wide range of off-vertical angles, each with different Doppler shifts, thereby leading to significant beating and hence more rapid signal variability. The fading time was therefore seen as a way to determine information about the scatterers.

It was found that generally the lower echoes (70–79 km) exhibit longer fading times than the higher ones, especially during the equinoxes. *Lindner* (1972, 1975b) also presented fading times for Adelaide, Australia. He found that above about 80 km, fading times are typically less than 2–3 seconds, whilst below, fading times as large as 20 seconds can be attained, at least for the more dominant scatterers. *Schlegel et al.* (1978) found a similar trend for Tromso, Norway, with fading times less than about 4 seconds at 90 km but quite large at the lower heights.

The use of fading times needs to be treated with caution also, since the fading time actually can depend on the mean wind speed and the width of the radar beam (*Hocking*, 1983a; *Hocking*, 1983b). This will be discussed in much more detail in Chapter 7. As an example, *Schlegel et al.* (1978) and *Cunnold* (1978) also tried to relate their fading times directly to turbulent energy dissipation rates, but this aspect of these papers should be ignored, since they did not take account of so-called "beam-broadening" (*Hocking*, 1985).

One interesting parameter that is related to some extent to the fading times is the echo phase. Just as with meteorological radars, MF radar observations before 1960 essentially recorded only amplitude. Then in the early 1960s, measurement of Doppler frequency offsets, and phases of the returned signal, became possible. As with the meteorological case, it was not a trivial measurement, and computerized measurements of complex data had to wait till the methods of *Woodman and Hagfors* (1969) were introduced. Nevertheless, some useful measurements were possible. Phase measurements are particularly useful if the returned signal comes from a single specularly reflecting entity. If the entity drifts slowly upward or downward, its phase will change, as the number of wavelengths from the radar to the reflector changes. If the phase drifts through say π radians, it means that the reflector has moved one quarter of a wavelength vertically. If the scatterer is not perfectly flat, then the phase can also be affected by horizontal advection through the radar beam.

Fraser and Vincent (1970); *Vincent* (1973) used phase information of the returned signals to gain a better feel for height fluctuations and echo "coherence." Results suggested that echoes from above 85 km were often quite incoherent (frequent random phase jumps and variations), whereas those from around 70 km altitude often showed very smooth phase variations over intervals of several minutes at times, and frequently showed quite slow fading rates. Thus it seemed that echoes from around 70 km (or at least the stronger ones selected for monitoring) could well be due to one or two Fresnel reflectors drifting across the beam, producing specular scatter (in some cases, *Vincent* (1973) measured the speed of drift of these reflectors via phase measurements; results appeared to compare favourably with measurements of wind speeds made by the spaced receiver technique). Above about 85 km, volume scatter appeared to dominate. The night-time 90 km echoes showed similar incoherence to the daytime ones. The lower echoes were, at times, found to be extremely stable in height, showing very small fluctuations over quite long periods, particularly during the equinoxes. At times, height fluctuations were less than a few wavelengths over several minutes. However, even these lower echoes became somewhat incoherent during the solstices.

Just as the development of small computers, and the procedures to digitize in-phase and quadrature components (*Woodman and Hagfors*, 1969) impacted other areas, it also impacted MF radars. Computers were introduced on-site at many MF observatories, and complex data acquisitions became common. Coherent integration was implemented (summing successive pulses to optimize the signal-to-noise ratio – see later for more details and warnings). However, useful coherent integration requires that the integration length be less than one quarter of the smallest fading cycle, and hence higher pulse repetition frequencies and data acquisition rates were necessary to get the best value out of coherent integrations. The concept of phase recording and coherent integration spread rapidly. The spaced antenna drifts method was adapted to deal with complex data, and coherent integration became common to improve signal-to-noise ratios. This development also allowed a new approach to wind measurement, namely the use of narrow beam Doppler techniques. This involved steering the beam of a large array of antennas to off-vertical angles, and then measuring the Doppler-shifted frequencies of backscattered radiation. This was similar to developments with meteorological radars around the same time. In the case of the MF radars, beam tilts of only 10 to 15 degrees off-vertical were most common. Horizontal winds could then be determined from these measurements, provided it could be assumed that vertical velocities were small in magnitude. Not all MF radar could be used in this way – large arrays were necessary, and only a few were available world-wide. Of particular note were the Buckland Park MF array (e.g., see *Reid and Vincent*, 1987; *Briggs et al.*, 1969, and references therein), and the Bribie Island array (*From and Whitehead*, 1984), both in Australia. A similar large system was also built much later on the Island of Andoya in Norway (*Singer et al.*, 2008). This so-called DBS (Doppler beam swinging) technique was more commonly applied with VHF radars than with MF and HF radars, because at MF the radars typically need to be of the order of a kilometer in width to be useful.

The ability to make Doppler measurements led to significant new capabilities, including the development of ways to measure gravity-wave momentum fluxes (e.g., *Reid and Vincent*, 1987; *Vincent and Fritts*, 1987; *Fritts and Vincent*, 1987). (This method was also developed roughly concurrently for under-ocean sonar studies (e.g., *Lhermitte*, 1983). The technique made use of differences of radial velocity variances on symmetric but oppositely directed beams.

Doppler capability also led to methods for measurement of turbulence strengths, which made use of measurements of the spectral width. As noted earlier in the section on meteorological/precipitation radars, some of the underlying theory for this had already been developed. For example, *Hitschfeld and Dennis* (1956), *Frisch and Clifford* (1974), and *Labitt* (1979) considered the relation between turbulence strength and spectral width, but without fully considering the effects of the buoyancy scale of turbulence. Additionally, a major point to be considered was that a large fraction of the measured spectral width was not due to the turbulence itself, but simply to the horizontal motion of the mean wind across the beam. Without this recognition, turbulence measurements were not possible (*Atlas*, 1964; *Sloss and Atlas*, 1968; *Atlas et al.*, 1969). This point was independently recognized, modeled and extensively utilized by *Hocking* (1983a, 1985) for D-region studies. The numerical model of *Hocking* (1983a) was also the first to recognize that wind-shears do not always widen the spectrum, and may also narrow it – previously it had been assumed that widening was the obvious result. *Hocking* (1996a) was also important in that it consolidated the separate theories of *Labitt* (1979) and *Hocking* (1983a), which had treated the relative roles of the buoyancy scale of turbulence, and the dimensions of the radar-volume, in quite different ways.

A follow-on from the introduction of spaced antenna methods and Doppler beam swinging was the development of the so-called imaging Doppler interferometer technique, which combined aspects of both spaced antenna and Doppler procedures. Based on earlier work by *Pfister* (1971) and *Farley et al.* (1981), *Adams et al.* (1985, 1986) and *Brosnahan and Adams* (1993) developed the method as a new way to measure mesospheric winds. The method is based on the assumption that scattering of radiowaves could be considered to be from many individual and identifiable scattering centers in the upper atmosphere, and that when the signal was spectrally analyzed, each spectral line corresponded to a distinct scatterer. The location of each scatterer could be found by comparing the phases on multiple receivers, and using the known range to the "target." The idea was controversial, because such a one-to-one correspondence between spectral lines and scatterers is not necessary – it is easy to produce a time series from random numbers which, when Fourier transformed, has spectral lines which do not correspond to any identifiable part of the original time series. Considerable debate ensued about the physical interpretation of the scatterers (*Franke et al.*, 1990; *Roper et al.*, 1993; *Briggs*, 1995; *Holdsworth and Reid*, 1995), and the argument revolved around whether scatter was truly volume scatter (with no identifiable scatterers), or whether distinct scatterers really did exist. Part of the resolution was that even volume scatter would not be uniform, but would have spatial statistical fluctuations which would appear to be point scatterers. In addition, simulations suggested that even if the point scatterers were not real, if an effective target could indeed be located for any spectral line (self-consistent phases), then it would behave as if it were a true target. In general, it seems that the IDI

Figure 2.10 Looking down on the Jicamarca radar.

estimated the size and power of a radar required to observe this scatter, and his calculations suggested an antenna several hundred meters across with peak powers of more than a megawatt. In fact it turned out that his calculations overestimated the size of the radar needed, resulting in construction of arrays that were larger than necessary. The result was all for the good, since it meant the instruments had lifetimes far in excess of their initial expectations and could be used for studies unenvisaged at the time of construction – including, ultimately, the MST-VHF technique.

Subsequently, two major facilities were constructed, one developed by Bowles at the Earth magnetic dip equator in Jicamarca, near Lima, Peru, and another developed under the guidance of Gordon at the subtropical location of Arecibo on the island of Puerto Rico. Figures 2.10 and 2.11 show the large antenna fields of these two radars. Each is about 300 m across. Whereas Jicamarca uses a phased array consisting of rows of coaxial-collinear dipoles, Arecibo uses a spherical dish which is illuminated by a feed in the focal area (see Chapter 5 for discussions about antenna design).

These two early incoherent scatter radars have, in the years since their development, contributed a great deal to the understanding of complex plasma processes in the Earth's ionosphere and near-space environment. They had been the forerunners of several newer such radars, particularly ones developed in high auroral latitudes. The techniques have become more sophisticated, and these radars can now be used to measure electron densities down into the D-region (e.g., *Evans*, 1969; *Mathews*, 1984a, b).

Figure 2.11 Looking down on the Arecibo radar with the Gregorian feed-system, installed in 1997, attached. The inset shows some of the earlier feed mechanisms, used prior to 1997. A 46.8 MHz VHF Yagi antenna is shown at the center of the inset, installed by the SOUSY group, and a log-periodic antenna was also available. The newer Gregorian dome was designed to give better performance on transmission and reception. (Photo credit, J. Röttger)

2.7.1 Coherent echoes seen with incoherent scatter radars

For the purposes of this book, these radars were most important for their roles in leading to the development of MST-VHF radar. The Jicamarca radar was of special importance. To a large extent, the history of MST-VHF radar begins with the Jicamarca radar.

Initially, these radars concentrated on scatter from the E- and F-regions of the ionosphere. Then, using measurements made in 1957, *Bowles* (1958) first reported "ionospheric scattering of the turbulence variety" from altitudes of 75–90 km, measured with the Illinois VHF radar in the USA. *Flock and Balsley* (1967) detected similar kinds of echoes with the Jicamarca radar. These echoes were different in form to the ones later used for upper ionospheric electron density calculations (e.g., *Mathews*, 1984b), since they had a coherent character quite different to the normal incoherent scatter.

This was not the first time that VHF echoes had been seen from the mesosphere. Some scatter from the ionosphere at VHF was observed in the early 1950s, and *Ellyett and Watts* (1959) mention that Bailey, Bateman and Kirby, and Pineo, did observe some VHF scatter from 75–90 km. *Booker* (1959) also discussed VHF scatter from around 90 km. *Villars and Weisskopf* (1955) also observed VHF scatter, as did *Friend* (1949) and *Saxton et al.* (1964).

Many of these examples were cases of oblique scatter, with transmitter and receiver well separated, so that the scatter came from Bragg scales many times the wavelength of the radiation (see Chapter 3). In fact, the scatter was most likely due to irregularities of electron density with scales similar in order of magnitude to those causing the previously discussed MF and HF backscatter. (*Villars and Weisskopf* (1955) suggested vertical scales of about 14 m, with scatter being from heights below 90 km.) Previous observations of backscatter had also been made in the lower atmosphere (e.g., *Friend* (1939) – see the review by *Gossard* (1990), for example, which proposes that the scatter seen by Friend was truly atmospheric scatter and not insect-scatter).

The unique features of the echoes seen by *Bowles* (1958) and *Flock and Balsley* (1967) were that they were D-region echoes observed with a monostatic system, so corresponded to Bragg scales of 3 meters from D-region heights. Observing the echoes was interesting, but of limited value. It was *Woodman and Guillen* (1974) who substantially improved the technique by implementing coherent detection in order to measure velocities, thereby turning the echoes from a novelty item to one of great practicality. Coherent integration refers to the practice of averaging returned echoes from several successive pulses together to improve the signal-to-noise ratio, and this will be discussed in more detail later in this book. It is only possible with these so-called "coherent echoes."

Nevertheless, as will be seen in Chapter 7, a warning needs to be sounded about coherent integration. It does improve the signal-to-noise ratio, but this is only relevant when using procedures like the autocorrelative method to determine the radial velocities. If a full spectral analysis is performed, then the important parameter is not the signal-to-noise ratio, but rather the so-called "detectability." This will be discussed further in Chapters 7 and 8. With the right spectral approach, coherent integration is often unnecessary (e.g., see *Hocking*, 1997a), except perhaps to save processing time.

Woodman and Guillen (1974) also reported echoes from the stratosphere, and recognized the great potential of this VHF radar technique for studying the mesosphere, stratosphere, and the troposphere as well. It is at that time that we can consider VHF-MST radars to have been "born."

2.8 MST radar techniques at VHF and some atmospheric science highlights

We can now discuss the development of MST-VHF radars. As noted earlier, the observations with the Jicamarca radar (see Figure 2.12), showed that the relative echo power of this newly detected type of backscatter from altitudes below 100 km was significantly stronger than the underlying incoherent scatter from the ionospheric D-region, and that the echo signals were much more coherent than the incoherent scatter signals.

A coherent signal is defined here to exhibit long temporal persistency over many interpulses. That means the coherent signals do not substantially change amplitude and phase over typically one millisecond. The coherency of this new type of radar signal from altitudes below 100 km allowed application of novel radar coding and data pre-processing techniques such as pulse-coding, coherent integration and specialized filtering techniques (see Chapter 4 on data processing).

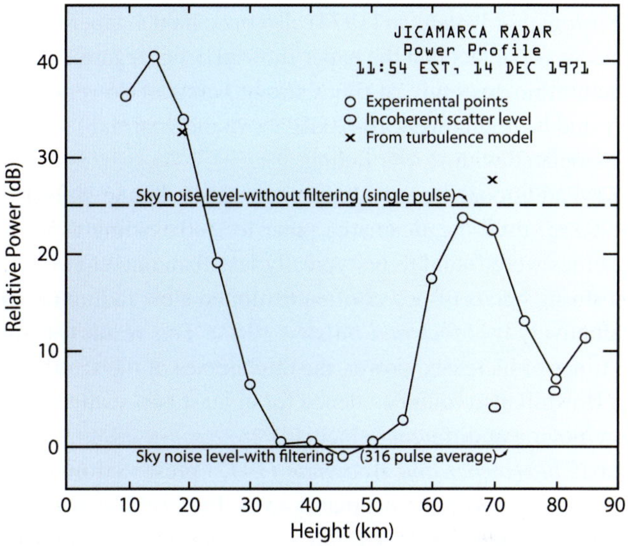

Figure 2.12 Early evidence of VHF backscatter from the D-region of the ionosphere using the Jicamarca VHF radar, from *Woodman and Guillen* (1974). (Reprinted with permission from the American Meteorological Society.)

At the time, it was already known that the radar echoes from the mesosphere are caused by scatter from irregularities of the radio refractive index, as discussed in previous paragraphs of this chapter (see the sections on D-region scatter at MF and HF frequencies). As seen there, these refractive index variations are caused by the neutral atmosphere turbulence mixing the electron density distribution of the lower ionosphere. Mixing could be due to turbulence, and also to other mechanisms that could produce stratified steps and sheets.

Detection of these coherent echoes not only opened the possibility of measuring mean and fluctuating wind velocities from the Doppler frequency shift and spread, but also yielded information on atmospheric turbulence. While in the mesosphere the scattering irregularities are electron density disturbances, in the stratosphere these are caused by temperature variations and in the troposphere by temperature and humidity variations (see Chapter 3 for details).

Interestingly, the discovery of these echoes actually arose from a lack of funding! At a time when support for the Jicamarca radar was low (the Americans and Peruvians were feuding over fish off the coast of Peru, so American support was withheld for projects like this), Woodman had insufficient funds to carry out important ionospheric studies and so turned to other, seemingly less important and curiosity-driven research. Among these studies was investigation of weak, slowly fading echo fluctuations that he had noticed from the lower regions of the atmosphere. It was at the same time that he recognized the potential to improve the detectability of the echoes by utilizing their similarity from pulse to pulse, thereby leading to the development of coherent integration.

Rastogi and Woodman (1974) also presented a paper on these VHF echoes. Of significance was the fact that the paper showed detail regarding the temporal variation of echo structure; interestingly, at times strong bursts of power occurred, lasting a few minutes only and having powers 10–20dB above the "normal" level. Such bursts also seemed to be associated with slower fading.

Key additional observations were made in the mid 1970s. *Rastogi and Bowhill* (1976a, b) used the Jicamarca radar to study fading times and power variability. Fading times were found to be typically less than one or two seconds and evidence existed that strong bursts of power often exhibited slow fading (a similar result had been noted qualitatively by *Flock and Balsley*, 1967). This result was interpreted to mean that during times of increased power, the thicknesses of the scattering layers decreased. Rastogi and Bowhill also found evidence for at least two scatterers contributing to their echoes, each moving at different velocities.

In 1977, *Harper and Woodman* (1977) presented more VHF results of mesospheric scatter – again with the Jicamarca array. Preferred scatter came particularly from around 75 km altitude, as seen by earlier authors. Detailed temporal analyses of the echoes were presented, similar to those of *Rastogi and Woodman* (1974). Results showed that frequent power bursts occurred, often 10–20 dB above the "normal" level, lasting around 2–5 minutes, and with a quasi-periodicity of around 2–5 minutes. This periodicity, they claimed, correlated with a 10 minute gravity wave observed in the Doppler measured winds at the same height. Thus emerged some of the earliest direct evidence of the possible effects of gravity waves on these VHF mesospheric scatterers. The authors also found a correlation between strong echo power and slow fading – but this did not always exist. At times, there was even an inverse correlation, with strong powers showing quite rapid fading. These results generally supported the findings of *Rastogi and Woodman* (1974).

Following these striking observations with the Jicamarca radar, a relatively explosive growth and development occurred in coherent backscatter radars for studying the structure and dynamics of the troposphere and stratosphere. These radars all operated in the low VHF band at frequencies around 50 MHz and their name, MST radars (standing for Mesosphere Stratosphere Troposphere radars), was cast during a small radar workshop at the University of Utah in the end of the 1970s. Because these all operate in the lower VHF band between 40 and 55 MHz, they are also called VHF radars. In this book, as noted at the beginning, we will consider MST radars to cover a wider class of radar, including meteor and MF, HF, UHF, and even higher frequencies, so long as they are used to investigate the MST region or have significant aspects in common with regard to technique. VHF radar used for MST work will be called VHF-MST. It is true to say, however, that VHF-MST radars are the only radars capable of observing all three regions (mesosphere, stratosphere, and troposphere) simultaneously with the same radar.

In the middle of the 1970s, two important new radars were developed almost in parallel. These were the Sunset VHF radar (operated on 40.5 MHz) near Boulder in Colorado (see *Gage and Balsley*, 1978; *Green et al.*, 1979, and references therein), and the SOUSY (SOUnding SYstem) VHF radar (53.5 MHz) in Germany (*Röttger et al.*, 1978). They both used the same novel data-processing techniques introduced at Jicamarca, but were designed primarily for the study of the height regions below 100 km, including the troposphere. Both improved on earlier methods. Both made extensive use of computers and digital processing. Each made their own new developments. For example, the Sunset radar applied a coaxial cable version of the Jicamarca antenna array and the SOUSY radar applied a new approach combining a large set of Yagi antennas (Figures 2.13, 2.14, and 2.15).

It was at this time (and shortly thereafter) that the separate paths of the meteorological and ionospheric radars began to merge somewhat, since both had led back to studies of the lower atmosphere and meteorology. Scientists who formerly had worked in the ionosphere now began to attend meteorological conferences, and joint publications were presented (e.g., *Atlas* 1990 and the associated volume of papers).

These MST radars, and all MST radars since, use antenna beams pointing to or close to the zenith direction, at least for Doppler studies. Their beam-widths are of the order of a few degrees, requiring their antenna array diameter to be at least ten wavelengths. The wavelengths of these VHF radars are typically between 7.5 and 5.5 m. This requires the antenna arrays of the VHF radars to be between at least 50 m and 100 m diameter. The radars therefore all concentrate most of their transmitted radiation in narrow beams and receive the backscattered radiation along the same beams, giving large increases in sensitivity (high gain). Antenna gains (see Chapter 3) of the order of 30 dB or more (30 dB corresponds to a power-density increase of 3 orders of magnitude over a simple radiating dipole). In order to enhance the signal-to-noise ratio of the backscattered signals further, such VHF radars apply peak transmitter powers of some 100 kW up to some 1000 kW.

Figure 2.13 The 40 MHz Sunset MST radar. This radar used Coco antennas, and had multiple fixed beams in various directions (from *Gage and Balsley*, 1978). (Reprinted with permission from the American Meteorological Society.)

(a)

(b)

Figure 2.14 The Sousy MST radar, circa 1982: (a) a general view of the array; (b) a close-up of some of the antennas. This radar used Yagi antennas and used multiple beams in various directions, which could be selected by the user.

The Sunset Radar, located in a narrow Rocky Mountain canyon to shield for scatter returns from distant mountains, operated with vertical and off-vertical beams in order to measure the Doppler shifts of the returned echoes. These measurements were then in turn converted to vertical and horizontal wind velocities. The Doppler shifts are produced by "beating" procedures similar to those discussed with regard to the spectrum analyzers earlier, but use digital computer methods based on the principles introduced by *Woodman and Hagfors* (1969). Now Doppler shifts of a fraction of a Hertz could be measured routinely, without the need for operator assistance. Specific details will be discussed in Chapter 4.

Figure 2.15 Aerial view of the Sousy MST radar

The Doppler capability of these radars was important not only for determination of winds, but also for application of coherent integration, although the warning at the end of Section 2.7 should be heeded here.

In earlier paragraphs, there was some discussion about so-called "preferred heights" of scatter, as observed with the Jicamarca and Arecibo radars. *Czechowsky et al.* (1979) used the (then) new SOUSY radar to carry out a comprehensive study of the heights of preferred scatter. (Figure 2.16, taken from that reference, shows the seasonal variation of echo heights for typical summer and autumn conditions – other seasons were included in the original paper.) The maximum height of echoes detected is about 75–80 km, in line with expectations for turbulent scatter; at greater altitudes, the scattering scales associated with VHF radar are in the so-called "viscous range" of turbulence, and are generally too suppressed by the effects of viscosity to produce measurable radar echoes. However, exceptions do exist, which will be discussed later in regard to polar mesosphere summer echoes.

Returning to discussions of the Doppler effect, we recognize that the Doppler frequency shift of the echo signal from the vertical beam results generally from the vertical wind velocity component. Thus, measuring Doppler frequency shift can immediately yield the vertical velocity, but some precautions regarding this interpretation must be borne in mind, as discussed in later chapters. The off-vertical beams, usually directed with an off-zenith angle in the order of 10–15 degrees, are used to measure the combination of the components of the vertical and horizontal wind velocity. By suitably combining these velocities, the three-dimensional wind velocity components are deduced – again, details will be discussed in later chapters.

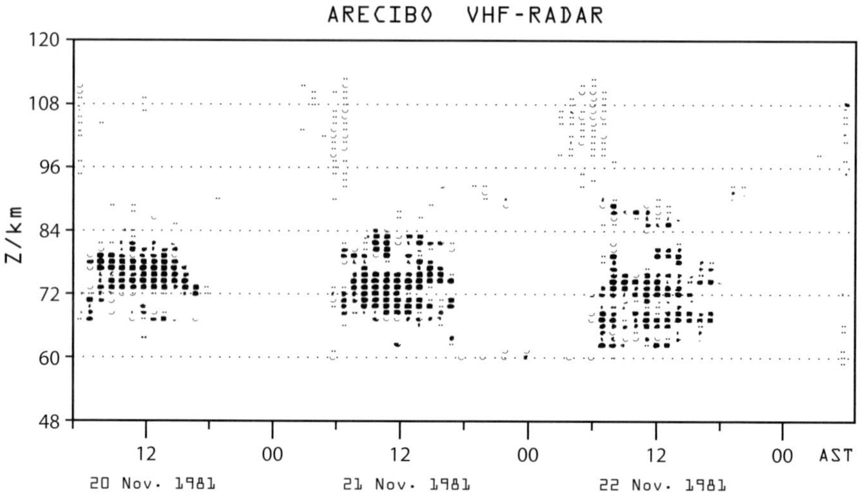

Figure 2.19 Some of the earliest mesospheric echoes observed with the Arecibo radar using the SOUSY 46.8 MHz system. The plots represent a primitive "density plot," where higher densities of dots represent higher powers. Echoes from altitudes between 90 and 100 km in the morning hours are the result of meteor backscatter (from *Röttger*, 1984a).

to 85 km, where the lower height is determined by the diminishing electron density as the altitude decreases, and the upper height is defined by the fact that the Bragg scale of this radar (3.2 m) becomes less than the inner scale of the turbulence at these heights. The echoes occurring between 90 and 110 km are from meteor trails, which can be used for wind and temperature observations, and will be discussed later in this book.

2.9 Newer-generation radars

Following these early developments, other MST-VHF radars were developed. The initial radar observations of the mesosphere, stratosphere and troposphere were performed at low and middle latitudes. The first high-altitude MST radar was built at the end of the 1970s in Poker Flat, Alaska. It was a high-power radar working on 49.9 MHz that consisted of 64 phase-coherent transmitters with a total peak power of 6.4 MW feeding a coaxial-collinear antenna array of aperture 40000 m^2 (*Balsley et al.*, 1979, 1980).

Another high technology radar was constructed in the early 1980s by the Radio Atmospheric Science Center of the Kyoto University in Japan (*Fukao et al.*, 1985a, b). This middle and upper atmosphere radar (MU radar) was the first active phased-array MST radar system. Each Yagi element is fed by a low-power transmitter and all of the 475 solid-state amplifiers are driven phase-coherently. This makes it possible to steer the beam electronically from pulse-to-pulse into 1657 directions between 0 and 30 degrees in zenith and from 0 to 360 degrees in azimuth. The maximum peak power is 1000 kW and the antenna area is 8330 m^2, yielding a beam half-width at half-power of 3.6 °.

This distributed, multi-transmitter MST radar design has flexible beam steering and the advantage that a failure of a single transmitter unit does not significantly compromise the system operation.

Figure 2.20 shows a bird's-eye view of the MUR, located in Shigaraki near Kyoto in Japan. It was inaugurated in November 1984 and has since contributed very substantially to research of the lower and middle atmosphere as well as of the ionosphere.

Another key radar was the University of Illinois radar, which was originally used by Bowles to observe some of the earlier VHF echoes, and has been discussed earlier.

Parallel to the development of these large radars was the development of many smaller so-called ST radars, which could reach into the stratosphere but could not obtain significant mesospheric scatter. They became key tools in meteorology, and will be discussed later. Two of these include the Chung-Li radar in Taiwan (*Chao et al.*, 1986), and the Adelaide radar (*Vincent et al.*, 1987).

(a)

(b)

Figure 2.20 (a) Aerial view of the Japanese MU radar (Photo credit, T. Sato). (b) The antennas of the MU radar from ground-level (Photo credit, T. Sato).

Many other radars have been built since these earlier ones, both large and small. The large Indian radar near Gadanki, India was commissioned in the 1990s (*Rao et al.*, 1995). In recent years, a major focus has been on systems employing many distributed transmitters, often with one transmitter (typically 1 or 2 kW peak power) on each antenna element. New large-scale radars with distributed transmitter systems include the PANSY radar in Antarctica (*Sato et al.*, 2014) and the MAARSY radar in northern Norway (*Latteck et al.*, 2012).

But there are too many new systems to discuss all of them here. Some design aspects will be discussed further in other chapters, especially Chapter 6.

2.10 Scattering and partial reflection

We will now turn to some of the essential understanding of the atmosphere that was gained from application of these radars. The first item relates to the nature of the irregularities producing the scatter.

In early interpretations of radar backscatter, it was assumed that it is essentially caused by turbulent fluctuations of the radio refractive index of the clear air, and that mixing of humidity temperature variations and electron density produces the necessary refractive index Bragg scales. Some of the relevant theory for describing turbulent backscatter was derived by *Ottersten* (1969b), and extra detail was provided by *Hocking* (1985). The turbulence refractive index structure constant then determines the scatter cross-section of the Bragg-scale irregularities causing the radar echoes. Indeed, turbulent scatter does turn out to be a (probably the) major contributor to atmospheric scatter. However, it is not the only form.

2.10.1 Specular and Fresnel reflectors

A comparison of the scatter cross-section deduced theoretically for conditions of reasonable turbulence intensity demonstrated that the measured cross-sections in the troposphere and lower stratosphere could sometimes be much larger than turbulence theory would allow. In addition, signals received on the vertical beam were often much more pronounced than when the antenna beams were pointing off-vertically. Investigations were therefore undertaken to determine whether the radar echoes seen on the vertical beam might be caused by some kind of partial reflection from horizontally stratified refractive index variations. This assumption is analogous to similar arguments discussed earlier with regard to D-region MF partial reflection (e.g., see Figure 2.7 and the discussions surrounding that figure). However, the difference in scales was significant – a step of depth say 25 meters (as required for MF scatter) could be considered reasonable, whereas a step of depth only 1–3 meters, as required for VHF partial reflection, which was also flat over a horizontal width of several hundred meters to a kilometer or more, seemed much more unlikely.

The requirement that the step must be horizontally stratified over such large distances arises from the fact that the layer needs to cover at least one "Fresnel zone" in horizontal extent, where a Fresnel zone is the plane area perpendicular to the wave propagation direction over which the distance to the transmitting antenna changes by a half-wavelength as one moves from the center to the Fresnel radius. Numerically, the Fresnel radius is given approximately by $r_f \simeq \sqrt{d\lambda}$ where λ is the radar wavelength and d is the distance from the radar to the scattering region. Typically this is of the order of 0.5 to 1.0 km at mesospheric heights, and several hundred meters at upper tropospheric heights for a VHF radar. Objects that have smaller horizontal extents appear more like point reflectors and do not have such substantial variation of back-scattered power with angle (e.g., *Ratcliffe*, 1956).

Another factor which worked against the existence of these specular reflectors was their likely lifetime. If the major diffusion mechanism was turbulence, then such a step would be expected to be destroyed by turbulent diffusion in a time $t \simeq L^2/K$, where $L \simeq 2$ m, and K is the expected diffusion coefficient, taken to be typically 100 m^2 s^{-1}, so $t \simeq .04$ seconds! Thus the step should be rapidly destroyed. Clearly, if the step was to survive, it must exist in a region where there is no turbulence – in a laminar region of the atmosphere – where the only form of diffusion is molecular diffusion.

Remarkably, observations and experiments did indeed show that such specular-type reflectors do actually occur – requiring the existence of vertical steps in the refractive index that occur within depths of less than (or at least on the order of) the radar wavelength. It needs to be noted that the step *cannot* simply be part of a larger scale structure – the step must start and finish within this depth. (See Chapter 7 for more details.) The existence of these specular reflectors captivated the scientific community.

Radar evidence for these specular reflectors came with publications by *Röttger and Liu* (1978), *Gage and Green* (1978), and *Röttger* (1978) for the troposphere, and *Fukao et al.* (1979) for the mesosphere. The publication by *Fukao et al.* (1979) came out in the same year that MF D-region was confirmed to be specular using the beam-steering method (Figure 2.7). Specular reflectors should also be associated with slow fading times, as discussed in the MF case, and *Hocking et al.* (1991) showed one case where the fading time for stratospheric scatterers was of the order of minutes.

Figure 2.21 shows some early profiles observed with 150 m height resolution, using vertically and obliquely directed antenna beams, which demonstrate the noticeable difference in received echo power in these two directions. (An early part-version of this plot appears in *Röttger* (1980b), but the full figure was taken from *Röttger and Larsen* (1990).) This difference in powers on different beams is attributed to the anisotropy of the irregularities responsible for the radar echoes. We note highest anisotropy in the stratosphere (above about 10–12 km). This might be expected, since its positive gradient of temperature means that it is very stable and stratified (as its name expresses). The increase in vertical echo power at about 12 km is due to the very steep gradient of temperature above the tropopause, and thus it can be used to determine the height of the tropopause. Since the troposphere is usually not as stable as the stratosphere, the anisotropy is less, as we note in Figure 2.21. It is reasonable to assume that on 20 June 1978, when the echoes from vertical and oblique directions were almost equal below 9 km, the scatterers were isotropic, possibly homogeneous, and probably due to

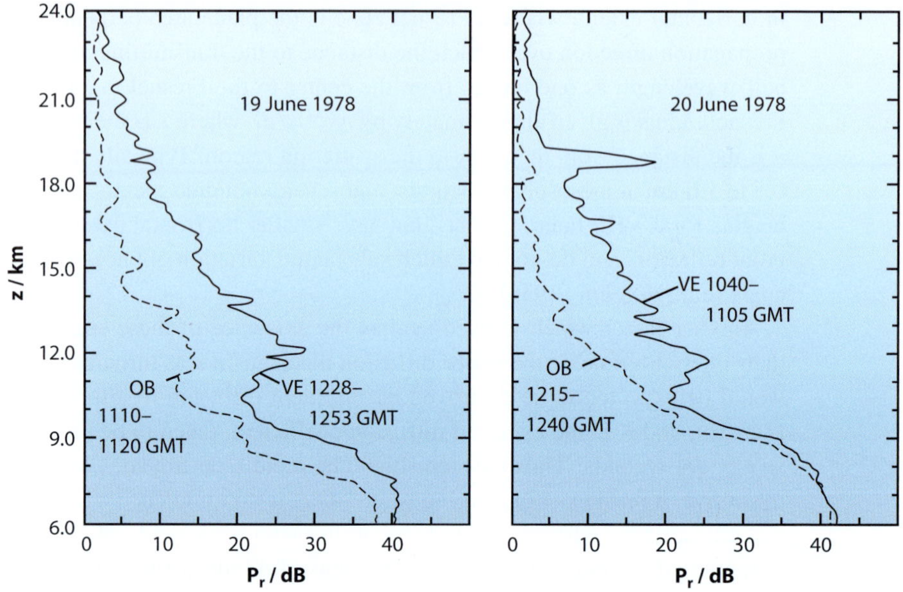

Figure 2.21 Some early profiles observed with 150 m height resolution, using vertical and oblique antenna beams, demonstrating the noticeable difference in received echo power in these two directions. The vertical beams are indicated with solid lines, and labelled "VE," while the oblique beams are labelled "OB" and are represented by broken lines (from *Röttger and Larsen*, 1990). (Reprinted with permission from the American Meteorological Society.)

turbulence. Figure 2.22 shows another example of this anisotropy; in this case, eight separate beams were used to allow the radar to look in multiple directions. Figure 7.19 in Chapter 7 shows another example of such anisotropy between vertical and off-vertical beams. Eventually studies of such anisotropy were not only continued in the vertical plane, but also examined for azimuthal variation (e.g. *Hocking et al.*, 1990; *Tsuda et al.*, 1986, 1997a, b; *Worthington et al.*, 2000, 1999a, b).

The reality of these observations was a subject of much debate – could these signals actually be an instrumental artifact? No definitive mechanism could be found that could explain such steps, although various suggestions have been made, and will be discussed further in Chapter 7. Nonetheless, the evidence was hard to refute.

In the case of *stratospheric* VHF reflections, any steps would have to be steps in temperature, since humidity concentrations in the stratosphere are very low, and refractive index in the non-ionized atmosphere depends primarily on humidity and temperature. While possible electron density steps had been measured with Langmuir probes in the mesosphere, resolution was generally too poor to resolve steps of the order of 2–3 m. Since temperature measurements are much easier to perform in the lower atmosphere than in the mesosphere, the opportunity existed for extremely detailed investigations of the form of the temperature profiles using balloon soundings. It was anticipated that if balloon measurements could be made, and if thin, sharp steps in temperature could be detected, then these results might shed light not only on the nature of the VHF scatterers

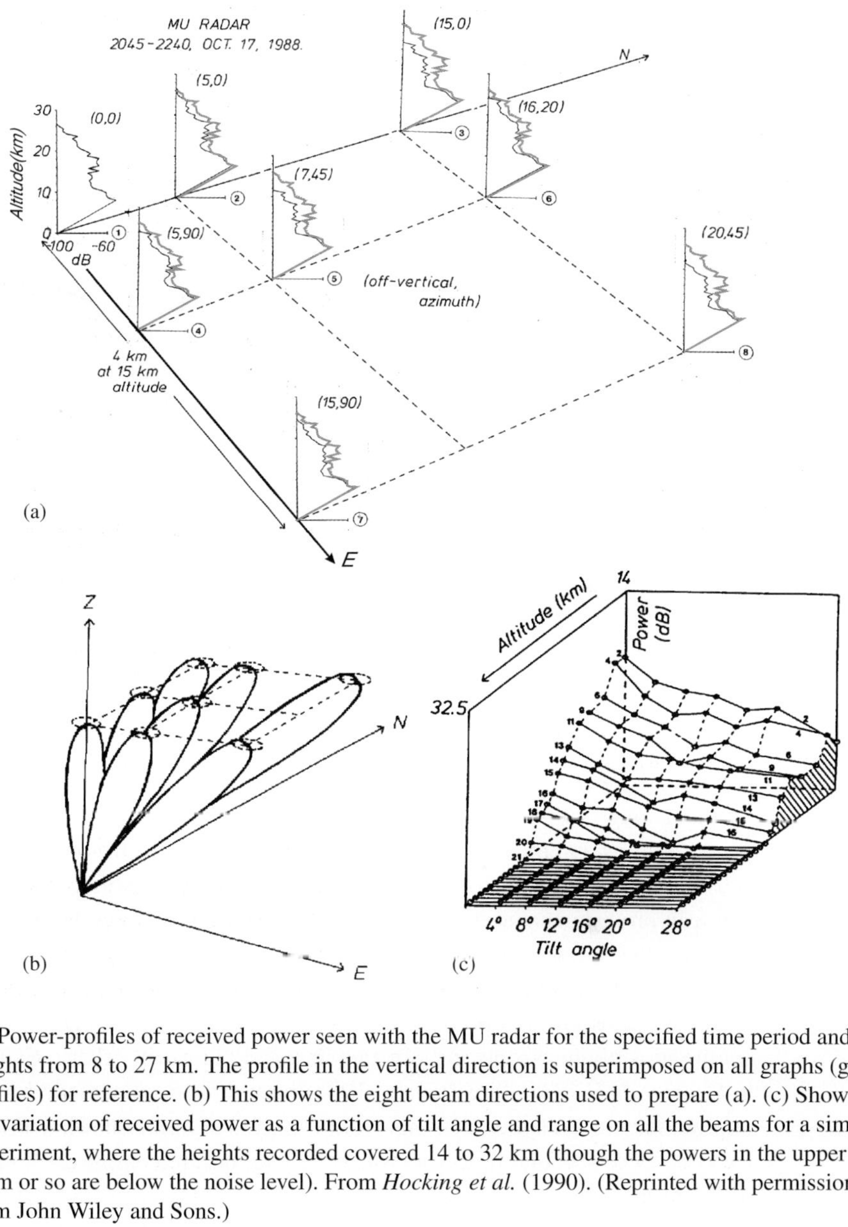

Figure 2.22 (a) Power-profiles of received power seen with the MU radar for the specified time period and for heights from 8 to 27 km. The profile in the vertical direction is superimposed on all graphs (grey profiles) for reference. (b) This shows the eight beam directions used to prepare (a). (c) Shows the variation of received power as a function of tilt angle and range on all the beams for a similar experiment, where the heights recorded covered 14 to 32 km (though the powers in the upper 6 km or so are below the noise level). From *Hocking et al.* (1990). (Reprinted with permission from John Wiley and Sons.)

in the mesosphere but possibly on MF and HF reflectors (and even VHF reflectors) in the mesosphere.

Despite the physically challenging nature of these steps, the "step" model was adopted as a viable explanation. As suggested above, balloon measurements were indeed attempted, and balloon-supported measurements with high resolution probes seemed to confirm the existence of these steps (e.g., see *Dalaudier et al.*, 1994; *Luce et al.*, 1996,

2001b, and references therein). The reflection/scattering process was called either Fresnel reflection or Fresnel scattering, where Fresnel reflection referred to a single step, and Fresnel scattering referred to several steps co-existing within the same range gate (see *Röttger and Larsen*, 1990, and references therein). Of course this definition is somewhat pulse-length dependent, because with a short enough pulse, a "Fresnel scatter" situation can be turned into a "Fresnel reflection" situation. But the definition is still useful in a general sense.

In reality, the truth turns out to be a little more complex. While specular reflectors do exist, they are never perfectly flat – small wrinkles and undulations on them broaden the spread of backscattered radiation, so that signals recorded with an off-vertical beam increase relative to those for a perfectly flat reflector. Some evidence for this can be seen in Figure 2.18, where spikes in the Doppler frequency spectra can be seen at various frequency offsets. This suggests that these sheets must exhibit some roughness, or, in other words, they are vertically and horizontally corrugated surfaces of refractive index structures. This has some analogy with Figure 7.21, where we will see evidence of near-simultaneous cases of specular reflection and quasi-isotropic (possibly turbulent) scatter.

Some warning needs to be sounded here. It was generally assumed that the MF/HF mesospheric reflectors and the VHF reflectors were one and the same, and that the characteristics of these sheets in the stratosphere could be assumed to be valid in the mesosphere. The possibility must be borne in mind, however, that they could each be reflections due to different mechanisms. Generally it is assumed that they are related – not unreasonably – but the possibility of different structures must be remembered.

With enough wrinkles, the power received on off-vertical beams can approach that for vertical beams. On the other hand, turbulence at 3 meter scales is not always isotropic, and if the scatterers are on average stretched out more horizontally than vertically, they can produce enhanced scatter on the vertical beams over the off-vertical beam. It becomes an important issue to be able to distinguish these different models, and there are times the two cannot be distinguished. Methods were developed to allow measurements of power ratios on oblique and vertical beams to be used to determine the length-to-depth ratios of anisotropic turbulent eddies, and these will be discussed in more detail in Chapter 7. *Hocking and Hamza* (1997) have developed an approximate protocol to differentiate these classes, and this will be discussed further in Chapter 7. Nevertheless, the concept of specular reflectors is real. Sometimes, to complicate things even further, tilted specular reflectors can be seen.

In distinguishing Fresnel reflection from Fresnel scatter, one test was quite useful – to look at the dependence of the backscattered power on the pulse length. Of course, the easiest way to distinguish multiple layers would be to make a radar with a very short pulse-length – maybe a few meters at most – so that each and every specular reflector could be independently identified. But such a proposal is unrealistic since it would require a bandwidth of the order of 20 MHz or more, which would be difficult to build and hard to obtain a radio license for. Failing that, the best test is to look at pulse length. *Hocking and Röttger* (1983) studied this behavior numerically, and showed that under normal circumstances the power should increase proportionally to the pulse length for the case of Fresnel scatter, but should be independent of the pulse length for the case of Fresnel reflection. This corrected earlier erroneous assumptions about the pulse length, and allowed *Gage et al.* (1985) to develop a newer model for Fresnel scatter from the atmosphere.

It was noted above, with regard to Figure 2.21, that there is often enhanced power close to the tropopause. This was noted by several authors (e.g., *Gage and Green*, 1979; *Larsen and Röttger*, 1982; *Rastogi and Röttger*, 1982; *Gage and Green*, 1982b; *Hocking et al.*, 1986, among others). *Rastogi and Röttger* (1982) also noticed longer fading times associated with the tropopause echoes, while *Gage and Green* (1982b) developed a semi-formal procedure for tropopause height determination by VHF radar. The ability of VHF-MST/ST radar to detect the tropopause height has become a powerful capability. Many processes require knowledge of this height, including satellite processes of stratospheric temperature retrieval. The method was also more recently useful for allowing studies of ozone transport from the stratosphere to the troposphere (*Hocking et al.*, 2007a). It should be noted, however, that the reasons for enhanced scatter at the tropopause are complex. Originally it was believed that this was because of an increase in specular reflections, but this is true only on some occasions. On other occasions, turbulence can occur just above the tropopause, as waves propagating from below break (like waves on a beach), giving rise to enhanced turbulence. So the increase in power at the tropopause is sometimes exclusive to measurements on the vertical beam when specular reflection dominates, and sometimes can occur on all beams when turbulence dominates. See Chapters 7 and 12 for more details.

2.10.2 Scattering by turbulence

Specular reflections are of course not the only reason for radar scatter from the atmosphere. It was fully realized that scatter from turbulent layers was probably the most common form of scatter. *Crane* (1980b) discusses various observations of turbulent layers. But even turbulent scatter needed further understanding. It was uncertain whether the scatter came from thick layers that covered the full extent of the radar pulse, or from multiple, thin layers of turbulence separated by much quieter regions. *Van Zandt et al.* (1978, 1981) performed important investigations into the nature of the turbulent scatterers and developed models for the distribution of layer thicknesses within a given height range, based on relatively low-resolution (kilometer-scale) wind and temperature profiles. Their work was important not only for physically understanding the nature of turbulent scatter, but also for determining the strength of the turbulence from measurements of backscatter cross-section – since the fraction of the radar volume occupied by turbulence was an important parameter in these determinations (e.g., *Hocking and Mu*, 1997). The turbulence does not always have to be isotropic, either. At scales of a few meters, it can be quite anisotropic, with average eddy shapes stretched horizontally (e.g., *Hocking and Hamza*, 1997). (The radar volume for a particular range is defined as the three-dimensional region of space (often considered as a cylinder) defined by a circle with a radius equal to the half-power half-width of the radar beam, and with a length equal to the full-width half-power length of the effective pulse.)

Much of the radar scatter was indeed due to turbulence, but the nature of the specular reflectors still captivated many researchers. The realization that Fresnel reflectors and scatterers seemed to be strongest when the atmospheric temperature gradient (with the tropopause and stratosphere, during times of strong specularity, being extreme examples) was most stable led *Gage and Green* (1982a) to develop a procedure for using the strength of scatter to estimate the local temperature gradient and then to integrate in height to produce a temperature profile in the troposphere. The method has been adopted in some form or other by subsequent authors (e.g., *Hooper et al.*, 2004), but has not become universally used. The reasons are not fully clear, but probably relate to the fact that while the method works in some situations, it is not always valid.

With regard to scatter by turbulence, Chapter 3 will begin to look in more detail at the ways that turbulence affects radio scatter, and measurement of turbulence strengths by radar will be a recurring theme throughout many of the subsequent chapters.

2.10.3 Amplitude distributions

In the section related to MF/HF D-region scatter, the concept of using amplitude distributions to learn about the scatterers was introduced. The same techniques were also introduced into VHF studies (e.g., *Hocking*, 1987b; *Kuo et al.*, 1987; *Rastogi and Holt*, 1981; *Röttger*, 1980a; *Sheen et al.*, 1985, amongst others).

The idea is that if scatter is due to an ensemble of roughly similar scatterers, as might occur in a turbulent patch, then the amplitudes of the resultant distribution will have a so-called "Rayleigh distribution" (*Rayleigh*, 1894). If, however, there is also a much stronger single scatterer in addition to these weaker scatterers, the distribution changes to a so-called "Rice distribution" (*Rice*, 1944, 1945). Figure 7.22 in Chapter 7 shows how these distributions change as the specular component is made larger. Each curve is parameterized by a parameter called the "Rice parameter" α, which is a measure of the strength of the specular component divided by the RMS "random" component. For a Rayleigh distribution, this parameter is zero.

In principle, by making histograms of the amplitudes of the received signal and comparing them to the above curves, it is possible to determine if there is a single dominant scatterer within the radar beam. More complex variations on this process exist, including looking at the phase distributions (e.g., *Röttger*, 1980a) and using more complex distributions such as the Nakagami-M distribution (e.g., *Sheen et al.*, 1985; *Kuo et al.*, 1987). The latter generalization is particularly useful if the specular component has undulations on it and causes focusing and de-focusing of the reflected radiation.

More specific details will be considered in Chapter 7. The major hurdle to successful implementation of this technique lies in the correct choice of data length, as discussed in the section on D-region scatter. Data sets that are too short are statistically unreliable, and data sets that are too long suffer from too much geophysical variability. Nevertheless, some useful results were obtained. An example is Figure 7.24 in Chapter 7, taken from *Hocking* (1987b), which shows the height profile of the mean Rice parameter ($< \alpha >$) as a function of height measured with the SOUSY radar using a vertical beam and two off-vertical beams, one directed at 7 °off-vertical to the north, and one at 7 ° off-vertical to the east. Note the increase in $< \alpha >$ just above the tropopause when observing with the vertical beam, indicating the presence of a few dominant reflectors within the radar volume in the stratosphere. Notice also that there is still a non-Rayleigh character to the scattering process on the north beam, but on the east beam the mean Rice parameter is fairly constant with height and consistent with a Rayleigh process. One possible interpretation of these results is that there could be specular reflections at off-vertical angles. Other evidence of specular reflection in directions other than vertical were discussed by *Röttger and Larsen* (1990), and references therein, who discussed tilting of the specular reflectors by gravity waves, tilted isopleths and frontal boundaries. Further discussions will appear in Chapters 7 and 12.

2.11 VHF-MST radar methods for measuring the horizontal wind velocity

One of the most pragmatic applications of the VHF-MST radar technique was measurement of atmospheric winds. The radar method provided the capability of continuous wind coverage from the ground to heights of 20 km and more, a somewhat unprecedented capability. Although meteorologists were slow to accept profilers, over the period from 1980 to 2000 they have become important instruments in the meteorological toolbox.

The earliest measurements of winds with VHF-MST radars were generally made using the Doppler method, in which a high-gain beam is pointed to off-vertical directions of the sky, then Doppler shifts are measured, and the information from various beams is assimilated to produce an estimate of the horizontal wind over the radar. Pointing directions for the beams are usually within 20 degrees off-zenith or less, with 10–15 degrees being the most common angles used.

Studies with the vertical beam also had the potential to measure vertical wind motions directly, and this was originally flagged as an important capability. However, even small errors in the vertical beam direction could lead to large contamination of the supposedly vertical wind measurements by horizontal winds, and so for long-term mean vertical wind determinations, care is needed. Even if the radar beam is truly vertical, off-horizontal tilts in the scatterers themselves could cause similar biases. VHF-MST radars therefore can be used to measure vertical mean winds, but care is needed. Further complications arise when it is recognized that at least some of the waves observed are ducted (e.g., *Isler et al.*, 1997), and this will be further discussed in Chapter 11.

One area that turned out to be ideally suited to the application of VHF-MST radar was measurement of gravity-wave activity. The Doppler shift on the vertical beam was a relatively simple measurement to make, and the resulting temporal oscillations thereby measured (on time scales of minutes to hours) proved to be excellent for studies of gravity wave motions. Early results of this type of measurement, and the ensuing theoretical

In the section on MF/HF radar, the so-called IDI (imaging Doppler interferometry) method was also introduced, and this method soon found application in VHF/MST studies. Its tropospheric applications will be discussed later.

2.12 Measuring momentum flux and turbulence

In the section on MF/HF scatter from the D-region, the ability to measure momentum fluxes (e.g., *Vincent and Reid*, 1983) was discussed, as was the ability to measure turbulence strengths. Turbulence strengths were also discussed earlier with regard to the early days of meteorological radar. Determination of both momentum flux and turbulence were also soon implemented for MST/VHF studies, and several papers appeared capitalizing on tropospheric measurements of these papers (e.g., *Fukao et al.*, 1994, among others). It is interesting that although the basic principles of turbulence measurement were known to radar meteorologists in the 1970s and 1980s, its routine application in the troposphere waited till much later, and could not really be properly implemented until the unifying paper of *Hocking* (1996a).

However, an alternative method for turbulence measurements was also applicable in the troposphere that was not available for D-region measurements. This was application of absolute power using a calibrated radar. It also required background atmospheric information like temperature and humidity profiles, which is why it was restricted to lower altitude studies. Further to this, the theory of *Van Zandt et al.* (1978, 1981) was required to estimate the fraction of the radar volume occupied by turbulence. Because of these complicating features, the method has not been used widely, but examples can be found in *Hocking and Mu* (1997) and references therein.

Another significant development relating to MST radar was studies of the mechanism of diffusion in the atmosphere, and especially the stratosphere. Although this topic is not restricted to radar work and has more general implications, its origins related to the radar observations of thin turbulent layers, and so were grounded in MST studies. Traditionally, turbulence has been considered to be quasi-homogeneous, with all parts of the atmosphere being turbulent to some extent. Diffusion was considered to be a spatially continuous process. The discovery that turbulence existed in stratified layers, separated by regions of little to no turbulence, altered the understanding of how diffusion could occur vertically over large scales (scales greater than 1 km). *Dewan* (1981) and *Woodman and Rastogi* (1984) proposed that while diffusion across an individual layer was straightforward, movement across non-turbulent regions was slow, occurring at the rate of molecular diffusion rather than turbulent diffusion. Therefore these authors proposed that turbulent layers randomly appear and disappear at different altitudes (especially due to the random superposition of gravity waves), and a particle which manages to cross one layer needs to then essentially wait until another turbulent layer happens to form on top of it before it can be further diffused. The rate of diffusion therefore depends on the rate of formation of these layers, and their spatial distribution and depths. The details will be discussed in more detail in Chapter 11. For now, we note that this new approach to understanding turbulent transport was another important advance in comprehending the physics of the atmosphere.

2.13 Radar meteorology and networks using MST radars

As discussed earlier in this chapter, atmospheric radar development diverged around and after World War II, with one group concentrating on meteorological applications, and another concentrating on ionospheric studies. The purpose of the latter studies was in part to assist the operation of world-wide communications systems which depended on ionospheric reflections. Meteorological systems tended to operate at frequencies of thousands of MHz (L-band, S-band and higher). The discovery that MST-VHF radars could make useful meteorological measurements was not at first regarded to be of much significance by the meteorological community.

At regular conferences on radar meteorology, such as the ones sponsored by the the American Meteorological Society, presentations were generally about these higher frequency systems. However, they did not have the height coverage of VHF systems, since they looked almost horizontally. They had limited clear-air capability, and it was not possible to easily determine atmospheric winds with them, especially in clear air conditions. It took until the early 1980s to convince this community about the merits of the MST radars, when reviews in the *Bulletin of the American Meteorological Society* (e.g., *Balsley and Gage*, 1982), *Larsen and Röttger* (1982) were accepted. Slowly, over time, the advantages of VHF radars were recognized. Smaller VHF radars (capable of seeing only into the troposphere and lower stratosphere (ST radars) were developed for purely meteorological applications. Such systems have been set up in networks in several countries, which will be described in Chapter 12. These include networks in Europe and the USA. In some cases (e.g., Europe) VHF frequencies were chosen, and in others (e.g., USA) frequencies around 400 MHz were selected (mainly with the hope of getting winds lower to the ground). Despite the different frequency, the 400 MHz systems still used all the new techniques developed for VHF radars, and so are still considered to be in the same class of radar-type. ST radars designed principally for tropospheric and stratospheric studies, and which used near-vertical beams, became known as "windprofilers," regardless of their frequency. The networks became very popular with meteorologists, being especially useful for assisting with forecasts 24–36 hours in advance and with detection of precursors of tornadoes (e.g., *Benjamin et al.*, 2004).

During the middle and late 1970s, a major international program was established, called The Middle Atmosphere Program, or MAP. The program was established since it was recognized that the altitude region between the tropopause and the lower thermosphere (called the middle atmosphere) was the least explored region of the Earth's atmosphere. This program, developed under the auspices of the Scientific Committee on Solar Terrestrial Physics (SCOTEP), boosted the funding for further MST radar systems, which were regarded as important for the research during MAP. The International Union of Radio Science (URSI) also recognized the importance of these new radar systems. Since the early 1980s, regular workshops were held which brought together radar engineering and scientific user communities. These workshops on technical and scientific aspects of MST radar were critical for the mutual understanding of the engineering and the scientific community, which was another benefit observed during the historical development of the MST radars. The tenth workshop of this kind, MST-10, was held in 2003 in Piura, Peru, the country where the MST radar technique was born.

Coordination at all levels was an important aspect of MST work during the 1980s and 1990s and even into the 21st century. Local networks were developed, and (especially for mesospheric studies) global-scale networks were also established. Cooordination with members of the modeling community helped unify results.

With regard to meteorology, VHF-MST/ST radars and windprofilers generally have become important parts of forecasting and nowcasting. Many individual features have been studied with these radar, including frontal systems, thunderstorms, typhoons, and other synoptic and mesoscale events. One of the earliest studies of frontal systems by VHF radar is highlighted in Figure 12.2, Chapter 12 (from *Larsen and Röttger*, 1982), and many others have followed. Studies of stratosphere–troposphere exchange have been possible. Working at VHF and UHF, it has also been possible to study water precipitation with these radars. Lightning channels also produce reflection of radiowaves, and it was possible to study these with VHF/ST radars as well. We will not discuss these many experiments here, but leave them for more detailed discussion in Chapters 10 and 12. ST/VHF profilers now work side by side with meteorological precipitation radars on a regular basis. It is interesting that MST radars have, to some extent, re-united the different branches of atmospheric radar (meteorology and ionospheric) that developed after World War II.

2.14 Strange scatterers in the polar upper atmosphere

In 1981, Figure 2.24 appeared in the literature, showing echo occurrence over the large Poker Flat VHF-MST radar in Alaska, USA. It will be seen that echoes appear at 82–85 km in summer. At the time, little was made of this data, but in subsequent years it came to be realized that this was an anomaly. Scattering eddies for 50 MHz radiowaves must be of the order of 2–3 meters in size (e.g., *Briggs and Vincent*, 1973; *Hocking*, 1987a). At these altitudes, molecular diffusion rates are relatively high, and an eddy formed at these scales would quickly be destroyed, as velocity fluctuations on one side of the eddy diffused across to the other side. This results in eddies of this size being part of the so-called "viscous range" of the turbulence spectrum. Therefore, there should be almost no eddies and therefore no scatter at these scales. There simply should not be any echoes from this height.

The anomaly in these observations appears to have been first documented by *Kelley and Ulwick* (1988) and *Kelley et al.* (1987). In fact, perhaps it should have been recognized earlier, since Figure 2.16 also shows echoes from around 85 km. In any case, the realization that these echoes should not exist using classical turbulence theories led to a flurry of investigation about their cause. The key appears to be recognition

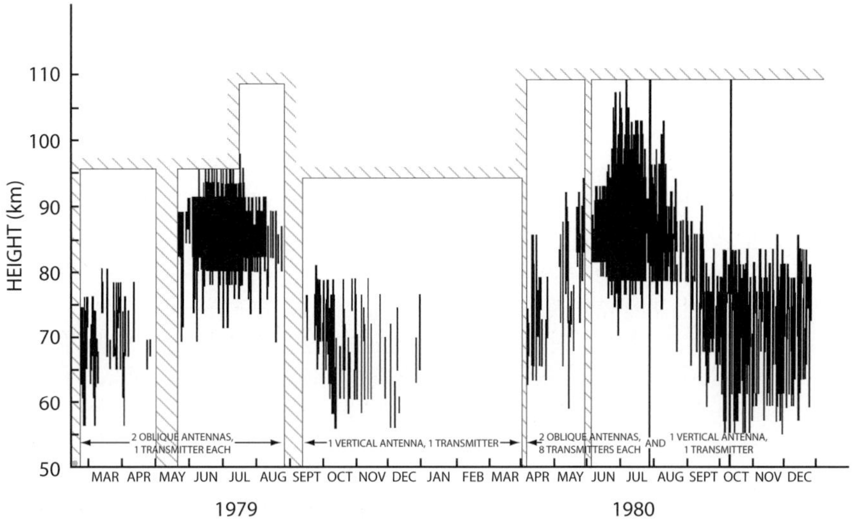

Figure 2.24 Evidence of backscattered power from above 80 km altitude observed with the Poker Flat VHF-MST radar (from *Ecklund and Balsley*, 1981). Occasions when the signal-to-noise ratio exceeded some (unspecified) critical value were marked with a small vertical line, so that regions of dark black correspond to strong echo occurrence. (Reproduced with permission from John Wiley and Sons.)

that the so-called "Schmidt number" at these heights may be quite large. The Schmidt number is the ratio of the molecular diffusion coefficient of the neutral gas to that of the electrons. In the case of the summer echoes at Poker Flat, it seems to be very large – of the order of 1000 or more. The reasons relate to the positively charged ions at these heights. If the ions are massive, they will move sluggishly, and so slow the diffusion of the electrostatically attracted electrons. *Kelley et al.* (1987) proposed that the positive ions were large water cluster-ions that had formed due to the very low temperatures that exist in the polar summer at 80–90 km altitude (see the section of Chapter relating to the mesospheric circulation, and especially Figure 1.23). Later studies recognized that the water-cluster ions would not have this effect, but electrons loosely bound to even larger positively charged aerosol particles (or even anti-correlated with negatively charged aerosols) might.

These echoes came to be known as PMSE (polar mesospheric summer echoes). The "aerosol model" was introduced by *Cho et al.* (1992) and came to be known as the "dressed aerosol" model because the aerosols behaved as if they were "dressed" in a layer of electrons. Subsequent results and reviews of PMSE were presented by, among others, *Röttger et al.* (1988), *Cho and Kelley* (1993), and *Cho and Röttger* (1997). *Woodman et al.* (1999) drew attention to significant differences in PMSE detection in the northern and southern polar regions, with northern hemisphere echoes seeming to be much stronger.

Studies of PMSE are still an ongoing area of research. More extensive discussion of these fascinating echoes will appear in Chapter 10.

2.15 Imaging, improving spatial resolution, and application of interferometry

2.15.1 Introduction

While producing data using a few discrete beams allows a lot of information to be determined about the atmosphere, some of this information is based on assumptions about the spatial variation of the signal. For example, it may be assumed that the atmosphere is statistically homogeneous over all the beams, or varies smoothly from beam to beam. However, in truth the spatial variation could be much more complex, and one beam may in fact probe a region of turbulence while another might point at a region of laminar flow, yet determinations of overall winds speeds assume that regions of space viewed by the beams have similar behavior. These problems could be bypassed if it was possible to look into every possible direction. Likewise when looking vertically, better resolution would unquestionably help in interpretation of the vertical structure. With newer and faster digitizers and computers, the possibility of obtaining higher resolution, and more complete coverage of the sky, has become a reality. Several methods can be used to do this, and some of these will be discussed below.

Resolution issues also relate to radar bandwidth. In order to use a radar, it is necessary to obtain a frequency allocation. Government departments usually control such allocations. Both a central frequency and a bandwidth are allocated. A user might receive a central frequency of 45.5 MHz and a bandwidth of maybe 600 kHz. The bandwidth limits the minimum pulse-length that can be transmitted, with wider bandwidths allowing shorter pulses and thus better height resolution. This is a consequence of Fourier theory: it can be shown that the product of the pulse length and the minimum associated bandwidth required to allow that pulse to be transmitted relatively undistorted should be of the order of 1. The rule is a parallel to the Heisenberg uncertainty principle (HUP) in quantum mechanics – indeed the reciprocity between widths of Fourier-transform pairs in the the spatial and inverse-wavenumber domains, and in the temporal and frequency domains, was the basis for development of the HUP, and was known in Fourier theory well before it was applied in quantum mechanics.

At frequencies of around 50 MHz, typical bandwidth allocations are of the order of 250 to 500 kHz, and special permissions are needed to procure allocations of 1 MHz and more. Wider bandwidths can be obtained if the central frequency is of the order of 1 GHz, but such allocations limit the radar's ability to see into the atmosphere to heights greater than about 15–20 km (due to fundamental limits associated with the smallest detectable turbulence scales), limiting the radar applications largely to the troposphere. A 1 MHz bandwidth limits the best available height resolution to about 150 m.

The beam-width available to a radar depends on the physical size of the radar. Larger radars have narrower beams and hence better angular resolution, again because the radar aperture and the beam-pattern are Fourier-transform pairs. A radar that is about 16λ wide (where λ = wavelength) has a half-power beam full-width of typically 3°. The beam-width limits the available angular resolution.

While the above theories about the reciprocity of functional widths in the Fourier domains are generally true, their determination of resolution limitations is based on the assumption that nothing at all is known about the scattering region. However, these limits can sometimes be bypassed if additional assumptions about the scattering region can be made. If, for example, it is known that all the scatterers are point scatterers, like stars in the sky, then these limitations can be bypassed. This is often done in astronomy. If signal is recorded from a star, the phase differences between antennas on the ground can be used to determine the angular location of the star in the sky. If the antennas are say 5 wavelengths apart, Fourier theory tells us that the location can only be accurate to about 10 degrees. But if it is known that the object under detection is a star – essentially a point source – then angular accuracy of a degree or better can be assumed. Likewise in atmospheric studies, if it can be assumed that the scatterers come in the form of discrete, well-defined scatterers of small physical extent, improved resolution is possible.

2.15.2 Resolution improvement

As will be seen in more detail later, the received amplitude versus range profile is a convolution between the refractive index gradient and the transmitted pulse. For readers unfamiliar with this process, the details are discussed in Chapter 4, but for now it is only necessary to know that the convolution processes can be inverted by applying special deconvolution procedures, which in principle can lead to improved resolution. However, it is a process that depends very much on how the noise is treated, and if done wrongly, can lead to very misleading results. It does require that the profiles are digitized at steps substantially smaller than the pulse resolution. In the late 1970s, digitization rates were too slow for it to be routinely applied, but *Röttger and Schmidt* (1979) did make the attempt. They used a strategy by which successive profiles on successive pulses were sampled at different sets of range-bins, each staggered in start-bin relative to the last, and were able to use these data, after application of deconvolution theory, to achieve 30 m resolution and show the existence of Kelvin–Helmholtz instabilities. This was a significant improvement in resolution over the more common 150 m resolution. *Reid et al.* (1987) also demonstrated cat's eye Kelvin–Helmholtz-like structures using high resolution studies.

The resolution improvements discussed above were within the expectations of standard Fourier theory. While applied as a demonstration in the above cases, the procedure was not used routinely until much later (*Hocking et al.*, 2014). The above procedure is also only valid to help with improvements in height resolution – it cannot help with angular resolution. In the meantime, the procedures used for improvement in resolution, particularly through the 1990s and into the 21st century, relied on interferometric procedures.

2.15.3 Interferometry

One important new method introduced in the 1980s was interferometry, where multiple receivers on the ground can be used to effectively produce an almost unlimited

It may be fair to say that there are really only three distinctive areas: (1) Wind determinations using spaced antennas; (2) Angular imaging (AIM); and (3) Range imaging (RIM). For this section, item (1) will not be considered so much, as it is not usually used in an imaging sense. Also in the context of this section, angle-of-arrival (AOA) techniques are not really imaging methods – their primary application is in Doppler vertical velocity correction. They may be considered loosely as subsets of either item (1) or (2). All other categories are subsets of these, albeit with different analysis techniques, numbers of frequencies, antennas, and so forth.

One of the nicest studies of imaging was presented by *H'elal et al.* (2001). The transmitter beam-width was kept quite wide – greater than 100 ° – but a large number of separate receivers were employed which could be used to either carry out interferometry or form multiple beams. The investigations concentrated on the troposphere, using up to 16 receiving channels. An example is seen in Figure 2.28. Layers are clearly seen, but the physical extent of the layers is an artifact. In truth they were much wider, but signal starts to disappear at the edges because anisotropy of the scatterers in the layers suppresses the signal received from lower elevations. The layers which appear to be widest correspond to layers in which the scatterers have greater isotropy. Layers which appear to be less broad in extent in the figure are actually ones which have greater aspect-sensitivity (more elongated horizontally). Further refinements were investigated using Capon and MUSIC algorithms, but the improvements were modest: most of the important details are apparent in Figure 2.28.

Mead et al. (1998), concentrating on boundary-layer studies, developed an interferometric 915 MHz radar at the University of Massachusetts, which they called the Turbulent Eddy Profiler (TEP) (also see *Pollard et al.*, 2000). The radar was

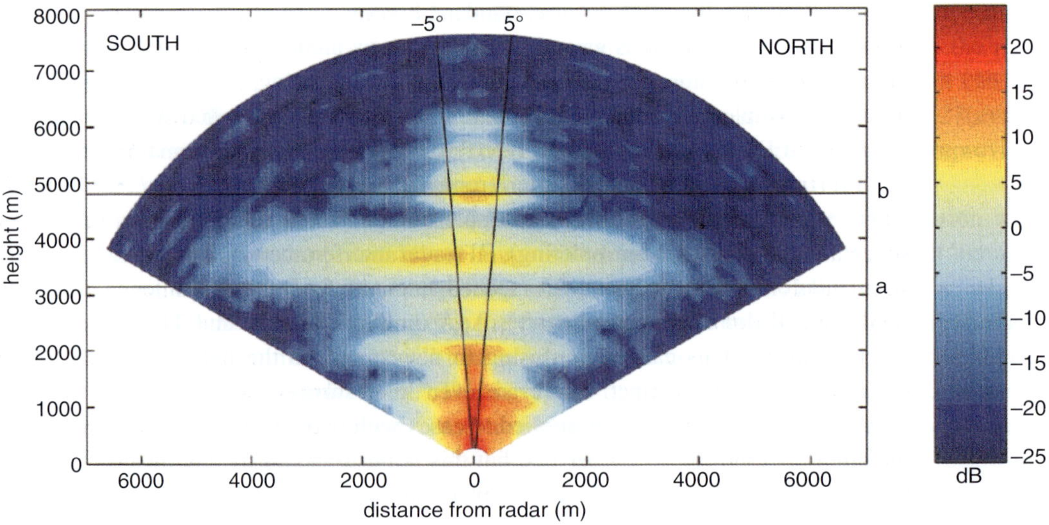

Figure 2.28 Scattered power as a function of both range and angle (from *H'elal et al.*, 2001). See the text for further details.

simultaneously able to use up to 64 independent receivers, and used a transmit beam with 25 ° width. High-resolution images with temporal resolution of about 2.5 s were possible (*Lopez-Dekker and Frasier*, 2004; *Cheong et al.*, 2004). The TEP system was also used to study the interaction of precipitation and clear-air turbulence, utilizing spectral sorting (*Palmer et al.*, 2005). Separate three-dimensional images of precipitation and clear-air echoes with good spatial and temporal resolution were also produced using the Capon CRI (or AIM) method. *Cheong et al.* (2006) used the adaptable TEP receive array to mitigate moving clutter (e.g., birds, aircraft) effects due to grating lobes, demonstrating the quality-control aspects of the system. Lastly, the TEP system was also used for a detailed study of the DBS method by comparing CRI results to various DBS beam configurations (*Cheong et al.*, 2008).

Other developments include a paper by *Chau and Woodman* (2001), who looked at 3-D coherent radar imaging, and *Yu and Brown* (2004), who simultaneously combined range interferometry and spaced-antenna methods in a technique they referred to as RIM-SA. The application by *Yu and Brown* (2004) employed three receivers and two distinct frequencies, with different frequencies being interleaved on a pulse by pulse basis. *Röttger* (2013) also looked at combined spatial and frequency domain interferometry, though with a modestly different emphasis.

It probably goes without saying that for any interferometric system to work well, great care is needed in ensuring that the phases of all the antennas and receive paths (including cables and receiver delays) are properly measured and accounted for. More detailed discussions about special cases of these various interferometric modes are given in Chapter 9.

2.15.7 The relation between IDI and FCA-type methods, and the validity of point scatterers

A few pages back, the issue of the reality of the scattering points utilized in interferometry was raised, just prior to the subsection on FDI. We have discussed in some sections of the chapter the possibility of so-called Fresnel reflectors, so if these are real, then they serve as the scattering points, at least for vertically directed beams and FDI. But if the scatter is due to turbulence, do such scatterers really exist? What about the case of angular interferometry, where we assume discrete scatterers exist at angles significantly offset from vertical? These cannot be specular reflectors, since our theory on specular reflectors is based on horizontally aligned reflectors. Such reflectors that are significantly off-vertical have never been seen, so they cannot function as our "point scatterers." Finally, if the point scatterers are not real, does this invalidate measurements made assuming their existence? Perhaps the method works even if the scatterers are some sort of artifact. Extensive work was done investigating these questions.

As part of such studies, *Franke et al.* (1990) compared winds deduced with the IDI method to the more traditional spaced antenna winds. These methods will be discussed more extensively in Chapter 9, but for now the reader does not need to know all the details about how they work – just that they are in principle different, and in contrast to interferometric methods, the spaced antenna winds method makes no assumptions about the existence of discrete points of scatter. The spaced antenna wind technique produces two estimates of the wind, denoted as the "apparent" and "true" winds. These author concluded that the so-called "apparent velocity" and the IDI winds were similar, while the so-called "true velocities" used in spaced antenna FCA (full correlation analysis) were underestimates of the actual physically real wind speeds, and were anything but "true winds."

Briggs (1995) proposed that if scatter was truly due to turbulence, artificial "scatterers" would result as artifacts of the very assumption that they should exist! If this were so, then quasi-sensible winds might be deduced using IDI, but they would overestimate the true winds. So these authors, in contrast with *Franke et al.* (1990), were of the opinion that the "apparent winds" from FCA, and the IDI winds, were both overestimates of the true winds. *Holdsworth and Reid* (1995) adopted a similar theme.

Adopting the opposite viewpoint, *Roper* (1996) and *Roper and Brosnahan* (1997) claimed that the scatterers were real, even for turbulent scatter. They argued that because the idea of smoothly formed eddies in turbulence was a myth, and that real turbulence comprises twisted and contorted structures – even string-like (e.g., *Hocking and Hamza*, 1997) – then glints and regions of enhanced scatter will arise naturally, appearing as "point scatterers." *Roper and Brosnahan* (1997) did note that due to the short-lived nature of these scatterers, more IDI scatterers were needed to make a reliable wind measurement than was the case for meteor trails.

Holdsworth and Reid (2004a, b) then used an MF radar in Australia to analyze three years of comparisons between FCA and IDI analyses. They showed that the IDI winds were statistically larger than the FCA winds (*Hocking et al.*, 2001c). These differences were attributed to both an overestimation of the IDI winds and underestimation of the FCA winds. They developed refinements to the IDI theory (mainly involving application of thresholds to limit the range of accepted values), and adopted a compromise between the two sets of measurements to produce optimal values.

One interesting side-effect of these discussions was a renewed interest within the research community in Spaced Antenna Wind methods, which had previously been considered of less value than Doppler methods. *Liu et al.* (1990) show analytically that when turbulence is neglected, some sort of equivalence between SDI/IDI and spaced antenna drift methods can be demonstrated. *Briggs and Vincent* (1992) developed a modified spectral version of full correlation analysis, and *Sheppard and Larsen* (1992) showed that a full-spectral analysis (FSA) applied to SDI data is equivalent to full correlation analysis (FCA).

The earlier FCA analysis by Briggs and colleagues concentrated primarily on the diffraction pattern on the ground, and did not concern itself with the nature of the scatterers that produced the scatter. *Doviak et al.* (1996) took a new approach, emphasizing the role of the scatterers themselves, with some evidence on turbulent scatter. *Chau and Balsley* (1998a) further extended studies of AOA contributions due to tilted layers, and also considered off-vertical transmitting beams and various geometrical effects. *Praskovsky and Praskovskaya* (2003) developed a structure-function approach to spaced antenna analysis, in contrast to the more common use of correlation functions. Overall however, some level of compromise was reached that both IDI and SA methods can be used for wind measurements, although some form of calibration may be needed. The details of these various techniques will be further explained in Chapter 9.

2.16 Temperature measurements and RASS

Another area of some historical interest in MST radar studies was measurements of atmospheric temperature profiles. While not a primary objective of the radar work, some interesting techniques were developed. None has become mainstays of the radar core research, but each has seen some periods of research interest.

In an earlier section of this chapter, the existence of "Fresnel reflectors" was recognized. *Gage and Green* (1982a) and *Gage et al.* (1985) suggested that in regions of stable air, Fresnel scatterers seemed more common. They proposed that the amount of Fresnel scattering might correlate with temperature gradients. Increased Fresnel scattering generally enhances the backscattered signal strength, so these authors attempted to develop a procedure for determining the temperature profile in the troposphere, utilizing the scattered signal strength to deduce stability (and so temperature gradients) as a function of height. (These authors originally did not recognize that the backscattered power is proportional to the pulse-length (see earlier), but this did not negate the general principle of the model.) Further developments recognized that it is not just the specular reflectors that can enhance the backscattered signal, but also turbulence (e.g., *Klaus*, 2008). This complicates matters, since tubulence has the odd feature that when it is very strong, the backscattered powers can be very weak. This is because very strong turbulence can mix the turbulent layers so severely that the internal temperature gradients within the layers become close to adiabatic, so that displaced parcels of air do not show much density contrast to the surrounding air. This results in weak backscatter. Hence it cannot be certain whether weak scatter is associated with very weak or very strong turbulence. Furthermore, regions of high stability, while having weak to zero turbulence, might have strong backscatter due to Fresnel reflectors. Due in part to these complications, and even paradoxes, the method has never really been further developed, and it has not been used much in the last 25 years.

Some extra discussion of these methods is provided in the chapter on meteorology later in this book (Chapter 12).

Another quite different way to measure temperatures is the use of gravity-wave spectra. Gravity waves will be discussed in more detail later, but one key point about them is that their spectrum has a sharp cutoff at high frequencies, greater than the Brunt–Väisälä (BV) frequency. The BV frequency depends on the local temperature gradient, so if the gravity-wave cutoff can be detected by suitable spectral analysis of the velocity fluctuations (usually measured with a vertically-directed beam – see the next sub-section), then the temperature gradient can be found and thence the temperature profile deduced by integration. The method is limited to conditions when winds are very weak – non-zero mean winds smear out the cutoff frequency and void the method. It has been applied on a few occasions (e.g., *Röttger*, 1986; *Revathy et al.*, 1996). A sample spectrum of gravity waves produced in application of this method can be seen later in Figure 12.30.

Each of the above methods is limited. The first is only approximate, the second requires generally still air (no wind). To supplement these methods, a method called "RASS" was introduced. RASS is an acronym for "Radio Acoustic Sounding System." It is a radar method for measurement of temperatures as a function of height in the troposphere and lower stratosphere.

The RASS method utilizes a combination of sound-waves and radar, and produces reliable results. However, it is limited in application because of the need for a sound-source, which can be annoying to nearby neighbors.

Early investigations of this method were introduced by *Marshall et al.* (1972) and *North and Peterson* (1973), although these had limited height range, and seem to have been based mainly around FM-CW and microwave radars. Other systems (also FM-CW and/or microwave) were discussed by *Bonino et al.* (1979) and *Peters et al.* (1985), with later studies presented by, for example, *Peters et al.* (1988). It was in the 1980s that the method began to be adopted more regularly by the MST community. The principle of the method is described below.

Consider a radar with wavelength λ_r. In RASS, a sound wave is transmitted into the atmosphere with a wavelength $\lambda_s = \lambda_r/2$. This produces periodic artificially induced wavefronts in the air with wavelength λ_s. A radar pulse is transmitted, and receives Bragg reflections from the propagating acoustic wave, which is of course moving at the speed of sound in air. The Doppler shift of the received signal is then measured. It is typically of the order of 100 Hz or so for a 50 MHz radar – much higher than the Doppler shifts due to natural atmospheric scatterers. The Doppler offset gives the speed of the acoustic wave, and since the speed of sound in air is temperature-dependent, the temperature of the air may be deduced.

The source of the sound waves does not need to be particularly special. It can be CW, or pulsed, or even white noise. The main requirement is just that the source is strong enough to produce useable refractive index variations in the air at the desired wavelength. The possibility of using environmental sources, such as traffic on a busy road, or a large waterfall, is probably worth pursuing in the future, although has not been done yet.

Having measured the Doppler offset due to the sound waves, the following relation may be used to determine the temperature:

$$c_s = \sqrt{\frac{\gamma p}{\rho}} = \sqrt{\gamma R T_v} \approx 20\sqrt{T_v}. \tag{2.5}$$

Here, T_v is the *virtual* temperature, which is not quite the true atmospheric temperature. In the troposphere, air is a mixture of dry air and water vapor. Because water vapor has a large value of latent heat, and water has a high specific heat, the presence of water has an effect on the thermodynamics of the air that belies its relative mass. So the existence of water vapor has an impact on the temperature of the air. The *virtual temperature* is the temperature that currently moist air *would have* if it were replaced by dry air at the same pressure and density as its current (moist) values. More specific details about how virtual temperature is calculated will be given in Chapter 12, but for the present, we simply indicate that the speed of sound in air given by the RASS method is a direct calculation of virtual temperature.

The RASS method is not without limitations, however. In windy conditions, the sound waves are moved around and become distorted. For useful radar backscatter, the sound wave-fronts must satisfy a specularity condition for the radar reflections. If the wind twists the sound wave-fronts sufficiently, the specularity condition can be lost, and so the radar signal reflected by the sound waves will no longer arrive at the radar receiver. This often limits the upper altitude to which useful temperature measurements can be made, which frequently is in the range 2 to 5 km.

One interesting exception is the Japanese MU radar. *Matuura et al.* (1986) were able to extend the height range of the method to previously unachievable heights, using the MU MST radar, and achieved a 21 km altitude – still the highest measurement achieved. This was done using pulse-to-pulse beamsteering, allowing the radar to scan the sky searching for remnants of the acoustic wave which still satisfied specularity conditions. Multiple sound sources scattered around the radar can also help. Subsequent applications with MST and windprofiler radars were developed in the late 1980s and the 1990s (e.g., *Peters et al.*, 1985; *May et al.*, 1988b, 1990; *Adachi et al.*, 1993; *May et al.*, 1996; *Tsuda et al.*, 1994; *Yamamoto et al.*, 1996; *Angevine et al.*, 1994, among others). Interesting developments include the use of multiple distributed acoustic sources with RASS (*Klaus*, 2008), and more routine measurements of temperature profiles up to 20 km altitude (*Alexander and Tsuda*, 2008). Researchers in India have also made interesting studies using the powerful Gadanki Indian radar (e.g., *Sarma et al.*, 2008).

2.17 Precipitation measurements with MST radar

In earlier sections of this chapter, the usefulness of C-band and S-band and higher frequencies were discussed for measurements of hail, rain, and snow. Generally the strength of MST radars is their ability to detect clear-air scatter, and indeed it is an advantage in this regard that they are not especially sensitive to precipitation. Hence they can measure reliable clear-air winds in zero, light, and moderate rainfall conditions. Traditionally the higher frequency radars have been the primary ones used for rainfall studies. But VHF radars have been found to have some use for studies of precipitation, since rainfall does produce a contribution to the spectrum, albeit often a smaller one.

Fukao et al. (1985c) were the first to study precipitation with MST/VHF radars. Subsequent papers showed how the neutral air and precipitation spectra could be distinguished, and also looked at the shapes of spectra, relating them to drop-size distribution studies (*Gossard*, 1988; *Sato et al.*, 1990; *Rajopadhyaya et al.*, 1993; *Maguire and Avery*, 1995). *Chilson et al.* (1993) used dual frequency studies, especially comparing results from radars working at around 50 MHz and 400–900 MHz, to better contrast the neutral air and precipitation spectra.

When a radar is absolutely calibrated in terms of signal power, quantitative rainfall rates can be determined. (Procedures for radar calibration will be discussed later in the text, especially in Chapter 5.) For example, *Sato et al.* (1996) used absolute calibration of the MU radar to supplement and help calibrate the precipitation satellite TRMM. *Campos et al.* (2007a, b) were able to use the calibrated McGill VHF radar for absolute rainfall measurements, and obtained accuracy of better than 1.5 dB. While precipitation measurements are not a key aspect of MST radar studies, useful work can be undertaken in this area. A slightly more extended discussion of recent developments in precipitation studies, and also with the developing field of humidity measurement by VHF radar, will be given in Chapter 10.

2.18 Additional applications

Of course, new developments are continually occurring, and in a book like this it is hard to draw a line between "history" and "new developments." A few more recent research topics include attempts to determine humidity profiles from radar backscatter data. *Tsuda et al.* (1997b), *Mohan et al.* (2001), and *Furumoto et al.* (2003) are representative of some of the earlier papers in this area.

As will be seen in later chapters, key terms in determining the strength of backscattered signal from the troposphere are the temperature gradient and the humidity gradient. If a radar is calibrated for absolute power, and if the temperature gradient can be deduced by independent means (either by RASS, or by radiosondes for example), then in principle the humidity gradients can be determined as a function of height. If the humidity gradient is known as a function of height, then the gradient-profile can be integrated from the ground upward, in principle leading to a final height-profile of humidity. The problem with the method is that the backscattered power depends on the *square* of the humidity gradient, so it is not easily possible to determine the sign of the gradient. This prevents the integration just described. A major focus in this type of research is to find ways to determine the sign of the humidity gradient.

Methods for determination of turbulence strengths have been discussed earlier. *Van Zandt et al.* (2000) developed an interesting technique, which utilized comparisons of backscattered power at two different frequencies. These authors used Hill's model of turbulent refractivity fluctuations (*Hill and Clifford*, 1978), together with simultaneous observations of radar reflectivity by S-band and UHF-band radars, to determine turbulent energy dissipation rates per unit mass (ε). The technique was self-calibrating, requiring only determination of the relative powers received by the two radars during rain for calibration.

Another area of some interest is the absolute calibration of radars. For wind measurements, absolute calibration is not needed, but it is increasingly becoming a necessity for other data. Determination of C_n^2 is one example where it is required, and the measurements of humidity discussed above require it. *Hocking et al.* (1983); *Green et al.* (1983); *Mathews et al.* (1988) and *Hocking and Röttger* (1997) discussed earlier calibration procedures, and *Campos et al.* (2007a) and *Swarnalingam et al.* (2009b) show more recent calibrations.

Other recent events include the development of new networks of windprofilers for meteorological studies, and use of MST radars/windprofilers to study hurricanes/cyclones/typhoons and other specialized mesoscale phenomena. New high-resolution radars have also been developed: the Turbulent Eddy Profiler was discussed earlier, although that is primarily a boundary-layer instrument. Other fine-resolution techniques have been developed, which will be discussed further in subsequent chapters, notably Chapters 5 and 6. We will consider developments beyond the above discussions as "recent" rather than historical, and consider them in more detail within subsequent chapters. Even many of the items already discussed above will be revisited in later chapters, with greater detail and better mathematical development.

We therefore now turn to more detailed investigations of the basics of radar, beginning first with a more thorough approach to the details of atmospheric refractive index (Chapter 3), and its many associated phenomena, and then moving on to more specifics about radar and antenna design (Chapters 4 to 6). Following that we will proceed to examine the methods used to extract information about atmospheric radar targets (Chapter 7), and then signal processing techniques (Chapter 8). Subsequent chapters will look at more subtle radar techniques, and real applications relating to waves, turbulence and meteorology of the atmosphere.

3 Refractive index of the atmosphere and ionosphere

3.1 Introduction

The most important atmospheric parameter for applications of MST radar is the refractive index. Refractive index variations define the propagation and scattering of radiowaves as they move around in the atmosphere.

Therefore, this chapter covers extensively the relation of refractive index to other parameters like electron density, pressure, neutral density, temperature, etc. The early stages are simple, but the chapter builds in level of difficulty as it develops. It is not crucial that this chapter be fully understood before progressing to Chapter 4 and beyond. Chapters 4 to 6 are developed from a more technical perspective. If, for example, the reader is new to the field, or maybe more focused on radar construction and engineering aspects, rather than on fundamentals of scatter, this chapter could be skipped, or read only partially, and perhaps revisited later. In the end, it is highly advised that the chapter is covered, of course.

The refractive index, n, is simply the ratio of the speed of an electromagnetic wave in a vacuum (c) divided by its speed in the current medium, viz.

$$n = \frac{c}{c_\phi}, \tag{3.1}$$

where c_ϕ is the phase speed of the wave in the medium of interest.

For most MST applications, the refractive index is close to unity, and generally the propagation direction can be considered to be in a straight line (i.e., little to no refraction). However, when a radiowave enters the atmosphere, it frequently encounters regions where the refractive index changes. These changes in refractive index cause small but not insignificant fractions of the incident radiowaves to be reflected. This reflected signal returns to the radar receivers, where it is amplified and stored (usually digitally), and this received signal becomes the basic data from which all of the necessary parameters for subsequent analysis are derived.

There are exceptions to the rule that there is no refraction. Once radiowaves enter the ionosphere (a plasma), the refractive index can deviate substantially from unity, and refraction can become important. Sometimes MST radars are used for ionospheric investigations, so it is useful to understand a little about index variation in a plasma. But it also happens that the refractive index of radiowave propagation through a simple plasma is a relatively easy function to derive, and so we will begin this chapter by deriving the refractive index for radiowave propagation through an electron gas. This derivation is not only relatively simple, but also is a useful basis for dealing with more complex media.

First we will give a general equation for the form of a propagating wave, and then we will consider the refractive index for some simple cases, beginning with the simple electron gas.

3.2 Wave representation

A wave is a function in space that changes position systematically as a function of time, and generally satisfies the wave equation

$$\nabla^2 \phi_D = \frac{1}{v_{\phi_D}^2} \frac{\partial^2 \phi_D}{\partial t^2}, \tag{3.2}$$

where ϕ_D could be any parameter like the displacement, velocity, electric field, magnetic field etc. This differential equation describes the motion of a pulse on a stretched string, sound waves, water waves, and so forth.

The most general solution in one dimension is

$$\phi_D(x, t) = f(x - vt), \tag{3.3}$$

where f is, in principle, any function, but generally we consider it as any bounded continuous function, and v is the velocity (either positive or negative) at which the waveform moves. A positive velocity corresponds to motion to the right along the x-axis. (Sometimes $\phi_D(x, t)$ is written as $f(x + vt)$, where in this case v is positive to the left; mathematically this is sometimes more convenient, but we will use the former definition.)

We can equally write Equation (3.3) in the form

$$\phi_D(x, t) = f_2(kx - kvt) = f_2(kx - \omega t), \tag{3.4}$$

where we have simply multiplied the term inside the brackets by a constant k, and ω is simply defined as kv.

Because any such function can be represented as a Fourier sum or Fourier integral, we often consider the solution to be of the form

$$\phi_D(x, t) = \sum_\ell \sum_j \xi_{\ell,j}, \tag{3.5}$$

where

$$\xi_{\ell,j} = \xi_{0(\ell,j)} \mathrm{e}^{\mathrm{i}(\vec{k}_\ell x - \omega_j t)}, \tag{3.6}$$

with $\mathrm{i} = \sqrt{-1}$.

We could equally use

$$\xi_{\ell,j} = \xi'_{0(\ell,j)} \mathrm{e}^{\mathrm{i}(\omega_j t - \vec{k}_\ell x)}, \tag{3.7}$$

or even

$$\xi_{\ell,j} = \xi''_{0(\ell,j)} \mathrm{e}^{\mathrm{i}(\omega_j t + \vec{k}_\ell x)}, \tag{3.8}$$

but we will persist with (3.6) as our standard.

This approach allows us to individually solve for separate Fourier components, which is relatively easy because of the ease with which a complex exponential function can be differentiated. The solution is also easily visualized as a set of sinusoidal oscillations in space and time. In addition, often the desired solution is in fact a single Fourier component.

When the solution is considered in this way, we see that for fixed time t, the function repeats in space at steps Δx, where $\Delta x.k = 2\pi$. This spatial repetition length is called the wavelength, denoted here as λ, so that $k = \frac{2\pi}{\lambda}$. k is called the wavenumber. Likewise, for fixed position x, the function repeats itself in time at steps of Δt, where $\Delta t.\omega = 2\pi$. The interval Δt is defined as the period, T, and we see that $\omega = \frac{2\pi}{T}$. This is called the "angular frequency." The quantity $f = \frac{\omega}{2\pi}$ is simply called the frequency, and represents the time for the wave to repeat itself. The wave speed is $v = \frac{\omega}{k}$.

In three dimensions we can write each Fourier component as

$$\xi = \xi_0 e^{i(\vec{k}\cdot\vec{r} \pm \omega t)}, \tag{3.9}$$

where

$$\vec{k}\cdot\vec{r} = k_x\vec{x} + k_y\vec{y} + k_z\vec{z}. \tag{3.10}$$

The choice of the sign (±) defines the assumed direction of positive propagation, and we will again use a minus sign, viz.,

$$\xi = \xi_0 e^{i(\vec{k}\cdot\vec{r} - \omega t)}. \tag{3.11}$$

The quantity ξ may be complex, and can be written as $\underline{\xi}_0 = |\underline{\xi}_0| e^{i\varphi}$.

As an example, in one dimension a wave can be written as

$$\underline{\xi} = \underline{\xi}_0 e^{i(kx - \omega t)} = |\underline{\xi}_0| e^{i(kx - \omega t + \varphi)}, \tag{3.12}$$

where φ is a phase offset.

As is normal for complex analysis, the phase φ can be found as

$$\varphi = \arctan\left(\frac{\mathrm{Im}(\underline{\xi}_0)}{\mathrm{Re}(\underline{\xi}_0)}\right). \tag{3.13}$$

The phase velocity is given by

$$\frac{dx}{dt} = \frac{\omega}{k} = v_p, \tag{3.14}$$

as expected from our original definition of k and ω.

When the wave is not a pure continuous wave with a single frequency, but comprises a collection of waves of closely-spaced frequencies and wavenumbers, the "collective" travels as a so-called "packet," and can be considered as a continuous carrier wave modulated by some amplitude and phase variation. The carrier wave travels at the phase speed of the wave, but the modulation may travel at a different velocity, called the "group velocity." Information can only be transferred at the group velocity.

The group velocity is given by

$$v_g = \frac{d\omega}{dk}. \tag{3.15}$$

We will not derive this expression – it is derived in many simple texts on wave theory, usually by determining the velocity of the beat pattern produced when two waves of frequencies ω_1 and ω_2 add together in a dispersive medium (a dispersive medium is one in which the phase velocity is a function of frequency).

We now turn to derivation of the phase speed of a radiowave in an electron gas.

3.3 Electromagnetic waves in a dielectric

We now consider the case of perturbed electric and magnetic fields in a dielectric, and show that these in fact satisfy a wave equation.

Maxwell's equations for a medium which contains no free charges (net charge density equals zero, no free current density) are

$$\vec{\nabla} \times \vec{E} = -\mu \frac{\partial \vec{H}}{\partial t}, \tag{3.16}$$

$$\vec{\nabla} \cdot \vec{E} = 0, \tag{3.17}$$

$$\vec{\nabla} \cdot \vec{H} = 0, \tag{3.18}$$

$$\vec{\nabla} \times \vec{H} = \epsilon \frac{\partial \vec{E}}{\partial t}, \tag{3.19}$$

where $\vec{E}$ is the electric field, $\vec{H}$ is the magnetic field (the magnetic induction is $\vec{B} = \mu \vec{H}$), ϵ is the electrical permittivity, and μ is the magnetic permeability. The operation "·" represents a dot-product, and "×" represents a cross-product.

We now take $\vec{\nabla}\times$ Equations (3.16), (3.17), (3.18), and (3.19). For example, applying $\vec{\nabla}\times$ to the left-hand side of (3.16) gives

$$\vec{\nabla} \times (\vec{\nabla} \times \vec{E}) = \vec{\nabla}(\vec{\nabla} \cdot \vec{E}) - \nabla^2 \vec{E}. \tag{3.20}$$

Taking $\vec{\nabla}\times$ of the right-hand side of Equation (3.16) gives

$$\vec{\nabla} \times \left(-\mu \frac{\partial \vec{H}}{\partial t} \right) = -\mu \frac{\partial}{\partial t} (\vec{\nabla} \times \vec{H}) = -\mu \frac{\partial}{\partial t} \left(\epsilon \frac{\partial \vec{E}}{\partial t} \right), \tag{3.21}$$

where we have used Equation (3.19) in the last step. Thus combining Equations (3.20) and (3.21), we have

$$\vec{\nabla}(\vec{\nabla} \cdot \vec{E}) - \nabla^2 \vec{E} = -\mu\epsilon \frac{\partial^2}{\partial t^2} \vec{E}. \tag{3.22}$$

But from (3.17), $\vec{\nabla} \cdot \vec{E} = 0$, so

$$\nabla^2 \vec{E} = \mu\epsilon \frac{\partial^2 \vec{E}}{\partial t^2}. \tag{3.23}$$

Similarly, we may do the same to Equation (3.19) to obtain

$$\nabla^2 \vec{H} = \mu\epsilon \frac{\partial^2 \vec{H}}{\partial t^2}. \tag{3.24}$$

Thus both $\vec{E}$ and $\vec{H}$ obey the wave Equation (3.2), where in this case ϕ_D is either $\vec{E}$ or $\vec{H}$. We see that both the $\vec{E}$ and $\vec{H}$ fields propagate at the same phase speed

$$c_\phi = \frac{1}{\sqrt{\epsilon\mu}}, \tag{3.25}$$

where we use c to represent speeds associated with an electromagnetic wave, and various subscripts to discuss different types of speeds. Here, we use the subscript ϕ to indicate phase speed.

Recall that $\epsilon = \kappa_e\epsilon_0$ and $\mu = \kappa_m\mu_0$, where ϵ_0 and μ_0 are the permittivity and permeability of free space, and κ_e and κ_m are the relative permittivity (also called the dielectric constant) and relative permeability respectively. For a plasma, and indeed many media apart from magnetic media, $\kappa_m = 1$. Then

$$c_\phi = \frac{1}{\sqrt{\kappa_e\epsilon_0\mu_0}} = \frac{c}{\sqrt{\kappa_e}}. \tag{3.26}$$

The refractive index is

$$n = \frac{c}{c_\phi} = \sqrt{\kappa_e}. \tag{3.27}$$

Determining κ_e for a plasma is relatively easy.

When an electromagnetic field impinges on a small region of the plasma, indicated by the star in Figure 3.1, it causes a movement of electrons and produces a polarization of the region. The electric displacement (or D-field) is

$$\vec{D} = \epsilon_0\vec{E} + \vec{P}, \tag{3.28}$$

where $\vec{P}$ is the induced polarization.

In many situations the polarization is proportional to the applied electric field, $\vec{E}$, i.e.,

$$\vec{P} = N\vec{p} = N\alpha\vec{E}. \tag{3.29}$$

Lower-case $\vec{p}$ is the polarization per particle and upper-case $\vec{P}$ is the polarization of N particles. The value α is a constant which we do not at this stage know. Then

$$\vec{D} = \epsilon_0\vec{E} + N\alpha\vec{E}. \tag{3.30}$$

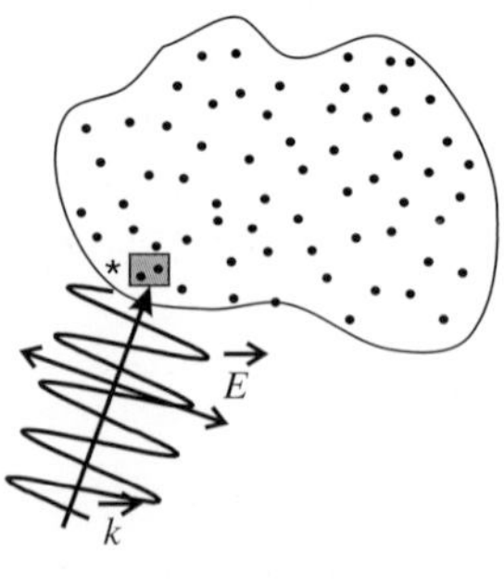

Figure 3.1 An electromagnetic wave incident on a region of air or plasma. The direction of propagation ($\vec{k}$) and the direction of oscillation of the electric field ($\vec{E}$) are shown. The small black dots represent electrons.

But

$$\vec{D} = \epsilon\vec{E} = \epsilon_0\kappa_e\vec{E}. \tag{3.31}$$

Equating Equations (3.30) and (3.31) gives

$$(\epsilon_0 + N\alpha)\vec{E} = \epsilon_0\kappa_e\vec{E}, \tag{3.32}$$

so cancelling the $\vec{E}$ terms gives $\kappa_e = 1 + \frac{N\alpha}{\epsilon_0}$. Hence the refractive index is then

$$n = \sqrt{\kappa_e} = \sqrt{1 + \frac{N\alpha}{\epsilon_0}}. \tag{3.33}$$

Hence if we can determine α, we can determine the refractive index for any electromagnetic wave passing through a medium, or at least for those media in which the polarization is proportional in strength to the electric field amplitude. Note that α could be complex, as may n be also.

3.3.1 Use of complex numbers

Throughout this text, use will be made of complex numbers in representing oscillatory motion. There are two distinct applications.

In the first case (referred to as type I), we represent the motion as a complex number, of the type $\underline{A}_0\mathrm{e}^{\pm i\omega t}$, and do a variety of calculations, and then in the end take *real parts* of our calculations. This is the case when motions are in one dimension, or for things like voltages in an electronic circuit, for cases where the relevant parameter might have both an amplitude and a phase. The procedure is something of a mathematically convenient way to speed up the calculations. There are certain additional rules that are needed, which we assume the reader is familiar with. For example, the power dissipated in an electronic circuit is $\frac{1}{2}\mathrm{Re}\{\overline{V^*I}\}$, where V is the voltage, * represents the complex conjugate, and I is the current.

In the second application (type II) of complex numbers, we often deal with motions that are *truly two-dimensional* or which have two orthogonal components – things like circularly polarized light or radiowaves spring to mind. In that case, we actually use *all* of the information embodied in the complex number, and the real component of the complex number represents a variable on one axis (perhaps displacement) and the "imaginary component" represents the other axis, and is in fact not imaginary at all, but a very real aspect of the problem. This second case is especially common in radio receivers, where the received signal is represented as "in-phase" (real) and "quadrature" (imaginary) components.

Both formats will be used in this book, and we assume the reader can recognize each case.

It should also be noted here that in the following text we often write vectors in the following way: $\vec{\underline{E}}$. Here, we are representing the electric field. The arrow overhead tells us it is a vector in 3-D or 2-D space, and the underline tells us that each component of this vector is complex. We use complex notation to represent each component because this allows us to represent the parameter with both an *amplitude* and a *phase*. In so doing, we are using the first application discussed above. There will be times when we might choose to represent a two-dimensional spatial vector as a complex number (second application above) but this can only be done if there is no phase variation in any of the components – we cannot use complex numbers to represent *both* the vector character and the phase behavior – we need to choose one or the other. So for an electric field, we may need up to six dimensions to describe it – three spatial dimensions, each with a real and imaginary component.

Another way to approach this is to set the equations up as matrix or tensor equations, which could be represented as say a 3×2 matrix. Some texts do this – or even have higher levels of complexity – we will avoid it, but will mention this alternative approach briefly on a couple of occasions.

In this chapter, we will mainly use type I complex number representation, so when we write a complex number we really mean we are only considering its real part, and the complex number representation is a mathematical convenience which makes the equations easier to solve. Type II applications will occur in other chapters.

3.4 Refractive index of an electron gas

We now consider one of the simplest cases – the refractive index of an electron gas. An electron gas is a collection of electrons in random motion, but in which we assume that the electrons themselves do not interact with each other. This is clearly unrealisitc, since the electrostatic forces between the electrons would drive them apart. Nevertheless, it is a reasonable model, in that it describes the propagation of an electromagnetic wave through a weakly ionized gas, such as the ionosphere. In that case we have a large number of neutral molecules, and a smaller number of positive ions and negative electrons. The mixture is electrostatically neutral, and the electrons are far enough apart that they do not interfere too greatly. We assume that the ions have no real effect, which is true at relatively high radiowave frequencies, since the electrons respond quickly to an applied electric field, and the ions are much more sluggish. Initially we do not even allow the electrons to have velocities of their own – we consider them as stationary, and they are only allowed to move under the influence of the incident electric field.

We start from Equation (3.33). The first step is to establish that the induced polarization is proportional to the applied electric field, and then to determine the value for α.

As a radiowave approaches an electron, it applies a force on the electron. We look at a fixed point in space, so following the Equation (3.6), but keeping x fixed, the temporal variation is of the form $e^{-i\omega t}$. For the following derivation it is also possible to assume a temporal variation of the form $e^{+i\omega t}$ and the correct formula will result, but for subsequent derivations involving collisional and magnetic field effects, it is very important to use the correct sign. If the form $e^{+i\omega t}$ is used, then the form for the wave equation must follow Equation (3.7) rather than (3.6). If the forms are mixed up, then strange things happen like the wave increasing in amplitude when damping is applied, and vectors rotating the wrong way in magnetic fields.

So we now write that the temporal variation of the electric field of the incident wave is given by

$$\vec{E} = \vec{E}_0 \mathrm{e}^{-\mathrm{i}\omega t}, \tag{3.34}$$

and consider that it acts on an electron. The motion of the electron obeys Newton's second law, or

$$\vec{F} = m_e \vec{a}, \tag{3.35}$$

where m_e is the electron mass of 9.110×10^{-31} kg. For our specific situation, this gives

$$q_e \vec{E} = -e\vec{E} = m_e \frac{\mathrm{d}^2 \vec{x}}{\mathrm{d}t^2}, \tag{3.36}$$

where

$$e = |q_e|, \tag{3.37}$$

and where q_e is the charge on an electron (which is of course negative). Numerically e equals 1.602×10^{-19} Coulomb. The vector $\vec{x}$ is the electron displacement.

Then $\vec{x}$ will obey a sinusoidal oscillation,

$$\vec{x} = \vec{x}_0 \mathrm{e}^{-\mathrm{i}\omega t}, \tag{3.38}$$

where $\vec{x}_0$ could in fact be negative or even complex.

So

$$-e\vec{E} = -m_e \omega^2 \vec{x}_0 \mathrm{e}^{-\mathrm{i}\omega t} \tag{3.39}$$

or

$$e\vec{E}_0 = m_e \omega^2 \vec{x}_0. \tag{3.40}$$

Since $\vec{E}_0$ and $\vec{x}_0$ are parallel, we will work in terms of scalars and write

$$x_0 = \frac{eE_0}{m_e \omega^2}, \tag{3.41}$$

or, more generally (multiplying both sides by $\mathrm{e}^{-\mathrm{i}\omega t}$),

$$x = \frac{eE}{m_e \omega^2}, \tag{3.42}$$

which is true for all t in the case that the electric field and the displacement are assumed to be along the same line of action. Note that the displacement is in fact in phase with the electric field oscillation so that the acceleration is in anti-phase with $\vec{E}$, consistent with the expectation that the electron accelerates against the electric field.

We think of an electron moving back and forth as producing an oscillating dipole in the following manner: + −, − +, + −, etc., where the "−" is the electron and the "+" is essentially the empty "hole" left behind in the electron-plasma after the electron has left. The dipole moment is orientated from the negative charge (where the electron currently sits) to the positive hole, so is oppositely directed to the displacement x. Hence the dipole moment is $-ex$, so

$$p = -ex = -e\left(\frac{+eE}{m_e\omega^2}\right) = -\frac{e^2E}{m_e\omega^2}. \tag{3.43}$$

For N electrons we multiply by N to produce the total polarization $\vec{P}$, and then comparison with (3.29) gives

$$\alpha = \frac{-e^2}{m_e\omega^2}. \tag{3.44}$$

Hence, we have

$$\begin{aligned} n &= \sqrt{1 + \frac{N\alpha}{\epsilon_0}} \\ &= \sqrt{1 + \frac{N\frac{-e^2}{m_e\omega^2}}{\epsilon_0}}, \end{aligned} \tag{3.45}$$

or

$$n = \sqrt{1 - \frac{Ne^2}{\epsilon_0 m_e\omega^2}}. \tag{3.46}$$

Sometimes, this is written as

$$n = \sqrt{1 - \frac{\omega_p^2}{\omega^2}}, \tag{3.47}$$

where $\omega_p^2 = \frac{Ne^2}{\epsilon_0 m_e}$ is called the plasma frequency.

In a plasma with collisions (i.e., a fluid with electrons surrounded by neutral molecules, in which the electrons frequently collide with the neutrals), the formula changes to

$$n = \sqrt{1 - \frac{Ne^2}{\epsilon_0 m_e(\nu_{ec}^2 + \omega^2)}}, \tag{3.48}$$

where ν_{ec} is the number of collisions that one electron experiences with neutral atoms or molecules per second. We will derive this expression later.

However, before doing that, attention needs to be drawn to a point of some note. No matter which of the above formulas are used, it will be seen that n is less than 1, so that c_ϕ in Equation (3.27) is greater than c. In other words, the phase velocity in the plasma exceeds the speed of light in a vacuum. It may seem that this violates the theory of special relativity. However, it turns out that the group velocity (denoted as c_g) is still less than c, and special relativity only restricts the rate of transfer of information to be less than c. Information is transferred at the group velocity, not the phase velocity.

To see this, we will now derive an expression for the group velocity. Let us begin with the expression for the phase speed of the wave. We will use the collisionless case, for simplicity. Then

$$c_\phi = \frac{c}{\sqrt{1 - \frac{Ne^2}{\epsilon_0 m_e \omega^2}}} = \frac{\omega}{k}. \tag{3.49}$$

Therefore,

$$k = \frac{\omega}{c}\sqrt{1 - \frac{Ne^2}{\epsilon_0 m_e \omega^2}}, \tag{3.50}$$

and we know that the group velocity can be calculated from Equation (3.15) as

$$v_g = \frac{d\omega}{dk} = \left(\frac{dk}{d\omega}\right)^{-1}. \tag{3.51}$$

Let us find $\frac{dk}{d\omega}$. Using the product and chain rules for differentiation gives

$$\begin{aligned}\frac{dk}{d\omega} &= \frac{1}{c}\sqrt{1 - \frac{Ne^2}{\epsilon_0 m_e \omega^2}} + \frac{1}{2}\frac{\omega}{c}\frac{\frac{2Ne^2}{\epsilon_0 m_e \omega^3}}{\sqrt{1 - \frac{Ne^2}{\epsilon_0 m_e \omega^2}}} \\ &= \frac{1}{c}\frac{1}{\sqrt{1 - \frac{Ne^2}{\epsilon_0 m_e \omega^2}}}\left[\left(1 - \frac{Ne^2}{\epsilon_0 m_e \omega^2}\right) + \frac{Ne^2}{\epsilon_0 m_e \omega^2}\right].\end{aligned} \tag{3.52}$$

Therefore

$$\frac{1}{c_g} = \frac{1}{c}\frac{1}{\sqrt{1 - \frac{Ne^2}{\epsilon_0 m_e \omega^2}}}, \tag{3.53}$$

or

$$c_g = c\sqrt{1 - \frac{Ne^2}{\epsilon_0 m_e \omega^2}}, \tag{3.54}$$

which is indeed less than c. Notice that

$$c_g . c_\phi = c^2. \tag{3.55}$$

We emphasize here that the above equations are not the most sophisticated forms for the refractive index – in reality, we also need to include the impact of the magnetic field, and the fact that electrons collide with other atoms. We will come to these considerations later in this chapter, but for the time being, the simple formulas given above allow us to study several important features of radiowave propagation. While the details may change when we introduce magnetic fields and collisions, the general principles remain the same.

3.4.1 Relevance of refractive index in MST studies

For many MST applications, especially for studies of events below the ionosphere, the refractive index remains close to unity. When the scattering is due to free electrons, the refractive index is only marginally less than unity in these cases. For scatter from the neutral atmosphere, especially the troposphere and stratosphere, scattering is from electrons that are bound to atoms, and this changes the scattering mechanism in such a way that the refractive index becomes greater than unity, but again only marginally so.

However, for radiowave propagation in the upper mesosphere and ionosphere, especially for radio frequencies of a few MHz, deviations of the refractive index from unity can become substantial. We could avoid discussion of these cases on the grounds that they are not relevant to most applications of MST radar, but it would leave this book somewhat incomplete. Therefore we choose to pursue the discussion of refractive index in a plasma further. If readers are not interested in cases where the ray paths of the radar may bend, and/or in cases where the refractive index deviates substantially from unity, then they should feel free to skip through to the sections on radar scattering later in this chapter. But for completeness, we will now examine the impact of significantly non-unity refractive indices on radio ray paths. There are significant lessons to be learned about the nature of refractive index in the ensuing section.

First, we will discuss just how the refractive index can be less than unity, since it gives a physical feel to the propagation process. Then we will consider bending due to refractive index variations, as well as look at the refractive index when collisions and magnetic field effects are included.

3.4.2 How can the phase speed be greater than *c*?

We noted above that the phase speed exceeds the speed of light in a vacuum, but that this does not violate the theory of special relativity because that only requires that the group velocity is less than c. But it is nonetheless useful to examine just how the situation arises that the phase speed can be so large.

The key point to bear in mind is that the wave propagates because each part of the advancing wave is the vector sum of Huygen's wavelets produced by oscillations in other electrons at an earlier stage.

To begin, consider a single electron that is accelerated from one point at time $t = 0$ to another point at a later time $t = 1$. Let us examine the electric field at times $t = 0$ and $t = 1$ units.

The situation is shown in Figure 3.2(a). Initially the electron is at A. The broken lines show the electric field at $t = 0$ and the solid lines show the electric field lines which the particle would have had if it were positioned at B at that time.

For simplicity of representation, consider that the electron is first at A, and then moves instantaneously to B at time $t = 1$. Of course this already violates the special theory of relativity, but we allow it only for purposes of illustration, and justify the procedure shortly.

The actual field lines at $t = 1$ will not be the solid lines shown in Figure 3.2(a), because points far from both A and B will not know that the particle has moved. Such information can only travel at the speed of light. A point which is a distance $\zeta = ct$ away from the region will only just be aware of movement. This region is represented as a grey semicircle in Figure 3.2(b).

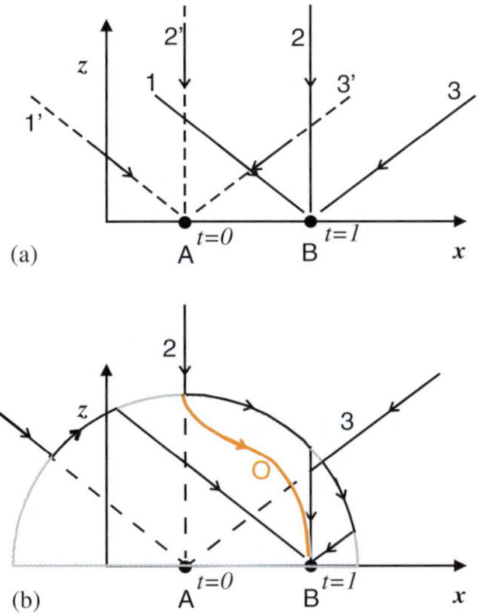

Figure 3.2 Representative electric field lines at two times for a particle instantaneously moved from A to B at time $t = 1$. See the text for a discussion.

Thus the true electric field lines at time $t = 1$ will be a combination of the field lines at $t = 0$ (far from the region) and the solid lines at $t = 1$, i.e., the field lines will look like the picture shown in Figure 3.2(b). There is a sharp change in the electric field on a circle of radius $\zeta = ct$ (where $t = 1$) centered on the point A. The electric field lines are shown by the solid lines. The fainter broken lines with shorter segments are simply left in as a reminder of the previous configuration, and do not form part of the $\vec{E}$ field.

It was mentioned that the assumption of an instantaneous jump at time $t = 1$ violates the special theory of relativity. If we allow the electron to move at a finite speed, less than c, then the electric field line denoted by the number 2 will actually follow the orange path labelled "O;" the impact of the change will be felt at successively further points from the origin as time progresses, resulting in a smoother curve. But the key point remains that the electric fields at $t = 0$ and $t = 1$ will look like the dark black lines for points beyond the circle, and like the black lines inside the circle for points close to B. Hence for diagrammatic purposes, we can still use our representation deduced on the basis of an assumed instantaneous jump.

Note that the induced electric field is strongest in the direction perpendicular to the direction of displacement, and the direction of the electric field at that point is to the right of the page. However, the electric field which caused the electron to displace to the right (from A to B) must have been orientated towards the left in order to cause the negatively charged electron to accelerate to the right. Hence, the applied field and the induced field are oppositely orientated. (This situation is unique for the electron – it would not be true for a proton, but the lighter electron is the particle responsible for defining propagation characteristics.)

It is also clear that the impact at a general point $\vec{r}$ at time t is due to the effect experienced by the electron at an earlier time $t' = t - |\vec{r}|/c$. The term $t - |\vec{r}|/c$ will

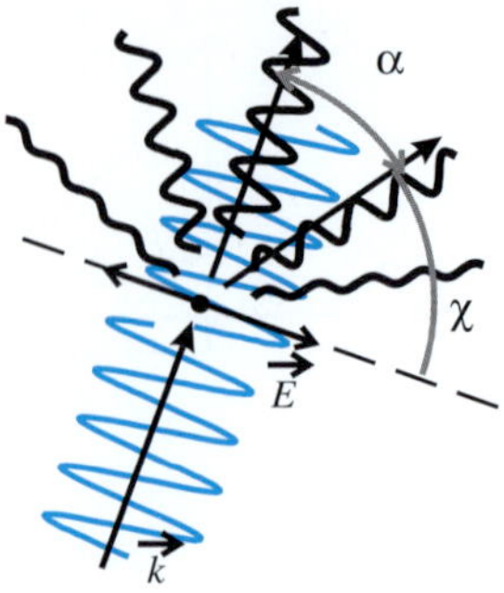

Figure 3.3 Incident electric wave field $\vec{E}$ impinging on an electron (black dot), showing the scattered electric field. See the text for discussion.

appear many times when we discuss electromagnetic waves, and is called the *retarded time*.

We now return to our thought-experiment. We have discussed an impulsive application of an applied electric field, but similar things hold true if the applied field is sinusoidal.

We will show later, when we discuss scattering, that the induced electric field is proportional to $-e^2 \sin\chi$, where χ is the angle of scatter relative to the electric field oscillations, but the key point is the existence of a negative sign, indicating that the scattered field and the incident field are out of phase by $180\,^\circ$. The situation is shown schematically in Figure 3.3. Note we have drawn two angles – χ is the angle of scatter relative to the electric field oscillation, and α is the angle relative to the direction of propagation of the wave.

The figure shows an incident oscillating field $\vec{E}$ (blue), and the field scattered in various directions (black sinusoidal lines). The polarization angle is given by χ, as shown representatively in the figure, and the radiation angle is represented by α. It is clear that $\alpha + \chi = 90\,^\circ$. Note that the forward radiation (black sine curve at $\alpha = 0\,^\circ$) is $180\,^\circ$ out of phase with the continuing incident radiation, as expected due to the fact that the electron has negative charge and so produces a radiating electric field in antiphase with the incident driving field. Even without any mathematics, it should be expected that the strongest scattered radiation will be at $\chi = 90\,^\circ$, where a viewer looking back at the electron sees it oscillating from side to side. At an angle of χ of $0\,^\circ$, a viewer looking back at the electron sees it end-on, and sees no movement. Hence it should not be surprising that the electric field in that direction should be zero. Field strengths at angles intermediate between $\chi = 90$ and $0\,^\circ$ should be intermediate in strength, so the $\sin\chi$ law predicted above should not be surprising. Scattered radiation also exists in backward directions ($\alpha \geq 90\,^\circ$), but is not drawn on the figure in order to reduce confusion. Hence the figure shows the main features that we need to know about for our ensuing discussions, without the need for mathematics – it shows the anti-phase between the scattered and incident waves, and the reduction in scattered field strength as the angle χ rotates from 90 to $0\,^\circ$.

We now wish to look at the impact of the total electric field at a general point P. The situation is shown in Figure 3.4(a). The contributions to the electric field experienced at

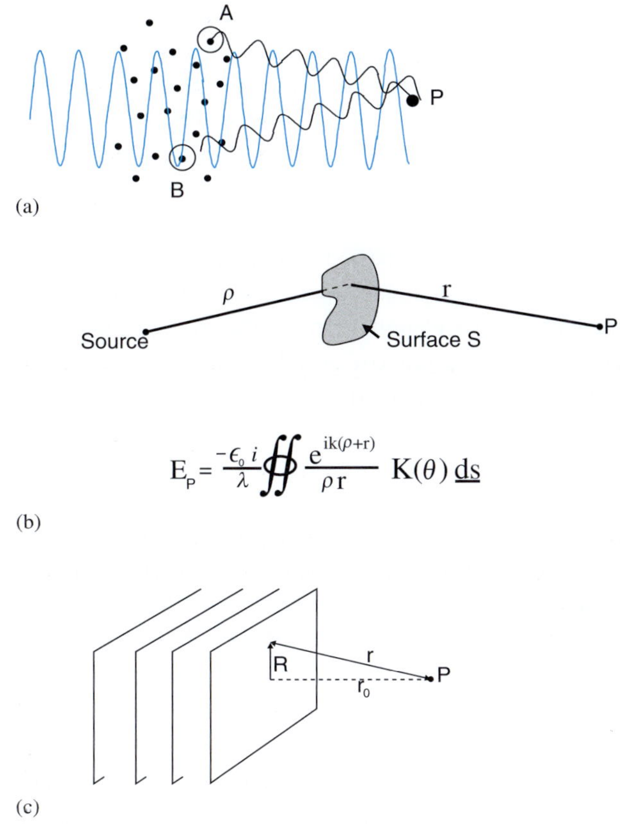

Figure 3.4 (a) The effect at a point P due to the original EM wave as well as the contribution from all the radiators in between the source and P. (b) The Kirchoff integral for the simple case that all Huygen's radiators have the same radiating efficiency. A more general form will be introduced later. (c) Representation of the original electron distribution as a series of successive planes. See the text for discussion. (Adapted from *Hecht and Zajac*, 1974.)

P are due to both the original field and the fields radiated by Huygen's wavelet sources (i.e., all the original electrons) prior to the point. Two such radiators are shown in the figure as A and B, but all electrons will radiate.

In order to determine the total field at the point P, we need to sum all of the effects of all of the waves and also the original wave. This can be done using the simplified Kirchoff integral (e.g., *Hecht and Zajac*, 1974), p. 391, which is shown in the figure as (b), with the time-varying component $e^{-i\omega t}$ removed for simplicity. Figure 3.4(b) shows the net effect of the radiation received from a surface S shown in the figure, for a source at finite distance. In our case, we may consider the source to be at an infinite distance, and the scattering entities to cover a 3-dimensional volume, rather than a plane. The term $K(\theta)$ is called the obliquity factor – in our case, it can be identified with the term $\sin\chi$ discussed in relation to Figure 3.3.

To allow for modification to a 3-dimensional volume of scatterers, we consider the volume to comprised a series of successive slabs, or successive planes of thickness much less than one wavelength, as shown in Figure 3.4(c). This representation will allow us to employ a well known construct known as the "Cornu spiral" in our explanation.

We will consider the impact of a plane L on the point P. The plane could be any one of those shown in Figure 3.4(c). In a qualitiative sense the impact of all of the planes will be similar. We will assume that the incident wave loses no energy as it propagates, which is of course false but sufficient for the moment.

To effectively determine the received electric field at P, we divide the plane L into half-period zones. Half-period zones are annuli of successively larger radii, with inner and outer perimeters defined by the fact that the distance from the circles to the point P is one half of a wavelength more for the outer one than for the inner.

The resultant amplitude received at P is obtained by vectorially summing the complex vectors from each annulus. To do this properly, we actually divide the half-period annuli into even finer thicknesses, and then add. The resultant geometrical construction is called a Cornu spiral.

To begin, consider Figure 3.5(a). It shows successive "half-wave zones." The first is considered to be the innermost one. The distance R_1 is chosen to be a half-wavelength greater than the distance R_0, where all references are relative to the point of interest, P. The distance R_2 is a further half-wavelength longer than R_1, and so forth.

Each half-wave zone may be further subdivided into smaller annuli (referred to as "sub-annuli" or "sub-zones"), where the phase difference across each sub-annulus is a small angle, much less than $180\,^\circ$. Examples are shown by the circles formed with broken lines in Figure 3.5(b). The solid circles in the figure divide successive half-wave zones.

Suppose that the very first (central) half-wave zone (which will be a disk, rather than an annulus, as shown in Figure 3.5(b)) is divided into sub-zones. Within each sub-zone, the phase is considered approximately constant. We now determine a net vector which represents the signal received from this zone at the point P. The length of the vector will be proportional to the area of the annulus, multiplied by the "obliquity factor." We consider the amplitude that reaches the point P, and take the phase as the phase at the point P relative to that which the incident wave would have when it arrives there. The phase from all radiators within a sub-zone will be the same, since the subzones are annuli and the distances to the point P are the same from all regions (to within the error defined by the sub-zone thickness).

It is assumed that the incident wave is a plane wave, so that all of the point radiators radiate in phase, and the only effect that causes phase differences is the different path lengths travelled to reach P. The signal from the very central sub-zone of the first half wave zone travels exactly parallel to the incident wave: both move along the line R_0 in Figure 3.5(a). Both therefore arrive at P with the same phase change. However, recall that the radiated wave will be $180\,^\circ$ out of phase with the incident wave.

We represent the vectors on an Argand diagram, as shown in Figure 3.5(c). Phase is defined to increase in an anticlockwise direction. Hence the vectors for the incident wave, and that radiated by the central sub-zone (indicated by the letter A in Figure 3.5(b)) will, when they arrive at P, be as shown by the "incident" vector and the vector "a" in Figure 3.5(c), i.e., they have a $180\,^\circ$ phase difference. This can be considered as

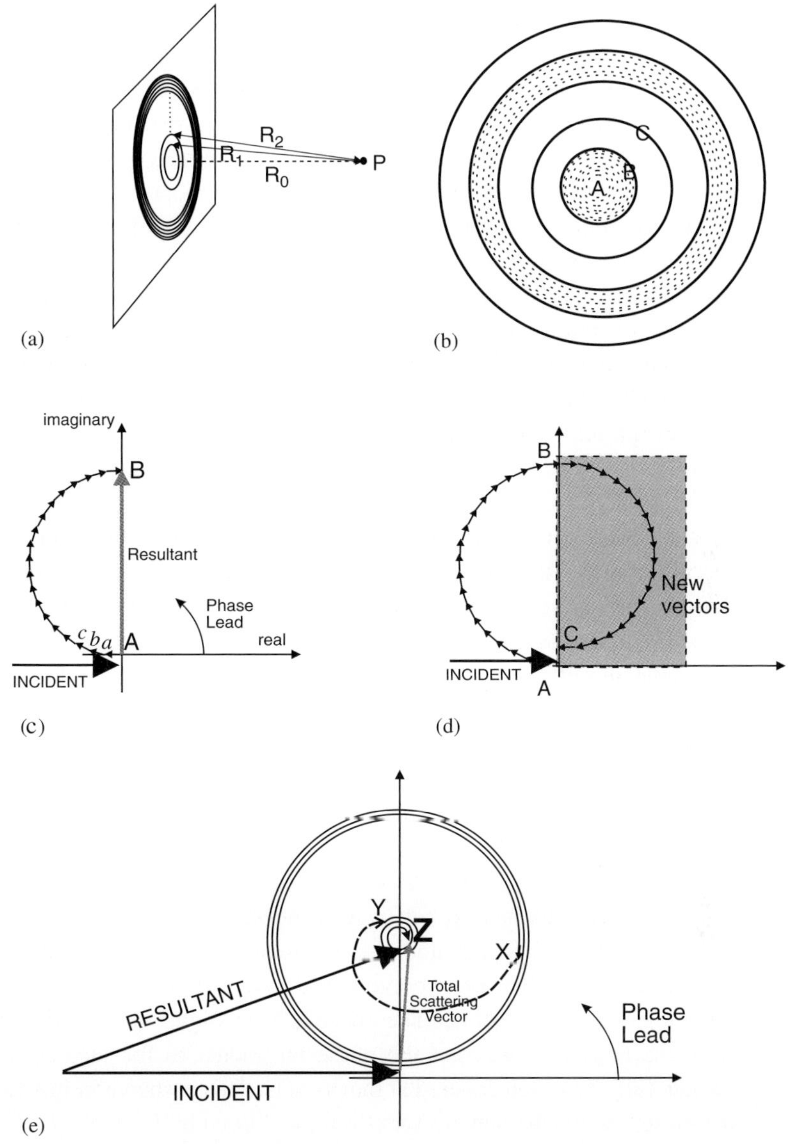

Figure 3.5 Demonstration of the net impact at a point P due to interference between radiated signals from multiple scatterers in a common plane. (a) Shows how the plane L is divided into concentric half-wave zones, with the distances R_i being of lengths that allow the phases from successive perimeters to increase in steps of 180° as i increases by unity. (b) Shows further subdivision of a half-wave zone into smaller zones (the sub-zones being marked by the broken circles). (c) Shows the addition of phasors from within these sub-zones for a single half-wave zone. (d) Shows the effect of addition of an extra half-wave zone. (e) Shows the net result of addition of all of the half-wave zones of a single plane out to some outer radius (which may or may not be at infinity). The broken line from X to Y is not a path, but simply reminds the reader that the circles due to successive zones keep spiralling inwards with successively smaller radii until we get to the final point Z. See the text for discussion.

either a 180 ° phase lead or a 180 ° phase lag – it is immaterial. The next sub-zone out from the center of the first half-wave zone is an annulus (as are all subsequent ones), and the radiation from radiators in that sub-zone must travel slightly further than that from the first sub-zone, so it will have a slightly larger phase delay than that from the central sub-zone, or a slightly lesser phase lead. Since in our 3.5(c), we have defined a phase lead to be anticlockwise, the vector due to this second sub-zone must be slightly clockwise of that due to the first sub-zone, as shown by the small vector "b" in 3.5(c). In order to determine the net effect of the two sub-zones plus the incident vector, we place them end to end. Now we move to the next sub-zone, which has an even greater phase lag, and it is represented by the small vector "c".

So we may continue to the outer limits of the first half-wave zone. The net summation is shown in 3.5(c) for addition of all the signals from all of the sub-zones in 3.5(b) i.e., from the point A to the circle B in 3.5(b).

Note that we assumed that each sub-zone contributed equal amplitudes. This is approximately true, as there are several counterbalancing effects. First, each successive sub-zone has a slightly larger radius, and therefore larger circumference, than the preceding one. This should lead to a larger area for the outermost sub-zones. But by defining the phase difference across successive sub-zones to be the same, this in fact means that the thickness of successive subzones becomes slightly less, almost exactly balancing the effect of the increase in circumference. Thus the net area covered by each sub-zone is close to equal, so that each successive vector in 3.5(c) has the same length.

The net vector at this point is the vector sum of the "incident" vector and the "resultant" vector in 3.5(c). Note that the "incident" vector was intended to have its tip at the origin, but has been displaced downward so that the other smaller vectors are not hidden.

We now move to the next half-wave zone, shown by the annulus between the points B and C in 3.5(b). We again subdivide it into sub-zones, and add successive vectors from each. Each successive vector continues to have a slightly greater phase lag (lesser phase lead) than the previous one. The phase at B is slightly less than that of the last vector from the first sub-zone, and the phase lag increases for successive sub-zones until a further 180 ° has been covered at the point C. This is shown in Figure 3.5(d); the shaded region represents the new vectors from the second half-wave zone.

Note, however, that the summation does not return quite to the origin, but finishes slightly above it. This is because we need to remember the impact of the obliquity factor. Although we ignored it in our discussion of the first half-wave zone, it was always present. By the time we consider the second half-wave zone, we can no longer ignore it. Because the radiation from this zone is no longer parallel to the incident radiation, there is a slight reduction in radiated amplitude, since χ (e.g. see Figure 3.3) is no longer 90 ° (or alternatively, α is no longer 0 °). Hence each vector from this half-wave zone is slightly smaller than those from the first half-wave zone.

We may then consider the contributions from the third half-wave zone, which will produce another semi-circle very similar to that in 3.5(c), but with an even further reduced diameter than that due to the second half-wave zone.

We may continue considering contributions from successive half-wave zones, and continue to add them. The net result is a spiral that spirals inwards towards the center, as shown in 3.5(e). The successive semi-circles have smaller and smaller diameters as the obliquity factor becomes stronger.

The final result is shown in 3.5 (e). The net effect of adding all the vectors due to all radiators in this one plane is to produce the vector labelled "total scattering vector," which is a vector from the origin to the final point Z. (Note that the broken line from X to Y simply indicates that the full process is one of ever decreasing circles that spiral inward and finish at Z.) The total vector experienced at P is then the vector sum of the "incident" vector and the "total scattering vector," giving rise to the vector labelled "resultant."

When we add the applied (incident) field and the effect of the plane L, the resultant is a vector $\vec{E}_p$ that leads the applied field.

This is the important point – the resultant *leads* in phase.

Of course, in reality the effective field at P is due to the sum of the applied field plus the fields due to many planes at different distances from P, but the fact remains that the resultant field will lead the applied field.

Because the resultant field leads the applied field, and because the phase difference gets successively larger as the wave progresses through the medium, this means the phase fronts travel faster than they would in a vacuum, so the phase velocity is greater than c and the refractive index is less than 1.

As discussed, if the oscillating particles were positively charged, the reverse would be true and the phase velocity would be less than c.

It is also true that if the radiating electrons are bound (as in dielectric materials), then the re-radiated field is also in phase with the incident field, so that the phase speed is less than c and the refractive index is greater than 1. This is why most materials with which you are familiar have a refractive index greater than unity. Bound electrons will be discussed later in the text.

The value of n for any medium is hence dependent on the phase of the induced dipoles relative to the incident field. Intermediate phases are also possible. When absorption occurs, the phases of the induced dipoles are neither in phase or in anti-phase with the incident radiation. A typical phase diagram is shown in Figure 3.6.

Here, the resultant wave ($\vec{E}_p$) lags in phase and is reduced in amplitude compared to the incident wave.

The fact that the refractive index can be less than unity has some interesting consequences, which we will discuss shortly in regard to the ionosphere. But one interesting

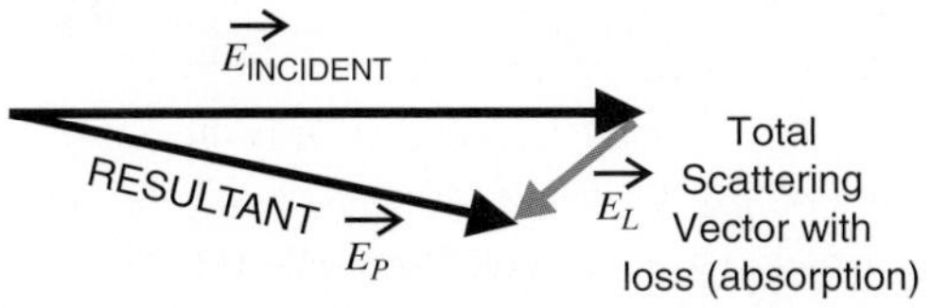

Figure 3.6 Resultant vector when the scattering centers also absorb some of the incident radiation.

consequence in experiments and equipment involving plasmas (e.g. fusion reactors, ion drives, etc.) is that a convex lens produces a diverging electric field and one must make a concave lens to focus a beam. This is the opposite of normal optics.

3.5 Radiowave refraction

We now move to the topic of refraction. Refraction concepts with radiowaves are similar to the optical case, except for the possibility that the refractive index may be negative. Again we will consider plasmas as examples, but the applications are broader. For VHF studies in the stratosphere and troposphere, refraction is not normally an issue, but it is not unheard of. Certainly long distance communication can suffer refractive effects due to layers of differing density in the atmosphere, and although locally the effects are minor, the accumulated effect over long distances can be quite substantial.

To first order, we can envisage the Earth's ionosphere as a series of successive shells of different electron density. Indeed in many cases where refraction is an issue (even tropospheric effects at VHF) the refracting phenomena are frequently layered, so that the ionospheric examples discussed below can also be considered applicable in part to other wavelengths and indeed even optical effects. Mirages and stellar scintillation, for example, are evidence of refraction at optical wavelengths.

We look to Figure 3.7(a), which shows a layered atmosphere or ionosphere. Consider what happens as an electromagnetic wave (denoted by $\vec{k}$ in the figure) impinges on such a sequence of layers. Investigation of the trajectory of such a radiowave through the ionosphere was an important scientific study in the early days of radio work, since refraction

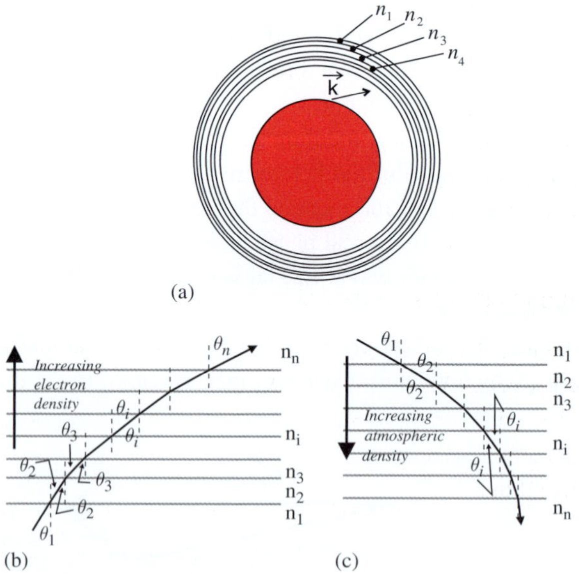

Figure 3.7 Various types of layered refractive index structures, at different scales in the ionosphere and atmosphere. See the text for details.

and reflection in the ionosphere was a primary method of global communication, particularly before the days of satellites. As discussed in Chapter 2, radio communication was one of the main branches of radio studies undertaken after World-War II.

To simplify things, consider the case of layering without curvature, as shown in Figure 3.7(b) and (c). We show two cases – one in which the wave moves through a plasma which has increasing density with increasing height (Figure 3.7(b)) and one where we consider the neutral atmosphere (Figure 3.7(c)). In case (b) we consider an upward-propagating wave, and in case (c) we consider radiowaves or light-waves incident from above (e.g. like starlight).

In case (b), as the wave moves up in height, the electron density (initially, at least up to the *E* region) increases. This means that the refractive index decreases! Thus, the ray path bends away from the normal. In contrast, the second case (c) is just the opposite, and is more similar to the normal case that might be expected as a ray of light enters a region of increasing density – for example, a slab of glass. In this case the ray bends towards the normal as the density (and hence the refractive index) increases.

However, the same mathematics can be used to describe the ray path in each case. Suppose that each successive layer has a different refractive index. At the *i*th boundary, Snell's law gives

$$\frac{\sin\theta_{i-1}}{\sin\theta_i} = \frac{n_i}{n_{i-1}}, \tag{3.56}$$

where θ_i is the angle of the wave-normal after it has passed through the interface and θ_{i-1} is the angle of incidence. Here n_{i-1} and n_i are the refractive indices either side of the junction. This equality applies at each interface, and the angle at which the ray in the *i*th layer enters the layer (at the bottom in 3.7(b), or the top in (c)) is also the angle at which it impinges on the next interface. Thus,

$$n_1 \sin\theta_1 = n_2 \sin\theta_2 = \cdots = n_i \sin\theta_i = \cdots = n_n \sin\theta_n. \tag{3.57}$$

Hence $n_i \sin\theta_i$ is constant for any given wave launch-angle. This expression holds in the limit as we let the layer thicknesses get narrower and narrower. Therefore, if we know n as a function of height, we can find θ as a function of height and trace the ray. We may also use this information to study refraction. This applies for both cases (b) and (c).

3.5.1 Refraction in the ionosphere

In Chapter 2 we discussed the so-called ionosonde. Here we can see mathematically how it works. We continue to use the refractive index for a non-magnetized, collisionless free electron gas – inclusion of such terms simply changes the details, but not the fundamental principle.

Recall that

$$n = \sqrt{1 - \frac{Ne^2}{\epsilon_0 m_e \omega^2}} \tag{3.58}$$

in a collisionless plasma. As a radiowave rises through the ionosphere, it turns more and more horizontal, as seen in the last section. Eventually it may turn completely horizontal.

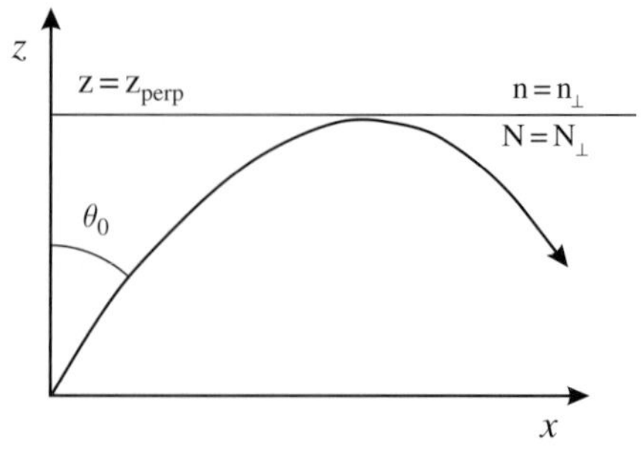

Figure 3.8 Total reflection of a radiowave by the ionosphere.

Let $N_\perp$ be the electron density at the point where a propagating electromagnetic wave turns horizontal, and therefore suffers total internal reflection.

Suppose that θ_0 is the angle at which the wave was launched at the ground (see Figure 3.8). Then if we take the refractive index at the ground to be unity,

$$1 \sin\theta_0 = n_\perp \sin 90°. \tag{3.59}$$

Then $n_\perp = \sin\theta_0$. Hence,

$$\sqrt{1 - \frac{N_\perp e^2}{\epsilon_0 m_e \omega^2}} = \sin\theta_0, \tag{3.60}$$

or

$$1 - \frac{N_\perp e^2}{\epsilon_0 m_e \omega^2} = \sin^2\theta_0, \tag{3.61}$$

or

$$\frac{N_\perp e^2}{\epsilon_0 m_e \omega^2} = \cos^2\theta_0. \tag{3.62}$$

Hence,

$$N_\perp = \frac{\epsilon_0 m_e \omega^2 \cos^2\theta_0}{e^2}. \tag{3.63}$$

Thus, as θ_0 increases, $N_\perp$ decreases. Radiowaves launched at angles closer to the ground (lower elevation) reflect from lower heights (see Figure 3.9). This was an important point when most world-wide communication took place using total reflection from the ionosphere. Radiowaves launched at one site would internally reflect from some point in the ionosphere, bounce back to the ground, reflect off the ground, return to the ionosphere, and so bounce their way around the world. The big disadavantage of the method was that the ionosphere was never perfectly flat, and suffered distortion due to waves and turbulence events, distorting the signal. Entire organizations were developed (and some still exist) for studying and forecasting ionospheric conditions, and for predicting optimum frequencies for ionospherically propagated communications.

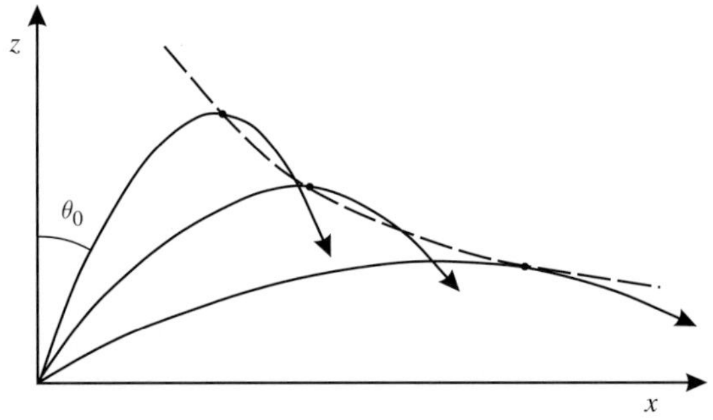

Figure 3.9 Total reflection of a radiowave by the ionosphere for various radiowave launch angles. Here, z is the height and x is horizontal displacement. It is assumed that the electron density increases as a function of height. The broken line simply guides the eye along the points of critical reflection.

3.6 Vertical incidence

Another special case arises, which has special relevance to the ionosonde. When $\theta_0 = 0\,^\circ$, we have vertically incident waves. The condition for critical reflection is then

$$N_\perp = \frac{\epsilon_0 m_e \omega^2 \cos^2 0}{e^2}. \tag{3.64}$$

The $N_\perp$ nomenclature was useful for the case of non-vertically incident waves in order to emphasize that the turning points occurred at different values of N for different angles θ_0. However, it is not needed for the case of vertical incidence, where we have only one launch angle, so we just refer to the electron density at the point of critical reflection as N in this section.

So we may write, from (3.64), that at the critical reflection point for vertical incidence,

$$N = \frac{\epsilon_0 m_e \omega^2}{e^2}. \tag{3.65}$$

This occurs when

$$n = \sqrt{1 - \frac{Ne^2}{\epsilon_0 m_e \omega^2}} = 0. \tag{3.66}$$

The radar angular frequency at which this occurs is

$$\omega_c = \sqrt{\frac{Ne^2}{\epsilon_0 m_e}}, \tag{3.67}$$

or, for frequency expressed in Hz, we write

$$f_c = \frac{\omega}{2\pi} = \frac{1}{2\pi}\sqrt{\frac{Ne^2}{\epsilon_0 m_e}}. \tag{3.68}$$

This is called the *critical frequency* for this particular value of N.

A special device has been used for many years to study the ionosphere which utilizes these principles of critical reflection. It is called the ionosonde, and was discussed in Chapter 1. A pulse of radiowaves with carrier frequency of say 1 MHz is sent into the ionosphere (vertically) and reflects back from some level where

$$N = \frac{\epsilon_0 m_e (2\pi \times 1\,\mathrm{MHz})^2}{e^2}. \tag{3.69}$$

Then a second pulse is sent up with a slightly higher frequency, and the same reflection process occurs. The system sweeps in frequency from about 1 MHz up to about 20 MHz. In each case we measure the time delay between pulse transmission and pulse reception. The critical frequency at each height can be converted to an electron density, thereby giving the electron density profile at that period of time. (In truth, it is a little more complicated than that, since the height is actually moderately difficult to find. If the radiowave travelled at close to the speed of light in a vacuum during its entire journey, then the height could be determined as $\frac{ct}{2}$ where t is the time delay from transmission to reception. However, because the group velocity approaches zero at the critical level, this increases the delay in returning to the ground, and produces an artificially high altitude (called the "virtual height"). Special additional corrections are needed to determine the true height.)

3.6.1 Evanescence

Note that if $\frac{Ne^2}{\epsilon_0 m_e \omega^2}$ is greater than 1, then n^2 is less than 0, so n is imaginary. If ω is greater than 0 (forced wave), then k must be imaginary, and we write it as $k = 0 + \mathrm{i}k_i$. The wave formula for a vertically propagating wave is then

$$\begin{aligned} \vec{E} &= \vec{E}_0 \mathrm{e}^{\mathrm{i}(kz - \omega t)} \\ &= \vec{E}_0 \mathrm{e}^{\mathrm{i}(\mathrm{i}k_i z - \omega t)} \\ &= \vec{E}_0 \mathrm{e}^{-k_i z} \mathrm{e}^{-\mathrm{i}\omega t}. \end{aligned} \tag{3.70}$$

In other words, the wave does not show a sinusoidal variation in x, but decays exponentially.

So consider a radiowave rising from the ground up into the ionosphere. If the electron density increases with height, then n approaches zero as the wave propagates upward. Hence the wavelength increases. Finally, the wave may reach a level where

$$N = \frac{\epsilon_0 m_e \omega^2}{e^2}. \tag{3.71}$$

At this point, it ceases to propagate and decays exponentially above this altitude. This decaying part of the wave is called an evanescent wave. Similar waves also exist at any reflection interface, even in glass during total internal reflection of light. In the ionospheric case, the wave cannot propagate beyond this height, and reflects back.

The process is shown in Figure 3.10, in which we see that the wavelength increases as the critical level is approached, and then turns to an evanescent wave above the critical level.

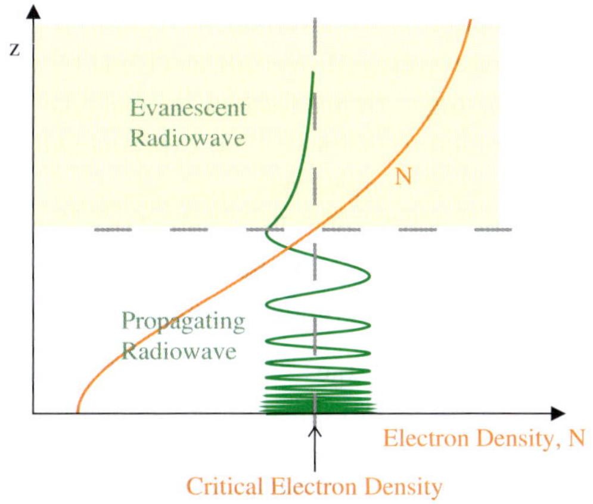

Figure 3.10 Behavior of an electromagnetic wave as it approaches a critical reflection level (denoted by the horizontal broken line). Note that the abscissa refers only to the profile denoted by "N" – the radiowave is drawn on in a representative way, but the wave amplitude is arbitrary and unrelated to the abscissa.

Note that $\omega = \sqrt{\frac{Ne^2}{\epsilon_0 m_e}}$, the condition for so-called critical reflection, is also equal to the plasma frequency which was discussed earlier.

3.6.2 Inclusion of collision rates in the expression for refractive index

Now, as promised, we turn to consideration of the impact of things like collision rates and magnetic fields on the refractive index.

Previously, in Equation (3.36), we used the formula

$$q_e \vec{E}_0 e^{\ i\omega t} = m_e \frac{d^2\vec{x}}{dt^2} \tag{3.72}$$

to describe the motion of an electron subject to an electric field. We also discussed the importance of using $\vec{E}_0 e^{-i\omega t}$, rather than $\vec{E}_0 e^{+i\omega t}$.

However, if the plasma is electrically viscous (or equivalently, the collision frequency of electrons with neutrals is non-zero), then the equation of motion for an electron becomes

$$m_e \frac{d^2\vec{x}}{dt^2} = q_e \vec{E}_0 e^{-i\omega t} - m_e \nu_{ec} \frac{d\vec{x}}{dt}, \tag{3.73}$$

where ν_{ec} is the collision frequency.

The collision frequency of an electron of speed v with its surrounding neutral molecules can be found in most elementary books on thermodynamics to be

$$\nu_{ec} \simeq \frac{v}{\sqrt{2}\pi r_\sigma n_N}, \tag{3.74}$$

where r_σ is the radius of a typical molecule, and n_N is the neutral number density. In particular, it is noted that the collision rate is proportional to the electron speed, and if

then for a phase-speed of c_ϕ, we can rewrite this as $e^{i(\frac{\omega}{c_\phi}x-\omega t)} = e^{i(\frac{\omega n}{c}x-\omega t)} = \exp\left\{i(\frac{\omega[n_R+in_I]}{c}x - \omega t)\right\}$. The part involving n_I appears now as $\exp\left\{-\frac{\omega n_I}{c}x\right\}$, which represents a decrease in amplitude as a function of x, as it should. If n_I were negative, then the wave would grow with distance, which is incorrect. So we need n_I to be positive, and we have in fact shown that it is. However, if the wave is treated as $e^{i(\omega t-kx)}$, then a negative value of n_I is necessary to ensure a decay in amplitude with increasing distance.

Some additional words of warning are needed here.

Firstly, we made the assumption earlier that the polarization is proportional to the applied electric field, or $\vec{p} \propto \vec{E}$, and that the two terms act along the same line. In fact, this is not really true in many cases, especially those involving bound electrons. In the bound case, the effective field is

$$\vec{E}_T = \vec{E}_i + \frac{\vec{p}}{3\epsilon_0}. \tag{3.87}$$

The second term is called the Lorentz term. It arises because of the need to consider the dipole impact of other electrons bound to other molecules and atoms in the immediate vicinity of the electron of interest. It turns out that the extra term is not needed for free electrons, but the reasons are subtle. This changes the equations substantially. *Fejer* (1985) discusses this in some detail. We will briefly consider the effect of these neighboring molecular dipoles when we discuss propagation through neutral gases later in this chapter.

Secondly, an additional assumption that has been made is that the collision frequency of the electrons with other particles is proportional to the particle velocity. This would seem intuitively reasonable, but experiments show it to be untrue, leading to further modifications of the equations, which will be discussed shortly.

3.6.3 Inclusion of the magnetic field

In this section we will examine the impact of allowing the EM wave to propagate through a plasma embedded in a magnetic field (anisotropic plasma). For now, we will ignore the effects of electron collisions.

As in all cases discussed so far, it is largely the electrons that re-transmit the received radiowaves, and this alters the refractive index of the medium because the original unimpeded waves and the re-radiated waves interfere to produce a new wave which moves at a speed different to that of light in a vacuum. However, the magnetic field introduces an extra complication: whereas previously the electron moved in a sinusoidal manner along a straight line, the magnetic field causes the electron to want to deflect perpendicular to the field. This complicates things, and in fact leads to the ionosphere becoming a double-refracting (bi-refringent) medium at radio-wavelengths. We will now examine how this arises.

When an electron enters a magnetic field at an angle perpendicular to the field lines, it experiences a force to the left of its motion (as viewed by a person standing parallel to the magnetic field line and looking along the electron's direction of motion). Mathematically, we write

$$\vec{F} = q_e \vec{v} \times \vec{B}, \tag{3.88}$$

where q_e, the electronic charge, is of course negative. The term $\vec{B}$ is formally called the magnetic induction, or sometimes the B-field (although many scientists call $\vec{B}$ the magnetic field, which formally it is not – the proper magnetic field is $\vec{H}$). The electron quickly establishes a circular motion, with force evB acting radially inwards. By Newton's second law, this equals the mass of the electron multiplied by its acceleration, or $m_e \frac{v^2}{r_{ce}}$, where m_e is the mass of the electron and r_{ce} is the radius of the circle. Equating the two gives the radius. Of course the radius is positive, so since we only need the magnitude, we replace q_e with e to give

$$r_{ce} = \frac{m_e v}{eB}. \tag{3.89}$$

The rotation period T_B is just the orbital circumference ($2\pi r_{ce}$) divided by the speed, or $T_B = \frac{2\pi m_e v}{eBv} = \frac{2\pi m_e}{eB}$. The corresponding angular frequency (called the *angular gyrofrequency*) is then just 2π divided by this, or

$$\Omega = \frac{eB}{m_e}. \tag{3.90}$$

The gyrofrequency in cycles per second (Hz) is $f_\Omega = \frac{\Omega}{2\pi}$. In the ionosphere, the gyrofrequency is a few MHz. For typical magnetic fields experienced in the solar wind, it is about 100 Hz, and for typical magnetic fields encountered in interstellar space, it is typically 10 Hz.

In the most general situation, the electron will have a component of motion parallel to the external B-field, and two orthogonal components perpendicular to it. If the electron motion is purely parallel to the B-field, there will be no impact due to the magnetic field. If the velocity is purely perpendicular to the B-field, the motion becomes circular. If it is intermediate, the motion is helical, with both a circular motion and a mean drift along the field lines – the electron traces out a spiral, with axis along the field line and tracing out a complete revolution in time T_B.

Simple mathematical treatment

In order to understand the principles involved, we consider the simple case of a sinusoidal wave propagating parallel to an externally imposed magnetic induction $\vec{B}$, where $\vec{B}$ defines the x-axis, and we also assume that the electric field of the wave oscillates in the y-direction. The electric field will be denoted E_y, although we recognize (for now) that it is the only component of $\vec{E}$. The situation is illustrated in Figure 3.12.

In this figure, the coordinates, directions of the electric and magnetic fields, and the relevant velocity components are shown. In visualizing the various forces, the reader needs to think about directions of both direct-force and cross-product terms. In order to make this just a little simpler, the figure has been drawn for a *positive* charge of magnitude e. So just for this paragraph we consider e as a positive charge, although throughout most of the text we consider e as purely a magnitude. In the absence of the externally applied magnetic field, a positive charge e moves along the y-axis due to the force eE_y. The velocity at any time is taken to be $v_y(t)$. Once the external magnetic field

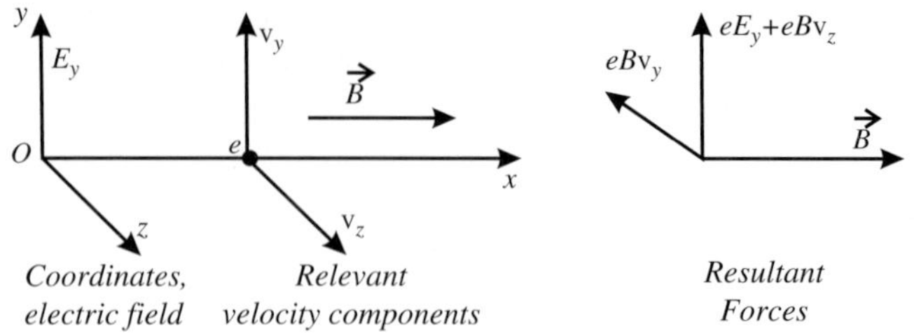

Figure 3.12 Coordinates, electric field, resultant velocities and resultant forces for a simple derivation of the refractive index in an externally applied magnetic field. The resultant forces have been drawn for a positive charge of magntiude e, since this case is a little less complicated to visualize, and adjustments for a negative charge are made in the text.

is applied, there is now an additional force due to the $\vec{v} \times \vec{B}$ force, equal to $-eBv_y$ along the z-axis *i.e., in the negative z-direction.* (Note we ignore the $\vec{v} \times \vec{B}$ force due to the wave's own magnetic field because this acts along the direction of propagation and so does not impact the transverse motion.) The force in the z-direction means that there must be an acceleration in this direction, which means we can no longer assume that the z-component of velocity is zero. If the z-component of velocity is non-zero, then the $\vec{v} \times \vec{B}$ force produces an additional force in the positive y-direction, in addition to the original electric-field force. Hence at any instant there is a force in the y-direction due to both the direct force of the electric field and the $\vec{v} \times \vec{B}$ force acting on the z-component of velocity, and there must also be a force in the z-direction due to the $\vec{v} \times \vec{B}$ force on the y-component of velocity.

We now apply Newton's second law in these two directions, but replace the positive charge e by the electronic charge q_e, which is of course negative. This gives

$$\begin{aligned} m_e \frac{\mathrm{d}^2 y}{\mathrm{d}t^2} &= q_e E_y + q_e B \frac{\mathrm{d}z}{\mathrm{d}t} \quad (y\text{-axis}) \\ m_e \frac{\mathrm{d}^2 z}{\mathrm{d}t^2} &= -\, q_e B \frac{\mathrm{d}y}{\mathrm{d}t} \quad (z\text{-axis}), \end{aligned} \tag{3.91}$$

where the last terms in each case are just the z- and y-components of velocity respectively.

Since the original wave is a sinusoid, the solution may be taken to be of the form $\propto \mathrm{e}^{-\mathrm{i}\omega t}$, (noting again the importance of using a negative exponent, for consistency with Equation (3.6)), so substituting into (3.91) gives

$$\begin{aligned} -m_e\omega^2 y &= q_e E_y - \mathrm{i} q_e \omega B z \\ -m_e\omega^2 z &= +\mathrm{i} q_e \omega B y. \end{aligned} \tag{3.92}$$

The second equation above gives

$$\begin{aligned} z &= \frac{-\mathrm{i} q_e \omega B y}{m_e \omega^2} \\ &= \frac{\mathrm{i}\Omega y}{\omega}, \end{aligned} \tag{3.93}$$

(where we have used $q_e = -e$, and e is now a magnitude once again), which may be substituted into the first equation of (3.92) to allow elimination of z and hence deduction of an expression for y.

Again replacing any q_e terms with $-e$ and solving gives

$$y = \frac{eE_y}{m_e\left(\omega^2 - \Omega^2\right)} \tag{3.94}$$

and

$$z = \mathrm{i}\frac{\Omega}{\omega}\frac{eE_y}{m_e\left(\omega^2 - \Omega^2\right)}. \tag{3.95}$$

The y- and z-components have different amplitudes. But how do we interpret the quantity i? Recall that we assumed that the applied field $\underline{\vec{E}}$ was of the form $\underline{\vec{E}_0}\mathrm{e}^{-\mathrm{i}\omega t} = \underline{\vec{E}_0}\{\cos\omega t - \mathrm{i}\sin\omega t\}$. We have also assumed (for now) that the total E-field is in the y-direction. Assume for demonstration purposes that $\underline{E_0}$ is purely real and equals E_0. Then to get the true motion, we take the real parts of y and z (see Subsection 3.3.1). Then $y \propto E_0\mathrm{Re}\{\mathrm{e}^{-\mathrm{i}\omega t}\} = E_0\cos\omega t$, and $z \propto \mathrm{Re}\{\mathrm{i}\mathrm{e}^{-\mathrm{i}\omega t}\} = E_0\sin\omega t$.

Hence the electron follows a cosine oscillation on the y-axis and a sine oscillation on the z-axis. So the two oscillations are in phase quadrature (i.e., 90 ° out of phase). Hence the motion is an ellipse with major and minor axes along the y- and z-directions.

But what if $\underline{E_0}$ is not purely real? Or what if the applied electric field $\vec{E}$ is not applied only along the y-axis? Recall from Subsection 3.3.1 that there are two different applications of complex numbers that will be used in this text (Subsection 3.3.1). In the above derivations, we have treated the y- and z-components separately, and coupled them through the equations given. So we have used the first type of complex number representation discussed in Subsection 3.3.1, but in fact have used it twice (once for the y- and once for the z-axis). To find the actual motion in the y- and z-planes, we need to take real parts of each expression, and we demonstrated this for a simple case in the last two paragraphs.

Now take real parts of both the y- and z-displacements separately to see how they relate. We will do this to see the shape of the overall trajectory. But this time we will allow for the possibility that the y-motions and z-motions might not be in phase-quadrature. This will prove useful later when we allow for an applied electric field that has components along both the y- and z-axes.

Suppose that the y-motion is represented by $A\ e^{-\mathrm{i}\omega t}$, so that its real component is $A\cos(\omega t)$. Let the z-component be represented by $\underline{R}y = R_0\mathrm{e}^{-\mathrm{i}\varphi}y$ (where $\underline{R} = R_0\mathrm{e}^{-\mathrm{i}\varphi}$ is a complex number). The z- and y-motions have a phase difference of φ, which may not be $\pi/2$ radians.

Then the z-motion is found as $\mathrm{Re}\left\{R_0\mathrm{e}^{-\mathrm{i}\varphi}y\right\}$, so that $z = AR_0(\cos\omega t\cos\varphi - \sin\omega t\sin\varphi)$. Using these real components we may readily write $z = R_0 y\cos\varphi - AR_0\sin\varphi\sin\omega t$, so that $\frac{z - R_0 y\cos\varphi}{AR_0\sin\varphi} = -\sin\omega t$. Then squaring each side of this, and adding to the equation $y^2 = A^2\cos^2(\omega t)$ gives $\left[\frac{z - R_0 y\cos\varphi}{AR_0\sin\varphi}\right]^2 + y^2 = \cos^2\omega t + \sin^2\omega t = 1$, which is a general equation for an ellipse tilted relative to the y- and z-axes.

Some special cases occur: if $\underline{R} = \pm iR_0$, then $\varphi = \mp\frac{\pi}{2}$ and $\cos\varphi = 0$, giving ellipses with major and minor axes along the axes, as discussed immediately following Equation (3.95). If $\underline{R}$ is purely real, then $\varphi = 0$, so the resultant is not an ellipse but a diagonal straight line with slope equal to the ratios of the peak amplitudes.

As noted in the earlier paragraphs, we considered only the real components of $e^{-i\omega t}$ in the above derivation, but an alternative perspective is to consider the various functions $\underline{K}e^{i\omega t}$ as a complete description of the vector motion. Each then represents a vector rotating in the Argand plane (or in our case, in the y–z plane). For $e^{-i\omega t}$ the rotation is clockwise. Two vectors with opposite senses of rotation, such as $e^{-i\omega t}$ and $e^{+i\omega t}$, add together to produce a plane wave ($e^{-i\omega t} + e^{+i\omega t} = 2\cos\omega t$). We will eventually move to this more general and elegant approach, and it even comes in partly useful now, when we consider a more general discussion about these complex exponentials in the context of propagating circularly polarized waves.

Impact of the polarization on the propagating wave

The previous discussion showed the response of the electron to an incident wave which was originally oscillating only in the y-direction. The resulting electron motion was elliptical, and so it re-radiated a new wave which was also elliptical. The resultant wave that was seen forward of this point was therefore the sum of the transmitted original wave plus the new wave radiated by the electron.

As a result, the form of the wave has changed – what started as a linear wave has now attained an elliptical component. This combined wave then drives the next electron, which adds even further to the elliptical component. Hence the waveform changes as it propagates, becoming more elliptical in nature.

We now ask the question: can we choose an incident wave with suitable ellipticity so that the resultant wave further down the path, even after contributions from scattering electrons along the way, retains the same elliptical shape? If so, the wave is called a *characteristic wave*. Our next goal is to find such waves. (In passing, we note that we will in fact find a pair of these waves, and any other wave (be it linear or elliptical) will be able to be represented as a sum of these two characteristic waves. In other words, the waves form a complete basis.)

We therefore need to consider a general incident wave with both E_y and E_z components, with a possible non-zero phase difference, and examine the induced polarization. If the phase difference is arbitrary, then the ellipse has a major axis that is at an angle to both the z- and y-axes. However, a simple rotation of the coordinates will enable alignment of the two axes along the major and minor axes of the ellipse, whereupon the phase difference between the y- and z-motions becomes $\pi/2$ radians. Hence we will consider only the case that the incident wave has ellipsoidal motion with oscillations along the y- and z-directions, which are either in phase-quadrature (ellipse), or in phase (linear-wave oscillation).

We now need to look at the oscillations induced by the applied electric field. We begin by determining the displacements due to the incident E_y field, which we may readily obtain from Equations (3.94) and (3.95), namely

$$y_1 = \frac{e\underline{E}_y}{m_e\left(\omega^2 - \Omega^2\right)} \quad ; \quad z_1 = \mathrm{i}\frac{\Omega}{\omega}\frac{e\underline{E}_y}{m_e\left(\omega^2 - \Omega^2\right)}, \tag{3.96}$$

where it is assumed that the actual values for y_1 and z_1 are attained by taking the real part of the right-hand side in the usual way.

For the case of an incident E_z field, the same equations occur as those in (3.91), but the term q_eE_y will disappear in the first equation, and a term q_eE_z will be added to the second term. The new solutions here are

$$z_2 = \frac{e\underline{E}_z}{m_e\left(\omega^2 - \Omega^2\right)} \quad ; \quad y_2 = -\mathrm{i}\frac{\Omega}{\omega}\frac{e\underline{E}_z}{m_e\left(\omega^2 - \Omega^2\right)}. \tag{3.97}$$

Obviously there is a great deal of symmetry between Equations (3.96) and (3.97), but a sign change occurs in the second part.

The overall displacements may be found by summing the respective y- and z-components. Notice z_1 and y_2 are imaginary, indicating a $\pi/2$ phase difference:

$$y = y_1 + y_2 = \frac{e}{m_e\left(\omega^2 - \Omega^2\right)}\left[\underline{E}_y - \mathrm{i}\frac{\Omega}{\omega}\underline{E}_z\right] \tag{3.98}$$

$$z = z_1 + z_2 = \frac{e}{m_e\left(\omega^2 - \Omega^2\right)}\left[\underline{E}_z + \mathrm{i}\frac{\Omega}{\omega}\underline{E}_y\right]. \tag{3.99}$$

The subsequent components of the induced polarization P_y and P_z are then

$$P_y = -Ney = \frac{-Ne^2}{m_e\left(\omega^2 - \Omega^2\right)}\left[\underline{E}_y - \mathrm{i}\frac{\Omega}{\omega}\underline{E}_z\right], \tag{3.100}$$

$$P_z = -Nez = \frac{-Ne^2}{m_e\left(\omega^2 - \Omega^2\right)}\left[\underline{E}_z + \mathrm{i}\frac{\Omega}{\omega}\underline{E}_y\right]. \tag{3.101}$$

As per usual we take the real part of the right-hand side to get the polarizations. Note that $\vec{P}$ and $\vec{E}$ are not necessarily in phase. This is not unusual in electromagnetism – indeed it is even possible that the polarization and the electric field are not even parallel or anti-parallel – and we could write

$$\vec{P} = \epsilon_0[\mathbf{M}]\vec{E}, \tag{3.102}$$

where **[M]** is a three-dimensional matrix and the (possibly complex) electric field is represented as a column vector. We will not use this tensor/matrix approach here, but it can be important in more complicated theory.

For the wave to travel with unchanging form, the re-radiated waves produced by the moving electrons must have the same form as the incident wave. In other words, the electron displacement ellipse must be similar in shape to the incident wave (indeed, since the radiated electromagnetic wave depends on the acceleration of the electron, we should really look at the acceleration vector, but since for a sinusoidal oscillation the displacement and acceleration are in anti-phase, the displacement and acceleration ellipses will be identical in shape and orientation – one is rotated by π radians relative to the other, which gives each the same appearance and orientation).

This requires that the electrons are displaced in the same direction as the overall electric field, so on a graph of y vs. x, or a graph of E_y vs. E_x, the position of the electron, and the total electric field vector, must have the same slope. In other words, the ratio of displacements must match the ratio of the electric fields for all time (and so must be in phase), viz.

$$\frac{y}{z} = \frac{E_y}{E_z}. \tag{3.103}$$

Hence by setting the ratio of displacements to $\frac{E_y}{E_z}$ using (3.98) and (3.99), (or, equivalently, by setting the ratio of polarizations (from (3.100) and (3.101) to $\frac{E_y}{E_z}$)), it is clear that the requirement is

$$\frac{y}{z} = \frac{E_y - \mathrm{i}\frac{\Omega}{\omega}E_z}{E_z + \mathrm{i}\frac{\Omega}{\omega}E_y} = \frac{E_y}{E_z}. \tag{3.104}$$

Solving gives

$$E_z\left[E_y - \mathrm{i}\frac{\Omega}{\omega}E_z\right] = E_y\left[E_z + \mathrm{i}\frac{\Omega}{\omega}E_y\right] \tag{3.105}$$

or

$$-E_z^2 = E_y^2. \tag{3.106}$$

Hence

$$E_z = \underline{R}E_y, \tag{3.107}$$

where $\underline{R}^2 = -1$, or $\underline{R} = \pm\mathrm{i}$.

In other words, for propagation of the wave along the magnetic field lines. the characteristic waves are circularly polarized, one with clockwise rotation and one with anticlockwise rotation. Hence we may think of *any* propagating wave as a sum of *two* circularly polarized waves with opposite senses of rotation. This can be used to describe circularly polarized waves, linear waves (represented as a sum of equal amounts of the two circular modes), and in the most general case, elliptically polarized waves.

Refractive indices of the characteristic waves

The above discussion has told us about the nature of the waves, but we also need to know their (complex) refractive indices, since this will describe how fast the waves propagate and how quickly they are absorbed. These quantities will now be derived.

We will consider $E_z = \pm\mathrm{i}E_y$. Where the symbols $\pm$ and $\mp$ appear, the upper "+" or "−" signs refer to the case $E_z = +\mathrm{i}E_y$ and the lower ones refer to $E_z = -\mathrm{i}E_y$.

Then from (3.100),

$$P_y = \frac{-Ne^2}{m_e\left(\omega^2 - \Omega^2\right)}\left[E_y - \mathrm{i}\frac{\Omega}{\omega}(\pm\mathrm{i}E_y)\right] \tag{3.108}$$

$$= \frac{-Ne^2E_y}{m_e\left(\omega^2 - \Omega^2\right)}\left[1 \pm \frac{\Omega}{\omega}\right], \tag{3.109}$$

or

$$P_y = \frac{-Ne^2E_y}{m_e\omega(\omega \mp \Omega)}. \tag{3.110}$$

Note that we established that for characteristic waves, the displacement and electric fields on each axis are in phase, so we have dropped the complex-number representation of E_y above.

The displacement field is given by Equations (3.28) and (3.29), viz. $\vec{D} = \epsilon_0 \vec{E} + \vec{P}$, and we also know by (3.30) that $\kappa_e = \frac{\vec{D}}{\epsilon_0 \vec{E}}$ provided that $\vec{D}$, $\vec{E}$, and $\vec{P}$ are all parallel or antiparallel, which we have shown is true for the characteristic waves (see Equation (3.110), where it is clear that P_y/E_y is purely real).

Then

$$\kappa_e = \frac{\epsilon_0 \vec{E} + \vec{P}}{\epsilon_0 \vec{E}} = \frac{\epsilon_0 E_y + P_y}{\epsilon_0 E_y}, \tag{3.111}$$

where the last term follows because $\vec{D}$, $\vec{E}$, and $\vec{P}$ are all parallel or antiparallel and in phase, so taking the ratios of their y components is the same as taking the ratios of their magnitudes.

Substituting from (3.110) gives

$$\kappa_e = \frac{\epsilon_0 E_y - \frac{Ne^2 E_y}{m_e \omega(\omega \mp \Omega)}}{\epsilon_0 E_y}. \tag{3.112}$$

But recall from Equation (3.27) that the refractive index $n = \sqrt{\kappa_e}$, so

$$n_{x,o}^2 = 1 - \frac{Ne^2}{\epsilon_0 m_e \omega(\omega \mp \Omega)}. \tag{3.113}$$

If Ω equals zero (no magnetic field present) this equation collapses to the expression developed earlier for zero magnetic field, as of course it should.

The two different modes are referred to as "O" (for ordinary, when rotation is in the sense that the electron would prefer to rotate) and the "X" (or sometimes "E") mode (for extraordinary).

We have assumed that the magnetic field is aligned along the x-axis, so this theory only applies strictly to the case that the radiowave phase vector is parallel to the external magnetic field. Nevertheless, the equations are still good approximations for θ less than about 30 degrees if we use the effective magnetic field as the component of the magnetic field along the wave's phase vector (i.e., use $B \cos\theta$ or $\Omega \cos\theta$ in place of B or Ω respectively).

Double refraction and Faraday rotation

The two characteristic waves have different refractive indices. As a result, they travel at different speeds and have different degrees of absorption, or in other words, different absorption coefficients. Both of these properties are utilized in experimental studies. The differences in the real part of n result in different phase-speeds, a property which leads to "Faraday rotation." The difference in the imaginary part leads to differences in absorption, which is utilized in the "differential absorption experiment" (DAE). Both may be used to measure electron densities in the ionosphere. Here, we will discuss Faraday rotation briefly – the DAE will be considered later in the text (Chapter 10).

Faraday rotation requires that the magnetic field is non-zero. In the case of a zero magnetic field, we may still consider any incident field to be a vector sum of two oppositely rotating circularly polarized waves, but once the correct combination is found to represent the wave in question, the two oppositely polarized circulating waves travel together with unchanging phase and so the total waveform travels without change.

However, in the case of a non-zero magnetic field, the different speeds result in a phase differential between the two characteristic modes.

Because the characteristic waves travel at different speeds, they will have different wavelengths. Assuming that the two waves start with the same phase, and that the refractive index is spatially constant, at a distance s from the origin each will have a phase

$$\varphi_{o,x} = 2\pi \frac{s}{\lambda_{o,x}} = 2\pi \frac{s\, n_{o,x}}{\lambda_v}, \tag{3.114}$$

where $\lambda_{o,x}$ are the wavelengths of the "O" and "X" modes and $n_{o,x}$ are the *real parts* of the relevant refractive indices. The term λ_v is the wavelength in-vacuo.

Then the phase difference is

$$\Delta\varphi = 2\pi \frac{s}{\lambda_v}[n_o - n_x]. \tag{3.115}$$

Consider the impact that this would have on a wave that was initially a linear wave. The wave can be represented as two oppositely rotating wave vectors, as shown in Figure 3.13(a), which shows the vectors at two distinct positions along the trajectory of the wave.

Since the two vectors corresponding to the characteristic waves rotate at different rates, so the vector sum (representing a linear wave) will also rotate. The vector sum is positioned at the line of reflectional symmetry between the two contributing characteristic-wave vectors, as can be seen in Figure 3.13(b) (i.e., the larger vector bisects the angle between the smaller vectors). The resultant vector oscillates back and

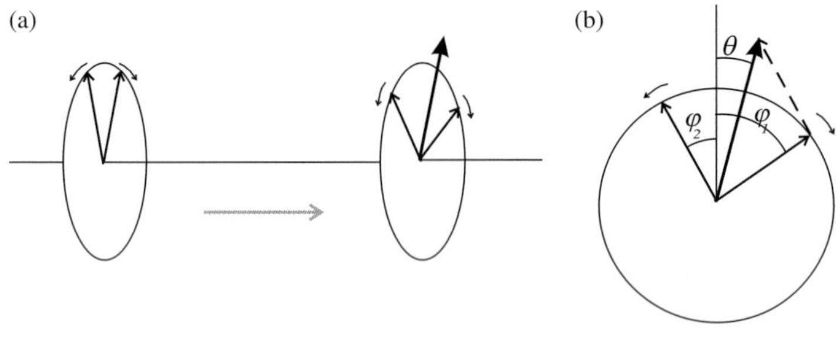

Figure 3.13 (a) The vectors corresponding to two oppositely rotating characteristic waves at two different positions. The circular motions are in a plane perpendicular to the page. In the right-hand part of (a), the vector sum of the two vectors is shown as the longer, darker line. (b) A "front-on" view of the right-hand part of (a), showing how the vector sum of the two contributing characteristic wave vectors produces the total (darker) vector. The large vector bisects the angle between the two smaller vectors.

forth in a straight line along the line defined by this larger vector, but also rotates very slowly (relative to the period of oscillation) along this line.

Note also that we define φ_1 and φ_2 in *opposite* rotational senses of rotation, to match their rotation directions.

As can be seen in the figure, the angle from the φ_1 vector to the bisecting line is $\varphi_1 - \theta$, and from the φ_2 vector to the bisecting line is $\varphi_2 + \theta$, and these two angles are equal, so $(\varphi_1 - \theta) = (\varphi_2 + \theta)$, giving

$$\theta = \frac{\varphi_1 - \varphi_2}{2}. \tag{3.116}$$

We will use this to represent the plane of the final vector of oscillation. Hence from Equation (3.115) we may write

$$\theta = \frac{\Delta\varphi}{2} = \pi \frac{s}{\lambda_v}[n_o - n_x]. \tag{3.117}$$

This has assumed that the two vectors each started at zero phase and rotated as the wave moved, and we have assumed that n_o and n_x are constants.

In reality we need to recognize that the electron density and magnetic field may vary with position, so the refractive indices may vary with position. However, over very short distances ds, Equation (3.117) will still apply, but in a differential form, viz.

$$d\theta = \pi \frac{ds}{\lambda_v}[n_o - n_x]. \tag{3.118}$$

This will be true irrespective of the starting phases at $t = 0$.

Our expressions for the refractive index given in (3.113) involved n^2, so we modify (3.118) to give

$$d\theta = \pi \frac{ds}{\lambda_v} \frac{n_o^2 - n_x^2}{n_o + n_x}. \tag{3.119}$$

If we are dealing with relatively high frequencies, we may take n_o and n_x as close to unity for purposes of division, so this produces

$$d\theta = \pi \frac{ds}{2\lambda_v}[n_o^2 - n_x^2]. \tag{3.120}$$

Substitution from (3.113) gives

$$d\theta = \pi \frac{ds}{2\lambda_v} \frac{2Ne^2}{\epsilon_0 m_e \omega} \frac{\Omega}{\left(\omega^2 - \Omega^2\right)}. \tag{3.121}$$

Finally, with $\omega\lambda_v = 2\pi c$, where c is the speed of light in a vacuum, we produce

$$\frac{\mathrm{d}\theta}{\mathrm{d}s} = \frac{Ne^2}{2\epsilon_0\, c\, m_e} \frac{\Omega}{\left(\omega^2 - \Omega^2\right)}. \tag{3.122}$$

The total rotation over an extended distance is just the integral over the path, viz.

$$\Delta\theta = \int_{\text{path}} \mathrm{d}\theta = \frac{e^2}{2\epsilon_0 c m_e} \int_{\text{path}} \frac{\Omega\, N\, \mathrm{d}s}{\omega^2 - \Omega^2}. \tag{3.123}$$

For high frequencies, where $\omega \gg \Omega$, this is

$$\Delta\theta = \frac{e^3}{2\epsilon_0 c m_e^2 \omega^2} \int_{\text{path}} NB \, \mathrm{d}s. \tag{3.124}$$

This equation can be used for experimental studies of the ionosphere. Suppose that a satellite emits a linear wave at a particular radio frequency, and that the plane of orientation of the signal is measured using a simple dipole or Yagi antenna at the ground. As the satellite moves closer to or further away from the dipole on the ground, the path length through the ionosphere will change, causing rotation of the orientation of the linear wave. The rate of rotation of the plane of the incident wave at the antenna, combined with knowledge of the satellite position, can be used to determine the integrated electron density (termed the "total electron content") between the satellite and the antenna. If the experiment is performed on two or more frequencies, some height information can also be achieved. Astronomers can use the same theory to determine the mean magnetic field in space, using quasars as a source of the radiowaves (estimates of the expected electron density being determined by other means).

3.6.4 Inclusion of both the magnetic field and collisional effects

Obviously we should next consider *both* the effects of collisions and the magnetic field. We can also allow the magnetic field to have a general direction, rather than being aligned close to the direction of the wave's phase velocity vector.

We will not carry out the full derivation, but just state the solution. The derivation can be developed from (3.91) by simply adding damping terms to the equations of motion – although the exact form of the damping turns out to be an issue for debate. The refractive index n obeys the following relation for the case that the collision rate is assumed proportional to the electron speed:

$$n^2 = 1 - \frac{X}{1 - \mathrm{i}Z - \left(\frac{Y_T^2}{2(1-X-\mathrm{i}Z)}\right) \pm \sqrt{\left(\frac{Y_T^4}{4(1-X-\mathrm{i}Z)^2}\right) + Y_L^2}}, \tag{3.125}$$

where $\mathrm{i} = \sqrt{-1}$, $X = \frac{\omega_N^2}{\omega^2}$, $Y = \frac{\Omega}{\omega}$, $Y_L = \frac{\Omega \cos\theta}{\omega}$, $Y_T = \frac{\Omega \sin\theta}{\omega}$, $Z = \frac{\nu_{ec}}{\omega}$, $\omega_N^2 = \frac{Ne^2}{\epsilon_0 m}$, and Ω is given by $\frac{eB}{m_e}$. Here, m_e is the electron mass and θ is the angle between the wave propagation direction and the Earth's magnetic field vector. The + and − modes correspond to the O and X modes of propagation, as outlined in the previous section. This equation is called the Appleton–Hartree equation (e.g., *Appleton*, 1932; *Ratcliffe*, 1959).

The equation simplifies if the special cases of θ parallel or perpendicular to the field lines are considered. These are called "quasi-longitudinal" and "quasi-transverse" cases, and correspond to $Y_T = 0$ and $Y_L = 0$ respectively. We will not give the specific formulas here, though some may be readily evaluated with these substitutions. However, a *word of warning* is needed here.

The Appleton–Hartree equations are *not* exact, and contain approximations. In particular, using the equation above for the special case that Y_L is close to (but not equal to) zero does not do a good job of producing the quasi-transverse refractive index, and some other aspects of the equation break down in special cases (e.g., *Whitehead*, 1952; *Davies and King*, 1961; *Benson*, 1964; *Melrose*, 1984; *Pacholczyk and Swihart*, 1975). The paper by *Whitehead* (1952) is especially concise (two pages only) but illustrates the issues for quasi-transverse propagation nicely. Even in the best cases, the quasi-transverse equation really only applies to angles of departure from perpendicularity to the magnetic field of the order of $\frac{\Omega}{\omega}$ radians and less. If the reader wishes to use any of these approximations, or even the complete Appleton–Hartree equation, some study of these papers is important. Another point to note is that the quasi-longitudinal approximation is valid out to quite large angles (e.g., *Benson*, 1964, who showed it can be valid out to 50 ° and more).

This formula needs further modifications when it is recognized that the collision rate is not proportional to the electron speed, as will be discussed below.

It should also be noted that in the most general case of propagation in a collisional medium with propagation at an arbitrary angle to the magnetic field, the characteristic modes of propagation are neither circular nor linear, but in fact pairs of ellipses with aspect-ratios and orientations dependent on the direction of propagation relative to the magnetic field lines. They are only circular for propagation along the field lines, and are linear for propagation directly perpendicular to the lines.

3.6.5 More sophisticated equations for refractive index

The previous sections introduced the reader to the fundamentals of radiowave propagation, with particular reference to the ionosphere and plasmas. Shortly, we will extend this work to consider refractive index variations due to irregularities in the neutral atmosphere. In the main, the latter expressions are empirically derived, which is in part why we have concentrated on the plasma situation – in the plasma case, some relatively simple but nevertheless instructive mathematics has been possible, with useful extensions to non-plasmas being evident in some circumstances.

However, our derivations have been modestly simple. We have considered largely the case that the induced dipole moment is parallel (or anti-parallel) to the applied electric field. This does not in fact have to be the case, and more generally the relation between the polarization and the electric field is

$$\vec{P} = \epsilon_0 [\mathbf{M}] \vec{E}, \tag{3.126}$$

where **[M]** is a three-dimensional matrix and the electric field is represented as a complex column vector. *Budden* (1965) discusses this more detailed process in considerable detail, e.g. see his Equation (71). Indeed our equations near Equation (3.101) could have been cast in this manner, as we discussed at that time.

In the later 1950s and early 1960s, an important further advance occurred. The earlier derivations had used a very classical assumption in regard to the collision frequency. In our simple derivation, we assumed that the collision frequency for an individual electron with the ions and neutrals was proportional to the electron velocity (e.g. the Appleton–Hartree formula). But experiments in the 1950s showed that in such a plasma, the collision rate varied as the *square* of the electron velocity, and that the distributions of velocity were not necessarily Maxwellian. This required a revision of the theory, especially in regard to absorption of the electromagnetic radiation. Early discussions by *Molmud* (1959), and a more complete derivation by *Sen and Wyller* (1960), were presented. These used a full matrix representation of the susceptibility matrix, and *Sen and Wyller* (1960) used a set of oblique axes to solve the refined expressions for the refractive index. Unfortunately, the paper by *Sen and Wyller* (1960) had several mathematical errors, necessitating revisions by *Manchester* (1965). A clearly written discussion of the new theory, using orthogonal axes, and adopting a tutorial approach, was presented by *Budden* (1965). Even today, that article is highly recommended to the reader with an interest in following this theory further. One result from the paper by *Budden* (1965) was that researchers could use *either* the equations presented by *Sen and Wyller* (1960) (with corrections by *Manchester* (1965)), *or* they could use the original Appleton–Hartree equations, but with the following replacement for the collision frequency (*Budden* (1965)):

$$\nu_{\text{eff}} = \frac{5}{2}\nu_{ec}. \tag{3.127}$$

This approximation is valid for frequencies greater than about 2 MHz and for heights above 90 km in the ionosphere.

A more complete formula is given by

$$n^2 = \frac{\epsilon_1\epsilon_2 \sin^2\theta + \frac{1}{2}\epsilon_3(\epsilon_1+\epsilon_2)(1+\cos^2\theta) \pm \sqrt{\sin^4\theta\{\epsilon_1\epsilon_2 - \frac{1}{2}\epsilon_3(\epsilon_1+\epsilon_2)\}^2 + \cos^2\theta\,\epsilon_3^2(\epsilon_1-\epsilon_2)^2}}{(\epsilon_1+\epsilon_2)\sin^2\theta + 2\epsilon_3\cos^2\theta}, \tag{3.128}$$

where

$$\epsilon_1 = 1 - X\left[(1+Y)\omega^2 C_{3/2}\{\omega(1+Y)\} + \frac{5}{2}\mathrm{i}\omega C_{5/2}\{\omega(1+Y)\}\right], \tag{3.129}$$

$$\epsilon_2 = 1 - X\left[(1-Y)\omega^2 C_{3/2}\{\omega(1-Y)\} + \frac{5}{2}\mathrm{i}\omega C_{5/2}\{\omega(1-Y)\}\right], \tag{3.130}$$

$$\epsilon_3 = 1 - X\left[\omega^2 C_{3/2}\{\omega\} + \frac{5}{2}\mathrm{i}\omega C_{5/2}\{\omega\}\right], \tag{3.131}$$

and where C is defined by

$$C_p\{\omega\} = \frac{1}{p!}\int_0^\infty \frac{u^p \mathrm{e}^{-u}}{u^2+\omega^2}du. \tag{3.132}$$

As usual, both O and X modes are represented, depending on the choice of + or − in Equation (3.128).

3.7 Electron backscatter cross-section

In the previous sections, we considered the refractive index of various plasma gases, and looked at the impact on radiowave propagation, refraction, and absorption. It is true that for some MST applications these effects are unimportant, especially for propagation through the troposphere and stratosphere at frequencies of 50 MHz or so. In those cases, the radiowaves can be considered to travel in straight lines, suffering no absorption as they propagate.

But there are still cases for which refraction and absorption are important, which is why we have discussed these issues. Measurements of electron density using radiowave absorption (the differential absorption experiment) is one example, and will be discussed in Chapter 10. Furthermore, understanding the details of propagation and scattering at a fundamental level, and over a wide range of circumstances, helps us to develop a better feel for the physics of the propagation process, an important goal in its own right.

Now we wish to turn to another important aspect of radiowave interaction with the atmosphere – namely that of scattering and reflection. This is important for all radio wavelengths, and all types of radar, for without it, no signal would be received at the ground, and no studies could be undertaken!

As before, we will begin by looking at scattering from plasmas. The theory for this is relatively well developed (e.g. see *Hagfors* (1975) and *Hafgors* (1989a) and references therein), whereas the scattering from neutral species in the atmosphere is more complex and often the expressions used are empirically derived. Understanding plasma scattering and reflection helps us to better understand all types of scattering.

Before we consider radiowave scatter from a plasma in any detail, we first need to derive the backscatter cross-section for a single electron. Some of this treatment overlaps with our discussions pertaining to refractive index, so some of the equations may look familiar. But the purpose here is different, so some repetition is warranted. We will study scattering and reflection in several ways. Some of the following discussion is loosely based on *Hagfors* (1975) and *Hafgors* (1989a), with additional references to *Tatarski* (1961), Section 4.2.

3.7.1 Cross-sections

Scattering cross-section formulas for a single electron, or a very low density electron gas, will now be considered from first principles. We will consider a plane wave with an electric field that oscillates in an abitrary direction, and assume that the frequency of oscillation is very large relative to the plasma frequency. This means that we can consider the speed of the wave to be the speed of light in a vacuum.

3.7.2 Scattering from a free electron gas

Suppose we consider a single electron, well separated from all of its neighbors (where, by well separated, we mean far enough apart that one cannot impact the other in any significant way; the exact definition of "far apart" will not be specified any more precisely). Imagine that at the location of this electron the electric field is given by

$$\vec{E}(t) = \mathrm{Re}\{\underline{\vec{E}_0}\, \mathrm{e}^{-\mathrm{i}\omega_0 t}\}, \tag{3.133}$$

where $\underline{\vec{E}_0}$ is in general complex (allowing it to have non-zero phase). This formula is consistent with our previous assumed forcings. As already discussed, we will do our working in complex number space and take real parts where needed to get the final results.

For now, we ignore the effects of external magnetic fields and collision frequencies (just as we did in the first instance for the derivation of the refractive index).

Then Newton's second law ($F = ma$) gives

$$q_e\underline{\vec{E}_0}\mathrm{e}^{-\mathrm{i}\omega_0 t} = -e\underline{\vec{E}_0}\mathrm{e}^{-\mathrm{i}\omega_0 t} = m_e\frac{\mathrm{d}\underline{\vec{v}}}{\mathrm{d}t} \tag{3.134}$$

(just as in Equation (3.36)), where the real velocity of the electron is

$$\vec{v}_{\mathrm{real}} = \mathrm{Re}(\underline{\vec{v}}). \tag{3.135}$$

In (3.134), q_e is the electronic charge (including the sign), and e is the magnitude of the elementary charge.

Substituting $\underline{\vec{v}}(t) = \underline{\vec{v}_0}\mathrm{e}^{-\mathrm{i}\omega_0 t}$ into (3.134) in the usual way produces

$$\underline{\vec{v}_0} = -\mathrm{i}\frac{e}{m_e\omega_0}\underline{\vec{E}_0}. \tag{3.136}$$

The current density associated with the motion of this electron at a general point $\vec{r}\,'$ is

$$\underline{\vec{j}}(\vec{r}\,', t) = -e\,\underline{\vec{v}}(t)\,\delta(\vec{r}\,' - \vec{r}_{pe}(t)), \tag{3.137}$$

where $\vec{r_{pe}}(t)$ is the position of the electron and where $\delta(\vec{r}\,')$ represents a spatial Dirac delta function. The vector potential due to this current at a radar receiver at some location $\vec{r}$ is then by definition (e.g. *Wangsness* (1986), p252)

$$\vec{A}(\vec{r}, t) = \frac{\mu_0}{4\pi}\int \frac{\vec{j}\,'}{|\vec{r} - \vec{r}\,'|}\,\mathrm{d}(\vec{r}\,'), \tag{3.138}$$

where the current density $\vec{j}\,'$ is to be evaluated at the retarded time $t' = t - \frac{|\vec{r}-\vec{r}'|}{c}$, to allow for the fact that the signal takes this amount of time to reach the receiver. In other words, $\vec{j}\,' = \underline{\vec{j}}(\vec{r}\,', t')$. The magnetic constant, μ_0, is defined as $4\pi \times 10^{-7}$ Henry/m. Then carrying out the integration in (3.138) using the usual rules for delta-function integration, and using (3.137) gives

$$\vec{A}(\vec{r}, t) = -\frac{\mu_0 e}{4\pi}\underline{\vec{v}}(t')\frac{1}{|\vec{r} - \vec{r}_{pe}(t')|}. \tag{3.139}$$

We now assume that relativistic effects are not important. That is, we assume $\frac{v}{c} \ll 1$, where v is the electron speed. If the electron were part of a larger collection of electrons in a plasma, its speed would be of the order of the RMS particle speed. This satisfies

$$\frac{1}{2}m_e v_{\mathrm{RMS}}^2 = \frac{3}{2}\,K_B\,T. \tag{3.140}$$

So

$$v_{\mathrm{RMS}} = \sqrt{\frac{3\,K_B\,T}{m_e}}. \tag{3.141}$$

If $T = 2000\,\text{K}$ (a large value), then $v_{\text{RMS}} = 300$ km/s. This is 1000 times less than the speed of light in a vacuum, which is $\sim 3 \times 10^5$ km/s. Hence, we are justified in using non-relativistic theory.

Thus far, we have seen how an applied electric field causes an electron to oscillate (as we also saw when we derived the refractive index), and have derived an expression for the resultant current. We have deduced the vector potential induced by this oscillating electron at a distant point $\vec{r}$. Our purpose is to find $\vec{H}$ and $\vec{E}$ at this point $\vec{r}$.

Now choose the origin as the mean position of the electron. This is valid as we assume that the observation point $\vec{r}$ is far removed from the electron, and the only dependence on $\vec{r_{pe}}$ in our equation is through the term $\frac{1}{|\vec{r}-\vec{r}_{pe}(t')|}$, which is purely a distance effect. So we may assume (to very good accuracy) that $\frac{1}{|\vec{r}-\vec{r}_{pe}(t')|} \approx \frac{1}{|\vec{r}- <\vec{r}_{pe}(t')>|}$, where $<>$ refers to the mean electron position. Setting our origin to $<\vec{r}_e(t')>$ gives

$$\vec{A} = -\frac{\mu_0 e}{4\pi}\vec{v}(t')\frac{1}{|\vec{r}|}, \tag{3.142}$$

where $\vec{r}$ is the vector from the mean electron position to the point of observation (which might, for example, be a radio receiver). We now substitute the retarded velocity as

$$\vec{v}(t') = \vec{v}_0 \mathrm{e}^{-\mathrm{i}(\omega_0(t-\frac{|\vec{r}|}{c}))}, \tag{3.143}$$

where c is the speed of the radiowave – which we take as the speed of light in a vacuum for now. Then use (3.136) to replace $\underline{\vec{v}}_0$ to give

$$\vec{A} = -\mathrm{i}\frac{\mu_0 e^2}{4\pi m_e \omega_0}\vec{E}_0 \cdot \mathrm{e}^{-\mathrm{i}(\omega_0(t-\frac{|\vec{r}|}{c}))}\frac{1}{|\vec{r}|}. \tag{3.144}$$

We now introduce the vector $\vec{k}_1$, defined by

$$\omega_0\frac{|\vec{r}|}{c} = \vec{k}_1 \cdot \vec{r}, \tag{3.145}$$

where $|\vec{k}_1| = \frac{\omega_0}{c}$. The wave-vector k_1 is assumed to be the wave-vector associated with a wave scattered *from* the electron, and k_1 is orientated along the line *from* the electron to the point of observation (e.g. a radio receiver), so $\vec{k}_1$ is parallel to $\vec{r}$. This choice ensures that $\vec{k}_1 \cdot \vec{r}$ remains positive. We then obtain

$$\vec{A}(\vec{r},t) = -\mathrm{i}\frac{\mu_0 e^2}{4\pi m_e \omega_0}\vec{E}_0 \mathrm{e}^{-\mathrm{i}(\omega_0 t - \vec{k}_1\cdot\vec{r})}\frac{1}{R_1}, \tag{3.146}$$

with $R_1 = |\vec{r}|$.

The magnetic field $\vec{H} = \frac{1}{\mu_0}\vec{B} = \frac{1}{\mu_0}\vec{\nabla}\times\vec{A}$ in the far field approximation, so we produce

$$\vec{H}(\vec{r},t) = \frac{1}{\mu_0}\vec{\nabla}\times\vec{A} \approx -\mathrm{i}\frac{e^2}{4\pi m_e \omega_0}\frac{\mathrm{i}\vec{k}_1\times\vec{E}_0}{R_1}\mathrm{e}^{-\mathrm{i}(\omega_0 t-\vec{k}_1\cdot\vec{r})}, \tag{3.147}$$

where we have used $\vec{\nabla}\times\underline{\vec{E}}_0\mathrm{e}^{-\mathrm{i}(\omega_0 t-\vec{k}_1\cdot\vec{r})} = \mathrm{i}\vec{k}_1\times\underline{\vec{E}}_0\mathrm{e}^{-\mathrm{i}(\omega_0 t-\vec{k}_1\cdot\vec{r})}$ (recognizing that $\underline{\vec{E}}_0$ is a (possibly complex) constant). The equality between these two terms can be easily seen if one writes out the $\vec{\nabla}\times\cdots$ expression long-hand and replaces occurrences of $\frac{\partial}{\partial x}$ by $\mathrm{i}k_{1x}$, $\frac{\partial}{\partial y}$ by $\mathrm{i}k_{1y}$, and $\frac{\partial}{\partial z}$ by $\mathrm{i}k_{1z}$, and then compares the result to an expansion of $\vec{k}_1\times\cdots$.

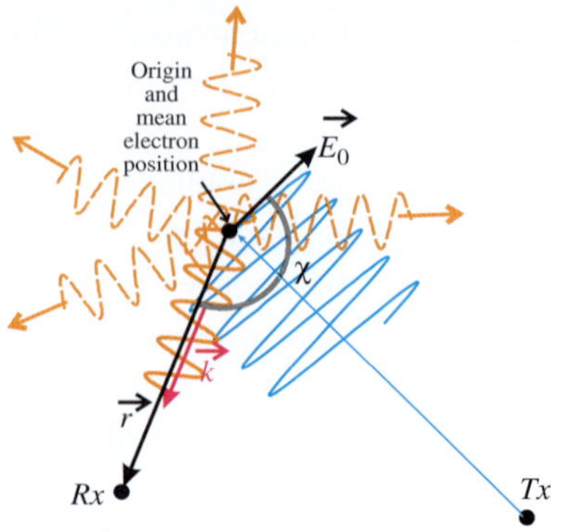

Figure 3.14 Coordinates and vectors relevant to the discussion about scattering in the text. The electron is forced to oscillate along the line $\vec{E}_0$. Waves are radiated in all directions (orange sinusoids), but only the solid one is of interest to us, since it is the portion that scatters to our observation point (a receiver at "*Rx*"). In the discussion, no transmitter is needed – it is only necessary that an electric field is present that causes the electron to oscillate. However, we have added a transmitter to help clarify the imagery, since the most probable way that an applied oscillating electric field will be produced is via signal from a transmitter. Note that the angle χ is defined to be between the line of electrical oscillation ($\vec{E}_0$) and the relevant scattered wave vector. See the text for discussion. The figure has some similarities to Figure 3.3.

Then

$$\vec{H}(\vec{r},t) \approx \frac{e^2}{4\pi m_e \omega_0} \frac{\vec{k}_1 \times \vec{E}_0}{R_1} \mathrm{e}^{-\mathrm{i}(\omega_0 t - \vec{k}_1 \cdot \vec{r})}. \tag{3.148}$$

The magnetic field and magnetic induction vectors are therefore in phase with the current density (after allowance for time retardation) and perpendicular in direction to both the driving electric field and the vector from the electron to the receiver. In Figure 3.14 the $\vec{H}$ vector would oscillate back and forth into and out of the page, being perpendicular to both $\underline{\vec{E}}_0$ and k_1. This is consistent with simple application of the right-hand rule typically used in first-year university physics when determining the direction of a magnetic field from a given current distribution. The same result could be obtained from the Biot–Savart law (with allowance for time-retardation).

The Poynting vector at the receiver due to the radiation from this electron becomes

$$S_r = \frac{1}{2}\eta|\vec{H}|^2 = \frac{1}{2}\eta\left(\frac{e^2}{4\pi m_e \omega_0}\right)^2 \frac{|\vec{k}_1 \times \vec{E}_0|^2}{R_1^2}, \tag{3.149}$$

where $\eta = \sqrt{\frac{\mu_0}{\epsilon_0}} = 376.7\ \Omega$ and $\epsilon_0 = 8.854 \times 10^{-12}$ F/m.

We now introduce the polarization angle χ through

$$\sin\chi = \frac{|\vec{k}_1 \times \vec{E}_0|}{|\vec{k}_1||\vec{E}_0|}, \tag{3.150}$$

and the power density incident on the electron, S_{in}, by

$$S_{in} = \frac{1}{2} \cdot \frac{|E_0|^2}{\eta}. \tag{3.151}$$

The angle χ is drawn in Figure 3.14 (and also in Figure 3.3). Since we are now working largely with magnitudes we will drop the vector and complex notations, and use $E_0 = |\underline{\vec{E}}_0|$, etc.

Then

$$\begin{aligned} S_r &= \frac{1}{2}\eta \left(\frac{e^2}{4\pi m_e \omega_0}\right)^2 \cdot \frac{k_1^2 E_0^2}{R_1^2} \cdot \sin^2\chi \\ &= \frac{\eta^2}{4\pi}\left(\frac{e^2}{m_e\omega_0}\right)^2 k_1^2 \cdot \sin^2\chi \left(\frac{1}{2}\frac{E_0^2}{\eta}\right)\left(\frac{1}{4\pi R_1^2}\right). \end{aligned} \tag{3.152}$$

Using $c = \frac{\omega_0}{k_1}$ and $\eta = \sqrt{\frac{\mu_0}{\epsilon_0}}$ then gives

$$S_r = \frac{\mu_0 e^4}{4\pi\epsilon_0 m_e^2 c^2} \cdot \sin^2\chi \cdot \frac{S_{in}}{4\pi R_1^2}. \tag{3.153}$$

Let us denote the first term as σ_e. Then

$$\sigma_e = \frac{\mu_0 e^4}{4\pi\epsilon_0 m_e^2 c^2}. \tag{3.154}$$

Using $\mu_0 = \frac{1}{\epsilon_0 c^2}$, we may also write

$$\sigma_e = \frac{e^4}{4\pi\epsilon_0^2 m_e^2 c^4} = 9.905 \times 10^{-29} m_e^2 \approx 10^{-28} m_e^2. \tag{3.155}$$

The backscattered power per unit steradian for one unit of incident power from a single electron is then

$$\sigma_e \sin^2\chi. \tag{3.156}$$

This has units of area. Sometimes, σ_e is also expressed in terms of the so-called "classical electron radius." This quantity is defined as the radius which an electron would have if all of its charge existed on a shell of radius r_e, and if the electrostatic energy associated with this distribution is assigned to be equal to the mass-energy, i.e.,

$$e \cdot \phi_e(r'') = m_e c^2, \tag{3.157}$$

where $\phi_e(r'')$ is the electrostatic potential at distance r'' from a point charge e. Note we have not concerned ourselves about the sign of the charge, since we are only interested in a magnitude of the radius. Then

$$e \cdot \frac{e}{4\pi\epsilon_0 r_e} = m_e c^2 \tag{3.158}$$

and

$$r_e = \frac{e^2}{4\pi\epsilon_0 m_e c^2} \approx 2.81 \times 10^{-15}\ \text{m}. \tag{3.159}$$

Then comparison with (3.155) shows that

$$\sigma_e = 4\pi r_e^2, \tag{3.160}$$

or in other words, σ_e is the surface area of a shell with radius r_e.

Thus (3.153) gives

$$\begin{aligned} S_r &= \sigma_e \sin^2\chi \cdot \frac{S_{in}}{4\pi R_1^2} \\ &= 4\pi r_e^2 \sin^2\chi \cdot \frac{S_{in}}{4\pi R_1^2}. \end{aligned} \tag{3.161}$$

Although σ_e is a useful quantity, it is not the one we seek. We now want to know the total amount of scattered energy. To find this, we must integrate S_r over all possible angles of scatter at some arbitrary radius R_1. This gives a total power, rather than a power density. Then we write this as

$$\Sigma_{\text{total}} = \oint S_r R_1^2 \sin\chi \; d\chi \; d\varphi, \tag{3.162}$$

where we have temporarily taken a new polar coordinate system with the $\vec{E}_0$ vector along the new z-axis, with χ being the zenithal angle and φ being the azimuthal angle. Substituting for S_r gives

$$\begin{aligned} \Sigma_{\text{total}} &= S_{in}\sigma_e \cdot \oint \frac{\sin^2\chi}{4\pi R_1^2} R_1^2 \sin\chi \; d\chi d\varphi \\ &= S_{in}\sigma_e 2\pi \int_{\chi=0}^{\chi=\pi} \frac{\sin^3\chi}{4\pi} d\chi \\ &= S_{in}\sigma_e \frac{1}{2}\left[\int_{\chi=0}^{\chi=\pi} \sin^3\chi \; d\chi\right] \\ &= \frac{1}{2} S_{in}\sigma_e \left[-\frac{2}{3}\cos\chi - \frac{1}{3}\cos\chi \sin^2\chi\right]_0^{\pi} \\ &= \left(\frac{2}{3}\sigma_e\right) S_{in}. \end{aligned} \tag{3.163}$$

We define $\sigma_T = \frac{2}{3}\sigma_e$ as the Thomson scatter cross-section, or simply the electron scatter cross-section. Thus

$$\sigma_T = \frac{\mu_0 e^4}{6\pi\epsilon_0 m_e^2 c^2} = \frac{e^4}{6\pi\epsilon_0^2 m_e^2 c^4} = \frac{8\pi}{3} r_e^2 = 6.6 \times 10^{-29} \text{ m}^2. \tag{3.164}$$

While our intent in this chapter is to concentrate only on the incident and reflected electric fields at the scattering region (in this case in the vicinity of the electron), it will be important later to relate this scatter cross-section to things like power emitted by a transmitter at the ground P_{Tx}, and the signal received by a nearby receiver antenna P_r. So Equation (3.161) is often generalized in radar studies to the following form:

$$P_r = \frac{P_{Tx} G_T A_e L}{(4\pi)^2 r_t^2 r_r^2} \sigma_A \qquad \text{Radar Equation}$$

Here, r_t and r_r are the distances from the target (in our case the electron) to the transmitter and receiver, respectively, and A_e is the effective area of the receiver antenna. The term L represents the transmitter system efficiency. As indicated, this equation is often referred to as the "Radar Equation." We will show these terms in more detail later, but it is appropriate to introduce the equation here, at least qualitatively. The term $P_{Tx}G_T/(4\pi r_t^2)$ equals the incident electric field power S_{in} as shown in (3.161), where G_T is a measure of the degree of angular concentration of the radar beam (referred to as the *gain*). The signal received at a receiver antenna of area A_e is the signal transmitted into a solid angle $\Delta\Omega = A_e/(4\pi r_r^2)$, which is just $S_r A_e/(4\pi r_r^2)$, which we write (using (3.161)) as $S_{in}\sigma_A A_e/(4\pi r_r^2)$, where we have used the substitution $\sigma_A = \sigma_e \sin^2\chi$. Using $S_{in} = P_{Tx}G_T/(4\pi r_t^2)$ leads to the Radar Equation shown above. We refer to σ_A as the cross-sectional area of the scatterer: in this case it is an electron, but in the general case it could equally be the area of a small reflecting object like a metal coin (though in that case there is no $\sin^2\chi$ dependence and A_e is essentially the surface area of the coin). More details are provided in Chapters 4 and 5. The case of scattering from a volume of space is also discussed further in Chapter 5.

3.8 Multiple electrons

Our previous discussion pertained to a single electron. But of course, a plasma consists of many electrons, and this will alter the power which a radio receiver will measure. Let us consider some options. We will start with the simplest case of a regular grid, often introduced in courses in crystallography, and develop the scenario from there, progressing to irregularly scattered electrons and then electrons with motion. The description loosely follows that presented by *Mathews* (1984a, b), although the approach is not uncommon and other references use similar developmental procedures.

3.8.1 A regular grid

First, consider a regular grid of electrons. This is analogous to crystallography. If electromagnetic waves (X-rays, optical waves, radiowaves, etc.) impinge on a regular grid of scatterers, it is found that for certain wavelengths and certain orientations, very strong back-scattered signal results. For example, waves which enter at $90°$ to the so-called "Miller planes," and for which the spacing between successive planes is a half-wavelength, produce strong backscatter. Such special cases are referred to as Bragg reflection. Waves which do not have one of these special orientations relative to a Miller plane produced very weak reflected signals. The resultant diffraction pattern in 3-D is called a Laue diffraction pattern. Examples can be found in any book on X-ray crystallography; such patterns appear as a series of dots scattered in a regular manner across the diffraction plane.

Figure 3.15 shows an example of such a regular grid. The wave vector indicated by $\vec{k}_1$ will produce strong backscatter, provided that the wave has a wavelength equal to twice the spacing between the successive planes (drawn as lines, but considered as planes

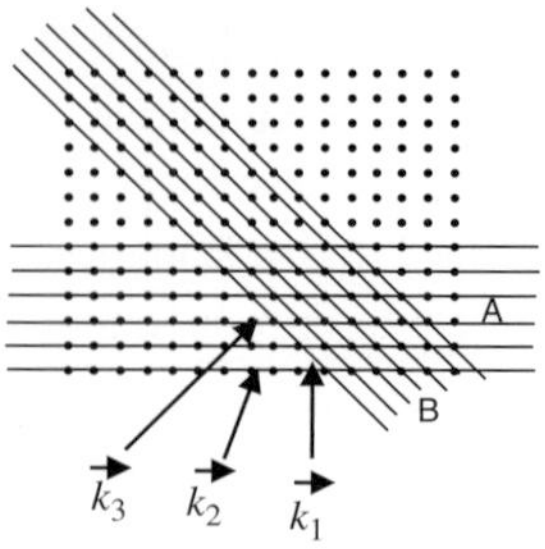

Figure 3.15 Scatter from a regular grid of electrons. The lines show examples of so-called "Miller planes" – which in the figure are lines, but can be considered to stretch back into the page as part of a three-dimensional grid, so we will refer to them as "planes."

in a 3-D image) in the cluster of planes denoted by "A." To see why this is, consider what happens when the wave strikes the first plane. It will be partly reflected, but most of the signal will move on to the next plane. Here, some further wave energy will be reflected. If the next plane is one half a wavelength beyond the first, then the wave travels an extra distance of a half wavelength from the first to the second plane, and then upon reflection travels a further half-wavelength from the second back to the first plane, giving a total round trip distance of one wavelength. Thus it arrives back at the first plane in phase with the next wavefront, which is just being reflected from that surface. The wave reflected from the second plane, and that from the first plane, travel back together toward the source as a wave that is stronger than either of the two individual contributing waves, since the two signals are in phase. Likewise, reflections from the third, fourth and successive planes also all add coherently with those from the other planes, leading to a strong coherent reflected signal.

Likewise the vector $\vec{k}_3$, being orientated perpendicular to a set of Miller planes (labelled "B") in the figure, also will produce a strong backscattered signal provided that the wavelength λ_3 is twice the spacing between the planes of the set "B."

The vector $\vec{k}_2$, having no corresponding Miller plane, will produce very little backscatter, although if the wavelength were correct, it could produce oblique scatter from one of the other planes. However, for now we concentrate only on backscatter.

3.8.2 Bragg scales

A key concept here is that of *Bragg reflection*. When the planes are separated by one half of the radar wavelength we get strong *constructive interference*. Figure 3.16 shows the concept in more detail. A wave is represented by the wavelike structure to the top right of the figure. The symbol "C" represents a crest, and "T" represents a trough. It is propagating towards the top left. Two Miller planes are labelled "A" and "B."

A plane-wave crest is assumed incident at the point a_1 in plane A. It will suffer partial reflection and may continue on to plane B. Similar processes will occur at all electrons situated in plane A. When these wave-fronts that pass through plane A to plane B reach B, they will suffer more partial reflection which will be reflected back towards A. Note that there is no scatterer at the point b_*, but it is representative of the plane B. Reflections will occur from all particles in plane A and all particles in plane B (and also from other planes which have not been annotated in the figure, of course). It is the accumulated reflections from all electrons in all planes that matters. So for the sake of description, we can consider that reflection did indeed occur from the point b_*.

The wave-front reflected from point b_* (or more exactly, from the plane B) will therefore have traversed one wavelength in its passage from plane A to B and back to A. Both the reflection from B and the reflection from A will produce identical phase changes (generally π^c) so that waves reflected from A and from B will be in phase once they meet up again at plane A, and therefore add constructively. Thus if planes are separated by $\frac{\lambda}{2}$, where λ is the radio wavelength, constructive interference occurs. This is called the *Bragg reflection* and leads to strong constructively interfering reflections.

Now turn to Figure 3.17. In it, we have replaced the Miller planes from Figure 3.15 with sinusoidally varying refractive index perturbations. Again, if electromagnetic waves with a wavelength of twice the perturbation wavelength (scale) impinge on the refractive index sinusoids, strong coherent reflections will occur. Indeed, this situation is actually a better one for consideration of Bragg reflection. When we considered the Miller planes, we discussed how an incident wave with a wavelength of twice the plane spacing would produce strong reflections – but we ignored some other important cases. If the incident radiation has a wavelength equal to the spacing between the planes, rather than double it, we will *still* produce strong reflection, since a reflected wave at the plane B in Figure 3.16 will now have a path difference of *two* wavelengths when it arrives back at plane A. Likewise if the wavelength of the incident radiation is 0.5 times the planar spacing, then the reflected wave will have a phase path of four wavelengths. All of these cases will produce strong reflection. However, in the case of the sinusoidal oscillation in Figure 3.17, strong reflection will *only* occur when the incident wavelength is twice the refractive index scale (a result best proved using Fourier theory, but one which we will not dwell on here). Thus a sinusoidal variation in refractive index is a *better* situation to employ when considering Bragg reflection. Indeed the case of the planes discussed in regard to Figures 3.15 and 3.16 can be considered as a series of delta-functions in the path of the wave, and a Fourier decomposition of these delta-functions will give Fourier

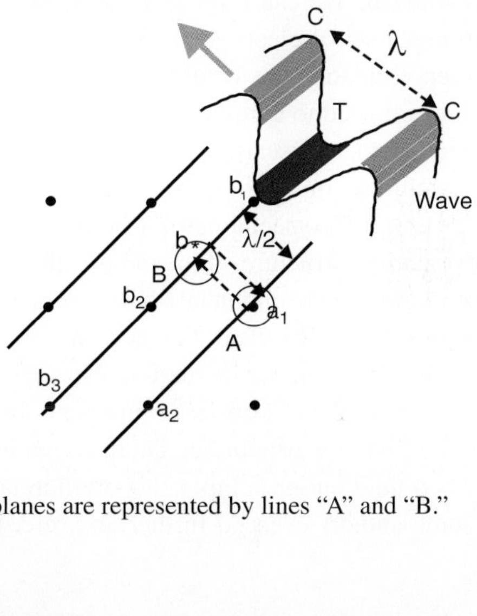

Figure 3.16 Bragg reflection: Miller planes are represented by lines “A” and “B.”

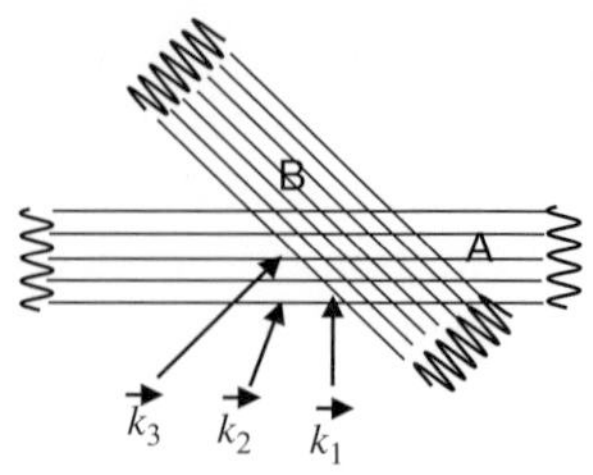

Figure 3.17 A structure composed of multiple Fourier harmonics in place of the Miller planes.

components with scales equal to the spacing, one half of the spacing, one third of the spacing, one quarter of the spacing, and so forth. Each of these Fourier components can be considered as a Bragg scale for a different wavelength wave.

Hence if we take *any* refractive index distribution, we can do a three-dimensional Fourier transform in space and thereby represent the distribution as a spectrum of Fourier scales. Each Fourier scale will act as a Bragg scatterer for a different incident wave, with differing orientations and wavelengths.

Conversely, if we probe the region with a single radiowave, it will in effect produce reflection *only* from those scales which match the Bragg condition, and will receive signal from no other scales.

It is common practice to use this visualization when considering scattering from any region of space, and we will employ it extensively in this book.

It may be noticed that we at first referred to the process as Bragg reflection, but more recently referred to it as Bragg scatter. This apparent confusion will now be clarified and expanded upon.

3.8.3 Random positions

In a real plasma, the electrons are never so organized as to form a grid. Nevertheless, our previous discussion makes a useful starting point and we will use the concept of Bragg scales shortly. In a real plasma, the electrons are quasi-randomly or even totally randomly positioned. We can represent the refractive index variations as either: (i) a randomly arranged set of electrons; or (ii) a randomly orientated spectrum of Fourier scales. Each representation has its uses.

At this point, we need to comment on some nomenclature. Many articles distinguish between different types of scatter, such as "incoherent scatter", "Fresnel scatter" (e.g. see Chapter 2, Sub-section 2.10.1), Bragg scatter and so forth. The term Bragg reflection is generally reserved for special cases where the Fourier scales discussed above have some sort of organized structure – for example, the region may be one which is stratified, so that all of the Fourier components have normal vectors that are vertical. A step in the refractive index in the vertical direction, which is uniform horizontally, is one such example. This is not an unusual situation, especially in MST studies. In this case, the phase and amplitude of the Fourier components of refractive index vary in a regular manner as a function of wavenumber. On the other hand, the term Bragg scatter is more often used when the Fourier components are unrelated in phase from one scale to the next – and some authors even go further and refer to this case as Rayleigh scatter, or even incoherent scatter, as if the concept of Bragg scales no longer applies. The addition of Fourier components for cases of random scatter was discussed in some detail by *Hocking and Röttger* (1983).

But use of the term Bragg scatter or Bragg reflection in this way can be misleading. *All* of the situations just discussed – including so-called "incoherent" scatter – involve some form of Bragg reflection. In the case of incoherent scatter, the Bragg scales are just randomly orientated and are all roughly equal in amplitude, so there is no preferred scale, and the phases of the Bragg scales change very rapidly in time (often on times scales less than a millisecond or so). *Bragg scatter is the underlying concept involved in almost all of the scattering processes involved in this book.*

3.8.4 Random electron position

We now return to the case of randomly positioned electrons. In the following sub-section we choose not to use a Fourier decomposition, but we will do so in later sections, which is why the topic of Bragg scales has been introduced at this point.

But for now, simply consider the scattering from a randomly distributed group of electrons, spaced far enough apart that they do not feel each other's presence. Each electron scatters incident radiowaves and each scattered radiowave arrives back at the receiver with a different phase.

Thus the resultant received signal must be found by vectorially adding a collection of randomly orientated vectors, as shown in Figure 3.18. It is because we visualize our electron gas as a random distribution of electrons which add randomly in phase, that we call this process incoherent scatter (but, as noted, a Bragg scales representation is equally valid).

3.8.5 Rayleigh distributions

Treatment of the vector sum of a randomly orientated group of N vectors in two dimensions is a classical problem in mathematics. If all the vectors all have length equal to a, and all are summed, the resultant vector follows a so-called *Rayleigh distribution.* There will be small probability that they will all cancel and sum to zero, and a small

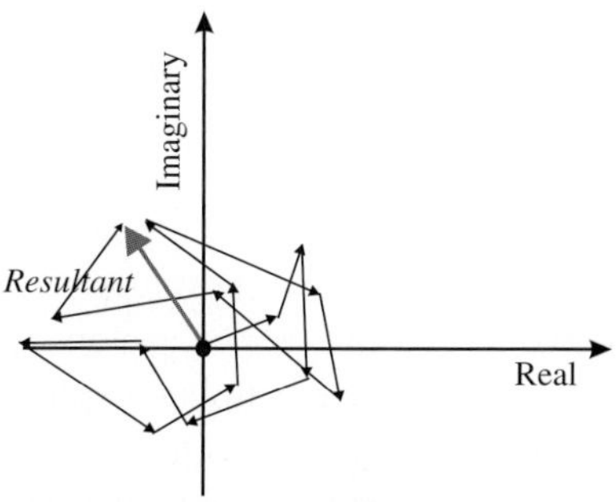

Figure 3.18 Resultant vector formed by adding a set of randomly phased vectors. If the process is repeated multiple times, for different phase orientations, and if all vectors have the same length, then the distribution of resultant vector amplitudes gives the Rayleigh distribution.

probability that they all happen to be aligned parallel and sum to an amplitude of Na. There will be higher probabilities that they will have a resultant vector of length greater than zero and less than Na. The actual shape of the distribution is not so important to us here, though the interested reader is referred to Figure 7.21 in Chapter 7 for a graph. The Rayleigh distribution was also discussed briefly in Chapter 2, Section 2.5.

[Special comment: It is worth noting here that in lidar studies of atmospheric constituents like sodium, there is a special scattering process called "Rayleigh scatter." This is *unrelated* to our discussion here about Rayleigh distributions, although both are named after the same great man.]

The contributing vectors do not actually have to be the same length in order to produce a Rayleigh distribution. Even if the contributing vectors have non-equal lengths, but the variation is not too large, the resultant distribution is still close to Rayleigh. Our primary interest in the Rayleigh distribution in this chapter will be in regard to its mean squared value, and we make the following hypothesis:

The mean squared length of the resultant vector produced by summing N equal-length vectors with random orientations is equal to N times the square of the length of an individual vector.

This is fairly readily proven, as will now be shown. We represent all the vectors as complex numbers in an Argand plane. In this case, the real component represents the x-component, and the imaginary component represents the y-component. The imaginary component is not considered as "imaginary," and is as real as the x-component (see the second method for using complex numbers discussed in Section 3.3.1).

Then the resultant vector produced in the sum is given by

$$\vec{R} = \sum_{k=1}^{n} a_k \mathrm{e}^{\mathrm{i}\varphi_k}. \tag{3.165}$$

The mean-square value of this resultant vector is then given by

$$|\vec{R}^2| = RR^* = \overline{\left(\sum_{k=1}^{n} a_k \mathrm{e}^{\mathrm{i}\varphi_k}\right)\left(\sum_{m=1}^{n} a_m \mathrm{e}^{\mathrm{i}\varphi_m}\right)^*}, \tag{3.166}$$

which is equal to

$$\vec{R}^2 = \left(\sum_{k=1}^{n} a_k^2 \mathrm{e}^{\mathrm{i}\varphi_k} e^{-\mathrm{i}\varphi_k}\right) + \left(\sum_{k=1}^{n}\sum_{m=1}^{n} a_k a_m \mathrm{e}^{\mathrm{i}(\varphi_k-\varphi_m)}\right), \tag{3.167}$$

where $m \neq k$.

For a finite number of vectors, the second term contributes various different values from case to case, and these different values give rise to the Rayleigh distribution.

However, as we let the number of vectors approach infinity (or if we average the results of a large number of repetitions of the process), the second term disappears, since the orientation of successive vectors is assumed uncorrelated (or equivalently, the phases φ_k and φ_m are assumed randomly distributed between 0 radians and 2π radians). For a large number of vectors, for every occurrence of one value of $a_k a_m \mathrm{e}^{\mathrm{i}(\varphi_k-\varphi_m)}$ there will exist another with the same magnitude and opposite sign, so the pair will cancel.

Indeed, it does not take much effort to extend this concept and recognize that the last term will disappear for *any* distribution that has a uniform phase distribution, regardless of the amplitude distribution, provided that the different phases are randomly scattered among the different amplitudes. So in this limit, only the first term in (3.167) is retained, which can be collapsed to

$$\vec{R}^2 = \sum_{k=1}^{n} a_k^2. \tag{3.168}$$

If a_k is of constant value, equal to a, say, then

$$\vec{R}^2 = Na^2. \tag{3.169}$$

Thus if we have a collection of N randomly positioned electrons in a volume V, the power flux received at a receiver will be

$$S_{Rx} = NS_r, \tag{3.170}$$

where S_r is given by (3.161), or

$$S_{Rx} = N\sigma_e \sin^2\chi \, \frac{S_{in}}{4\pi R_1^2}. \tag{3.171}$$

This is the key result that we need from these considerations.

Remember again – this formula only applies if the electrons are truly randomly distributed and are large in number. Later, it will be seen that this is sometimes not the case, so although this formula serves as a useful starting place, it is not the end of the story by far.

3.9 Backscatter cross-sections and reflectivities for a radar

3.9.1 Introduction of the spectrum

We now return to Equation (3.161), viz.

$$S_r = 4\pi r_e^2 \sin^2\chi \cdot \frac{S_{in}}{4\pi R_1^2}. \tag{3.172}$$

Since many MST radars use backscatter situations, with coincident transmitters and receivers, we will concentrate on the case $\chi = 90\,^\circ$, so that $\sin\chi = 1$. Although bistatic situations do occur, we can demonstrate the main points that we wish to make quite adequately by using a monostatic situation.

We crudely consider a radar that transmits a pulse of some length Δr, and transmits its signal over a limited angular extent. For simplicity suppose that the angular coverage is represented by a cone of radius $\Delta\theta$, so that at any instant the radar receives scatter from an approximate cylinder of length Δr and radius $r \cdot \Delta\theta$, or a volume

$$\Delta V = \pi(r \cdot \Delta\theta)^2 \Delta r, \tag{3.173}$$

where r is the mean range from which the scatter occurs. This region is called the *radar volume*. The value of r can be determined from the time taken for the pulse to travel from the transmitter to the scatter region and back to the receiver. In later chapters we will discuss range determination, and the shape of the transmitter and receiver beams in more detail, but this will be adequate for now. We will assume that $\Delta\theta$ is small (typically less than $10\,^{\circ}$).

In Equation (3.171), we adapted Equation (3.172) for the case of randomly positioned electrons. In a more general case, the electron gas will not be completely disordered, nor completely ordered. Its refractive index structure needs to be described by a spectrum in k-space. We replace the total number of electrons N in a special way. We recognize that although all electrons will scatter radiation, the total signal depends on how the phases of the reflected signals add. As discussed, by far the strongest signal will come from reflection from Bragg scales embedded within the structure. These are sinusoidal oscillations with "wavefronts" perpendicular to the line joining the region to the transmitter/receiver system, and with spatial wavelength equal to half of the radar wavelength.

In a real radar experiment, a pulse of radiowaves is used, and it has various Fourier components. Each Fourier component will reflect off its own Bragg scales within the region of interest. The total backscattered power depends on the integrated power across the spectral band of interest, as shown in Figure 3.19.

The basic idea embedded in this figure is as follows. The received amplitude as a function of range is a convolution between the transmitted pulse and the refractive index variability (e.g., *Hocking and Röttger*, 1983, and references there-in). These authors looked at the gradients of the refractive index and convolved them with the pulse, but we can also consider qualitatively convolving the pulse with the refractive index profile itself, since the Fourier transform of the fluctuations, and the Fourier transform of their gradient, are identical except for a rescaling factor proportional to the wavelength, (a well known and standard result in the Fourier domain when a function is differentiated (e.g., *Champeney*, 1973) – in fact some evidence of this rescaling term was already shown in Figure 3.4, where a term $\frac{-i}{\lambda}$ appears, although this particular aspect was not discussed at the time). We will look more carefully at the convolution of the pulse with the backscatter function, and its equivalent action in the Fourier domain, later in this and successive chapters, but this simple discussion is adequate for now.

As seen in the figure, the resultant Fourier transform is a product between the Fourier transform associated with the pulse, and that associated with the refractive index variations (shown as part (d)). The Fourier transform of the pulse is smoothly varying as a function of wavenumber, while the Fourier transform of the refractive index variations varies significantly in phase from one wavenumber line to the next, but could (although may not) maintain a broadly uniform amplitude. When integrating across the wavenumber domain, the problem becomes one of adding successive vectors which are the product of the refractive index Fourier transform and the Fourier transform of the pulse (represented by the vertical lines in part (d)). This amounts to vectorially adding a series of vectors with slowly changing amplitude, but with phases that change from wavenumber to successive wavenumber. In other words, we again have a Rayleigh sum,

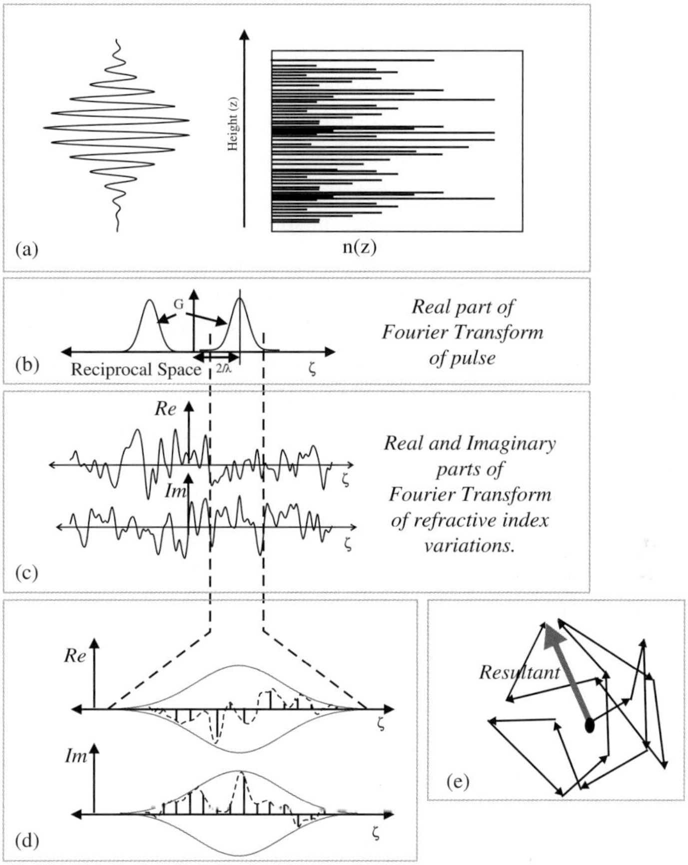

Figure 3.19 (a) One dimensional representation of the interaction of the radar pulse with the scatterers. The strengths of individual scatterers are shown as horizontal lines of various strengths, though in reality it could also be a continuum. The "wavelength" in the pulse is also not shown properly to scale – in reality it should be much less than the typical spacing between scatterers. (b) The Fourier transform of the pulse (real part only). (c) The Fourier transform of the refractive index variations, drawn as a continuum. (d) The product of (b) and (c). This figure concentrates only on the region defined by the positive wavenumbers of the spectrum of the pulse (expanded as shown by the broken lines descending from (b) and (c)). Vertical lines indicating representative spectral lines are shown in order to give the spectrum a discrete character, to allow it to be consistent with our discussion in terms of discrete scattering entities. A similar diagram would be produced if we had included the negative frequencies. (e) Vector sum of the Fourier components in (d) (positive and negative frequencies included), where the summation is performed in complex-number space (Argand diagram).

as in Figure 3.18, but this time the sum is over wavenumber space. The result is shown conceptually in 3.19(e).

Volume dependence of the backscattered power

Our next goal is to examine how the backscattered power depends on the electron density spectrum and on the radar volume. The most important aspect is to demonstrate that the backscattered power is proportional to the radar-volume of scatter. In light of Equation (3.171), this may seem obvious. Surely if the backscattered power is proportional to the number of electrons, is it not obvious that the number of electrons is proportional to the volume and so the backscattered power must be proportional to the radar volume?

However, it is important to look at how this same proportionality can be developed from the Bragg-scale and spectral perspective. One clear example of the need for this was a paper presented by *Gage et al.* (1981). Recall that the radar volume was given by $\Delta V = \pi(r \cdot \Delta\theta)^2 \Delta r$. Notice in particular that the radar volume is proportional to Δr. But *Gage et al.* (1981) proposed that the backscattered power should depend on the *square* of the pulse length. This clearly would violate the concept that the backscattered power should depend on ΔV. As it turned out, these authors had mistakenly developed their model based on an earlier paper which dealt with radar reflection from artificially generated refractive index structures which produced a highly coherent spectrum within the radar volume. The error was reported by *Hocking and Röttger* (1983) and the correct formula was developed, as well as a more general formula which could be applied in cases where the backscatter varied on larger scales as a function of height.

So to see how this volume-dependence arises, consider what happens if we keep the transmitted pulse at a fixed amplitude (thereby maintaining the same peak incident Poynting flux arriving at the scattering volume), but double the pulse length. This doubles the radar volume, but in wavenumber space it halves the width of the Fourier transform of the pulse, thereby halving the number of vectors available to add. However, since the amplitude of the pulse is fixed, the amplitude of the spectral lines in the Fourier transform of the pulse doubles. This is a well-known result from Fourier theory. In general, if the pulse width increases by £ times, then the bandwidth increases by $\frac{1}{£}$ times, but the amplitudes in the wavenumber domain all increase by £ times. This is required because the value of the pulse at zero time-lag equals the sum of the complex amplitudes in wavenumber space, so if the width in wavenumber space halves, and the sum remains fixed, then the wavenumber amplitudes must all double. This means that the number of vectors increases by $\frac{1}{£}$ times, but the amplitude of each vector increases by £ times. The total received power will be proportional to the number of vectors multiplied by the mean square amplitude of each individual vector, as discussed in Equation (3.169) (although in this case we are applying the summation over the wavenumber domain rather than across a group of randomly positioned electrons). Hence the power increases by $\frac{1}{£} \times £^2$ times, or a total factor of £. In the case considered by *Gage et al.* (1981), the spectrum could not be assumed to have quasi-random variations in spectral amplitude across the width of the Fourier transform of the pulse (as in Figure 3.19(c)), but rather all had the same amplitude and phase. Hence the Rayleigh sum shown in Figure 3.19(e) would not apply, and all the vectors would add end-to-end in a long straight line. Then the resultant vector is proportional to $£^2$, rather than £.

Hence we have now verified that the power should be proportional to the ΔV, and the pulse length dependence for normal atmospheric scatter should be proportional to the pulse length.

The role of the spectrum in determining the backscattered power

We now turn to examination of how the spectrum enters the picture. Suppose that the electron density is Fourier transformed over all space, with no regard for a limiting volume defined by the radar. The spectrum is then formed by finding the product of this Fourier transform and its own complex conjugate. Call this spectrum $\Phi_N(k)$. The values of the spectral lines shown in Figure 3.19(c) now no longer contain any phase information and all quantities are positive. Across the width of the band defined by the Fourier transform of the pulse, the powers (or more strictly, the power densities) are roughly the same, with some statistical variation. Then the total power is found by summing all these spectral contributions. Φ_N may have a spectral form, but we assume for now that this spectral variation occurs on larger scales than the spectral width of the pulse. For example, if the spectral width of the pulse (in units of inverse wavelength) is given by 0.001 m^{-1}, then we expect the natural variation in the atmospheric spectrum (due perhaps to turbulence) might only be significant on scales of the order of perhaps 0.1 m^{-1}. So we have assumed that the spectral densities are approximately constant across the region of interest.

The above result, when combined with the volume dependence developed in the previous sub-section, says that the total received spectral contribution integrated across the whole pulse is given by

$$\mathcal{P}_{s,\mathrm{tot}} \propto \Phi_N(k_B)\,\Delta V, \tag{3.174}$$

where k_B is the Bragg scale for the central frequency of the pulse. We did not have to choose the value at the central frequency scale – it is similar for all wavenumbers contained in the pulse, but it makes sense to choose the central one as most representative.

The astute reader may wonder why there is no multiplication by the spectral width of the radar pulse – surely the total received power depends on the band-width as well as the powers? The answer is that it *is* already incorporated through the pulse-length dependence which is embodied in the term ΔV – as discussed in the previous sub-section on pulse-length dependence.

Equation (3.174) has been confirmed numerically by *Hocking and Röttger* (1983). Hence we may modify Equations (3.172) and (3.171) to write the power at the receiver as

$$S_r \propto \sigma_e \frac{S_{in}}{4\pi R_1^2}\Phi_N(k_B)\,\Delta V. \tag{3.175}$$

The backscatter cross-section per unit volume (as distinct from the total back-scattered power) is given as follows. We calculate the power that would be backscattered, per unit steradian, for an incident Poynting vector of unity, and normalize by dividing by the scattering volume. This means we multiply Equation (3.175) by R_1^2 (to work out the total backscattered power per unit steradian), divide by S_{in} (to normalize with respect to the incident power), and divide by ΔV (to normalize with respect to the scatter volume). The result is

$$\sigma_s \propto S_r \frac{R_1^2}{S_{in}\Delta V} \propto \sigma_e \Phi_N(k_B)/(4\pi) \propto r_e^2 \Phi_N(k_B), \tag{3.176}$$

where we have expressed σ_e in terms of the electron radius from (3.160). For now we have left this as a proportionality, since we have not formally defined Φ_N quantitatively at this time. We will revisit this equation later.

The quantity σ_s represents the backscatter cross-section per unit volume, often referred to as the reflectivity (e.g., *Ottersten*, 1969a). However, this can be a confusing nomenclature, since a reflection coefficient should be unitless, whereas σ_s has units of m^{-1}. We will use the term "reflectivity" with reservation. Note it includes the spectral component of the electron density variations at the Bragg scale of the central carrier wavelength of the pulse.

3.9.2 The spectrum of refractive index variations

The previous sub-section developed an expression for the backscatter cross-section per unit volume in terms of the electron density spectrum. However, it turns out that this expression is limited in applicability. To see this, we will revisit the derivation of σ_s, but do it independently of electron density. We will instead consider the refractive index as the key parameter. In order to do this, we will first reproduce our expressions for backscattered power in terms of the permittivity, and then in terms of the *relative permittivity*, which is also referred to as the dielectric constant. We emphasize at this point that there is something oxymoronic about this process – we refer to a "dielectric constant" but in fact we will assume that it is anything *but* constant – indeed its variability is the source of all scatter! But the convention has been in existence for many years – so we need to accept the concept of a "constant" that is spatially and temporally variable. We will generally talk of the permittivity, or the relative permittivity, to avoid referring to a variable quantity as a constant.

In the previous derivation, we considered that all of the scattered electric field could be attributed to the oscillations induced in the free electrons. Now we allow a more general scenario. We still allow for the fact that the permittivity variations can be due to the electron oscillations, but we consider that there might also be ions involved, and that the electrons could move in a coordinated way, which might modify our simple assumption that the electrons act independently. One possibility is that the electrons move in a coordinated way which might partially cancel the effect of the incident electric field.

Important vectors, locations and other pertinent parameters which will be used in the following discussions are shown in Figure 3.20.

Consider the signal received at time t at the receiver. Assuming that the speed of propagation from the transmitter to the scatter point is c, and that the scatter is very small compared to the incident signal (Born approximation), then the pulse arrives at the scatter point "O" (roughly the center of the region) at time $t' = t - |\vec{R}_s|/c$, and was transmitted at time $t'' = t' - |\vec{R}_i|/c$, or $t'' = t - |\vec{R}_s|/c - |\vec{R}_i|/c$.

The transmitted pulse can be written as

$$g_p(t'') = g(t'')\mathrm{Re}\{\exp\{-i\omega t''\}\}, \tag{3.177}$$

where the function g describes the envelope of the pulse.

In order to keep things simple, we will ignore the envelope effect, and treat the wave as a continuous wave and consider g as unity over all time, but its presence should be recognized. However, even now we need to recognize that the pulse length often (indeed usually) partly defines the shape of the perimeter drawn in Figure 3.20, and we will return to this point later.

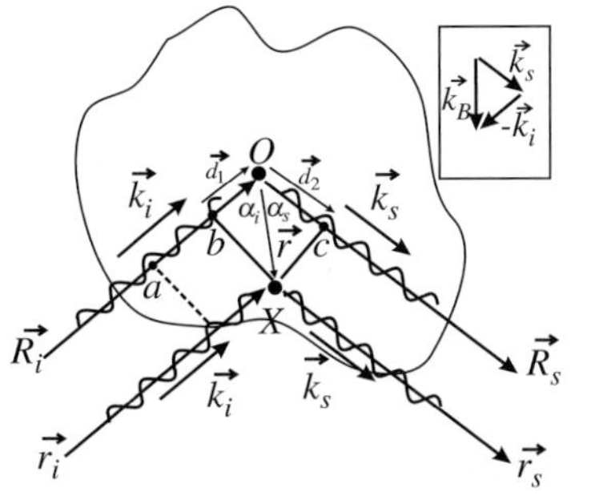

Figure 3.20 Major aspects of bistatic scattering from a region of generalized permittivity permutations. The perimeter outlines the region of interest, which may be defined by natural limits (e.g. a region of turbulence) or (*most commonly*) may be defined by the radar beams and pulse-length. The point O is an arbitrarily chosen origin, but we select it to be somewhere near the center of the region of interest. Point X is a representative region of scatter, which could be anywhere within the volume. The vectors $\vec{R}_i$ and $\vec{r}_i$ represent vectors from the transmitter to the points O and X respectively. The transmitter is considered far enough away that the two vectors may be considered as parallel. The incident signal is represented by the two sinusoidal lines on the left (with wave-vector $\vec{k}_i$). These two wave rays are in phase as they approach the points O and X, as indicated by the broken line at a. The vectors $\vec{R}_s$ and $\vec{r}_s$ represent vectors from the points O and X to the receiver (respectively). The receiver is considered far enough away that the two vectors $\vec{R}_s$ and $\vec{r}_s$ may be considered as parallel. The scattered signals (represented by the two sinusoidal lines on the right, and the vector $\vec{k}_s$) are *not necessarily* in phase after they leave the points O and X. The inset at the top shows the subtraction of $\vec{k}_i$ from $\vec{k}_s$ to produce $\vec{k}_B$. Although $\vec{k}_B$ appears vertical here, this is only because we have drawn the transmitter and receiver to be roughly symmetrically placed about the scatter point – if this were not so, $\vec{k}_B$ would be skewed in one direction or the other. Note that we have defined $\vec{k}_B$ as predominantly downward, broadly towards the receiver.

It should also be recognized that terms like $\omega_0|R_i|/c$ can be written as $ck_0|R_i|/c = k_0R_i = |\vec{k}_0\,\vec{R}_i|$, and we will use this construction frequently; indeed it was already used in Equation (3.145). The length $|R_i|$ also corresponds to a *phase difference* of $\frac{2\pi R_i}{\lambda}$.

Consider a general point X, as shown in Figure 3.20. Consider that relative to some (unspecified) origin, the location of this point can be represented by a vector $\vec{X}$. Then we may write the total permittivity at time t' and general point $\vec{X}$ as

$$\epsilon(\vec{X}, t') = \epsilon_0 + \Delta\epsilon(\vec{X}, t'), \tag{3.178}$$

where ϵ_0 is the mean value for the entire volume of interest averaged over the time that the radio pulse typically exists in the volume.

The signal received at the receiver at time t from a general point X will be due to the electric and magnetic fields produced by the polarization (and associated permittivity) at the point X at time t'_X.

The next step is to find the polarization induced in the medium (just as done previously). Adapting Equation (3.28), we write that for scatter from the point X

$$\underline{\vec{P}}(\vec{X}, t) = \Delta\epsilon(\vec{X}, t'_X)\underline{\vec{E}}_X e^{-i(\omega_0 t'_X)}, \tag{3.179}$$

where $\underline{\vec{E}}_X$ is the complex amplitude (i.e., includes a phase term) of the wave that drives the electron motion.

At this point in the derivation we pay no heed to the transmitter signal that drives the oscillation at X – we simply concern ourselves with the fact that an oscillation exists at X, and it has some complex amplitude.

We now need to look in more detail at the signal from X that is received at the receiver at time t.

Consider that the electron is forced to oscillate along a vector $\vec{\xi}$. The current density at any point X at the relevant time t'_X (which will be slightly different for each point $\vec{X}$), is $\vec{j} = q_e\vec{v} = q_e\frac{\mathrm{d}\vec{\xi}}{\mathrm{d}t} = \frac{\mathrm{d}(q_e\vec{\xi})}{\mathrm{d}t} = \frac{\mathrm{d}\vec{P}}{\mathrm{d}t}$, so that

$$\begin{aligned}\underline{\vec{j}}(\vec{X}, t'_X) &= -\mathrm{i}\omega_0\underline{\vec{P}}(\vec{X}, t'_X)\\ &= -\mathrm{i}\omega_0\Delta\epsilon(\vec{X}, t'_X)\underline{\vec{E}}_X\mathrm{e}^{-\mathrm{i}(\omega_0(t-\frac{r_S}{c})}\\ &= -\mathrm{i}\omega_0\Delta\epsilon(\vec{X}, t'_X)\underline{\vec{E}}_X\mathrm{e}^{-\mathrm{i}(\omega_0 t-\vec{k}_s\cdot\vec{r}_s},\end{aligned} \tag{3.180}$$

where r_s is the distance to the receiver (magnitude only) and we have used $\frac{\omega_0}{c} = |\vec{k}_s|$ and taken $\vec{r}_s$ and $\vec{k}_s$ as parallel.

Note that the direction of the induced current density is along the same vector as the applied electric field, because the induced polarization is parallel to it. This term is similar to Equation (3.137), except here there is no $\delta(\vec{r} - \vec{r}_e(t))$ term in the equation, since we allow ϵ to vary throughout the volume.

We may then calculate the vector potential at the receiver due to a volume of space dV surrounding the point X at time t as

$$\underline{\vec{A}}_{Rx}(\vec{X}, t)\mathrm{d}V = \frac{\mu_0}{4\pi}\frac{\vec{j}'(t'_X)}{|\vec{r}_s|}\mathrm{d}V, \tag{3.181}$$

where $\underline{\vec{A}}_{Rx}$ can be considered as the vector potential at the receiver per unit volume of dielectric surrounding X. This is similar to Equation (3.138), although here we consider the contribution due only to the small volume element dV. In principle each $\vec{r}_s$ is different for each different scatter point, and the time delays for the propagation from the transmitter to the scatter points, and from the scatter points to the receivers, are all different as well.

Using our explicit expression for $\vec{j}$ from (3.180), we may write

$$\underline{\vec{A}}_{Rx}(\vec{X}, t) = -\mathrm{i}\omega_0\Delta\epsilon(\vec{X}, t'_X)\frac{\mu_0}{4\pi}\frac{\underline{\vec{E}}_X\mathrm{e}^{-\mathrm{i}(\omega_0 t-\vec{k}_s\cdot\vec{r}_s)}}{|\vec{r}_s|}. \tag{3.182}$$

Since we will need to consider the collective effect of all parts of the scattering region, we need to somehow unify the various terms $\underline{\vec{E}}_X$; at present these are all different for different locations of X. In order to supply this unification, we recognize that the electric field at all points in the region is driven by the same source – namely the transmitter. The electric field at a general point X at time t'_X was initiated at the transmitter at time $t'_X - |\vec{r}_i|/c$.

The introduction of the transmitter pulse is just a little complicated. An idealized transmitter is a point source (zero surface area), producing a finite energy flux, so it must have an infinite electric field amplitude at the point of transmission. Of course in reality a transmitter antenna has a physical size, and there are all sorts of near-field effects – the far-field approximation is only set up some distance from the antenna. But we will assume that the transmitter is a point radiator, and we will consider the magnitude of the amplitude at a distance of 1 unit distance (e.g. 1 meter) from the source as being E_T. However, we will associate a phase with the transmitted wave which applies at the point of transmission, so will consider the radiated wave as $\underline{\vec{E}}_T$, where the phase is that at the source. In this way, we can write that the modulus of the amplitude at a distance R from the transmitter is $\frac{E_T}{R}$, provided that R is expressed as a multiple of a unit distance (typically meters).

Then using $t'_X - |\vec{r}_i|/c$, and this new definition of $\underline{\vec{E}}_T$, we write

$$\underline{\vec{E}}_X e^{-(i\omega_0 t'_X)} = \frac{\underline{\vec{E}}_T}{|\vec{r}_i|} e^{-i\omega_0 (t'_X - |\vec{r}_i|/c)} = \frac{\underline{\vec{E}}_T}{|\vec{r}_i|} e^{-i(\omega_0 t'_X - \vec{k}_i \cdot \vec{r}_i)}. \tag{3.183}$$

Replacing $\underline{\vec{E}}_X$ with $\frac{\underline{\vec{E}}_T}{|\vec{r}_i|} e^{i(\vec{k}_i \cdot \vec{r}_i)}$ from (3.183) into (3.182) and separating out the temporal and spatial parts of the exponent of e leads to

$$\underline{\vec{A}}_{Rx}(\vec{X}, t) = -i\omega_0 \Delta\epsilon(\vec{X}, t'_X) \frac{\mu_0}{4\pi} \frac{\underline{\vec{E}}_T e^{-i\omega_0 t}\, e^{i(\vec{k}_s \cdot \vec{r}_s + \vec{k}_i \cdot \vec{r}_i)}}{|\vec{r}_i|\, |\vec{r}_s|}. \tag{3.184}$$

With these changes the phases of all single-scatter signals by all paths are referenced against the transmitter signal.

We will now make a change of coordinates, and give more detail about our origin (which has not as yet been specified). We will choose the point O in Figure 3.20 as our new origin.

Then Equation (3.184) applies at the new origin as well (since the new origin is just another point in the scattering volume), and we may adapt (3.184) for the case of the point O to be

$$\underline{\vec{A}}_{Rx}(\vec{O}, t) = -i\omega_0 \Delta\epsilon(\vec{O}, t'_O) \frac{\mu_0}{4\pi} \frac{\underline{\vec{E}}_T e^{-i\omega_0 t}\, e^{i(\vec{k}_s \cdot \vec{R}_s + \vec{k}_i \cdot \vec{R}_i)}}{|\vec{R}_i|\, |\vec{R}_s|}. \tag{3.185}$$

Now (3.184) must be modified and expressed in terms of (3.185).

First, we recognize that the vector $\vec{d}_1$ in Figure 3.20 has length $|\vec{r}| \cos\alpha_i$, and is orientated parallel to $\vec{k}_i$. So it can be written as

$$\vec{d}_1 = -\left[\vec{r} \cdot \frac{\vec{k}_i}{|\vec{k}_i|}\right] \frac{\vec{k}_i}{|\vec{k}_i|} = -\left[\vec{r} \cdot \vec{k}_i\right] \frac{\vec{k}_i}{|\vec{k}_i|^2} = -\left[\vec{r} \cdot \vec{k}_i\right] \frac{\vec{k}_i}{\vec{k}_i \cdot \vec{k}_i}, \tag{3.186}$$

where the minus arises because $\vec{r} \cdot \vec{k}_i < 0$, yet $\vec{d}_1$ must be parallel to $\vec{k}_i$ according to Figure 3.20.

Likewise

$$\vec{d}_2 = \left[\vec{r} \cdot \vec{k}_s\right] \frac{\vec{k}_s}{\vec{k}_s \cdot \vec{k}_s}. \tag{3.187}$$

Note there is no minus this time as $\vec{r} \cdot \vec{k}_s \geq 0$.

Then we may write $\vec{k}_i \cdot \vec{r}_i = \vec{k}_i \cdot \vec{R}_i - \vec{k}_i \cdot \vec{d}_1$ and $\vec{k}_s \cdot \vec{r}_s = \vec{k}_s \cdot \vec{R}_s - \vec{k}_s \cdot \vec{d}_2$.

Hence

$$\vec{k}_i \cdot \vec{r}_i = \vec{k}_i \cdot \vec{R}_i - -\left[\vec{r} \cdot \vec{k}_i\right] \frac{\vec{k}_i \cdot \vec{k}_i}{\vec{k}_i \cdot \vec{k}_i} = \vec{k}_i \cdot \vec{R}_i + \vec{k}_i \cdot \vec{r}, \tag{3.188}$$

and

$$\vec{k}_s \cdot \vec{r}_s = \vec{k}_s \cdot \vec{R}_s - \left[\vec{r} \cdot \vec{k}_s\right] \frac{\vec{k}_s \cdot \vec{k}_s}{\vec{k}_s \cdot \vec{k}_s} = \vec{k}_s \cdot \vec{R}_s - \vec{k}_s \cdot \vec{r}. \tag{3.189}$$

We use (3.188) and (3.189) to re-write (3.184) as

$$\underline{\vec{A}}_{Rx}(\vec{X}, t) = -\mathrm{i}\omega_0 \Delta\epsilon(\vec{X}, t'_X) \frac{\mu_0}{4\pi} \frac{\underline{\vec{E}}_T \mathrm{e}^{-\mathrm{i}\omega_0 t}\, \mathrm{e}^{\mathrm{i}(\vec{k}_s \cdot \vec{R}_s + \vec{k}_i \cdot \vec{R}_i)}\, \mathrm{e}^{\mathrm{i}\{(\vec{k}_i - \vec{k}_s) \cdot \vec{r}\}}}{|\vec{r}_i|\ |\vec{r}_s|}. \tag{3.190}$$

We now recognize that the term $\mathrm{e}^{\mathrm{i}(\vec{k}_s \cdot \vec{R}_s + \vec{k}_i \cdot \vec{R}_i)}$ is independent of X.

Further, we assume that $\epsilon(\vec{X}, t'_X)$ is slowly varying relative to the duration of the pulse, so although t'_X is different for different positions of X, we can take it as the same from the point of view of ϵ, so we set $t'_X = t'_O$ in the argument of $\Delta\epsilon$, which we will then simply write as t'. In addition, the terms $|\vec{r}_i|$ and $|\vec{r}_s|$ in the denominator are simply distances, and if the typical width and depth of the scattering region are much less than the distances to the transmitter and receiver (as is normal), then all occurrences of $|\vec{r}_s|$ for purposes of division can be approximated to $|\vec{R}_s|$, and likewise $|\vec{r}_i| \approx |\vec{R}_i|$. Finally, we have referenced the electric field relative to the amplitude of the transmitted pulse, $\vec{E}_T$, which was convenient while we unified our time delays, but for purposes of determining the cross-section per unit volume, it makes better sense to use the strength of the incident electric field at the scattering region. So we replace $\frac{\underline{\vec{E}}_T}{|\vec{r}_i|}$ by $\underline{\vec{E}}_O$, which is the electric field strength arriving at the scattering region (and in particular at O). Note it may have a different phase to $\underline{\vec{E}}_T$, but this will not matter for determinations of power. This choice of amplitude will make it easier to determine cross-sections later on. We also define the Bragg wavenumber as

$$\vec{k}_B = \vec{k}_s - \vec{k}_i, \tag{3.191}$$

(see the inset of Figure 3.20). Note we could have adopted the opposite expression and written $\vec{k}_B = \vec{k}_i - \vec{k}_s$ (which would have pointed generally upward), but (3.191) is better suited to development of Fourier spectra later on. With all these adjustments, and writing the position X as the vector $\vec{r}$ relative to O, Equation (3.190) becomes

$$\underline{\vec{A}}_{Rxr}(\vec{r}, t) = -\frac{\mu_0 \mathrm{i}\omega_0 \underline{\vec{E}}_O}{4\pi R_s} \mathrm{e}^{-\mathrm{i}\omega_0 t}\, \mathrm{e}^{\mathrm{i}(\vec{k}_s \cdot \vec{R}_s + \vec{k}_i \cdot \vec{R}_i)} \left[\Delta\epsilon(\vec{r}, t') \mathrm{e}^{-\mathrm{i}\{\vec{k}_B \cdot \vec{r}\}}\right], \tag{3.192}$$

where the Rxr subscript reminds us that we are now using a new coordinate system which employs the vector $\vec{r}$ and has its origin at O, and where $R_s = |\vec{R}_s|$. All of the terms outside of the square brackets are now constant for our chosen scattering region.

The next step is to determine the total vector potential at the receiver by integrating over the radar volume. All current density vectors in the scattering region are close to parallel, so all vector potentials due to all current densities at the receiver will be parallel, so we simply need to add all the $\vec{\underline{A}}_{Rxr}(\vec{r},t)\mathrm{d}^3\vec{r}$ vectors algebraically i.e., integrate (3.192), giving

$$\begin{aligned}\vec{\underline{A}} &= \int\int\int_{Vs} \vec{\underline{A}}_{Rxr}(\vec{r},t)\mathrm{d}^3\vec{r} \\ &= -\frac{\mu_0 \mathrm{i}\omega_0}{4\pi R_s}\,\mathrm{e}^{\mathrm{i}(\vec{k}_i\cdot\vec{R}_i)}\left[\int\int\int_{Vs}\Delta\epsilon(\vec{r},t')\mathrm{e}^{-\mathrm{i}\{\vec{k}_B\cdot\vec{r}\}}\mathrm{d}^3\vec{r}\right]\vec{\underline{E}}_O\mathrm{e}^{-\mathrm{i}(\omega_0 t-\vec{k}_s\cdot\vec{R}_s)}, \quad (3.193)\end{aligned}$$

where we have defined the scatter volume as V_s, and written that a differential unit of volume is $\mathrm{d}^3\vec{r}$. The integral contained in the square brackets has a special significance, which will be considered shortly. Note that we have, for now, let $\vec{k}_B$ be fixed, and treated $\vec{k}_B$ and $\vec{r}$ as independent – which is strictly wrong, since $\vec{k}_B$ may at times vary as $\vec{r}$ varies. However, we will re-address this assumption shortly.

The terms $\vec{\underline{E}}_O$, $\mathrm{e}^{-\mathrm{i}(\omega_0 t)}$ and $\mathrm{e}^{\mathrm{i}(\vec{k}_s\cdot\vec{R}_s)}$ have been combined as a single term to the right of the bracketed integral. Everything to the left of the square brackets, and including the square brackets, can be considered as a complex constant.

The term $\vec{\underline{E}}_O\mathrm{e}^{-\mathrm{i}(\omega_0 t-\vec{k}_s\cdot\vec{R}_s)}$ is proportional to the temporal and spatial variability of the vector field at the receiver. In order to move to the next step, it needs to be recognized that the field at the receiver is part of a more general equation of the form $\vec{\underline{E}}_O\mathrm{e}^{-\mathrm{i}(\omega_0 t-\vec{k}_s\cdot\vec{\xi})}$, where $\vec{\xi}$ is a vector originating at O and extending parallel to $\vec{R}_s$ through the receiver and on to infinity. The receiver corresponds to the special case $\vec{\xi}=\vec{R}_s$. It is important to recognize this spatial variability since we will shortly determine $\vec{\nabla}\times\vec{\underline{A}}$, and that requires knowledge of the spatial behavior of $\vec{\underline{A}}$ around the receiver.

Another change is also needed. It is most common in this work to now change from $\Delta\epsilon$ to

$$\Delta\epsilon' = \Delta\frac{\epsilon}{\epsilon_0}, \quad (3.194)$$

where $\epsilon' = \frac{\epsilon}{\epsilon_0}$ is called the *relative permittivity*, or, equivalently, (though more confusingly) the dielectric constant. This change will also be incorporated into the next equation.

We now calculate $\vec{\underline{H}}$ as

$$\vec{\underline{H}} = \frac{1}{\mu_0}\vec{\nabla}\times\vec{\underline{A}} = -\frac{\mathrm{i}\epsilon_0\omega_0}{4\pi R_s}\,\mathrm{e}^{\mathrm{i}(\vec{k}_i\cdot\vec{R}_i)}\left[\int\int\int_{Vs}\Delta\epsilon'(\vec{r},t')\mathrm{e}^{-\mathrm{i}\{\vec{k}_B\cdot\vec{r}\}}\mathrm{d}^3\vec{r}\right]\vec{\nabla}\times\vec{\underline{E}}_O\mathrm{e}^{-\mathrm{i}(\omega_0 t-\vec{k}_s\cdot\vec{\xi})}. \quad (3.195)$$

But $\vec{\nabla}\times\vec{\underline{E}}_0\mathrm{e}^{\mathrm{i}\vec{k}_s\cdot\vec{\xi}} = \mathrm{i}\vec{k}_s\times\vec{E}_0\mathrm{e}^{\mathrm{i}\vec{k}_s\cdot\vec{\xi}}$, which simply equals $\mathrm{i}k_s E_0\mathrm{e}^{\mathrm{i}\vec{k}_s\cdot\vec{\xi}}\sin\chi$, with direction perpendicular to both E_0 and R_s, just as in Equation (3.150) and Figure 3.14. We evaluate this at $\vec{\xi}=\vec{R}_s$, i.e., the location of the receiver.

At this stage we become mainly interested in the Poynting vector, which we can get directly from the magnitude of $\vec{\underline{H}}$, so we concentrate on amplitudes only. The magnitude of $\vec{k}_s$ is just the radar wavenumber k_0. Then (3.195) becomes

$$\underline{H} = \frac{k_0\epsilon_0\omega_0\sin\chi}{4\pi R_s}\mathrm{e}^{\mathrm{i}(\vec{k}_i\cdot\vec{R}_i)}\left[\int\int\int_{Vs}\Delta\epsilon'(\vec{r},t')\mathrm{e}^{-\mathrm{i}\vec{k}_B\cdot\vec{r}}\mathrm{d}^3\vec{r}\right]\vec{\underline{E}}_O\mathrm{e}^{-\mathrm{i}(\omega_0 t-\vec{k}_s\cdot\vec{R}_s)}. \quad (3.196)$$

Our next step is to obtain the received electromagnetic flux (the Poynting vector) given by $S_r = \frac{1}{2}\eta \underline{H}\underline{H}^*$, or

$$S_r = \frac{k_0^2\epsilon_0^2\omega_0^2\sin^2\chi}{16\pi^2R_s^2}\frac{1}{2}\eta E_O^2\left[\int_{Vs}\Delta\epsilon'(\vec{r})\mathrm{e}^{-\mathrm{i}\vec{k}_B\cdot\vec{r}}\mathrm{d}^3\vec{r}\right]\left[\int_{Vs}\Delta\epsilon'^*(\vec{r}')\mathrm{e}^{+\mathrm{i}\vec{k}'_B\cdot\vec{r}'}\mathrm{d}^3\,\vec{r}\,'\right], \tag{3.197}$$

where H is the magnitude of $\underline{H}$ and E_0 is the magnitude of $\underline{E}_0$. We have abbreviated the integral to use only one integral symbol, although a three-dimensional integration is still intended. We have also dropped the time-dependence term t' in order to simplify the notation – the temporal dependence is inferred. In addition, the term $\mathrm{e}^{\mathrm{i}(\vec{k}_i\cdot\vec{R}_i)}$ and $\mathrm{e}^{-\mathrm{i}(\omega_0 t-\vec{k}_s\cdot\vec{R}_s)}$ have disappeared because they have been multiplied by their own complex conjugates.

It is tempting to look at the two integrations and consider them as Fourier transforms, and their product as a power spectrum. However, it is not so simple. While the functions look a lot like Fourier transforms, they apply only at a fixed wavenumber $\vec{k}_B$, and so even if they are considered as a special case of a Fourier transform, that transformation produces a delta-function in $\vec{k}$-space. Taking power spectra in cases involving delta-functions requires care, and special normalizations, as outlined by *Champeney* (1973) on pages 60–61. Furthermore, the Fourier transform for a quasi-random function does not always exist, as we will show later. So we proceed a little more cautiously for now, and will follow the approach presented by *Tatarski* (1961). We will revisit the discussion in this paragraph later in this chapter.

We now need to deal with the double-integral, so we collapse (3.197) to

$$S_r = \frac{k_0^2\epsilon_0^2\omega_0^2\sin^2\chi}{16\pi^2R_s^2}\eta^2\frac{1}{2}\frac{E_O^2}{\eta}\left[\int_{Vs}\int_{Vs}\Delta\epsilon'(\vec{r})\Delta\epsilon'^*(\vec{r}')\mathrm{e}^{-\mathrm{i}(\vec{k}_B\cdot\vec{r}-\vec{k}_B\,'\cdot\vec{r}')}\mathrm{d}^3\vec{r}\,\mathrm{d}^3\vec{r}\,'\right]. \tag{3.198}$$

We also write $\frac{1}{2}\frac{E_0^2}{\eta}$ as S_i, the incident flux, and use $\epsilon_0^2\omega_0^2\eta^2$ as $\epsilon_0^2c^2k_0^2\frac{\mu_0}{\epsilon_0} = \epsilon_0^2\frac{1}{\mu_0\epsilon_0}k_0^2\frac{\mu_0}{\epsilon_0}$, which contracts to k_0^2, and rewrite

$$S_r = \frac{k_0^4\sin^2\chi}{16\pi^2R_s^2}S_i\left[\int_{Vs}\int_{Vs}\Delta\epsilon'(\vec{r})\Delta\epsilon'^*(\vec{r}')\mathrm{e}^{-\mathrm{i}(\vec{k}_B\cdot\vec{r}-\vec{k}_B\,'\cdot\vec{r}')}\mathrm{d}^3\vec{r}\,\mathrm{d}^3\vec{r}\,'\right]. \tag{3.199}$$

At this point, we follow *Tatarski* (1961), pages 65–67, with some changes in nomenclature and normalization.

First, it needs to be recognized that for the case shown in Figure 3.20, the vector $\vec{k}_B$ is largely independent of the position $\vec{r}$, so we can replace $\vec{k}_B\,'$ with $\vec{k}_B$. There are cases where this is not true, especially for the case of backscatter, but we can safely ignore these digressions for now.

Secondly, we need to recognize that S_r is not a constant. The signal scattered from each part of the scattering region depends on the value of $\Delta\epsilon'$ at that point. If the function is homogeneous (though not necessarily isotropic), then the value of $\Delta\epsilon'(\vec{r})\Delta\epsilon'^*(\vec{r}')$ will depend (at least in a long-term average sense) only on the separation vector between $\vec{r}$ and $\vec{r}'$. For any particular pair of values $\vec{r}$ and $\vec{r}'$, there will be many others with the same separation vector. We could imagine that for any chosen vector separation, we can store all the instantaneous measured values of $\overline{\Delta\epsilon'(\vec{r})\Delta\epsilon'^*(\vec{r'})}$ and then form a histogram of them. We could then find a mean value, $\overline{\Delta\epsilon'(\vec{r})\Delta\epsilon'^*(\vec{r'})}$.

Since the received signal will contain many contributions (one from each such pairing), it will also be an average quantity. So it makes sense that the same result would be achieved if we replaced *each* occurrence of $\Delta\epsilon'(\vec{r})\Delta\epsilon'^*(\vec{r'})$ with its associated mean value – we expect that the mean value of the received signal will be correctly reproduced.

With these two sets of changes, (3.199) becomes

$$\overline{S_r} = \frac{k_0^4 \sin^2\chi}{16\pi^2 R_s^2} S_i \left[\int_{Vs}\int_{Vs} \overline{\Delta\epsilon'(\vec{r})\Delta\epsilon'^*(\vec{r'})} e^{-i(\vec{k}_B\cdot\{\vec{r}-\vec{r'}\})} d^3\vec{r}\, d^3\vec{r}\,' \right]. \tag{3.200}$$

Now we need to deal more carefully with the term $\overline{\Delta\epsilon'(\vec{r})\Delta\epsilon'^*(\vec{r'})}$. We discussed this in the context of collecting pairs of points with identical vector separations, and forming a histogram, but we can take another approach. Instead of randomly accessing every pair of points $(\vec{r}, \vec{r'})$ for a fixed vector separation and forming the average, we can accumulate different separations simultaneously by starting at a point $\vec{r}$, then tracking systematically away from this point along a line in ever increasing steps, successively adding the contribution from each new pair of points as we move away from $\vec{r}$ to the sums for other pairs with the same vector separation. Then we repeat for different directions of the line of tracking, and then start again at a new reference point. In other words, the autocovariance function is formed. Assuming this function is only dependent on the vector distance between any pairs of scatterers (i.e., we assume homogeneity, as discussed above), we may write

$$\begin{aligned} \overline{\Delta\epsilon'(\vec{r})\Delta\epsilon'^*(\vec{r}\,')} &= \frac{1}{V_s}\int_{Vs} \Delta\epsilon'^*(\vec{r}\,')\Delta\epsilon'(\vec{r}\,' + \vec{\xi}) d^3\vec{r}\,' \\ &= \rho(\vec{\xi}), \end{aligned} \tag{3.201}$$

which is the spatial autocovariance function. At this point we could substitute (3.201) into (3.200), which would produce a triple integral. However, there is an easier way to achieve our objectives. This is to return to (3.199) and, following the ideas expressed in our last few paragraphs, with $\vec{r} = \vec{r'} + \vec{\xi}$, write it as

$$\overline{S_r} = \frac{k_0^4 \sin^2\chi}{16\pi^2 R_s^2} S_i \left[\int_{Vs}\int_{Vs} \Delta\epsilon'(\vec{r'} + \vec{\xi})\Delta\epsilon'^*(\vec{r'}) e^{-i(\vec{k}_B\cdot\vec{\xi}} d^3(\vec{r'} + \vec{\xi})\, d^3\vec{r'} \right]. \tag{3.202}$$

Since $\vec{\xi}$ and $\vec{r'}$ are independent variables, $d^3(\vec{r'} + \vec{\xi})$ and $d^3\vec{r'}$ are independent, so we may separate the integrals thus:

$$\overline{S_r} = \frac{k_0^4 \sin^2\chi}{16\pi^2 R_s^2} S_i \left[\int_{Vs} \left[\int_{Vs} \Delta\epsilon'(\vec{r'} + \vec{\xi})\Delta\epsilon'^*(\vec{r'}) e^{-i\vec{k}_B\cdot(\vec{\xi})} d^3(\vec{r'} + \vec{\xi}) \right] d^3\vec{r'} \right]. \tag{3.203}$$

For each application of the inner integral (i.e the term in square brackets), $\vec{r'}$ is fixed, so $d^3(\vec{r'} + \vec{\xi}) = d^3\vec{\xi}$, which we may use to write

$$\overline{S_r} = \frac{k_0^4 \sin^2\chi}{16\pi^2 R_s^2} S_i \left[\int_{Vs} \left[\int_{Vs} \Delta\epsilon'(\vec{r'} + \vec{\xi})\Delta\epsilon'^*(\vec{r'}) e^{-i\vec{k}_B\cdot(\vec{\xi})} d^3\vec{\xi} \right] d^3\vec{r'} \right]. \tag{3.204}$$

The term in square brackets now needs to be examined more closely. For any choice of $\vec{\xi}$, there will be many pairs of vectors $(\vec{r'} + \vec{\xi}, \vec{r'})$ which have the same separation $\vec{\xi}$, and for such groupings the term $e^{-i(\vec{k}_B \cdot \vec{\xi})}$ will be a constant. So the final contribution to the overall integral will be the same as if we replaced $\Delta\epsilon'(\vec{r'} + \vec{\xi})\Delta\epsilon'^*(\vec{r'})$ by its average (as argued in the preceding paragraphs), and multiplied by the number of values of $\Delta\epsilon'(\vec{r'} + \vec{\xi})\Delta\epsilon'^*(\vec{r'})$ which contributed to the average. The average is of course just $\overline{\Delta\epsilon'(\vec{r'} + \vec{\xi})\Delta\epsilon'^*(\vec{r'})}$, which, by (3.201), is just $\rho(\vec{\xi})$.

Hence we write

$$\overline{S_r} = \frac{k_0^4 \sin^2 \chi}{16\pi^2 R_s^2} S_i \left[\int_{Vs} \left[\int_{Vs} \rho(\vec{\xi}) e^{-i(\vec{k}_B \cdot \vec{\xi})} d^3\vec{\xi} \right] d^3\vec{r'} \right], \tag{3.205}$$

and since $\vec{\xi}$ and $\vec{r'}$ are independent, this can be written as

$$\overline{S_r} = \frac{k_0^4 \sin^2 \chi}{16\pi^2 R_s^2} S_i \left[\left[\int_{Vs} \rho(\vec{\xi}) e^{-i(\vec{k}_B \cdot \vec{\xi})} d^3\vec{\xi} \right] \int_{Vs} d^3\vec{r'} \right]. \tag{3.206}$$

The reader may recall that we should multiply $\overline{\Delta\epsilon'(\vec{r'} + \vec{\xi})\Delta\epsilon'^*(\vec{r'})}$ by the number of contributing terms, which we appear not to have done. But this number would be, in principle, just the volume V_s divided by d^3r', so is taken care of through the outer integral, i.e., the last integral is just effectively the sum of $d^3\vec{r'}$ over the whole volume (since the main integrand contains no mention of $\vec{r'}$), or simply the volume V_s. Hence we are left with

$$\overline{S_r} = \frac{k_0^4 \sin^2 \chi}{16\pi^2 R_s^2} S_i V_s \int_{Vs} \rho(\vec{\xi}) e^{-i(\vec{k}_B \cdot \vec{\xi})} d^3\vec{\xi}. \tag{3.207}$$

Having developed the expression in terms of the autocovariance function, we next need to convert the equation to an expression in terms of the power spectrum. To do this, we need a small digression regarding Fourier transforms.

Some notes on Fourier transforms

It is assumed that the reader is familiar with the process of forming Fourier transforms and Fourier integrals, and most of the associated theorems, but one of the more annoying things about Fourier theory is that there are quite a few different formulations, used by a variety of authors. In order to proceed with our analysis, it is necessary to become familiar with the main ones.

A reasonably common procedure is to avoid using the wavenumber $k = \frac{2\pi}{\lambda}$, but instead use an inverse scale $\zeta = \frac{1}{\lambda}$. This latter formulation is well suited to computer simulations, and many commercial FFT algorithms employ this approach. But is it best for us?

In the equations below, four different versions of one-dimensional complex Fourier transform pairs are shown. We use the one-dimensional case for simplicity – the extensions to 3-D (as we need) will be obvious to any experienced student of Fourier theory:

$$
\begin{aligned}
\underline{F}_\zeta(\zeta) &= \int_{-\infty}^{\infty} \underline{f}(x)\mathrm{e}^{-\mathrm{i}2\pi\zeta x}dx &&: \quad \underline{f}(x) = \int_{-\infty}^{\infty} \underline{F}_\zeta(\zeta)\mathrm{e}^{\mathrm{i}2\pi\zeta x}d\zeta \\
\underline{F}_{c1}(k) &= \int_{-\infty}^{\infty} \underline{f}(x)\mathrm{e}^{-\mathrm{i}kx}dx &&: \quad \underline{f}(x) = \frac{1}{2\pi}\int_{-\infty}^{\infty} \underline{F}_{c1}(k)\mathrm{e}^{\mathrm{i}kx}dk \\
\underline{F}_{c2}(k) &= \frac{1}{2\pi}\int_{-\infty}^{\infty} \underline{f}(x)\mathrm{e}^{-\mathrm{i}kx}dx &&: \quad \underline{f}(x) = \int_{-\infty}^{\infty} \underline{F}_{c2}(k)\mathrm{e}^{\mathrm{i}kx}dk \\
\underline{F}_{c3}(k) &= \frac{1}{\sqrt{2\pi}}\int_{-\infty}^{\infty} \underline{f}(x)\mathrm{e}^{-\mathrm{i}kx}dx &&: \quad \underline{f}(x) = \frac{1}{\sqrt{2\pi}}\int_{-\infty}^{\infty} \underline{F}_{c3}(k)\mathrm{e}^{\mathrm{i}kx}dk.
\end{aligned}
\tag{3.208}
$$

The subscripts "c" in the last three rows are intended to indicate that these particular Fourier transforms have radian (i.e., ^{c}) wavenumbers (i.e., involve $k = 2\pi\zeta$) in their arguments.

Each form of Fourier transform has been used by different authors. The second form was used by *Champeney* (1973) (see his page 40), while *Batchelor* (1953) (pages 25–26, Equations (2.4.2) and (2.4.3)) and *Tatarski* (1961, 1971) (e.g., page 7, Equations (16) and (17) in the 1971 version) used the third form – but used it in regard to the autocovariance function and the spectrum, which is somewhat different, as we will see shortly. Both Batchelor and Tatarski were significant contributors to the development of turbulence theory. The fourth term has been used less, though its evident symmetry is considered an advantage by some authors and researchers.

A more thorough comparison of the different Fourier transforms can be found in *Champeney* (1973), pages 74–75.

Which of these different forms are best suited to represent our integral in Equation (3.207)?

All four forms ensure that if we apply the first operation on $f(x)$ to obtain the appropriate Fourier transform ($\underline{\mathcal{F}}$ say), and then apply the second operation to $\underline{\mathcal{F}}$, we return to $f(x)$.

The equations clearly differ, but in regard to our current theory, there is one important difference that stands out. This is in regard to Parseval's theorem (e.g., *Bracewell*, 1978). In part, this says that for a zero-mean function $f(x)$,

$$\int_{-\infty}^{\infty} \underline{F}_\zeta^* \underline{F}_\zeta d\zeta = L\sigma_f^2, \tag{3.209}$$

where σ_f is the standard deviation of the data string $\underline{f}(x)$ and L is its length. Another way to write this is

$$\frac{1}{L}\int_{-\infty}^{\infty} \underline{F}_\zeta^* \underline{F}_\zeta \; d\zeta = \sigma_f^2. \tag{3.210}$$

In other words, the area under the graph of the modulus of F_ζ squared (i.e.,the total energy) per unit spatial length is equal to the variance of the original function $f(x)$.

The three-dimensional equivalent statement (most relevant to our work here) would be

$$\frac{1}{V_s}\int\int\int_{Vs} \underline{F}_{\vec{\zeta}}^* \underline{F}_{\vec{\zeta}} d^3\vec{\zeta} = \sigma_f^2, \tag{3.211}$$

where V_s is the volume of scatter.

However, this equality applies only to the first form. It does not necessarily apply to the other forms.

Comparing the two terms on the left of the first two rows in (3.208), it is clear that they represent the same integral, but they are assigned to their relevant Fourier transform functions differently. The integration that is assigned to a wavenumber ζ_p in F_ζ is assigned to a radian wavenumber $k = 2\pi\zeta_p$ in F_c. Hence the same values of $\underline{F}$ – either $\underline{F}_\zeta$ or $\underline{F}_{c2}$ – (and by extension the same values of $\underline{F}^*\underline{F}$) appear in each case, but in the second case they are more widely spaced, by a factor of 2π. Hence the integrated area under $\underline{F}^*_{c1}\underline{F}_{c1}$ is 2π times greater than that under $\underline{F}^*_\zeta\underline{F}_\zeta$. Hence

$$\int_{-\infty}^{\infty} \underline{F}^*_{c1}\underline{F}_{c1} dk = 2\pi \int_{-\infty}^{\infty} \underline{F}^*_\zeta \underline{F}_\zeta d\zeta = 2\pi\sigma_f^2 L. \tag{3.212}$$

In the third expression of (3.208), the same rationale results in multiplication by 2π to adjust for the scale change from ζ to k, but the $\frac{1}{2\pi}$ term out the front is multiplied twice, squaring it. The net result is that

$$\int_{-\infty}^{\infty} \underline{F}^*_{c2}\underline{F}_{c2} dk = \frac{1}{2\pi} \int_{-\infty}^{\infty} \underline{F}^*_\zeta \underline{F}_\zeta d\zeta = \frac{1}{2\pi}\sigma_f^2 L. \tag{3.213}$$

In a similar way it can be shown that

$$\int_{-\infty}^{\infty} \underline{F}^*_{c3}\underline{F}_{c3} dk = \int_{-\infty}^{\infty} \frac{2\pi}{(\sqrt{2\pi})^2} \underline{F}^*_\zeta \underline{F}_\zeta d\zeta = \sigma_f^2 L. \tag{3.214}$$

Is any of these more suitable for our analysis of scattered radiation?

Fourier transforms for random data

The formulas shown in the last subsection assumed that the functions $\underline{f}$ were well defined. But in the previous analysis on scattering, it was assumed that the permittivity fields in our previous derivations were random – or at least quasi-random. In dealing with quasi-random processes, a different strategy to that developed for standard Fourier integrals has evolved.

The reasons are in part historical and in part practical.

To begin, Fourier theory in its purest sense requires functions that can be defined to exist between $-\infty$ and ∞. But consider a continuous quasi-random function that extends from $-\infty$ to ∞. Then its formal Fourier transform looks something like

$$F(k) = \mathcal{B} \int_{-\infty}^{\infty} f(x) \mathrm{e}^{-ikx} dx, \tag{3.215}$$

where the choice of $\mathcal{B}$ depends on the relevant transform chosen in Equations (3.208) i.e., 1, 2π, etc. We assume f exists as a randomly varying function over all space (or time).

Then, as discussed by *Champeney* (1973), on page 79 of his book,

$$\int_{-X}^{X} f(x) \mathrm{e}^{-ikx} dx \to \infty \text{ as } X \to \infty, \tag{3.216}$$

so the function becomes unusable. Even if an attempt is made to normalize the function, for example by dividing by $2X$, this is often also unsuccessful, since

$$\frac{1}{2X}\int_{-X}^{X} f(x)\mathrm{e}^{-ikx}dx \rightarrow 0 \text{ as } X \rightarrow \infty \quad [\text{for } k \neq 0]. \tag{3.217}$$

So the standard Fourier transform becomes largely meaningless. In addition, historically the fast Fourier transform was only introduced in 1965 (*Cooley and Tukey*, 1965), and prior to that computers were slow anyway. So much Fourier and spectral work was done using autocorrelation functions and spectra, since these were considered more useful for interpretation: these functions have generally smoother shapes, are differentiable and interpretable, while the raw data series are generally erratic and hard to interpret from visual inspection. (In more modern times, very long data-sets with even millions of points can be plotted quickly with computers, and sometimes this can be instructive, but it was simply impossible in those early days.)

So for all these reasons, a different protocol for Fourier analysis was developed compared to that in the last subsection. In contrast to (3.208), the appropriate conversions were defined in terms of the the autocovariance function ρ and the spectrum Φ. The relationship is defined in both *Tatarski* (1961) and *Batchelor* (1953); shortly we will demonstrate the relation using one dimensional real-number space for simplicity.

Before doing so, though, we need to recognize that another complication with random data is that they cannot be specified for all space or all time – usually only short data-sets are available. So the spectra and autocovariance functions need to be defined in terms of limits.

The autocovariance function is defined as

$$\rho(\xi) =^{\lim}_{X\rightarrow\infty} \frac{1}{X}\int_{X1}^{X1+X} f^{*}(x)f(x+\xi)dx, \tag{3.218}$$

and the power spectrum is defined as

$$\Phi(k) =^{\lim}_{X\rightarrow\infty} \frac{1}{X}F_X^{*}F_X, \tag{3.219}$$

where F_X is the appropriate Fourier transform of the spatially limited function. In practice, we only determine these functions on data-sets of length X that actually exist – being able to find the limit as X goes to infinity is usually a luxury that cannot be realistically achieved.

Then in this type of analysis, the main functions used are the autocovariance function and the spectrum, and the relevant conversions between them are *defined* as follows:

$$\Phi(k) = \frac{1}{2\pi}\int_{-\infty}^{\infty} \rho(x)\mathrm{e}^{-ikx}dx \quad : \quad \rho(x) = \int_{-\infty}^{\infty} \Phi(k)\mathrm{e}^{ikx}dk, \tag{3.220}$$

e.g., see *Batchelor* (1953), Equations (2.4.3) and (2.4.2). Here Φ is the power per unit length, which differs from the energy spectra described in Equations (3.209) to (3.214).

So in this regard the equations look a bit like F_{c2} in (3.208), except here the relations are in terms of autocovariance functions and spectra. However, the analogy is not complete. It can be seen from the second equation in (3.220) that if we take $x = 0$, we produce

$$\int_{-\infty}^{\infty} \Phi(k) dk = \rho(0) = \sigma_f^2, \tag{3.221}$$

for zero-mean functions f. This contrasts to Equation (3.213) which essentially says $\int_{-\infty}^{\infty} \Phi dk = \frac{1}{2\pi}\sigma_f^2$. So although the transformation formula between ϕ and ρ matches F_{c2}, the normalization is quite different.

For studies of quasi-random data, use of autocovariance functions and spectra is the most common approach. It differs from the four other transformations shown in (3.208).

Interestingly, if the user requires a properly normalized spectrum, the form F_{c3} in (3.208) is the most useful, for if its magnitude is squared, then it is normalized according to (3.221).

The three dimensional version of (3.220) is

$$\Phi(k) = \frac{1}{(2\pi)^3} \int_{-\infty}^{\infty} \int_{-\infty}^{\infty} \int_{-\infty}^{\infty} \rho(\vec{x}) e^{-i\vec{k}\cdot\vec{x}} d^3\vec{x} \quad : \quad \rho(\vec{x}) = \int_{-\infty}^{\infty} \int_{-\infty}^{\infty} \int_{-\infty}^{\infty} \Phi(\vec{k}) e^{i\vec{k}\cdot\vec{x}} d^3\vec{k}. \tag{3.222}$$

Note that the limits in these integrals *are* $\pm\infty$.

These new definitions of autocovariance and spectrum will form the basis of the rest of this section on backscatter analysis. They will be used until we have developed this theory. However, they will not be a standard throughout the book – different strategies will be used to represent Fourier transform pairs, depending on the most common usage of frequencies, wavenumbers, and normalizations employed in the relevant field.

Determination of the radio scatter cross-section

We now return to Equation (3.207), and apply Equation (3.222, second part) as our next step. Then we write

$$\overline{S_r} = \frac{k_0^4 \sin^2 \chi}{16\pi^2 R_s^2} S_i V_s \int_{Vs} \left[\int_{-\infty}^{\infty} \Phi_{\epsilon'} e^{+i\vec{\kappa}\cdot\vec{\xi}} d^3\vec{\kappa} \right] e^{-i(\vec{k_B}\cdot\vec{\xi})} d^3\vec{\xi}. \tag{3.223}$$

Merging the exponentials and swapping the orders of integration, this can then be written as

$$\overline{S_r} = \frac{k_0^4 \sin^2 \chi}{16\pi^2 R_s^2} S_i V_s \int_{-\infty}^{\infty} \Phi_{\epsilon'} d^3\vec{\kappa} \int_{Vs} e^{i(\vec{\kappa} - \vec{k_B})\cdot\vec{\xi}} \, d^3\vec{\xi}. \tag{3.224}$$

At this point in the derivation, it is necessary to examine the function $\int_{V_S} e^{i(\vec{\kappa} - \vec{k_B})\cdot\vec{\xi}} \, d^3\vec{\xi}$.

It is evident that this is a delta function in κ-space at the wavenumber $\vec{\kappa} = \vec{k_B}$ if the volume V_s is infinite. If the volume is not infinite, then further examination is required. The integral will now be a function of $\vec{\kappa}$. However, the fact that it is a delta-function for infinite volume suggests that even for finite volumes, the integral may be a function that is quite narrow in $\vec{\kappa}$-space. In addition, it needs to be remembered that the incident signal is probably a pulse – or, at the very least will be coded – and so contains a range of spectral components k_0. So in a sense the k_0^4 term should be inside the integral, alongside $\Phi_{\epsilon'}$.

So now we make another approximation. We look at the combined function $k_0^4 \Phi_{\epsilon'}$. We then ask whether this combined function is substantially broader in width in $\vec{\kappa}$-space than the integral function $\int_{V_S} e^{i(\vec{\kappa}-\vec{k_B})\cdot\vec{\xi}} d^3\vec{\xi}$. As an example, we will see later that for turbulence, the function $\Phi_{\epsilon'}$ varies as $|\vec{\kappa}|^{-11/3}$, and when combined with the k_0^4 term (remembering $k_0 \propto |\vec{k_B}|$ for any particular geometry), the overall function varies as $|\vec{\kappa}|^{1/3}$. This can be considered to be slowly varying. Hence we will take the step of moving $\Phi_{\epsilon'}$ *outside the integral*, ascribing it to its value at $\vec{k_B}$, and treating it as a constant. Then (3.224) becomes

$$\overline{S_r} = \frac{k_0^4 \sin^2 \chi}{16\pi^2 R_s^2} S_i V_s \, \Phi_{\epsilon'}(\vec{k_B}) \int_{-\infty}^{\infty} \left[\int_{Vs} e^{i(\vec{\kappa}-\vec{k_B})\cdot\vec{\xi}} d^3\vec{\xi} \right] d^3\,\vec{\kappa}. \tag{3.225}$$

The function of interest is enclosed by the square brackets, but the new formulation makes it clear that once we have found its functional representation, it then needs to be further integrated over $\vec{\kappa}$.

To keep life simple, we will derive this integral for an equivalent one-dimensional case. Extensions to three dimensions will be obvious.

The equivalent process is to first calculate the following one-dimensional integral

$$\mathcal{I} = \int_{-\mathcal{X}/2}^{\mathcal{X}/2} e^{i\kappa x} dx = \int_{-\mathcal{X}/2}^{\mathcal{X}/2} [\cos \kappa x + i \sin \kappa x]\, dx, \tag{3.226}$$

and then further integrate it over κ. Equation (3.226) will produce a peak at $\kappa = 0$, instead of at $\vec{k_B}$, but otherwise will be similar to the original 3-D function that we seek. Clearly the imaginary part is zero by symmetry, so we only need to find $\int_{-\mathcal{X}/2}^{\mathcal{X}/2} \cos \kappa x \, dx$. Hence

$$\mathcal{I} = \frac{1}{\kappa} \sin \kappa x \,|_{-\mathcal{X}/2}^{\mathcal{X}/2} = \frac{2}{\kappa} \sin \kappa \frac{\mathcal{X}}{2} = \mathcal{X} \frac{\sin \kappa \frac{\mathcal{X}}{2}}{\kappa \frac{\mathcal{X}}{2}}. \tag{3.227}$$

When considered as a function of κ, this is a "sinc" function. It is a function with a peak value of $\mathcal{X}$ at zero lag, and which falls to zero at $\kappa_0 = \frac{\pm 2\pi}{\mathcal{X}}$. Beyond these first zeros, it oscillates from positive to negative values as κ increases, with the amplitudes of oscillation becoming vanishingly small as κ approaches infinity.

It was noted above that we will eventually need to integrate this function over κ-space. So let us do this now. The area under the curve $\mathcal{I}$ between $\pm\infty$ in κ-space is 2π. To see this, substitute $\alpha = \kappa \frac{\mathcal{X}}{2}$ into the function, and recognize $d\kappa = \frac{2}{\mathcal{X}} d\alpha$ to give

$$\int_{-\infty}^{\infty} \mathcal{X} \frac{\sin \kappa \frac{\mathcal{X}}{2}}{\kappa \frac{\mathcal{X}}{2}} d\kappa = 2 \int_{-\infty}^{\infty} \frac{\sin \alpha}{\alpha} d\alpha = 2\pi, \tag{3.228}$$

where we have used the well known property that $\int_{-\infty}^{\infty} \frac{\sin \alpha}{\alpha} d\alpha = \pi$. (If the reader is unfamiliar with this expression, write $\frac{1}{\alpha} = \int_0^{\infty} e^{-\alpha t} dt$, and then find $\int_0^{\infty} \left[\int_0^{\infty} e^{-\alpha t} dt\right] \sin \alpha d\alpha = \int_0^{\infty} \left[\int_0^{\infty} e^{-\alpha t} \sin \alpha d\alpha\right] dt$. This will be one half of $\int_{-\infty}^{\infty} \frac{\sin \alpha}{\alpha} d\alpha$. The integral of $e^{-\alpha t} \sin \alpha$ with respect to α is $-e^{-\alpha t} [t \sin \alpha + \cos \alpha] / \left[1 + t^2\right]$, which is $-1/(1+t^2)$ at $\alpha = 0$ and zero at $\alpha = \infty$. Taking the difference of the two gives $\frac{1}{1+t^2}$. Then one half of $\int_{-\infty}^{\infty} \frac{\sin \alpha}{\alpha} d\alpha$ is $\int_0^{\infty} \frac{1}{1+t^2} dt$. The indefinite integral of this is arctan(t), so the definite integral between 0 and ∞ is $\frac{\pi}{2}$, giving $\int_0^\infty \frac{\sin\alpha}{\alpha}\mathrm{d}\alpha = \frac{\pi}{2}$, or $\int_{-\infty}^{\infty} \frac{\sin\alpha}{\alpha}\mathrm{d}\alpha = \pi$.)

As expected, the function is narrow, and is largely confined between the zeros at $\kappa = \frac{\pm 2\pi}{\mathcal{X}}$. If the integral over κ-space is evaluated numerically, it is found to be $1.18 \times 2\pi$, and if integrated between $\frac{\pm 4\pi}{\mathcal{X}}$, it is $0.90 \times \pi$. So most of the contribution of the integral of (3.226) comes from the main peak, and only about 10% comes from portions beyond the second zeros. The main lobe contributes about 80%. Hence, as expected, the function is relatively narrow, and most of its contribution comes from the main peak. So we make the approximation that

$$\int_{-\infty}^{\infty} \mathcal{I}\,\mathrm{d}\kappa \left\{ = \int_{-\infty}^{\infty} \left[\left[\int_{-\mathcal{X}/2}^{\mathcal{X}/2} \mathrm{e}^{\mathrm{i}\kappa x}\mathrm{d}x \right] \mathrm{d}\kappa \right\} \approx 2\pi, \tag{3.229}$$

and recognize that the function $\mathcal{I}$ is largely concentrated in the region between $k = \pm\frac{2\pi}{\mathcal{X}}$, or alternatively has a coverage of $\Delta\kappa \approx \frac{4\pi}{\mathcal{X}}$.

Note in particular that the coverage $\Delta\kappa$ is inversely proportional to the width $\mathcal{X}$ in x-space.

In the three dimensional equivalent, we find $\mathcal{I}$ to be a three-dimensional sinc function of $\vec{\kappa}$ centered around $\vec{\kappa} = \vec{k_B}$, and we expect from (3.229) that

$$\int_{-\infty}^{\infty} \mathcal{I}\mathrm{d}^3\kappa \approx (2\pi)^3. \tag{3.230}$$

The main lobe is confined to a *volume* in $\vec{\kappa}$-space of

$$\Delta V_\kappa \approx (2\pi)^3/V_s. \tag{3.231}$$

It may occur to the reader to ask why we are using approximations here. We found that the area under the 1-D sinc function in the main lobe was $1.18 \times 2\pi$, so why not use that, for example? Even the volume occupied by the 3-D sinc function's main lobe is only given as an approximation. Why not be more precise?

The answer is that here we have only looked at a special case – namely the situation that the scattered intensity is equally weighted over all of the physical volume V_s. In essence, we have assumed a boxcar function, with a weighting of unity inside, and zero outside. But in real life this is not always true – indeed it rarely is. For example, in most radar experiments, a pulse is transmitted and scattered, and this is not always a boxcar pulse. It may, for example, be a Gaussian function. Furthermore, the radar will probably use a narrow beam, and this beam itself will have some tapering, being strongest in the middle and falling away to the edge. Again, a broadly Gaussian shape is not unusual.

So in this case, the integral $\mathcal{I}$ will be different, and additional weighting terms will appear inside the expressions like (3.225) and (3.226). Hence the integrals will not be 2π, and the relation between the volume in κ-space and V_s will not be quite the same as in (3.231).

However, we do expect all practical weighting functions to give values *similar to* the values given above. So we expect (3.230) and (3.231) to be *approximately* true for all weightings, to within a factor of 2 or so. Hence we leave these two equations as

Great care is needed in the correct choice of Φ. For example, for scatter from the neutral atmosphere, it is normal to assume that the refractive index fluctuations are themselves proportional to a classical Kolmogoroff turbulence spectrum, and Φ_n above is taken to be Kolmogoroff in form (e.g., *Ottersten*, 1969a). This is valid because scatter is due to electrons that are bound to atoms, and the atoms vary in density according to a Kolmogoroff spectral form, while the electron backscatter coefficients are largely independent of radar frequency. However, for scatter from the mesosphere and lower ionosphere, the scattering entities are *free* electrons, in which case the scattering efficiency is strongly wavelength dependent. In this case the density perturbations are still largely determined by the neutral molecules (with some exceptions, such as polar mesosphere summer echoes, see later), which force the ions and electrons to follow suit. Because of the strong wavelength dependence of the scattering coefficients for free electrons, and because the electron densities have a Kolmogoroff spectrum, the refractive index variations are therefore significantly non-Kolmogoroff in form. It is necessary to start with the electron density spectrum and then produce the refractive index variations (e.g. *Hocking and Vincent*, 1982a).

In the next few sections we will briefly introduce some of these special cases; additional detail will be presented in Chapter 5.

3.10 Impact of electron motions and plasma waves in radiowave scattering

We now return to the case of plasmas, because there is more that they can teach us. This section is not directly relevant to MST VHF radars per se, but understanding the subtle points of this section will help us understand backscattering more generally. The issues *are* relevant to studies of the D-region with incoherent scatter (e.g., *Mathews*, 1984b).

Up until now, we have assumed that the electrons are approximately randomly distributed, and fixed in space. Both of these assumptions need to be re-considered. If the electrons are fixed in space, all of the received signal will come back at exactly the transmitted frequency. However, if the electrons are moving, each with different velocities, and we imagine each one backscatters incident radio signals, then each electron will produce a different Doppler shift. The Doppler shift will depend on the component of the electron speed in the direction of the radar. Therefore each will return to the receiver at a slightly different frequency. If the electrons have a Maxwellian distribution, as in a neutral gas, then the component of the velocity in the direction of the radar will have a Gaussian distribution, and a plot of the intensity of returned signal as a function of frequency would look roughly like Figure 3.22. The function is broad, and may have a small offset f_0 which is associated with a mean radial drift of the plasma, but generally the offset will be small compared to the width of the distribution.

But is this what is seen in a real situation? Now we need to reconsider some earlier assumptions – the assumption of random independent particle positions, and the assumption that the free electron gas has very low density. If the electron density is very low, the above scenario works in principle. But we need to remember that the gas also contains ions, in order to assure overall charge neutrality. When the electron density

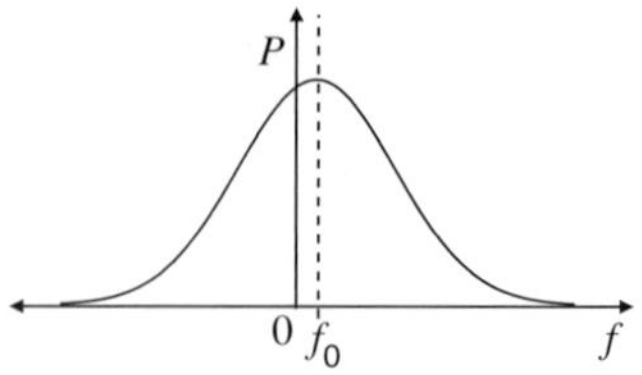

Figure 3.22 Distribution of Doppler shifted frequencies received by a radio receiver if the electrons have a Maxwellian distribution. A large fraction of the electrons will have radial components with values close to zero, and a smaller percentage will have larger radial velocity components. The resultant distribution is Gaussian.

is high, it can no longer be assumed that the electrons act independently – they cluster, sometimes guided by the effects of the larger ions, and produce electron density variations that depend on the interactions between neighbors. This alters the scattering properties of the medium.

We have used the terms "high density" and "low density," but how low does the density have to be before the electrons can be considered as independent? In order to understand where the dividing line between low density and high density lies in the context of radiowave scatter, it is necessary to understand the concept of the Debye length. This quantity can be understood as follows. Consider a gas comprising ions and electrons, each type of charged particle being uniformly distributed, so that in any localized region of space, the overall mean charge density is close to zero. Now imagine introducing a single charged particle – perhaps a positively charged one for argument's sake. The positively charged ions will be repelled by it, and the electrons attracted to it, resulting in non-zero charge densities locally. Because the electrons are smaller and more mobile, they will move quickly, while the sluggish ions can be considered as stationary on the time-scales of interest. Thus there will be a build up of electrons around our test charge. But it would be intuitively expected that the further one goes from the introduced charge, the less impact the test charge would have, so we expect no change in charge distribution at larger distances. In essence, these more distant charges are shielded from the effect of the introduced charge by the ones closer in. Hence the electrons redistribute into a charge distribution which is high near the introduced charge and falls to "average" values further away. The distance from the introduced charge at which the electron densities are essentially unchanged relates to the Debye length. More specifically, the electron density *perturbation* from the mean value is large at the introduced charge, and decays exponentially (no proof given, but the interested reader is referred to *Chen* (1984)) to zero in a manner proportional to e^{-r/λ_D}. λ_D is the Debye length.

Concisely, then, the Debye length is the distance within plasma at which a newly inserted electron or ion is pretty much invisible (i.e., has no impact, due to being overpowered by the shielding effects of nearby electrons and ions at the observer's location).

The Maxwellian assumption above is only valid for electron gases in which the radar wavelength is much less than the Debye length. In the case of higher density electron gases, when the spacing between electrons is reduced and the Debye length is smaller than one wavelength, the interactions between the charged particles result in an organized charge distribution which actually partly cancels the radiative effects of the individual electrons.

The second point is that many types of free waves exist in a plasma, generated by resonance phenomena. These exist as naturally as waves exist on the surface of the ocean. Examples include electron plasma, ion waves, and upper hybrid and ion cyclotron waves and lower hybrids. We will not discuss the details of derivations of all these waves; the interested reader is referred to, for example, *Kelley* (1989) and *Chen* (1984).

For our purposes, it is simply necessary to know that these waves exist in the ionosphere. The tiniest perturbation will set them oscillating. Thus, when our radiowave impacts the ionosphere, it will encounter all sorts of organized motion of these types.

If such an oscillation exists, and is aligned with a wave vector parallel to that of the incident radiowave, and the wavelength of the oscillation is one half of that of the incident radiowave, we satisfy the Bragg condition, as already demonstrated in Figures 3.16 and 3.17.

There will be multiple such plasma waves, all damped to some degree, moving in all directions. Each will have a phase velocity. A backscatter radar will only see waves with wavefronts aligned perpendicular to the radial direction from the radar to the point of scatter, and so the radar will only detect plasma waves moving towards it or away from it. Thus the reflected radiowave will be Doppler shifted in frequency. Hence the signal reflected from such a wave will look like Figure 3.23 as a function of frequency. At first glance this looks a little similar to Figure 3.22, but here the width is less since the plasma wave has (in principle) a well-defined frequency. The frequency offset will also be very different to that in Figure 3.22.

So what plasma waves might be involved? One example is electron waves. These are essentially pressure waves (acoustic waves) in an electron gas. They are often called *Langmuir waves*. Their phase velocity is given by (e.g. *Chen* (1984))

$$c_{l\phi} = \sqrt{\left(\frac{\omega_p^2}{k_p^2} + \frac{3}{2}v_{th}^2\right)}, \tag{3.260}$$

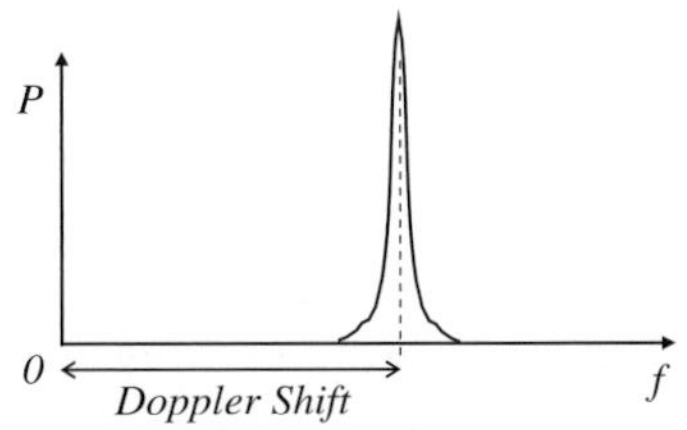

Figure 3.23 Doppler spectrum associated with plasma waves in the ionosphere. Only the positive frequencies have been shown – there will be a matching negative frequency as well, with positive and negative frequencies corresponding to movement towards and away from the radar. There may also be multiple such lines, each pair associated with different plasma processes.

where k_p is the plasma wavenumber and $v_{th} = \sqrt{\frac{3K_BT}{m_e}}$ is the root mean square electron speed. Here, ω_p is given by

$$\omega_p = \sqrt{\frac{n_o e^2}{\epsilon_0 m_e}}. \tag{3.261}$$

Thus the Doppler shift of such a wave will be

$$\Delta f = 2f_o \frac{v}{c}. \tag{3.262}$$

Here we have used a standard formula for the Doppler shift of signal emanating from a moving source, but have multiplied by 2 because the process involves reflection of an incident wave. Put simply, there is a Doppler shift as the scatterer "receives" the wave and a further identical Doppler shift as it retransmits it back to the source. Then we may write

$$\Delta f = 2\frac{c}{\lambda_o}\frac{v}{c} = \frac{k_\lambda}{\pi}v. \tag{3.263}$$

In this case, we use $v = c_{l\phi}$ and we get

$$\Delta f = \frac{k_\lambda}{\pi} \cdot \sqrt{\frac{\omega_p^2}{k_p^2} + \frac{3}{2}v_{th}^2}\ . \tag{3.264}$$

We also know that the Bragg condition tells us that $k_\lambda = \frac{1}{2}k_p$.

Let us consider some typical values. We know $v_{th} \approx \sqrt{\frac{3kT}{m_e}}$. For $T = 2000$ K, $v \approx 3 \times 10^5$ m/s. Typically, $\omega_p = \sqrt{\frac{n_o e^2}{\epsilon_0 m_e}}$ with $n_o \approx 10^{12}$ m^{-3}. Therefore, $\omega_p \geq 5.6 \times 10^7$ radians/second.

For a radar wavelength of 1 m to 0.3 m, which corresponds to a frequency of 300 to 900 MHz, we have $k_\lambda \approx \frac{2\pi}{\lambda}$, so $k_p \approx 12$–40 m^{-1}. Therefore $\frac{\omega_p}{k_p} \approx 1.4 \times 10^6$ to 6×10^6. This is at least five times larger than v_{th}. Thus,

$$\frac{\omega_p^2}{k_p^2} > 25v_{th}^2 \tag{3.265}$$

and

$$\frac{\omega_p^2}{k_p^2} + \frac{3}{2}v_{th}^2 \approx \frac{\omega_p^2}{k_p^2}. \tag{3.266}$$

Thus (3.264) is approximately

$$\Delta f \approx \frac{k_\lambda}{\pi}\frac{\omega_p}{k_p}, \tag{3.267}$$

so

$$\Delta f \approx \frac{\omega_p}{2\pi}. \tag{3.268}$$

The spectral line is therefore offset by

$$\Delta f \approx \frac{1}{2\pi}\sqrt{\frac{n_o e^2}{m_e \epsilon_0}}. \tag{3.269}$$

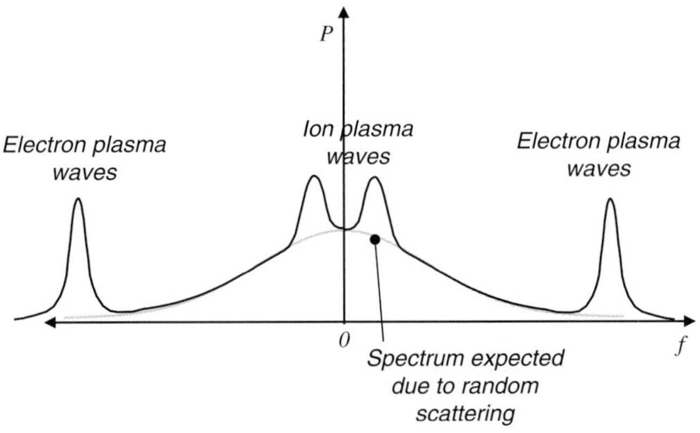

Figure 3.24 Expected spectrum for scatter from a plasma, combining the concepts of Figures 3.22 and 3.23 and including both Langmuir waves and ion waves. The ion wave spectral lines are generally much broader than those due to the electron waves, due in part to Landau damping.

In the case of larger k_p (shorter wavelengths), the thermal term may begin to contribute.

Of course, the atmosphere contains other plasma waves. Ion waves exist, for example. These are essentially sound (acoustic) waves propagated by the *ions* rather than the neutrals, and of course the movement of the ions affects (drags) the motions of the lighter surrounding electrons. These waves are also damped by a process called Landau damping, which has the effect of broadening the spectral lines associated with them. Other types of waves have also been mentioned. So let us consider what we might expect for the spectrum of received signals. We will concentrate primarily on the electron and ion waves. (Ion waves are also called ion-acoustic waves.)

It may be noted here that all of these waves are damped, especially by diffusive processes. In addition, a "wave picture" is not the only way to describe them. Some authors develop the relevant formulas in terms of a process called "dressed ions," in which the ions are considered to be surrounded by a dressing of electrons which they drag along with them. The wave description is the most common, however. The very earliest researchers expected a broad spectrum, based on the ideas of free electron scattering discussed in relation to Figure 3.22. We might therefore expect a combination of such a spectrum plus extra maxima due to other waves, as drawn schematically in Figure 3.24.

But in reality, we find that we do not get this. The broad spectrum in fact does not appear. This was a huge surprise to early researchers in the field. Only the wave-contributions were seen. The actual spectrum recorded in ionospheric backscatter experiments generally looks like Figure 3.25.

Only at radar wavelengths much less than a Debye length (less than typically $\frac{1}{2}$ cm) does the broad spectrum appear. At that stage, the electron and ion lines disappear. As briefly discussed earlier, the reason for the lack of a broad component like that shown in Figure 3.22 is that the scattering due to free electrons, and the scattering due to the fact that the electrons organize themselves due to the radiowave, partly self-cancel.

It also turns out that the power received in a spectrum of the type shown in Figure 3.22, when it does appear, is about one half of that which was originally expected from spectral broadening theory.

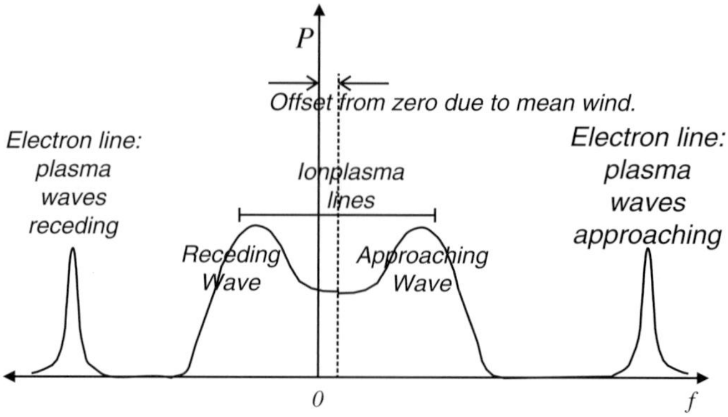

Figure 3.25 The actual spectrum produced for scatter from a plasma. The contribution from Figure 3.22 does not generally appear.

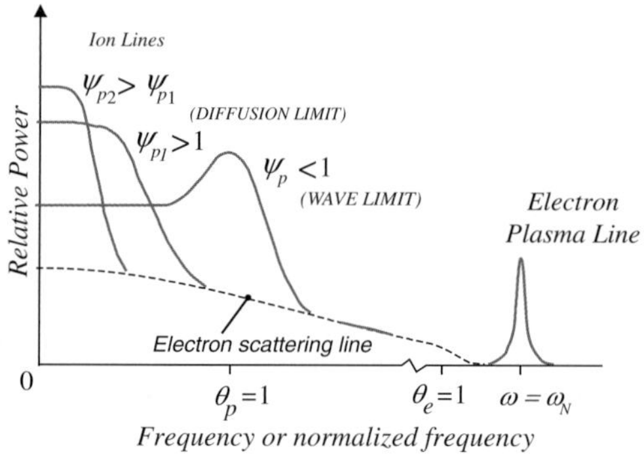

Figure 3.26 A more detailed plot of an incoherent radar spectrum identifying some of the different lines. The parameters θ_p and θ_e refer to the ratios of the radar frequencies to ion and electron plasma frequencies, and the quantities ψ_{pj} refer to positive-ion collision rates normalized relative to RMS thermal positive-ion speeds; specifically, $\psi_p = \nu_{+\text{ion}}/(\sqrt{2}k_B U_p)$, where $\nu_{+\text{ion}}$ is the collision rate of a typical positive ion with the neutrals, k_B is the Bragg wavenumber and $U_p = \sqrt{\frac{K_B T_p}{m_p}}$ is proportional to the thermal ion speed, K_B being Boltzmann's constant and T_p being the ion temperature. The graph labelled ψ_{p2} refers to large values of collision rates, which make the spectra narrower, while ψ_{p1} refers to more intermediate collision rates. The graph labelled ψ_p refers to even smaller collision rates where the ratio is less than unity.

In reality, the shapes of the spectral lines and regions are also modified due to the effects of diffusion and electron collisions. A better diagram of the spectrum is shown in Figure 3.26 (adapted from *Mathews*, 1984b). The situation can become even more complicated; for example, *Mathews* (1978), and *Mathews and Tanenbaum* (1978) discuss the inclusion of negative ions into the mix, although these effects are not shown in Figure 3.26. We will not further discuss these plasma phenomena – the interested reader is referred to *Mathews* (1984b), *Kelley* (1989), and *Chen* (1984), among others.

3.10.1 Further theory pertaining to scattering

At the beginning of Section 3.6.5, we discussed developments which involved matrix representations of the susceptibility matrix, but which were largely beyond the scope of this book. Here, we again need to remind the reader that our discussions to date have not been complete, and we have not fully considered issues like the susceptibility matrix.

Once again, the paper by *Budden* (1965) is worth examining. Other very useful papers relating to these more generalized theories include *Booker* (1956) and *Flood* (1968). The latter presented a revised theory for differential absorption of radiowaves (compared to Chapter 2), which turned out to be of limited value (see discussion paper and reply by *Holt* (1969) and *Flood* (1969)), but the initial introduction section in *Flood* (1968) is relatively complete and well worth reading.

Booker (1956) also discussed such so-called anisotropic media. However, despite the potential complexity added by these equations, one of the most important points in regard to MST studies is that Equation (3.256) remains robust, and we will draw on this fact several times in this book. This is of great importance for MST work, especially in regard to turbulent scatter, and we will consider this shortly.

3.11 Refractive index and scattering in the neutral atmosphere

In previous sections, we have seen the principles of refractive index, and have applied examples of various types. We have especially concentrated on plasmas, partly because the mathematical development is well understood, and partly because some component of it is also relevant to MST scatter, particularly for scatter from the lower ionosphere and D-region.

We will now move on to discussions about scatter from the neutral air. This is important for all aspects of MST scatter, and is the dominant process associated with scatter from the troposphere and stratosphere. While it is still the electrons that do the scattering, they are now bound to atoms, which leads to additional complications and the need for a somewhat different approach.

In addition, in the earlier sections of this chapter, we have concentrated on refraction and scattering by individual electrons. In dealing with the neutral atmosphere, we introduce an additional focus, namely that of diffraction theory. This is especially important when the scatterers have sizes that are physically constrained, perhaps limited to a few wavelengths or less. Of course the same size-constraints could equally apply for plasma scatter too, but we deal with the issue in this section because it seems suitable to give emphasis here, neutral scatter being a major aspect of this book.

We also view the scattering process slightly differently to the plasma case, and are less concerned about individual electrons. Both viewpoints can be important in both areas, but the emphasis is different. Further discussion about diffraction theory will appear in later chapters, especially Chapter 7.

3.11.1 Expressions for the refractive index in the neutral air

In order to consider the refractive index of radiowave propagation in air, we need to return to Equation (3.73), viz.

$$m_e \frac{\mathrm{d}^2 \underline{\vec{x}}}{\mathrm{d}t^2} = q_e \underline{\vec{E}}_0 \mathrm{e}^{-\mathrm{i}\omega t} - m_e \nu_{ec} \frac{\mathrm{d}\underline{\vec{x}}}{\mathrm{d}t}. \tag{3.270}$$

In that earlier equation, ν_{ec} was the collision frequency of the electrons with other particles (and especially the neutral molecules). Here, we consider it purely as a damping term, which may in part relate to the collision frequency between molecules when dealing with a gas, but has other associations if we consider molecules bound in a solid, for example.

In Equation (3.91) an extra term was included to account for magnetic fields. Here we will not (yet) consider the effect of magnetic fields, but will add an extra term of a different type. Recognizing that the electrons are bound to atoms with very large mass (relative to the electrons, at least), the electron can be considered to be bound to the atom by a spring-like force, of the type $-\kappa_{sp}x$, where κ_{sp} is a suitable spring constant. (A more thorough treatment uses quantum mechanics, but for our purposes this "spring-like" electronic oscillator approximation will do for now.)

In order to best demonstrate the concept, the vector notation will be dropped and only one dimensional motion along the x-axis will be considered. However, the variables may be complex.

Then Newton's second law becomes

$$m_e \frac{\mathrm{d}^2 \underline{x}}{\mathrm{d}t^2} = q_e \underline{E}_{0x} \mathrm{e}^{-\mathrm{i}\omega t} - m_e \nu_{ec} \frac{\mathrm{d}\underline{x}}{\mathrm{d}t} - \kappa_{sp}\underline{x}. \tag{3.271}$$

If we assume $\underline{x} = \underline{x}_0 \mathrm{e}^{-\mathrm{i}\omega t}$, then

$$m_e(-\mathrm{i}\omega)^2 \underline{x}_0 \mathrm{e}^{-\mathrm{i}\omega t} = qe\underline{E}_{0x}\mathrm{e}^{-\mathrm{i}\omega t} - m_e\nu_{ec}(-\mathrm{i}\omega)\underline{x}_0\mathrm{e}^{-\mathrm{i}\omega t} - m_e\omega_0^2\underline{x}_0\mathrm{e}^{-\mathrm{i}\omega t}, \tag{3.272}$$

where we have replaced κ_{sp}, the spring constant, by $m_e\omega_0^2$, ω_0 being the undamped angular frequency of oscillation for the spring.

Then collecting terms in $\underline{x}_0$, and simplifying, gives

$$-m_e(\omega^2 - \omega_0^2)\underline{x}_0 = q_e\underline{E}_{0x} + \mathrm{i}\omega m \nu_{ec}\underline{x}_0, \tag{3.273}$$

so

$$\underline{x}_0 = \frac{-q_e\underline{E}_{0x}}{m_e(\omega^2 - \omega_0^2) + \mathrm{i}m_e\omega\nu_{ec}}, \tag{3.274}$$

or

$$\underline{x}_0 = \frac{-q_e\underline{E}_0}{m_e[(\omega^2 - \omega_0^2)^2 + (\omega\nu_{ec})^2]} \cdot [(\omega^2 - \omega_0^2) - \mathrm{i}\omega\nu_{ec}]. \tag{3.275}$$

Subsequent developments follow the same methodology as that leading to Equation (3.82), but with the extra ω_0^2 term included, so the refractive index is

$$n = \sqrt{\left(1 - \frac{Ne^2(\omega^2 - \omega_0^2)}{\epsilon_0 m_e[(\omega^2 - \omega_0^2)^2 + (\omega\nu_{ec})^2]}\right) + \frac{\mathrm{i}Ne^2\omega\nu_{ec}}{\epsilon_0 m_e[(\omega^2 - \omega_0^2)^2 + (\omega\nu_{ec})^2]}}, \tag{3.276}$$

where we have replaced q_e with e since all charge terms are squared, so we need to only consider the magnitude. In all cases, ω^2 in Equation (3.82) is replaced by $\omega^2 - \omega_0^2$. This means the term has a maximum at $\omega = \omega_0$ and falls away on either side, giving in essence a spectral line.

Separate real and imaginary parts can be developed in the same way as the procedure leading to (3.85) and (3.86). This gives the real part of n, (which defines the propagation characteristics), as

$$n_R = \sqrt{1 - \frac{Ne^2(\omega^2 - \omega_0^2)}{\epsilon_0 m_e[(\omega^2 - \omega_0^2)^2 + \omega\nu_{ec})^2]}}, \tag{3.277}$$

and the imaginary part (which defines the absorption) as

$$n_I = \frac{1}{2}\frac{e^2\nu_{ec}\omega N}{\epsilon_0 m_e[(\omega^2 - \omega_0^2)^2 + (\omega\nu_{ec})^2]}. \tag{3.278}$$

Hence for bound electrons, absorption lines and bands develop, depending on the characteristics of the molecules. Scattering by bound electrons in this way is generally referred to as *Rayleigh scatter*.

A major radiowave absorber is water vapor, which has particularly strong absorption (and scattering) in the microwave region, specifically in the K-band at about 23 GHz (wavelength of 1.3 cm). In reality, the process is more complex than we have described, and that particular absorption band is not associated directly with electron energy levels but with rotational transitions associated with the water molecule. A detailed analysis requires a proper quantum-mechanical description, which is beyond the scope of this work, but the mathematics described above gives a physical feel for the origins of the absorption peak in a general sense. The absorption bands of water and many of the other atmospheric gases are very numerous and in general quite complex.

(The region around the 23 GHz water-vapor peak is referred to as the K-band. At frequencies above and under this – specifically 26.5–40 GHz (Ka) and 12–18 GHz (Ku) – absorption is weaker and so these bands are used for satellite transmissions, etc. (Ka – means "Kurtz-above" and Ku means "Kurtz-unten".)

At MF, HF, and VHF frequencies, spectral peaks are generally less likely, and the absorption is largely a continuum.

A similar theoretical framework can be used to look at Faraday rotation in cases where the electrons are bound to molecules, such as in solids, liquids, and gases. In our case, the gases of the neutral air are the most important.

Ignoring the effects of absorption (since simultaneous calculation of refractive indices accounting for bonded electrons, absorption, and magnetic fields is messy), but now considering the impact of the magnetic field, Equation (3.271) becomes

$$m_e\frac{\mathrm{d}^2y}{\mathrm{d}t^2} = q_eE_y + B\frac{\mathrm{d}z}{\mathrm{d}t} - \kappa_{sp}y, \tag{3.279}$$

where we have considered propagation along the x-direction with electron motions in the y- and z-directions. This equation can also be seen by adapting (3.91). We recognize that there should be a matching z-component like the second term in (3.91), viz.

$$m_e \frac{d^2z}{dt^2} = -q_e B \frac{dy}{dt} - \kappa_{sp} z. \tag{3.280}$$

Assuming solutions proportional to $e^{-i\omega t}$ leads to

$$\begin{aligned} -m_e(\omega^2 - \omega_0^2)y &= q_e E_y + i\omega B q_e z \\ -m_e(\omega^2 - \omega_0^2)z &= -i\omega B q_e y, \end{aligned} \tag{3.281}$$

where we have replaced κ_{sp} by $m_e\omega_0^2$, as we did above for the damped case. Once again the term $\omega^2 - \omega_0^2$ replaces occurrences of ω^2. As before, terms involving q_e are all squared terms, so we can substitute e in place of q_e. The refractive indices for the two characteristics modes are then modifed from (3.113) to become

$$n_{x,o}^2 = 1 - \frac{Ne^2}{\epsilon_0 m_e \left[(\omega^2 - \omega_0^2) \pm (\omega\Omega)\right]}. \tag{3.282}$$

This may also be written as

$$n_{x,o}^2 = 1 - \frac{Ne^2}{\epsilon_0 m_e \left[(\omega^2 - \omega_0^2)\left(1 \pm \frac{(\omega\Omega)}{\omega^2 - \omega_0^2}\right)\right]}. \tag{3.283}$$

For the case that $\omega\Omega \ll (\omega^2 - \Omega^2)$ (i.e., when the frequency is well removed from the resonance at $\omega = \omega_0$), the identity $\frac{1}{1-\xi} \approx 1 + \xi$ for $\xi \ll 1$ gives

$$n_{x,o}^2 = 1 - \frac{Ne^2}{\epsilon_0 m_e \left(\omega^2 - \omega_0^2\right)} \mp \frac{Ne^2\omega\Omega}{\epsilon_0 m_e \left(\omega^2 - \omega_0^2\right)^2}. \tag{3.284}$$

The first two terms give the refractive index in the absence of a magnetic field, while the last term distinguishes the two characteristic modes.

For applications of Faraday rotation, the same principles apply as in Equation (3.119), viz.

$$d\theta = \pi \frac{ds}{\lambda_v} \frac{(n_o^2 - n_x^2)}{(n_o + n_x)}, \tag{3.285}$$

but in this case we cannot assume n_o and n_x are unity – rather, their sum is twice the refractive index for the case of no magnetic field.

Hence (3.285) remains as it is, but $\frac{n_o^2 - n_x^2}{n_o + n_x}$ becomes

$$\frac{n_o^2 - n_x^2}{n_o + n_x} \approx \frac{Ne^2\omega\Omega}{\epsilon_0 m_e \left(\omega^2 - \omega_0^2\right)^2}. \tag{3.286}$$

The MF to UHF Band in the neutral atmosphere

At MF, HF, VHF, and even UHF radio frequencies in the un-ionized atmosphere, there are no real absorption lines, and the refractive index is largely a continuum. The following expression is sufficient for most work in this frequency range. The expression has been in part derived empirically, and unfortunately even some of the measurements were made at optical wavelengths. But the best estimate of the refractive index currently in use is:

$$n = 1 + n' = 1 + 77.6 \times 10^{-6}\frac{p}{T} + 3.73 \times 10^{-1}\frac{e_{wp}}{T^2}. \tag{3.287}$$

Plasma terms have been ignored. Here, p is the atmospheric pressure in units of millibars (hPa), T the absolute temperature (K), and e_{wp} is the water-vapor pressure. (We have used e_{wp} to represent the water vapor because in many meteorological texts it is represented by e, but here we need to differentiate it from electric charge.) Typical values for n' are of the order of 0–500 $\times$ 10^{-6}, with largest values occurring in the lower troposphere. In some texts n' is denoted as N and is called the "refractivity" of the air.

While the relation is, as stated, partly empirical, some modest amount of theoretical work has been carried out, particularly by *Owens* (1967) and *Hoenders* (2008). A key aspect of the derivation was recognition that the effect of an electromagnetic wave on an electron bound within a molecule is also affected by the presence of neighboring molecules and atoms. Estimates of the dipole moment of the molecule are needed, as well as the Debye length of molecular dipoles within the gas.

Perhaps more important from a practical perspective is the so-called "potential refractive index gradient," given by

$$M_n = \frac{-77.6 \times 10^{-6}}{T^2} p \left[1 + \frac{15500}{T} q_{wp}\right]\left[\frac{dT}{dz} + \Gamma_a - \frac{7800}{1 + \frac{15500}{T} q_{wp}} \cdot \frac{dq_{wp}}{dz}\right]. \tag{3.288}$$

Here, p is the pressure in millibars (hPa), T is the absolute temperature (K), Γ_a is the adiabatic lapse rate, and $q_{wp} = e_{wp}/(1.62p) \approx 0.62e_{wp}/p$ is the specific humidity (which is very similar to, but not exactly equal to, the mixing ratio), e_{wp} being the water vapor pressure. Some discussion of the derivation of this equation can be seen in *Tatarski* (1961) and *Hocking* (1985), and the equation is also briefly summarized in Appendix A, Section A.6. We will not go through the entire derivation here, but the basic idea is to look at the difference in refractive indices of an adiabatically displaced parcel and the refractive index of the surrounding air.

In most work on turbulent scatter, the potential refractive index gradient is the primary refractive-index term that is needed, as we will now demonstrate.

In radar work, the recorded power received at the receiver will be

$$P_R \propto \sigma_s, \tag{3.289}$$

where σ_s is the backscatter cross-section per unit volume at the point of radar backscatter, per unit steradian, per incident Poynting vector, as discussed in Equation (3.256), viz.

$$\sigma_s = \frac{\pi}{8}\, k_B^4\, \Phi_n(\vec{k_B}). \tag{3.290}$$

Note that here we have slightly modified (3.256) by using k_B instead of k, which results in division of (3.256) by $2^4 = 16$ on the right-hand side. The proportionality constant in (3.289) contains terms related to transmitted radar power, antenna directivity, receiver gains, range, efficiencies, etc., which will be discussed later.

The units of σ_s and η_s are inverse length (e.g. m^{-1}), so the nomenclature is a little confusing, as we have discussed. The electron backscatter cross-section had units of area, so can justifiably be called a cross-section. Neither σ_s nor η_s have units of area, so are not really a cross-section, though they are often referred to as such. They are also referred to as reflectivities, as discussed earlier, though this can also be a confusing name because the process of backscatter is not really a reflection. Even if we consider it as a reflectivity, a reflection coefficient has no units. The most common (slightly unsatisfactory) solution is to regard σ_s and η_s as *reflectivities*, but to consider "reflectivity" and "reflection" as separate processes.

In order to determine σ_s, we use the expression for the refractive index spectrum as given by *Tatarski* (1961, 1971), and also summarized in Appendix A, namely

$$\Phi_n(k_x, k_y, k_z) = 0.033 C_n^2 |\vec{k_B}|^{-\frac{11}{3}}. \tag{3.291}$$

(Here a normalization has been chosen such that $\int\int\int_{-\infty}^{\infty} \Phi(\underline{k}) d\underline{k} =< n^2 >$.) C_n^2 is a parameter which indicates the level of refractive index fluctuation.

Then combining (3.291) and (3.290), plus using $k_B = \frac{4\pi}{\lambda}$ (λ being the radar wavelength) gives

$$\sigma_s = 0.033 \frac{\pi}{8} (4\pi)^{\frac{1}{3}} C_n^2 \lambda^{-1/3} = 0.00655\pi C_n^{24/3} \lambda^{-1/3} = 0.03014 C_n^2 \lambda^{-1/3}. \tag{3.292}$$

The equivalent expression for η_s is just σ_s multiplied by 4π, or

$$\eta_s = 0.3787 C_n^2 \lambda^{-1/3}, \quad \text{often rounded as} \quad 0.38 C_n^2 \lambda^{-1/3}. \tag{3.293}$$

Hence the received power is

$$P_R \propto C_n^2 k_B^{\frac{1}{3}} \propto C_n^2 \lambda^{-1/3}. \tag{3.294}$$

As in (3.289), the constants of proportionality relate to geometrical and radar-beam related issues, and discussion of these will be left to later chapters.

While C_n^2 is a useful parameter, it is more often the strength of turbulence, represented by the turbulent kinetic energy dissipation rate ε, that is required. The conversion from C_n^2 to ε is carried out through the relation (*Hocking and Mu* (1997), Equation (3))

$$C_n^2 = \frac{1}{\gamma} \varepsilon^{\frac{2}{3}} F_t^{\frac{1}{3}} M_n^2 \omega_B^{-2}. \tag{3.295}$$

The term F_t refers to the fraction of the radar volume filled by turbulence, and ω_B is the Brunt–Väisälä frequency, which was discussed briefly at the end of Chapter 1. The term γ is a constant with contributions from a variety of terms. There is even some evidence that γ is not exactly a constant but may be dependent on other parameters like the Richardson number R_i, which is a measure of instability (see end of Chapter 1).

The details of (3.295) will be discussed later – for now our interest is in the relation of the power received by the radar to C_n^2 and thence to ε through this last equation. Of particular importance has been the potential refractive index gradient, M_n.

The Equations (3.294) and (3.295) are the primary equations needed for studies of turbulence in the neutral atmosphere using MST radars. Some further elaboration will be presented in Chapters 5 and 7.

Backscattered power and specular reflectors

In the previous sub-section, the basic equations required for using radars for studies of turbulence were introduced. But turbulence is not the only scattering mechanism for MST radars, and in this section we introduce the mathematical formalism to deal with another type of scatterer – the extended horizontally aligned reflector.

We have seen some reflectors when we dealt with the ionosphere, and especially critical-level reflections. In scatter associated with the neutral or weakly ionized air, effects like critical level reflection do not occur. Refraction occurs, but is less dramatic than in the upper ionospheric case. Different modes of the waves (O and X, for example) show only modest differences in behavior. However, neutral atmosphere scatter and reflection has special challenges of its own. Mixing of the air produces refractive index inhomogeneities which can act to cause weak reflectors (as distinct from scatterers), and MST radars are sufficiently powerful that they can detect the weak backscattered signal. While turbulence is considered the main mechanism for mixing of the air, it is not the only contender. Occasions also exist when the refractive index shows sudden steps in value, which are also horizontally stratified. The reason for these steps has not been fully understood, but explanations relate to small scale waves (viscosity waves), horizontal intrusions of air, and sharp ledges left after turbulence has died away, and even highly stretched structures embedded in the outer layers of active turbulence.

Potential mechanisms for their creation will be discussed in more detail later, particularly Chapter 11. For now we will consider simply that they may exist, and will represent them by a model of a sharp Heaviside step in refractive index as a function of height.

We begin by looking at Figure 3.27. This shows such a step, and also shows incident, transmitted, and reflected waves at the step. The horizontal axis is labelled z, and refers to height, but is plotted horizontally to save space. The reflection point is at $z = z_0$. We will assume for the present that the incident wave is a plane wave, with wavefronts perpendicular to the z-axis. The reflected and transmitted waves have the same characteristic.

In order to calculate the amplitude of the reflected signal relative to the incident signal, two seemingly different approaches may be adopted.

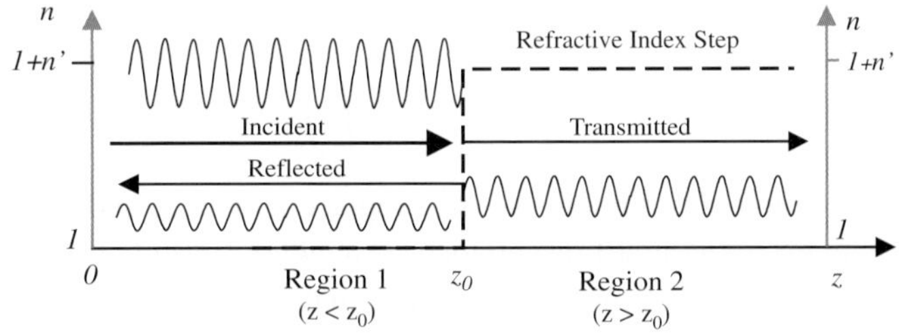

Figure 3.27 Incident, reflected, and transmitted waves as they occur at a Heaviside step-function in refractive index. An example would be light entering a slab of glass. However, in our case we will consider it as a radiowave entering a region of slightly enhanced refractive index.

the air around them. As a result, a long trail of plasma (typically a few km in length) exists behind the meteoroid, and this so-called "meteor trail" can reflect radiowaves. Yet another cause of refractive-index variability is turbulence, which can mix the refractive index of the medium in which it exists. For example, in the ionosphere, the refractive index is often quite different from unity, and events like turbulence can mix an initially uniform region into one in which the refractive index varies in both time and space. Such perturbations can then be the cause of radiowave scatter. The same thing can happen with regard to the neutral air. The air has a refractive index which is slightly different from unity, the exact value depending on temperature, pressure, density, humidity, and free-electron density (see Equation (3.287), Chapter 3). Typically, the variations from unity are a few parts to a few hundred parts in a million, which is quite small. Despite the small values, however, the effects of this non-unity refractive index can be profound. Systematic variations in refractive index are responsible for the formation of mirages, for example, and can cause deviations in the path of starlight as it enters the atmosphere. Turbulence, and even small-scale atmospheric wave phenomena, can cause the refractive index to vary as a function of position (and time), and these small perturbations can also be a source of radiowave reflection and scatter. Despite the very weak refractive-index perturbations produced by this process, a radar with suitable sensitivity can still detect scatter from regions of refractive-index inhomogeneity. Indeed, scatter from turbulence-induced refractive-index perturbations in the air is often one of the main processes employed with high-gain atmospheric radars.

As discussed in Chapter 3, the backscattered power relates to the amplitude of the Bragg scale vector (one half of the radar wavelength for a monostatic radar) aligned with wavefronts perpendicular to the line joining the radar and the point of scatter. Examples like turbulent scatter and scatter from atmospheric irregularities are therefore generally referred to as "Bragg scatter." Scatter from particulates like water droplets is referred to as "Rayleigh scatter," and is often considered to be non-Bragg scatter. In fact, this definition is misleading. Even for scatter from particulates, the particulates can each be considered as small "delta-function" scatterers, and if we Fourier-analyze the entire field of delta functions, we will again obtain various Fourier components. Even in this case, the backscattered power depends on the amplitude of the Bragg scale, so it is wrong to think of Rayleigh scatter as different to Bragg scatter – it is simply a different class of Bragg scatter/reflection. Likewise, ionospheric researchers often refer to "incoherent scatter." This is essentially a scattering process which decorrelates quickly from pulse to pulse. Again, the scatter is still from Bragg scales of refractive-index variations embedded in the atmosphere. Despite the similarities, each of these different types of scatter do have unique features associated with them, and it is important to understand the differences, because it helps define the most apt methods of radar design, data acquisition, and data analysis.

The primary purpose of an atmospheric radar is to probe the air, and to interpret the signals received so as to better understand the motions and dynamics of the atmosphere. In order to do that, we first need to understand the basic features and principles of atmospheric radar. The purpose of this chapter is to do just that.

4.3 A simple radar

Figure 4.1 shows a block diagram of a very simple radar. It has only five components, these being: (i) a radar controller, (ii) a transmitter, (iii) a transmit antenna, (iv) a receiver antenna, and (v) a receiver. Usually some sort of recording device would be attached to the receiver, but that has not been added here. In some cases, the same antenna is used for transmission as well as reception, although this requires a special fast-acting switch called a TR switch (or transmit-receive switch) to alternate between the transmitter and receiver. We have avoided this design for now by considering separate antennas. Multiple targets have also been drawn in the figure.

In Figure 4.2, we demonstrate the principle of operation of the radar. In this case, we have drawn the radar pulse to propagate horizontally, although for MST studies, near-vertical propagation is more common. In the figure, time increases as one moves down the page. We have also assumed that the same antenna is used for transmission and reception. A burst of radio-frequency electromagnetic radiation is produced by the transmitter and propagates away from the antenna. The pulse can be seen approaching the target in the first two figures, and then, in the third figure, the target is reached. The fourth figure shows that most of the pulse continues on right through the target, while a small portion is reflected. Typical reflection coefficients $\mathcal{R}$ for MST studies can vary between 10^{-9} and 10^{-3}, but in all cases, we can consider that almost all of the original pulse continues to propagate, and only a very small component is scattered or reflected. We often assume that all of the pulse continues unattenuated. Although energetically this is, of course, impossible, the approximation is often a very useful

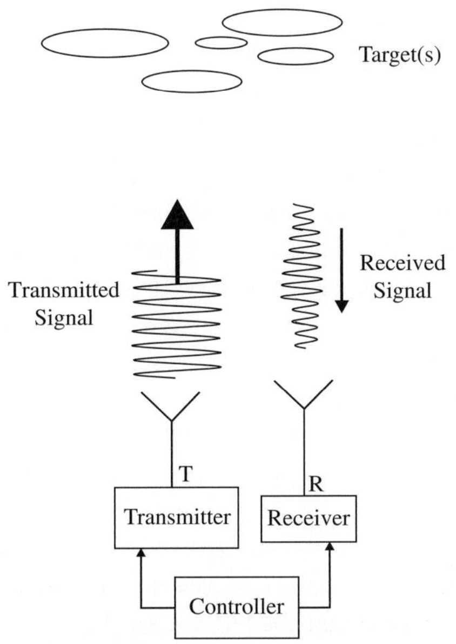

Figure 4.1 The simplest possible radar, showing the transmitted signal, the targets, and the receiver.

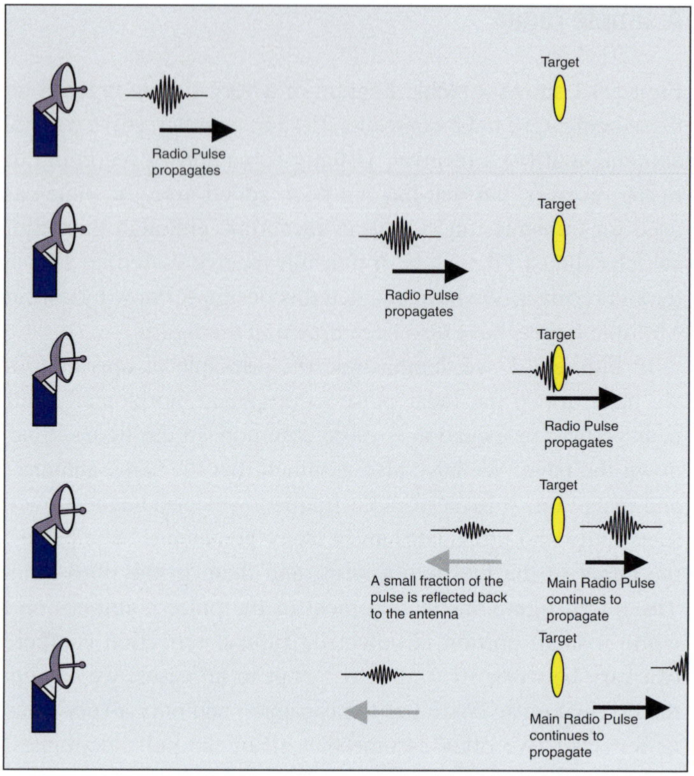

Figure 4.2 A sequence showing the principle of radar. In the upper figure, a radar pulse is transmitted, and it is then shown at later and later times as we move down the page. In the third figure, the pulse encounters a target, and then some portion is reflected back to the left while the bulk of the pulse proceeds further to the right. The reflected pulse is subsequently detected by the radar antenna.

one, and is called the "Born approximation." The fifth figure shows the reflected pulse on its way back to the antenna, while the original transmitted pulse has moved on to the right. The pulse that is moving to the right may subsequently encounter other targets, and suffer similar reflections. It is also possible that the reflected component of the pulse might encounter other targets on its way back to the antenna, but the scattered component will be attenuated significantly relative to the original pulse, which is already exceedingly weak because it was created by a weak reflection. Hence we ignore pulses which are multiply reflected, and consider only single-reflection (or scatter) pulses.

If we examine the received signal on a cathode-ray oscilloscope connected to the radar receiver, a figure something like Figure 4.3(a) will appear (adapted from *Hocking*, 2003a). In these cases, we consider amplitude only. Later we will talk about in-phase and quadrature components, but not yet. Figure 4.3(a) shows the expected variation in signal strength for three clusters of radiowave scatterers. A returned pulse can be seen from each cluster. These returned pulses are often called "echoes." An echo from the

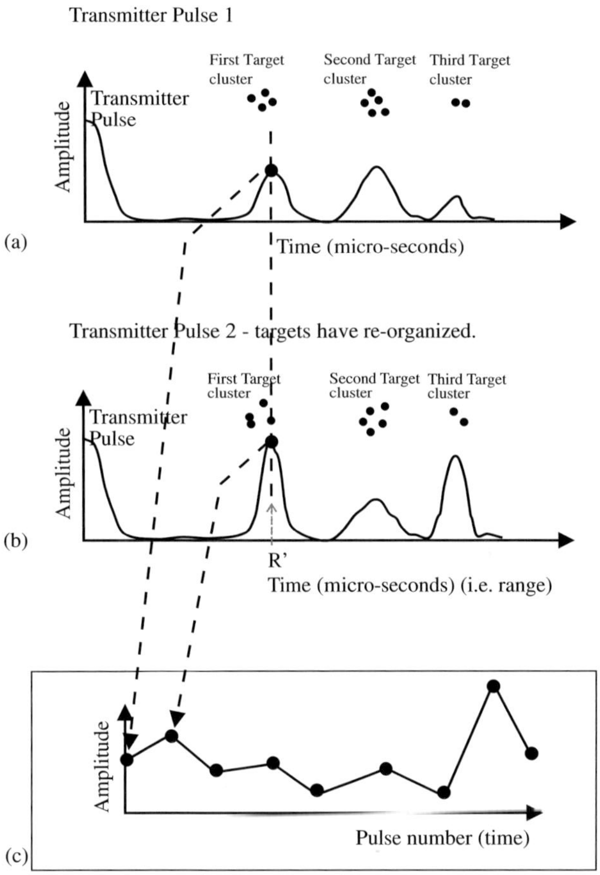

Figure 4.3 Typical scan of the amplitude detected by a radar for two successive pulses, (a) and (b). The targets producing the scatter are also shown schematically above each reflected pulse, demonstrating how they have moved from pulse to pulse. (c) By sampling at a particular time delay on each successive pulse (represented by R' in the figure), a time series is built up. This can be done at a multiplicity of different time delays (different ranges), so that similar time series can be constructed at a number of different effective ranges.

transmitted pulse can also be seen at zero lag. Typical delays are generally of the order of microseconds.

Figure 4.3(b) shows the result when a second pulse is transmitted at a slightly later time. The scatterers in the cluster have moved their position and perhaps changed their strength, so the returned pulses (echoes) have different amplitudes to the cases seen in Figure 4.3(a).

If we now imagine transmitting multiple pulses in succession, then we can measure the signal strength at a fixed lag (denoted by R' in the figure) after pulse transmission, and record the amplitude to a recording device, as shown in Figure 4.3(c). From pulse to pulse, the amplitude changes, and this evolution tells us about the behavior of the scatterers in the first target-cluster. Because all the recorded signals have the same time-lag

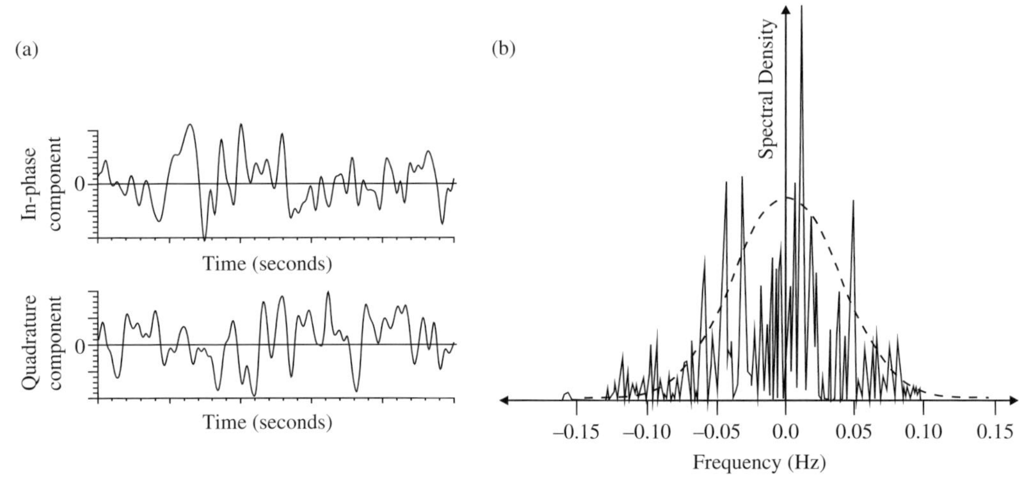

Figure 4.8 (a) Typical in-phase and quadrature components received with an MST radar. The abscissa is not specifically shown with actual times, since for oscillations like this, the length of the time axis would be different for different radar frequencies. At VHF, this total time axis might cover 5 or 10 seconds, whereas at medium frequencies (MF), it might cover 60 or 90 seconds. (b) Typical power spectrum for the time-series shown in (a). In this case, we have added a frequency scale, but the actual values will depend on the duration and sampling rate of the time-series in (a). The frequency scale is only added to give a rough idea of the sorts of frequencies typically involved, and could be up to an order of magnitude smaller or larger, depending on sampling details. Both figures were adapted from *Hocking* (1983b).

of the received signals, and will be discussed in considerable detail later in this book. For the present, the different spectral lines can each be considered to represent different radial velocities, and although we will see later that this representation is perhaps a little too simplistic, it makes a useful starting point.

The above discussion dealt with "radar" for continuous wave transmission. In the strictest sense, this is not really a radar, since it really has no useful ability to resolve range. Nevertheless, it has served as a useful starting point. We now turn to a more detailed consideration of pulsed radar.

4.6 Pulsed radar

We now have a simple representation of our received signal. However, we need to recognize that the radar does not normally transmit a continuous wave, but rather a sequence of pulses, as already discussed in regard to Figure 4.3. (In some cases a coded CW wave can be used, with variations in frequency being used to encode signal, but we will not consider these cases here.) The received amplitude and phase will differ from pulse to pulse. This is the information that we require and wish to diagnose.

In order to proceed, we will modify Figure 4.6 to allow it to transmit pulsed RF. The new circuit is shown in Figure 4.9. Our local oscillator reference remains as $\cos(\omega_0 t)$,

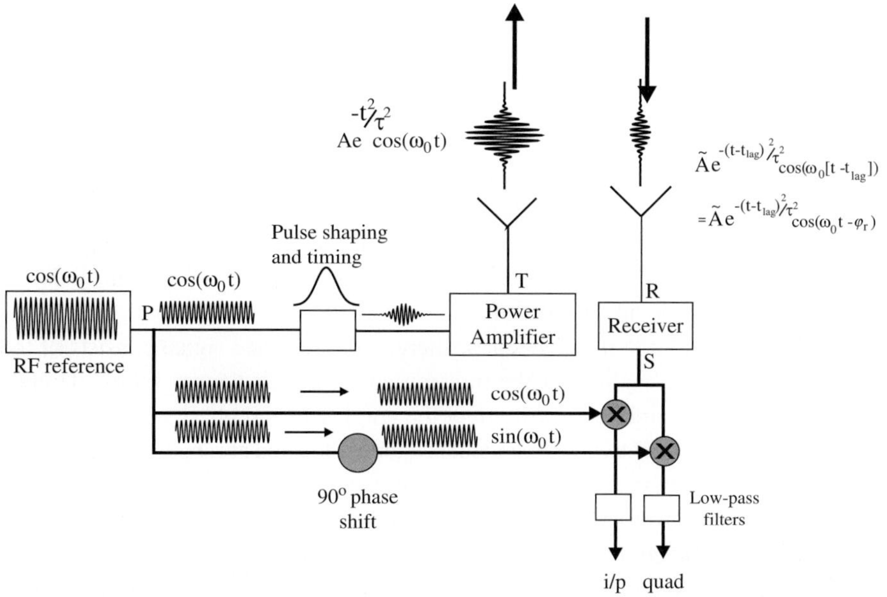

Figure 4.9 A further development of a block diagram for the radar, this time with pulse-shaping (pulse modulation) included. In this case, we have assumed that the pulse has a Gaussian form.

and again part of the signal is "sniffed" off to act as a reference for producing in-phase and quadrature signals later on. The rest of the signal proceeds to a pulse shaping and timing unit, where it is multiplied by a suitable pulse-shaping function. This unit is also called a pulse modulator or simply a pulse shaper. The envelope of the pulse could take a variety of forms. It could be a square wave, in which the reference is simply turned on and then off again, or it could be a Gaussian function, or some other form. Square pulses are quite common, although highly undesirable, since they produce significant Fourier harmonics that can be a serious nuisance to other users of the nearby frequency spectrum, and could result in the system being closed down by government radio-frequency-monitoring authorities.

Despite these reservations, we will begin our discussion here by referring to a square pulse. We will discuss a more carefully shaped radar pulse shortly. The simple square pulse shape is described mathematically as

$$p(t) = \begin{cases} 1 & 0 \leq \mathrm{t} \leq \tau \\ 0 & \text{elsewhere.} \end{cases} \tag{4.10}$$

A typical pulse length τ is 1–5 μs for many atmospheric MST radar applications at frequencies of typically 50 MHz. At medium frequencies, a pulse length of 10–15 μs is more common, and for meteor studies, a pulse length of 7–15 μs is frequently used. Using this pulse as a modulating signal, the received signal from a single-point scatterer will have the following form, where we have assumed that the transmitted frequency is f_0:

$$E_R(t) = \tilde{A}e^{i\{\omega_0 t-(2k_0 r+2k_0 v_r t)+\varphi\}}\underline{p}(t-2r/c), \tag{4.11}$$

where we are using complex representation now, and where φ represents various system offsets related to cable delays and the receiver. $\underline{p}$ may be complex, which means it may have embedded phase variations within the pulse, although we will mainly consider the simpler case of $\underline{p} = p$ i.e., a purely real function.

From (4.11), we see that the range information is not only embedded in the exponent of e, but is also present in the delay of the pulse $p(t)$. The fact that the time delay is embedded in the pulse allows us to use the temporal delay of the pulse to determine the range, as already discussed.

For multiple independent scatterers, the returned signal consists of the superposition of multiple replicas of the transmitted signal. Of course, the scatterers' physical characteristics and range will affect the amplitude and phase of the returned signal. For the simple case of two independent scatterers, located at ranges r_1 and r_2, the returned signal would consist of the sum of two pulse-modulated carrier signals. After coherent detection, the envelope of the signal present at the output of the receiver would have the shape of the transmitted pulse as illustrated in Figure 4.10. At this point, we have ignored the pulse-shaping effects of the low-pass filters (LPFs). It is also important to recognize that any filter also causes additional delays, producing a temporal delay of the order of the inverse of the filter width. These important topics will be addressed later.

From Figure 4.10, it is obvious that if the two scatterers are separated by less than the pulse length τ, we will not be able to resolve their signals by time sampling. Therefore, τ dictates the range resolution of the radar and is denoted by Δr,

$$\Delta r = \frac{c\tau}{2}. \tag{4.12}$$

We now return to Figure 4.9. We will no longer assume that the pulse is a square function, and rather will let it be quite general in form, but assume it is purely real for now. The transmitted waveform is then of the type

$$p(t) = A(t)\cos(\omega_0 t), \tag{4.13}$$

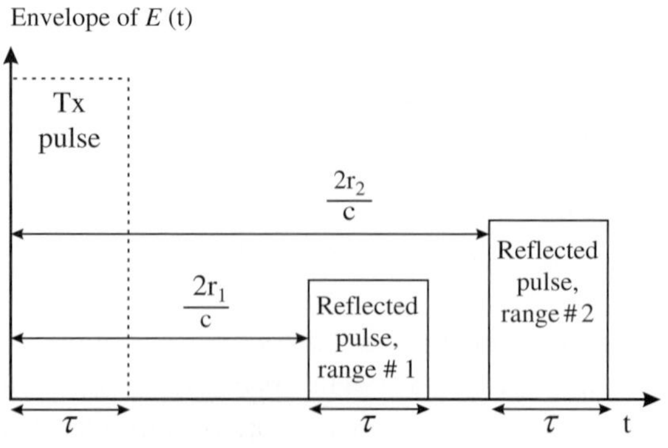

Figure 4.10 Idealized envelope of received signal $E_R(t)$ for two point scatterers located at ranges r_1 and r_2 for a square pulse of temporal length τ. Note that the envelope replicates the transmitted pulse with varying amplitude and time delay depending on the scatterers' reflectivity and range respectively.

where $A(t)$ is the envelope function, and we have assumed that the cosine function peaks at zero lag, for simplicity. We will demonstrate the application with a Gaussian function, $A(t) = Ae^{-t^2/\tau^2}$, although in practise Gaussian functions are not optimum either. The best realistic pulse shape is a square function with some sort of smooth tapering at the start and finish. The tapering should be designed to minimize frequency side-lobes. However, Gaussian functions are easy to deal with mathematically.

In general a radar transmits a continuous sequence of such pulses, at regular intervals, called the inter-pulse period. The purpose of the radar receiver is to determine the amplitude $\tilde{A}$ (or at least a quantity proportional to it) and the total phase $\varphi_0 = 2k_0r_0 + 2k_0v_rt - \varphi'$. The phase can also vary within one pulse, but for most atmospheric radar applications, we consider that the phase within a single pulse is invariant.

The pulse-shaping unit would normally require timing pulses, which would also be used to drive the digitizer and other aspects of the radar that need to know about the time of pulse transmission. We have not shown these timing signals here.

The modified radar pulse (still at low levels) then moves to a power amplifier unit, where the signal is amplified to high power, just as we discussed in regard to Figure 4.6. The pulse then moves out from the transmit antenna to the targets. Some of the signal is scattered back, and received by the receive antenna. From this point on, the analysis is similar to that discussed in relation to Figure 4.6, except that the signal must be sampled separately at distinct "range gates" and stored separately for each different delay, as shown in Figure 4.3. However, Figure 4.3 discussed amplitude only, whereas for our Doppler radar, we need to record both in-phase and quadrature components.

As mentioned before, the phase is made up of system-dependent parameters and the desired range information embedded in the term $2r/c$. The recorded signal again looks like Figure 4.8(a), and multiple sets of such time series are recorded, each at different time delays (or ranges). These delays are user-specified, and it is common to use a time interval Δt between successive samples equal to about one pulse length. As mentioned earlier, the sampling points are called gates, and the delay is commonly expressed as a range Δr, where $\Delta r = c\Delta t/2$. For example, if a transmitter were to transmit a pulse with a half-power half-width of 2 μs, ($\frac{c\Delta t}{2} = 300$ meters), and we wanted to sample from 1.2 to 12 km range, it would be necessary to sample at temporal delays of 8 μs, 10 μs, 12 μs, etc., up to 80 μs. These would be referred to as 1.2 km, 1.5 km, 1.8 km, ..., 12 km range gates. Figure 4.11 shows some typical in-phase and quadrature time series, sampled at range gates of 84.30 km, 84.45 km, etc. up to 84.90 km. Note that the two oscillations are seen to have a horizontal offset of about 90° relative to each other, as expected according to Figure 4.7.

With faster digitizers, it is becoming more common to "oversample" the data, or in other words, to sample at intervals much less than the pulse length. This method has the advantage that the user can later apply deconvolution procedures to improve the range resolution of the system. *Röttger and Schmidt* (1979) were the first to do this, though they used a slow computer and had to adopt a type of staggered interleaved sampling and do the deconvolution later. More modern systems can do this in real time (e.g., *Hocking et al.*, 2014).

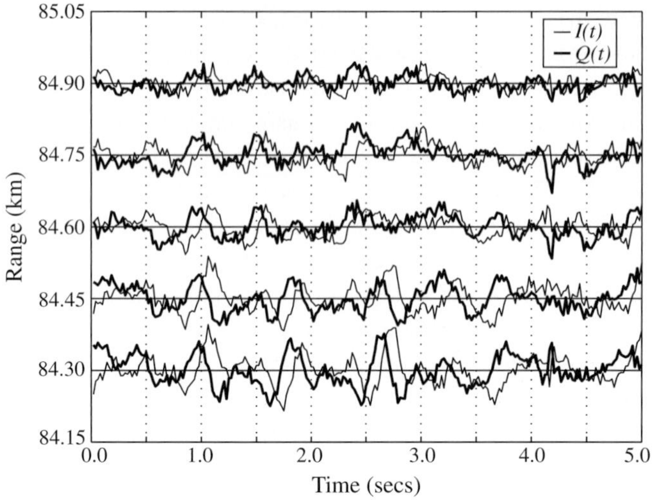

Figure 4.11 Example of time series ($I(t)$ and $Q(t)$) data from the EISCAT Observatory in Tromsø, Norway. Data are shown for a 5 s segment from five contiguous gates corresponding to the altitude range 84.3–84.9 km.

4.6.1 Backscatter as a convolution

In Chapter 3, Section 3.11.1, we qualitatively introduced the fact that the backscattered profile from the atmosphere is a convolution between the pulse and the scattering/reflecting profile. Here, we prove this statement.

Imagine a situation in which there are multiple scatterers in the atmosphere, at different altitudes. To keep things simple, we envisage the scatterers as horizontally aligned mirror-like reflecting planes, with different reflection coefficients, defined as $r(z)dz$. This is referred to as "Fresnel scatter." Consider also a pulse of some specified shape with a suitable carrier frequency, with amplitude of the form $p(t)$ specified by Equation (4.13). Let $t = 0$ be defined as the time at which some suitably designated part of the pulse (typically the time it achieves maximum amplitude, or the midpoint of the pulse: we will assume the value of the peak) is emitted from the antenna. We wish to look at the signal strength measured at the receiver at some later time t_*, after reflection from the layers in the atmosphere. A monostatic radar will be assumed.

Now consider the reflection of the pulse from a reflector at a height z_0. Different parts of the pulse will reflect from z_0 at different times, and we will consider that the peak of the pulse reflects from z_0 at such a time that the peak of the pulse arrives back at the receiver at our selected time t_*. Then, ignoring for now attenuation in signal-strength due to range effects, the amplitude received at time t_* will be

$$S_p(t_*) = r(z_0)p(0)dz, \tag{4.14}$$

where S_p refers to the signal due *only* to the peak of the pulse. For *other* parts of the pulse to arrive at the receiver at the same time t_*, they *cannot* be reflected from the

height z_0, since they had a delay or a lead in transmission time from the antenna relative to the peak. If the portion of interest of the pulse was transmitted *before* the peak, then it will have to travel to a slightly higher altitude than z_0 before reflection in order to reach the ground at t_*, in order to add some extra delay to compensate for its early start. Conversely, if the portion of interest was transmitted *after* the peak, the portion that arrives back at the receiver at time t_* will need to be reflected at a height lower than z_0 in order to "catch up" lost time. In general, we write

$$S_{t'}(t_*) = r(z)p(t')dz, \tag{4.15}$$

where $S_{t'}$ refers to the signal due *only* to the portion of the pulse transmitted at time t'. t' is negative if the associated part of the pulse was transmitted before the peak. *All* of these portions arrive at the receiver at time t_*. The quantities t_*, t', and z are related by

$$t' + \frac{2z}{c} = t_*, \tag{4.16}$$

i.e., in order to reach the receiver at time t_*, we need to sum the delay due to the travel to the reflecting element and back ($\frac{2z}{c}$), plus compensate for the delay or advance in transmission relative to the peak. Then Equation (4.15) can be written as

$$S_{t'}(t_*) = r(z)p(t_* - \frac{2z}{c})dz. \tag{4.17}$$

This equation is a mixture of spatial and time variables, so in order to simplify things we convert all items to distance. We do this by using $t_* = \frac{2z_0}{c}$, and we define new functions S_z and p_z by $S_z(z_0) = S_{t'}(t_*)$ and $p_z(z_0) = p(t_*)$. Thus we interpret each as a function of altitude, which is exactly what is done by experimentalists and observers – the delay seen on a cathode-ray oscilloscope is considered not as a time delay, but rather as a distance to the target assuming that the pulse travelled at speed c: it is often called the "virtual range." Then (4.17) becomes

$$S_z(z_0) = r(z)p_z(z_0 - z)dz. \tag{4.18}$$

All signals S_z from each height z arrive together at the receiver at the same time $t_* = \frac{2r_0}{c}$, so the total signal strength (expressed as an amplitude) is given by summing all terms, or

$$S_{\text{tot}}(z_0) = \int_{-\infty}^{\infty} r(z)p_z(z_0 - z)\text{d}z. \tag{4.19}$$

This is a *convolution* between the reflection coefficient profile r and the pulse described in spatial coordinates, $p_z(z)$. We often write

$$S_{\text{tot}}(z_0) = \int_{-\infty}^{\infty} r(z)p_z(z_0 - z)\text{d}z = r \otimes p_z. \tag{4.20}$$

The above discussion is missing one item, and that is the range effect. If the amplitude at a distance of 1 meter (ignoring near-field effects) is A_0, the amplitude at z meters is A_0/z, and for cases of reflection, the amplitude back at the ground is $A_0/(2z)$, so this effect needs to be integrated into the convolution. If the scattering is not due to mirror-like reflectors, but isotropic (as in turbulent scatter) the amplitude at z meters is still used. First, the received signal (with frequency f_R, which may be slightly different to the original transmitted frequency (f_0), due to Doppler shifting effects) is weakly amplified and then mixed with the local oscillator frequency f_L. This produces a frequency $f_L - f_R$, which will be similar to the IF frequency, but Doppler shifted with respect to it. At this stage, most of the amplification of the received signal takes place at a frequency well removed from both f_R and f_0. It is *this* signal which is then beaten down to near-DC by mixing it with the intermediate frequency f_I. Then the final signal is used to produce in-phase and quadrature components, just as in the non-superheterodyne case.

In reality, care must be taken to ensure that the in-phase component really does lead the quadrature component in phase. Whether this is true or not can depend on whether the local oscillator has a higher or lower frequency than the main radar frequency. In practice, the radar builder measures the outputs from the in-phase and quadrature components and selects the one with leading phase as the in-phase component.

In our example here, we have generated the carrier frequency by mixing signals from a local oscillator and an IF oscillator. This is not mandatory. It would be equally possible to generate the IF by mixing a local oscillator signal with another local oscillator signal which generates the RF directly. Another strategy that is often used is that of direct digital synthesis (DDS). In this process, a digital repetitive waveform is generated with a frequency much higher than the desired LO, IF, and RF frequencies. The digital signal is then sampled at suitable points along this reference waveform to produce the desired LO, IF, and RF. This employs a frequency-aliasing strategy to produce frequencies considerably smaller than the reference signal. The digital signal is then converted to analog form, and passed through some simple filters in order to remove spurious harmonics, giving ultimately three reference signals with frequences and phases which are very stable with respect to each other. If such a procedure is used, Figure 4.12 still applies but the frequencies f_L, f_I, and f_0 can be considered as three separate, but commonly derived, inputs at the top left-hand corner of the figure.

Figure 4.12 also shows some other important aspects of a radar that we have not yet discussed. First, it shows a computer, which in this case is used both to control the timing of the devices in the circuit, and also to record the received signal. Second, it shows the digitizers, which convert the received analog in-phase and quadrature components to digital signals. Third, it shows some trigger signals. These are required to make sure that each part of the radar switches synchronously. The trigger is produced from a master controller (in this case the computer), and designates when pulse shaping should take place. At the same time, it resets the digitizers to start sampling as the pulse is transmitted. The trigger signal also may control the transmit-receive switch, which will be discussed shortly.

As noted, some more modern radars do not convert to in-phase and quadrature components, but rather digitize directly at the IF level, or in some cases even digitize the RF directly. This requires very fast digitizers, but has various advantages in regard to system calibration and equalization, and halves the number of digitizers needed. This concept will be discussed further in Chapter 5.

4.6.3 Transmit-receive switches

There is one remaining device that we have not discussed in regard to Figure 4.12, and that is the transmit-receive (TR) switch (e.g., *Skolnik*, 2002). This unit is required whenever we wish to use the same antennas for reception as well as transmission. For narrow-beam MST radars, this arrangement is generally mandatory. The alternative would be to have separate transmitter and receiver arrays. Since the antenna array often comprises hundreds of antenna elements, and represents a major financial investment, to produce a second set of identical antenna elements would be a huge cost.

The name TR switch is actually something of a misnomer, since some types do not even involve physical switches at all. A more rigorous term is TR duplexer. Sometimes the term "circulator" is also used.

Figure 4.13(a) shows the idea behind a TR switch. The transmitter pulse passes through a switch to the antenna, and at this time, the switch has no contact with the receiver (broken arrow). Only after the transmitter pulse has left the switch is the contact point moved to the right, as shown by the broken arrow in Figure 4.13(a). Now the received signal may pass directly to the receiver. This arrangement prevents the transmitter pulse from flowing directly into the receiver, which would easily destroy the sensitive receiver. In a so-called "active" TR switch, the switching is controlled by the radar trigger; the system is designed to wait for a pre-specified length of time after the trigger pulse, and then allow the switching to the receiver to occur. The wait time should be long enough that the transmitted pulse has passed completely though the switch. Usually, fast-switching devices are required, like PIN diodes, because the user wants to start recording signals as soon as possible after the pulse has left. This is especially true if the radar's purpose is to record signals close to the ground, e.g. 500 m or 1 km in range. PIN diodes will be discussed in more detail in the next chapter.

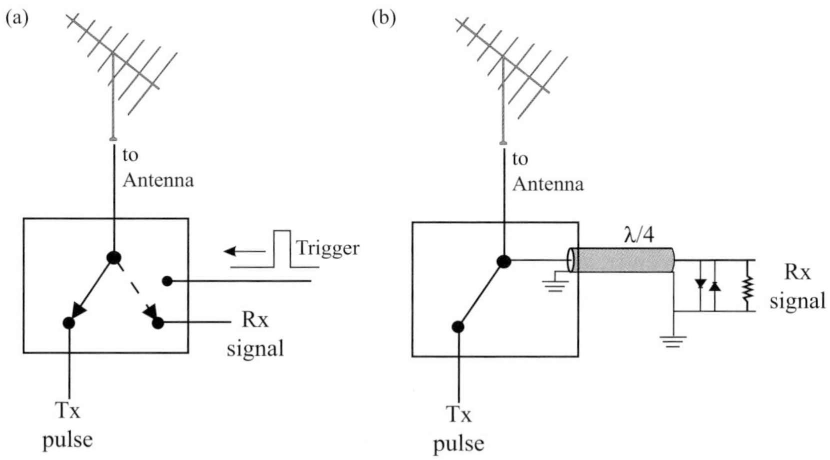

Figure 4.13 (a) The concept of a transmit-receive (TR) switch. (b) Design of a simple passive TR switch. The details will be discussed further in the next chapter.

of the pulse and time-lagged scatter. In this section, we address briefly the issue of how we combine these aspects. However, we will keep the discussion descriptive, and move to a more thorough discussion in the next chapter.

The returned signal depends in part on the radar transmitter power, the polar diagram of the radar, the distance to the radar, and the shape of the scatterer. Four of the key parameters measured with a radar are: (i) absolute signal power, (ii) noise power, (iii) spectral offset, and (iv) spectral width. These relate to: (i) reflectivity, or scatter cross-section, (ii) noise limitations as well as absolute calibration, (iii) radial velocity, and (iv) radial velocity variability (Doppler spread).

The term "reflectivity" strictly only applies to "mirror-like" reflectors, which are stretched out horizontally and involve relatively sharp steps in refractive index in the vertical direction. The reflection coefficient is defined as the ratio of the reflected amplitude to the incident amplitude at the plane of reflection. More isotropic scatterers are described in terms of their backscatter cross-section (see Chapter 3, Equations (3.255) to (3.258)). The backscatter cross-section σ_s is equal to the power backscattered per unit steradian per unit volume per incident power level. In some cases, an alternative definition is used which considers not the power scattered per unit steradian, but the power scattered if the radiation is assumed to scatter isotropically into a sphere. It is denoted η_s and equals 4π times σ_s (see Equations (3.255) to (3.258)). The scattering cross-section can be related to atmospheric characteristics, such as the intensity of fluctuations in temperature and humidity and the strength of turbulence (e.g., see *Hocking and Mu*, 1997). The Doppler shift, or radial velocity, provides the projection of the three-dimensional motion of the atmosphere onto the pointing direction of the radar. Combined with radial velocity estimates from other directions, the three-dimensional wind field can be estimated. Doppler spread, or *spectral width*, is a measure of the power-weighted distribution of radial velocities within the resolution volume of the radar. This distribution is directly affected by beam width, turbulence, and wind-shear (among others), and can therefore be used to study atmospheric velocity variability.

It should be emphasized that new methods are continually being developed to estimate additional atmospheric parameters. For example, methods based on multiple receivers and frequencies are providing insight into many atmospheric phenomena from the planetary boundary layer to the thermosphere (e.g., *Woodman*, 1991; *Kudeki and Surucu*, 1991; *Palmer et al.*, 1999; *Luce et al.*, 2001a)). It is one role of the MST radar scientist to develop new techniques that make use of the fundamental parameters provided by the radar.

We have already developed a physical picture of the signal received by a radar. We have seen that it is weighted in angle by the polar diagram of the radar, and that the received pulse suffers a temporal retardation due to its distance of travel to the target and back. When we combine this information, we can say that the electric field component along the polarization direction of the receive antenna for a continuous-wave EM field from a scatterer at range r can be found by adding polar-diagram terms to Equation (4.7), giving the following equation (e.g., among others, *Doviak and Zrnić* 1993, plus see Chapter 3):

$$E_T(r,\theta,\phi,t) \propto \frac{A(\theta,\phi)}{r} e^{i\{2\pi f_0(t-\frac{2r}{c})+\varphi''\}}, \tag{4.27}$$

where

θ	angular distance from bore-sight of radar
ϕ	azimuthal angle from the x-axis (generally, but not always, east)
$A(\theta,\phi)$	position dependent amplitude due to polar diagram and scatterer characteristics
r	range, which also relates to the pulse time delay
f_0	transmitter frequency
φ''	unknown, constant transmitter initial phase
c	speed of light (3×10^8 ms^{-1})

with of course the speed of light being related to the frequency by $c = \lambda_0 f_0$, where λ_0 is the EM wavelength.

The amplitude $A(\theta,\phi)$ is a function of the antenna patterns used for transmission and reception, and also depends on the scattering strength and the shape of the scatterers. It is essential to note that the EM wave is time-delayed by twice the distance away from the source divided by the propagation speed, so that the delay is $\Delta t = 2r/c$. For a moving target, we can add velocity terms as in Equation (4.7), and for a pulsed radar we need to recall that the backscattered signal is a convolution between the radar pulse and the range-dependent backscatter cross-section (or the reflectivity in the case of reflectors).

For our purposes, the polarization of the EM wave has been largely ignored. It should be emphasized, however, that the polarization will affect how the wave interacts with the scattering object. For example, dual-polarization methods have been well developed in the weather radar community, where significant advantages have been demonstrated for observations of precipitation (*Bringi et al.*, 2002). Further extensions of these concepts will be discussed in future chapters, but for now we recognize that we have been able to combine the concepts of polar diagram and the radar equations. We now turn to aspects relating to data optimization.

Equation (4.27) is only a proportionality. In the next chapter, we will develop this equation further and produce a quantitatively useful expression for backscattered power in terms of the important atmospheric and radar parameters.

4.8 Optimizing the signal

One of the principal requirements of good radar design is the need to optimize the signal. Optimization may, however, mean different things to different users. In one case, optimization may mean optimizing the signal relative to the noise, and in another case it might mean obtaining the shortest pulse length possible in order to produce optimum resolution. In this section, we will consider some of these types of optimization.

4.8.1 Matched filter

Here, we will first consider the simple case of a square transmitter pulse (Figure 4.10). The transmitted pulse is scattered from the targets, and then passes through various stages of reception. The final stage in the Doppler radar shown in Figure 4.12 is an LPF (low pass filter), which is used to eliminate the higher frequency signal present at the output of the final mixer (i.e., frequencies at and above f_I). Any finite bandwidth LPF will alter the shape of the input signal. An obvious question is: how should we choose the response of this LPF to optimize the performance of the radar system? As briefly mentioned earlier, noise (cosmic and electronic) is always present in the radar signals. We will now discuss a means by which the effects of this noise can be minimized by proper design of the LPF.

Let the signal present at the input to the LPF be given by

$$\underline{s}_i(t) = \underline{S}_i(t) + \underline{n}_i(t). \tag{4.28}$$

Although frequencies f_I and above are removed *after* passage of the signal though the LPF, we assume for convenience that they were removed prior to reaching the LPF, so $\underline{S}_i$ does not contain these higher frequencies. The term $n_i(t)$ is white noise resulting from receiver electronics and cosmic sources, where "white noise" refers to noise which has little to no variation in intensity as a function of frequency over the band of interest.

Of course, the output of this filter is given by the convolution of the input signal with its impulse response denoted by $\underline{h}(t)$:

$$\underline{s}_o(t) = \underline{s}_i(t) \otimes h(t). \tag{4.29}$$

Assuming the input white noise $\underline{n}_i(t)$ has a power spectral density of $N_0/2$, the signal-to-noise ratio (SNR) at the output of the filter can be shown to have the following form:

$$\mathrm{SNR}(t) = \frac{\left|\int_{-\infty}^{\infty} \underline{S}_i(t-\gamma)\underline{h}(\gamma)\,\mathrm{d}\gamma\right|^2}{\frac{N_0}{2}\int_{-\infty}^{\infty} \left|\underline{h}(\gamma)\right|^2\,\mathrm{d}\gamma}. \tag{4.30}$$

The numerator can be rewritten using the Schwartz inequality,

$$\left|\int_{-\infty}^{\infty} \underline{S}_i(t-\gamma)\underline{h}(\gamma)\,\mathrm{d}\gamma\right|^2 \le \int_{-\infty}^{\infty} \left|\underline{S}_i(t-\gamma)\right|^2\,\mathrm{d}\gamma \int_{-\infty}^{\infty} \left|\underline{h}(\gamma)\right|^2\,\mathrm{d}\gamma.$$

The maximum instantaneous SNR is obtained when the equality holds in the above equation, which occurs when $\underline{h}(\gamma) = \underline{S}_i^*(t-\gamma)$. In other words, the optimal receiver LPF, in terms of output SNR, has an impulse response which is *matched* to the input signal, viz.

$$\underline{h}(t) = \underline{S}_i^*(-t). \tag{4.31}$$

It is important to note that any filter causes additional delays in the received signal, a point we have not discussed here, but which should always be borne in mind.

It is interesting to note that the receiver filter should be matched to the input signal $\underline{S}_i(t)$. In an experimental scenario, however, the input signal is not known a priori since it results from atmospheric scatter. As a first approximation it is common to assume that the received signal has the same shape as the transmitted pulse (albeit with reduced amplitude), but when atmospheric scatter effects are considered, it has been demonstrated that moderately better SNR performance may be obtained by slight modifications to the receiver filter impulse response (*Johnston et al.*, 2002).

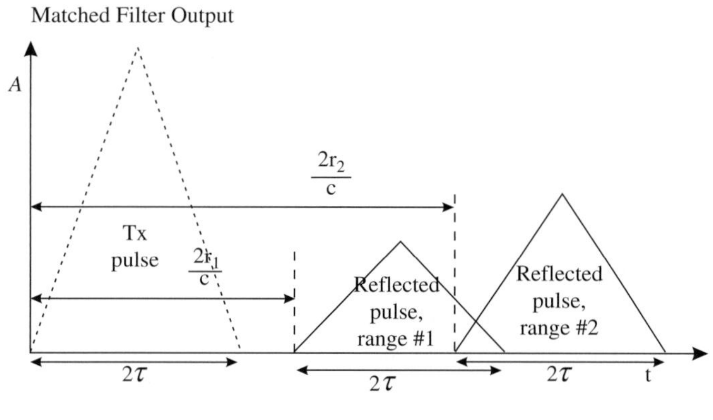

Figure 4.14 Output of matched filter for two point scatterers and a pulse-modulated transmission, where the transmitter pulse has been assumed to be a square function and the low-pass filter has been assumed to have a square impulse response. The resultant pulse is the self-convolution of the transmitted square pulse. Additional delays may also occur in realistic filters, but have been left out.

As mentioned previously, the signals shown in Figure 4.10 were obtained by largely ignoring the effects of the LPF. With the matched filter concept, we can now more precisely describe the output of the receiver. We assume in this case that the filter has a boxcar impulse response to match the transmitted pulse. Of course, such a filter cannot be achieved in practice, nor would we want to produce such a filter, since it would follow a $[\sin(f)/f]^2$ variation as a function of frequency and would pass frequencies all the way out to infinity! A more reasonable filter would have low and high pass cutoffs.

After detection, the signal $\underline{s}_i(t)$ is passed through the matched filter. Given the impulse response in Equation (4.31), the output of the matched filter will have a triangular shape, as illustrated in Figure 4.14. Note that its base is now twice as long as the boxcar function shown in Figure 4.10. The effects of noise have been ignored for this new figure, although they have been minimized by the matched filter.

After matched filtering, the resolution is still dictated by Equation (4.12), since we do not require complete separation between the signals in order for them to be resolved. Assuming the scatterers have equal returned power, Δr will determine the range at which the output of the matched filter has decreased by 3 dB in power. It is this amount of power separation which is customarily assumed to be needed to resolve two targets.

4.8.2 Filters and resolution

The calculation in the previous section showed how we might optimize the signal-to-noise ratio. However, we need to be careful here – if the filter is too narrow, the resultant pulse width can be wider, and the system resolution worsened. In fact, if a Gaussian pulse is used, and the above criteria are followed, the pulse lengthens by 25%. For some experiments, this might be unacceptable, and it might be desirable to use a wider filter.

To see that this is true, we will develop an expression for the pulse width as a function of filter width. In this case, we will not use a square pulse, but rather a Gaussian one, and similarly we will assume that the final stage low-pass filter also has a Gaussian form.

First, we define the transmitted electric field of the pulse as

$$\mathcal{E}(t) = A\mathrm{e}^{-t^2/\tau^2}. \tag{4.32}$$

We have ignored the carrier portion of the signal (i.e., the $\mathrm{e}^{\mathrm{i}(\omega t+\varphi'')}$ terms), though they should be considered to be present. We will concentrate on the simplest case of a purely real pulse amplitude.

If we let τ_H be the half-power full-width, then we can write this as

$$\mathcal{E}(t) = A\mathrm{e}^{-2(\ln 2)t^2/\tau_H^2}. \tag{4.33}$$

We shall also consider the receiver filter to have a Gaussian profile, and define it as

$$\mathcal{F}_R(f) = \mathcal{F}_{R0}\mathrm{e}^{4\ln 2 f^2/f_F^2}, \tag{4.34}$$

where f_F is the 3 dB full width. $\mathcal{F}_R$ specifies the *power density* as a function of frequency.

The system response can best be evaluated by determining the signal received from a single stationary point target or a single reflector. The final pulse received after the pulse has been transmitted, scattered, and passed through the filter has a Fourier transform specified by the product of the amplitude-response of the receiver and the Fourier transform of the pulse. The Fourier transform of the pulse is given by

$$A_f(f) = A_{0f}\mathrm{e}^{-\pi^2\tau_H^2 f^2/(2\ln 2)}, \tag{4.35}$$

where A_{0f} is equal to $\beta A\tau_H$ where β is a constant, the exact value of which is unimportant for subsequent calculations.

The Fourier transform of the final pulse is then given by

$$F(f) = A_f(f) \times \sqrt{\mathcal{F}_R(f)} = \exp\left\{-f^2\left[\frac{\pi^2\tau_H^2}{2\ln 2} + \frac{2\ln 2}{f_F^2}\right]\right\}, \tag{4.36}$$

which we will write as

$$F(f) = F_0\mathrm{e}^{-f^2/b^2}, \tag{4.37}$$

where

$$\frac{1}{b^2} = \frac{\pi^2\tau_H^2}{2\ln 2} + \frac{2\ln 2}{f_F^2}. \tag{4.38}$$

The effective received pulse, after passing through the receiver, is just the Fourier transform of this, or

$$E_{\mathrm{eff}}(t) = E_{\mathrm{eff}0}\mathrm{e}^{-\pi^2 b^2 t^2}. \tag{4.39}$$

We may write this as

$$E_{\mathrm{eff}}(t) = E_{\mathrm{eff}0}\mathrm{e}^{2\ln 2\, t^2/(\tau_{H(\mathrm{eff})}^2)}. \tag{4.40}$$

Comparing (4.39) and (4.40) shows us that

$$\tau_{(H_{\text{eff}})} = \sqrt{\tau_H^2 + \frac{4(\ln 2)^2}{\pi^2 f_F^2}}, \tag{4.41}$$

or

$$\tau_{(H_{\text{eff}})} = \sqrt{\tau_H^2 + \frac{1}{5.136 f_F^2}}. \tag{4.42}$$

This is an important expression, which we will utilize shortly. Before we do, we need to look at the signal-to-noise ratio.

Simultaneous optimization of signal-to-noise and pulse length

The total received power is of course just $\int P(f)\mathcal{F}_R(f)\mathrm{d}f$, where $P(f)$ is the square of A_f, and $\mathcal{F}_R(f)$ is already defined by (4.34). The received noise will be proportional to $\int \mathcal{F}_R(f)\mathrm{d}f$. We will use the general result that $\int_{-\infty}^{\infty} \mathrm{e}^{-x^2/(2\sigma^2)}\mathrm{d}x = \sqrt{2\pi}\sigma$ in the following calculations. Hence the signal-to-noise ratio is given by

$$S_N \propto \int_{-\infty}^{\infty} A_{0(\text{eff})}^2 \tau^2 \mathrm{e}^{-(\pi^2\tau_H^2 f^2)/\ln 2} \times \mathrm{e}^{-(4\ln 2 f^2)/f_F^2}\mathrm{d}f / [\mathcal{F}_{R0}\sqrt{2\pi} f_F/(\sqrt{8\ln 2})]. \tag{4.43}$$

The upper integral may readily be evaluated to give

$$S_N \propto \frac{\tau_H}{f_F} \frac{1}{\sqrt{1 + \frac{1}{5.136 f_F^2 \tau_H^2}}}. \tag{4.44}$$

The optimum combination for τ_H and f_F can be evaluated by holding τ_H as a constant and differentiating with respect to f_F, and finding the point where the derivative is zero. In our case, this can easily be shown to satisfy

$$f_F \simeq 0.624 \frac{1}{\tau_H}. \tag{4.45}$$

We now need to look back at the effect of the pulse length. Substitution of this choice for f_F into (4.42) shows us that

$$\tau_{H(\text{eff})} \simeq 1.225\tau_H. \tag{4.46}$$

Hence we see that our "optimum" choice of signal-to-noise ratio gives a somewhat non-optimum effect on the pulse length, with a 22.5% increase in pulse length. Of course, the effect may vary depending on our choice of pulse shape and filter shape, but the concept remains – the user often has to make a choice in regard to optimizing various aspects of the radar. A common choice is to use $f_F \simeq 1.0/\tau_H$.

4.8.3 Pulse compression

As shown in Chapter 3, the radar equation relates the transmitted and returned powers via numerous parameters. In general, we would like to increase the returned power, since this results in better performance of the radar. It was observed that

pulse with itself. Another way to think of this is to consider the correlator output which would result if the pulse-train were reflected from a perfect reflecting sheet somewhere above the radar and returned to the correlator. Then the returned signal will be an exact duplicate of the transmitted pulse, and so we expect the output of an ideal correlator to be a narrow spike at zero lag, indicating a reflecting region of near-zero depth.

The correlator self-output (also loosely referred to as the ACF) is shown in the bottom panel of Figure 4.15. For a Barker code, the side-lobes alternate between zero and unity and the peak of the ACF has a value equal to the code length. Since the code is a function of time, the side-lobes in the ACF correspond to *range* side-lobes. In other words, if a sample was taken at the output of the matched filter at a range corresponding to the peak of the ACF, the signal would also be contaminated by the side-lobes of scatter from other ranges. A solution to this problem is provided by the use of *complementary codes* (e.g., *Golay*, 1961; *Schmidt et al.*, 1979; *Farley*, 1985), which will now be described.

Given the long coherence time of typical MST radar signals, it is possible to employ a sequence of codes transmitted on a pulse-to-pulse basis. By transmitting different codes and summing, it is theoretically possible to eliminate range side-lobes and any residual DC component in the signal. Referring to Figure 4.15, it is obvious that if we could devise a code that had range side-lobe levels which were opposite in sign, the sum of the two decoded signals would result in zero range side-lobes. Complementary codes have just such a property and are often used in MST radar applications. The simplest 2-bit complementary code set can be constructed as $++$ and $+-$, where $+$ and $-$ correspond to 0 and π phase shifts respectively. By calculating the ACF of these two codes, it can be seen that the range side-lobes have opposite signs. It is also interesting to note the relationship between the codes. The first half of the codes (one bit in this case) is the same, while the latter half is opposite. A 4-bit code can be constructed by a concatenation of the two 2-bit codes, resulting in $+++-$. The complementary code is obtained by retaining the first half of the code and complementing the latter half $++-+$. This process can be repeated for any length code that is a power of 2. Typically, the complementary codes are referred to as codes A and B. An example of a 16-bit complementary code set is provided in Figure 4.16 along with their ACFs and summed ACF. As expected, the range side-lobes are canceled by summing the two ACFs.

Given the expected coherence time, signals from several consecutive pulses are often summed, i.e., integrated. This process is called *coherent integration* and results in increased SNR. By alternating complementary codes and coherent integration, it is possible to eliminate range side-lobes as a by-product of the integration procedure. Coherent integration will be discussed in more depth later in this chapter.

A residual DC signal in the receiver can cause severe quantization errors during the digitization process. These can easily be eliminated by simply *flipping* (complementing) the code from pulse to pulse. For example, a complementary code set could be used, with flipped codes inserted on alternating pulses, such as $A\bar{A}B\bar{B}$, where the bar operator represents the complement. With such a process, any DC signal present would have an opposite sign from pulse to pulse and would thus be canceled during coherent integration. Since four codes are used in the complete code set, the number of coherent integrations should be set to an integer multiple of four.

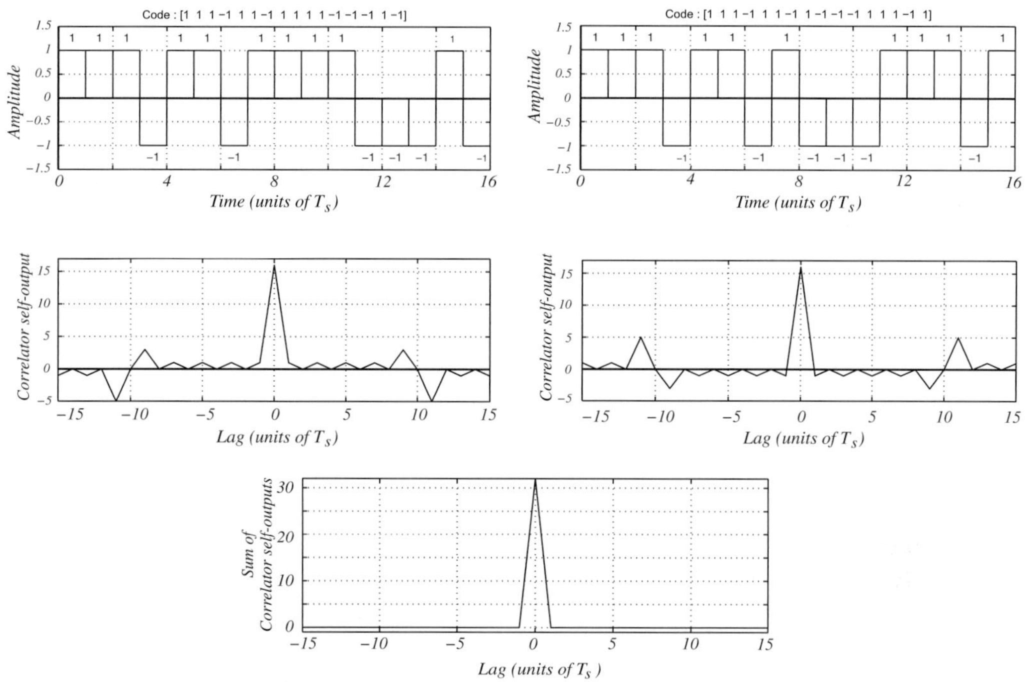

Figure 4.16 Similar to previous figure, but for both codes comprising a complementary set. The two codes *A* and *B* are shown in the upper row, and are designed to have opposite range side-lobes, as shown with the correlator self-output functions in the second row. Thus, the addition of the two partially decoded signals results in zero range side-lobes, as seen in the third row of the figure.

An actual experimental example of the complementary code process is shown in Figure 4.17. The data were obtained with the EISCAT VHF radar in Tromsø, Norway, from the mesopause region. Polar mesosphere summer echoes (PMSE) are extremely strong echoes observed in this region and are often confined in narrow layers of high reflectivity. Therefore, PMSE provide a good example of the coding process. The left-most panel of the figure provides the raw returned signal power as a function of altitude. A 64-bit coded pulse was transmitted with a 300 m subpulse resulting in a total pulse length of 19.2 km. A thin PMSE layer exists at approximately 85.5 km. Notice how the wide-coded pulse smears the thin PMSE layer over the width of the code. After decoding with only code *A* of the complementary code set, the power profile provided in the center panel of Figure 4.17 results. As expected, the decoded, or compressed, signal possesses significantly higher power. Since we have decoded with code *A* only, however, range side-lobes appear and have been marked with arrows in the figure. After full decoding with codes *A* and *B* and coherent integration, the power profile in the right panel is generated. The PMSE layer is now obviously observed at 85.5 km and no range side-lobe effects are present. One might notice a slight drop in noise level power at altitudes higher than 100 km. This is due to the upper limit on the sampled range gates. Normally, only fully decoded gates are presented as valid data.

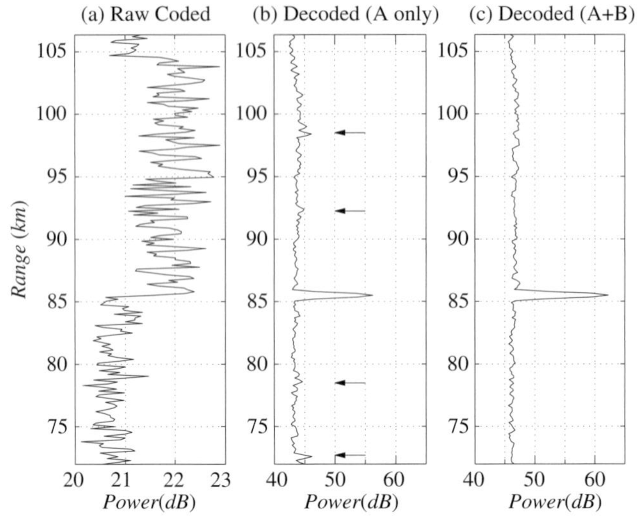

Figure 4.17 Example of complementary coded PMSE data from the EISCAT Observatory in Tromsø, Norway. A code length of 64 bits was used with a subpulse width of 300 m. The power profile of the raw coded signal is presented in the left panel. Partial decoding using only code *A* results in the power profile shown in the center panel. Note the range side-lobes indicated with arrows. The fully decoded power profile is provided in the right panel where a strong PMSE layer is observed at an altitude of 85.5 km.

For observations of the atmospheric boundary layer (ABL), it is desirable to have range gates as close to the surface as possible. Two problems limit the minimum range gate altitude. First, the far-field of the radar is determined by the aperture of the antenna and radar wavelength. Large antennas possess a more distant near/far-field transition. As a result, boundary layer radars (BLR) are typically designed with a shorter wavelength and a smaller antenna aperture. The second factor which affects the minimum range gate is the overall length of the coded pulse. Decoding can be thought of as a matched filtering process. By filtering the signals from the sampled range gates, it is possible to produce a decoded, or compressed, signal. The first fully decoded range gate, however, is not produced until after the filter has passed completely into the sampled gates. This does not occur until the range gate corresponds to the length of the code. Therefore, many of the first range gates have to be ignored since they were not fully decoded. Theoretical work in the generation and implementation of complementary codes subsequently resulted in an important breakthrough (*Spano and Ghebrebrhan*, 1996). Through the use of so-called *truncated* codes, it is possible to partially decode these initial range gates, allowing boundary layer (ABL) measurements at lower altitudes than were previously possible. It is expected that most future MST radar systems will exploit these codes.

More sophisticated codes have also been developed in recent years, and these are dependent on the faster computing power of modern computers. These codes can be very long, and often are pseudo-random in structure (e.g., *Rastogi and Sobolewski*, 1990;

Huang and MacDougall, 2005). Other more advanced procedures exist which utilize inversion techniques (e.g., *Lehtinen et al.* (1996)). We will not discuss these at this time.

4.9 Doppler radial velocity and coherent integration

We saw in Equation (4.9) that the received signal is represented by "in-phase" and "quadrature" components, which we denoted as $I(t)$ and $Q(t)$. That equation considered only a CW signal, but if we include the effect of a transmitted pulse, then the total received signal is given by

$$\underline{s}(t) = I(t) + \mathrm{i}Q(t) = \tilde{A}\mathrm{e}^{\mathrm{i}(\omega_r t - \varphi_0)} p\left(t - \frac{2r_0}{c}\right), \tag{4.47}$$

where $\omega_r = -2k_0 v_r$, v_r being the radial velocity, and where radial velocities away from the transmitter are defined as positive. The phase term depends on $\frac{2r_0}{c}$ as well as various instrumental effects and delays. The signal $\underline{s}(t)$ can be thought of as a phasor which rotates as a function of time, as shown in Figure 4.7. An example of actual experimental data was shown in Figure 4.11 for a 5 s record. The $I(t)$ and $Q(t)$ data were obtained with the EISCAT VHF radar in Norway from the mesopause region. It was also shown in Figure 4.8(b) that the spectrum produced by Fourier transforming the time series has significant structure if there are multiple scatterers in the radar beam. We now wish to turn to some preliminary examples of how to extract dynamical atmospheric data in a practical sense.

4.9.1 Radial velocity

Arguably the most important MST application occurs in the determination of Doppler velocity, since supply of wind information is one of the primary applications of MST and wind profiler radars. This information is contained in the returned radar signal, and it is the frequency content (rate of change of phase) of $\underline{s}(t)$ that provides the Doppler velocity estimate. For a pulsed radar, this manifests itself in two ways.

First, there is a small change in phase from the start to the finish of the scattered (or reflected) pulse, since the scatterer will move slightly during the time that the pulse is in contact with it. Nevertheless, for the majority of MST applications, the phase of $\underline{s}(t)$ does not change significantly over the pulse duration. For a 6 m wavelength MST radar, a scatterer with a 40 ms^{-1} radial velocity would produce less than 10^{-4} rad phase change over the duration of a typical pulse (1 μs). Such a small phase change would be difficult to detect. In cases of much larger velocities, such as the entrance speed of a meteor, phase change across the pulse length can be usefully employed (e.g., *Sato et al.*, 2000), but this is the exception rather than the norm.

A more realistic solution is to monitor the changes of phase from pulse to pulse for the case that multiple pulses are transmitted, and track the phase change over these pulses. This idea has already been conceptually addressed, but we now wish to be more qualitative. We will denote the inter-pulse period (IPP) (also called the pulse repetition time (PRT)) by T_s.

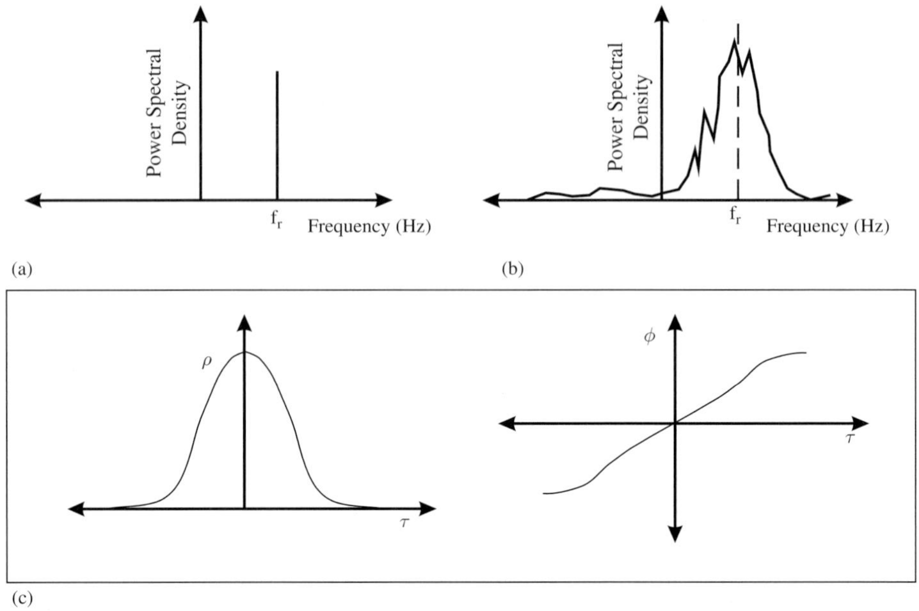

Figure 4.18 Examples of spectra for: (a) a single scatterer; and (b) a more realistic situation involving multiple scatterers. In part (c), we show a schematic of the autocovariance function. The graph on the left shows the magnitude, and the graph on the right shows the phase (which may vary between $-\pi$ and π).

In order to be able to use the phase changes between successive pulses as a method to determine radial velocity, it is first necessary that the scatterer move by much less than a half wavelength in the time interval T_s. If this condition is not satisfied, the speed is said to be "aliased." This special case will be discussed later in this chapter. For now, we will assume this condition is satisfied.

If the received signal is produced by a single scatterer, the in-phase and quadrature components will both be sinusoidal oscillations, with a 90° phase offset between each other. The corresponding power spectrum will have a single peak with a frequency offset of $-2v_r/\lambda$, where v_r has been taken to be positive when moving away from the receiver. Such a spectrum is shown in Figure 4.18(a).

In a more realistic situation, there will generally be multiple scatterers in the radar beam, or perhaps even a continuum of scatterers. In this case, the spectrum will be more complex. Such a spectrum has been drawn schematically in Figure 4.18(b). There will be a range of spectral offsets because the scatterers will each be moving at different speeds, depending on their relative location in the beam, the wind speed in their direct vicinity, the existence of turbulence and the existence of wind-shear. The spectrum will have a width, and a mean offset from zero Hz. The area under the spectrum is proportional to the total backscattered power, and the mean offset of the spectrum is a measure of the mean radial velocity measured in the radar volume. "Radar volume" refers to a region of space defined by the beam width and the pulse length at the height of scatter. In order to determine the wind speeds, it is first necessary to determine the radial velocity. The horizontal wind can then be determined through the relation

$$v_H = v_r / \sin(\theta), \tag{4.48}$$

where v_H is the horizontal wind-speed component in the vertical plane in which the radar beam is tilted, θ is the tilt of the beam from vertical, and v_r is the radial velocity, which equals $-\frac{\lambda}{2} f_r$. We have made several assumptions here, which will be scrutinized more carefully in a later chapter. We have first assumed that the vertical velocity component is zero, and we have assumed that the scatterers are all isotropic.

The spectrum must be recorded at each gate (range) at which data are recorded. It is often required that this determination be done in real time, since storage of all the point-by-point data recorded can quickly consume large quantities of storage space. An example of a series of "stacked" spectra from one of the earlier papers in the field is shown in Figure 4.19.

In order to determine v_r, we need a method to estimate f_r. Several procedures exist to do this, some of which we will describe shortly. However, before doing so, we make one more comment. Sometimes the spectrum is not well defined and is hidden in large levels of noise. To improve this situation, various averaging procedures exist. One common process is coherent integration, and we will describe this procedure shortly. A second is incoherent spectral summation. In this process, several successive data sets are acquired with the radar. Each might typically be 10 or 15 seconds in length. Then the spectra for each of these separate time series are formed and then averaged together. This produces a much cleaner spectrum with higher detectability. It should be emphasized that incoherent integration does not increase the SNR, but simply reduces the variance of the spectral estimate and therefore improves subsequent processing. Not all radars employ this process, but it is not uncommon. We now return to determination of the offset f_r in Figure 4.18(b).

One of the earliest (and still commonly used) techniques is to use *weighted moments* to calculate spectral offsets and spectral widths. For example, the mean offset is found as

$$\bar{f} = \int_{f_{\min}}^{f_{\max}} P_N(f) f \, df, \tag{4.49}$$

where P_N is the power spectral density function normalized to have unit area. Determination of the RMS spectral width σ_f is often found through the relation

$$\sigma_f^2 = \int_{f_{\min}}^{f_{\max}} P_N(f) (f - \bar{f})^2 \, df. \tag{4.50}$$

Application of these formulas requires an assumption that the noise level is negligible relative to the signal, or else serious biases can result. A more detailed discussion of these formulas occurs in later chapters, especially Chapter 7.

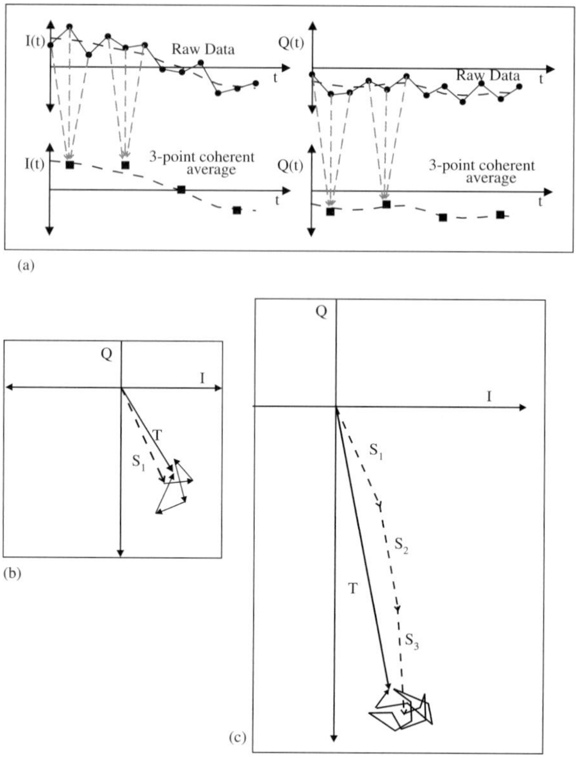

Figure 4.20 (a) Demonstration of the process of coherent integration. (b) Vector sum of a signal component and noise. (c) The coherent integration process seen from the perspective of a phasor diagram.

of Figures 4.20(a). The variance about the broken line has decreased, indicating an improvement in SNR. There are now also one third as many points. This procedure is termed coherent integration. The word "integration" is used because in order to save computation time in earlier computers, the three successive values were simply summed, and no division by the number of points was undertaken. The sum is obviously proportional to the average, so the relative fluctuations are the same in each case. It should be noted that in reality a coherent integration over three points would be unusual, and many radars do integrations over numbers of points that are powers of 2. However, we have used three for convenience in drawing the diagram, and the concept is the same regardless of the number of points.

Figures 4.20(b) and 4.20(c) show a different perspective to the averaging process. In this case, we have produced the plots as phasor diagrams. Figure 4.20(b) shows the signal, S_1, at one selected time, drawn as a straight broken vector. On the tip of this vector, we have added five successive smaller arrows, randomly oriented. These represent noise. Finally, the resultant vector is shown as the vector T (for total). The I and Q components can be extracted and plotted as in Figure 4.20(a) if desired. Each successive vector in the time series would be different – the sequence shown in Figure 4.20(a) would suggest a slow rotation of the signal vector in a clockwise sense, and each new point would have a different combination of noise vectors. A detailed treatment of the amplitude distributions can be found in various texts (e.g., *Rayleigh*, 1894; *Rice*, 1944, 1945; *Whalen*, 1971; *Hocking*, 1987b), but we do not need this level of detail at this time.

Figure 4.20(c) shows the effect of a coherent integration. The vectors S_1, S_2, and S_3 represent successive signal vectors, and it will be noted that they have rotated slightly with time. Each of these signal vectors would have had some noise vectors associated with them. However, for convenience, we have added all of these *after* we have summed the signal vectors, since order of addition is unimportant. There are now 15 noise vectors, although we have not placed arrows on their tips simply in order to make visualization easier.

The noise vectors constitute a two-dimensional random walk (*Rayleigh*, 1894), and the mean square displacement from the center position is proportional to the number of vectors. Thus the magnitude of the mean displacement is proportional to the square root of the number of vectors. Hence the vector $\vec{T}$ will be offset from the tip of $(\vec{S}_1 + \vec{S}_2 + \vec{S}_3)$ by about $\sqrt{3}$ times the offset between $\vec{T}$ and $\vec{S}_1$ in Figure 4.20(b). Since the signal has increased in length by 3 times, the signal-to-noise ratio has increased by $3/\sqrt{3} = \sqrt{3}$ times. Thus the vector $\vec{T}$ more closely follows the true signal in the second case.

The process of coherent integration certainly is an important one in VHF studies. Coherent integration of thousands of points is not unusual. For example, if the PRF is 4096 Hz, coherent integration over 512 points would be quite common, giving a new temporal resolution of 0.125 seconds, but an SNR improvement of about 23 times in amplitude, and 512 times in power (26 dB). Indeed, coherent integration was one of the prime reasons that VHF radars were able to be developed in the first place, because without it, the signal was often buried too deeply in noise to be useful.

If using coherent integration, care is needed with regard to sampling rates. An important process can occur called frequency aliasing. This refers to the undesirable situation, in any sampling process, when the sampling rate is slower than twice the highest frequency embedded in the signal. It is even more of a serious issue if the sampling is slower than the frequency of most interest. This can happen in situations of high wind speeds with corresponding large radial velocities. Velocity aliasing causes severe distortion in the frequency content of the sampled radar signal, and generally should be avoided. It can even cause spectral lines to appear in completely the wrong position. Specifically, if the sampling period is δt_s, then any frequency greater than $\frac{1}{2\delta t_s}$ will be improperly sampled. (In some special cases aliasing can be turned to advantage (e.g., *Hocking et al.*, 2014), but the user must understand the aliasing process very clearly in order to do this.) Aliasing can also occur if the data are sampled fast enough, but excessive coherent integration is applied. In general aliasing should be avoided.

Recent developments have allowed us to move even beyond coherent integration, and simultaneously reduce the impact of aliasing, as will be discussed in the next section.

4.9.3 An alternative to coherent integration

With the rapid increase in computational power of digital signal processing (DSP) integrated circuits and the vast improvement in high-speed memory, it has now become

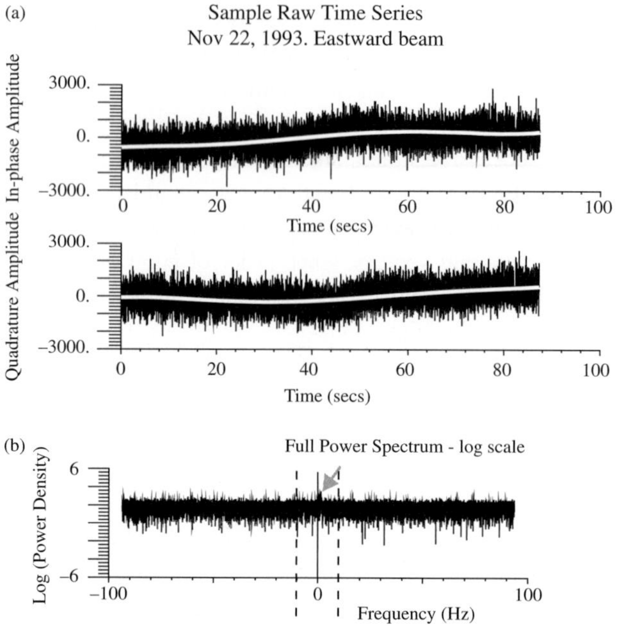

Figure 4.21 (a) Typical in-phase and quadrature time series detected with a VHF radar without coherent integration, and using an effective PRF of 190 Hz. The solid white line shows a 7th-order polynomial fit. (b) Power spectrum of the time series shown in (a).

realistic to sample, store, and process much more data than previously possible. As a result, recent MST radar systems have been designed with fast sampling rates and a minimized number of coherent integrations (e.g., *Hocking*, 1997a). By doing so, digital filtering methods can be employed to remove various types of interference, such as echoes from aircraft and interference from nearby radio transmitters.

Figure 4.21 shows a typical time series recorded with limited coherent integration (*Hocking*, 1997a). In this case, we can consider that the transmitter pulse was (effectively) transmitted at 190 Hz PRF, and so the aliasing frequency is at 95 Hz. (The actual PRF was 1520 Hz, with an 8-point coherent integration, but for our purposes in the ensuing discussions, we can consider this as a 190 Hz PRF system with no coherent integration. The important point is that we retain an aliasing frequency of 95 Hz, whereas more traditional processes leave an aliasing frequency closer to 5 or 10 Hz.) The white line through the center shows the result of a 7th-order polynomial fit, which specifies the mean behavior of the signal. This line will not be discussed at this point. The plot clearly looks much noisier than Figure 4.8(a), but this should not be taken to indicate that it contains fewer useful data. The corresponding spectrum is shown in Figure 4.21(b). The bulk of the spectrum appears to be at a constant mean level, which is primarily noise. However, a small peak can be seen, as indicated by the grey arrow.

It would be easy to believe that this peak is insignificant. However, the region between −10 and 10 Hz is indicated by the vertical broken lines, and this region has been expanded in the lower portion of Figure 4.22. Now the peak is more evident.

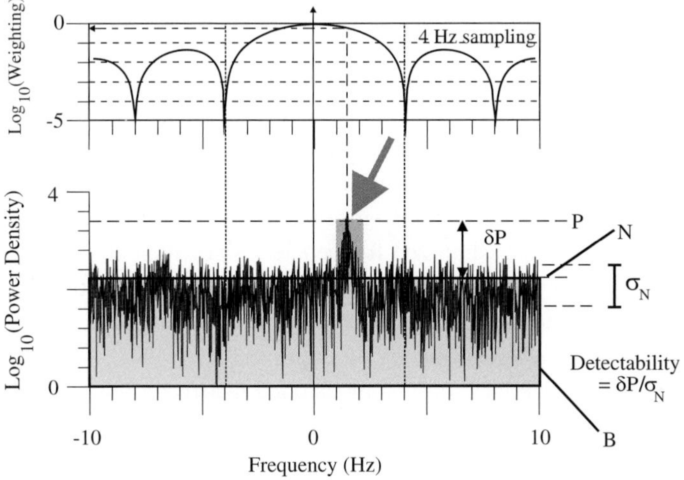

Figure 4.22 The lower curve shows an expanded view of the spectrum from Figure 4.21(b), while the upper curve shows the effective weighting produced by a coherent integration process that utilizes an effective sampling rate of 4 Hz. δP and N refer to peak power and mean noise levels respectively. A traditional coherent integration process would only produce spectra to ±2 Hz in this case.

The important point to note in regard to these two figures is that the parameter which best describes the ability to extract useful data from the spectrum is a diagnostic called the *detectability*. This is essentially the ratio of the peak power (without noise included) in the spectral signal divided by the standard deviation of the noise level. Often the SNR is considered to be a useful guide to the relative contributions of signal and noise, but for spectral calculations it is somewhat meaningless. The signal would be the integrated power in the spectral peak region, and the noise would be the integrated power in the box denoted by B in Figure 4.22. If we increase the frequency limits to ±95 Hz, as in Figure 4.21, then the SNR decreases by a factor of almost 10. The detectability, however, remains unchanged. The only reason that the signal appears to be less attainable in Figure 4.21(b) is that it is normal to think in terms of this SNR parameter. However, once we recognize that the detectability is what defines our ability to extract useful information, then it is clear that the fact that we have digitized out to large frequencies does not affect our ability to retrieve spectral information. Indeed, the function of coherent integration is little more than to remove the high frequency spectral contributions. For example, if we had applied a 10-point coherent integration to the data in Figure 4.21(a), then the spectrum would now only exist out to ± 9.5 Hz. The spectral peak would largely be unaffected.

An additional point arises here. While coherent integration for a *fixed* PRF simply removes the higher frequencies, increasing the PRF does substantially improve the detectability. So a system that runs with a PRF of 100 Hz has reduced detectability compared to a system that runs at 1600 Hz PRF but uses a 16-point coherent integration – even though each has the same frequency limits. To see how this arises (apart from the ability to suppress high frequency interference peaks, which we have already discussed), recall that the noise recorded by the system is limited by the IF bandwidth – which might be say 100 kHz. Since the system samples at perhaps 1000 Hz, then the true noise bandwidth is much wider than the digitized bandwidth – for this simple example, 100 times wider. Noise outside the limits of ±PRF/2 Hz is (upon digitizing) frequency aliased in to the recorded band between −PRF/2 and +PRF/2. The higher the PRF, the wider is the digitized bandwidth. This means that the mean noise level per unit frequency is less for a system with higher PRF, since the total noise (defined by the IF filter) is spread over a wider range of digitized frequencies. Since the detectability depends on the signal's peak spectral values relative to the standard deviation of the noise fluctuations (in the frequency domain), the detectability improves when the mean noise level as a function of frequency is reduced, giving improved detectability for higher PRFs.

To date, we have only demonstrated that recording a spectrum out to large aliasing frequencies is no worse than performing coherent integration. Now we wish to show that it is actually a superior procedure. Foremost among the rationale for this is the fact that the coherent integration procedure leads effectively to application of a biased filter. If we apply a coherent integration, it is equivalent to performing a running mean across the data with a boxcar weighting of width Δt, but then only sampling the new function at steps of Δt (where Δt is taken to be much larger than the inter-pulse period). In the Fourier domain, this is equivalent to multiplying the Fourier transform of the signal by the Fourier transform of the boxcar function. The square of the Fourier transform of a boxcar is shown in the upper graph in Figure 4.22. In effect, the spectrum we would record is the product of the two functions shown in Figure 4.22. In this case, we show the effect if we were to apply a sampling period (after coherent integration) of 0.25 s (effective PRF of 4 Hz). Different parts of the spectrum are biased in different ways.

But the situation is even worse than this. The coherent integration also maps frequencies from outside the aliasing frequency into the central region. For example, suppose it is known that real atmospheric signals should have a Doppler offset of say less than 12 Hz, and suppose that spectral interference exists at a frequency of +15.6 Hz. Suppose we transmitted at 190 Hz PRF, but then applied a 10-point coherent integration. Then data are recorded at a digitization frequency of 19 Hz, and the aliasing frequencies will be ± 9.5 Hz. However, our RF interference at 15.6 Hz will be mapped into a new frequency of −3.4 Hz. There will be some suppression due to the filter function (see the upper curve in Figure 4.22), but the signal will nonetheless be there. If the interference were strong, it could seriously affect the determination of spectral peaks. Since it appears with a frequency less than 12 Hz, any analysis software would consider that it should be treated as an atmospheric signal.

It is worth noting that the idea that the aliasing frequency (in our case 9.5 Hz) is a "folding frequency," as presented in some earlier text books, is in error. The interfering frequency (in our case 15.6 Hz) does *not* get reflected about this frequency (which would produce 15.6 − 9.5 = 6.1 Hz), but rather is transformed in a "modulo" manner – i.e., we repeatedly add or subtract 19.0 Hz until the new frequency lies in the region between −9.5 and +9.5 Hz. Hence 15.6 Hz maps to 15.6 − 19.0 = −3.4 Hz. More details about the frequency mapping process that occurs when data are undersampled or coherently averaged can be found in *Hocking et al.* (2014).

Thus when we use coherent integration, we can introduce undesirable frequencies into our region of interest. In addition, if the true spectral peak were to lie slightly outside the aliasing frequency, it gets mapped to a negative frequency. For example, in our case, if the true frequency were 9.8 Hz, it would appear at −9.2 Hz, giving a completely false picture of radial velocity direction.

The procedure of recording all of the data at 95 Hz results in the interfering signal truly appearing at 15.6 Hz, where it belongs. The user will know that such a high frequency could not be due to real atmospheric signal (since we already noted that our maximum useful atmospheric signals should have offsets of less than 12 Hz in magnitude in this case), and knows to ignore it. In contrast, when coherent integration was used, the peak appeared at −3.4 Hz and so appeared as a true atmospheric signal. In Chapter 8, we will show examples of meteor and lightning contamination, which can hide the atmospheric signal when coherent integration is used, but permits useful determination of the winds when higher sampling rates are used. In addition, *Hocking* (1997a), Figure 5, shows the effect of aircraft on the spectrum. It shows cases where the spectral lines due to aircraft occur at typically 40 or 50 Hz, which suffer severe aliasing when coherent integration is used, but are well separated from the atmospheric signal when faster sampling is used with limited coherent integration.

In addition, examination of Figure 4.22 shows that when we multiply by the weighting function, we actually diminish the spectral peak by about 2 dB in this case, whereas no such effect occurs when we record at higher PRFs.

There are many other advantages to recording at high data rates, which will be discussed in more detail in Chapter 8. However, it is also important to note that there can be limits to the ability to record all the available data. For example, if a PRF of 10 000 Hz is used, then it can become prohibitive to record all data at such a high PRF. A suitable compromise might be to use 50-point coherent integration, leaving an effective PRF of 200 Hz – still large enough to eliminate aliasing effects, but small enough that data storage does not become a serious issue. In reality, coherent integration is frequently performed over numbers of points which are a power of 2.

One final additional comment is required in regard to noise aliasing, which is often misunderstood. This concept applies both to data recorded with coherent integration, and to data without it. The final stage low pass filter (see Figures 4.9 and 4.12) has a typical bandwidth in the range 100 kHz to perhaps 1 MHz. The width needs to be fairly large in order to accommodate the full spectral range of the radar pulse. With regard to the digitization rate, even in extreme circumstances, the highest digitization frequency might be 1000 Hz or perhaps 10 000 Hz. This is at least 10 times lower than the filter width, and in some cases the ratio can be 100 or even 1000 or more. Hence high frequency noise is invariably under-sampled. All this noise is mapped into the region between the aliasing frequencies. Thus it is important to remember that the effective filter width for the system is still the final stage filter, and is *not* defined by the sampling rate or the aliasing frequency. This is an important point to bear in mind in regard to radar calibration, which is discussed elsewhere in this book.

4.10 Range and velocity ambiguities: ambiguity function

The ambiguity function is fundamental to the design of radar systems (e.g., *Barton*, 1988), providing both range and Doppler resolution limitations. In addition, ambiguities in both range and velocity can be determined from this important design step. The ambiguity function can be thought of as the power of the cross-correlation of the transmitted waveform $\underline{s}(t)$ (normally a train of pulses) with a time and Doppler-shifted version of itself $\underline{s}(t + t_0)e^{i2\pi f_d t}$. It is obtained by calculating the magnitude squared of the cross-correlation function and is given by the following equation:

$$|\chi(t_0, f_d)|^2 = \left| \int_{-\infty}^{\infty} \underline{s}^*(t)\underline{s}(t + t_0)e^{i2\pi f_d t}\, dt \right|^2 . \tag{4.53}$$

The usefulness of the ambiguity function becomes apparent when its dependence on f_d and t_0 is investigated. For a returned signal with zero Doppler shift, the ambiguity function reduces to the cross-correlation of the transmitted signal. For a train of rectangular pulses, each of width τ, the result would be a train of triangular functions that repeat at intervals of the IPP T_s. As shown previously, the output of the matched filter for a rectangular pulse input is a triangular function with width 2τ. Thus, the ambiguity function and the matched filter concept are closely related. A similar argument can be made for a returned signal with zero time delay and varying Doppler shift in order to retrieve resolution and ambiguity information for Doppler measurements. However, for most MST radar applications, both range and velocity ambiguities can be more easily understood through a study of general sampling theory.

As mentioned previously, Doppler phase varies relatively slowly across one pulse-width for most MST radar applications. Therefore, the phase change is tracked over a sequence of pulses (often up to hundreds and even thousands of pulses), which is inherently a sampling process. As such, we are constrained by the *Nyquist Sampling Theorem*, which states that the sampling frequency $(1/T_s)$ must be at least twice the highest frequency in the signal in order to have the possibility of fully recovering the original signal. Therefore, the highest observable frequency would be $1/2T_s$, which is termed the Nyquist frequency. Converting this Doppler frequency into units of velocity results in an expression for the so-called *aliasing velocity*, denoted v_a. Any target moving with a speed greater than v_a will be mapped to an incorrect frequency when the time-series is Fourier analyzed. The value of v_a is

$$v_a = \frac{\lambda}{4T_s}. \tag{4.54}$$

The term aliasing was described just prior to Section 4.9.3. One could produce an extremely large v_a by simply minimizing T_s, which we discussed in Section 4.9.3. However, a limitation exists which will now be introduced.

If the IPP is set to be too short, an effect called *range aliasing* can occur. This concept is illustrated in Figure 4.23.

In this example, the IPP was set to 1 ms and a high-reflectivity region exists at a range from 50–100 km. Unfortunately, a communication tower is in the field of view of the

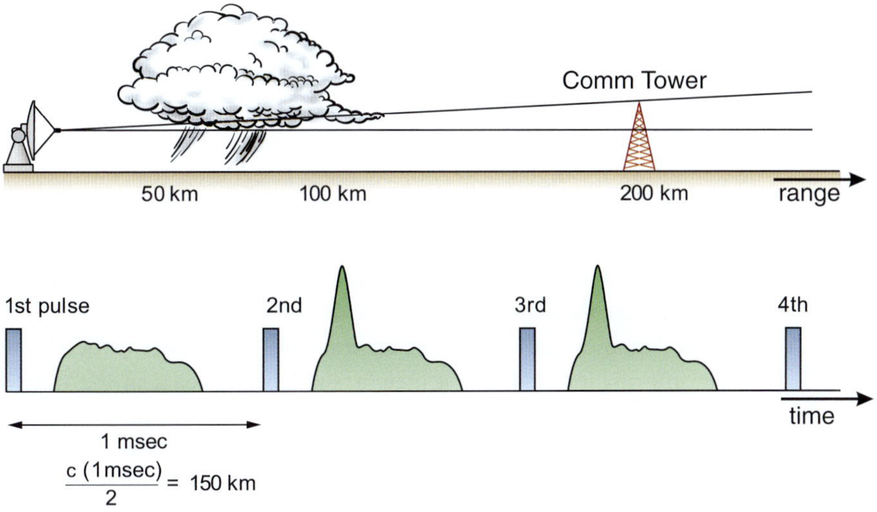

Figure 4.23 Depiction of the range-aliasing effect. The top panel illustrates the radar beam with a high-reflectivity region from 50–100 km, with a communications tower located at a range of 200 km. The bottom panel provides a representation of the transmitted pulse with returned atmospheric signals. The reflectivity from the tower due to the first pulse does not arrive at the receiver until after the second pulse was transmitted. Such range aliasing effects can cause misinterpretation of the data and should be avoided.

radar at a range of 200 km. As seen in the figure, the desired signal is received at times corresponding to ranges of 50–100 km from the first pulse. The echo from the tower, which will typically be relatively strong, will not arrive at the radar until after the second pulse was sent. Of course, this is a result of the 1 ms IPP. Although the tower is actually at a range of 200 km, the echo would be interpreted as being from a range of 50 km, based on the assumption that it was an effect of the second pulse. The problem is that we do not know from which pulse the echo was received. For this situation, the tower echo is called a *2nd-trip echo*. Of course, it is possible to have 3rd- or even 4th-trip echoes depending on the strength of the signals. The maximum unambiguous range, or aliasing range, is determined by the IPP and is given by the following equation:

$$r_a = \frac{cT_s}{2}. \tag{4.55}$$

Since r_a and v_a have opposite proportionality to the IPP, we have a dilemma as to whether to increase the aliasing range or the velocity, a problem often referred to as the "Doppler dilemma." We cannot do both.

It should be mentioned that methods to mitigate both range and velocity aliasing have been conceived (see references cited in Chapter 7 of *Doviak and Zrnić*, 1993). Phase coding and the use of varying IPP are examples of such methods. However, the fundamental trade-off between v_a and r_a remains as a major experimental design criterion.

4.10.1 Deliberate range aliasing

In the previous section, we considered the issue of range aliasing, and indicated that it is often an undesirable effect. However, there are times that it can be turned to advantage. Generally, a small IPP, or high PRF, helps improve the signal relative to the noise, since it allows for higher levels of coherent integration (or equivalently, better numerical filters). Hence if a scientist were planning an experiment to study say the lower troposphere, then a PRF as high as 15 kHz might seem reasonable. But any signal from beyond 10 km will be "range aliased" and care will be necessary to ensure that this does not contaminate the signal.

Let us suppose that our researcher transmits the signal vertically, recording tropospheric information, but that a meteor suddenly appears in the radar beam. Meteors produce very strong signal. A meteor at a range of say 94 km will appear in the recorded signal at a range of 4 km, due to this aliasing effect. Of course, this is undesirable, and one might conclude that a PRF of only 1500 Hz should be used in this case, so that meteors can be avoided. But this will result in an effective loss of 10 dB in signal power. One alternative might be to use 1500 Hz and a 10-bit pulse code (as discussed earlier in this chapter), but then signal is lost at the lowest few km of altitude. The best solution is in fact to keep the high PRF of 15 000 Hz and design analysis software that can locate and remove meteors. This is best done in the time domain, and a more specific illustration will be considered in Chapter 8. It is clear, however, that good experimental design requires a sound knowledge of the atmosphere, knowledge about likely contaminants (like meteors), a willingness to be flexible and inventive with choice of radar parameters, and detailed understanding about analysis processes. We will further discuss design aspects in Chapter 5.

Another interesting example, related in some sense to the previous one, is the study of tropospheric processes in the polar regions in summer. Here again, high PRFs are desirable, but unfortunately in summer there is a layer of strong radar scatterers at about 84 km altitude called PMSE (polar mesosphere summer echoes). Unlike meteors, which are very transient events and can be removed by suitable signal processing, PMSE signals are very persistent and hard to separate from tropospheric signals. So a PRF of 15 kHz is impractical. But if the operator uses a PRF of 10344 Hz, then the aliasing range is 14.5 km, so a PMSE echo from 84 km appears at 11.45 km and is therefore outside the sampling range of interest. In this case, we have been able to use the fact that the PMSE layers are very stable in height.

Early versions of the SKiYMET meteor radar (*Hocking et al.*, 2001a) also used a very high PRF to optimize detections, but used the knowledge that the vast majority of meteors observed by VHF radars occur between 70 and 110 km altitude to determine the correct ranges. Improved and faster digitizers eventually allowed this process to be bypassed.

Good radar experimental design requires a sound knowledge of atmospheric processes, often allowing special "tricks" to be enacted, like the ones above, which allow the user to get more information out of the radar than might classically be considered possible. Progress in MST studies has frequently capitalized on such innovative thought processes. With this in mind, the next chapter looks at a more detailed understanding of the internal workings of a radar.

4.11 Radar calibration

We have discussed at length the ways in which we extract information from the radar signal, but in all cases we have considered the received power only as a relative term. We have dealt with it largely qualitatively, and have concentrated on determinations of radial velocity, spectral width, and relative variations in power. Often users consider power in terms of signal-to-noise ratio, but this idea can be quite misleading. Power depends on the choice of filters used, for example, and even for fixed filters the noise can vary throughout the day. A primary source of noise for VHF radars is galactic sky noise, and this can show significant daily variability. For example, sky noise increases when the galactic center passes through a beam or side-lobe of the radar, and there are other strong astronomical radio sources like Cassiopeia A.

It is therefore a good idea to calibrate the radar, usually using a noise source. This allows the digital units recorded by the system to be converted to more useful products like power in μWatts or dBm, and this in turn allows the received powers to be converted to backscatter cross-sections and reflection coefficient (e.g., *Green et al.*, 1983; *Hocking and Vincent*, 1982a; *Cohn*, 1994; *Hocking and Mu*, 1997; *Campistron et al.*, 2001).

The process can be somewhat complicated, due to unknown inefficiencies in the radar, and variation of sky noise due to ionospheric absorption. Nonetheless, for meaningful comparisons between different radars and different frequencies, an absolute calibration is essential. Specific details about radar calibration will be given in later chapters, especially Chapter 5.

systems, can also be applied to monostatic systems. For now, we simply present the relevant formula – a more thorough derivation will come later. In the bistatic case, the radar equation is expressed as follows:

$$P_r = \frac{P_{Tx} G_T A_e L}{(4\pi)^2 r_t^2 r_r^2} \sigma_A, \tag{5.7}$$

where r_t and r_r are the distances from the target to the transmitter and receiver, respectively, and A_e is the effective area of the receiver antenna (or antenna array), after consideration of receiver antenna losses. The term L represents the transmitter system efficiency (i.e., relates to system Losses, but is 1.0 for no losses) and σ_A refers to backscatter cross-section (see Chapter 3). Here, backscatter cross-section σ_A is a little different to that discussed in Chapter 3. It is considered as the integrated backscatter effect across the entire scattering volume. It can be seen that dimensionally, σ_A has dimensions of area, and we could consider it as the surface area of a piece of perfectly reflecting flat metal at the range of scatter (also see the last equation of Section 3.7).

This definition of σ_A differs from our uses of σ_s and η_s in Chapter 3, Equations (3.256) to (3.258), where those terms referred to backscattered power *per unit volume* per unit incident Poynting vector per specified solid angle (either 1 or 4π steradians). As discussed there, they are referred to as "reflectivities," but even this notation can be a bit confusing. We will combine these different backscatter concepts later in this chapter.

The quantity G_T is referred to as the *gain* of the transmitter antenna or antenna-array. We will introduce a more formal definition of the gain shortly, but in essence it is a measure of the degree to which the radiated signal is concentrated into a narrow beam, with narrower-beam radars having higher gain. Typical gains are of the order of 1 to 100 000 and more, and are usually expressed in decibels (dB). Many MST radars have gains in the range 20 to 40 dB.

It is essential in a bistatic system to keep an accurate synchronization between the transmitter and the receiver. While this synchronization can be established by providing a separate communication channel between the transmitter and the receiver, the accurate timing provided with GPS receivers can also be used to obtain synchronization with an accuracy of the order of 100 ns (e.g., *Sahr*, 2008). Highly stable atomic clocks are also now commercially available which can be used as references at different sites.

The Doppler velocity measured with a bistatic radar is the rate of change of the total path delay from the transmitter to the receiver via the target. It is the velocity component of the target motion projected onto the direction of a vector given by

$$\vec{n'_{tr}} = \vec{i_t} + \vec{i_r}, \tag{5.8}$$

where $\vec{i_t}$ and $\vec{i_r}$ are the unit vectors in the direction of the target as seen from the transmitter and the receiver, respectively. By measuring the Doppler velocity at three different receiver locations which do not form a line, the three components of the wind velocity vector can be determined for a single scattering volume.

In contrast to the monostatic radar, where the entire range profile can be measured for each pulse, only the common volume illuminated by the transmitter and receiver antenna beams can be measured at any one time. It is thus necessary to steer at least one of the beams in order to cover a wide height range, which reduces the time resolution. One way to avoid this problem is to employ a so-called "fan beam" on transmission, which is obtained by using a linear array of small antennas aligned perpendicular to the baseline between the transmitter and the receiver. This produces a beam which has a narrow width in a vertical plane aligned along the row of antennas, and is wide in the vertical plane which is orientated along the line from the transmitter to the antenna (e.g., *Amayenc et al.*, 1973).

Even with this configuration, the common volume is limited to a finite range when the transmitter antenna beam is steered to a direction perpendicular to the baseline.

An alternative approach to producing a wide range coverage at moderate to good resolution is to transmit with a broad beam, but on reception to use an array of receivers with separate digitizers for each antenna element. The receiver array could be separate to the transmitter one, or a single array could be used for both transmission and reception. By sampling and recording data for each antenna element separately, a sharp receiver antenna beam focused to any desired direction can be formed in software by adding the output of each receiver with appropriate phase shifts in the off-line data processing. This method is called post-beam steering or digital beam forming. Although such configurations can be costly for atmospheric radars which use a large number of antennas, recent progress and reduced costs in digital signal processing techniques are making this approach feasible.

One advantage of employing multistatic configurations is that the isolation between the transmitter and the receiver is usually much higher than in the monostatic case, and it is even possible to analyze data recorded on the receiver before the pulse has finished transmission (e.g., *Hocking and Hocking*, 2010). It is thus possible to use a long pulse, or even a continuous wave. As the range resolution provided by the size of the common volume between the transmitter and the receiver antenna beams is sometimes not sufficient, pulse compression is often (but not always) required.

The most typical pulse compression code applicable only to the CW signal is the M-sequence (maximum-length sequence) code (*Cohn and Lempel*, 1977; *Golomb*, 1981). This code is generated by feeding back the exclusive OR of pairs of outputs of an N-bit shift register as shown in Figure 5.4. Its name comes from the fact that a

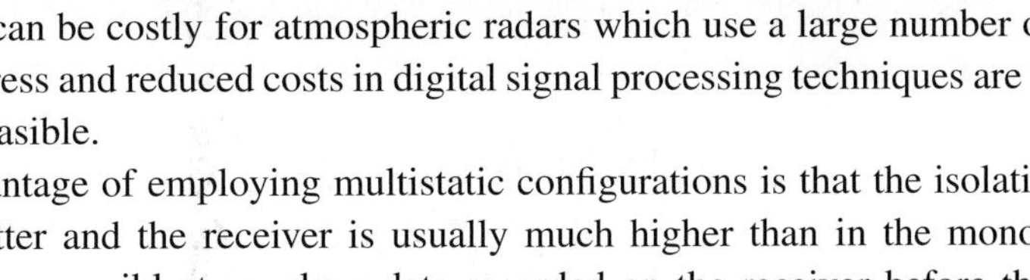

Figure 5.4 Generation scheme of an M-sequence for $N = 4$ ($M = 15$), and its autocorrelation function. The 4-bit shift register has a feedback equal to the exclusive-OR of the output of the 3rd and 4th bits, which is then fed back as an input to the first bit at the left. This sequence repeats as each new output of the OR product of bits 3 and 4 is fed back into bit 1, forcing all bits to move to the right. The output on the right is then the sequence of bits used to create the code. This produces a sequence of period 15, the autocorrelation function of which is shown on the right.

maximum possible period of $M = 2^N - 1$ bits is achieved with an N bit register by properly selecting the feedback bits. The autocorrelation function of the binary phase code given by this sequence has a maximum of value M at a period of M bits, and the value of -1 for the rest.

An efficient pulse compression is achieved by making the code length M large enough that its corresponding distance in space is larger than the size of the common volume.

5.2 Radar antennas

5.2.1 Basic theory

An antenna is a device that receives a signal from a transmitter, and then turns this received signal into radiated electromagnetic waves. Conversely, it can receive radio waves and turn them into electromagnetic currents within the antenna, which may then be amplified and detected by electronic software. Specific examples will be given shortly, but to begin we need to introduce some general topics. The first of interest is the antenna *gain*.

Previously, the concept of antenna gain was introduced in a descriptive sense. We now introduce the concept more formally. We will consider the case of antenna transmission, although the same formulas apply for an antenna used for reception as well.

Performance of a general antenna during transmission is expressed in terms of the *absolute gain*, which is expressed as (e. g. *Balanis*, 1997; *Drabowitch et al.*, 2005)

$$G_a(\theta, \phi) = \frac{L \ |E(\theta, \phi)|^2}{\frac{1}{4\pi} \int |E(\theta, \phi)|^2 d\Omega}, \tag{5.9}$$

where E is the electric field generated by the antenna at the location of the target (assumed to be at a distance r from the radar), (θ, ϕ) denotes the direction of interest from the radar to the target, Ω is the solid angle, and L is the loss factor including the ohmic loss and reflection due to impedance mismatch. The absolute gain G_a can be seen qualitatively to be a measure of power density (power per unit area) radiated in the direction (θ, ϕ) relative to the total power radiated in all directions and multiplied by 4π. More specifically, the numerator represents a quantity proportional to the Poynting vector at angle (θ, ϕ) per unit steradian, while the demominator represents the total power radiated in all directions, divided by 4π radians, and so represents the mean power per unit steradian that would result if all of the power fed into the system were radiated isotropically. To see this equation derived from first principles, it is normally evaluated at a fixed radius r, but then terms involving the radius r all cancel. Normally the gain is expressed in decibels (dB), or more precisely dBi, where this means "gain in dB relative to an isotropic antenna."

The term L deserves some comment. Although often referred to as a loss factor, it is more accurately an efficiency factor. Its largest value is unity, which means perfect efficiency or zero losses. We will use the symbol L to represent it, although later we will distinguish between the cases of transmission (L_{Tx}) and reception (L_{Rx}). The terms efficiency and loss factor will be used interchangeably throughout the text.

An alternative definition of gain is the directional gain, or *directivity*. This is the same as (5.9) but does not consider the loss term, so is given by

$$G_d(\theta,\phi) = \frac{G_a(\theta,\phi)}{L} = \frac{|E(\theta,\phi)|^2}{\frac{1}{4\pi}\int |E(\theta,\phi)|^2 \mathrm{d}\Omega}, \tag{5.10}$$

which can be determined from the measured directional pattern.

In the case of atmospheric radars, we usually take θ as the zenith angle, and ϕ as the azimuth angle. We also generally take the z-axis to the zenith, and the x-axis corresponds to $\phi = 0$. It is assumed that the target is located sufficiently far from the antenna that a far-field approximation may be used.

A graph of the radiated power (proportional to E^2) as a function of zenith angle θ and azimuth ϕ is referred to as the polar diagram of the aperture. For cases in which the aperture is several wavelengths wide or more, the polar diagram comprises a large central value (referred to as a beam), then various smaller side-lobes, as shown in Chapter 4, Figure 4.5.

As discussed, the gain without the L factor included is called the directional gain. The absolute gain G_a is also referred to as the isotropic gain (dBi). Sometimes the gain is given relative to a reference antenna, usually taken as a half-wave dipole. As a half-wave dipole has a gain of 2.16 dB, G_a is higher than the gain relative to a dipole by this amount.

Although the above definitions give G as a function of angle, it is common to refer to its maximum value, which is usually at the center of the beam (referred to as being along the bore-sight), as being the *gain*). This is a constant for any particular radar antenna.

On reception, the performance of an antenna is generally represented by the effective area

$$A_e = P_r/S_r, \tag{5.11}$$

where P_r is the total power received by the antenna, and S_r is the power density of the incident wave per unit area. The effective area indicates how much power the antenna collects from a given incident wave field in terms of the collecting area. For a large array or reflector antenna with uniform illumination, A_e is roughly equal to its physical aperture A_{ant} times the loss (or efficiency) factor. However, losses at the edge of a radar dish, for example, may reduce the effective area. For an array of antennas, the effective area is comparable to but usually smaller than the physical area. From this point on we will distinguish between the loss factors on transmission and reception and and refer to them as L_{Tx} and L_{Rx} respectively.

The reciprocity theorem of the electromagnetic field allows us to show that the polar diagrams on transmission and reception are the same. Essentially, if a set of voltages are applied to the terminals of the antennas in an array, then a particular current distribution results, which produces a particular polar diagram. If the same array is radiated with a set of plane waves which had the same functional form as the polar diagram just discussed for transmission, then the same current distribution results as that discussed for transmission, and then the voltages produced at the terminals are the same as for the case of transmission, leading to the identical natures of the polar diagrams for transmission and reception.

This is in essence a special case of the so-called *Lorentz reciprocity condition*, a fundamental result that exists whenever sinusoidal electric field oscillations are involved.

A universal relation between G_a and A_e exists, given by

$$G_a = \frac{4\pi A_e}{\lambda^2}, \tag{5.12}$$

where λ is the radar wavelength. This relation only includes losses in G_a associated with the antenna design itself, and does not consider losses in the feed cables.

This relation shows that the requirements for a narrow beam for transmission and a high sensitivity for reception can be simultaneously satisfied by a large aperture area.

Shortly we will consider the antenna pattern, which determines the shape of the main beam, the level of side-lobes, and so on. Before that, however, we will derive (5.12), and some other associated equations.

5.2.2 Relation between gain, effective area, and beam-width

In the following pages, we will look in more detail at the relation (5.12). A variety of ways exist to show this relation, from quite different perspectives. One method relies on electric circuit theory, another on Fourier diffraction theory. Each approaches teaches us something new about antennas.

Gain and radiation resistance for an infinitesimal dipole

Here, we consider a radiating element comprising an extremely short section of wire of length l. Such an antenna is rarely used in practice, but the theory is relatively simple and such a short section can be used as a building-block for more realistic antennas. In this subsection, we will derive its gain and effective area as a prelude to a proof of the relation (5.12).

The electric field generated by a current in this short piece of wire at a range $r\,(\gg \lambda)$ from the antenna and angle θ from the z-axis has a component in the θ direction only, and is given by (e.g., *Balanis*, 1997)

$$\underline{E}_\theta = \mathrm{i}\frac{IlZ_o}{2\lambda r} \sin\theta \, \mathrm{e}^{-\mathrm{i}kr}. \tag{5.13}$$

Where Z_o is the characteristic impedance of the space, $k = 2\pi/\lambda$ is the wavenumber and $\mathrm{i} = \sqrt{-1}$. This is a donut-like pattern symmetrical around the z-axis.

The absolute gain of an antenna was defined by Equation (5.10). Combining this with Equation (5.13) and carrying out the integral gives

$$G_a = 3/2. \tag{5.14}$$

The total radiation power is also given by using Equation (5.13) as

$$P_r = \oint_S \frac{|\underline{E}_\theta|^2}{2Z_o} \mathrm{d}S = \frac{\pi Z_o}{3}\left(\frac{Il}{\lambda}\right)^2, \tag{5.15}$$

where S is a closed surface surrounding the dipole.

The radiation resistance is given by solving $P_r = \frac{1}{2}I^2 R_r$, by analogy with power dissipated by a resistor in a circuit, except that for an antenna the power is not lost as heat but is radiated to space. Then

$$R_r = \frac{P_r}{\frac{1}{2}I^2} = \frac{2\pi Z_o}{3}\left(\frac{l}{\lambda}\right)^2. \tag{5.16}$$

Effective area of infinitesimal dipole

We now need the effective area. Consider an infinitesimal dipole placed in vacuum where a planar wave exists, and examine the impact on the dipole (which we regard as a receiver antenna). Let voltage $\underline{V}_o$ be excited at the dipole. If a load $\underline{Z}_L$ is connected, the current flowing on it is given by

$$\underline{I} = \frac{\underline{V}_o}{\underline{Z}_a + \underline{Z}_L}, \tag{5.17}$$

where $\underline{Z}_a = R_r + \text{i}X$ is the radiation impedance of the antenna. The power consumed at the load becomes maximum when the matching condition

$$\underline{Z}_L = \underline{Z}_a^* = R_r - \text{i}X \tag{5.18}$$

is satisfied. This result can be seen in most elementary text books on electrical circuit matching theory. For this case, the power consumed at the load is given by

$$P_r = \frac{V_o^2}{8R_r}. \tag{5.19}$$

For an infinitesimal dipole of length l, the electric field is regarded as constant over this length, and the voltage is given by

$$V_o = E\, l, \tag{5.20}$$

where E is the magnitude of the applied electric field.

Using the Poynting flux given by

$$S_r = \frac{E^2}{2Z_o}, \tag{5.21}$$

we obtain

$$P_r = \frac{Z_o l^2}{4R_r} S_r. \tag{5.22}$$

The effective area of an antenna is defined by Equation (5.11), viz.

$$A_e \equiv \frac{P_r}{S_r}. \tag{5.23}$$

Using Equation (5.16) we obtain

$$A_e = \frac{Z_o l^2}{4R_r} = \frac{3\lambda^2}{8\pi}. \tag{5.24}$$

Combining Equations (5.14) and (5.24) then leads to Equation (5.12), at least for this special antenna. We now need to extend this to a general antenna.

Relation between gain and effective area for a general antenna

Our purpose here is to show that (5.12) works for more general antennas than a simple infinitesimal dipole.

The Lorentz reciprocity condition

In order to do that, it is necessary to recall the Lorentz reciprocity condition. This was discussed earlier, but we re-iterate it here. More complete discussions can be found at *http://en.wikipedia.org/wiki/Reciprocity_(electromagnetism)*, and in *Born and Wolf* (1999), *Mahan* (1943), and *Rayleigh* (1900), for example.

For our purposes, we quote a relatively general form of the law (from the Wikipedia reference given above), namely

$$\int\int\int_V \left[\vec{\underline{j}}_1 \cdot \vec{\underline{E}}_2 - \vec{\underline{E}}_1 \cdot \vec{\underline{j}}_2\right] dV = \oint_S \left[\vec{\underline{E}}_1 \times \vec{\underline{H}}_2 - \vec{\underline{E}}_2 \times \vec{\underline{H}}_1\right] \cdot d\vec{A}, \tag{5.25}$$

where $\vec{\underline{E}}_k$ and $\vec{\underline{H}}_k$ are electric and magnetic fields at location k, and the j_i are current densities. For an electric circuit involving (sinusoidal) electromagnetic waves, this becomes a statement that if a voltage $\underline{V}_a$ is applied at point 1 in a circuit and this induces a current $\underline{I}_b$ at a point 2, then the same voltage $\underline{V}_a$ applied at point 2 produces a current $\underline{I}_b$ at point 1. It also means that if any particular current distribution is produced on an antenna, and it produces a particular distribution of radiating waves, then that same distribution of *incoming* waves produces the original current distribution on the antenna. We have already used this fact to indicate that the transmission and reception polar diagrams of an antenna are identical in shape.

In an antenna array which is lossy, the law can also be used to prove that for a pair of antennas in which one transmits and one receives (case 1) and the case that the second tramsmits and the first receives (case 2), see Figure 5.5, we may write

$$P_{Rx} = P_{Tx}(G_{a1}L_{Tx1})(A_{e2}L_{Rx2}) = P_{Tx}(G_{a2}L_{Tx2})(A_{e1}L_{Rx1}), \tag{5.26}$$

where the L_k are loss terms. This says that transmission and reception loss terms are related, but are not necessarily equal. We will therefore distinguish between the terms L_{Tx} and L_{Rx} throughout this text.

For a lossless case,

$$G_{a1}A_{e2} = G_{a2}A_{e1}, \tag{5.27}$$

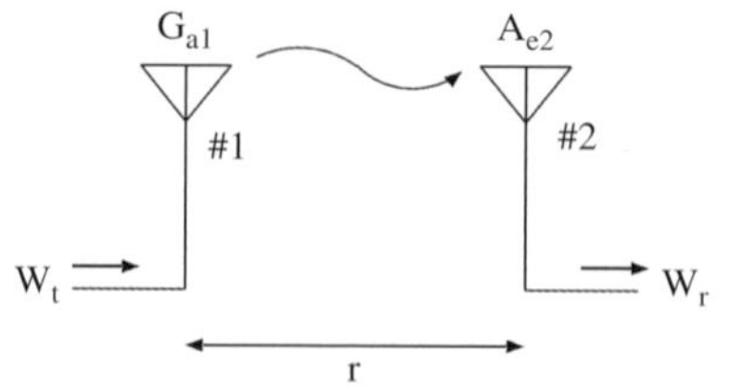

Figure 5.5 Two antennas used to transmit signals between each other.

so

$$L_{Tx1}L_{Rx2} = L_{Tx2}L_{Rx1} \tag{5.28}$$

is the requirement.

Physically, even a large array does not create a planar wave on transmission, but it is placed in a planar wave on reception. So the electromagnetic field around the antenna may, of course, be different on transmission and reception, which is the reason that the transmitter and receiver efficiencies may be different.

Application of the Lorentz condition to antennas

We now consider transmission between two antennas as shown in Figure 5.5. We transmit power W_t from antenna #1 with absolute gain G_{a1}, and receive the signal with an antenna of effective area A_{e2} at a distance r. We assume that both antennas have perfect matching with the transmission line connected to each antenna, and that the antennas have no loss. Hence our equations may not apply for lossy systems.

The power density at antenna #2 (for $r \gg \lambda$) is given by

$$S_r = \frac{W_t G_{a1}}{4\pi r^2}. \tag{5.29}$$

Then the received power is given by

$$W_r = S_r A_{e2} = \frac{W_t G_{a1} A_{e2}}{4\pi r^2}. \tag{5.30}$$

Lorentz reciprocity shows that if we exchange the transmitter and the receiver, and transmit power W_t from antenna #2, then we receive power W_r by antenna #1.

We thus obtain the relation

$$G_{a1}A_{e2} = G_{a2}A_{e1}. \tag{5.31}$$

As this relation holds for an arbitrary pair of antennas, we have

$$\frac{A_{e1}}{G_{a1}} = \frac{A_{e2}}{G_{a2}} = K, \tag{5.32}$$

where K is a universal constant for *any* lossless antenna. Using Equation (5.14) and Equation (5.24), we have

$$\frac{A_e}{G_a} = K = \frac{\lambda^2}{4\pi} \tag{5.33}$$

for an infinitesimal dipole. However, in our discussion above, we may consider one antenna to be an infinitesimal dipole, and the other a more complex antenna. The theory still applies, and the constant K is unchanged. Now we can introduce another more complex antenna in the place of the remaining infinitesimal dipole, and the theory still remains valid, but now we have no infinitesimal dipoles in the picture. Hence the constant K deduced from the case of an infinitesimal pair of dipoles also applies for any pair of lossless antennas, making (5.33) quite general.

A Fourier approach to the gain/effective-area relation

The above derivation used an approach based on electromagnetic theory, but other proofs exist which can also be illuminating. We now use such an alternative approach. Although the proof can be done more generally, we will concentrate here on the sub-class of antennas which have moderately large size and some level of rotational symmetry.

When the radar aperture width is several to many wavelengths across, (e.g., 5–10 or more), the central lobe takes an elliptical shape, and for the case that the aperture has some degree of broad rotational symmetry, such as a circle, or even a square, the main beam is circular. Any non-circular nature of the aperture is most evident in the side-lobes of the polar diagram. In such cases we specify the main beam as having a "half-power-half-width" $\theta_{\frac{1}{2}}$, which represents the angle from the bore (the line of maximum gain) at which the power falls to half of the value along the bore. The half-power full-width, which is two times $\theta_{\frac{1}{2}}$, will be denoted as θ_h.

In this subsection, we will derive relationships between the isotropic gain G, the effective area A_e, and the values $\theta_{\frac{1}{2}}$ and/or θ_h using a Fourier approach, rather than the electrical-circuit approach used in the previous derivation. We will not distinguish between G_a and G_d here, but consider a situation without losses, so both are represented by G.

We start with the recognition that the diffraction pattern of a two-dimensional aperture is its 2-D Fourier transform in wavenumber-adapted direction-cosine space (*Champeney* (1973)), where the specific meaning of this type of space will become evident shortly. The function $f(x, y)$ will be used to describe the aperture. We take the function to have value C inside the perimeter of the aperture, and zero outside. If we wish to adjust the strength of transmission within the aperture, we can reduce any part of it to a value less than C. Very often we take C as unity. In principle the transmission coefficient can even be complex, representing phase variations. We will assume only real transmission coefficients for now.

Then the radiated electric field is given by

$$\underline{E}(\nu_x, \nu_y) = \kappa_Z \int_{-\infty}^{\infty} \int_{-\infty}^{\infty} e^{(2\pi i \vec{\nu}\cdot\vec{r})} f(\vec{r}) d\vec{r}, \tag{5.34}$$

where $\nu_x = \frac{\cos\theta_x}{\lambda}$ and $\nu_y = \frac{\cos\theta_y}{\lambda}$, with $\cos\theta_x$ and $\cos\theta_y$ being the direction cosines relative to the x and y axes respectively and λ the radar wavelength. The constant κ_Z includes various normalization constants and impedance matching values, but since it will self-cancel in our calculations we will not specify it in detail.

If the aperture function is spherically symmetric, then we may use a suitable Hankel transform, given by

$$\underline{E}(\nu) = \kappa'_Z \int_0^{\infty} f_r(r) J_0(2\pi\nu r) 2\pi r dr, \tag{5.35}$$

where $f_r(r)$ represents the variation of the 2-D function f along a radial line from the center of the aperture in the x–y plane, and J_0 is the zeroth order Bessel function. In this case ν is given by $\frac{\sin\theta}{\lambda}$, where θ is the angle from the aperture normal vector (originating at the center of the aperture). The polar diagram is proportional to $|\underline{E}|^2$.

For the case where the main beam is circularly symmetric, we can define a half-power half-width, as discussed above. Determining this angle is simply a matter of knowing the Fourier transforms. For example, the diffraction pattern for a uniform circular aperture of radius a is an Airy function, with first minimum at $\theta_{\min} \approx 2.44a/\lambda$, and half-power angle at

$$\theta_{\frac{1}{2}} \simeq \frac{0.255\lambda}{a}. \tag{5.36}$$

The *full width* of the beam at half-power is (e.g., *Balanis*, 1997)

$$\theta_h \simeq \frac{1.02\lambda}{D}, \tag{5.37}$$

where $D = 2a$ is the aperture diameter. This is often written in terms of the area of the aperture by squaring the above expression and producing

$$\theta_h^2 \simeq \frac{1.04\frac{\pi}{4}\lambda^2}{A_e} \simeq 0.82\frac{\lambda^2}{A_e}, \tag{5.38}$$

where $A_e = \pi a^2$ is the area of the aperture.

For a square aperture, the Fourier transform is the product of two sinc functions orientated along the x- and y-axes. The main beam is close to circularly symmetric, and the half-power half-width is given by

$$\theta_{\frac{1}{2}} \simeq 0.44\lambda/D, \tag{5.39}$$

where D is the width of the square. Squaring and rewriting in terms of the area we produce

$$\theta_h^2 \simeq 0.77\frac{\lambda^2}{A_e}. \tag{5.40}$$

For most reasonable apertures, to an accuracy of about 5% or better for θ_h, we can write

$$\theta_h^2 \simeq \kappa\frac{\lambda^2}{A_e}, \tag{5.41}$$

where κ is about 0.8. Hence we consider this as a useful relation for simple calculations involving antenna areas and beam-widths.

We now turn to proof of Equation (5.12). To do this, we will represent the radiation patterns as the Fourier transforms of the aperture, just as we did in the preceding paragraphs. The aperture function will be described by $f(\vec{r})$, where $\vec{r}$ lies wholly in the plane of the aperture and is zero outside the aperture. In the case we consider, f is taken to be a constant, C, inside the aperture and zero outside. We write

$$G = \frac{\Upsilon(0,0)}{\frac{1}{4\pi}\int_{\theta=0}^{\pi}\int_{\phi=0}^{2\pi}\Upsilon(\theta,\phi)\sin\theta d\theta d\phi}, \tag{5.42}$$

where Υ is proportional to $|E|^2$, as described by (5.34). We will now write $|E|^2$ as E^*E, where the superscript $*$ means complex conjugate.

Any constants of the system apply equally in the numerator and the denominator, and self-cancel. Then the numerator is simply proportional to $\left[\int E(x,y)dxdy\right]^2$.

We will consider the case that $E(x,y) = C$ is a constant all over the aperture, and zero outside it. Then the numerator for the case $\vec{\nu} = \vec{0}$ becomes

$$C^2 \left[\int dxdy\right]^2 = C^2 A_{\text{ant}}^2, \tag{5.43}$$

where A_{ant} is the physical area of the aperture.

We therefore write (5.42) as

$$G = \frac{4\pi C^2 A_{\text{ant}}^2}{\mathcal{I}}, \tag{5.44}$$

where

$$\mathcal{I} = \int_{\theta=0}^{\pi} \int_{\phi=0}^{2\pi} E^*(\theta,\phi) E(\theta,\phi) \sin\theta d\theta d\phi. \tag{5.45}$$

Note that the volume element for integration here is not the usual $r^2 \sin\theta d\theta d\phi$, but contains no r^2 term since it was cancelled by division by $4\pi r^2$ earlier in the definition of G.

Using $E = \int\int f(\vec{r}) e^{2\pi i \vec{\nu}\cdot\vec{r}} d\vec{r}$, where $\vec{r}$ is a vector in the two-dimensional plane of the aperture from the origin to a point in the aperture, and $\vec{\nu}$ is a unit vector from the origin in the direction of interest, (θ,ϕ), the following result is produced:

$$\mathcal{I} = \int_{\theta=0}^{\pi} \int_{\phi=0}^{2\pi} \left[\int\int_{\text{all } \vec{r}} f(\vec{r}) e^{-2\pi i \vec{\nu}\cdot\vec{r}} d\vec{r}\right] \left[\int\int_{\text{all } \vec{r}'} f(\vec{r}') e^{2\pi i \vec{\nu}\cdot\vec{r}'} d\vec{r}'\right] \sin\theta d\theta d\phi. \tag{5.46}$$

Letting $\vec{r}' = \vec{r} + \vec{\xi}$, and collapsing the two terms in square brackets into one, we produce

$$\mathcal{I} = \int_{\theta=0}^{\pi} \int_{\phi=0}^{2\pi} \left[\int\int_{\vec{\xi}} \left\{\int\int_{\vec{r}} f(\vec{r}) f(\vec{r} + \vec{\xi}) d\vec{r}\right\} e^{-2\pi i \vec{\nu}\cdot\vec{\xi}} d\vec{\xi}\right] \sin\theta d\theta d\phi. \tag{5.47}$$

The term in curly brackets is simply the autocovariance function of the aperture function, which we will denote as $\rho(\vec{\xi})$, producing

$$\mathcal{I} = \int_{\theta=0}^{\pi} \int_{\phi=0}^{2\pi} \left[\int\int_{\vec{\xi}} \rho(\vec{\xi}) e^{-2\pi i \vec{\nu}\cdot\vec{\xi}} d\vec{\xi}\right] \sin\theta d\theta d\phi. \tag{5.48}$$

The term inside the square brackets is simply the Fourier transform of ρ, which we will denote as R, leading to

$$\mathcal{I} = \int_{\theta=0}^{\pi} \int_{\phi=0}^{2\pi} R(\nu_x, \nu_y) \sin\theta d\theta d\phi, \tag{5.49}$$

where R is of course the square of the Fourier transform of the aperture function, confirming that the polar diagram is the square of the modulus of the Fourier transform of f, which is of course not really a surprise.

To evaluate $\mathcal{I}$, we now assume that over the region in which most power is radiated (within the main beam), the function is circularly symmetric, so is independent of direction, allowing us to consider only the radial form of the function, which we will denote as $R_r(\nu)$, where ν is $\frac{\sin\theta}{\lambda}$. In addition, we express $\sin\theta$ in terms of ν as follows:

$$\sin\theta = \lambda\nu \qquad \text{and} \qquad \mathrm{d}(\sin\theta) = \lambda \mathrm{d}\nu. \tag{5.50}$$

The right-hand expression can be written as $\cos\theta d\theta = \lambda d\nu$, and we use $\cos\theta = \sqrt{1-\sin^2\theta} = \sqrt{1-\lambda^2\nu^2}$, so that $d\theta = \frac{\lambda d\nu}{\sqrt{1-\lambda^2\nu^2}}$. Recognizing that R_r is independent of ϕ, and that the ϕ integral therefore becomes simply 2π, we may write

$$\mathcal{I} = \lambda^2 \int_{\text{all } \nu} 2\pi R_r(\nu) \frac{\nu}{1-\lambda^2\nu^2} d\nu. \tag{5.51}$$

At small ν, the denominator is close to unity, and at large ν, $R_r(\nu)$ falls rapidly to zero, at least for the case that the aperture is say 5–10 wavelengths wide or more, i.e. when the beam half-power half-width is less than 5 degrees. The R_r function falls to zero more rapidly than the denominator, and the entire integrand approaches zero at large ν. Hence we can take the denominator as unity everywhere in the integrand (at small ν it *is* close to 1, and at large ν its value is immaterial), to produce

$$\mathcal{I} \simeq \lambda^2 \int_{\text{all } \nu} R_r(\nu) 2\pi \nu d\nu. \tag{5.52}$$

The integral is simply the area under the function $R(\vec{\nu})$, which is just the value of its Fourier transform at zero, and its Fourier transform is the function $\rho = \int f(\vec{r}) f(\vec{r}+\vec{\xi}) d\vec{r}$ evaluated at $\vec{\xi} = 0$. Since f has value C everywhere within the aperture, this integral is $C^2 A_{\text{ant}}$, which differs from Equation (5.43) in that here we have the autocovariance peak value whereas there we had the square of the integrated transmission function. Then from (5.44) we produce

$$G \simeq \frac{4\pi C^2 A_{\text{ant}}^2}{\lambda^2 C^2 A_{\text{ant}}}. \tag{5.53}$$

Replacing A_{ant} with A_e leads to Equation (5.12), viz.

$$G = \frac{4\pi A_e}{\lambda^2}, \tag{5.54}$$

as required.

The above derivation made some approximations, although the result is actually quite accurate, as seen in the previous proof using the Lorentz reciprocity condition. Another fairly accurate proof for the special case of a circular aperture is shown in Appendix B.

Other consequences of the gain/effective-area relationship

The following two results will also be of use in our work; first, from Equation (5.41),

$$\theta_h^2 \simeq \kappa \frac{\lambda^2}{A_e}, \tag{5.55}$$

where κ is about 0.8, and secondly, by combining (5.55) and (5.54), we produce

$$G = \frac{4\pi\kappa}{\theta_h^2}. \tag{5.56}$$

There is one other special type of beam that deserves mention. Although we have considered beams in general above, and particularly took interest in square and circular apertures, it is not uncommon to consider a Gaussian beam for mathematical purposes. However, in the cases just discussed, we were able to take the effective area of the antenna as being pretty much the physical area covered by the antenna, In the case of a Gaussian beam, "effective area" does not have such a direct interpretation, since the aperture of a Gaussian beam would also be Gaussian as a function of radius and so would be infinite in extent. However, an effective area can be defined of course by forming some sort of weighted average across the antenna field, or from knowledge of the gain and then inverting Equation (5.54). Equations like (5.54) and (5.55) now become a definition of effective area rather than a derivation. However, the equivalent case for (5.56) (i.e. the relation between beam half-power full-width and gain) can be obtained without recourse to the effective area when dealing with a Gaussian beam. If the Gaussian beam (assumed fairly narrow – less than 10 degees half-power half-width) is described by

$$P(\theta) = P_0 e^{-\frac{\theta^2}{2\theta_{\mathrm{RMS}}^2}}, \tag{5.57}$$

then for narrow beams the gain is

$$G = 4\pi \frac{P_0}{P_0 \int_{\theta=0}^{\pi} e^{-\frac{\theta^2}{2\theta_{\mathrm{RMS}}^2}} 2\pi\theta d\theta}. \tag{5.58}$$

Substituting $\chi = \theta^2$, so that $d\chi = 2\theta d\theta$, and letting the upper limit of the integral tend to infinity (since the contribution from $\pi/2$ to infinity is negligible), and recognizing that $e^{-\infty}$ is zero, we produce

$$G = \frac{4\pi}{2\pi\theta_{\mathrm{RMS}}^2} = \frac{2}{\theta_{\mathrm{RMS}}^2}. \tag{5.59}$$

Finally, we need to relate θ_{RMS} to $\theta_{\frac{1}{2}}$, which can be achieved by solving $\exp\left(-\frac{\theta_{\frac{1}{2}}^2}{2\theta_{\mathrm{RMS}}^2}\right) = \frac{1}{2}$. We produce $\theta_{\mathrm{RMS}} = \theta_{\frac{1}{2}}/\sqrt{2\ln 2}$, leading to

$$G = \frac{16\ln 2}{\theta_h^2}. \tag{5.60}$$

The constant in this equation is $16\ln 2 = 11.09$, which is moderately similar to $4\pi\kappa = 10.05$ in (5.56) for $\kappa = 0.8$. A choice of $\kappa = 0.88$ would be better for a Gaussian beam, but the general form of the equation persists, and so is useful for many simulations. Distinction between Gaussian polar diagrams and more realistic ones will later be discussed further.

One final point should be made here. If a radar is used in monostatic mode, then the received power is proportional to the product of the transmitter gain and the receiver area. Since the receiver effective area is proportional to the gain, this means that the receiver power arriving from zenithal angle θ is proportional to the square of the polar diagram function. If the transmitter polar diagram has a shape of the form given by Equation (5.57), then after reception the system can be considered as the equivalent of a combined two-way polar diagram given by the *square* of the normal one-way polar diagram, so in this case it would be given by

$$P(\theta) \propto \left[e^{-\frac{\theta^2}{2\theta_{RMS}^2}} \right]^2, \tag{5.61}$$

which can be written as

$$P(\theta) \propto \left[e^{-\frac{\theta^2}{2\theta_{RMS2way}^2}} \right], \tag{5.62}$$

where

$$\theta_{RMS2way} = \theta_{RMS}/\sqrt{2}. \tag{5.63}$$

In other words, the effective beam is narrower by $\sqrt{2}$ times relative to the original. The original is called the *one-way* beam, while the combined one is referred to as the *two-way* beam, since it involves radiowave propagation from the radar to the scatterers and back again, thus traversing the same region of atmosphere twice.

5.2.3 Radiation patterns for simple antennas

We now begin to look at antenna radiation patterns in more detail. To start, consider the simplest case of a small electric dipole. We consider the radiation field from two oscillating electrical charges with opposite sign $\pm Q \exp \mathrm{i}\omega t$ separated by a small distance $l(\ll \lambda)$ along the z-axis. From the continuity of current, this situation is equivalent to considering a constant current $I = I_o \exp \mathrm{i}\omega t$ over the length l, where $I_o = -\mathrm{i}\omega Q$ and $\mathrm{i} = \sqrt{-1}$.

Here, we return again to Equation (5.13), viz: The electric field generated by this current at a range $r(\gg \lambda)$ from the antenna and angle θ from the z-axis has a component in the θ direction only, and is given by

$$\underline{E}_\theta = \mathrm{i}\frac{IlZ_o}{2\lambda r} \sin\theta \, e^{-\mathrm{i}kr}, \tag{5.64}$$

which, as discussed earlier, is a donut shape with the antenna located in the central hole of the donut and aligned perpendicular to the plane of the donut.

The radiation pattern of a linear antenna having arbitrary current distribution $I(z)$ $(-L \le z \le L)$ can be considered as a summation of small current elements as shown in Figure 5.6. Its far-field pattern is given by

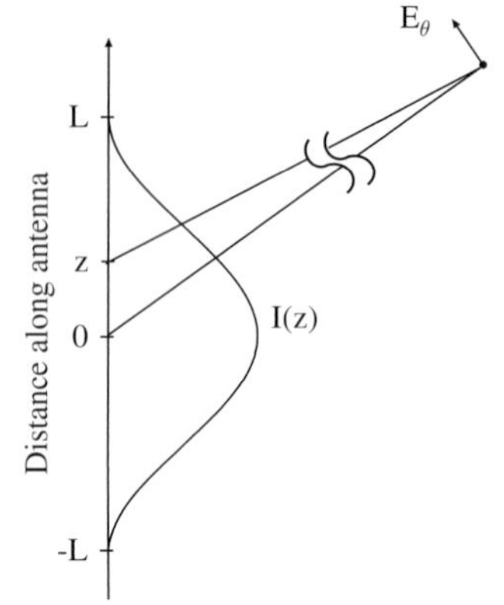

Figure 5.6 Radiation field generated by a linear current source.

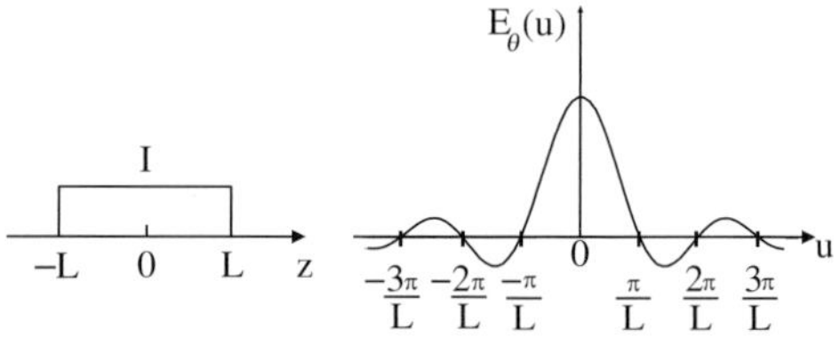

Figure 5.7 A uniform linear current source and its radiation pattern.

$$\underline{E}_\theta = \mathrm{i}\frac{Z_o}{2\lambda}\frac{\mathrm{e}^{-\mathrm{i}kr}}{r}\sin\theta \int_{-L}^{L} \underline{I}(z)\mathrm{e}^{\mathrm{i}kz\cos\theta}\,\mathrm{d}z\,. \tag{5.65}$$

Replacing θ by $u = k\cos\theta$, we obtain

$$\underline{E}_\theta(u) \propto \int_{-L}^{L} \underline{I}(z)\mathrm{e}^{\mathrm{i}zu}\,\mathrm{d}z\,, \tag{5.66}$$

which is the inverse Fourier transform of the current distribution $I(z)$.

The radiation pattern is thus easily computed if the current distribution is given. Figure 5.7 shows an example for the case of a uniform distribution.

As clearly seen in the figure, the radiation field has a maximum (main lobe) in the direction perpendicular to the current. Multiple smaller maxima in other directions are termed side-lobes.

5.2.4 Reflector antenna

The discussion in the previous section can be readily expanded into a two-dimensional case. If the distribution of the electric field or the magnetic field $f(x, y)$ is given for a two-dimensional aperture on the ground, the directional pattern of the radiated field is given by

$$\underline{F}(\theta, \phi) = \underline{F}'(\theta_x, \theta_y) = \int\int f(x, y)\mathrm{e}^{\mathrm{i}k(x\cos\theta_x + y\cos\theta_y)}\,\mathrm{d}x\mathrm{d}y, \tag{5.67}$$

where θ_x and θ_y are the angle of the observed direction measured from the x and y axes, respectively, and the cosines of these angles are the direction cosines (also see (5.34)). (The two functions F and F' are the same physical function but depend on different variables and so have different dependencies on their respective arguments.)

The radiation pattern is thus given by a two-dimensional Fourier transform of the source distribution, as already discussed.

For example, the half-power full-width (angular distance between the two points where the radiated power becomes half of the central point of the beam) of a circular aperture with diameter $D(\gg \lambda)$ illuminated uniformly is given by (see Equation (5.37))

$$\theta_h \simeq \frac{1.02\lambda}{D} \tag{5.68}$$

in units of radians (1 radian $\approx 57.3\,°$). This relation shows that the beam-width of a large aperture antenna is roughly given by the inverse of the antenna diameter measured in units of wavelengths.

The level of side-lobes is also determined by the source distribution. While the maximum side-lobe level of a linear and square aperture with uniform illumination is 13.2 dB down from the peak of the main lobe, it is 17.6 dB for a circular aperture. As easily understood by the relation of the Fourier transform, the main cause of side-lobes is the sharp edge of the source distribution. It is thus possible to reduce the side-lobe levels by gradually reducing the intensity of illumination at the outer edge of the aperture. However, the beam-width is increased as the effective size of the source is reduced.

One way of configuring a large antenna (large compared with the wavelength) is to use a reflector surface, such as the paraboloid. In the case of a parabolic reflector antenna, the transmitted waves from the initial feed located at the focus point of the paraboloid are reflected by the main reflector surface, and then form a plane wave at the aperture plane. The radiation pattern of this type of antenna is determined first by computing the illuminated field at this aperture plane from the initial feed, and then by substituting this distribution as $f(x, y)$ in Equation (5.67).

In this type of antenna, it is necessary to reduce the weight of $f(x, y)$ at the outer edge of the illuminated field in order to prevent the emission of the initial feed outside the main reflector (called spill over). The effective area is thus usually somewhat smaller than the physical aperture of the main reflector.

For large antennas like those used for atmospheric radars, it is not an easy task to support the heavy structure consisting of the initial feed and the waveguide above the main reflector. This structure also blocks part of the reflected waves, and the scattering caused by them significantly increases the side-lobe level. To avoid this problem, a sub-reflector configuration is commonly used. Figure 5.8 shows cross-sectional views of various types of reflector antennas.

The Cassegrain feed uses a hyperboloid, one of whose foci matches the focus of the main reflector, and the initial feed is located at the other focus. The Gregorian feed uses an ellipsoid instead.

As the problem of increased side-lobe level due to blocking of the reflected waves by the sub-reflector still persists with these configurations, offset antennas are used in

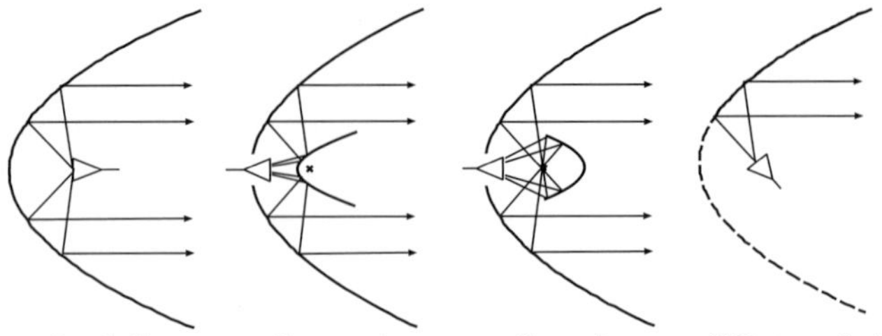

Figure 5.8 Cross-section of reflector antennas.

applications where the side-lobe level is the major concern. Because of the anisotropic configuration of the reflector, offset antennas have the problem of poorer polarization characteristics, which means that the radiated field contains polarizations other than that excited by the initial feed.

The inherent limitation of the reflector antenna in atmospheric radar application is the beam steering. The antennas discussed above have a single focus point and correspondingly a single beam. It is thus necessary to mechanically steer the entire structure to change the beam direction. In communications applications, especially those for satellite communications, the configuration of multiple beams using a single reflector surface has been intensively studied. It has been found that effective allocation of beams is usually limited to beams tightly arranged in a narrow angular region, which is not useful for atmospheric radars.

One unique exception is the spherical reflector antenna used for the Arecibo incoherent scatter radar system. As a plane wave incident upon a spherical surface produces a focal line (instead of a focal point as for parabolic reflectors), a slotted waveguide (called a line feed) with properly controlled inside phase speed can collect the energy (*La Londe*, 1979; *Kildal*, 1986). By moving this line feed, the beam direction can be changed within a wide angular range without moving the reflector surface, although the illuminated region is limited to a part of the entire reflector surface. A more sophisticated initial feed system consisting of three sub-reflectors, which converts the focal line into a single focus point, has been introduced at Arecibo to cover a very wide frequency range. It is called a Gregorian feed because the sub-reflectors are placed above the focusing area of the main reflectors, but in reality it has a more complicated design than the original Gregorian feed for a parabolic reflector (*Kildal et al.*, 1994).

5.2.5 Array antenna

Another useful method of realizing a large effective area is to use an array of small antennas. Each antenna, which is called an element antenna, can be of any type. In the case of atmospheric radar, half-wave dipoles, Yagi–Uda antennas, and CoCo (coaxial co-linear) antennas are commonly used, as discussed later.

The phase relation is of crucial importance in arranging and connecting the element antennas. If the electromagnetic fields generated by two elements at the target location have the opposite phase, they cancel each other. In other words, the desired beam is generated in a direction where fields from element antennas arrive in phase. Figures 5.9 and 5.10 schematically show wavefronts generated by a linear array.

Figure 5.9 shows the case where all elements are excited with the same phase. In this case a wavefront parallel to the array propagates perpendicular to the array. On the other hand, the wavefront is generated in a tilted direction in Figure 5.10, and propagates in an oblique direction, acting as if the entire array were tilted. Although these figures illustrate situations for a transmitting antenna, the same relation applies to a receiving

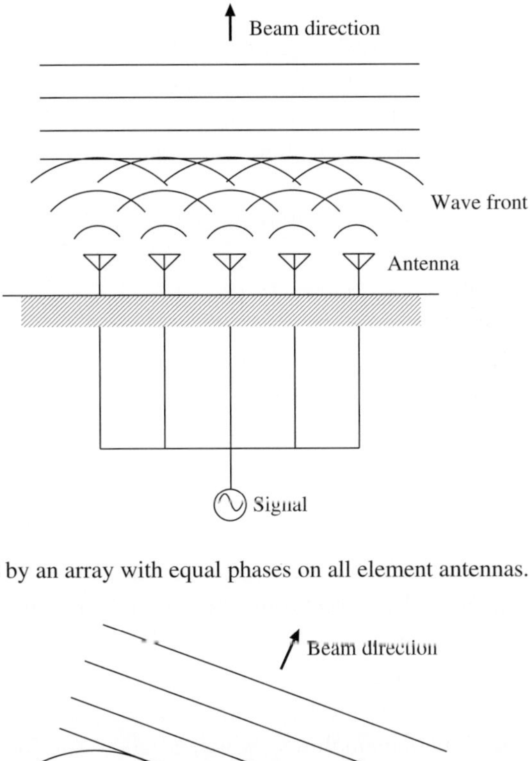

Figure 5.9 Wavefront generated by an array with equal phases on all element antennas.

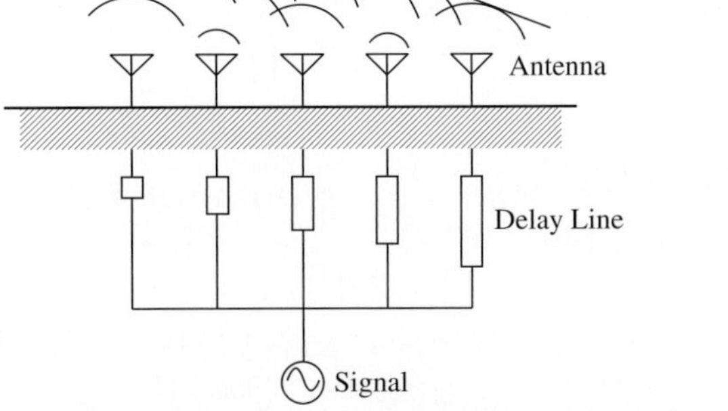

Figure 5.10 Wavefront generated by an array with constant phase progression across the element antennas.

antenna. An array antenna with the capability of controlling the phase of individual elements is called a phased array antenna. Phased arrays with electronic phase control can change the beam direction very rapidly, and thus are useful for atmospheric observations.

It should be noted in regard to Figures 5.9 and 5.10 that these diagrams give the impression of the formation of perfect plane waves, but due to the limited extent of the element antennas, these plane waves will be diffraction-limited. Nevertheless our arguments are still largely valid.

Next we examine the characteristics of the array antenna more precisely. For simplicity, we consider the case of linear array with equal elements arranged along the x-axis. Planar arrays and arrays on a curved surface can be treated in a similar manner, but with some complications. If the directional pattern of a linear array consisting of N elements is, for the case of no phase offsets applied to any of the antennas, given by $F_e(\theta, \phi)$, then the directional pattern once phases are applied is given by

$$\underline{F}(\theta, \phi) = F_e(\theta, \phi) \sum_{i=1}^{N} a_i e^{i\psi_i}, \tag{5.69}$$

where a_i is the amplitude of each element, and ψ_i is the phase, which is given as

$$\psi_i = kd_i \cos\theta_x + \delta_i, \tag{5.70}$$

where d_i is the location of each element, and δ_i is the phase of excitation, which determines the direction of the main beam. This equation shows that the characteristics of an array antenna are expressed as a product of the characteristics of the element and a term expressing the effect of the array, which is called the array factor.

The array factor indicates the characteristics of the array independent of the type and characteristics of its elements. In an actual case, however, the current induced at an element in the midst of the array is different from that of the elements at its outer edge because of the mutual coupling between elements, and thus Equation (5.69) does not hold in a rigorous sense.

We consider the simplest case where the elements are evenly spaced at a distance d with equal excitation amplitude, which is called a uniform array. We further assume that the phase of excitation linearly progresses from one end of the array to the other with a constant increment δ between adjacent elements:

$$\psi_i = \left(n - \frac{N-1}{2}\right)\psi, \tag{5.71}$$

where

$$\psi = kd\cos\theta_x + \delta. \tag{5.72}$$

The normalized array factor is expressed as

$$A_{\mathrm{arr}}(\psi) = \frac{1}{N}\sum_{i=1}^{N} e^{i(n-\frac{N-1}{2})\psi} = \frac{\sin\left(\dfrac{N\psi}{2}\right)}{N\sin\left(\dfrac{\psi}{2}\right)}. \tag{5.73}$$

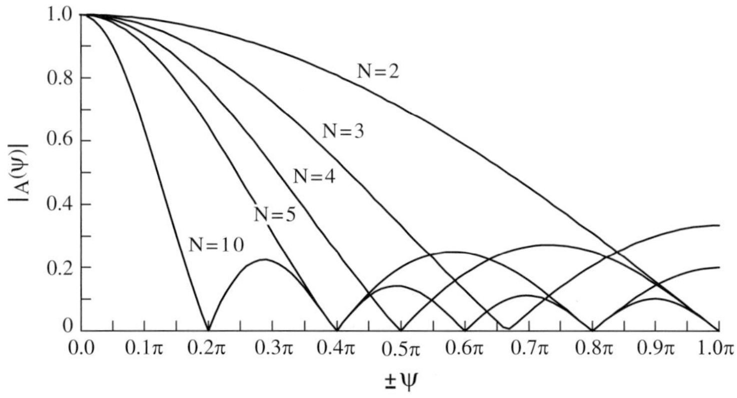

Figure 5.11 Array factor of a uniform array with a constant interval.

Figure 5.11 shows $A_{arr}(\psi)$ for several N values.

The peak at $\psi = 0$ shows the main lobe, and other smaller peaks correspond to side-lobes. The main beam direction $\theta_{x\max}$ measured from the x-axis is expressed as

$$\theta_{x\max} = \cos^{-1}\frac{-\delta}{kd}. \tag{5.74}$$

When N becomes large, the array factor approaches the characteristics of a linear current source with uniform current distribution. As in the case for reflector antennas, application of tapering to the amplitude weight of the elements is effective in reducing the side-lobe levels. In a large array, reduced weight at the outer edge can also be achieved by a scheme called thinning, which is the random removal of elements to reduce the density of elements.

As is clear from Equation (5.73), the array factor is a periodic function of ψ with a period 2π. However, owing to the condition $-1 \leq \cos\theta_x \leq 1$, only a region

$$-kd + \delta \leq \psi \leq kd + \delta \tag{5.75}$$

appears in a real antenna pattern. This region is called the visible range.

It can be seen from this equation that multiple main lobes are included in the visible range when $kd \geq 2\pi$ holds. This is because radiation from all elements is added in phase in multiple directions when the interval between elements is larger than one wavelength. In this case, main lobes appear in a direction other than the desired one, and these are called grating lobes. On the other hand, no grating lobe appears when $kd < \pi$, which means that the interval is less than half of one wavelength. For an intermediate value of $\pi \leq kd < 2\pi$, the behavior of the grating lobe depends on the value of δ. The condition for d at which the grating lobe does not appear is given by

$$d < \frac{\lambda}{1 + \cos\theta_{x\max}}. \tag{5.76}$$

The grating lobe appears at a smaller value of δ as kd is increased.

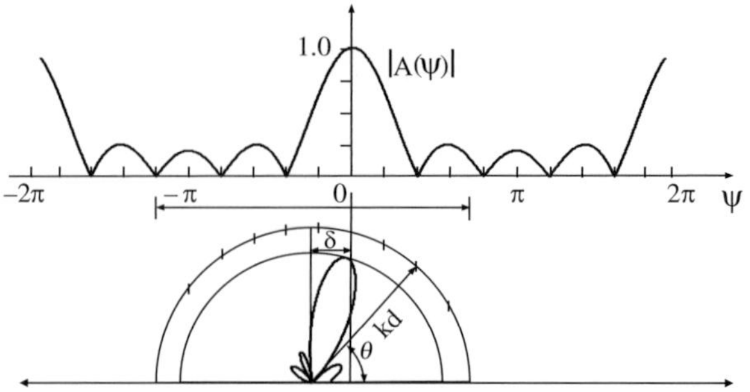

Figure 5.12 Relation between the array factor and the radiation pattern.

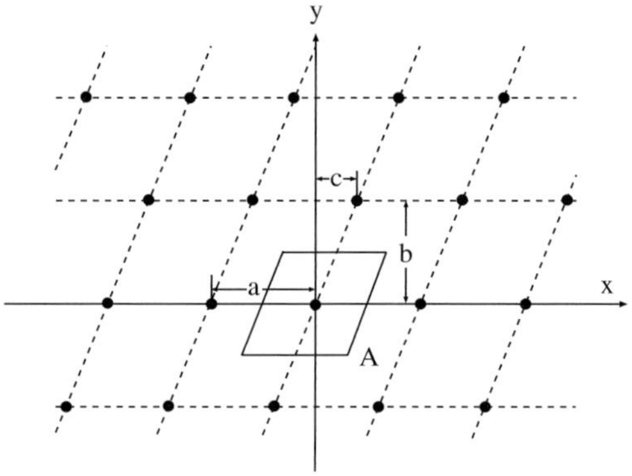

Figure 5.13 General triangular arrangement of array elements (dots) on the antenna plane. The area A in the figure denotes an element unit cell.

Figure 5.12 schematically shows the relation between the array factor and the actual antenna radiation pattern (*Weeks*, 1968). It can be seen that the visible range increases as kd is increased, and that the grating lobe(s) appears when the visible range includes $\psi = \pm 2\pi, \pm 4\pi$, etc. In a phased array antenna, the maximum beam steering angle, and thus the maximum value of δ, should be controlled so that the grating lobe does not appear in the visible range. It should be noted that the condition in Equation (5.76) is for the center of the grating lobe, so the actual limit should be set more conservatively so that high-level side-lobes of the grating lobe are kept outside the visible range.

The discussion of the visible range can be readily expanded to a two-dimensional case. The phase term ψ_i in the directional pattern of Equation (5.69) now becomes

$$\psi_i = k(x_i \cos\theta_x + y_i \cos\theta_y) + \delta_i, \tag{5.77}$$

where (x_i, y_i) gives the location of each element. When the elements are arranged on a general triangular grid shown in Figure 5.13, Equation (5.69) can be written as (e. g. *Mailloux*, 2005)

$$F(\theta, \phi) = F_e(\theta, \phi) \sum_{n=-\infty}^{\infty} \sum_{m=-\infty}^{\infty} w_{nm} \mathrm{e}^{ik(na+mc)(\cos\theta_x - \theta_{x0}) + ikmb(\cos\theta_y - \theta_{y0})}, \tag{5.78}$$

where w_{nm} is 1 for the N elements of the aperture and 0 for all other locations, where a, b, and c denote the intervals between the elements, and where $(\theta_{x0}, \theta_{y0})$ is the direction of the main beam.

The array factor of Equation (5.78) takes maxima in the directions

$$\begin{aligned}\theta_x &= \cos^{-1}(\cos\theta_{x0} + \lambda n/a)\\ \theta_y &= \cos^{-1}(\cos\theta_{y0} + \lambda m/b - \lambda cn/ab),\end{aligned} \tag{5.79}$$

where $n = m = 0$ corresponds to the desired main beam, while others are grating lobes if such combinations of n and m fall into the visible range. The condition for which no grating lobe appears is given by

$$\max\left[b, \frac{a}{2}, \frac{a}{\sqrt{1+c^2/b^2}}\right] < \frac{\lambda}{1+\sin\theta_{\max}}, \tag{5.80}$$

where $\theta_{\max}$ is the zenith angle of the main beam, and $-a/2 < c \le a/2$.

Here we derive the shape of the grid that allows the largest value of $\theta_{\max}$ for a given size of the unit cell $S = ab$. The problem is to find the minimum of the function

$$f(a) = \max\left[\frac{S}{a}, \frac{a}{2}, \frac{a}{\sqrt{1+a^2c^2/S^2}}\right]. \tag{5.81}$$

Clearly the third term takes its minimum value for the maximum value of c, which is $a/2$. Then $f(a)$ takes its minimum value $\sqrt[4]{3/4}\sqrt{S}$ when $b = \sqrt{3}a/2$ and $b = a/2\sqrt{3}$. These values correspond to an equilateral triangle, which gives the widest scanning angle for a given density of elements. For example, for a square grid array of $a = b = 2\lambda/3$, the grating lobe appears when $\theta_{\max} = 30\,°$. The equilateral triangular grid with the same element density is $a = 0.72\lambda$ and $b = 0.62\lambda$, which allows $\theta_{\max} = 37.8\,°$.

For a large two-dimensional array of omnidirectional elements arranged on the x–y plane, the total effective area as a function of angle is approximately given by (e. g. *Mailloux*, 2005)

$$A_e(\theta,\phi) = \frac{N\lambda^2}{4\pi}|A_{\text{arr}}(\psi)|^2\cos\theta. \tag{5.82}$$

The $\cos\theta$ term expresses reduction of the effective area as seen from an oblique angle, which is called scan loss. It is found from this equation that each element in the array can be regarded to have an effective area of

$$A_{ee}(\theta,\phi) = \frac{\lambda^2}{4\pi}\cos\theta, \tag{5.83}$$

which is contrary to the assumption that the element has an omnidirectional pattern. This is explained by the fact that the mutual coupling between nearby elements in the array modifies the element pattern. Such a pattern is called an array element pattern. When the antenna pattern is modified, the radiation impedance of the element also changes. It should be noted that the radiation impedance is a function of the main beam direction, although the variation is usually not large.

5.2.6 Element antenna for array

As Equation (5.69) indicates, the total gain of an array antenna is proportional to the element antenna gain, G_{ae}, and N^2. A high gain is thus achieved by increasing G_{ae} or by increasing the number of elements N. As Equation (5.12) shows, however, increased gain means an increased effective area. For the case of the two-dimensional arrays used for atmospheric radars, the effective area of the element antenna cannot exceed its physical area determined by the interval of elements. If the former exceeds the latter, the mutual coupling between the elements reduces the effective area.

For example, for a square array with $N(\gg 1)$ elements uniformly arranged at an interval d in two orthogonal directions, the total effective area is apparently limited by the physical area of Nd^2. Even if the effective area A_{ee} of each element in the free space is larger than d^2, each element can receive the energy entering into an area of no more than d^2. The reduction of the effective area takes place as the result of interference between adjacent elements through mutual coupling. On the other hand, if $A_{ee} < d^2$, the total effective area of the array is roughly given by NA_{ee}. The optimum choice in terms of the best use of the antenna area is to employ the element whose effective area is roughly equal to d^2. From Equation (5.76), d should be less than λ for a phased array with the beam direction near $\theta_{\max} = \pi/2$, and $\lambda/2$ for a fully steerable beam. Thus, from Equation (5.12), it is found that the element gain should not exceed 4π (= 11 dB) even for a very limited scan angle, and 2π (= 8 dB) if a wide scanning angle is required.

The gain of a half-wave dipole in free space is 1.6 (= 2.16 dB). If it is horizontally set at a height of $\lambda/4$ above a perfectly conducting ground, the radiated field toward zenith is roughly doubled, and thus the gain increases by four times to about 8 dB, which is already sufficient for a dense array of $\lambda/2$ intervals (e.g., *Fukao et al.*, 1986a).

Some reduction in the number of elements while maintaining the total area is obtained by using the Yagi–Uda antenna commonly used for reception of VHF/UHF television (e. g. *Balanis*, 1997). This type of antenna is a dipole antenna equipped with slightly shorter unfed sub-elements, called directors, on its front, and a slightly longer unfed sub-element, called a reflector, on its back. The gain of a 3-to-4-element Yagi is 7–10 dB, depending on its design. The interval between the antennas is chosen to be around 0.7λ. Although the maximum beam steering angle from the zenith is reduced, it is not a serious limitation because the $\cos\theta$ term in Equation (5.82) limits the steering angle of large arrays anyway.

The above restriction for the maximum beam steering angle applies to the case where the beam is to be directed to an arbitrary azimuth direction. When the beam is steered only in one azimuth plane, no restriction is set on the arrangement in the direction perpendicular to this plane. One extreme case is the so called CoCo (coaxial co-linear) antenna (e.g., *Balsley and Judasz*, 1989). This antenna consists of a coaxial cable with its inner and outer conductors alternately reversed at every half wavelength. While the inside of the coaxial cable serves as a transmission line, the outside of the outer conductor works as the radiating element. As a result this antenna works as a linear array of half-wave dipoles fed serially at its end. Usually a CoCo antenna has a length of several to a few tens of wavelengths and is fed at its center point as a very long dipole. Naturally, all of the half-wave dipoles are excited in the same phase, so arrays of CoCo antennas can be steered in a single plane perpendicular to their axis. The major limitation of the CoCo antenna is its inherently narrow bandwidth. As the length of the line increases, the condition for reversing at the half-wavelength can be satisfied only in a narrower bandwidth. Reduction of the bandwidth can also lead to extra losses as more frequency components of the pulse are suppressed. Similar antennas using a coaxial cable coupled with radiation elements at constant intervals are designed and used for atmospheric radars.

We have so far disregarded the issue of polarization. Scattering from atmospheric irregularities is in general co-polar, which means that the scattered waves have the same polarization as the incident wave. Thus any antenna that transmits and receives the same polarization can be used for MST applications, for which the medium is transparent. However, the ionospheric plasma, which experiences the effect of the Earth's magnetic field, causes rotation of the polarization plane, which is called Faraday rotation (e.g., *Farley*, 1969; *Hysell and Chau*, 2001). Faraday rotation was also discussed extensively in Chapter 3. In this case the received wave has a different polarization plane from the transmitted one. In order to adapt to this problem, a pair of crossed antennas, such as crossed dipole or crossed Yagi's, is used. By exciting the two orthogonal elements with a phase difference of $\pi/2$, a circular polarization can be transmitted and received. A helical antenna is commonly used to generate circularly polarized waves for microwave applications, but these are not a preferred choice for atmospheric radars of VHF or UHF bands, where the physical structure of each element becomes large.

5.2.7 Antenna impedance and matching

The radiated power from a transmitting antenna is computed by integrating the Poynting flux $\underline{\vec{E}} \times \underline{\vec{H}}^*/2$ over the surface of a sphere surrounding the antenna. For an infinitesimal electric dipole, it is given by using Equation (5.13) as (e.g. *Balanis*, 1997)

$$P = \oint \frac{|E_\theta|^2}{2Z} dS = \frac{\pi Z_o}{3} \left(\frac{Il}{\lambda} \right)^2 . \tag{5.84}$$

From the viewpoint of an electrical circuit, this power can be considered as being consumed by an imaginary resistance, which is called the radiation resistance, although of course in truth it is being usefully radiated into the space surrounding the antenna.

The radiation resistance of an antenna is given by the relation

$$P = \frac{1}{2} I^2 R_r, \tag{5.85}$$

and is computed for the small dipole as

$$R_r = \frac{2\pi Z_o}{3} \left(\frac{l}{\lambda} \right)^2 . \tag{5.86}$$

The total radiation impedance $\underline{Z} = R_r + iX$ can be defined in a similar manner, but by integrating the flux over the volume covering the antenna, which is usually more complicated. For a linear antenna, it is given by

$$\underline{Z} = -\frac{1}{|I_o|^2} \int_{-L}^{L} \underline{E}_z(z)\underline{I}(z)\mathrm{d}z, \tag{5.87}$$

where I_o is the current at the feed-point of the antenna. In the general case, the current distribution $I(z)$ is unknown, and should be computed by solving an integral equation for $I(z)$. For a thin dipole antenna with length $2L \simeq \lambda/2$, it is approximated by

$$I(z) = I_o \sin\left\{k\left(\frac{l}{2} - |z|\right)\right\}, \tag{5.88}$$

where $k = \frac{2\pi}{\lambda}$. For a half-wavelength dipole, the radiation impedance is computed to be $Z = 73.13 + \mathrm{i}42.55\,\Omega$.

In order that the antenna efficiently radiates the power provided from the transmitter, the radiation impedance should be well matched to that of the feed line. The matching is usually measured in terms of the voltage standing wave ratio (VSWR), which is the ratio of the highest to the lowest amplitude of the voltage observed along the feed line, and is given by

$$\mathrm{VSWR} = \frac{1 + |\underline{\mathcal{R}}|}{1 - |\underline{\mathcal{R}}|}, \tag{5.89}$$

where $\underline{\mathcal{R}}$ is the voltage reflectivity given as

$$\underline{\mathcal{R}} = \frac{\underline{Z} - \underline{Z}_l}{\underline{Z} + \underline{Z}_l} \tag{5.90}$$

and where $\underline{Z}_l$ is the characteristic impedance of the feed line. For the case of linear antennas, the reactance component X varies significantly as the length of the element is changed. For example, it becomes zero for a dipole of length 0.47λ. On the other hand, this large variation with length limits the bandwidth of linear antennas, because the length in terms of the wavelength varies as the frequency changes.

The intrinsically narrow bandwidth of linear antennas can also be interpreted as the result of resonance. This type of antenna has a high radiation efficiency because it makes use of a resonance at around half-wavelength. It is often desirable to reduce its Q-value in order to broaden the bandwidth. One way is to employ thicker elements, a method used at UHF or higher frequencies. The biconical antenna, which can be regarded as a dipole whose diameter increases from its center to the outer end, also has a wideband characteristic. Folded dipoles also have wider bandwidths than unfolded ones.

For the case of a Yagi antenna, the bandwidth can be broadened by making the feed element into a folded dipole. The radiation impedance can also be controlled by changing the distance between the sub-elements, as well as their length, through mutual coupling between the sub-elements as described below. By this approach, it is possible to obtain a relative bandwidth of more than 10% with this type of antenna, which is usually sufficient for atmospheric radar applications.

A parasitic element called a stub with one end open or short-circuited is also commonly used to adjust the radiation impedance of a linear antenna to that of the feed line. Another way of impedance matching is to use a quarter-wavelength transformer. As is well known, two lines with characteristic impedance of $\underline{Z}_1$ and $\underline{Z}_2$ can be matched by an inserted transmission line $\lambda/4$ in length with characteristic impedance of $\sqrt{\underline{Z}_1\underline{Z}_2}$.

When two antennas are located close to each other, the voltages induced at the feed points are given by

$$\begin{aligned} \underline{V}_1 &= \underline{Z}_{11}\,\underline{I}_1 + \underline{Z}_{12}\,\underline{I}_2 \\ \underline{V}_2 &= \underline{Z}_{21}\,\underline{I}_1 + \underline{Z}_{22}\,\underline{I}_2, \end{aligned} \tag{5.91}$$

where $\underline{Z}_{12}$ and $\underline{Z}_{21}$ are called the mutual impedances, as they represent an induced voltage on a second element due to the current flowing on a first element.

The reciprocity principle of the electromagnetic field assures $\underline{Z}_{12} = \underline{Z}_{21}$.

For an array antenna consisting of N elements, Equation (5.91) becomes

$$[\underline{V}] = [\underline{Z}][\underline{I}], \tag{5.92}$$

where $[\underline{Z}]$ is an $N \times N$-dimensional symmetrical matrix called the impedance matrix, and $[\underline{V}]$ and $[\underline{I}]$ are column matrices (vectors) holding voltages and currents respectively for the N antenna elements. The impedance matrix determines the relation between the current flowing on each element and the voltage at the antenna terminals. The ratio of the voltage to the current of each element in an array is called the active impedance. For a phased array with uniform illumination, the active impedance is given for the jth element as

$$\underline{Z}_j = \underline{V}_j/\underline{I}_j = \sum_{i=1}^{N} \underline{Z}_{ij} e^{i(\phi_i - \phi_j)}, \tag{5.93}$$

where ϕ_i is the excitation phase of the ith element. Obviously, the active impedance is a function of the phase relation between elements, which is determined by the beam direction.

As it is undesirable that the active impedance varies as the beam direction is changed, it is important to keep the mutual impedance sufficiently smaller than the self-impedance $\underline{Z}_{ii}$. One partial solution is to make the interval between elements as large as possible, although this introduces problems for the requirement for a wide range of beam-steering angles. The triangular grid arrangement discussed in the previous section is beneficial because it helps to enlarge the distance between parallel antenna elements, whose mutual impedance is larger than that in the case of other orientations. The power amplifier that feeds the antennas should be designed to be robust against the expected variation of load impedance. For a three-element Yagi antenna in a large array with equilateral triangular grid with 0.7λ interval, the total of the mutual impedance between the surrounding elements is about 10% of the magnitude of the self-impedance (*Fukao et al.*, 1986b).

It is tempting to think that the most important mutual coupling effects associated with any antenna are due to its nearest neighbors, but this is not entirely true. Although the effect of more distant antennas is individually less, there are more of them. Suppose an array is assembled from a grid of Yagi antennas, with each Yagi tuned at its theoretical stand-alone best value. Once the array is assembled, all impedances are altered due to mutual coupling. Suppose that now the most central antenna is tuned to 50 Ω. Now suppose we move to the outermost antennas, and tune all of these. A new measurement of the central antenna will now show a change from 50 Ω, because even though the outer antennas are further away, there are more of them, with numbers roughly proportional to the distance from the central antenna (since the outermost antennas at radius r form a circle of perimeter $2\pi r$). Therefore when tuning an array in which mutual coupling is an issue, the user should always tune the outer antennas first and work in towards the center.

A reflector type antenna has a large advantage as far as the impedance characteristics are concerned, because its radiation impedance is determined by that of the initial feed, and in principle is not related to the configuration of the reflector surface.

Another issue in matching the antenna to the feed line is the problem of unbalanced current. The most commonly used feed line, namely a coaxial cable, is an unbalanced transmission line, which means that the potential of the outer conductor is kept equal to that of the ground, and only the voltage of the inner conductor varies. On the other hand, a dipole antenna is a balanced load, for which both sides of the element have equal and opposite voltages. Connecting a dipole antenna directly to a coaxial cable results in a current flowing on the outside of the outer conductor of the coaxial cable, which contributes to undesired radiation.

A device called a balun (balanced-unbalanced converter) is used for such a case (e.g., *Balanis*, 1997). One common type of balun is gamma match. In this case, the radiating element is a single rod of about a half-wavelength without a feed-gap, with its center point connected to the outer conductor of the coaxial cable. The inner conductor is connected to an offset point on the radiating element through a short parasitic element and a capacitor. By properly choosing the length and capacitance, the current on the outside of the outer conductor can be cancelled. The gamma match also works as an impedance transformer.

Another common type is the $\lambda/2$ coaxial balun, which is often called a U-balun. One feed-point of a dipole is directly connected to the inner conductor, while the other is connected through an extra coaxial cable of length $\lambda/2$, which reverses the sign of the signal. The U-balun converts the antenna impedance seen from the feed line to four times its original value, because the current on the radiating element is half of that in the coaxial cable, while the voltage is doubled.

5.2.8 Effect of random errors in an array antenna

Here we consider the effect of random phase error Δ_ℓ of element antennas. In this case the array factor of a uniform array becomes

$$\underline{A}_{\mathrm{arr}}(\psi) = \frac{1}{N}\sum_{\ell=1}^{N} \mathrm{e}^{\mathrm{i}\psi_\ell} \cdot \mathrm{e}^{\mathrm{i}\Delta_\ell}, \tag{5.94}$$

where we have for now allowed the array factor to be complex. The difference of the pattern from that without error $A_{0\mathrm{arr}}(\psi)$ is expressed as

$$\underline{\Delta}_A(\psi) = \underline{A}_{\mathrm{arr}}(\psi) - \underline{A}_{0\mathrm{arr}}(\psi) \simeq \frac{\mathrm{i}}{N}\sum_{\ell=1}^{N} \mathrm{e}^{\mathrm{i}\psi_\ell}\Delta_\ell, \tag{5.95}$$

where we have assumed that the Δ_ℓ are all small so that $e^{i\Delta_\ell} = \cos\Delta_\ell + i\sin\Delta_\ell \simeq 1 + i\sin\Delta_\ell \simeq 1 + i\Delta_\ell$. Assuming that Δ_ℓ are mutually independent random variables with mean 0 and standard deviation σ, the variance of $\underline{\Delta}_A(\psi)$ is given by (e.g. *Skolnik*, 2002)

$$< \underline{\Delta}_A^2(\psi) > = \frac{1}{N^2}\sum_{\ell=1}^{N}\sum_{m=1}^{N} < \Delta_\ell \Delta_m > e^{i(\psi_\ell - \psi_m)} = \frac{1}{N^2}\sum_{\ell=1}^{N} < \Delta_\ell^2 > = \frac{\sigma^2}{N}, \tag{5.96}$$

where $< >$ denotes an ensemble average. Note that the complex exponential terms have disappeared for similar reasons to those discussed in regard to Equations (3.166) to (3.169).

This result means that the side-lobe level is expected to increase from its designed level by σ^2/N, and that the level to which the random phase errors must be reduced depends on the level of side-lobes that is permitted. For example, if the side-lobe level of -40 dB from the peak of the main lobe is of concern for an array of $N = 500$, σ should be sufficiently smaller than 0.22 radian (13 °).

The same relation applies also to random amplitude errors by defining the amplitude of each element as

$$A_\ell = A_{\ell 0}(1 + \Delta_\ell), \tag{5.97}$$

where $A_{\ell 0}$ is the designed amplitude of each element. Errors in positioning can be treated in the same manner.

5.2.9 Digital beam forming (DBF) antennas

As we have seen in previous sections, the total pattern of an array antenna during reception is determined by adding the received signals of individual elements. Thus different patterns can be synthesized from the same received signals if they are recorded before adding. In this manner, multiple directions can be observed simultaneously.

This idea has been realized by recent advances in digital signal processing, such as the digital beam forming (DBF) antenna (e.g., *Drabowitch et al.*, 2005; *Mailloux*, 2005; *Frank and Richards*, 2008). Here the outputs from each antenna element or groups of elements are pre-amplified, demodulated, and then converted to digital signals, which can then be used to generate arbitrary receiving patterns during off-line processing. It should be noted that coherent integrations can be applied (if necessary) to individual receiver outputs, so that the off-line processing can be made on signals which were recorded at a relatively slow rate.

However, as this technique can be applied only for reception, it is not of much use to steer the receiving beams outside the illumination region of the transmitted beam. One possible way to expand the scope is to transmit a broad beam that covers the entire angular region of interest, and then observe it with multiple receiving beams. Let us briefly compare this method (hereafter referred to as the DBF method) with conventional beam steering (BS) using the same beam pattern both for transmission and reception. We assume that the number of beams is N and that the beam is switched sequentially at every transmitted pulse in a sufficiently short time in the latter case. In the BS method, the same beam direction can be observed only $1/N$ times compared with the DBF method. If signals can be added coherently over multiple cycles of beam scanning, the DBF method can improve the SNR by N times over the BS method. On the other hand, the transmitted beam for the DBF method should have a maximum gain of $1/N$ compared with the BS method. As a result, the SNRs for both methods are equal.

In this sense, the DBF method does not have an essential advantage over the conventional BS method in terms of sensitivity in atmospheric radar applications. If the transmitter beam illuminates regions that are not sampled by the receiving beams, additional loss factors take effect. Its advantage is in the fact that different directions can be measured exactly simultaneously, which is essential in interferometric observations. Another merit is that undesired echoes due to side-lobes of the transmitted antenna can be cancelled by designing the receiving pattern to have nulls in such directions. An adaptive signal processing approach is easily implemented for clutter suppression, for example.

As for the hardware, the merit of DBF is that no mechanism for beam steering, such as a phase shifter, is required for the transmitter if a wide beam is transmitted using a small antenna equipped with a high-power transmitter. This may lead to a large reduction in the cost of the transmitter compared to the BS configuration with an active phased array, which requires an array of N amplifiers with phase control capability. The negative aspect is of course the need for a large number of receivers, but fast A/D converters and low-cost digital signal processing boards are rapidly mitigating this problem, as discussed later.

5.2.10 The feed system

The mechanisms by which the transmitter feeds the antennas, and by which the antennas feed the receivers, are an important element of the antenna system, especially for an array antenna, where design factors such as transmission loss, impedance matching, and phase variation have a serious effect on its performance. The feed is often called a transmission line or occasionally a transmission feed. Technical details of transmission lines can be found in handbooks such as *Wadell* (1991).

As already discussed, a coaxial cable is the most popular transmission line for VHF and UHF array antennas. Its advantage is that the shielding by the outer conductor provides good isolation of the transmitted signal from other lines or antennas as long as the potential of the outer conductor is kept to that of the ground. However, cables have a relatively high dielectric transmission loss, in addition to possible problems related to balanced/unbalanced mode conversion. For example, a typical coaxial cable of 10 mm diameter with polyethylene insulator (10D-2V) has a loss of about 3 dB/100 m at a frequency of 50 MHz, and 10 dB/100 m at 430 MHz. The loss can be reduced to roughly half of this value by replacing the insulator with highly foamed polyethylene.

A Lecher wire system (two conducting wires in a parallel line) has a lower loss than the coaxial cable, and can be directly connected to dipole antennas since it is a balanced-mode transmission line. On the other hand, it is subject to interference between other lines and antennas owing to its open-air structure.

In a passive phased array antenna, it is necessary to divide the transmitted power to feed individual antenna elements. One convenient means is to use a $\lambda/4$ hybrid. A bifurcated transmission line followed by two bifurcated lines after a length of $\lambda/4$ constitutes a 4-port divider. From the same consideration as for the case of the $\lambda/4$ transformer, it is found that a perfect match is obtained when all parts of the lines are made with the same cable. It should be noted, however, that the matching is obtained only for the forward waves and for the received waves from the antennas with equal phase for all of the four ports. For other phase combinations of the received signals, the matching conditions differ, varying the receiver efficiency, which is why we have discriminated between L_{Tx} and L_{Rx}.

5.2.11 Beam steering and phase shifting

In a phased array antenna, the phase shifter is an essential component in determining the performance of the array (e.g. *Frank and Richards*, 2008; *Barton*, 2005). When the phase of a high-power signal must be changed, as is the case for a passive phased array in which the output power of the transmitter is subdivided to excite each antenna element, a ferrite latching phase shifter is commonly used. This device makes use of a particular property of the ferrite, namely that its magnetic permeability, and thus its electrical length, change according to the value of an applied DC magnetic field. The phase shifter consists of a cascade connection of switches, each of which changes the phase between its input and output by π, $\pi/2$, $\pi/4$, and so on, depending on the on/off switching of the DC bias.

This type of digital phase shifter can generate an arbitrary phase value with a fixed quantum error corresponding to one half of the minimum phase difference. The effect of this phase error on the antenna pattern can be evaluated by Equation (5.96). For this case of discrete errors, the effect can be roughly evaluated by replacing σ^2 with $h^2/12$, where h is the minimum step size of the phase shifter. Thus the necessary number of bits is determined by the required level of the side-lobe degradation.

Another type of digital phase shifter employs a series of extra delay lines of the length of $\lambda/2$, $\lambda/4$, $\lambda/8$, and so on, each of which is inserted or skipped by the control of a diode switch. For a frequency of UHF or higher, this configuration is readily realized with microstrip lines, and is commonly used for a solid-state active phased array.

An example of phased shifters with continuous phase control is a varactor diode, which works as a voltage-controlled capacitor when used under a reverse bias voltage.

5.2.12 Adaptive clutter rejection

Side-lobe cancellation or, in a more general sense, the adaptive antenna technique, has been an important issue in various fields of antenna engineering (e.g., *Fenn*, 2008). It is now gathering wide attention since the speed of signal processing devices has reached a level that enables real-time implementation of the technique.

In atmospheric radar applications, where a well-defined antenna beam is usually configured by a large array or aperture antenna, strong clutter echoes from surrounding mountains are the major source of interference. As the desired echoes from the atmosphere are so weak, even very weak echoes entering low-level side-lobes can become a serious problem, especially when they contain fluctuating components (*Sato and Woodman*, 1982a). The fluctuating components may arise, for example, from atmospheric fading. It has been a common practice to remove such undesired echoes in the offline signal processing after the data are recorded.

A clear advantage of the adaptive antenna is that it makes use of extra information about the direction of arrival in discriminating undesired echoes, in contrast to other clutter rejection schemes based on signal processing of the received echoes, such as adaptive filtering. As the adaptive filtering technique tries to remove the clutter after it is mixed with the desired echo, it cannot completely remove the clutter when the clutter has a fading component. It also tends to over-kill the DC component of the desired echo, as is the case of observing the vertical direction. It is thus a superior process to cancel the clutter echo before it is mixed with the desired echo by modifying the antenna pattern adaptively.

Here we describe an algorithm called DCMP (directionally constrained minimization of power) (*Takao et al.*, 1976). The procedure is described as a problem of finding optimal weights which are used in weighted sums of signals received at each antenna element. We assume there are N antennas, and that the signal arriving at the nth antenna at time t_ℓ is $x_n(t_\ell)$. We assume that the received signal is discrete and is sampled only at times $t_1, t_2, \ldots t_\ell, \ldots$. Then the signal arriving at all antennas at time t_ℓ is given by $\underline{\vec{x}}_n(t_\ell)$. Since we will use matrix notation, we will write $\underline{\vec{x}}$ as a column matrix $[\underline{x}]$, and denote the matrix element at time step ℓ as

$$[\underline{x}]_\ell = \left[\underline{x}_1, \underline{x}_2, \ldots, \underline{x}_N\right]_\ell^{\mathrm{T}}, \tag{5.98}$$

where the superscript T represents a matrix transpose, and in this case converts the row matrix (presented in the equation) to the desired column matrix. We also define a set of weightings for the N antennas by $[\underline{w}]$ viz.

$$[\underline{w}] = \left[\underline{w}_1, \underline{w}_2, \ldots, \underline{w}_N\right]^{\mathrm{T}}. \tag{5.99}$$

The optimal weight vector is given as a solution that minimizes the resulting average power

$$P_{\mathrm{out}} = \sum_\ell \mid [\underline{w}]^{\mathrm{H}}\, [\underline{x}]_\ell \mid^2 = \frac{1}{2}[\underline{w}]^H[\underline{R}][\underline{w}], \tag{5.100}$$

under the condition of constant gain to waves coming from the target volume, which is given by

$$[\underline{c}]^{\mathrm{H}}[\underline{w}] = \text{constant}. \tag{5.101}$$

Here, the superscript H represents the Hermitian conjugate, and

$$[\underline{c}] = \left[e^{-i\phi_1}, e^{-i\phi_2}, \ldots, e^{-i\phi_N}\right]^{T}, \tag{5.102}$$

where the ϕ_n terms are the phases required to point the mean beam of the antenna ensemble in the direction of interest.

The optimal weight $[\underline{w}]_{opt}$ is given by

$$[\underline{w}]_{opt} = [\underline{R}]^{-1}[\underline{c}]. \tag{5.103}$$

In the above equations, $[\underline{R}]$ is an averaged covariance matrix given by

$$[\underline{R}] = \begin{bmatrix} \Sigma \underline{x}_1(t_\ell)\underline{x}_1^H(t_\ell) & \Sigma \underline{x}_1(t_\ell)\underline{x}_2^H(t_\ell) & \ldots & \Sigma \underline{x}_1(t_\ell)\underline{x}_N^H(t_\ell) \\ \Sigma \underline{x}_2(t_\ell)\underline{x}_1^H(t_\ell) & \Sigma \underline{x}_2(t_\ell)\underline{x}_2^H(t_\ell) & \ldots & \Sigma \underline{x}_2(t_\ell)\underline{x}_N^H(t_\ell) \\ \vdots & \vdots & \ldots & \vdots \\ \Sigma \underline{x}_N(t_\ell)\underline{x}_1^H(t_\ell) & \Sigma \underline{x}_N(t_\ell)\underline{x}_2^H(t_\ell) & \ldots & \Sigma \underline{x}_N(t_\ell)\underline{x}_N^H(t_\ell) \end{bmatrix}. \tag{5.104}$$

Here, $\underline{x}_n(t_\ell)$ represents the signal received at the nth antenna at time t_ℓ, and all summations are over ℓ.

We have not proven all the statements above, since our intent here is to give only an overview. But in words, the idea is that by minimizing P_{out} (by optimizing the antenna weightings w_n), the method effectively breaks the signal into components from various directions and then essentially tries to cancel out those that are not close to the bore. However, it never specifically identifies the interfering signals, but simply removes them. Like any method, it has limitations. In principle, with N antennas we can deal with $N - 1$ sources, out of which one is the desired signal and the rest of at most $N - 2$ undesired signals can be cancelled. If there are more sources than antennas, the model may start to break down. In addition, errors can occur. For example, if a multipath signal reflected from a planar wall were to exist, it becomes a perfectly correlated interference to the desired signal. In this case DCMP gives the reflected signal exactly the same amplitude as the true signal, but with opposite phases for the two signals. Then the output power becomes zero while keeping the gain in the desired direction, which is the perfect answer, but has no use. However, this is a somewhat contrived example, and in general the method works quite well.

The method is especially well suited to removal of ground clutter, since undesired ground-echoes may be mutually correlated with each other, since fading along the path to the ground echoes is slow, and echoes from different ground echoes show similar behavior. However, these ground echoes do not correlate well with the desired atmospheric echo, especially if the target is moving. This is a good situation for DCMP, since it is well suited to removal of the partially correlated ground echoes, but retention of the atmospheric echo.

Figures 5.14 and 5.15 show altitude–velocity spectra before and after the adaptive clutter suppression process, respectively. Other similar graphs, and further discussion about the procedure used, can be found in *Nishimura et al.* (2006). The data were

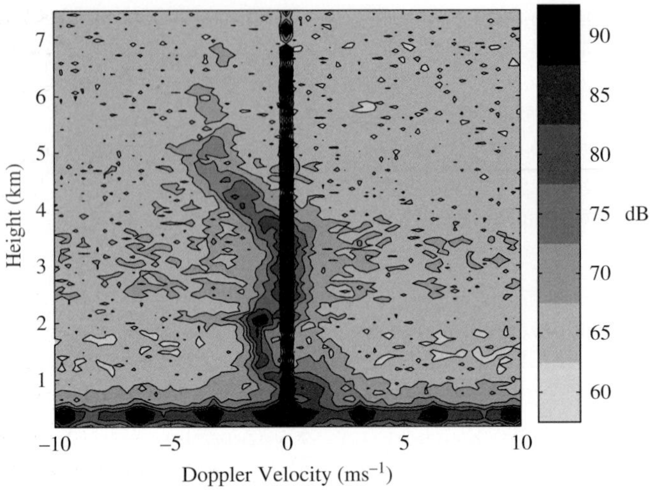

Figure 5.14 An altitude–velocity spectrum observed after a coherent sum of eight antennas out of ten. (Due to a stability issue with the equipment, only eight antennas worked during this period.) (More details about this graph and other similar graphs can be found in *Nishimura et al.*, 2006.)

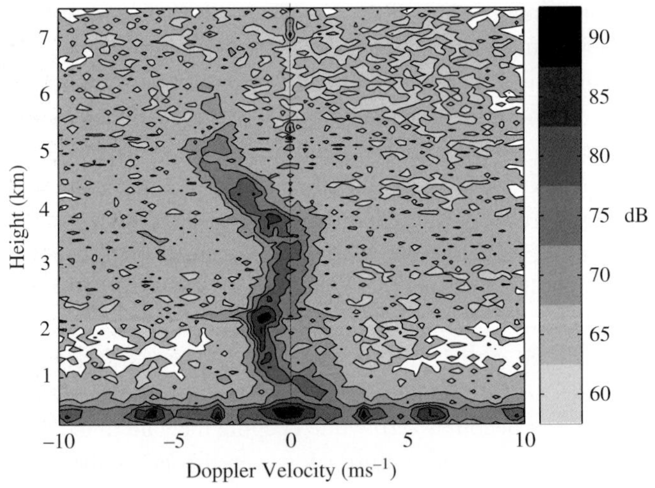

Figure 5.15 As for Figure 5.14 but with the adaptive spatial filtering procedure. (Other similar graphs can be found in *Nishimura et al.*, 2006.)

obtained with digital receiver systems consisting of ten Yagi antennas placed at a distance of 1.3 km from the Equatorial Atmosphere Radar, West Sumatra, Indonesia. The strong ground clutter component that is dominant at and around zero Doppler component in Figure 5.14 is almost completely suppressed in Figure 5.15. It should be noted that the atmospheric echo components with zero Doppler frequency seen at 2.5–3.5 km height are conserved in this processing. Additionally, broad spectral components spread over the entire frequency range seen in Figure 5.14 at 2.5–4 km height are clearly suppressed in Figure 5.15. These components are most likely due to fading ground clutter, which is spread over a wide frequency region by the process of windowing of the time series before FFT operation. There is a slight depression of the SNR as seen for the echoes at 6–7 km, which is the side effect of controlling the phase of individual antenna elements.

The original DCMP algorithm does not guarantee that the main lobe of the antenna keeps the desired pattern. For example, the bore of the main beam may be tilted after adaptation, especially when strong interference exists in a direction near to the main beam. This is undesirable for atmospheric radar. DCMP-CN is an algorithm to make sure that the bore direction is not altered by the adaptation process. A modified algorithm DCMP-CN (DCMP with constrained norm) is introduced to give a constraint that the main lobe is not altered by the adaptation (*Kamio et al.*, 2004). The principle of DCMP-CN is expressed as

$$\min_{[\underline{w}]} \left(P_{\text{out}} = \frac{1}{2}[\underline{w}]^H [\underline{R}][\underline{w}] \right), \tag{5.105}$$

$$\text{subject to } [\underline{c}]^T [\underline{w}]^* = \text{constant} \qquad \text{and} \qquad [\underline{w}]^H [\underline{w}] \leq U, \tag{5.106}$$

where U is the upper limit of the weight norm, which is directly related to the accepted side-lobe level. Note that, although expressed differently, the term P_{out} in these last equations is the same P_{out} as appears in Equation (5.100). This minimization problem with an equality constraint and an inequality condition is solved by using the penalty function method. Our discussion of this topic has been essentially an overview, and lacks some detail: more details can be found in the cited references.

5.3 Transmitter and receiver systems

We now leave the general area of antennas and move to consideration of the transmitters and receivers attached to the antennas.

5.3.1 System configuration

Although a variety of systems are employed in radars, depending on their use and designs, a general configuration is shown in Figure 5.16.

Since the purpose of an atmospheric radar is to detect the Doppler shift of the received signals, the receiver must maintain a constant phase relationship with the transmitter. Such a system is called a coherent radar, in order to discriminate it from a non-coherent radar, which can only measure the echo power. It should be noted that the incoherent scatter radar, which observes random scattering from ionospheric free electrons, is named after the incoherent nature of the target, but it is still a kind of coherent radar with Doppler capability.

The radio frequencies (RF) used in atmospheric radar studies ranges from around 30 MHz (lower VHF) to about 1 GHz (L-band). It is common to modulate and demodulate the transmitter and receiver signals at an intermediate frequency (IF) of 1–10 MHz,

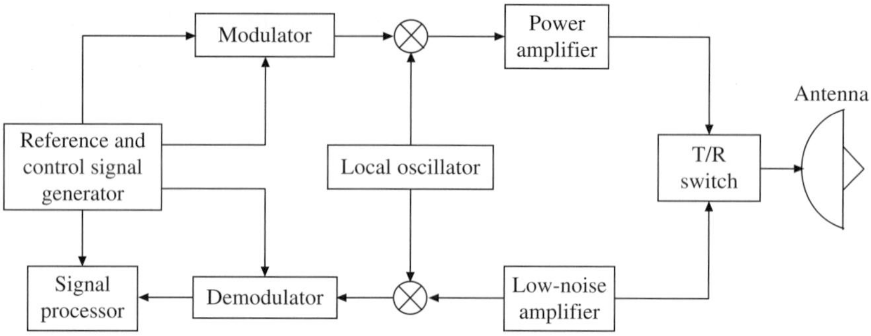

Figure 5.16 A general configuration of the radar system.

and then up convert to RF in the transmitter and down-convert from RF in the receiver. This was discussed extensively in Chapter 4. The IF oscillator and the local oscillator of a coherent radar system are customarily called COHO (COHerent Oscillator) and STALO (STAble Local Oscillator), respectively, for historical reasons.

5.3.2 Transmitter

Transmitters used for radars are classified according to the oscillator type and the amplifier type. One representative oscillator is the magnetron. It can directly generate the high power at RF with a single-stage unit, and thus enables a simple system configuration. The peak output power of magnetrons ranges from about 1 kW to several MW. They also have high efficiency, small size, and low cost. The main limitation of the magnetron for use in atmospheric radars is the low frequency stability. As it is a self-excited unit, signals generated by the unit always have frequency jitter, which is not acceptable for the purpose of detecting the small Doppler shifts associated with atmospheric echoes. One way of circumventing this problem is to sample and record the transmitted waveform of each pulse, and to correlate it with the received signals. This technique has been utilized with a 35 GHz cloud profiling system (*Hamazu et al.*, 2003).

Figure 5.16 is an example of an amplifier type in which the final output power is obtained by amplifying the IF signal from a reference signal generator and the RF signal after mixing with the local oscillator signal. The advantages of using the amplifier are high frequency stabilty and freedom in designing the transmitted waveform. Since the total gain of the transmitter may exceed 90 dB, it is crucial to separate the input and output signals. Amplifying at multiple stages with different frequencies has the effect of avoiding oscillations due to RF leakage of the final output to earlier stages.

The type of the final power amplifier depends on the configuration of the radar system. One kind may be called the concentrated system, with which the output of a single (or a small number of) power amplifier(s) is fed to a reflector-type antenna, or to an array antenna through a power-dividing network. For this kind of system, a high-power amplifier with peak output power of 10 kW–10 MW is required. Commonly used amplifier units for this category are vacuum tubes such as multi-cavity klystrons, travelling-wave tubes (TWTs), and cross-field amplifiers (CFAs) (e.g., *Skolnik*, 2002; *Weil and Skolnik*, 2008). Among these units, the klystron provides the highest peak output power of more than 10 MW, but it has a limited use in atmospheric radars because of its narrow bandwidth, low efficiency, and large size. While the peak output power of TWTs is limited to several 100 kW, they have a relatively wide bandwidth, and allow high duty cycles. The CFA has high efficiency and high output power, but has a low gain, and thus is used only for the final-stage amplifier.

Solid-state amplifiers are characterized by their long life and wide bandwidth, although the peak output power is limited to about 10 kW even in the case where the outputs of several units are combined (e.g., *Borkowski*, 2008). They are thus most suited for use in active phased array systems, in which the output power of many elements is combined in space. The element for the final power amplifier is usually a silicon bipolar transistor at the frequency of S-band or lower, and GaAs FET for C-band and higher. In contrast to the duty factor of a few per cent for vacuum tube amplifiers, solid state amplifiers can be used even with CW operation. The actual duty factor of solid state amplifiers is usually limited by the design of thermal radiator and power supply. The high duty factor of solid state amplifiers allows for the use of long pulses which are pulse-compressed with a high compression ratio, as discussed in detail in Chapter 4, Section 4.8.3.

One problem associated with the solid state amplifier has been its relatively low power efficiency. In order to have precise control of the transmitted waveform, a linear amplifier is required. A traditional design is to use a class-B push-pull amplifier, which has a theoretical upper limit of power efficiency of $\pi/4$ (78.5%). Actual amplifiers have much lower power efficiency, because the above limit does not include losses in the power supply.

A promising means to improve the efficiency is to employ switching amplifiers such as class-E and class-F (*Snider*, 1967; *Sokal and Sokal*, 1975). The basic idea is to keep the current on the transistor as a 50%-duty sinusoid as is the case for the class-B amplifier, but make the voltage rectangular so that there is no overlap between the current and voltage waveforms, as shown in Figure 5.17 (e.g., *Raab*, 1978).

As the power comsumption in the transistor is the product of current and voltage, the theoretical upper limit of the power efficiency is 100%. The class-E amplifier realizes this operation by adding matching capacitive and inductive elements to the

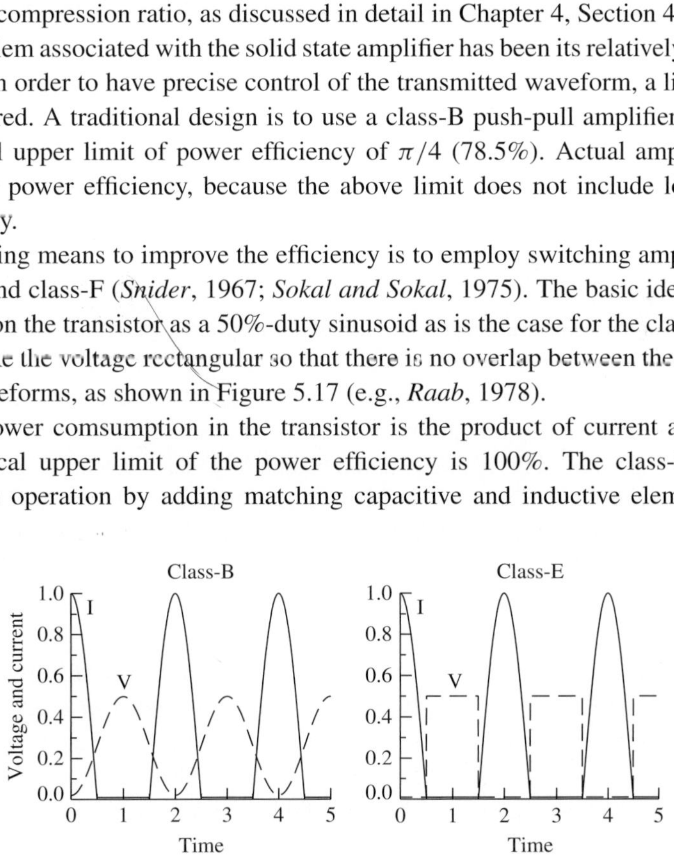

Figure 5.17 Voltage and current of class-B and class-E operations.

load impedance, and the class-F amplifier makes use of a distributed-element circuit for operating at higher frequencies.

Although the principle of class-E operation is old, it has become popular in recent years because of the development of high-speed and high-power switching devices. It was first utilized as an efficient power supply, and is increasingly applied to RF power amplifiers at higher frequencies as the available switching speed becomes faster. It is commonly used as the final-stage amplifier of mobile phones (e.g. *Sowlati et al.*, 1995; *Tsai and Gray*, 2002). The maximum output power is roughly inversely proportional to the operating frequency, with a typical value of 100–200 W at 50 MHz.

A critical issue of class-E amplifier design in applying it to atmospheric radars is its sensitivity to the load impedance variability. For the class-E operation to be realized, precise impedance matching is required. The active impedance of a phased array antenna is variable because the phase of the mutual coupling changes as the beam direction is steered. It is thus important to provide some means of isolating the power amplifier stage from the load impedance variability, such as insertion of a circulator which directs the reflected signal from the antenna to a dummy load.

Another issue is the output signal waveform design. A switching amplifier provides sharp rectangular pulses owing to the strong nonlinearity of the device. The relatively wide bandwidth associated with such pulses can be reduced by inserting a bandpass filter, but with an appreciable power loss. The output waveform can be modified to have slower rise and fall times by controlling the drain (or collector) via voltage waveform, although an optimized design is required depending on the characteristics of the device used.

5.3.3 The receiver

As the scattering echo power of atmospheric radars is very low, substantial gain is required for the receiver. The basic details of a receiver were discussed in the previous chapter: here we add some extra hardware-related comments, particularly pertaining to receiver noise and optimization.

An atmospheric radar should be able to at least detect the galactic background noise, because the minimum detectable signal before coherent integration is much lower than that noise level. The galactic noise power is discussed in Section 5.6.3. It should be possible to digitize the galactic noise to typically say 10 to 14 bits (depending on digitizer word-size), so that the weaker atmospheric signal embedded in the noise can also be digitized in the process.

Noise is a major consideration in receiver design. It is important to consider not only the galactic noise, but also the noise due to the receiver itself, noise due to the ambient temperature, noise from the antenna and feed lines, and also the amplification of noise as it progresses through the receiver. The noise characteristics of a receiver are determined by the noise figure, which is defined by the ratio of the input and output signal-to-noise ratios as

$$\mathcal{F} = \frac{(\mathrm{S/N})_{\mathrm{in}}}{(\mathrm{S/N})_{\mathrm{out}}}. \tag{5.107}$$

For a multi-stage amplifier whose gain and noise figure at stage i is expressed as G_i and $\mathcal{F}_i$, respectively, the noise figure is given by

$$\mathcal{F} = \mathcal{F}_1 + \frac{\mathcal{F}_2 - 1}{G_1} + \frac{\mathcal{F}_3 - 1}{G_1 G_2} + \cdots, \tag{5.108}$$

which is determined mostly by the noise figure of the front end.

At frequencies of less than 200 MHz or so, the galactic noise temperature significantly exceeds the environmental temperature, as we will see later. At these frequencies the noise figure of the receiver amplifier is not a critical factor. Noise levels are often expressed as a temperature, such that the noise power across a frequency bandwidth B_f (in Hz) equals Boltzmann's constant K_B multiplied by B_f and multiplied by the temperature. In a well-designed VHF receiver, the receiver noise should be less than the skynoise. Typical skynoise temperatures at 50 MHz should be of the order of 3000 K to 12 000 K, while typical receiver noise should be around 1000 K or less. On the other hand, at UHF and higher frequencies, skynoise is at a much lower temperature, and can be comparable to the receiver noise. Then the front-end amplifier is required to have low-noise characteristics. HEMT (high electron mobility transistor) amplifiers are most suited for this purpose.

It is difficult to design an amplifier with a very high gain of about 60 dB or more at a single frequency, because coupling between the output and input signals may cause instability. This can be mitigated by placing attenuators between successive stages to reduce feedback, although this is not often done. However, it was successfully employed by *Hocking et al.* (2014) to produce a stable gain of over 95 dB. More commonly, multi-stage amplifiers with high gain usually employ a superheterodyne configuration, with which amplification is made at two (or more) frequencies. GaAs Schottky barrier diodes are often used for the frequency conversion mixer because of their low-noise characteristics.

MMIC (monolithic microwave integrated circuit) is an effective technique to reduce the size and power consumption of a receiver. It is also useful to achieve the uniform phase and amplitude characteristics that are required for a large number of the devices used in an active phased array system.

An important part of a receiver is the filter. We discussed these in some detail from a general perspective in the previous chapter. Hardware filter design is heavily engineering orientated, and many books have been written about this topic alone. We will not pursue this aspect further here. However, we will comment that the future of receiver design will be more and more in software, with filters applied inside computer code. This will allow much wider variety in the types of filters that can be used for radar work.

5.3.4 TR switch

The concept of a TR switch is an important one, and such duplexers are a key component of most monostatic radars. We have discussed these devices briefly in the previous chapter, and refer the reader to that section for introductory material. Here, we give more details about hardware specifics.

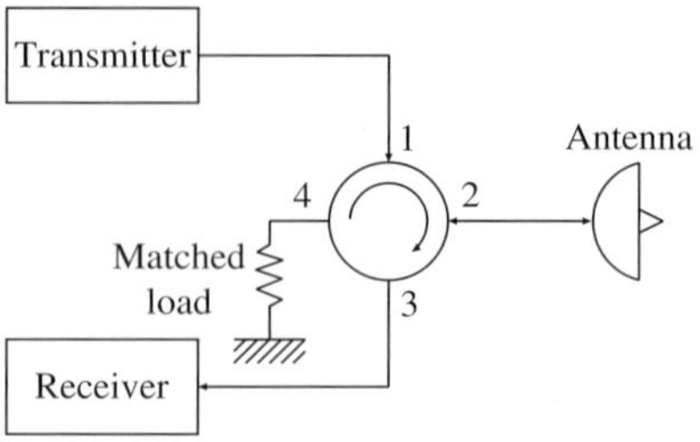

Figure 5.18 TR switch with a 4-port circulator.

The basic devices used for the TR switch are the directional coupler and the diode switch. One method of achieving directionality is to make use of the nature of ferrite, which becomes an anisotropic medium under a DC magnetic field. Figure 5.18 schematically shows a TR switch consisting of a 4-port ferrite circulator.

The signal enters from port 1 into the circulator and is output only to port 2. Similarly, the signal from port 2 is sent to port 3, and so on. A matched dummy load is attached to port 4 in order to absorb the signal reflected from the receiver and achieves a high separation between the transmitter and the receiver. The circulator has the advantage that the receiver can receive signals even while the transmitter is sending signals to the antenna, thus enabling CW operation. However, the isolation between the adjacent ports is relatively low, and it is aggravated when the transmitted signal is reflected at the feed point of the antenna because its radiation impedance is not well matched to the transmission line.

Actively controlled TR duplexers

An active TR duplexer is one in which the switching is actively controlled by electronic triggers from the radar controller. A diode is the principal component. The impedance between its two terminals under forward voltage bias mostly determines the isolation. Besides the conductance of the diode itself, the parasitic impedance, due to items such as the lead line inductance and the capacitance between the terminals, is a major factor which lowers the isolation. The open (reverse bias) and closed (forward bias) states of a diode can be reversed by connecting a diode through a $\lambda/4$ transmission line, which inverts the load impedance seen from the other end. In order that the open or closed state be ensured when a strong RF signal is applied, the bias voltage should be higher than the peak RF voltage, which is difficult or inefficient for switching the high-power signal from radar transmitters.

A PIN diode is a device that can switch high-power RF signals with a small DC bias voltage. It consists of three layers: Positive, Intrinsic, and Negative semiconductors. A charge carrier injected from the P (or N) layer into the I layer keeps its direction of motion even though the sign of the applied electric field is altered while it traverses the I layer. Thus the forward/reverse bias state is determined by the DC bias voltage regardless of the RF voltage.

Figure 5.19 shows an example of a TR switch consisting of 3 dB hybrid directional couplers and PIN diodes.

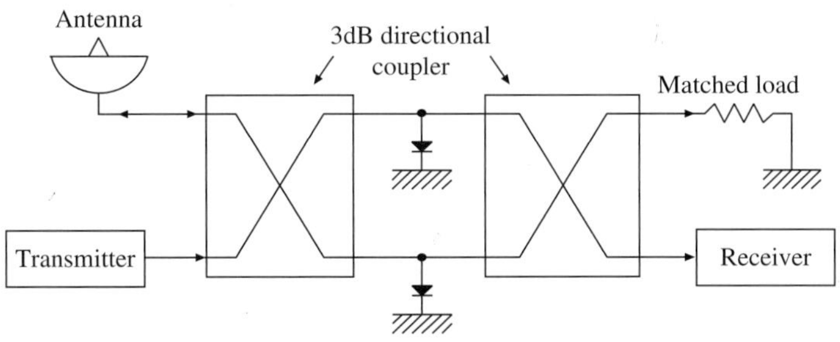

Figure 5.19 TR switch with 3 dB directional coupler and PIN diodes.

The 3 dB coupler splits the input signal into two outputs with equal amplitude, but with a phase difference of 90 °. This device can be realized by two parallel strip lines aligned at a small distance over a length of $\lambda/4$ at microwave frequencies, but also by lumped constant circuits at VHF and UHF.

The two PIN diodes are forward-biased during transmission. The reflected signals at the two diodes are synthesized with equal phase at the antenna port, but cancel at the transmitter port. The leaked signal is directed to the dummy load. During reception, the diodes are reverse-biased. The received echo from the antenna is synthesized at the receiver port, and the noise from the transmitter is directed to the dummy load.

Passive TR switches

In some cases, active switches are too slow. Usually, the pre-set delay at which switching occurs is set to be quite a bit longer than the pulse length in order to ensure that there is no possibility of damage to the receiver. A cleverer system could actually determine when the pulse strength has fallen to sufficiently low values itself, and then allow switching. This could, in principle, allow the user to start recording information much earlier than for the case of an active switch. This concept is the principle behind the design of so-called "passive TR switches."

Figure 5.20(b) (also shown in the previous chapter, but given again here for ease of reference) shows one example of a passive TR switch. Instead of using a trigger to motivate the switching, the system is self-switching. It works in the following way. The receiver is located at the end of a quarter-wavelength of coaxial cable (or an equivalent circuit), and sets of back-to-back diodes are placed across the end of the cable. When the pulse transmits, high voltages are generally created, often involving powers of the order of kilowatts. This causes the back-to-back diodes to break down and short-circuit, resulting in a load impedance of approximately zero. From transmission line theory, it is known that the impedance looking into a quarter-wavelength cable is the characteristic impedance squared divided by the load impedance (receiver impedance). As a result of the short-circuited diodes, the impedance at the input of the coaxial cable is essentially infinite, thus blocking the transmitted signal from entering the receiver. As soon as the transmitter voltages drop below 0.7 volts, which is the breakdown voltage of typical diodes, the back-to-back diodes become open circuit. Hence the impedance seen looking into the cable is now no longer infinite, and equals the receiver impedance transformed

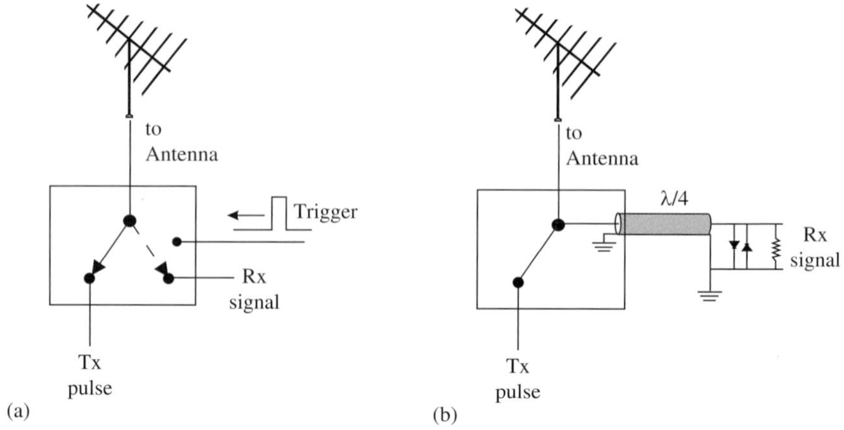

Figure 5.20 (a) The concept of a transmit-receive (TR) switch. (b) Design of a simple passive TR switch. (This figure was also shown in the previous chapter, but is discussed here in greater detail.)

by a quarter-wavelength cable. Normally, both the cable characteristic impedance and the receive impedance will both be 50 Ω. Hence any signal remaining at the front end of the cable (which will include the received signal coming back from the antenna) will now be able to pass through to the receiver. In this way, the user can start recording the signal just as soon as the transmitter signal falls below 0.7 volts, and there are no system-related delays. This procedure can be of enormous benefit for radars, which need to sample as soon as the pulse has left the transmitter. It is a common device used in many modern MST radars. Practical application involves some subtle extra points which have not been discussed nor drawn in the figure. Here, we have tried to concentrate on the concept rather than on the details.

5.4 Radar signal acquisition system

Radar signal reception requires a receiver. In the past, these have been predominantly hardware orientated, but as time goes by, more and more of the receiver design is being incorporated into software. In the next few paragraphs, we will discuss practical aspects of various types of receivers, from purely hardware system through to modern software controlled systems.

The radar signal acquisition system consists of an A/D (analogue to digital) converter and a further digital signal processing unit. As explained in Chapter 4, Figures 4.6 and 4.9 and the associated text, we need to obtain two orthogonal baseband signals: in-phase (I) and quadrature (Q) components which together constitute a complex time series. Traditionally, this has been achieved by mixing the IF signal with local signals with orthogonal phases, and then the signal of each channel is separately A/D converted.

However, this method has two drawbacks. Firstly, it is not easy to keep the gain of the two channels equal. Secondly, it is difficult to keep the phase difference between the channels to exactly 90°. If one of the two channels has a small amplitude error ϵ ($\ll 1$)

and/or phase error δ ($\ll 2\pi$) relative to the other channel, the complex baseband signal can be written as (e.g. *Richards*, 2005)

$$\underline{s}(t) = I(t) + \mathrm{i}Q(t)(1+\epsilon)\mathrm{e}^{\mathrm{i}\delta}. \tag{5.109}$$

For the case of a single Doppler frequency signal, where $I(t) = \mathcal{A}_s \cos\omega_0 t$ and $Q(t) = \mathcal{A}_s \sin\omega_0 t$, the output signal becomes

$$s(t) \simeq \mathcal{A}_s\left(1 + \frac{\epsilon + \mathrm{i}\delta}{2}\right)\mathrm{e}^{\mathrm{i}\omega_0 t} - \mathcal{A}_s\frac{\epsilon + \mathrm{i}\delta}{2}\mathrm{e}^{-\mathrm{i}\omega_0 t}, \tag{5.110}$$

which contains a "ghost" Doppler component at frequency $-\omega_0$ with relative magnitude of $(\epsilon^2 + \delta^2)/4$.

These problems can be avoided by converting the IF signal into a digital form, and performing the above operations by digital signal processing. Such systems are often referred to as digital receivers (e.g., *Yeomans*, 2008; *Meikle*, 2008). In the case of VHF atmospheric radars, it is even possible to directly sample and convert the RF signal to a digital form, which significantly simplifies the configuration of the receiver.

5.4.1 Digital receiver systems

Here we show an example of a digital receiver system developed for the multistatic operation of the Equatorial Atmosphere Radar in Indonesia (*Nishimura et al.*, 2006). In general, digital receiver systems with full capability can be quite expensive. For this observation, a simple and highly cost-effective digital receiver system was developed using personal computers (PCs) and Echotek ECDR-214PCI PCI boards with A/D converters and digital down-converters on-board.

Figure 5.21 shows the system. The PCI board has two analog inputs, A/D converters, and four independent signal processing units. Hence each board is used as two receiver systems with I/Q channels. Each receiver system consists of an antenna, a band-pass filter with the bandwidth of 10 MHz, and a low noise amplifier with a gain of approximately 25 dB, followed by the board.

The sampling clock is generated by a GPS receiver system with a reference clock output of 10 MHz followed by a digital direct synthesizer (DDS) which up-converts

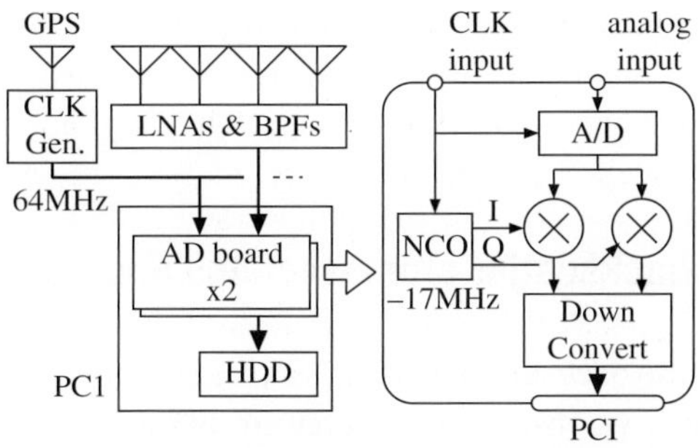

Figure 5.21 Example of a digital receiver system with a commercial digital signal processing board.

the clock signal to 64 MHz. The 47 MHz RF signal received by the antenna is directly sampled and quantized by the A/D converter on-board at this clock frequency with a resolution of 14 bit. The digitized signal is filtered and down-converted to the baseband with a bandwidth of 2 MHz by three-stage CIC (cascaded integrator-comb) and FIR (finite impulse response, see Chapter 8) filters and a down-sampler in a 4-channel digital down-converter (DDC) chip on the ECDR-214PCI board. The down-converted signal is thus read from a CPU via PCI bus, and is averaged over 30 times to compress the data with a repetition interval of 1 msec.

5.4.2 Fully digital systems

Modern technology allows flexibility not previously possible. Digitization rates of 1 GHz and more are possible, and digitizers working at frequencies of a few tens of MHz are quite cost-effective ($1000 per channel and less). This allows extremely simple systems to be built. There is no need to apply an IF, nor to beat the signal down to baseband. In-phase and quadrature components do not need to be created in hardware, and no mixers are required. This simplifies operations and control, and reduces the noise level of the receiver. It does require that a single frequency direct conversion receiver can be built with a gain of typically 90 to 100 dB, but this can be achieved by paying special attention to reduction of feedback between successive stages of the amplifier. Such receivers may be a little nonlinear, but because there is no frequency conversion, nonlinearities can be easily corrected in software. Once digitized, the signal can be subject to any type of filters, decoding, or integration techniques that are desired. The process at this time (2015) is relatively new, but has been tested on some systems and found to be quite reliable. The only limitation of such systems is data-download speed and processing speed, but use of computers with multiple CPUs (eight or more), programmed with threads so that each new pulse is directed to a new CPU, allows the computer to keep pace with the data acquisition (e.g., *Hocking et al.*, 2014). Digital systems also allow simple generation of pulses and radar control – often in the past the pulse generation and control has been one of the more difficult aspects of radar electronics.

5.4.3 Pulse-coding, coherent integration, and software issues

Items like pulse-coding, coherent integration, signal-to-noise, detectability, etc. are considered as primarily software issues for the purposes of this book, and so have already been discussed in the previous chapter. They will not be reconsidered here.

5.5 Relating backscatter cross-sections and reflectivities to received power

We now turn back to consideration of items involving the antennas and backscatter. In Chapter 3, we discussed the concept of backscatter cross-section and reflectivity at some length. In the earlier parts of this chapter, we have discussed antenna gain and polar diagrams. But to most effectively utilize these terms, we need to be able to link the concepts together in order to allow measurements of power received at the receiver to be used to work back to determination of atmospheric parameters through the backscatter cross-section.

In order to do this, we return to Chapter 3, and specifically Equation (3.171), viz.

$$S_{Rx} = NS_r, \tag{5.111}$$

where S_r is given by (3.171), or

$$S_{Rx} = N\sigma_e \sin^2 \chi \, \frac{S_{in}}{4\pi R_1^2}. \tag{5.112}$$

5.5.1 An example: naive determination of electron density

Our purpose is to show how appropriate measurements can be used to determine atmospheric parameters. We will demonstrate the concept first for an ionospheric plasma, where we will show how the electron density can, in principle, be measured. Then we will demonstrate the process for a turbulent neutral atmosphere. However, we will use the simple idea that the plasma comprises a random distribution of electrons, and no coherent structures or Debye shielding. As discussed in Chapter 3, Section 3.10, this is an oversimplification – indeed, it is totally in error, due to the presence of plasma waves – but nevertheless it is a useful way to demonstrate the ideas involved without too much complication. We can then move on to more realistic discussions.

Let us consider a monostatic radar in which the same transmitter sends the signal and then receives the backscattered component. Assume that the electron number density is n_e. Then

$$N = n_e V, \tag{5.113}$$

where V is the volume of the region of the ionosphere which the radar probes.

Figure 5.22 shows a crude representation of the radar volume. The pulse is assumed to be a square pulse of length Δr and the beam is assumed to have sharp edges at

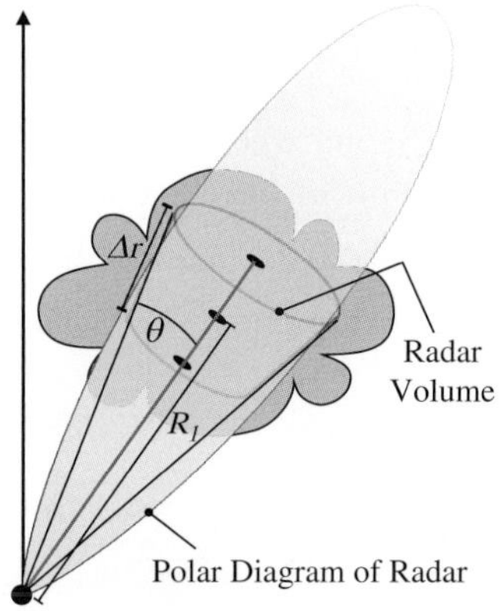

Figure 5.22 The "radar volume" normally associated with a radar beam at range R_1 using a pulse-length Δr.

the angle $\theta_{\frac{1}{2}}$ from the bore. Neither approximation is accurate, but they make a useful starting point for our discussions.

Using the notation shown in Figure 5.22, the volume is approximately given by

$$V = \pi (R_1 \theta_{\frac{1}{2}(2)})^2 \cdot \Delta r, \tag{5.114}$$

where $\theta_{\frac{1}{2}(2)}$ is the effective *two-way* half-power half-width (see Equation (5.63)) and Δr is the pulse length. Therefore

$$N \approx \pi (R_1 \theta_{\frac{1}{2}(2)})^2 \Delta r \cdot n_e. \tag{5.115}$$

The received power is then equal to

$$P_{Rx} = A_{\text{eff}} \cdot S_{Rx}. \tag{5.116}$$

Combining Equation (5.116) with Equations (5.115) and (5.112) gives

$$P_{Rx} \approx \pi R_1^2 \, \theta^2_{\frac{1}{2}(2)} \Delta r \, n_e \cdot \sigma_e \sin^2 \chi \frac{S_{in}}{4\pi R_1^2} \cdot A_{\text{eff}}. \tag{5.117}$$

For a monostatic radar, $\sin \chi = 1$. Finally,

$$S_{in} = \frac{P_{Tx} \, G}{4\pi R_1^2} \tag{5.118}$$

is the power flux at range R_1 where G is the radar gain and P_{Tx} is the transmitted power. Thus

$$P_{Rx} \approx \frac{P_{Rx} \, G}{16\pi R_1^2} \, \theta^2_{\frac{1}{2}(2)} A_{\text{eff}} \, n_e \, \Delta r \, \sigma_e. \tag{5.119}$$

Two relations can be used to remove $\theta^2_{\frac{1}{2}}$ and A_{eff}. These were given by Equations (5.54) and (5.56). They are repeated below in slightly modified form:

$$A_{\text{eff}} = \frac{G\lambda^2}{4\pi} \tag{5.120}$$

and

$$\theta^2_{\frac{1}{2}(2)} = \frac{\frac{\pi}{2}\kappa}{G}, \tag{5.121}$$

where κ is around 0.8, and where λ is the wavelength of the radar, G is the one-way gain and $\theta_{\frac{1}{2}(2)}$ is the *two-way* beam half-power half-width. We have used $\theta_h = 2\theta_{\frac{1}{2}}$ and $\theta_{\frac{1}{2}(2)} = \theta_{\frac{1}{2}}/\sqrt{2}$, where $\theta_{\frac{1}{2}}$ is the one-way half-power half-width.

For a Gaussian polar diagram, (5.121) can be written as

$$\theta^2_{\frac{1}{2}(2)} = \frac{2 \ln 2}{G} \tag{5.122}$$

to good accuracy. This can also be seen from Equation (5.60). If we use (5.122) and (5.120), then this leads (5.119) to become

$$P_{Rx} \approx \frac{P_{Tx} \, G}{64\pi^2 R_1^2} \, \lambda^2 2 \ln 2 \, n_e \, \Delta r \sigma_e. \tag{5.123}$$

At this point, some remarks are needed. First, note that σ_e is the electron backscatter cross-section. However, the term

$$\sigma_s = n_e \sigma_e \tag{5.124}$$

represents a backscatter cross-section per unit volume, and so represents a reflectivity. This is the link between Equation (5.7) and Equations (3.256) to (3.258). Note that we use σ to represent each term, but the meanings of the terms are quite different and even have different dimensions – so the subscript used is very important.

The other key point is the term $2\ln 2$ in (5.123). Note that Equation (5.7) contains no such term. This is because Equation (5.7), even when used for a monostatic radar, was valid for a uniform aperture, whereas in the derivations above, a Gaussian polar diagram was assumed. It may seem better then to drop the term $2\ln 2$, and simply use 1.0. Indeed, if we use (5.121) we have the number 1.26 in place of $2\ln 2$, which is even closer to unity. Then

$$P_{Rx} = \frac{P_{Tx} G \lambda^2}{64\pi^2 R_1^2} \Delta r \, n_e \sigma_e, \tag{5.125}$$

where

$$\sigma_e = \frac{\mu_0 e^4}{4\pi m_e^2 c^2}. \tag{5.126}$$

However, often the term $2\ln 2$ in (5.123) is retained. There is good reason for this. In many radar backscatter situations, scatter is from horizontally aligned layers. If a vertical beam is used, then the side-lobes are offset from vertical. Scatter is received at fixed range, so in the direction of these side-lobes, the point of scatter at the range under study may be below the height of the layer, and so produce no scattered signal. In addition, as we will see later (notably in Chapter 7), the scatterers themselves can be considered as ellipsoidal, which can confine the scatter to points closer to overhead. The net result of these effects is that there are times when the impact of side-lobes for scatter is diminished. Versions of (5.123) which do not involve $2\ln 2$ often (effectively) use the full side-lobe structure in their calculations, which can therefore be inappropriate due to the reasons outlined above. So the exact form of (5.123) varies between applications. In truth, the variations between the terms are of the order of 40% or less, which is often considered a small error in radar theory.

Now return to (5.125). Using $\lambda = \frac{c}{f}$, where f is the frequency of the radar, we can write

$$P_{Rx} = \frac{P_{Tx} G}{256\pi^3 R_1^2} \Delta r \frac{\mu_o e^4}{m_e^2 f^2} n_e. \tag{5.127}$$

A somewhat equivalent expression can be found by modifying (5.123), but retaining the $2\ln 2$ term. Thus a radar may be used to determine the electron density as a function of range if the transmitter power and radar parameters are known, at least to an accuracy better than 40% with this simple theory.

This derivation has been approximate. A more thorough derivation should include efficiency and loss factors, plus apply a 3-D integral. But the expression (5.127) is a useful one to work with. A more detailed derivation was presented by *Hocking* (1985) for a Gaussian and produced similar results. That derivation will also be reproduced in the next section on turbulence studies.

It can be seen that the received power is inversely proportional to the radar frequency squared, so it may be assumed that it is better to use low-frequency radars to optimize the received power. However, this is not true.

Firstly, for a radar dish of fixed diameter (or fixed area A_{eff}), the gain increases as the frequency increases,

$$G = \frac{4\pi A_{\text{eff}}}{\lambda^2} = \frac{4\pi A_{\text{eff}} f^2}{c^2}. \tag{5.128}$$

Secondly, we must consider the effects of noise. The major source of noise for a well designed radar is galactic radiation. (Astronomers, of course, consider this as signal!) Skynoise decreases as f increases. In fact, the noise is proportional to $f^{-\beta}$, where β is generally of the order of 2.2 to 2.5. One of the more detailed studies (*Swarnalingam et al.* (2009a)) gives a value of 2.5.

Thus, at higher frequencies, noise levels are much less. Therefore the signal to noise ratio is superior at higher frequencies if the backscatter cross-section is constant. In general the cross-section has a wavelength dependence, so the ratio of signal to noise depends on the joint wave-number dependence of the signal spectrum and the skynoise. We will see an example of this in the next section.

5.5.2 Determination of turbulence parameters

In this section, we will somewhat repeat our discussions above, but this time will consider turbulent scatter from the neutral atmosphere. However, instead of considering the radar volume described by Figure 5.22, this time we will do a full integration using an assumed Gaussian polar diagram. We will follow in part the derivation in *Hocking* (1985).

First, recall the expressions developed for backscatter cross-section per unit volume developed in Chapter 3. These were given by Equations (3.256) and (3.258). As discussed there, care was needed with regard to the normalizations. If we ensure that the normalization is chosen correctly, then the two most commom forms of reflectivity are given by

$$\eta_s = \frac{\pi^2}{2} k_B^4 \Phi_n \quad \text{and} \quad \sigma_s = \frac{\pi}{8} k_B^4 \Phi_n. \tag{5.129}$$

Remember that σ_s and η_s, being reflectivities, are quite different conceptually to the cross-section σ_A discussed in Equation (5.7).

Chapter 11 is devoted to a more detailed discussion about the physics of turbulence. For now, we will not look in detail at the derivation of the spectral form for neutral atmospheric turbulence, but simply use Φ_n from Equation (A27) in Appendix A, which gives

$$\Phi_n = 0.033 C_n^2 k^{-11/3}, \tag{5.130}$$

so that, as seen in Equations (3.292) and (3.293),

$$\eta_s = 0.3787C_n^2 \qquad \text{and} \qquad \sigma_s = 0.03014C_n^2, \tag{5.131}$$

as also discussed by *Ottersten* (1969a, b). It should be noted that this simple form only arises because the refractive index in the neutral atmosphere is largely independent of the wavelength at VHF frequencies, as discussed in Chapter 3. In the ionsophere, this expression no longer applies because the scattering is from free electrons, and so has a more complex dependence of refractive index on electron density and wavelength. This matter was discussed at some length in Chapter 3.

Now it is necessary to show how η_s (or σ_s) relates to the power received in a backscatter experiment. Here, we will use a full integral, rather than the simpler radar-volume approach discussed in the last section.

Consider scatter from a range r, at a zenith angle θ from vertical and azimuth angle ϕ from north. Then the peak power per unit area incident at (r, θ, ϕ) is

$$P_r(\theta, \phi) = \frac{L_{Tx}\, G_T(\theta, \phi)\, P_{Tx}}{4\pi r^2}, \tag{5.132}$$

where P_{Tx} is the transmitter peak power, $G_T(\theta, \phi)$ is the transmitter array directional gain at angle (θ, ϕ), and L_{Tx} is the efficiency of transfer of power from the transmitter to the transmitting antenna. Note that L_{Tx} is less than 1. The power backscattered per unit steradian from a volume $\delta V = r^2 \sin\theta\ \mathrm{d}\theta\ \mathrm{d}\phi\ \mathrm{d}r$ is given by $\sigma_s \delta V P_r(\theta, \phi)$, and the power returned to the receiving antenna back at the ground per unit area is simply this quantity divided by r^2. The power received by the receiver from the volume δV is therefore

$$P_{Rx}(\delta V) = \frac{P_{Tx}\, G_T(\theta, \phi)\, L_{Tx}\, \sigma_s\, \delta V\, L_{Rx}\, A_R(\theta, \phi)}{4\pi r^4}, \tag{5.133}$$

where $A_R(\theta, \phi)$ is the effective area of the receiving antenna for the direction (θ, ϕ), and L_{Rx} is the efficiency of transfer of signal from the antenna to the receiver. Note that it has been assumed that the transmitter and receiver antennas are very nearly collocated (or could be one and the same), so that the same (θ, ϕ) can be used for each.

The total power received at the receiver can now be found by integrating (5.133) over all values of (θ, ϕ) and over the depth of the atmosphere defined by one pulse. Assuming a square pulse of length $c\tau$ for simplicity (or a range resolution of $\ell_{rr} = \frac{c\tau}{2}$), assuming σ_s to be constant everywhere within the radar volume, and assuming that $r \gg \ell_{rr}$, then

$$P_{Rx} \approx \frac{P_{Tx}\, L_{Tx}\, L_{Rx}\, \sigma_s\, \ell_{rr}}{4\pi r^2} \int_{\theta=0}^{\pi} \int_{\phi=0}^{2\pi} G_T(\theta, \phi)\, A_R(\theta, \phi) \sin\theta\ \mathrm{d}\theta \mathrm{d}\phi. \tag{5.134}$$

Suppose further that the transmitter and receiver polar diagrams are both phased such that their maxima point in the same direction (θ_o, ϕ_o), and that the directional gain and effective areas at (θ_o, ϕ_o) are G_{Tm} and A_{Rm} for the transmitter and receiver respectively (where the extra "m" in the subscript reinforces that this is the maximum (bore-sight) gain). If further, the polar diagrams are axially symmetric about (θ_o, ϕ_o), and both can be described by a Gaussian form, and the polar diagram of at least one has a half-power half-width of less than about $10\,^\circ$, then (5.134) becomes

$$P_{Rx} \approx P_{Tx} \frac{\ell_{rr}\, L_{Rx} L_{Tx} G_{Tm} A_{Rm}}{4\pi r^4} \cdot \sigma_s \frac{\pi \theta^2_{\frac{1}{2}c}\, r^2}{\ln 2}, \tag{5.135}$$

where $\theta_{\frac{1}{2}c}$ is the half width of the *combined* transmitter-receiver polar diagram (i.e. the product of the two polar diagrams). This equation is strictly only valid for a Gaussian polar diagram, but is similar for other forms but with ln 2 changed in value (see prior to (5.127)).

This can also be written as

$$P_{Rx} \approx \frac{P_{Tx} L_{Rx} L_{Tx} G_{Tm} A_{Rm}}{4\pi r^4} \cdot \sigma_s \cdot \frac{V}{\ln 2}, \tag{5.136}$$

where V is the volume defined by the locus of the half-power half-width of the polar diagram at the height of scatter, and the pulse length, namely $V = \pi(r\theta_{\frac{1}{2}})^2\, \ell rr$. Combining (5.131), (5.136) and the formula for V leads to

$$C_n^2 \approx \frac{92.0 P_{Rx} r^2 \lambda^{\frac{1}{3}}}{P_{Tx} G_{Tm} A_{Rm} L_{Tx} L_{Rx} \ell_{rr} \theta^2_{\frac{1}{2}c}}, \tag{5.137}$$

where the constant 92.0 is $\frac{16\pi \ln 2}{0.3787}$ (from (5.131)). If the same array is used for transmission as for reception, we may take $A_{Rm} = \frac{G_{Tm}\lambda^2}{4\pi}$. In this case, if the one-way polar diagram has half-power half-width of $\theta_{\frac{1}{2}}$, then $\theta_{\frac{1}{2}c} = \theta_{\frac{1}{2}}/\sqrt{2}$. If the *one-way* gain of the polar diagram is α times the directivity G_d, we can make use of the relation $G_d\theta^2_{\frac{1}{2}} = 4 \ln 2$ (valid for Gaussian polar diagrams, see Equation (5.60), with $\theta_{\frac{1}{2}} = \theta_h/2$), to obtain

$$C_n^2 \approx 66.4 \cdot \frac{P_{Rx} r^2 \lambda^{\frac{1}{3}}}{P_{Tx} A_{Rm} L_{Tx} L_{Rx} \alpha \ell_{rr}}. \tag{5.138}$$

Either (5.135), (5.136), (5.137), or (5.138) can be used to estimate C_n^2 from absolute measurements of received power.

In this derivation we have assumed a Gaussian polar diagram. Assuming a square or circular aperture will change the constant 66.4 to a slightly larger value of about 73.3. These differences were discussed just prior to Equation (5.127).

Finally, it is worth recognizing the term $P_{Tx}A_{Rm}$ in Equation (5.138). This term is often called the *power-aperture product*, and represents a measure of the effectiveness of the radar at target detection, especially for monostatic radars, where the area is the area of the whole array. Both peak and average power-aperture radars are used. Typical values of peak power-aperture products are in the range 10^6 to 10^{10} W m^2, and mean values are typically 5% of this.

5.6 Calibration

As described in earlier sections, calibration of the radar system is desired. This can take several forms; range calibration, polar diagram verification, calibration of the received powers and radar efficiencies.

We will consider each in turn.

5.6.1 Range calibration

Proper recording of the range of the scatterers may seem a trivial process, but various complications exist. First, there may be delays within the transmitter between the input and output signal, and the time taken for the signal to move from the transmitter to the antennas is another factor to be considered. Likewise the time taken for the received signal to return to the receiver from the antennas must be recognized, and finally the receiver itself has a finite bandwidth which must result in further delays. A target at a range of 1 km should have a delay of 6.67 microseconds between original triggering of the transmitter and receipt of the scattered signal, but in reality this can easily be as high as 8 microseconds or more. This difference must be incorporated into the data acquisition so that any measured delay corresponds to the true range.

Calibration of the ranges can be done in a variety of ways. One is simply to determine all likely delays, measure them independently, and add them up. This is usually complicated, since it is easy to miss terms or make calculation errors in receiver delays. Another method is to transmit on the main beam, and collect signal on a small secondary antenna, which is then sent directly to a cathode-ray-oscilloscope, bypassing the receiver entirely. The signal recorded on the special receiver antenna has travelled out to the transmitter array, across the ground to the receiver, and back through the cable from the receiver antenna to the oscilloscope. Hence it has a delay equal to all of the various cable delays in the system plus an additional delay of a few tens of meters due to the travel across the ground from the main array. (For a distributed transmitter antenna, the received signal may be a little smeared out because signal is coming from all antenna-elements in the array, but the center of the received pulse should be a reasonable representation of the center of the array.) Usually it is possible to see the transmitter pulse on the scope, and the delay between the strong transmitter pulse and the much weaker receiver pulse can be measured. This usually requires quite narrow pulses so that the two signals can be separated. Once the delay has been found, this becomes part of the intrinsic delay parameter of the system.

The receiver delay then needs to be added. This is usually of the order of the inverse of the receiver bandwidth. Many systems have separately generated "range-markers" that can be displayed on the scope at the same time as the signal, simplifying measurements.

If this practice is not possible, another common strategy is to insert a delay line of known length between the input to the transmitter and the receiver, bypassing all elements (including the transmitter). A typical delay line might have a delay of 15 km (100 microseconds). The recorded signal is then digitized, and then the offset of the recorded peak from the 100 μsec marker gives the desired correction. This method has the advantage that it includes the delays associated with the receiver, though it misses delays associated with the feed cables to the antennas.

Commercial delay lines can be expensive. An alternative is to make up a delay line from lengths of coaxial cable used to feed the antennas. However, such a structure needs to be several km long for most effective use, and the losses through several km of coaxial cable can be substantial. Amplifiers often need to be inserted at regular intervals, and these need to be wide-band so as not to introduce delays of their own.

If a suitable reflecting structure exists nearby, such as a tall tower, then it often produces a signal from reflection of the transmitter pulse, provided that the transmitter leakage and receiver pickup along the ground is sufficient. If the distance to the tower is known, offsets in range calibration can be calculated. Aircraft flying near the radar can also be used in a similar way, provided that they have navigational equipment on board capable of giving exact locations at the time of radar reflection.

Another method of range calibration includes measuring winds as a function of height and comparing to radiosonde winds, either instantaneously or over extended periods. This generally has poorer accuracy than the methods discussed above, but can be used as a last resort.

5.6.2 Calibration of the polar diagram

Ensuring that the polar diagram achieved by the antennas matches expectations is important for many calculations, including determination of absolute powers and turbulence measurements. Correct determination of the polar diagram shape also means that the system gain and beam-width can be properly determined. Such measurements can also be used to verify that the beam points in the expected directions – errors in beam directions can lead to substantial errors in wind calculations. Polar diagram tests are often difficult and not always done, though should be carried out where possible.

First, we recall that there are a few slightly different definitions of the antenna gain, and we especially recall the absolute gain G_a and the directivity G_d. Equation (5.9) defines the absolute gain G_a, which includes the loss of the antenna, such as the ohmic loss and reflection loss at the antenna terminal due to impedance mismatch. The directional gain without this loss term is defined by (5.10).

Although radio stars are safely regarded as a point target with precisely known direction, the weak signal makes it difficult to use these sources for measuring the antenna beam pattern. Signal intensity fluctuations due to ionospheric scintillation aggravate the situation. Nevertheless, the method has been used successfully by some authors (e.g., *Carey-Smith et al.*, 2003; *Vincent et al.*, 1986). Often interferometric procedures are used. The method has the advantage of being relatively cheap to implement, and, in contrast to some other techniques, does not require great beam-pointing flexibility.

Another (and perhaps more reliable) means is to use satellites or the moon (e.g., *Sato et al.*, 1989; *Mathews et al.*, 1988), as the radar target. Some satellites have been placed in Earth-orbit specifically for this purpose. These are perfectly reflecting spheres, for which the cross-section is well-known, and this method will be discussed shortly.

For now, we will concentrate on using the moon as a target. In spite of the fact that it has an angular size of about 0.5 °, most of the echo power returns from a narrow region where the surface is close to perpendicular to the line of sight. It is thus possible to regard the moon as a point target as well. Its reflectivity relative to a perfectly conducting sphere of the same size is not accurately known at VHF and UHF bands, but is estimated to be less than 0.1. The signal-to-noise ratio measured with the MU radar at 46.5 MHz, using 1 MW peak power, employing the full 100 m circular antenna, and using a 512 μs pulse width, is slightly higher than 50 dB. Figure 5.23 compares the antenna pattern of the

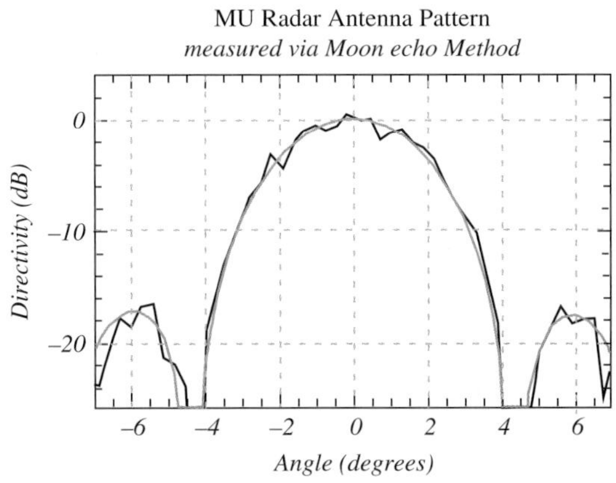

Figure 5.23 Antenna pattern of the MU radar as measured by the moon reflection method (*Fukao et al.*, 1990). The darker line shows the measurements, while the lighter line shows the results of a numerical simulation. (Reprinted with permission from John Wiley and Sons.)

MU radar antenna measured by moon echo, when it traversed a stationary beam, with that computed theoretically (*Fukao et al.*, 1990).

In order to apply coherent integration, the beam was pointed to a direction where the moon has no radial velocity with respect to the radar. Note that the measured two-way pattern is converted to a regular one-way pattern, so the dynamic range in this figure is about 25 dB. The background noise is subtracted. The fluctuations in the measured pattern are due to ionospheric scintillation. The major limitation of the moon as a radar target is that its declination is limited to $\pm 30\,^{\circ}$, and thus it is not visible to radars located at high latitudes with limited antenna steerability.

The directional gain G_d without loss factors is difficult to evaluate in reality, because the denominator in Equation 5.10 cannot be measured for most of the angular region outside the main beam. The directivity in Figure 5.23 is thus normalized by its unknown peak value.

It is also possible to use artificial radio sources or radar targets to measure the antenna beam-width. The antenna pattern is usually measured in the antenna far-field, which occurs at a distance r satisfying

$$r > \frac{D^2}{\lambda}, \tag{5.139}$$

where D is the diameter of the antenna. For $D = 100$ m and $\lambda = 6$ m (50 MHz), for example, it is 1.7 km. As most of the antennas used for atmospheric radars point to near the zenith, we need to employ spacecraft or aircraft with relatively high altitude.

It is technically possible to use an airplane or helicopter with GPS route control for this purpose, but these methods are in general quite cost inefficient. In order to study the antenna pattern of the MU radar including the region of low elevation side-lobes, a special purpose receiver was installed on a scientific satellite OHZORA (*Sato et al.*, 1989).

Artificial satellites with known radar cross-section can be used for the overall calibration as well as for measuring the beam-width. As of 2008, ten satellites launched for the purpose of radar calibration are listed on the catalog of artificial space objects provided by the National Space Science Data Center (*Kelso*, 2009). Among these satellites, the largest is LCS-4, which is an aluminum sphere with physical cross-section of 1 m^2 (*Prosser*, 1965). It should be noted that its radar cross-section is slightly different from this value at UHF and VHF because it falls in the Mie scattering regime.

The major difficulty of using satellites such as LCS-4 lies in the fact that most atmospheric radars do not have the function of tracking moving objects. The estimated orbit computed from the so-called two-line element, which is provided weekly by NORAD, naturally contains some error due to their orbital perturbations. It is thus necessary to arrange at least a few antenna beams overlapping each other across the expected track of the satellite, and then to determine the actual path relative to the center of each beam direction by comparing the echo power received on different beams (*Sato et al.*, 1994).

5.6.3 Power calibration

Although windprofiler radars are so-named because of their ability to measure winds in the atmosphere, they can be used for many other things as well. For some of these extra applications, it is important to be able to convert the received digital records to some absolute measure of received signal strength, usually measured in Watts. This requires additional calibration procedures. One example is determination of the atmospheric refractive index structure constant, C_n^2 (e.g., *Hocking and Mu*, 1997). Such a calibration is achieved by determining all parameters in the radar equation. If we take Equation (5.7), we can consider the case of a monostatic radar by setting $r_t = r_r = r$ and $A_e = \frac{G_a \lambda^2}{4\pi}$ (valid since the transmission and reception beams are identical in form), and we produce

$$P_{Rx} = \frac{P_{Tx} G_a^2 \lambda^2 L_{Tx} L_{Rx}}{(4\pi)^3 r^4} \sigma_A, \tag{5.140}$$

which gives one form of the radar equation for a monostatic radar. Here P_{Rx} is the received power, P_{Tx} is the transmitted power, G_a is the antenna gain, λ is the wavelength, and L_{Tx} and L_{Rx} are the one-way external loss factors for transmission and reception respectively. These losses lie outside the antenna and include cable and matching losses. The term r is the range, and σ_A in this case is the radar scattering cross-section and not the reflectivity.

Equivalent expressions, perhaps more directly related to measurable parameters, were given as Equations (5.137) and (5.138). The same calibration constants need to be determined in all equations.

The most important requirements are: (i) an accurate measure of the transmitted power; (ii) determination of the loss terms L_{Tx} and L_{Rx}, and hence determination of the radar efficiency; (iii) accurate determination of the polar diagram G_a and the gain (already discussed); and (iv) a suitable way to convert the received power to an absolute value.

Measurement of the transmitted power is relatively straightforward. Many transmitters are specified at certain power levels, so as long as the manufacturer accurately reports the specifications then the transmitted power (item (i)) should be known. Power meters exist which can be used for verification of the specified values, or for general longer-term measurements and status-monitoring, so we will not dwell on this aspect. Item (iii) has already been discussed.

Hence the key remaining items are (iv) receiver system calibration and (ii) system efficiency and losses. We will now deal with these.

Measuring received echo power: system gain and efficiencies

While many previous publications have presented power measurements in terms of a signal-to-noise ratio, this term is rather unhelpful in an absolute sense, as noted by e.g., *Hocking* (2011).

Our purpose here is to develop a less arbitrary descriptor of receiver input power. We will take this to be the power from the antenna within the bandwidth of the receiver. Our discussion will follow earlier publications and internal reports, notably *Hocking et al.* (1983) and *Green et al.* (1983), who calibrated the SOUSY and SUNSET radars respectively. The former report was particularly thorough, and formed the basis for several subsequent articles, including *Campos et al.* (2007a), *Swarnalingam and Hocking* (2007), *Latteck et al.* (2007), *Kirkwood et al.* (2007), *Swarnalingam et al.* (2009a) and *Swarnalingam et al.* (2009b).

Figure 5.24 shows the situation under discussion. A signal is input to the first pre-amplifier (if it exists) and then passes through the receiver, filters, mixers, and digitizer, finally being stored as a digital value in memory or on some form of storage medium. Assuming that the receiver has been designed to be linear, a particular input voltage will appear at the output amplified by a gain factor g_S. The units of the gain factor will depend on the units of the input signal (generally volts) and the output signal. It does not matter exactly what form the output signal takes – it could be volts, or digital numbers produced by a digitizer, for example. The only requirement is that the output units, once established, are maintained as the same for all time. Typically nowadays they will be digital units, and the units of gain will be digital units per volt.

The easiest way to determine g_S is simply to remove the antenna and feed a pure sinusoid signal into the input on the left of the figure, and digitize the output. If the

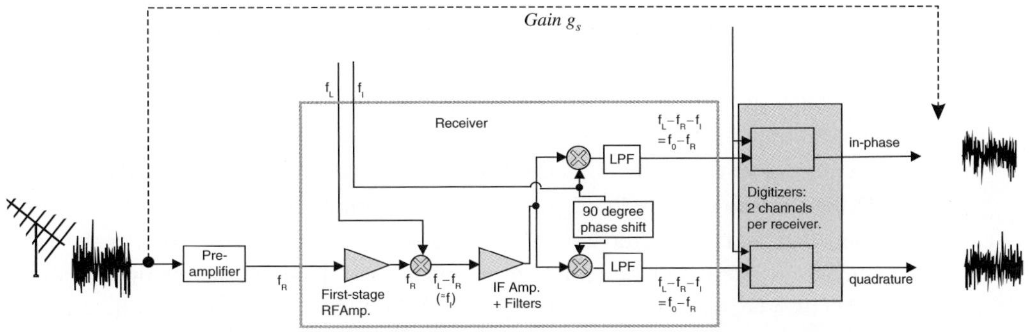

Figure 5.24 Passage of received signal from the receiver antenna through to the digitizers. The figure is adapted from Figure 4.12, with just the receiver portion extracted. Minor changes of a cosmetic nature have been made for ease of interpretation in regard to this section.

input signal is slightly offset from the radar central frequency, both the in-phase and quadrature signals will show sinusoidal variations in time, and the gain g_s can be found as the ratio of the peak of the digitized sinusoid divided by the peak of the input signal. Both in-phase and quadrature signals will have the same peak values, but displaced in time by a quarter cycle, and so will show the same gain. Essentially,

$$\sqrt{\overline{D^2_{\text{ip[out]}}} + \overline{D^2_{\text{quad[out]}}}} = g_s u_{\text{in}}, \tag{5.141}$$

where $D_{\text{ip[out]}}$ and $D_{\text{quad[out]}}$ are the instantaneous digital in-phase and quadrature signals and u_{in} is the instantaneous input signal in volts. g_s is a ratio in digital units per volt.

This should be easy, but complications exist. Typical receiver gains are of the order of 90–100 dB, so for a 5 volt output, an input of a few tens of microvolts is typical. Any input signal must be generated in a wave-form generator, then suppressed to values of this order, then passed to the receiver. For the method outlined above, which involves injecting a pure sinusoid into the receiver, this requires a good and reliable attenuator, and careful choice of the amplitude of the input signal. It needs to be small enough that it does not saturate the receiver, or enter an amplitude regime where the receiver becomes nonlinear, but large enough that it is significantly stronger than the receiver noise. Sometimes this is not possible, and as a result, the output signal can be noisy, non-sinusoidal in appearance and hard to read. Suitable averaging can be applied to smooth out the noise, but it still remains a nuisance. The gain is also dependent on frequency, depending on its location within the filter bandwidth, and so the frequency should not be too far from the center of the band.

An alternative approach is to "sniff" a small portion of the transmitter pulse (with an attenuation of perhaps 40 or 50 dB), and feed this directly into the receiver. The effective gain can then be calculated by comparing digital output values to the known input amplitude. If the user sniffs a small part of the transmitter pulse, the signal should also be passed through a suitable delay line, which delays the signal by several kilometers, so that it is not mixed up with indirect receiver pickup from the transmitter.

Despite the complications,the above methods are still valuable. They are particularly useful if the user knows that the receiver is linear, as supplied by the manufacturer. Then the signal fed into the receiver can be moderately large – large enough for it to dominate over the receiver noise, but small enough that the receiver does not saturate. Then the gain can be measured quite easily. However, if system performance is required at very low input levels as well, then other methods are sometimes needed.

So while the above procedure is a possible calibration method, it is not always used. Quite often, the calibration is done using an input signal which is not a pure sinusoid, but a mixture of randomly phased sinusoids – in other words, a noise source.

In the case of noise, assuming no coherent integrating or pulse-to-pulse averaging, we assume that we record over some short time period T_p (maybe a second or so) and then calculate the mean square sum of the in-phase and quadrature components. Equation (5.141) then becomes

$$\overline{D^2_{\text{ip}}} + \overline{D^2_{\text{quad}}} = g_s^2 \left\{ \sum_j u_{0j}^2 \right\}. \tag{5.142}$$

Here, we have assumed that each different value of j is a different frequency, and have assumed all frequencies differ from the carrier frequency, so that all rotate over a non-zero number of cycles during the averaging period. The jth contributing harmonic oscillation has an amplitude u_{0j}. We have also used the fact that during an integral number of rotations, a cosine or sine function of amplitude a_0 has a mean-squared value of $a_0^2/2$, so that the sum of the in-phase and quadrature mean square values will be a_0^2. Special consideration is needed for any frequency component with exactly the carrier frequency, as it will contribute fixed values to each of the in-phase and quadrature components, since the vector will not rotate in the Argand diagram relative to the carrier. However, these contributions will disappear when the means are subtracted, and in any case make only a negligible contribution to the left-hand side since they represent only a small fraction of the contributing frequencies.

It could be argued in regard to Equation (5.142) that because the input signal must be divided into two halves in order to generate in-phase and quadrature components, then each of the in-phase and quadrature components should be divided by $\sqrt{2}$, therefore altering g_s. This argument is, however, irrelevant. The only thing we need to know is that the digital values and the noise terms are proportional, and the constant g_s can then be found from experiment. Whether there should be $\sqrt{2}$ terms or not might affect the absolute determination of the receiver gain, but as far as determining an absolute measure of the input power from a suitable digital value, it is immaterial. It is only necessary that the calibration of g_s is done in a manner consistent with the way in which the data are to be used.

It should also be noted in regard to Equation (5.142) that the relation between the digital values and the input power is only applicable in this form if the receiver generates no internal noise. If it does generate internal noise (which it must), then Equation (5.142) is valid only when the right-hand side includes *both* radio noise *and* receiver noise. For now we will consider a noiseless receiver, and introduce the noise term when it is appropriate.

Multiplying and dividing in (5.142) by twice the receiver impedance $2R$ on the right-hand side then gives

$$\overline{D_{\text{ip}}^2} + \overline{D_{\text{quad}}^2} = 2g_s^2 R \left\{ \sum_j \frac{u_{0j}^2}{2R} \right\}, \tag{5.143}$$

or

$$\overline{D_{\text{ip}}^2} + \overline{D_{\text{quad}}^2} = 2g_s^2 R \left\{ W_{Rx} \right\}, \tag{5.144}$$

where W_{Rx} is the mean power which enters the pre-amp and receiver block. The symbol W is used to represent Watts.

Hence if g_s is known, the measured digital values allow determination of the received noise power, which can be used as P_{Rx} in Equation (5.140).

Determination of g_s is performed by feeding known noise into the receiver input and recording the output digital readings.

However, the process is complicated by two factors. First, the receiver produces noise of its own, and second, many systems use various types of inter-pulse averaging (e.g., coherent integration) which complicates the interpretation. We now discuss these aspects.

First, we now replace W_{Rx} by $W_{Rx} + W_{Rxnse}$ where W_{Rxnse} represents internal receiver noise plus the ambient system temperature. The term W_{Rxnse} is the equivalent receiver noise at the entrance to the pre-amp/receiver system.

Second, we need to compensate for coherent integrations and pulse coding. This means that an input noise signal has its output digital values suppressed by the averaging process (we assume that resulting coherent integrations are an average, rather than a sum – if a sum is used, then the following theory must be adjusted for that fact, see shortly). Application of an N-point coherent integration reduces the digital variance by N, and application of an M-element Barker code reduces the recorded digital variance by a further M. If an M-element complementary code is used, then the reduction is by $2M$, where M is the number of elements in a single pulse of the two-pair code. We therefore write that the power reduction is NLM, where $L = 1$ for a Barker code and $L = 2$ for a complementary code.

We assumed an averaging process here. Care must be taken to establish exactly how coherent integration is achieved. Some radars simply sum the successive in-phase and quadrature data-point amplitudes within an N-point coherent integration, and some average the values (i.e. they divide by N after the summation is complete). In the first case, the signal power increases by N^2 times, but the noise power increases by N times. In the second case, the signal power stays unchanged, and the noise level actually decreases by N times. If the radar under examination uses sums rather than averages, the discussion above needs to be adapted to suit.

We will let g_s be the maximum receiver gain. We will also assume that the receiver contains an attenuator which can be adjusted to alter the gain (e.g., this might be necessary to properly record stronger echoes like meteors, or to record lower tropospheric signals), and allow that the system may change its gain by remote control to say g_x. Then we may finally write that (5.144) becomes

$$\overline{D_{\rm ip}^2 + D_{\rm quad}^2} = \frac{2g_s^2 R 10^{-\frac{X}{10}}}{N \cdot M \cdot L} \{[W_{Rx} + W_{Rxnse}]\}, \tag{5.145}$$

where $X = 10\log_{10} g_s - 10\log_{10} g_x$ is the adjustment applied to the gain in dB. If coherent integrations and pulse coding are stored as sums, rather than averages, the term $N \cdot M \cdot L$ appears in the top line of (5.145) rather than in the denominator.

Equation (5.145) forms the basis of the most common calibration method. In practice, the most straightforward and reliable way to perform the calibration using noise is to connect a known noise source such as a noise diode to the receiver input. The spectrum of the noise is flat (white noise) and the noise across a width B of the spectrum is given by

$$W_{\rm in} = K_B\, T\, B_f, \tag{5.146}$$

where K_B is Boltzmann's constant (1.38×10^{-23} W s/K), T is the equivalent temperature, and B_f is the bandwidth in Hz. The bandwidth B_f should include not only the effect of the analog filter used in the receiver, but also that of digital filtering including the effects of coherent integration and pulse-coding, as discussed in regard to (5.145).

Note that this process is independent of the pulse repetition frequency (PRF) used. While the spectrum of the time series will only be available between frequencies of $\pm PRF/2$, this spectrum contains noise from frequencies right out to the edges of the final stage filter. Frequencies beyond $\pm PRF/2$ will be aliased, into the region between $-PRF/2$ and $+PRF/2$, but will appear at full strength. Hence the filter bandwidth is what defines the noise level, not the digitization rate.

If the filter is not flat, then an effective bandwidth B_f should be found that corresponds to a flat-topped (boxcar) filter of width B_f that has the same area under the graph of response vs. frequency as the true filter.

The result (5.146) was shown experimentally by *Johnson* (1928) and then derived theoretically by *Nyquist* (1928). It is called "Johnson noise" by experimentalists, and "Nyquist noise" by theoreticians.

Our problem is now to deduce the gain g_s and the receiver noise W_{Rxnse}. This requires several measurements, as described below.

Suitable, calibrated noise sources are available commercially. Most are based around a diode as the noise source. The antenna is removed from the pre-amp or receiver, and then the noise signal from the noise generator is fed directly into the receiver. Such a noise source can be adjusted in equivalent temperature, and has a scale which allows the noise level to be adjusted. Such a scale will allow the noise level to be altered by a factor $\mathcal{F}$, so that the output noise temperature will be $\mathcal{F}T$, where T is the physical temperature of the instrument, which would normally be about the ambient temperature in the room. The signal power passed into the receiver is then

$$W_{\text{in}} = K_B\,(\mathcal{F} + 1)\,T\,B_f, \tag{5.147}$$

where the factor $\mathcal{F} + 1$ arises because we also need to recognize that the receiver itself, and the noise generator, have a true ambient thermal temperature which adds to the noise produced by the diode. For most MST radars, the receiver's physical temperature is the room temperature, though in higher frequency devices it might be less if the receiver is specially cooled (such as by liquid nitrogen). It should be noted here that we refer to *two* temperatures of the receiver, one being the actual temperature, and the second being the noise temperature, which relates to W_{Rxnse}.

By increasing the values of $\mathcal{F}$, different noise powers are fed into the receiver, and the corresponding digital data are recorded. Suitable variances are then found, and $\overline{D_{\text{ip}}^2} + \overline{D_{\text{quad}}^2}$ can be plotted as a function of W_{in}. The result will be a straight line that will *not* pass through (0,0). Rather, if a straight line is fitted using a suitable regression algorithm, and the best-fit line is extrapolated back towards zero, it will cross the W_{in} axis at some negative value. The distance from this cross-over point to (0,0) can be used to determine the receiver noise, W_{Rxnse}.

The slope of the best-fit line is given by

$$S_w = \frac{2g_s^2 R 10^{-\frac{X}{10}}}{N \cdot M \cdot L}, \tag{5.148}$$

from which g_s can be found (at least for the case that averaging is used to determine the coherent integrations and pulse codes; if these are sums, again NML appears in the numerator).

Swarnalingam et al. (2009a) have confirmed this process experimentally using multiple experiments with different types of pulse codes and different levels of coherent integration.

Once the receiver noise and gain constant g_s have been found, the remaining task is to find the radar efficiency. The simplest way to do this is to use the galactic skynoise as a reference.

As a point of interest, if a pure frequency harmonic is fed into the system, rather than noise, and coherent integration and pulse-coding are performed using averaging (rather than sums), then the terms N, M, and L disappear in Equation 5.145). However, no matter how the calibration is performed, whether by using a pure input signal or noise, the same value of g_s should be found, and it represents the amplification of a pure siuusoid through the receiver–digitizer system.

Radar efficiency and loss factors

The losses in a radar system may differ for transmission and reception. This is especially true for a system comprising multiple antennas. Antennas are often fed from one or a few transmitter ports, which then branch out like a tree to more and more antennas (e.g., see *Campos et al.*, 2007a, for design details). This is sometimes called a "Xmas tree" arrangement. At each point where one cable splits into two, the input and output impedances need to be carefully matched. This is often achieved by carefully choosing the lengths of the cables – typically a combination of quarter-wave and half-wave cables is used. Alternative strategies involve special tuning circuitry at cable junctions: in this case it is possible for one cable to split into a higher number, such as three or four or more. Highest priority is given to optimizing radar pulse transmission, since it is critical that as much of the transmitter pulse is radiated, and as little reflected back to the transmitter from the antennas, as possible. Hence efficiencies of transmission and reception may differ. Generally if a choice is needed, the efficiency of transmission is made higher. This is especially true at VHF, where the main form of noise is skynoise, so extra losses on reception attenuate the signal and skynoise equally, maintaining the ratio of the two. As long as the skynoise (after losses are considered) exceeds the receiver noise, the reception efficiency is not seriously compromised.

It is a difficult task to evaluate loss factors L_{Tx} and L_{Rx} on a component-by-component basis, because the loss produced by each component can be quite small. For example, a single connector may cause a loss of $\sim$ 0.1 dB, but if several exist in series, they may have an appreciable cumulative effect. Also, corrosion due to long-time use is not negligible.

For transmission, there is little other way to determine efficiency than by measuring the transmitter power at the transmitter and then measuring the received signal at each antenna. This ignores losses within the antenna, but other than that can be quite efficient. It works because transmission mode deals with large signals which are easily measured without concern about internally generated noise. *Swarnalingam et al.* (2009b) showed efficiencies varying from 35 to 80% for different types of radars.

However, for reception, direct measurement is more difficult, particularly because we deal with weak signals in the presence of various types of noise. Rather than seeking out each possible source of power loss, an easier procedure is to use the background galactic skynoise as a reference. In order to understand this process, it is necessary to have a short discussion about skynoise.

Galactic (cosmic) skynoise

Noise is radiated from multiple sources in and outside of our galaxy, producing a broad background of radiowaves emanating from all directions. Astronomers use this signal to learn about stars and other entities, but in MST work this is simply considered as noise. Some strong individual sources exist, but in many places in the sky the signal is a broad background with little angular variation.

Figure 5.25 shows an example of the skynoise variation as a function of time of day for a site near Montreal, Canada. Notice strong peaks at 0500 and 2000, but a weaker section with no outstanding peaks between 1200 and 1800. There is clearly a strong dependence of T on the direction, being stronger towards the center of the galaxy, and in the direction of certain discrete sources.

When the galactic noise power is measured in a direction with no outstanding radio stars present (such as overhead between 1200 and 1800 in Figure 5.25), the measured power can be found by adapting (5.117). The received power is given by that expression, but S_{iii} is replaced by the galactic radiowave flux, and the Δr term disappears because there is no pulse involved. All range-dependencies disappear as well:

$$P_{Rx} \propto \theta_{\frac{1}{2}}^{2} P_{sky} A_e, \tag{5.149}$$

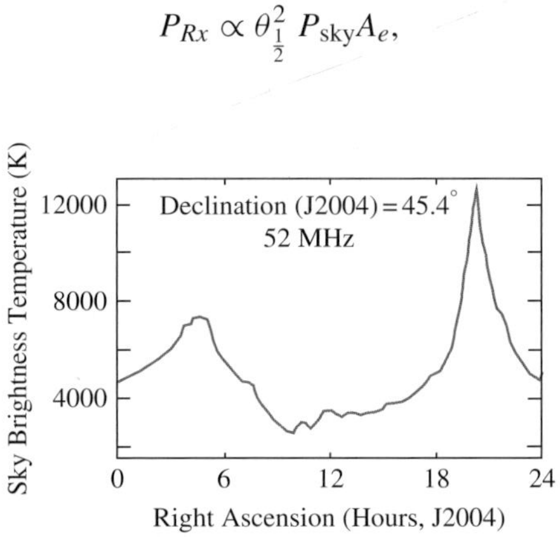

Figure 5.25 Typical skynoise variations as a function of time of day at 52MHz (adapted from *Campos et al.*, 2007a)). In this case the skynoise is shown for a vertically pointing radar beam on the McGill radar near Montreal, Quebec, Canada, at 45.4 °N. The original figure (*Campos et al.*, 2007a) showed three separate estimates of the 52 MHz skynoise based on rescalings from frequencies of 22, 38, and 45 MHz, but we have shown just one representative plot based on weighted averages of all three of these estimates.

We may now combine Equations (5.145) and (5.152) to give:

$$T_{\text{rec}} = L_{Rx}\left[T_{\text{sky}} + T_{\text{loc}}\right] + (1 - L_{Rx})\,T_{\text{lin}} + T_{Rx}, \tag{5.153}$$

where

$$T_{\text{rec}} = \frac{\overline{D_{\text{ip}}^2} + \overline{D_{\text{quad}}^2}}{K_B B_f \left(\frac{2g_s^2 R \cdot 10^{\frac{-X}{10}}}{N \cdot M \cdot L}\right)}, \tag{5.154}$$

and where we have expressed the noise power in terms of equivalent temperature in Kelvin by dividing by $K_B B_f$ on both sides of the equation. By comparing the received signal and the expected skynoise, the efficiency of the system can readily be determined. Often it is necessary to assume that T_{lin} and T_{loc} are small, although sometimes these assumptions can be relaxed. An example would be the case that T_{loc} shows no diurnal variation, in which case comparisons of the variance of T_{rec} and T_{sky} can allow L_{Rx} to be found.

Figure 5.26 shows the diurnal variation of skynoise for a radar at Resolute Bay in Northern Canada, taken from *Swarnalingam et al.* (2009a). The measured and model curves match well, and there are no major galactic point-sources to contaminate the data, so the efficiency could be found from the full day of data by fitting the experimental curve to the model, with the efficiency left as a free variable.

Campos et al. (2007a) showed that the efficiency of the McGill VHF radar was 63% on transmission and 44% on reception. *Swarnalingam et al.* (2009a) showed that the efficiency of the Resolute Bay radar was 35% on transmission but only 12% on reception, in part because of the long cables required to feed the array. A smaller array of 16 antennas at Resolute Bay, closer to the main building and using very low loss cable, had a transmission efficiency of 90% and a reception efficiency of 59%. Other radar efficiencies are presented, for example in *Swarnalingam et al.* (2009b).

However, the method described above is not realistic in the case of active phased array radar, where a large number of transmitter/receiver antennas are used. In this case, a natural noise source such as a radio star can be used. The strongest isolated radio source in the northern hemisphere is Cassiopeia A, whose declination is 58.82 °. At lower latitude, Cygnus A (40.73 °), Taurus A (22.02 °), and Virgo A (12.38 °) are the possible alternatives. It should be noted, however, that the strength of these stars relative to the galactic background is a function of the frequency and the antenna gain G_T. At lower VHF, even the strongest, Cassiopeia A, has a contribution only roughly equal to the galactic background when seen by an antenna with a gain of 30 dB, which corresponds to diameter of about 100 m at 50 MHz.

Unusual calibration methods

An interesting power-based method for radar calibration was introduced by *Van Zandt et al.* (2002). In this method, two radars are used side-by-side. Although both radars should be absolutely calibrated, it is possible to use the method even if the two radars are only calibrated in a relative sense. Powers received by the two radars are then compared, and the theoretical framework developed by *Hill and Clifford* (1978) is used to allow the power ratio to be converted to turbulent kinetic energy dissipation rates. In their case, they used an S-band radar and a UHF radar, and the radars were "calibrated" by comparing the strengths of the signals received by each in the presence of rain.

Phase calibration

One other form of calibration is important for many imaging and interferometric location procedures. This is the calibration of phases when multiple receivers are used. The simplest and most common method is to feed small signals into the receivers and measure the phase of the output signal relative to the input, though this does not take into account phase delays through the antennas and cables. Another strategy is to use a small transmitter located near the array and measure phase offsets after this signal has passed through the receiver antennas, feeds, and receivers. This needs to be then adjusted for the phase delay from the transmission point to the relevant antenna. *Chau et al.* (2008) discussed various aspects of this process, and offered several alternatives for calibration protocol, including phase determinations using known galactic sources.

Ionospheric calibration

When working with MF radars, it is possible to use another interesting trick to measure the efficiency of a radar. Indeed, because the main source of noise at MF and HF is not skynoise but rather man-made noise, lightning and other earth-bound sources, the skynoise approach is not very effective.

At low and medium frequencies, radiowaves are totally reflected from the E-region, or from sporadic E layers. At night, when absorption is low, it is often possible to see so-called "multiple-hop" reflections. These are echoes which occur when the vertically transmitted pulse reflects from the E-region, returns to the ground, then reflects off the ground and returns to the E-region, where it is again reflected back to the ground. This can happen multiple times, so an echo is seen at a range equal to the height of the E-layer, twice its height, three times its height and so forth. During times of very stable E-layers or stable sporadic E, the reflected pulses can be very steady and show little variation in amplitude over time scales of tens of seconds. By comparing the strength of the first and second hops, it is possible to deduce the combined signal losses due to range effects, absorption in the D-region, and reflective losses at the E-region and at the ground. Then, by comparing the transmitted power to the strength of the first hop echo, and removing the effects of these various losses deduced using the second hop, the radar efficiency can be determined.

In MF and HF work, it is not uncommon to measure effective reflection coefficients R_{io}, rather than C_n^2, and we will perform our discussions in terms of R_{io}. Conversions between effective reflection coefficient and turbulent backscatter cross-section are relatively straight forward, with the backscatter cross-section being proportional to the effective reflection coefficient squared (e.g., *Hocking and Vincent*, 1982a).

To see how this theory is applied, let us write that the amplitude $\mathcal{A}_R$ of the signal received from a reflecting layer at height z (generally the E-region for calibration purposes), and of reflection coefficient R_{io}, is

$$\mathcal{A}_R = {K_*}^{-1} a_D R_{io} z^{-1} \mathcal{A}_T, \tag{5.155}$$

where $\mathcal{A}_T$ is the transmitted amplitude at unit range from the transmitter, K_* is a calibration constant, and a_D is the total accumulated absorption (less than unity) through the D-region and anywhere else along the path. a_D will be unity if there is no absorption.

The parameter $\mathcal{A}_R$ does not have to be an absolute measure like the received amplitude in volts, but can be any parameter that is proportional to the received voltage. It is only important that the same parameter is used consistently. Likewise, the transmitter amplitude does not need to be a direct measure of the transmitted power, but only needs to be a consistently chosen parameter known to be proportional to the transmitter amplitude. The dependence on range is proportional to z^{-1} because the reflector is assumed to act like a perfect mirror, so the signal received at the ground after transmission to the reflector and back (total distance covered being $2z$) is proportional to $(2z)^{-1}$. The factor 2 is absorbed into K_*.

If K_* can be found, then the reflection coefficient R_{io} can be found for any reflector, as long as $\mathcal{A}_R$ and $\mathcal{A}_T$ are always measured in the same way, through the relation

$$R_{io} = K_* {a_D}^{-1} z \frac{\mathcal{A}_R}{\mathcal{A}_T}. \tag{5.156}$$

For cases in which absorption is small, an absolute value of R_{io} is therefore readily determined.

Then, if a second hop echo exists, it reflects first off the E-layer, then off the ground, and then off the E-layer again. It also travels through the absorbing medium twice, so the term a_D must appear twice in a multiplicative way. The received amplitude of the second hop can therefore be found as

$$\mathcal{A}_{2R} = K_*^{-1} R_{io}^2 a_D^2 (2z)^{-2} r_g \mathcal{A}_T, \tag{5.157}$$

where r_g is the reflection coefficient at the ground. The factor $2z$ arises because the second hop needs to travel up to the E-region and back twice, and so travels twice as far as the first hop.

Then squaring (5.155) and dividing through by (5.157) allows elimination of R_{io} and a_D and so

$$K_* = \frac{2\mathcal{A}_T \mathcal{A}_{2R}}{z \mathcal{A}_R^2 r_g}. \tag{5.158}$$

Provided that the reflection coeffcient at the ground is known (and it is often of the order of unity for MF systems), K_* can be determined as above, and then is known for any echo, whether it be strong enough to have a second hop or not. Further discussion can be found in *Piggott et al.* (1957).

The list of calibration methods described above is not exhaustive, and with sufficient ingenuity others can be developed, but we have covered the most commonly-used approaches.

6 Examples of specific atmospheric radar systems

6.1 Introduction

As discussed in Chapter 2, the MST technique began in part with *Woodman and Guillen* (1974) after discovery of atmospheric echoes from the troposphere using the Jicamarca incoherent scatter radar in Peru. Following this, recognizing the potential for meteorological applications, several groups set about building specialized radars for low altitude (less than 20 km) application, based broadly on the Jicamarca system. Primary groups who followed this course included a NOAA group in Boulder, Colorado, and a group at the Max Planck Institut für Aeronomie in Northern Germany. The NOAA Aeronomy group developed the so-called “Sunset radar” which was installed close to Boulder, and later a larger system at Poker Flat in Alaska. The Poker Flat system was in part also designed for mesospheric studies. The Max Planck group built a radar in the Harz Mountains, near Katlenburg-Lindau. These were the first VHF instruments designed specifically for meteorological studies.

Later, similar radars were developed by other groups in the UK, Japan, Australia, and various other countries, eventually leading to large networks of such radars. The term “windprofiler” was adopted to describe such radars when used for tropospheric and lower stratospheric (meteorological) wind measurements. One notable development was the construction of the large MU (middle-upper) radar near Shigaraki in Japan. At the time this was a state-of-the-art instrument, and had many important developments incorporated into it.

In this chapter, we will describe some of the details of three radars. One will be the German SOUSY radar in the Harz Mountains, one the MU radar, and the third a low-cost radar (CLOVAR) built in the 1990s in Canada. The objective is not so much to discuss the history of these radars (that was considered in some detail in Chapter 2), but to give more detail about technical developments through the course of evolution of the MST technique. The SOUSY radar was built at a time when personal computers were just starting to be developed, but were of very slow speed. Instruments like the Data General PDP-8 and NOVA mini-computers were just under development; the PDP-8 was developed around 1965, and the NOVA came into being in the late 1960s and early 1970s. While slow and awkward compared to modern computers, with limited memory (a few kB was often the limit), their introduction, and their ability to allow direct control of the radars and permit on-line coherent integration (usually by controlling a hardware processor) was considered a major breakthrough at the time. Programs were often stored on floppy disk, and it was generally not possible to do operations like fast Fourier transforms in real time. Look-up tables were needed for such processing, and often spectra were not formed at all. Rather, the first three moments of the autocovariance function were usually the parameters used to determine powers, radial velocities, and equivalent spectral widths. As time passed into the 1980s, personal/mini computers became more common, and specialized cards capable of relatively fast Fourier-processing were developed. The MU radar was introduced in the mid-1980s, and took advantage of these computer developments. However, computer disk space was limited, and data were generally stored to magnetic tape, with detailed processing often being left to post-processing on larger computers at a later time.

By the 1990s, personal computers were becoming even faster, and disk space larger, so the possibility of real-time processing was becoming available, including real-time FFT. Computer–computer communications were improving, allowing relatively fast transfer of data files between adjacent machines. Due to these developments, one set of data could be analyzed on one computer while the next set was recorded on another. Disks with storage capability of several tens of megabytes – impressive at the time – were also available, so data could be processed and stored on hard disk. These developments made it possible to build very low-cost radars with good real-time functionality, and the CLOVAR radar will be used to demonstrate these advances.

Of course since the late 1990s, computer speeds and storage capabilities have grown further, to the extent that the entire software suite for a radar could fit onto a simple transportable cell phone, allowing much more sophisticated processing techniques to be applied, including massive phased-arrays, which process data on all antennas independently. We will not discuss these here. The basic principles of the signal-processing required to produce useable results from a profiler radar were largely developed before the 21st century: newer developments have concentrated on incorporating more elegant processing methods, and allowing processing on multiple antennas simultaneously (compared to earlier systems which generally could process only one or two channels at a time). Other developments since the year 2000 have worked towards movement of many functions that were previously performed in hardware to software, reducing production costs. Digital receivers have been very common, for example. Examples include a special new design developed by Mardoc Inc., (*Hocking et al.*, 2014) and a system developed by *Yamamoto et al.* (2014).

Because the fundamentals were largely developed prior to the year 2000, we will concentrate initially on the SOUSY, MU, and CLOVAR radars. Of course several radars can be used to demonstrate these early developments, and their evolution, but we have chosen these three because the authors of this book have most familiarity with the three selected instruments. Two more recent designs will be considered at the end of the chapter.

6.2 The SOUSY radar

The SOUSY radar was first conceived conceptually in the early 1970s. The Aeronomie group at the Max Planck Institut in Katlenburg-Lindau (J. Röttger, R. Rüster,

P. Czechowsly, G. Schmidt, and J. Klostermeyer) had been studying traveling ionospheric disturbances, which were assumed to be gravity waves (see Chapter 11), and it was assumed that they were generated at ground-level and propagated upward. They applied various ground-based instruments such as microbarographs for these studies, but after learning about the work of *Woodman and Guillen* (1974), and also after being informed of plans by the NOAA Aeronomy group (B. Balsley, J. Green, W. Ecklund, and T. E. Van Zandt) to build the Sunset radar (also following on the developments of *Woodman and Guillen*, 1974), it was resolved in 1973/4 that the SOUSY group would also build a similar system.

About ten scientists and engineers at MPAE worked together to lay the foundations of the new radar. J. Röttger was nominated as group spokesman, and funding was approved, first through the director Prof. Walter Dieminger and, after his retirement, through the new director, Prof. Ian Axford. The name SOUSY (for SOUnding SYstem) was proposed at the time by J. Röttger; the VHF radar was intended to be part of a larger collection of instruments designed to allow monitoring over a significant depth of the atmosphere, up to 100 km and more, but the VHF radar in time evolved into the primary instrument. Parts for transmitters and receivers were ordered, and some member of the group (notably J. Röttger and G. Schmidt), visited John Green at NOAA in the USA (August 1974), and then the Jicamarca radar in Peru (in October–November 1974), in preparation for building the new system.

Serious construction occurred through 1975–1977, with all team members involved, where the main technical parts were mostly guided by K.H. Geisweid (analog system and RF) and G. Schmidt (digital system). Ron Woodman also made some contributions: during a stay with the SOUSY group, he designed a dedicated digital decoding processor, although this was later replaced by an updated version dedicated for complementary coding, as will be discussed later. Over the next few years the SOUSY group contracted to five people (Czechowsky, Klostermeyer, Röttger Rüster, and Schmidt), but by employing international contacts, and by being active participants in the Middle Atmosphere Program (MAP), the group remained viable and the radar was completed. First details were published in *Czechowsky et al.* (1976). Ron Woodman also visited and worked with the SOUSY group from January 1976 to July 1977, writing several technical reports during the stay.

The original SOUSY system was located at the MPAE institute, comprising eight Yagi antennas with a mean transmitter power of 250W. First echoes were detected on 3 February 1977. By 16 March 1977, observations of a frontal system had been made, and early results were published in *Röttger and Czechowsky* (1977), with many more to follow (e.g., *Röttger et al.*, 1978). Interestingly, the first echoes detected on 3 February 1977 were simply recorded on film, and the radar control was implemented using signal generators. More sophisticated computer control and preprocessing followed only after the system had been moved to the Harz mountains.

Following these preliminary tests at the institute in Katlenburg-Lindau, the radar was then moved to the Harz mountains and expanded: first measurements were made in July and August 1977. The antenna field site was in the Sperlutter valley (10.5 °E, 51.7 °N),

Figure 6.1 The process of moving the SOUSY radar from the institute in Katlenburg-Lindau to the Harz mountains. (a) A container being removed from the Max Planck Institut on 25 March 1977. (b) A truck towing the container to the final site in the Harz mountains. (c) and (d) Installation of the radar equipment at the site. (e) A view of some of the installed antennas.

about 25 km east of Katlenburg-Lindau. The mountains surrounding the valley provide considerable shielding, up to elevation angles of 5–25 ° up to to distances of about 2.8 km, which was the maximum ground-clutter range.

Some photographs showing the transfer of the radar to the Harz mountains are shown in Figure 6.1.

The concept of coded pulses was introduced to the radar on 29 July 1977, and complementary codes were introduced in 1978. The codes were introduced by both G. Schmidt and R. Woodman. The implementations of complementary codes on the SOUSY radar were the very first introduction of these codes to the MST radar community, and demonstrated the value of them to MST measurements. Such coded experiments allowed detection of stratospheric layers, and especially showed some of the earliest detections of enhanced tropopause echoes. It was proposed then that these echoes were due to a mixture of specular reflection from stable layers, diffuse reflection from rough layers, and turbulence.

By 1980, the SOUSY radar was known world-wide as a leader in the field.

6.2.1 Technical details

The SOUSY radar, while adapting many ideas from the Sunset and Jicamarca radars, also had many unique features. In contrast to the Jicamarca and Sunset radars, which used dipole-antennas for transmission and reception, the SOUSY radar used Yagi antennas. Photographs were presented in Figures 2.14 and 2.15 in Chapter 2. The Jicamarca and Sunset radars actually used coaxial-colinear (Co-Co) antennas (e. g. *Balsley and Judasz*, 1989). These are good for remote site installations, since long strings of Co-Co antennas can be assembled in advance and simply rolled out and suspended at the side. However, Yagi antennas are more efficient and have wider bandwidths.

A moderately detailed report about the SOUSY radar was presented by *Czechowsky et al.* (1976), and the discussion below is based to some extent on that document. That paper was presented prior to completion of the radar, so some of the items were projections of future plans, while others had already been enacted. A few were even implemented a few years later. In our report below, we will not comment too much about the timing of when the various features were implemented, but rather simply recognize that the plans presented were eventually (in the main) implemented, and that the radar was a true leader in the field for many years.

Choice of wavelength

One item discussed in some detail in that article was the rationale for the choice of frequency. Briefly, the authors did not consider frequencies greater than about 1 GHz because this limited the upper altitude at which they could make useful measurements. Figure 11.25 in Chapter 11 shows the expected limiting scales as a function of height expected for Kolmogoroff turbulence (controlled especially by the atmospheric kinematic viscosity), and an inner scale of about 0.15 m (one half-wavelength at 1 GHz frequency) occurs around 20–30 km altitude. A 1 GHz radar would not be expected to receive useful turbulent scatter from above that height. Since the authors desired signal from a wide altitude range, a lower frequency was desirable. A frequency below 20 MHz was also not desirable – lightning and man-made noise can be quite high in this frequency band, and construction of a high-gain antenna at such frequencies would require a large land area.

The SOUSY group also considered signal-to-noise issues, and noted that the cosmic noise varies roughly as $\lambda^{-\beta}$, where β is of the order of 2.2 to 2.5 (see Chapter 5). At the same time, Equation (5.138) in Chapter 5 suggests that the received signal varies proportionally to $P_T C_n^2 \lambda^{-1/3} A_{\text{eff}}$, where P_T is the transmitter power, A_{eff} is the effective area, λ is the radar wavelength, and C_n^2 is the refractive index structure constant (which is wavelength-independent in the inertial range of turbulence). This suggests that for a fixed effective area, longer wavelengths will have slightly less backscattered power. However, the more important parameter is the signal relative to the noise, and so the signal-to-noise ratio should change proportionally to roughly $\lambda^{-(\beta+0.33)}$, being therefore much stronger at shorter wavelengths (higher frequencies). However, if it is assumed that the antenna arrays are to be made from discrete antennas such as Yagis, then the limiting factor might not be the size of the array, but the number of sub-element antennas. Since most radar designs place the antenna sub-elements a fixed multiple of a wavelength apart (typically say 0.7 wavelengths), a fixed number of antennas means that the area increases proportionally to the wavelength squared. Maintaining a fixed number of antennas in such a way also maintains the gain as independent of wavelength. The signal-to-noise ratio then varies as $\lambda^{-(\beta-1.67)}$, or typically $\lambda^{-0.5 \text{ to } -0.8}$. So a system working at 500 MHz would typically have a better signal-to-noise ratio than a system of the same gain working at 50 MHz by 3–6 times. In the view of the SOUSY group, this was a small sacrifice relative to the loss of data above 20 km altitude. In addition, it was felt that suitable application of coherent integration and pulse-coding (see earlier chapters) would alleviate this loss of signal-to-noise. It was decided that a frequency of 30–100 MHz would be optimum for the experiments, and after seeking frequency licencing, 54 MHz was chosen.

Block diagram

We now turn to technical details associated with the SOUSY radar. Figure 6.2 shows a block-diagram of the system.

The system can be seen to be largely controlled by a computer, a radar controller and a master oscillator. The radar was designed as a pulsed system. It needs to be remembered that at that time, computer memory was limited, and speeds were slow. The initial controlling computer was a Hewlett-Packard HP-100-RTE, which was soon upgraded to a HP21-MX. This latter unit had a capacity of 16 384 × 16-bit words. By modern standards, this is very small, so the amount of processing possible was limited. Generally, raw data were written to magnetic tape or disk, and analyzed later on a larger computer, whereupon spectral analysis or correlative analysis could be applied. However, some limited real-time capability was incorporated, as will be seen below. (The Sunset radar in Colorado, built around the same time, was configured to allow quasi-realtime spectral processing, but used large banks of look-up tables to achieve this.) The limited computer capability also meant that many other actions needed to be performed in hardware. Key aspects of all the new MST radars, including the SOUSY system, were the recording of in-phase and quadrature signals (as distinct from amplitude-only, or amplitude and phase recordings, which had been the earlier norm), plus the use of coherent integration and pulse-coding. This allowed significant improvements in signal-to-noise calculations, but again, the pulse-decompression and coherent integrations (actually summation) required special dedicated hardware (see the items pulse-decoder and hardware adder in the block-diagram). In modern systems, these processes can often be performed in software.

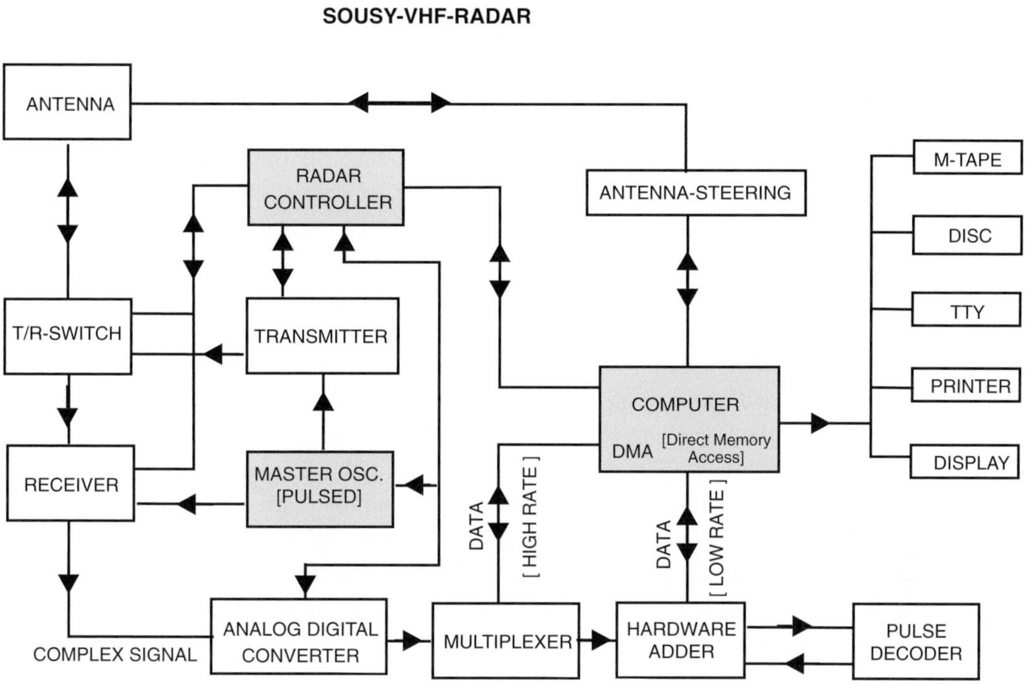

Figure 6.2 The SOUSY radar (from *Czechowsky et al.*, 1976).

Computer controller

The computer represented an essential unit of the radar facility. Prior to this time, extensive computer control of a radar was rare, so this was something new. The computer was used for system control, plus data acquisition and preprocessing. This ensured that the system operations could be optimally adjusted to compensate for changing characteristics of the backscatter returns.

After coherent detection at the receiver output, the two quadrature signals (real and imaginary parts of the video signal) underwent an analog-to-digital conversion. The two-channel analog–digital converter had a high (for the time) conversion rate of more than 2×10^6 words per second, which matched the 1 MHz video bandwidth. Each word comprised 10 bits, adapted to match the linear dynamic range of the receiver. The digital data were either directly transferred to the computer memory via the high-rate channel, or were sent to the hardware adder. This adder was designed to integrate the data coherently, as discussed above.

Data acquisition was perfomed at a rate of 1.2 MHz, giving a typical sampling interval of 0.8 microseconds (120 m resolution), with the capability to record sweeps of up to several tens of km per pulse. The system was planned so that the signal coherence time could be checked by the computer as a function of height so that the integration time of the hardware adder could be set to an optimum value.

Hardware pulse-decoding could be directly accomplished after hardware addition. The data were then transferred via the low-rate channel into the computer. After further pre-processing, the data (e.g. power and velocity as function of range) could be displayed for quick looks and were used for real-time decisions about radar control. As noted above, the data were also written to magnetic tape for subsequent post-processing.

The computer also controlled the steering of the antenna beam. The radar controller, either under manual or computer control, generated the transmitter pulse sequences (which are then passed to the master oscillator), as well as the control pulses for the transmitter and receiver amplifier, the transmit-receive switch and the analog-to-digital converter. The computer could also be used to check things like excessive VSWR, excessive duty cycle and receiver overload.

The transmitter

Experiments at Jicamarca had suggested that scattering layer depths were of the order of 100 m, so the radar was designed to achieve resolutions of the order of this value. The transmitter was designed as a linear amplifier with a bandwidth of 5 MHz, allowing (in principle) pulse lengths down to about 30 m, although the final pulse width chosen was about 0.8 to 1 μsecond. The peak power was 600 kW. The transmitter was based on vacuum tubes and valves; although solid-state transmitters were becoming available by the late 1970s, they were still something of an unknown, and vacuum systems were considered more reliable. Vacuum systems required higher internal voltages to operate, and required replacement of the tubes on a regular basis, while solid-state systems required lower voltages (but higher internal currents), and (in principle at least) could last indefinitely. However, solid-state systems really only started becoming the norm in the 1980s and 1990s. Views of the SOUSY transmitter are shown in Figures 6.3 and 6.4, while a valve transmitter with its metallic shields removed is shown in Figure 6.5.

The reasons for the use of such a wide-band amplifier were two-fold: first, it allowed optimal suppression of harmonics, and secondly, the pulse shaping depended only on the input from the pulse shaping master oscillator, so a wider bandwidth allowed less distortion of the pulse. Strong suppression of harmonics was necessary because of administrative regulations in Germany. The second condition also assured that the radar had the advantage of easy pulse-shape matching in the receiver and transmitter path.

The transmitter could handle pulse-coded signals up to a maximum duty cycle of 4%, so that a considerable improvement of the signal-to-noise ratio of at least 10 dB could be expected compared to single pulse operation. The power supply was capable of handling a maximum length of a coded pulse-train of 100 μs, which was required for mesospheric sounding. Phase and amplitude jitter caused by the high power amplification, which could have influenced the accuracy of the power spectral analysis of the received scattered signals, was kept at a minimum. The driver and the final amplifier were screen-grid modulated in order to keep shot noise generation during the receiving phase at low levels, so that the limiting receiver sensitivity was not influenced.

The transmitter final stage operated at grounded grid using two water-cooled tetrodes (Siemens PS 1054) in push-pull mode. Both the radar transmitter and the receiver could be tuned in a wide frequency range – from 48 MHz to 67 MHz – without any complicated mechanical reconstruction.

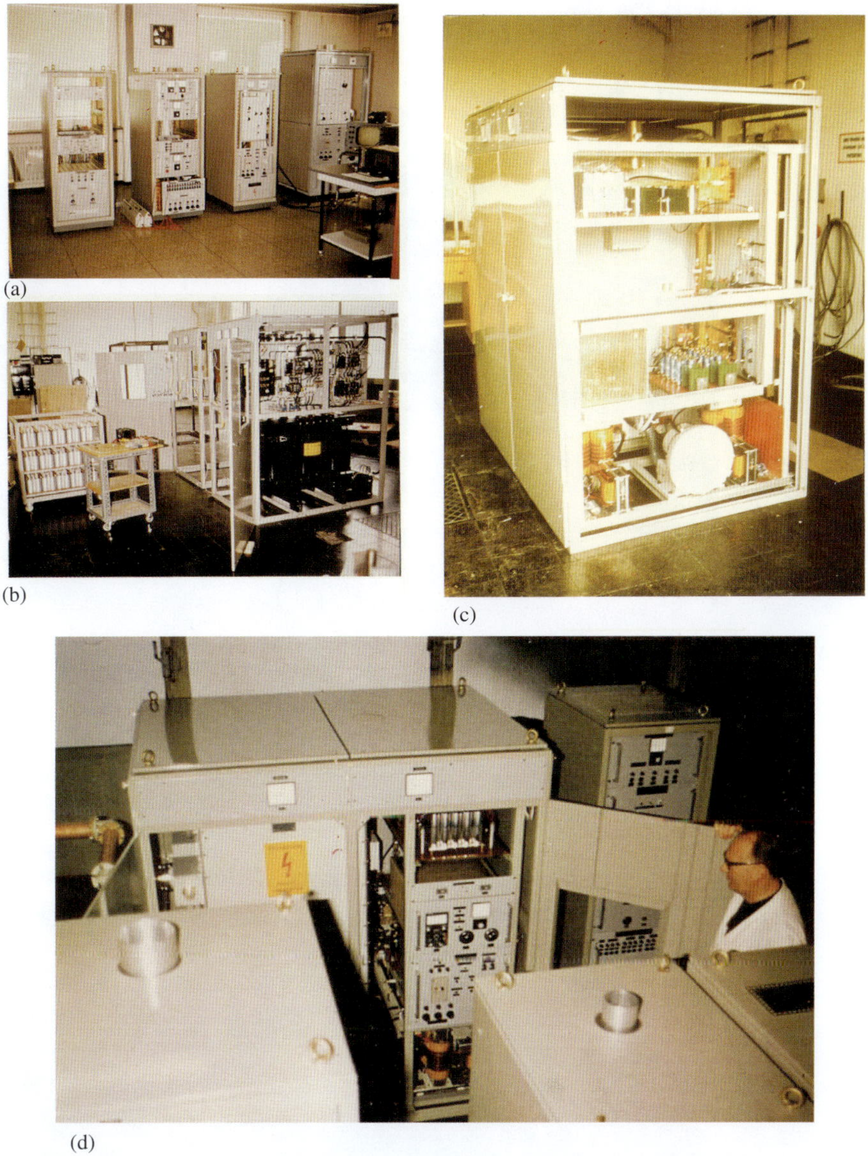

Figure 6.3 Views of the SOUSY high-power transmitter. (a) Transmitter driver stages, including the Endtreiber (final driver stage), which delivered 20 kW and had a bandwidth of 7 MHz. (b) High voltage power portion of the transmitter. (c) The *steuerteil de endstufe*, which, literally translated, means *the control unit of the final power amplifier*. (d) The final output stage of the transmitter, which produced 600 kW peak power and had a bandwidth of 3 MHz.

The receiver

The receiver used a solid-state design and had a noise level of 4 $K_B\,T_0$ under optimal conditions (noise temperature of about 1200 K). It was capable of linear amplification within a dynamic range of 70 dB at 1 MHz video-bandwidth. The deviation from

Figure 6.4 The assembled transmitter.

Figure 6.5 A valve transmitter similar to the SOUSY one. In this case, it is actually one of the Poker Flat transmitters, taken around a similar time in history (from *Balsley et al.*, 1980). (No such clear photos of the SOUSY transmitter with the shields removed were available to the authors, so this serves as a suitable substitute, although the SOUSY vacuum tubes were somewhat larger; the Poker Flat radar used multiple 100 kW transmitters, while the SOUSY radar was a single 600 kW unit.) (Reprinted with permission from John Wiley and Sons.)

linearity throughout the 70 dB range was smaller than ±1 dB. This high linearity was necessary because of the hardware coherent integration procedure which had to be applied to improve the signal-to-noise ratio during the signal processing (as discussed above and in earlier chapters).

The output signal amplitude of the receiver was planned to be controlled by fast-switched attenuators which could be controlled by the computer; in such a scenario, the computer could in principle alter the attenuation according to measured strengths of recorded signal in real-time, allowing range-dependent gain control. Thus, an overloading of the receiver and the analog-digital converter could be avoided, and the dynamic range was increased considerably. Because of this flexibility, ground-clutter problems can be reduced, for example. However, this feature was never implemented: before coherent integration, the signal is largely invisible in the skynoise, and the optimum receiver gain is usually defined by the level of the skynoise, so the advantages of automatic gain control are largely restricted to reducing the gain in situations with large ground-echoes. Once oscillations associated with ground echoes die out at the higher altitudes, the gain at higher altitudes is largely set at the same levels.

The receiver IF-band-width was 2 MHz, and was adjustable to different low-pass filters in the video path (i.e. the final receiver stage), which could be chosen to properly match the transmitted pulse shape. All reference frequencies in the transmitter and receiver were coherent, in order to ensure that Doppler measurements of the echoes could be reliably determined. The complex video signals from the quadrature receiver outputs generally had a phase stability better than 1° throughout the entire dynamic range. These were then fed to the analog-to-digital converter for subsequent processing, as discussed above and as shown in the block diagram.

The transmit-receive switch

A transmit-receive switch (TRS) was necessary to successively connect the antennas to the transmitter and receiver, depending on whether the pulse was being transmitted, or whether transmission was complete. Transmit-receive switches have been discussed in some detail in Chapters 4 and 5 of this book. The switches had to not only handle a maximum power of 600 kW transmitted in the VHF-band, but also allow rapid switching to receiver mode in a time less than a few μs. The SOUSY radar used an active TR switch, since passive units were only developed some years later (see Chapter 5 in particular). The transmitter and receiver were tunable in a rather broadband fashion, which added some extra complication to development of a good TR switch. A broad-band balanced duplexer from *Unitrobe* (1968) was applied to the switch, based on pin-diode operation. In this duplexer, two hybrids were used and the switching was performed by 2 × 6 high-power pin-diodes. The only frequency-dependent part was a tunable stub which was used to compensate for the diode capacities. It was estimated that the decoupling of transmitter and receiver during the transmitting phase was at least 66 dB. The hybrids performed an additional decoupling between the transmitter and receiver even during the receiving phase, so that shot noise generated in the transmitter would not limit the receiver sensitivity. A fast diode-switching control assured a maximum recovery time

(a) (b)

Figure 6.7 The SOUSY antennas. (a) A module comprising 8 × 4-element Yagis, used for test purposes at the Max Planck Institut. (b) The antennas in the array in the Harz mountains. The module shown in (a) also had some additional interesting history, which is discussed in the main text.

sampled, since only one receiver channel existed. The analysis showed that the spaced antenna method can be applied in MST work in addition to the standard method using beam steering. In addition, the first interferometric applications were introduced with this set-up (e.g., *Röttger and Vincent*, 1978; *Vincent and Röttger*, 1980).

6.2.2 Summary of the SOUSY radar

The SOUSY radar was one of the very first radars designed and built specifically for MST studies. All of the electronics except the antennas were housed in three large transportable containers, and could be moved. Subsequent work by the SOUSY group saw experiments in Arecibo and Andenes, Norway, using similar mobile systems, and the SOUSY group and radar were world leaders for many years in the area of MST Studies. (In the early 21st century, the radar was moved to the Jicamarca radar in Peru, but we will not discuss that move here: the interested reader is referred to *Woodman et al.* (2007).)

6.3 The MU radar

6.3.1 Introduction

The next radar for discussion is the MU radar, see Figure 6.8. This radar was completed in November 1984, and was able to take advantage of many advances in computing and engineering technology that took place in the late 1970s and early 1980s. For example, solid-state transmitters were now more common, and the design of this radar capitalized on that. Computers were also improving rapidly. In addition, the Japanese scientists who built it had a high commitment to producing a cutting-edge instrument, and invested more money in this radar than any previous MST radar. The result was a highly efficient and adaptable radar that stood as a world leader for many years to come. The discussion

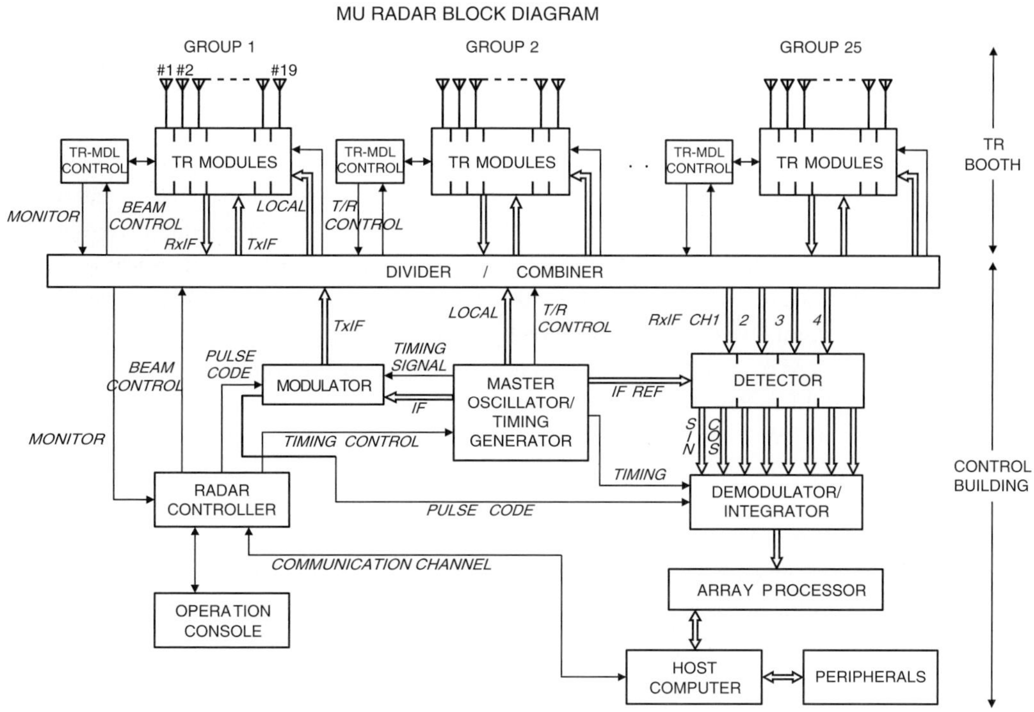

Figure 6.8 The MU radar.

that follows will be given in the past tense, since it largely refers to the conditions at the time of construction, but at the time of writing (2016) the radar is still in existence and still functioning well. Indeed, having gone through several upgrades, it is still a leader in the field even at this time.

The design plans were started in the late 1970s and early 1980s. One early limitation was a restriction on purchase of arable land. As a general rule, flat land should only be used for agriculture in Japan, although exceptions could be made upon application to relevant authorities. But even if such a case could be made for non-agricultural use, flat land was very expensive – the area required to build the radar would have cost over US\$10 M (approximately US\$1000 per m^2). This problem was resolved by buying a mountain, which had a purchase price of only about US \$1 M, and then digging it out to form a hollow. The soil removed from the mountain could not be moved from the site, and was built up around the edges, allowing it to act to some extent as a radio-shield. (Earlier plans to take excess earth to the coast and use it to build up the coastal area never eventuated due to the large amount of trucking that would have been required, disturbing the local community.) Total costs of the exercise broke down roughly as follows: construction = 9 M US\$ (including land movement, antenna installation, fencing, roadways etc – all construction except the building), radar = 18 M US\$, building = 3 M US\$, land cost = 1M US\$ (taking the conversion at the time of 1 US\$ = 100Yen).

The radar itself was built by Mitsubishi Corp., who have also supported the system over the past 30 years.

The antenna array, both from above and on the ground, is shown in Figure 2.20.

The details of the radar are discussed in *Fukao et al.* (1985a, b). Here, we will give a summary of those reports; for more specifics, the reader is referred to those articles.

The MU radar was the first MST system to employ solid-state transmitters on a large scale. This had advantages and disadvantages. Valve transmitters are capable of high power transmission, while solid-state systems are usually limited to a few kW. The MU system used transmitters that were only capable of 2.4 kW of peak power. However, this was compensated for by using 475 of them, giving a total peak power of over 1 MW. Solid-state transistors are capable of higher duty cycles, and the MU radar could function comfortably with a 5% duty cycle, allowing a mean power output of 50 kW. Forcing 475 separate transmitters to operate in unison of course had issues of its own, but was accomplished by carefully designed circuitry and tight timing control.

In fact the design employed turned the issue of having 475 separate amplifiers to advantage, by ensuring that each amplifier fed a different antenna. By applying different phases to the separate transmitters, the beam could be steered to almost any angle in the sky, as described in Chapter 5. This gave the system a flexibility that was simply unavailable on any other MST radar. Indeed the beam could be steered on a pulse-to-pulse basis, allowing unprecedented experiments to be performed. This capability is referred to as an "active phased array system," whereas a system in which a single port feeds all the antennas is referred to as a "passive phased array system."

As a further advance over most earlier systems, the radar used crossed antennas, allowing the users to transmit and receive either circular or linear polarizations.

The basic parameters of the radar are shown in Table 6.1.

6.3.2 Computers

The radar had three large computers integrated into its design. The primary instrument was a desktop HP9835A, which was the main component of the radar controller. In addition to the larger computers, each TR module controller had a microprocessor, based on an Intel 8085 chip. If these are considered as computers, it might be considered that the system had a total of $3 + 25 = 28$ computers of various types. The HP9835A was linked to the 25 transmitter-receiver (TR) module controllers. Various timing signals necessary for real-time system control were generated according to instructions from the HP9835A. Software in the radar controller made a variety of flexible operations possible. For instance, it was possible to steer the antenna beam within each interpulse period (IPP), i.e., up to 2500 times every second, virtually to any direction within 30 degrees of the zenith. It was also possible to excite only a portion of the antenna array and receive the echo with other portions.

The system also incorporated two other computers, namely a superminicomputer (VAX-11/750) and an array processor (MAP-300) with a 2 Mbyte random access memory (RAM). These were used for post-recording processing.

Table 6.1 Basic parameters of the MU Radar at the time of its commissioning (adapted from *Fukao et al.*, 1985a). (Reprinted with permission from John Wiley and Sons.)

Parameter	Value/description
Location	Shigaraki, Shiga, Japan (34.85N, 136.10E)
Radar sytem	monostatic pulsed radar; active phased array system
Operational frequency	46.5 MHz
Antenna	
Antenna description	circular array of 475 crossed Yagis
Aperture	8330 m^3 (103 m in diameter)
Beam-width	3.6 ° (one way half-power half-width for full array)
Steerability	Steering is completed in each IPP
Beam directions	1657 directions; 0–30° off-zenith angle
Polarizations	linear and circular
Transmitter	
Transmitter design	475 solid-state amplifiers
TR modules	Each module has output power of 2.4 kW peak and 120 W average
Peak power	1 MW (maximum)
Average power	50 kW
Duty cycle ratio	5% maximum
Tx bandwidth	1.65 MHz (maximum)
Pulse width	1–512 μs (variable)
IPP	400 μs to 65 ms (variable)
Pulse compression	binary phase coding up to 32 elements; Barker and complementary codes
Receiver	
Receiver bandwidth	1.65 MHz (maximum)
Dynamic range	70 dB
IF	5 MHz
A/D converter	12 bits × 8 channels

6.3.3 The antenna array

The antenna array is shown in Figure 6.9. It clearly has a circular perimeter. The diameter is 103 m. Within the array, individual antennas are arranged in a triangular grid, which optimizes the number of beam directions that can be properly formed. Antennas were three-element crossed Yagis, and the antenna spacing was 0.7λ. The antennas, which are crossed, are each aligned in the N-S and E-W directions.

The driven element of each Yagi antenna was folded in order to cancel the unbalanced current on it. This folding works as an impedance transformer, and the input impedance becomes four times that of an unfolded antenna. (Folded dipoles also usually have a wider bandwidth than unfolded ones.) Therefore another impedance transformer, or balun, which reduces the input impedance to a quarter, was inserted at the feed point (see Figure 6.10). We will not fully describe the operation of the balun here – the interested reader is referred to *Fukao et al.* (1985a), and references therein, for further details. The Yagis were also designed to optimally balance the gain, the bandwidth, and also

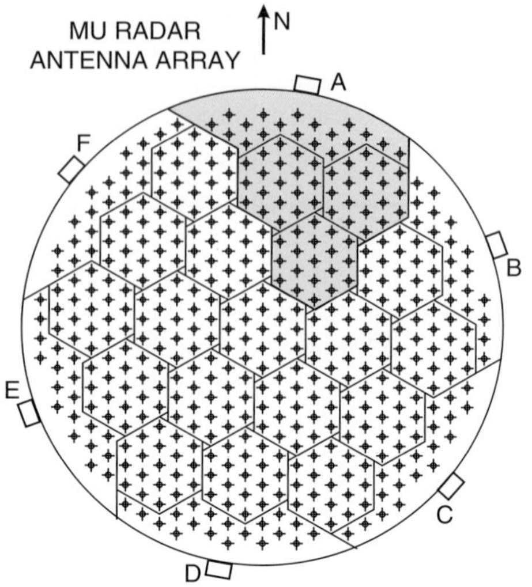

Figure 6.9 Arrangement of the antennas of the MU radar within the main array. Each cross is a 3-element crossed antenna. The antennas are grouped into hexagons, as shown, with additional antennas arranged around the outside to complete the circle. The six boxes outside the antenna array, indicated as A–F, represent the booths which accommodated the TR modules. All booths except F accommodate the TR modules for three hexagonal sub-arrays and one peripheral sub-array. The shaded area shows groups accommodated by booth A. The TR modules for the central hexagon are housed in booth F.

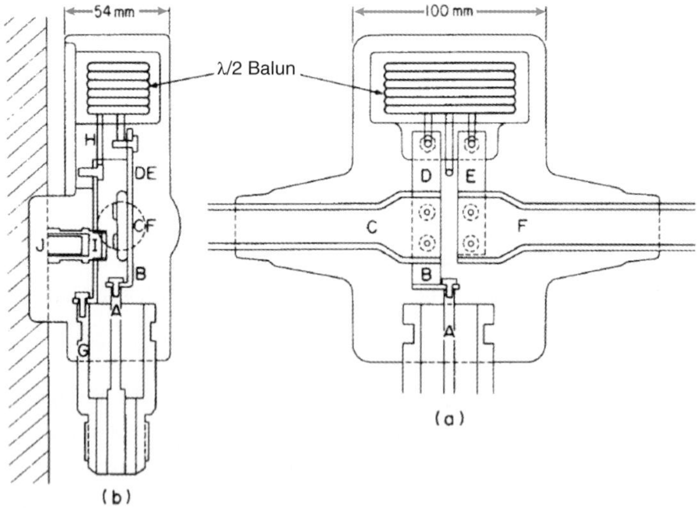

Figure 6.10 Feed-point circuit and balun for the antennas. C and F are the terminals of the driven element (folded dipole) of a three-element Yagi, while D and E are the ports of the half-wavelength balun. A and G are the inner and outer coaxial conductors, respectively, while H is the outer conductor of the balun. See *Fukao et al.* (1985a) for further details.

to produce a good front-to-back ratio, thereby minimizing the effect of the underlying ground.

The gain of the three-element Yagis was chosen so that the effective area of the antennas roughly matched the available area which could be considered as "occupied" by each antenna (i.e., an equilateral triangle with sides of length 0.7λ), so that the whole array could be considered as a "filled" dish (i.e., with no holes). This gain is about 7 dB for a single three-element Yagi.

Whereas the SOUSY radar used tapering of the signal sent to the outer antennas in order to reduce the side-lobes, the MU radar used no tapering, either spatial or electronic.

The impedance matching is generally very good, with the VSWR of the Yagis always being below 1.3 for beam-pointing angles less than 30 ° from zenith. The frequency dependence of the element properties is fairly small within the permitted bandwidth, and the VSWR is kept below 1.25 within this range. A polar-diagram for a typical MU Yagi antenna can be found in Figure 6.11. The theoretical beam-width was approximately 17.3 ° for a single hexagonal sub-array, and 9.0 ° for a combination of three hexagonal sub-arrays.

The polar diagram of the entire array of the radar was extensively simulated using numerical procedures, and also verified using satellite tests. The moon was also used as a target for calibration purposes (see Chapter 5). The beam-width of the main lobe was approximately 3.6 °, corresponding to a gain of 34 dB. The first side-lobe was

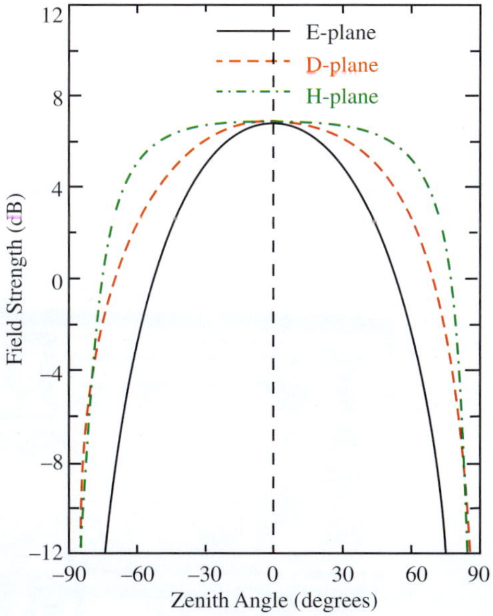

Figure 6.11 Polar diagrams of a typical stand-alone linear Yagi used in the array. The field strength has been normalized relative to the radiation field of an isotropic antenna matched to the feed-line impedance. Solid, dashed, and dot-dashed lines show the vertical plane patterns in the directions parallel (E-plane), diagonal (D-plane) and perpendicular (H-plane) to the excited dipole, respectively.

suppressed by more than 18 dB with respect to the main lobe. The spillovcr radiation at elevation angles below 20 ° was reduced by more than 40 dB, even when the antenna beam was tilted 30 ° from zenith. It was found that in the azimuthal directions, the sidelobes were distributed almost symmetrically and uniformly, without any outstanding peak in any particular direction.

6.3.4 The transmitter-receiver system

As mentioned, the MU radar made extensive use of solid-state transmitters for pretty much the first time in MST work. For this reason, we will spend a little time discussing this new feature. Some of these points were discussed in the introduction to this section, and in Figure 6.9. Figures 6.12 and 6.13 show a typical transmitter printed circuit board and also a view inside one of the booths that housed the 475 transmitter and receiver modules.

The following section describes some of the details of the TR unit. Readers who are less familiar with some of the electronics may wish to refer to Figure 4.12 in Chapter 4 as we go through the details. The TR modules were some of the more expensive parts of the radar, costing US$30 000 each.

As seen in Figure 4.12, the system starts with a local oscillator (LO) and an intermediate frequency (IF), which then is used to develop the main radio frequency (RF). This is referred to as up-conversion. Conversely, the received RF needed to be beaten down to the IF upon reception (down-conversion). With the MU radar, both up-conversion and down-conversion were performed in the remote booths discussed in Figures 6.9 and 6.13. The IF (5 MHz) and LO (41.5 MHz) signals were transferred between the booths and the control building; the RF was generated within the booths. The signals were then sent to each antenna via AF coaxial cables of equal electronic length. (AF is an abbreviation used for coaxial cables which employ an aluminum outer conductor

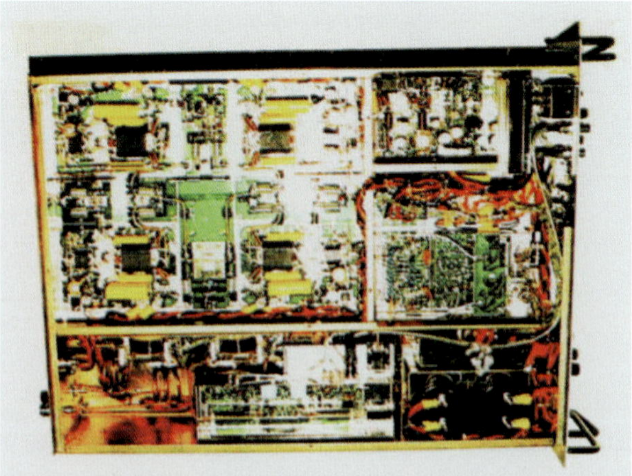

Figure 6.12 A single solid-state transmitter unit of the MU radar.

Figure 6.13 Inside one of the transmitter buildings of the MU radar.

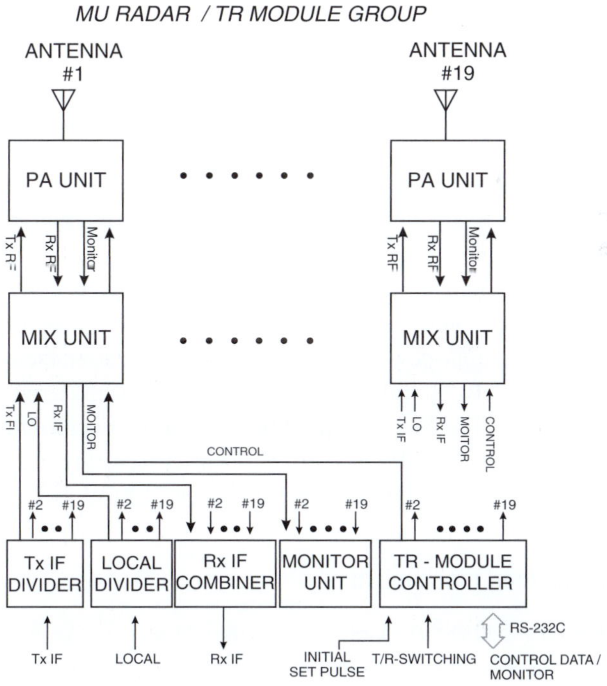

Figure 6.14 The transmit-receive configuration used at the MU radar.

and are insulated with foamed polyethylene.) Upon arrival at the booth, the signals were divided into four (or five) groups by the TR module divider (denoted TX IF divider and local divider, respectively) and distributed to the 19 TR modules in each of the groups. Upon reception, the received RF was down-converted to an IF signal, and the received IF signals from the 19 TR modules were combined into one by the TR-module combiner (RX IF combiner) and sent directly to the control building.

Figure 6.14 is a block diagram of one group of TR modules. The timing signals for T/R switching and initial set pulse (ISPL) were split in the same way. The main constituents of the TR module were a mixer (MIX) unit and a power amplifier (PA) unit (not shown – see *Fukao et al.*, 1985a, for more details). Frequency conversion was performed in the MIX unit. The 41.5 MHz LO was fed to the MIX unit, passing through a digital phase shifter for beam steering and TX/RX phase correction. The MIX unit also contained a buffer amplifier, a gate for transmission, and preamplifier for reception.

The PA unit amplified the RF (46.5 MHz) signal supplied with the MIX unit by up to 63.7 dBm (2350W), and fed it to an array element. An exciter consisting of a three-stage amplifier operated in A class with a gain of 39.5 dB. The final power amplifier stage was composed of four push-pull circuits operating in parallel mode with a gain of 12 dB. Eight high-power transistors (TH-430) were employed.

A transmit-receive switch was used which consisted of a combination of two 3 dB hybrids and high-power PIN diodes, similar to the type used by the SOUSY group, and discussed above. Three additional diode switches were inserted in the RX channel in order to obtain a total isolation of 100 dB between TX and RX signals. Switching required 10 μs, limiting the time between cessation of the pulse transmission and the start of sampling.

A bandpass filter was inserted after the TR switch to prevent transmission of unnecessary harmonics. The second and third harmonics were reduced by 85 dB and more than 90 dB, respectively, in comparison with the fundamental frequency.

Phase and intensity of the TX signal are monitored by the TR module controlled via a directional coupler. During observation, the radar controller polled the TR module controller for monitor data, and the ouput VSWR was detected by the hardware in order to protect the high-power transistors. The polarization (selection) switch comprised two relays and a 3 dB hybrid, and gave two linear as well as right-hand and left-hand circular polarizations. The overall gain of the PA unit was approximately 50 dB.

The output of each PA is 2350 W, but after factoring in cable losses, the final radiation power from each antenna element becomes 2100 W.

6.3.5 Antenna feed mechanism

Fukao et al. (1985a) discussed at length the ways in which the signal was fed from the power amplifier units to the antennas, and how the received signal was then fed back from the antennas to the TR unit. The main points are: (i) each TR unit feeds a separate antenna with a separate cable; and (ii) the cables were all of different lengths, complicating the phase relationships between transmitted signals. The cable-paths were assured to be equal to integral numbers of half-wavelengths, and then further phase adjustment was applied by adding additional delay lines to the shorter cables. Delay lines were adjusted in steps of 50 ns, which is considered to be a fine enough unit that it should not deteriorate the beam pattern. More details can be found in *Fukao et al.* (1985a).

6.3.6 Summary of the MU radar

The MU radar was unquestionably a state-of-the-art instrument, which set a standard for many years to come. Scientific output from this system has been enormous, and the system has seen several upgrades, keeping it at the forefront of MST research. The radar has also led to several subsequent radars, including an equatorial radar in Indonesia (EAR), a system in Antarctica (PANSY), and a newer equatorial radar currently in the planning stages. We have only summarized its features here – the interested reader is referred to *Fukao et al.* (1985a) for more details, as well as the companion paper *Fukao et al.* (1985b) for more details about the electronics and the early results determined with the radar.

6.4 The CLOVAR radar

6.4.1 Introduction

The third radar for discussion looks not at further, more sophisticated radars but in fact regresses to a simpler system. While large, expensive radars are obviously impressive, many scientists do not have access to millions of dollars of funding. So we turn here to examine a radar that cost under $100 000 to build in the mid-1990s, demonstrating that even modest research groups can build their own radars. To be fair, it is not an MST radar, and is designed for tropospheric and lower stratospheric work, but it worked sufficiently well that it allowed several important publications, and allowed the developers to test and improve various software tools that were new to the field. A radar of similar design was later built at Resolute Bay (*Hocking et al.*, 2001b), where it could in fact be considered as an MST radar because it was able to observe polar mesosphere summer echoes (PMSE). Of course there are several other radars which were built at modest cost; this one is chosen because of its familiarity to one of the authors (WKH). The radar to be described is the CLOVAR (<u>C</u>anadian [<u>L</u>ondon, <u>O</u>ntario] <u>V</u>HF <u>A</u>tmospheric <u>R</u>adar). While only a small radar, it led to the design of many other subsequent radars. The radar was first reported in the refereed literature in *Hocking* (1997a). In the following pages, we will describe some of the key aspects of the design and implementation, and discuss some of the more pragmatic features that were not discussed in the formal publication.

The requirements for the CLOVAR radar were as follows:

(i) The cost should be under Ca$100 000.
(ii) It should be able to observe above the middle troposphere, and at times reach into the lower stratosphere.

(iii) It should have a narrow beam, to enable measurements of turbulence using the beam-broadening spectral method (see Chapter 7). This requirement arose because turbulence studies were an area of great interest to the designer.
(iv) It should be possible for the designer to be able to build it alone, or with modest amounts of help from students and, only on rare occasions, additional technical support (if and when available).
(v) All calculations of powers, radial velocities, spectral widths, and other useful parameters, had to be performed in real time, since availability of support personnel for post-acquisition processing was not an option.
(vi) Ground-level radiation and reception had to be kept to small values to reduce interference with neighbors and to reduce noise levels in the radar.
(vii) Aircraft contamination was likely to be a major form of interference, so the design needed to be able to eliminate the impact of aircraft.
(viii) The array should be an active one (i.e. all antennas joined to the central building by separate cables, like the MU radar). This was needed to simplify beamsteering, and also to allow the health of the antennas to be checked regularly without visiting each one.
(ix) It should be possible to do interferometry with the radar, which necessitated that each antenna could be accessed separately (or at least small groups of antennas should be accessible independently).

6.4.2 The antenna array

A radar frequency of 40.68 MHz was the only one available from Industry Canada Spectrum Management, so this was the frequency selected. Figure 6.15 shows the radar antenna array. The design was inspired by the Mill's cross array used in astronomy, but instead of two perpendicular lines of antennas, the "lines" (or "arms") had a width of just over two wavelengths. This design gave a relatively narrow main beam for the polar diagram, with a main-beam half-power half-width of very close to 2°. The two-way beam half-power half-width was 1.4°. Of course, while this unique cross-structure led to a narrow main beam, the first side-lobes were only 7 dB down from the maximum in one-way mode, and 14 dB down in two-way mode. This limitation had to be dealt with – it was not possible to use weighted moments to calculate radial velocities, and the radar relied heavily on spectral-fitting procedures in order to produce accurate wind estimates.

A small wooden building and a small live-in trailer were built at the center of the array, to minimize cable losses. The trailer cost Ca$900, and the hand-built wooden building cost under Ca$2000.

The design shown in Figure 6.15 also incorporates several other special features. First, in order to keep costs low, the antennas were simple center-fed dipoles, rather than Yagis or Co-Co antennas. The dipoles were suspended one quarter of a wavelength above a ground-plane, the ground-plane comprising sections of wire stretched across the array, with spacing between them of about 60 cm (wide enough that a person could walk between them, allowing the grass to be kept low using a weed-eater or similar cutting device). The ground plane was not laid on the ground, but instead was suspended about 70 cm above it. This was because it was not possible to level the land with earth-moving equipment, and since the land was slightly undulating, with height variations of up to a meter, the ground-plane was suspended as a horizontal plane above the ground. The entire cost of the wire for the ground plane was Ca$600. One draw-back of the ground-plane was that it led to substantial after-pulse ringing due to resonance effects in the wires: this was removed by earthing the ground-plane wires to stakes driven deep into the ground at multiple points around the array.

The layout of the anntennas within the array was not regular. Dipoles were grouped into sets of four (called a quartet), each quartet having four dipoles at the corners of a square of length one half of a wavelength. Feeding every antenna with separate cable was too expensive so, as a compromise, each quartet was fed by a separate cable, giving 32 feed cables in all. Initially Belden 9913 cable was used, at a cost of Ca$700 per 1000 feet (305 m). A total of about 900 m of cable was needed, at a total cost of Ca$2100. (This is not too expensive, but we are aware of one interesting case, namely the VHF radar at the University of Canterbury in New Zealand. In that case, in order to reduce costs even further, the designers did not use coaxial cable at all, but in fact built all of their own feeds as transmission lines. This had the added advantage that the lines were not stolen – any coaxial cable at the radar disappeared very quickly due to theft (Graham Fraser, personal communication). However, in the case of the CLOVAR radar, the extra expense of coaxial was deemed worthwhile because of its flexibility and the ease of working with it.) The feed cables were housed in long wooden boxes that ran the length of the array. As an additional note, in later years the 9913 cable was replaced with lower-loss Andrews half-inch Heliax (LDF4).

As can also be seen in Figure 6.15, the quartets are not located on a simple grid. Rather, they were arranged to be 1.5 wavelengths apart in a diagonal direction. This means that the distances of the quartets from the center were multiples of 1.06 wavelengths. The reason for these distances is as follows. Since the individual quartets had lengths of a half-wavelength on each side, each quartet radiated very little in directions along its edges, since each edge comprises two antennas one-half of a wavelength apart, so the radiation from these two antennas cancels when we consider radiation along the ground (at least if we ignore coupling effects). The spacing of 1.5 wavelengths diagonally was chosen so that quartets which were diagonally adjacent also cancelled each other with regard to ground radiation. Almost complete cancellation of ground-radiation was to be expected in all directions which were an azimuthal multiple of 45 ° from the arms of the array. This tightly constrained the ground-level interference, and when the polar diagram was finally determined numerically, the ground level suppression in all directions was over 60 dB below the power transmitted along the bore-sight of the main beam. In two-way mode, the supression was over 120 dB. This is better than both the SOUSY and MU radars, and proved to be a valuable feature, even allowing similar radars to be built in large cities like Montreal, and near major airports, in later years. A partial view of the one-way polar diagram is shown in Figure 6.16, and we will return to that figure again later.

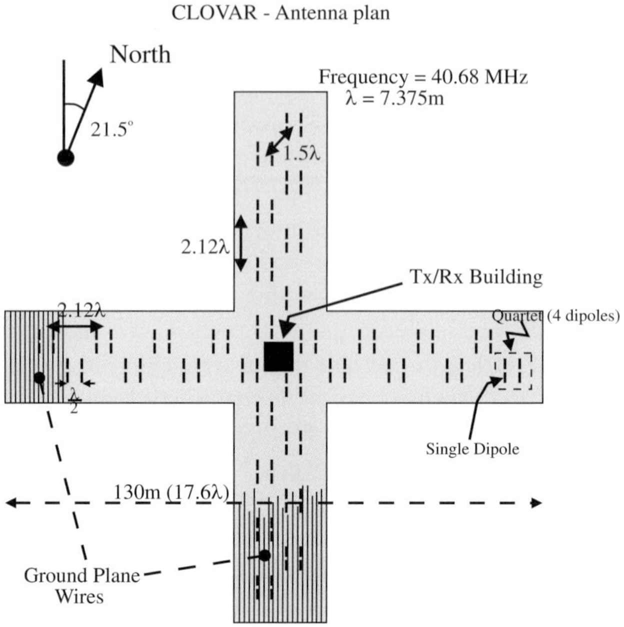

Figure 6.15 Plan view of the CLOVAR radar, showing antennas, antenna spacing, orientation and location of the ground-plane.

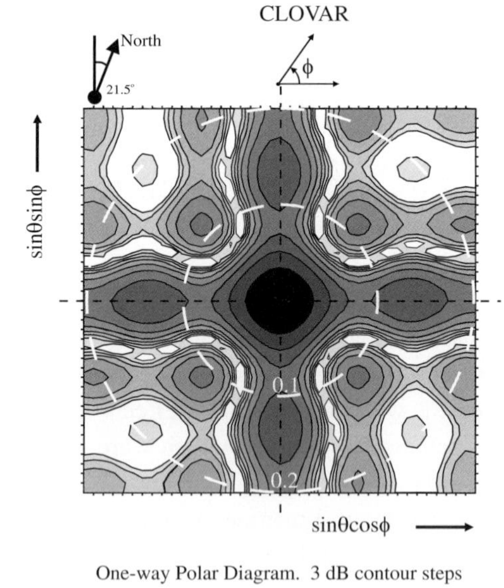

Figure 6.16 One way polar diagram of the CLOVAR radar, with shading in 3 dB steps.

Quartet design

Schematics of the quartets are shown in Figure 6.17. These had several cost-saving features. While it is true that Yagis have much better performance, they can be expensive to build, often costing over Ca$500 each at the time. This would have required a figure of

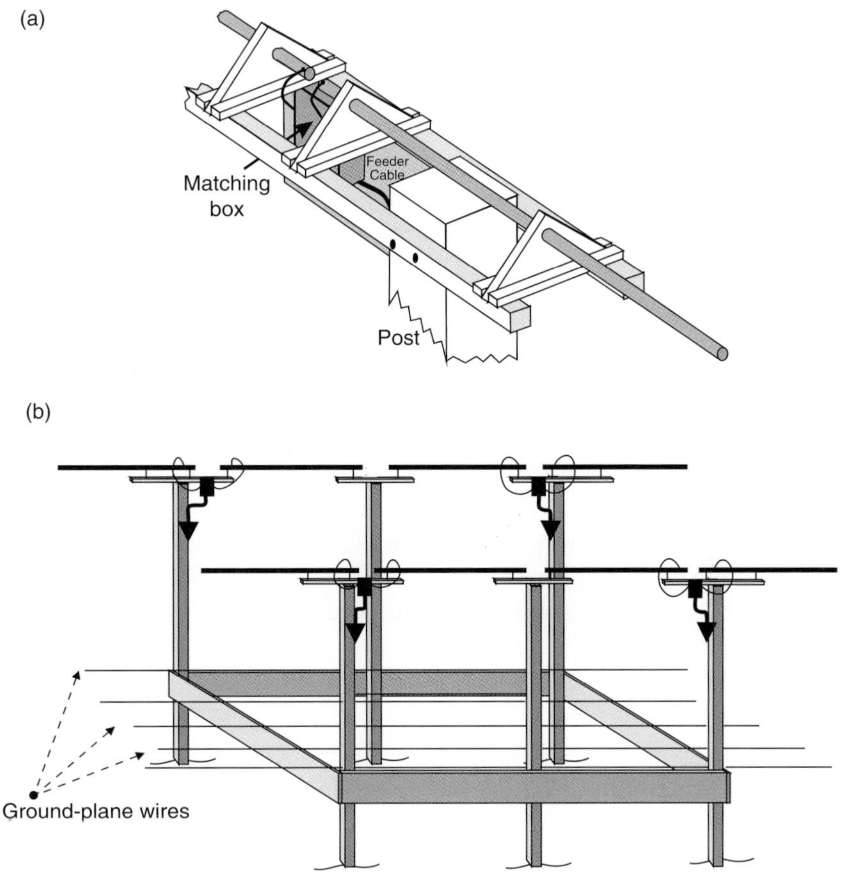

Figure 6.17 (a) Antenna support mechanism. (b) A single quartet. Some features have been adjusted from the real situation for ease of display, such as the length of the jumper wires in (b) and the symmetry of the support mechanism in (a). See the text for clarification of these issues.

over Ca$60 000 just for antennas. The cost could be reduced by building them in-house, but even so, the expense was too much.

Most of the support mechanisms of a quartet were made of wood, which is cheap in Canada relative to aluminum, and only the dipoles themselves required any metal. Figure 6.17(a) shows the feed-point of a dipole, and Figure 6.17(b) shows the physical layout of a quartet. All posts were wooden, and the horizontal beams shown in (b) act both to strengthen the support structure and also to support the ground plane (as discussed earlier).

Figure 6.17(a) is a little misleading, in that it suggest that the horizontal wooden support structures were asymmetric about the main support post. In reality, the portion that the reader sees in front of the post was somewhat longer than shown, so that the sections were supported at their center. However, the matching boxes shown in the figure were offset from center, for ease of construction and ease of access for the supply cables. To compensate, the main posts were positioned slightly more than a half-wavelength apart, so that the matching boxes and the dipole feedpoints could be exactly a half-wavelength apart.

The dipoles were fed by a balanced feed, so that the matching box had to include a balun, transforming the balanced antenna impedance to an unbalanced cable. Further details of the antenna design will appear in the next section. As discussed above, a ground plane was supported across the horizontal planks of the quartet, such that it was horizontal to better than 2 cm height over the whole array, while the earth itself undulated across the array. The ground-plane wires ran parallel to the dipoles.

The antenna elements themselves were simple half-inch aluminum tubes bought as second-hand surplus materials (essentially bought at a yard-sale). Figure 6.17(a) shows how the tubes were supported by triangular-shaped units, which were made of clear perspex (acrylic glass). These sheets had holes drilled through them so that the aluminum tubes could be supported, but still have good electrical isolation from any other parts of the structure. The tubes were held in place by simple gear-clamps, one placed either side of the perspex structures. The antennas were then coupled to the matching boxes using short sections of wire. (The lengths of these wires were kept very short – Figure 6.17(b) exaggerates their length for display purposes.)

Figure 6.18 shows the completed antennas.

Dipole design

The radar array was designed as an active array, like the MU radar, but instead of having separate access to each of the 128 antennas, access was only available to each quartet. Hence in many ways, the quartet can be considered as the fundamental antenna unit of the array. Each quartet could be monitored from the main building, and the phases

Figure 6.18 The CLOVAR antennas at sunset.

and amplitudes delivered to each quartet could be independently controlled, allowing beam-pointing and interferometry.

The design of the dipoles required significant experimentation in order to produce optimum impedances. Less use was made of numerical simulation, and more use of experimentation, in determining optimum parameters. The paper by *Proctor* (1950) was of great assistance in developing the final design, and discussions with S. Dillon of Tomco (Australia), plus the manual *Amidon* (1992), were of great help in developing the balun design.

Figures 4 and 5 from *Proctor* (1950) show that for a reflecting ground plane with spacing of about one tenth of a wavelength, as was planned for this array, the maximum impedance occurs for a dipole length of about 0.92 to 0.93 times a half-wavelength in length, and for a height of 0.25 to 0.3 wavelengths above the ground plane. The plan was then to tune the antennas to close to 100 ohms, then couple a pair of antennas together using cables of length equal to one wavelength. This in principle produces 50 ohms. These outputs could be further attached in parallel, producing nominally 25 ohms, which could then be transformed up to 50 ohms. The final matching was changed a little from this design, as will be discussed shortly, but the decision to use a dipole tuned to approximately 100 ohms remained. While the measurements by *Proctor* (1950) suggested that maximum impedances occurred when the height of the dipoles above the ground-wires was 0.3λ, the final design used a vertical separation of 0.25 wavelengths, which was adequate.

A good balun was required which produced approximately a 100 ohms impedance. After extensive investigation, the design shown in Figure 6.19 was adopted.

This was relatively cheap to make – the ferrite ring (Amidon FT240-61) cost about $8, and the other components required only transformer wire, some stand-offs and a mounting box. Coaxial cable was soldered directly to the transformer to reduce connector costs.

One of these units was built, and although the input to output impedance ratio was close to optimum, the system was calibrated by plotting graphs of input versus output impedance. Then the unit was connected to a 3-wavelength cable, connected to a dipole, and measurements were made as function of dipole length, tube radius, spacing of the feed-points etc. All measurements were made with a complete quartet, though three of the dipoles were left open-circuit and measurements were made on the fourth. This was to try and include coupling effects in the determinations.

Measurements showed that the balun introduced some capacitance, so a wire coil made from transformer wire was placed in series to cancel this effect. The optimum dipole had a length of 5 feet 6.5 inches (168.9 cm), and the jumper wires were placed at the ends of each half of the dipole, near the center, with the halves having a gap of 2.0 cm between them. The final impedance of each dipole when present in a quartet was 100 ohms, with a phase of $\sim 2°$.

It was also found that by gently spreading or squeezing the separation between the wires AA' and CC' in Figure 6.19, the transformed impedance could be altered by typically $\pm$ 15 degrees, and any capacitance could be removed by squeezing the spacing of the in-series wire transformer coil. This turned out useful in the final tuning stage,

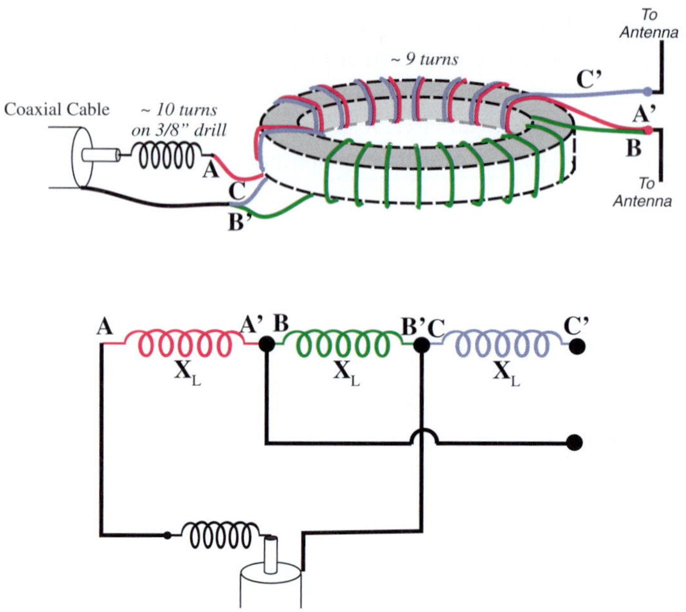

Figure 6.19 The matching balun used in the CLOVAR radar, and the equivalent circuit. Note that the equivalent circuit misses one point: wires A and C were wound in parallel, and actually functioned in part as a transmission line. Some degree of tuning was possible by slightly adjusting the spacing of this transmission line.

allowing tuning out of minor effects due to the underlying ground and to compensate for antenna coupling.

Antennas were built using the dimensions discussed above, and then each was connected to a balun like that shown in Figure 6.19, and then fed through a one-wavelength cable, where pairs were coupled. Then the outputs of the two coupled pairs were intended to be fed through a 71 Ω cable of length 0.75 wavelengths, which should ideally map 50 Ω to 100 Ω, but 71 Ω cable was not readily available. In the end, using a Smith chart for calculations, the final design involved changing the dipoles to have an impedance of 112 Ω, coupling pairs of antennas, then using two 75 Ω RG6 cables of length 0.785 wavelengths (one from each antenna pair), to produce a final output impedance with a real part of 50 Ω. The small capacitative component was tuned out using a small four turn hand-wound coil in series, of diameter 3/8 in (9.5mm). The final matching was accurate to better than ±2 Ω, and at no stage of the circuit did the VSWR exceed 2.0, even within matching cables, thus keeping radiative losses from the cables to a minimum. The total cost of all the materials for the baluns, including FT-240-61 ferrites, 256 insulating stand-offs, 600 feet of thermalized #12 transformer wire, and various types of insulating tape, was Ca$2045.

6.4.3 The controller computer

Computing showed remarkable progress in the time from the 1970s (when the SOUSY radar was developed) through the 1980s and into the early 1990s. The MU radar had a fast and capable computer, but it was not cheap, and was certainly not a personal computer. But by the 1990s, personal computers had become good enough to act as radar controllers. They were even fast enough that most on-line processing could be done in something like real time. For example, whereas in earlier years a look-up table was used for calculating Fourier transforms in reasonable time, by the 1990s it was possible to do full mixed-radix fast Fourier transforms (*Singleton*, 1969) of a few thousand points in times of a second or so. The CLOVAR radar capitalized on this by running two computers in parallel, linked by a Novell serial network. One computer controlled the radar, recorded the data, and stored the data in RAM (random access memory), and then a second computer copied the data across and performed all the post-acquisition analysis while the first began recording a new set of data. The analysis computer was able to perform calculations fast enough (including full spectral analysis) that it could keep ahead of the acquisition computer. This was a necessary requirement for the radar: because of the small size of the research group, and limited staff, it was necessary that all processing be done in real-time, since post-processing on other machines was too time-intensive to be practical.

The analysis computer had the following features – considered "top-of-the-line" in 1993: (i) a 486 DX2 66 MHz processor, (ii) 16 MB of memory, (iii) 250 MB hard drive, (iv) 1.2 and 1.44 MB floppy drives, (v) one parallel port, (vi) three comm (serial) ports, (vii) SVGA graphics, and (viii) up to eight expansion slots, plus keyboards, mouse, tape backup and other capabilities. The price was Ca$3900 at the time. The acquisition computer had similar types of capabilities. The computers ran under the DOS5 operating system.

This combination of computers, although modest by modern standards, was sufficient for the scientists who used them to make several important contributions to the literature, which were especially introduced in *Hocking* (1997a). Particular items of interest were the reduced use of coherent integration (no longer needed to the same extent as in earlier years), greater use of spectral fitting procedures, and extensive on-line quality control. Several figures in this book were prepared using data from this system. The acquisition computer and much of the low-power supply system and digitizer hardware were provided by Genesis Software Pty. Ltd. of Australia, and the cost of their contribution was of the order of Ca$30 000.

6.4.4 Beam-pointing

One necessary requirement for a Doppler ST-radar is the ability to steer the beam. This is often done using high-power vacuum relays. These can typically each cost a hundred dollars or more. However, because the radar was an active one, with 32 separate cables feeding 32 quartets, the power delivered in each cable was quite modest. The transmitter power was initially only about 5 kW peak, and even later, when upgraded, was 10 kW. It was capable of 10% duty cycle, so the mean power in any cable was only 37.5 W, and the peak power never exceeded 375 W.

As a result, it was possible to use smaller, much less expensive relays. Hermetically sealed units were used, but there was no need for vacuum relays except in particularly high-power areas. Such switches showed no worse loss on contact than vacuum relays (less than 0.1 dB typically). (It should be noted that the typical loss in a connector is 0.1 dB as well, but the system used very few connectors, with all joints being soldered where possible; this reduced losses and costs.) The relays typically cost \$5 to \$10 each – the ones used for CLOVAR were produced by Omron, although on subsequent systems other brands were used. These relays can handle up to 16 amps, have high break-down voltages, (more than enough to handle the peak voltages expected), and have lifetimes of typically 10^7 clicks. The main requirement for effective use was that they never be switched while the transmitter is active – a condition easily satisfied by proper software control of the transmitter and relays. The same design, using similar relays, worked even for later radars with up to 50 kW of peak power.

The details of the switching unit are shown in Figure 6.20. These will be referred to as "beam-pointing" (BP) boxes. Each requires only four power supply inputs of either 12V or 5V (depending on the relay design). The design is based on the fact that the antenna positions are rotationally symmetric in 90 degree steps: examination of Figure 6.15 shows that if the figure is rotated by 90, 180, or 270 degrees, the positions of the antennas appear unchanged. Application of a suitable DC voltage to the relays labelled BP0 rotates the transmission port by 90 degrees: application of a supply voltage to the relays labelled BP1 rotates the beams by 180 degrees, and application of a voltage to both sets of relays rotates the beams by 270 degrees. There were eight such beam units in all: each applied to a different set of antennas. One BP unit would rotate the transmitter signals between the innermost quartets, another would apply to the quartets second from the center (one on each arm), another to the four quartets spaced three positions from the center, and so forth.

The diagram shows coils labelled Delay Cables. Although they look like inductors, they were in fact lengths of coaxial cable that were of differing lengths, designed to produce the required beam. Each set was different from one beam-pointing box to the next, depending on the antennas that they controlled. This beam could be at any chosen zenith and azimuth in the sky, depending on the choice of delay cables. The beam-pointing boxes could change the beam direction to three other azimuthal positions, in 90 degree steps. Assuming that the main cross-structure is aligned in a north-south east-west direction (not mandatory, but it simplifies the discussion), then typically the first beam would be a northward beam, and application of the beam-pointing unit rotates the beam to the east, south and west. A new basic beam direction could be established by using a new set of cables, which could then be rotated in 90° azimthal steps. Most commonly the "basic beam direction" was a chosen zenithal tilt along one of the radar arms, but this was not mandatory.

It was also necessary to be able to generate a vertical beam. The path through the BP box is shown in Figure 6.20, in the portion labelled BP2. A suitable DC relay voltage was applied to these relays, and no voltage was applied to any of the relays in sections BP0 or BP1. When the BP2-relays flip to the normally open position, the delay cables are bypassed, and the path goes through the lower pins of the BP2-relays in the figure. The circuit between the lower pins has been drawn to appear as a short-circuit, with only a short piece of wire joining them, although this is a little misleading: in truth a section of carefully cut coaxial cable was used here, with length chosen to optimize impedance

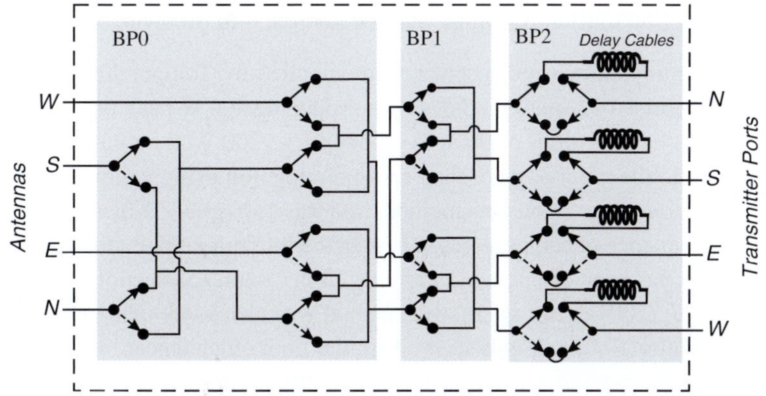

Figure 6.20 Switches within a single beam-pointing unit of the CLOVAR radar. Solid arrows represent normally closed positions, and broken ones indicate "normally open" positions. Application of a suitable DC voltage flips the relay from the normally closed position to the normally open one. All path lengths within the unit are set to be equal, so that the box introduces no extraneous delays.

(a) (b)

Figure 6.21 A typical beam-pointing unit. (a) Internal wiring, where the coils of cable are the paths for a vertical beam (they are of non-zero length in order to optimize impedance matching). (b) The back of the casing: the beam-pointing delay cables are attached to these N-type jacks. The transmitter and antenna jacks are on either side of the box; they are not visible in (b), but can be seen in (a) to the left and right looking from above. The DC-power leads can just be seen on the left of (b), and are also visible from above to the right (at the top of the photograph) in (a)).

matching through the box. All cables used for this purpose were of equal length, so that the paths followed between any pair of ports in the vertical arrangement were equal.

Figure 6.21 shows a typical BP box, and items of interest are described in the caption. The cables that can be seen coiled up here are the ones for generating the vertical beam – the delay cables are not shown, but are normally attached to the eight N-type jacks shown in Figure 6.21(b).

6.4.5 The transmitter, transmit-receive switch, and receiver

The transmitter and receiver were supplied by Tomco, Australia. The system initially comprised 12 small modules each producing 400 W peak power, and these were coupled together in groups of three to produce 4×1200 Watt output. These four ports were then fed to the input ports of the beam-pointing unit using various matching methods to map the four ports to 32. One method used was a so called "Xmas tree" arrangement, in which a sequence of cables are used to map the four output ports to eight, then 16, then 32. This unit has been discussed in more detail in Chapter 5, Section 5.6.3, as well as in *Campos et al.* (2007a). An alternative method was also used, using simple inductive-capacitive circuits. Both methods showed similar high efficiciences.

The transmit-receive switch used PIN diodes, just as the SOUSY and MU radars did: passive TR switches only became available a few years later. The receiver had a unique capability in that it had a multiplexer inserted at the front end, permitting the input to be cycled between multiple supply ports: this was invaluable for the subsequent development of a meteor radar attachment. The use of a multiplexer was required as a cost-saving measure, so that only a single receiver was needed. If funds are available, it is best to have separate receivers for each receiver port, but at the time of development of CLOVAR, this was too expensive. The downside of such a design is that each received signal is only sampled once every cycle, so if five received signals are sampled, the sample rate for a single receiver drops by a factor of five, reducing the detectability. The upside is of course the ability to sample multiple input signals at low cost.

The system was capable of producing usable echoes from 1.5 to to about 6 km in altitude, and these components cost about Ca$25 000. A year or so later, monies became available to upgrade the system to 10 kW, for a price of about Ca$20 000, which increased the range coverage to about 8–9 km maximum altitude.

The system was also used for meteor studies, when the full 10 kW peak power was funnelled with a 4:1 combiner into a single four element Yagi for transmission, and the multiplexer on the receiver was used to allow cycling between five receive antennas. This system served as the prototype for the future successful SKiYMET radar (*Hocking et al.*, 2001a), which also introduced many new concepts to the area of meteor research.

A summary of the costs associated with the radar is given in Table 6.2.

6.4.6 System tests and usefulness

The radar, while certainly not of the status of the MU radar, was still nonetheless a productive system. It was used to show the benefits of reducing the amount of coherent integration, for example. Due to computing limitations, it was common prior to the 1990s to use typically 1024 coherent integrations for say a pulse-repetition-frequency of 4096 Hz, giving one point every 0.25 seconds. This was considered standard procedure, with the argument being that large amounts of coherent integration were required to optimize the signal-to-noise ratio. Using the CLOVAR radar, it was shown that this was unnecessary, and in fact the need for coherent integration was a requirement only because it reduced the number of points that needed to be processed. Newer, faster

Table 6.2 Costs associated with construction of the CLOVAR radar. All prices are Ca$ as of 1993. Most prices are accurate, based on accounting records: a small number are based on memory-recall, but are believed to be relatively accurate. Labour costs were minimal – construction was mainly done by the principal investigator with occasional help from technical personnel and students.

Item	Cost, Ca$
Buildings and sheds	
Used trailer for holding equipment	900
Wooden side-shed	2000
Wooden troughs to hold cables	1000
Electrical installation	3000
Air-conditioning	900
Support pipes, springs for ground-plane wires	2500
Post-hole digger rental	500
Antennas	
Aluminum tubing for antennas	2000
Ground plane wires	600
Wooden posts and beams for quartets	3840
Antenna support brackets incl. perspex	1280
Antenna matching units (incl. ferrites, transformer wire, etc.)	2045
Hammond weatherproof boxes to house matching units	1024
Cable (Belden 9913)	2100
Cable (RG6 75 Ω)	450
Meteor antennas and cabling	3000
Beam pointing	
Beam pointing rack	0 (surplus supplies)
Beam pointing boxes	1600
Beam pointing relays	1224
Beam pointing power supply	200
Computers, digitizers, and transmitter	
Acquisition computer and digitizer system	30 000
Transmitter (5 kW initial system)	25 000
4:1 combiner for use in meteor studies	5000
Analysis computer	3900
Software (Novell network, operating system, etc.)	600
Transmitter upgrade to 10 kW	25 000
Total Cost	
Cost before Tx upgrade	94 663
Cost after upgrade to 10 kW	119 663

computers could handle larger numbers of points easily. Older systems used weighted moments to determine powers and radial velocities, which were indeed affected by the signal-to-noise ratio, but since newer systems could calculate FFTs and perform on-line spectral fitting, the important parameter for determination of the limits of capability became the *detectability*. This has been discussed in Chapter 4, Section 4.9. By reducing the coherent integrations, so that the final data were recorded at say 100 Hz (after coherent integration), there were also advantages in removing aircraft contamination, meteors, and various other types of noise (also see Section 4.9). The system also used five beams, whereas some others at the time used only three; use of five beams gives better reliability about the radial velocities in that the user can check for anti-symmetry between opposite beams. Use of spectral fitting also allowed the users to compensate for the impact of the unusually large side-lobes shown in Figure 6.16. The system had all of the features introduced by the NOAA Casper system (*Wilfong et al.*, 1999), but the CLOVAR system pre-dated that system by at least 2 years.

Tests of the data showed good accuracy, as shown in *Belu et al.* (2001). Figure 6.22, from that reference, shows examples of accuracy checks. The radar was the forerunner of the so-called O-QNet radar network in Canada, which has been used for many years as input to large-scale numerical forecast models produced by NOAA and the UK Met. Office.

6.5 More recent radars

To close out this chapter, we now discuss very briefly two more recent radars. These are the PANSY radar in Antarctica (with first full-scale operation in 2015) and the MAARSY radar recently completed ($\sim$ 2011) in northern Norway.

6.5.1 The PANSY radar

Introduction

One newer radar of considerable interest is the PANSY radar, constructed in Antarctica and funded by the Japanese government. The acronym stands (loosely) for "Program of the ANtarctic Syowa MST/IS", where MST refers to mesosphere-stratosphere-troposphere and IS refers to incoherent scatter. The project aim was to construct the first full-scale MST/IS radar in the Antarctic region at Syowa Station (69 °00′S, 40 °35′E) (*Sato et al.*, 2014).

The scientific goals of the instrument were: (i) clarification of the role of atmospheric gravity waves at high latitudes in the MST region; (ii) exploration of the dynamical aspects of polar mesospheric radar echoes, polar mesospheric clouds (PMCs), and polar stratospheric clouds (PSCs); and (iii) other studies of dynamical processes unique to the polar regions. Like the MU radar, the system is a VHF monostatic pulsed phased array Doppler radar. It is designed to operate at 47 MHz and the antenna comprises 1045 three-element crossed Yagi antennas with a total peak output power of 520 kW. Major parameters of the radar are given in Table 6.3.

It was intended that the radar should attain a sensitivity comparable to that of the MU radar described in previous sections of this chapter. This was ambitious, as the availability of power at the site is limited. For example, the MU radar consumes 230 kW of power at its maximum duty ratio of 5% because it employs class AB power amplifiers, which helps suppress spurious radiation. In contrast, the maximum available power for the PANSY radar at Syowa Station is less than 100 kW, because fuel for power generation is one of the most precious resources.

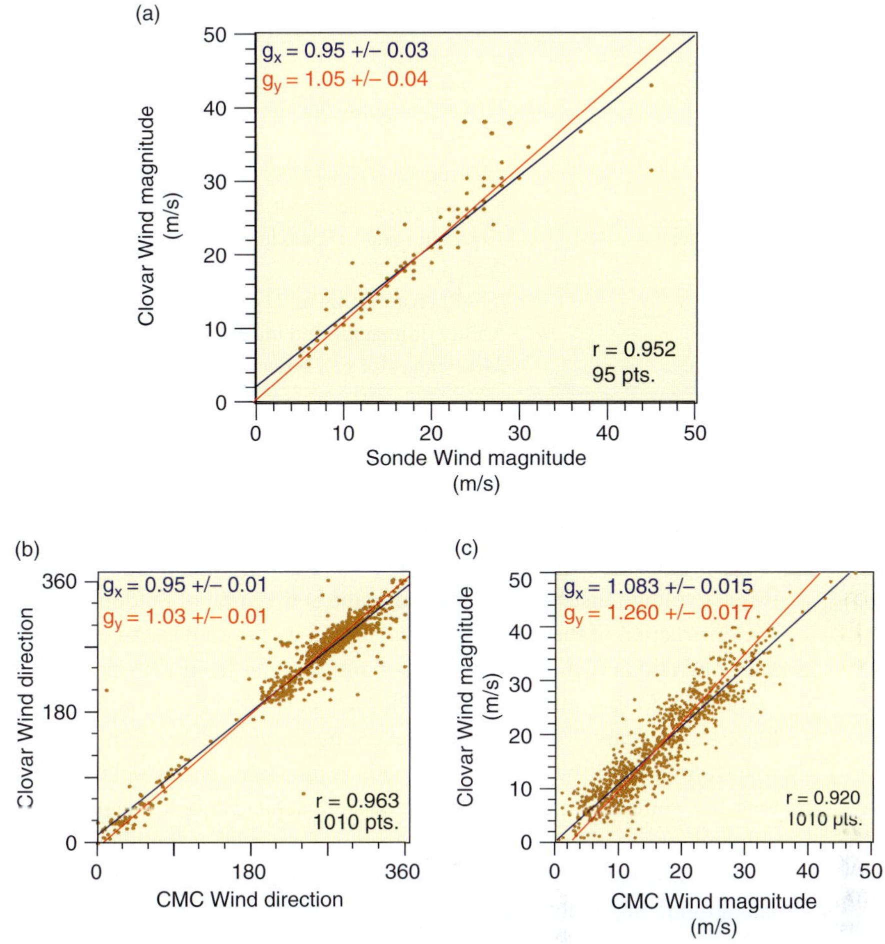

Figure 6.22 Verification of the CLOVAR winds against radiosondes and zero-hour computer forecast data. (a) A scatter-plot of comparisons of radar wind magnitudes against radiosondes launched at the CLOVAR radar site (b) and (c) Azimuthal wind directions and magnitudes compared to Canadian Meteorological Centre zero-time analysis models. This means that the CMC took all the data available at the time (except the radar data) and coordinated it into a self-consistent representation of weather conditions present at the time across all of Canada: these "maps" could then be used for predictive forecasts for typically 6, 12, 18, or 24 hours ahead. However, the data also stood as a record of conditions present at the time of preparation. Hence data used for the comparisons in this figure had no predictive element: they simply incorporated and interpolated the best available data at the time, allowing retrospective comparisons between (in this case) the radar and the analysis model.

Since the sensitivity of an atmospheric radar system is determined by the product of the total output power and the antenna's effective area (the so-called power-aperture product, see Chapter 5, Section 5.5.2), the output power was reduced to half of that of the MU radar, and the antenna area was doubled, maintaining the same power-aperture product for the two radars. In order to further reduce the power consumption, a new

Table 6.3 Basic parameters of the PANSY radar (adapted from *Sato et al.* (2014)).

Parameter	Value/description
System	Monostatic pulse Doppler radar with active phased array
Center frequency	47 MHz
Antenna	An irregular-shaped array consisting of 1045 crossed Yagi antennas. Effective diameter about 160 m (18 000 m^2)
Transmitter	1045 solid-state TR modules Peak power: 520 kW
Receiver	(55+8) channel digital receiver system Ability of imaging and interferometry observations
Peripheral	24 antennas for FAI observation

class-E amplifier with a total power efficiency exceeding 50% was developed (the principle of class-E amplifiers is explained in Chapter 5, Section 5.3.2). Subsequently the total power consumption, including that of the digital signal processing system, was reduced to less than 75 kW. One bi-product of the high efficiency in the power amplifier is that cooling fans could be avoided in the entire suite of outdoor units.

Construction of the radar at Syowa Station was started in December 2010. Major outdoor operations such as assembly of buildings and facilities were confined to the short two-month long summer season. Also, sea ice around Syowa Station was exceptionally thick in the years 2011 to 2012, preventing the icebreaker ship *Shirase* from reaching Syowa Station in that time frame. This resulted in serious delays of transportation. It took five summer seasons before the system achieved its full capability in March 2015, although observations with various forms of a partial system have been on-going since March 2011. Figure 6.23 shows a field view of part of the antenna system during winter operation in September 2012.

System configuration

The system configuration of the PANSY radar follows closely that of the MU radar, except that the number of antenna sub-arrays was increased from 25 to 55. Another major difference is that the use of intermediate frequency (IF) was avoided. The pulse waveform for transmission was generated and directly converted to RF of 47 MHz and then distributed to the 55 antenna sub-arrays.

In contrast to the configuration of the MU radar, where the TR modules are stored in six container houses around the antenna field, each TR module of the PANSY radar is mounted on the antenna mast. This approach is becoming more common on many large radars nowadays, since it reduces RF losses due to long runs of cables to the antennas. The strategy does, however, require very rugged construction of the transmitters, and makes servicing them more difficult.

As discussed, the power savings of the PANSY radar are significant. Each of 1045 TR modules transmits an average output power of 25 W. With its power efficiency of 50%, it radiates only 25 W as heat, for which natural air cooling is sufficient. In contrast, the MU radar has 475 TR modules, each of which transmits an average output power

Figure 6.23 A part of the antenna array of the PANSY radar (courtesy of R. Ito).

of 120 W. As its power efficiency is about 25%, each module consumes 360 W as heat, requiring a forced air cooling system for each container house.

On reception, RF signals are combined at 55 divider/combiner modules located near the center of each antenna subarray, then transferred to the operations building, and subsequently directly converted to baseband IQ signals. While routine observations employ only a single digital receiver to save power consumption, the radar is actually equipped with 55 digital receivers which can be used for specialist applications like imaging studies.

The antenna array

As noted above, an important issue in designing the radar was the anticipated complications of construction in the harsh Antarctic environment. Additionally, alteration of the natural topography is prohibited by the Antarctic Treaty, which excludes the use of heavy construction vehicles in the antenna field. In order to fulfill these requirements, the weight of the antenna elements and the TR modules had to be minimized so that they could be eaily moved around. On the other hand, the antenna elements must withstand maximum continuous wind speed of up to 65 ms^{-1} during severe snow storm conditions that often last for a few days. The final design of the three-element crossed Yagi antenna, which has a weight of 12 kg, was established after examining several test antennas at Syowa Station. Each individual Yagi antenna has an isotropic gain of 7.2 dB, a bandwidth of 2 MHz, and a voltage standing wave ratio (VSWR) of less than 1.25 in this band. Its half-power beam-width is 68 ° in the E-plane, and 126 ° in the H-plane.

The antennas follow the natural topography, and so vary in height above sea-level. This means that the geopotential heights of the antennas are variable in a range of approximately 3 m. The phase of the radiated signal from each antenna is electronically adjusted to compensate for the path difference for each antenna beam direction.

Inside an antenna subarray, the horizontal projections of antenna locations are aligned to a grid of equilateral triangles with 4.5 m intervals, which corresponds to 0.7 wavelengths. A subarray of hexagonal shape consists of 19 crossed Yagi antennas, and the entire antenna array consists of 55 sub-arrays. The arrangement of these 55 sub-arrays was originally designed to form a quasi-circular shape 160 m in diameter, as shown at the upper-right part of Figure 6.24. During the Antarctic winter of 2011, however, the level of snow fall at Syowa station was the highest on record since the foundation of the station in 1956. Since the wind direction at Syowa station is predominantly from the northeast, the effect of snow was most significant at the central and southwestern parts of the circular area, where the maximum snow accumulation exceeded 3 m. While the antenna elements buried in the piled snow were severely damaged, major outdoor components, including TR modules mounted on the antenna masts, suffered only minor damage. After examining the snow record during the winter, 30 antenna subarrays out of 55 were moved from the original location to four blocks on surrounding

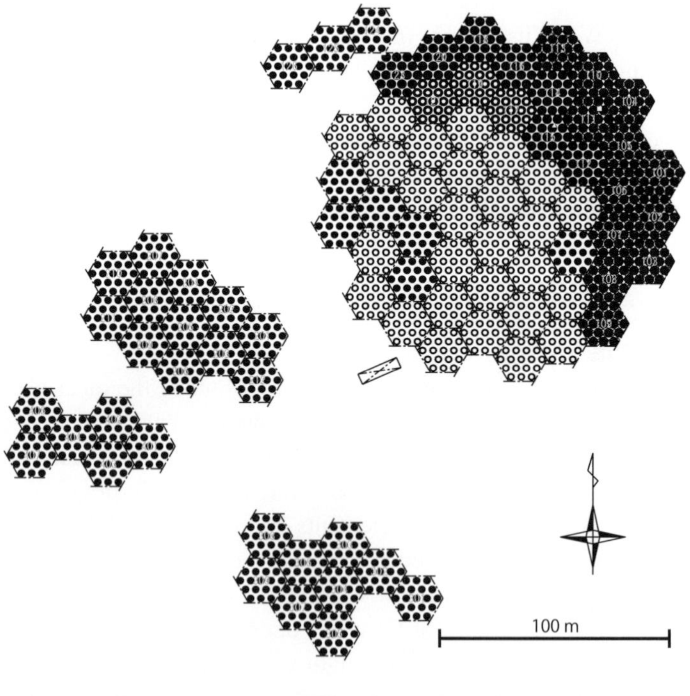

Figure 6.24 Arrangement of the PANSY antenna array. White circles show the modules in the original circular array, and black dots indicate the modules after relocation. Each hexagon denotes a subarray consisting of 19 crossed Yagi antennas. The white rectangle indicates the radar operations building.

hilly areas, where snow accumulations were relatively low. Black dots in Figure 6.24 show the arrangement of the relocated antennas.

The guiding principle for this relocation was maintenance of the hexagonal arrangement of each sub-array, retention of the basic triangular grid, and minimization of the distance between the sub-arrays, while simultaneously avoiding areas prone to deep snow. In this manner, the antenna patterns of the sub-arrays, which determine global properties such as the behavior of grating lobes, remains unchanged. On the other hand, the diameter of the entire antenna area was roughly doubled to about 320 m compared to the original circular array. This enlargement had the twofold effects of sharpening the antenna main lobe and increasing the level of near-axis side-lobes. Although the total output power and the effective antenna area remained the same as that of the original circular array, this sharpening of the antenna beam, which means a very small κ in (5.41), reduced the effective volume illuminated by the main lobe, and thus resulted in a reduction of the system sensitivity. A numerical computation of (5.134) showed that the sensitivity of a monostatic radar using the relocated array is 4.56 dB less than that using the original circular array. A long-term comparison of the echo power by the full system with that by a partial system consisting of a dense array of 12 groups, located west of the operating building in Figure 6.24, roughly confirmed this estimate (see Figure 6.26).

The upper panel of Figure 6.25 shows the computed two-way antenna pattern of the rearranged array with uniform excitation. The half-power width of the main lobe was reduced to about 1° compared to 2.2° for a 160 m diameter circular array, but the maximum *two-way* side-lobe level was increased to −7.5 dB. This contrasts with a

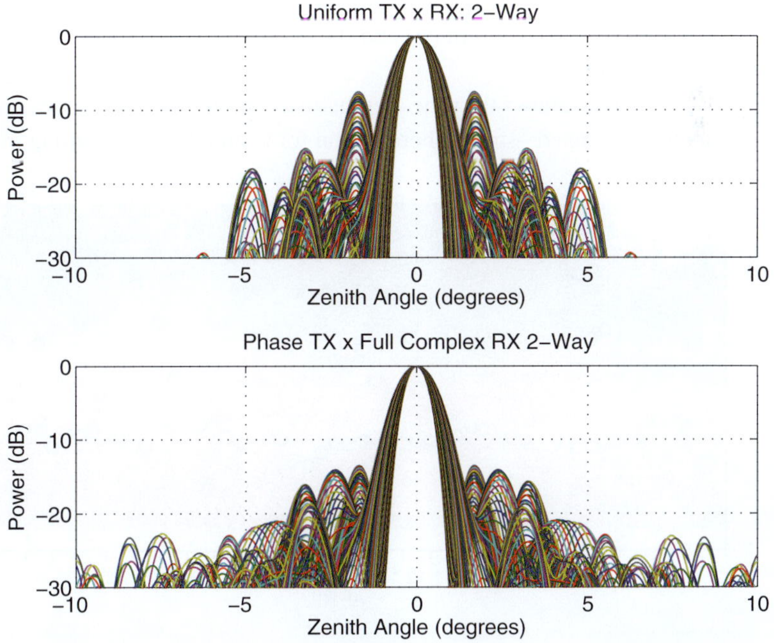

Figure 6.25 Computed two-way antenna pattern of the relocated array with uniform excitation (upper panel), and with an optimization of phase in transmission plus phase and amplitude in reception (lower panel). In each case there are 36 graphs, corresponding to polar diagrams at 10° steps in azimuth.

value of −35 dB for the original uniform circular array. Since this level is too high to disregard the contribution of echoes entering into side-lobe regions, it is not appropriate to consider the beam width only at its half-power level. The beam-width measured at −10 dB level including the first side-lobes is about 4.0 °, which seems to be a more realistic value for an effective beam-width for an atmospheric radar.

Levels of these high side-lobes can be reduced by adjusting the phase and amplitude of each antenna module. The lower panel of Figure 6.25 shows an example of computed patterns with such adjustment. For transmission, only the phase of each module is adjusted because the amplitude is hard to control for class-E amplifiers. On reception, both the amplitude and phase of all elements are adjusted. Suppression of the maximum side-lobe level was achieved by using a simulated annealing algorithm commonly used for optimization problems (*Kirkpatrick et al.*, 1983). The maximum two-way side-lobe level was reduced to −13.5 dB, which is tolerable for most atmospheric observations. In this case, the cost of this reduction is loss of sensitivity of about 6 dB, due to reduced amplitude and cancellation of signals with variable phases. Conventional uniform excitation will be used for routine observations that require maximum sensitivity. The optimized pattern will be effective for imaging observations that require sharp beams.

Preliminary results

Full-system continuous observations began on 5 April 2015, and continued for over a month. The observation was made with five antenna beams pointing to the zenith and the north, east, south, and west directions with 10 ° zenith angle. The observational altitude was switched between the troposphere–stratosphere region and the mesosphere region alternately every one minute, and the echo power spectra were averaged for each one minute period.

Figure 6.26 shows the time–height distribution of the echo power averaged for 30 min in the troposphere–stratosphere region for zenithal beam-pointing in this period.

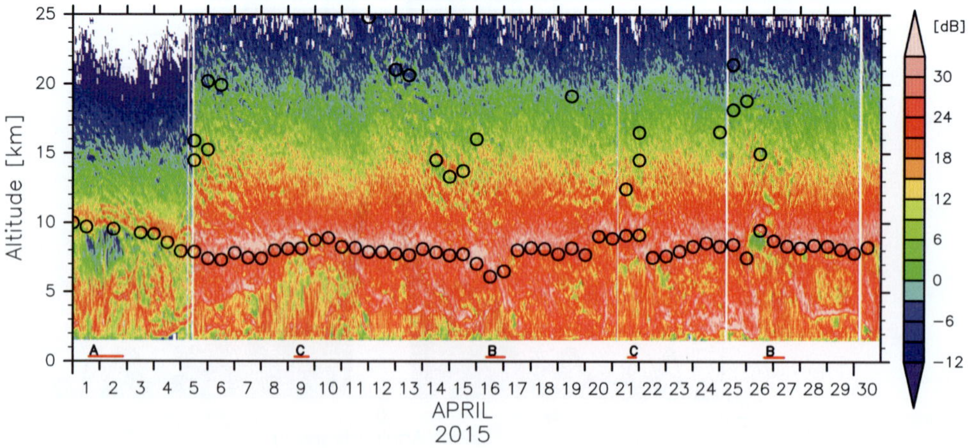

Figure 6.26 Time–height distribution of the echo power in the troposphere–stratosphere region for a zenithal beam observed by the PANSY radar over one month of April 2015.

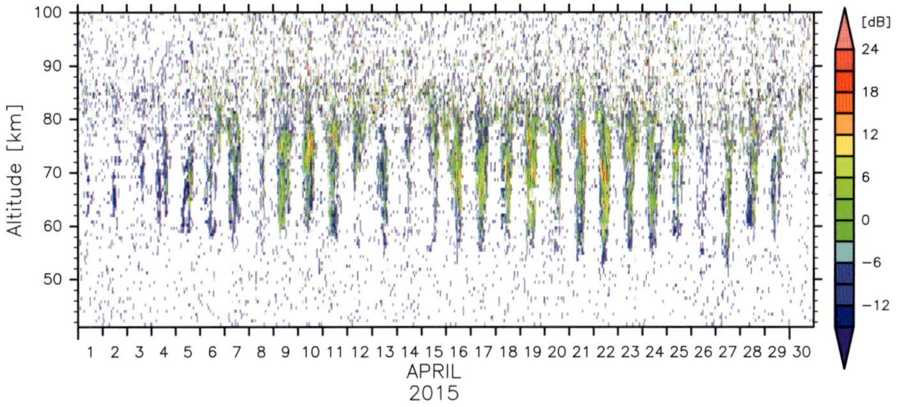

Figure 6.27 Same as Figure 6.26, but for the mesosphere region.

Figure 6.28 The MAARSY radar (from *Latteck et al.*, 2012). (Reprinted with permission from John Wiley and Sons.)

The radar operated with 12 groups until 5 April, then subsequently utilized all 55 groups. Black circles indicate the tropopause height determined from radiosonde observations. Red bars at the bottom of the figure indicate occurrence of severe snow storms categorized as Class-A to -C. As a strong aspect-sensitivity exists in the echo power at this height region, the echo power at off-zenith beam directions is 5–10 dB lower than that of the zenith direction.

Figure 6.27 shows the echo power in the mesosphere region. Strong polar mesosphere winter echoes (PMWE) are observed throughout the period during daylight hours, sometimes from around 53 km altitude up to about 85 km.

6.5.2 The MAARSY radar

MAARSY stands for Middle Atmosphere Alomar Radar SYstem. The radar is located on the island of Andoya in Northern Norway (69.30 °N, 16.04 E), and an overview is presented in *Latteck et al.* (2012). The radar is shown in Figure 6.28. First results with the full array were produced in May 2011.

scattering and radar engineering, and so has links both to the preceding chapters as well as Chapter 8. Chapter 8 will repeat some of the items in here, but generally from a different perspective.

This chapter is organized in such a way that the simplest parameters are discussed first, and parameters which are more difficult to extract, or perhaps less commonly determined, are considered later.

7.2 Wind vector determination

One of the simplest and yet most important parameters to calculate is the wind speed, so we shall begin with a brief discussion of its determination, examining in detail some of the assumptions made in this evaluation. Of course we have already seen this parameter several times already, but here we focus in more detail on the physical relationship between the scatterer motions and the spectra produced with the radar.

There are at least two different approaches to the determination of the mean wind, as briefly introduced in Chapter 2. One approach utilizes large antenna arrays with correspondingly narrow radiation patterns, and with the beams pointed in various directions to measure wind speeds. The Doppler shift of the backscattered signal is utilized for this calculation, and such techniques are called "Doppler beam swinging" (DBS) techniques. The second class of method, called spaced antenna methods, utilizes systems of separated (spaced) arrays; wind speeds are determined by using phase and time differences between signals received with the arrays. The sets of spaced antenna arrays usually have smaller physical dimensions than the antenna arrays used in the DBS mode. In some senses, the techniques which use time delays and the techniques which utilize phase delays can even be regarded as distinct techniques, and they are often described independently. They are both considered as "spaced antenna" techniques. Spaced antenna techniques are discussed more extensively in Chapter 9.

7.2.1 Doppler measurements

The principle of Doppler determination of the wind speed is to utilize the change in the phase of scattered radio waves as a function of time. It is probably the most common procedure currently in use, so some time will be devoted to discussion of this technique. The method was discussed to some extent from the perspective of engineering issues, such as the way that Doppler shifted signals are sampled, in previous chapters, but here we look more carefully at the atmospheric and mathematical aspects of the process.

Basic principles of the Doppler method

As the scatterers move, the path length between the transmitter, scatterer, and receiver changes, and this shows as a change in phase (Figure 7.1(a)). We can think of this as a rotation of the vector in the Argand plane (Figure 7.1(b)). This idea was also discussed in Chapter 3 in regard to discussions about application of complex numbers. For

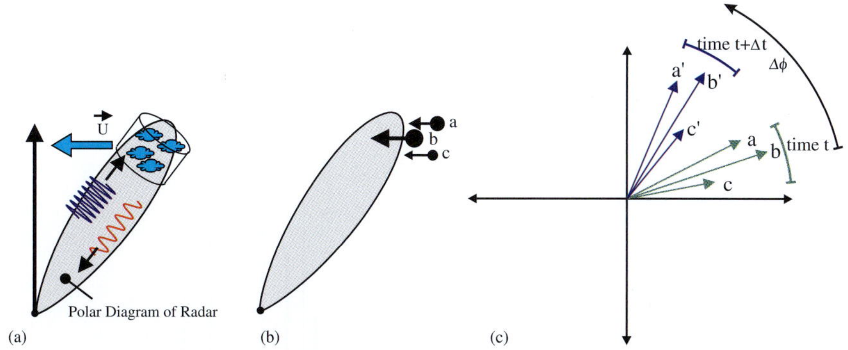

Figure 7.1 Principle of the Doppler method of wind measurement. (a) Shows some scatterers being blown along through the beam and illustrates the change in radar frequency produced by the motion of these scatterers. (b) Recognizes that the scatterers may have different speeds, illustrated with three scatterers near the radar beam. (c) Shows the phasor rotations of each scatterer from (b) from time t to $t + \Delta t$.

a monostatic radar, the mean rate of change of phase is a measure of the mean radial component of the velocity of the scatterers, viz.

$$< v_{\mathrm{rad}} >= \frac{\lambda}{4\pi} < \frac{\mathrm{d}\varphi}{\mathrm{d}t} > . \tag{7.1}$$

Here, $<>$ represents the average value, averaged over the radar volume and the sampling time, v_{rad} is the radial component of the velocity, λ is the radar wavelength, φ is the phase, and $d\varphi/dt$ is the rate of change of phase. Each scatterer causes its appropriate vector to rotate at a slightly different frequency, depending on the component of the wind away from the radar (radial velocity), and we can represent this on a spectrum where each line corresponds to a different scatterer (Figure 7.2). Of course it should be borne in mind that in a real spectrum, it may not be physical to think of each separate spectral line as due to a separate scatterer (a single scatterer that changes in amplitude as it moves, and even has some variation in radial speed with time, will produce a spread of spectral lines by itself), but for our purposes this is an adequate analogy.

We shall assume for simplicity that the spectral density and frequency of each line are invariant, so just the phase of the signal from each scatterer changes. If a single phasor at time t can be described by a complex number $\underline{a}_0 = a_0 \mathrm{e}^{\mathrm{i}\omega t}$ ($a_0 =\mid \underline{a}_0 \mid$), then at time $t + \Delta t$ it is given by $\underline{a}'_0 = a_0 \mathrm{e}^{\mathrm{i}\omega(t+\Delta t)}$. Given the two phasors, the phase difference can be found by calculating $\underline{a}_0{}^* \underline{a}'_0$, where (*) means complex conjugate. This calculation gives $a_0^2 \mathrm{e}^{\mathrm{i}\omega\Delta t} = a_0^2 \mathrm{e}^{\mathrm{i}\Delta\varphi}$. We actually seek $<\Delta\varphi>$ averaged over all the scatterers in the radar volume, but if the phasors all have equal amplitude, or even more generally the spectrum has a symmetric shape, then we can say that $<\Delta\varphi>= \arg\{<a^2 \mathrm{e}^{\mathrm{i}\Delta\varphi}>\}$, even for large values of $\Delta\varphi$. In other words, summing the phasors and finding the rotation of the resultant vector gives the same result as averaging the angles of rotation of each phasor. This is also true even in the presence of a moderate amount of noise. Hence we

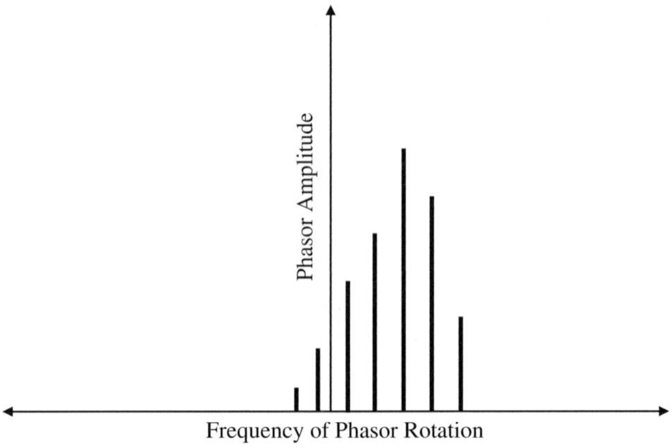

Figure 7.2 Idealized spectrum produced by six scatterers of different cross-sections, each moving at different radial speeds.

will consider averages of $a_0^2 e^{i\Delta\varphi}$. This mean value calculated for n phasors $a_j e^{i\Delta\varphi_j}$ ($j = 1$ to n) is

$$<a_0^2 e^{i\Delta\varphi}> = \frac{1}{n}\sum_{j=1}^{n} \underline{a}_j^*(t)\underline{a}_j(t+\Delta t). \tag{7.2}$$

To improve the accuracy, imagine averaging over a reasonable length of time, say at N time steps $t_1, t_2, \ldots, t_k, \ldots, t_N$, where $t_k = k\Delta t$. Then

$$<a_0^2 e^{i\Delta\varphi}> = \frac{1}{N}\sum_{k=1}^{N}\left[\frac{1}{n}\sum_{j=1}^{n} \underline{a}_j^*(t_k)\underline{a}_j(t_k+\Delta t)\right]. \tag{7.3}$$

We can write this as

$$<a_0^2 e^{i\Delta\varphi}> = \frac{1}{N}\sum_{k=1}^{N}\left[\frac{1}{n}\sum_{j=1}^{n} \underline{a}_j^*(t_k)\sum_{j'}^{N} \underline{a}_{j'}(t_k+\Delta t)\right], \tag{7.4}$$

since the cross terms in the square brackets of (7.4) all involve terms like $e^{-i\omega_j t}e^{i\omega_{j'}(t+\Delta t)}$, where $\omega_j \neq \omega_{j'}$, and such terms sum to zero when summed over a period substantially longer than their beat period. In fact, in the case that the time series is Fourier transformed by a discrete Fourier transform, all frequencies are harmonically related and so these cross terms summed over the data length are all exactly zero. But the term $\sum_{j=1}^{n} \underline{a}_j(t_k)$ is simply the value of the (complex) time series which would be recorded by the radar at time t_k, which we will denote as $\underline{s}(t_k)$, and so (7.4) can be written as

$$<a_0^2 e^{i\Delta\varphi}> = \frac{1}{N}\sum_{k=1}^{N} \underline{s}^*(t_k)\underline{s}(t_{k+1}), \tag{7.5}$$

which is simply the autocovariance function at the first lag, $\underline{\rho}(\Delta t)$, say. Thus the mean rate of change of phase can be found as

$$< \frac{\Delta\varphi}{\Delta t} >= \frac{1}{\Delta t} \tan^{-1} \left\{ \frac{\mathrm{Im}(\underline{\rho}(\Delta t))}{\mathrm{Re}(\underline{\rho}(\Delta t))} \right\}, \tag{7.6}$$

where Re means "the real part of" and Im means "the imaginary part of." This estimator of the rate of change of phase was introduced by *Woodman and Guillen* (1974). Note that since, by the Wiener–Kintchine theorem (e.g., *Bracewell*, 1978), the autocovariance and the spectrum are a Fourier transform pair, the value of the autocovariance $\underline{\rho}(\tau)$ at $\tau = \Delta t$ can also be found from the power spectrum $P(f)$ as

$$\underline{\rho}(\Delta t) = \sum_{j=1}^{n} P(f_j) \mathrm{e}^{2\pi \mathrm{i} f_j \Delta t} \Delta f, \tag{7.7}$$

where $f_j = (j-1)/T$, T being the data length of the time series $\{\underline{s}(t_k)\}$, and $\Delta f = \frac{1}{T}$.
Then

$$v_{\mathrm{rad}} = \frac{\lambda}{4\pi} < \frac{\Delta\varphi}{\Delta t} >= \frac{\lambda}{4\pi\,\Delta t} \tan^{-1} \left\{ \frac{\mathrm{Im}\{\sum_{j=1}^{n} P(f_j)\mathrm{e}^{2\pi \mathrm{i} f_j \Delta t}\}}{\mathrm{Re}\{\sum_{j=1}^{n} P(f_j)\mathrm{e}^{2\pi f_j \Delta t}\}} \right\}, \tag{7.8}$$

where the term Δf in (7.7) divides out since it appears in both the numerator and the denominator. Since $P(f)$ is real (by definition),

$$v_{\mathrm{rad}} = \frac{\lambda}{4\pi} < \frac{\Delta\varphi}{\Delta t} >= \frac{\lambda}{4\pi\,\Delta t} \tan^{-1} \left\{ \frac{\sum_{j=1}^{n} P(f_j)\sin(2\pi f_j \Delta t)}{\sum_{j=1}^{n} P(f_j)\cos(2\pi f_j \Delta t)} \right\}. \tag{7.9}$$

In the limit that the term in the right-hand brackets {} is << 1 and the $P(f_j)$ values are small for the larger $| f_j |$, this approximates to

$$v_{\mathrm{rad}} \simeq \frac{\lambda}{4\pi\,\Delta t} \frac{\sum_{j=1}^{n} 2\pi f_j \Delta t P(f_j)}{\sum_{j=1}^{n} P(f_j)} \simeq \frac{\lambda}{2} \frac{\sum_{j=1}^{n} f_j P(f_j)}{\sum_{j=1}^{n} P(f_j)}. \tag{7.10}$$

This last expression is one commonly employed as an estimator of the radial component of the velocity (e.g., *Gage and Balsley*, 1978; *Zrnić*, 1979). Nevertheless, note it is only an approximation of the more exact expressions (7.6) and (7.9), and breaks down when the argument of the $\tan^{-1}\{\}$ expression in (7.9) becomes comparable to 1. This can happen particularly when the signal is noisy or when the spectral peak is close to the Nyquist frequency. In these cases, the approximation (7.10) can give erroneous estimates. In the case of high noise levels, the true radial velocity is underestimated. This can be seen by recognizing that the noise itself will produce a zero-mean radial velocity, so the final estimate will lie between the value determined from the spectrum of the true signal and zero. The more exact expressions (7.6) and (7.9) will work well in these cases, however.

Some workers extract the radial velocity from the spectrum not by using expressions like (7.6)–(7.10), but by fitting a particular spectral shape to the data. Usually a Gaussian function of the type

Liu, 1978). Various models have been proposed, including sharp edges to turbulent layers and viscosity waves (*Hocking*, 2003a; *Hocking*, 1996b; *Hocking et al.*, 1991; *Luce et al.*, 2001a, b). We will consider experimental ways to study these layers and discuss them in greater depth later in this chapter.

For now we will simply recognize that scatterers may be anisotropic, without addressing the cause, and examine the impact on radiowave scatter in such cases. Radiowaves incident vertically will be more efficiently backscattered than those incident obliquely. Thus for an off-vertical beam, stronger scatter will be received from angles nearer to the zenith than from the mean direction of tilt of the beam (e.g., *Hocking et al.*, 1986; *Röttger*, 1981). We might parameterize the backscatter as

$$B(\theta) \propto \mathrm{e}^{-\frac{\sin^2\theta}{\sin^2\theta_s}}, \tag{7.17}$$

and then θ_s is a parameter typifying the nature of the scatterers. For example, $\theta_s = 90°$ corresponds to almost isotropic scatter, provided we are only interested in angles of θ out to $20°$ or so; $\theta_s = 0°$ is for the case when reflection only occurs from overhead. (Some authors use the form $\mathrm{e}^{-\frac{\theta^2}{\theta_s^2}}$ for $B(\theta)$.)

It can be shown that θ_s relates to the ratio of the length to the depth of the scatterers (*Briggs*, 1992; *Hocking*, 1987a; *Hocking and Hamza*, 1997). We will not reproduce this derivation here in detail: the interested reader is referred to *Briggs* (1992) for a very thorough discussion. We will come back to this point in due course.

For the present, we will only concern ourselves with the fact that θ_s can take a variety of values. In such cases one should replace the angles θ_T in Equations (7.12) to (7.16) with the parameter θ_{eff} where

$$\sin\theta_{\mathrm{eff}} = \sin\theta_T \left[1 + \frac{\theta_0^2}{\theta_s^2}\right]^{-1}. \tag{7.18}$$

Here, it has been assumed that the radar two-way polar diagram has a Gaussian shape of the form $\exp\{-\sin^2\theta/\sin^2\theta_0\}$ (when aligned vertically), so that the half-power half-width of the two-way beam is $\theta_{\frac{1}{2}(2)} = \sqrt{\ln 2}\,\theta_0$.

We will now prove this statement.

Correction for θ_s

As pointed out by *Röttger* (1981), anisotropy in the scattering mechanism alters the effective pointing angles for an off-vertical radar. Such anisotropy also alters the effective beam-width. The principle is illustrated in Figure 7.4.

Let the polar diagram of backscatter for the scatterers be

$$P_s(\theta) \propto \mathrm{e}^{-\frac{\sin^2\theta}{\sin^2\theta_s}}, \tag{7.19}$$

and the two-way polar diagram for a vertically pointing radar be

$$P_R(\theta) \propto \mathrm{e}^{-\frac{\sin^2\theta}{\sin^2\theta_0}}. \tag{7.20}$$

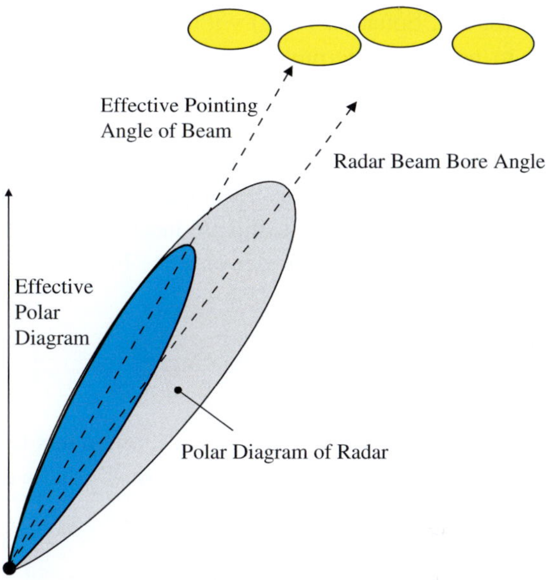

Figure 7.4 Polar diagram of the radar, and modified polar diagram after the effects of the scatterer anisotropy are included.

Then for a radar pointed off-vertical by angle θ_T in the azimuth direction $\varphi = 0$, the polar diagram at angle (θ, φ) is

$$P_{RT}(\theta, \phi) \propto e^{-\left[\frac{(\sin\theta\sin\phi)^2 + (\sin\theta\cos\phi - \sin\theta_T)^2}{\sin^2\theta_0}\right]}. \tag{7.21}$$

(Note that the expression $\exp\left\{-\sin^2(\theta - \theta_T)/\sin^2\theta_0\right\}$ (which has in the past been used to represent a tilted beam) is NOT a good approximation, as it describes an annulus around the zenithal point at a mean angle θ_T.) When the effects of the polar diagram of the scatterers are included, the effective polar diagram is the product of (7.19) and (7.21). This is a maximum when the derivative of the exponent with respect to $\sin\theta$ is zero, or at

$$\sin\theta_{\text{eff}} = \sin\theta_T\left[1 + \frac{\sin^2\theta_0}{\sin^2\theta_s}\right]^{-1}. \tag{7.22}$$

For θ_0, θ_s less than about $10\,^\circ$, this approximates to

$$\sin\theta_{\text{eff}} \approx \sin\theta_T\left[1 + \frac{\theta_0^2}{\theta_s^2}\right]^{-1}. \tag{7.23}$$

Thus the effective pointing angle is given by (7.22), and horizontal wind speeds will be underestimated by the factor

$$R_1 = \left[1 + \frac{\theta_0^2}{\theta_s^2}\right], \tag{7.24}$$

if one uses, say, Equation (7.12) without any correction. This is in fact only approximate – to properly determine the actual measured radial velocity, equation (35) of *Hocking* (1983a) should be integrated (including an aspect sensitivity for the scatterers) to produce the expected power spectrum; this will not have a maximum at the exact point described by (7.22), but it will be close.

The half-width of the effective beam can be found by finding the angles (θ, ϕ) where the effective polar diagram (i.e. the product of (7.19) and (7.21)) falls to one half of the value at $(\theta, \phi) = (\theta_{\mathrm{eff}}, 0)$. Consider only the line $\varphi = 0$. Then the product of (7.19) and (7.21) gives

$$e^{-\left[\frac{(\sin\theta - \sin\theta_T)^2}{\sin^2\theta_0} + \frac{\sin^2\theta}{\sin^2\theta_s}\right]}. \tag{7.25}$$

We seek the two zenithal angles $(\theta_{\frac{1}{2}(2)})_{1,2}$ where this falls to one half of the value at $\theta = \theta_{\mathrm{eff}}$. Some algebraic manipulation soon shows that a quadratic in $\sin\theta$ results, which has two roots at

$$(\theta_{\frac{1}{2}(2)})_{1,2} = \sin\theta_{\mathrm{eff}} \pm \sqrt{\ln 2}\,\sin\theta_0 \left[1 + \frac{\theta_0^2}{\theta_s^2}\right]^{-\frac{1}{2}}, \tag{7.26}$$

(for θ_0, θ_s less than about $10\,°$), and this shows that the effective half-power half-width is

$$R_2 = \left[1 + \frac{\theta_0^2}{\theta_s^2}\right]^{-\frac{1}{2}} \tag{7.27}$$

times the half-width of the radar alone. Note that this ratio is independent of the radar tilt direction, at least out to angles of $10\,°$–$15\,°$.

Equivalently, we can write that the effective half-power half-width $\theta_{\mathrm{eff}\frac{1}{2}}$ can be found as

$$\frac{1}{\sin^2(\theta_{\mathrm{eff}\frac{1}{2}})} = \frac{1}{\sin^2(\theta_{\frac{1}{2}(2)})} + \frac{1}{\sin^2(\theta_{s\frac{1}{2}})}, \tag{7.28}$$

where $\theta_{\frac{1}{2}(2)}$ is the two-way half-power half-width of the radar beam and $\theta_{s\frac{1}{2}}$ is the half-power half-width of the backscatter polar diagram of the scatterers.

Now let us address the issue of how the power received by the radar changes as a function of tilt angle θ_T. The power received by the vertical beam can be found by integrating over the beam. For a Gaussian polar diagram, this integral is proportional to $(\sin\theta_{\mathrm{eff}\frac{1}{2}})^2$ where $\theta_{\mathrm{eff}\frac{1}{2}}$ is the effective half-power half-width of the *combined* polar diagrams of the radar and the scatterers. For small $\theta_{\mathrm{eff}\frac{1}{2}}$, the power is then proportional to $(\theta_{\mathrm{eff}\frac{1}{2}})^2$.

When the radar beam is phased to look at an off-vertical angle θ_T, the peak power will be reduced by a factor

$$f_1 = e^{-\frac{(\sin\theta_{\mathrm{eff}} - \sin\theta_T)^2}{\sin^2\theta_0}}, \tag{7.29}$$

because the peak returned power is returned from θ_{eff} and not θ_T, and then reduced by a further factor

$$f_2 = e^{-\frac{\sin^2\theta_{\text{eff}}}{\sin^2\theta_s}}, \tag{7.30}$$

because of the reduction in power due to the polar diagram of backscatter of the scatterers.

Thus the total received power will be proportional to the product of f_1 and f_2, and then multiplied by the effective beam half-power half-width squared. Thus the received power on the off-vertical beam divided by that received on the vertical beam is equal to

$$\frac{f_1.f_2.(\theta_{\text{eff}\frac{1}{2}})^2}{(\theta_{\text{eff}\frac{1}{2}})^2}, \tag{7.31}$$

or

$$\frac{P(\theta_T)}{P(0)} = \exp\left\{-\left[\frac{(\sin\theta_{\text{eff}} - \sin\theta_T)^2}{\sin^2\theta_0} + \frac{\sin^2\theta_{\text{eff}}}{\sin^2\theta_s}\right]\right\}. \tag{7.32}$$

Note that this final expression corrects an error in the original derivation of *Hocking et al.* (1986), in which the factor f_2 was neglected.

Inversion of Equation (7.32) gives

$$\theta_s = \sin^{-1}\left[\sqrt{\frac{\sin^2\theta_T}{\ln[P(0)/P(\theta_T)]} - \sin^2\theta_0}\right], \tag{7.33}$$

(e.g., *Hooper and Thomas*, 1995).

More effects of the shapes of the scatterers

Even beyond the above issues, there are still potential problems with Doppler determination of wind speeds. If there are a variety of shapes, for example, the simple theory of the previous sections is not valid. If stratified reflecting steps exist as well as isotropic and anisotropic scatterers, then complications also arise.

The shapes of the scatterers can also affect determination of the vertical velocity. If, for example, the atmospheric scatterers are not aligned exactly horizontally, but have a small tilt, then the direction of preferred scatter will not be immediately overhead, but off to one side. The result is that the small vertical velocities will be contaminated with a small component of the horizontal wind. For example, if the effective tilt is only 1°, and the beam half-power half-width is, say, greater than about 3 °, a horizontal wind of 50 ms^{-1} results in a contribution to the "vertical" velocity of $\sim$ 1 ms^{-1} . This is why many analyses of vertical velocity are made only by using long-term means; it is hoped that such tilts average out to zero when averaged over long times, but even then some caution must be exercised. Indeed, a better radar configuration for estimates of the vertical velocity is a bistatic radar, with the transmitter and the receiver well separated (e.g., *Waterman*, 1983).

Other problems also exist; for example it is possible that erroneous wind speeds and wind-shears will result if the scattering layers are much thinner in depth than the radar pulse-length (e.g., *Fukao et al.*, 1988a, b; *Hocking*, 1983a; *May et al.*, 1988a).

Despite all these potential problems, the Doppler method still remains a good way to measure mean winds in the atmosphere, but any user must be aware of these limitations and bear them in mind during any experimental study.

7.2.2 Spaced antenna methods: FCA and interferometer techniques

There are alternative approaches for determining atmospheric wind speeds, and these are spaced antenna methods (e.g., see reviews by *Briggs*, 1984; *Hocking*, 1983c; *Hocking et al.*, 1989). In them, one uses separate groups of antennas, spaced apart on the ground, to determine the wind speed. There are two main approaches; the first uses cross-correlation techniques to determine the time it takes for the diffraction pattern of the irregularities to cross between groups of antennas; in its most sophisticated form it is called full correlation analysis, or FCA (*Briggs*, 1984). The second approach, originally introduced by *Pfister* (1971) and later by *Farley et al.* (1981), *Röttger and Ierkic* (1985) and *Adams et al.* (1985), involves using the groups of antennas to form an interferometer. Briefly, such interferometer methods use phase differences between signals received at the groups of antennas to determine angles of arrival. Cross-spectral techniques are used for such angle-of-arrival determinations. Doppler methods are then used to determine the radial velocities associated with each separate scatterer. With such methods, it is possible to calculate the positions of the main scatterers in the sky, hence enabling more accurate determination of horizontal and vertical winds. The major disadvantage of this technique is that it requires that preferred regions of scatter, of narrow angular extent, do indeed exist, so that a direction can be determined. If scatter is diffuse, from a wide range of angles, the method breaks down; even though apparent directions of preferred scatter might still seem to result from the analysis in this case, they are not meaningful (e.g., *Briggs*, 1980).

These two techniques are discussed in much greater detail in Chapter 9 of this book.

7.2.3 Brief comments on the various wind-measurement techniques

There are advantages and disadvantages to all these methods. For example, correlation analysis techniques often use small groups of antennas, with corresponding wide polar diagrams. As a consequence, they often produce winds that are averaged over a large area of the sky. On the other hand, there is the advantage that both components of the wind speed are measured in the same volume, directly above the radar. Furthermore, even if the atmospheric scatterers have non-isotropic backscatter polar diagrams, correct estimates of the wind speed still result. The vertical wind speed is not measured, and Doppler methods must be used to determine this. A negative aspect of the FCA method is that the correlation functions often show oscillations, making it hard to locate the true cross-correlation maxima, and the FCA method often has a low acceptance rate of good quality data.

If isotropic scatterers are the main type of scatter, then the spaced antenna and Doppler methods are equivalent (*Briggs*, 1980). If there is a significant contribution from horizontally aligned, mirror-like partial reflectors (called specular reflectors), the result can be enhanced scatter from the vertical, an advantage for the spaced antenna technique, since that method uses only vertically aligned beams. However, in the extreme that these specular reflector regions form a continuous blanket across the sky, with buoyancy waves causing undulations in this blanket, then the FCA and other simpler versions of the spaced antenna method can break down and effectively measure the phase speeds of the gravity waves. (Gravity waves (or buoyancy waves) are propagating atmospheric oscillations that move through the air. They are discussed in much greater detail in Chapter 10.) This is a problem for E-region studies using MF and HF frequencies, but in the middle atmosphere it is rarely a problem (e.g., *Hocking et al.*, 1989). It is worth noting that DBS methods seem to be able to work at lower values of signal-to-noise ratio (SNR) than correlation methods. At low SNR values, the cross- and auto-correlation functions often develop large secondary peaks that can severely limit the application of the technique, whereas it is often easier to identify a single Doppler peak even at low SNR with DBS methods.

As discussed, the major disadvantage of interferometer techniques is that they require that there be preferred regions of scatter in the sky, of narrow angular extent, so that a direction can be determined. If scatter is diffuse, from a wide range of angles, the method breaks down completely. On the other hand, if such discrete scatterers are present, interferometer methods enable high resolution studies of the scatterers.

It is clear from the preceding discussions that while the principles of estimation of wind speeds are simple, in practice there are many complicating features. Determination of perhaps the simplest target parameters – their speeds – is quite complex for the atmospheric case. To first order, all the methods are sound; but if one is interested in details about wind fluctuations, it is clear that it is necessary to know other features of the target, such as aspect sensitivity, shape, the spatial distribution of the scatterers, and perhaps even the cause of the scatterers. In due course, we will address methods to determine such target parameters.

7.3 Spectral width estimates

So far we have concentrated on determination of mean winds. In the Doppler method, this relates largely to the mean frequency offset of the spectrum, whilst in the FCA method it relates to the time delay of the peak of the cross-correlation function. There is still more information in the signal. In the spaced antenna method, the width of the auto- and cross-correlation functions holds extra information about the targets; in the Doppler method, the width of the spectrum contains the information. In some ways, the second case is easier to visualize, so let us concentrate on that one. It should be remembered throughout these discussions that the spectrum and the autocorrelation functions are Fourier transforms of each other.

A variety of methods can be used to determine the spectral width. One can utilize either the width of the autocorrelation function where it falls to one half of its value at zero lag, or the second lag of the autocorrelation function, or the second moment of the spectrum itself (see the discussion by *Woodman*, 1985). In all cases, one must be careful about the effects of noise, since noise can cause systematic errors. For example,

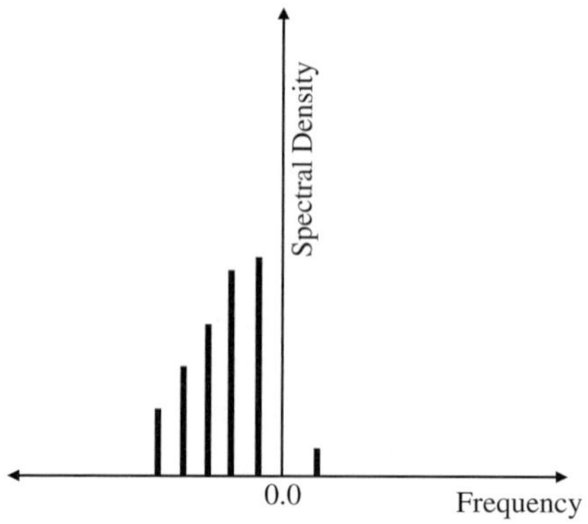

Figure 7.5 Spectral lines produced by a group of scatterers with a non-zero mean radial velocity and a dispersion of radial velocities.

noise produces a narrow spike at zero lag of the autocorrelation function, and this spike should be eliminated before proceeding with analysis. A procedure commonly used to determine the spectral width is least-squares fitting of a Gaussian-like function as in Equation (7.11). In some cases, it is necessary to remove excessively large spikes from the spectra, a procedure which is especially necessary when there are mirror-like partial reflectors in the radar volume (e.g., *Hocking*, 1983b). The details of these procedures will not be considered here; we are more concerned with the interpretation of the spectral width.

What then can cause the broadening of the spectrum? Perhaps the most obvious cause is random motion of the scatterers. If each scatterer has a velocity superimposed upon the mean speed, then each produces a line in the spectrum with a different frequency, as illustrated in Figure 7.5.

If the scatterers have, for example, a Maxwellian distribution, then the vertical component of velocity (w) must have a Gaussian distribution that is proportional to $\exp\{-w^2/(2w_{RMS}^2)\}$. Since for a vertical beam the Doppler shift from any scatterer is $f = \frac{2}{\lambda} w$, the spectrum will have a shape of the form $\exp\{-f^2/(2f_{RMS}^2)\}$, where $f_{RMS} = \frac{2}{\lambda} w_{RMS}$.

For some years in the early period of VHF middle atmosphere studies, it was assumed that this was the major cause of spectral broadening. However, for most VHF radars, this is not in fact the case. There are other causes of spectral broadening, which, while understood by a few (e.g., *Atlas*, 1964; *Atlas et al.*, 1969; *Hocking*, 1983a; *Hocking*, 1983b; *Sloss and Atlas*, 1968), were not generally appreciated in the early days of radar studies of turbulence. However, these effects are now well recognized, and will be discussed below.

For a vertically pointing beam, probably the main cause of the non-zero spectral width is the so-called "beam broadening," which is illustrated in Figure 7.6(a).

Even if all the scatterers are moving horizontally at the same velocity, each scatterer will produce a different Doppler shift because of its different zenithal angles. The net

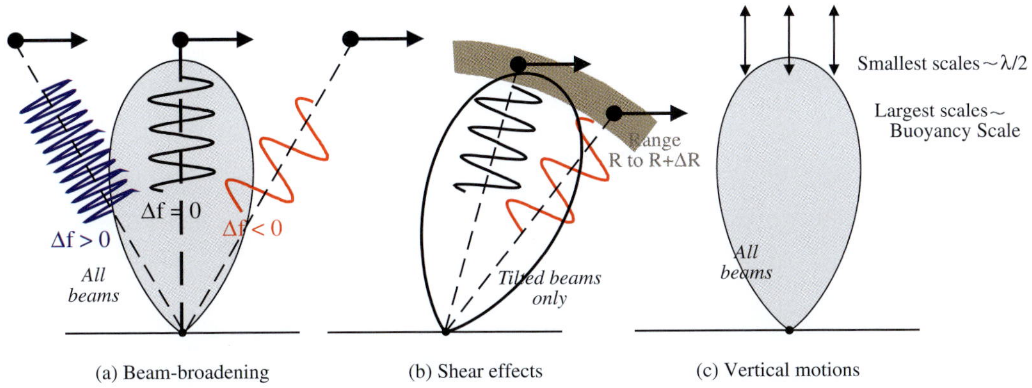

Figure 7.6 Doppler effects due to scatterers in different positions of various beams.

result is a spectrum of finite width. This spectral broadening has been modeled by several workers (e.g., *Atlas*, 1964; *Atlas et al.*, 1969; *Hitschfeld and Dennis*, 1956; *Hocking*, 1983a; *Hocking*, 1983b; *Sloss and Atlas*, 1968). For relatively narrow beams ($\leq$ about 5 ° half-power half-width), the spectral half-power half-width $f_{\frac{1}{2}b}$ obeys the approximate relation (in units of Hz)

$$f_{\frac{1}{2}b} = \frac{2}{\lambda}(1.0) \mid \underline{V}_{\text{hor}} \mid \theta_{\frac{1}{2}c}, \tag{7.34}$$

where $\theta_{\frac{1}{2}c}$ is the two-way half-power half-width of the polar diagram (including both the transmitter and receiver polar diagrams – for a monostatic Doppler radar it is the two-way value, $f_{\frac{1}{2}(2)}$) in radians, and $\underline{V}_{\text{hor}}$ is the *total* horizontal wind vector. The "1.0" has been added to ensure that it is recognized that this is an accurate equation and not an approximation, and the constant of proportionality truly is 1.0.

The same approximation is also fairly accurate even for off-vertical beams. It is important to note that the *total* wind speed should be used, and *not* just the component parallel to the tilt direction of the beam. This formula is based on the assumption that the scattering is statistically isotropic, an assumption which we will relax shortly. When one compares the spectral half-widths due to the non-fluctuating components of the wind-field to the experimental spectral half-widths measured with the vertical beam, one frequently finds that the two are very similar. For example, Figure 7.7(a) (from *Hocking*, 1986) shows an almost 1:1 relationship between the two parameters when spectra produced from 11s data sets were used. Normally for turbulent scatter, the experimental half-widths exceed the theoretical values (where "theoretical" is used to describe the non-turbulent contributions), as illustrated in Figure 7.7(b).

There are other effects which alter the spectral width, particularly if the beam is tilted from the vertical. Horizontal fluctuating motions will alter the spectral width (see Figure 7.8) and so will changes of the mean wind with height, as occurs, for example, in a wind-shear (e.g. Figure 7.6(b)). The former effect always broadens the spectrum, whilst the latter can either reduce or increase the spectral width depending on the sign of the wind-shear. These points are discussed in more detail by *Hocking* (1983a), for example, and later by *Nastrom* (1997).

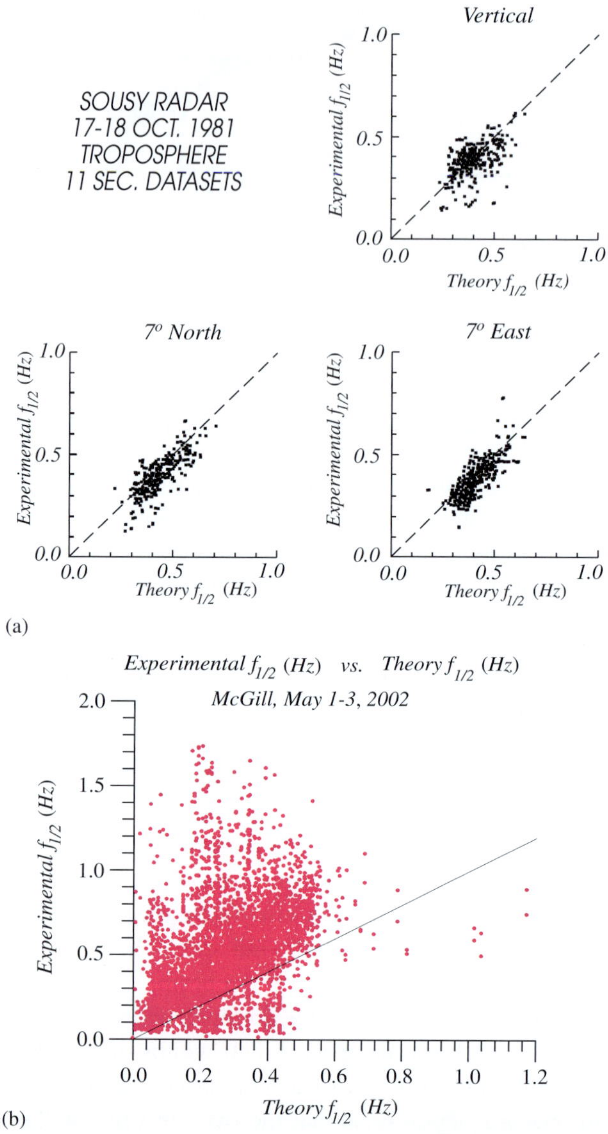

Figure 7.7 (a) Scatter plots of experimental half-power half-widths as a function of beam-broadened half-width for various beam tilts, recorded with the SOUSY radar (*Hocking et al.*, 1986). (b) Scatter plots of experimental half-power half-widths as a function of beam-broadened half-width for a vertical beam, recorded with the McGill VHF radar. Generally the experimental values slightly exceed the theoretical values, with a few exceptions which are discussed further in the text.

Of course the target parameter which is desired is the RMS fluctuating velocity of the scatterers, but this often contributes only a small fraction to the total spectral width. To determine the RMS fluctuating velocity, one should first use the measured mean wind speeds as a function of height and the known polar diagram (radiation pattern) of

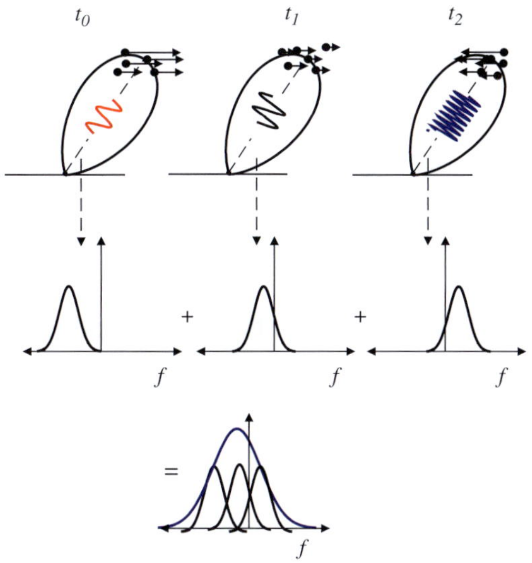

Figure 7.8 Showing how horizontal fluctuating motions can broaden the observed spectrum.

the radar to determine the spectral half-power half-width $f_{\frac{1}{2}nf}$ contributed by the non-fluctuating effects. Then the contribution from the fluctuating component f_{fluct} can be found through the relation

$$f_{\text{fluct}}^2 = f_{\frac{1}{2}\text{expt}}^2 - f_{\frac{1}{2}nf}^2. \tag{7.35}$$

This arises because the experimental spectrum is approximately a convolution between the spectrum that would be produced if there were no fluctuating components, and the spectrum due to the fluctuating components alone, at least for very narrow beams ($\leq$ about 5 ° half-power half-width; the more general case has been modeled by *Hocking* (1983a)).

To properly consider all the contributions from the mean wind including wind-shear, a more accurate computer model needs to be used (e.g., *Hocking*, 1983a; *Hocking*, 2003b).

In some cases, Equation (7.34) serves as a useful approximation to obtain $f_{\frac{1}{2}nf}$, but generally it is advised to incorporate wind-shear effects, especially in view of the speed of modern computers, which makes this easily possible. *Nastrom* (1997) has performed a study of the effects of shear-broadening. It is worth bearing in mind that these effects were known 15 years before that publication, and indeed the model presented by *Hocking* (1983a) recognized this fact. *Chu* (2002) has also performed studies of beam-broadening with tilted beams, and again repeats the earlier results of *Hocking* (1983a) (although Chu assumed that the contours of constant radial velocity were ellipses, which is only true if there are substantial vertical velocities; for small vertical velocities, these contours are in fact hyperbolas). Indeed, the model of *Hocking* (1983a) was quite general, incorporating a full representation of wind-shears, and still represents the most general model published to date. These different models will be discussed in more detail shortly.

(vii) No allowance was made for dependence of backscattered power on range within the pulse.
(viii) The receiver was to have an infinite band-width.
(ix) The beam-width was unchanged as the beam moves to off-zenith angles.

Many of these assumptions are quite unreasonable, but they formed the basis for *Nastrom* (1997) to develop an approximate formula which showed different terms separately (although this approach could not deal with things like non-linear wind profiles). One interesting feature was that the formula developed involved the beam one-way half width where the two-way beam-width should have been. However, calibration against the full numerical model allowed fine-tuning of the various constants, and the analytical expression developed has found application in various research groups.

In 2011, *Dehghan and Hocking* (2011) were concerned about the fact that on some occasions, the calculated beam-broadened spectral width *exceeded* the measured width, which should not, in principle, happen. They therefore re-examined all of the available literature on this topic, to ascertain if there could be missing features in the model. The reasons for this seemingly unphysical result are discussed in the next subsection, and, as it turns out, are easily understood. However, in this section, the relevance of this paper is in regard to Nastrom's analytical model.

Dehghan and Hocking (2011), using Figure 7.10 as a starting point, began by independently developing an expression like that of Nastrom's, but allowed the various terms and cross-terms that they expected to have power-law formats, with the coefficients and power laws being left as variables. Then they performed calculations of the spectral widths using the full and exact model of *Hocking* (1983a) and performed a massive least-squares fit to optimize the coefficients and power laws of the various terms. They were able to reproduce the expression developed by *Nastrom* (1997), plus found one extra term. This established the equivalence of the different methods.

The most detailed analytical expression for the mean-square fluctuating velocity is the following:

$$v_{\mathrm{RMS}}^2 = \frac{\nu^2}{\kappa} u_0^2 \cos\alpha - a_0 \frac{\nu}{\kappa} \sin\alpha \left(u_0 \frac{\partial u}{\partial z} \zeta \right) + b_0 \frac{2\sin^2\alpha}{8\kappa} \left(\frac{\partial u}{\partial z} \zeta \right)^2 + c_0(\cos^2\alpha \sin^2\alpha) \mid u_0 \xi \mid + d_0(\cos^2\alpha \sin^2\alpha)\xi^2, \tag{7.36}$$

where (following to some extent the notation of Nastrom), ν is the *one-way half-power half-width*, α is the zenithal tilt angle, u_0 is the wind speed, $(\partial u/\partial z)$ is the vertical shear of horizontal wind speed, R_0 is the range, ΔR is the range resolution, $\kappa = 4\ln 2$, $\zeta = 2R_0\nu \sin\alpha$ (drawn in Figure 7.10), $\xi = (\partial u/\partial z)\Delta R/\sqrt(12)$, $a_0 = 0.945$, $b_0 = 1.500$, $c_0 = 0.030$, and $d_0 = 0.825$.

However, minor functional adjustments were also recommended at certain points and for some coefficients, which optimized the equation even further. The same is true even for Nastrom's expression, which can be improved with modest adjustments. *Dehghan and Hocking* (2011) also give a simple Fortran program that calculates this expression, and this includes these corrections. The reader is recommended to at least use, or adapt, that subroutine, to obtain optimum results. An even better recommendation is to use a full integral numerical calculation, but the above equation (as presented with corrections in *Dehghan and Hocking* (2011)) is usually adequate. The slightly less accurate expression from *Nastrom* (1997) could also be used, though with less reliability.

7.3.2 "Negative" energy dissipation rates

Equation (7.35) implies that the experimental spectral width must always exceed the beam-broadened theoretical one, but Figure 7.7 shows that this is not always so in practice. If we apply (7.35) then, (using procedures described in more detail in the next subsection) the net result will be an apparent "negative" value for the square of the spectral width due to fluctuating velocities, implying a physically unreal negative turbulent kinetic energy dissipation rate. How do we interpret such a negative value? Should we simply ignore negative values, as some authors (e.g., *Nastrom*, 1997, among others) have done? (We shall discuss the conversion of f^2_{fluct} in Equation (7.35) to ε shortly; for now, it is simply necessary to know that a negative value for f^2_{fluct} will lead to a negative estimate of turbulence strength.)

Dehghan and Hocking (2011) have investigated this question in some detail. They concluded that these negative values should *not* be ignored, and indeed ignoring them can lead to over-estimates of the mean turbulence strengths. These authors investigated three possible causes of apparent negative turbulent energy dissipation rates. The first was related to scatterer anisotropy, another was the accuracy of determination of the experimental spectral width, and the last was the accuracy of the mean wind determination. Each of these will now be discussed in turn.

Equation (7.34), and even the more advanced Equation (7.36), are still only estimates of the spectral half-width due to the non-fluctuating component, since they (and even a full numerical calculation) assume that the scatterers scatter isotropically. If the scatterers are anisotropic, as may be the case and as has been discussed previously, then the true contribution from non-fluctuating components will be less than that calculated with (7.34). That equation can still be used, but $\theta_{\frac{1}{2}}$ must be replaced by $\theta'_{\frac{1}{2}} = R.\theta_{\frac{1}{2}}$ where

$$R = \left[1 + \frac{\theta^2_{\frac{1}{2}}}{\theta^2_{s\frac{1}{2}}}\right]^{-\frac{1}{2}}, \qquad (7.37)$$

$\theta_{\frac{1}{2}}$ being the true half-power half-width of the radar beam, and $\theta_{s\frac{1}{2}}$ the half-power half-width of the polar diagram of backscatter due to the scatterers (i.e. $\theta_{s\frac{1}{2}} = \sqrt{\ln 2}\,\theta_s$, θ_s being defined by Equation (7.17)). Hence in this case the experimental spectral width may be less than the beam-broadened one. This assumes Equation (7.18) has been already applied to properly determine the true wind speed. If, on the other hand, the mean winds were determined without correction, then they will be an underestimate by a factor given by Equation (7.18), which is an even larger reduction than that given by (7.37). Hence the anisotropy will in fact still produce results with a measured spectral width greater than the beam-broadened one. However, neither case is ideal, and the user should correct both the wind speed and the spectral width before anisotropy.

However, *Dehghan and Hocking* (2011) concluded that as long as turbulence measurements were made using an off-vertical beam in the troposphere, the contribution of anisotropy to "negative turbulence" measurements should be relatively small, and they turned to other explanations. Notice, however, that once again it is important to be aware of anisotropy, and know the backscatter polar diagram due to the scatterers. It is becoming more and more important as we proceed through this text to know this parameter.

We now turn to consideration of other factors. The previous explanation (anisotropy) represented an asymmetric error, with the potential to systematically produce small experimental spectral widths. The next two explanations do not do this – rather, they produce explanations by which *either* the experimental spectral width *or* the beam-broadened width may be significantly under-estimated *or* overestimated. Each scenario is equally likely, so the effect is symmetric, producing both underestimates (and potentially negative values) of turbulence strengths, and also overestimates of turbulence strengths. The reasons relate to statistics. Cases in which the experimental width is over-estimated, or the beam-broadened width is underestimated, will produce overly large values of turbulence strength, and will generally go unnoticed by most users. Cases of extreme underestimates of the energy dissipation rates will be especially noticed when they become negative. However, because both the over- and under-estimates arise from the same root cause, *both types of data must be included in any estimate of the mean strengths of turbulence*: elimination of the negative values without complementary elimination of the over-estimates (usually near-impossible to identify) leads to a bias in the positive sense. Cases where the negative values reach high percentages (e.g. 30% or more) should be discarded entirely (including both positive and negative values) since they probably contain little useful information.

The first source of this type of error to consider relates to the accuracy with which the experimental spectral width can be determined. There are various reasons why this width may be in error. Noise naturally contributes, and is especially bad if the width of the spectrum is determined by using a weighted integral for the variance (see Equation (7.10) – this shows how to do a weighted estimate of the mean of the spectrum – the variance is found in a similar way but the weighting is not f, but rather $(f - \bar{f})^2$).

A far superior method is spectral fitting, as discussed earlier. Unfortunately, if the spectral width is over-estimated (as in the case of using weighted spectral moments), the percentage of negative estimates of energy dissipation rates will be reduced, giving the appearance of improved data. Nothing could be further from the truth. Hence spectral fitting should *always* be used for determinations of turbulence strengths.

In the case of no noise, or the case where spectral fitting is applied, the width is still limited by the sampling rate of the data. The spectral resolution will be proportional to the frequency resolution, which will be proportional to the inverse of the data length. We will not go into mathematical detail about this here, and refer the reader to *Dehghan and Hocking* (2011) for more specific details. However, we will summarize the effect in a general sense.

The accuracy of a determination of the width of a spectrum is dependent on its resolution, and the resolution is dependent on the data length. Many radars use data-lengths of 10 seconds or so, so the resolution is 0.1 Hz. However, if we apply Equation (7.34) as an approximation, and assume a horizontal wind of 10 ms^{-1} and a half-power two-way beam half-width of 2 degrees, then the spectral half-power half-width is close to 0.1 Hz – i.e., comparable to the resolution. So a single spectrum might have meaningful powers at maybe two and probably at most four points, the others being buried in noise. As the wind speed increases, the spectra do get wider, but then the relative contributions of beam-broadening and turbulence to the spectrum increase, making extraction of turbulence strengths harder once a certain critical wind speed is achieved (e.g., *Hocking*, 1986).

Hocking (1997a) proposed use of data sets of length 30–40 s. Although this runs some risk of allowing the scatterers to change statistically in this time (e.g. the turbulence may grow or die out), it provides a much better frequency resolution, and so allows a larger number of points in the spectrum. Indeed, it is worth remembering that those researchers that do use 10 s data bins in fact often record several 10 s bins in succession, form spectra, and then incoherently average the spectra. So their total temporal coverage is still several tens of seconds anyway.

Dehghan and Hocking (2011) discuss the dependence of the errors produced by these processes in a quantitative way, and demonstrate that the spectral resolution and fitting procedures do indeed contribute to the likelihood of negative turbulence estimates. They very much recommended the use of data-lengths of the order of 30–40 s for spectral studies, to reduce these errors, but some degree of error is always present.

The next cause of errors in determination of turbulence estimates (and subsequently increased likelihood of negative turbulence estimates, as well as over-estimates of the turbulence strength) is the most important one. It arises not because of any fitting issues, but rather as an intrinsic problem associated with the Doppler method itself. The problem can be seen in Figure 7.3(b). As seen, two orthogonal beams are required in order to measure two components of the wind. The total wind vector is formed by vectorially adding these two components. The main issue is that the two components were measured in physically distinct regions of space, separated by typically 1 to 3 km in the troposphere and lower stratosphere, and even larger horizontal distances in the mesosphere. So if, for example, the spectrum is measured using a northward pointing beam, then the northward component of the wind is measured correctly for the region of scatterers, but the eastward component is measured in the other beam, some distance away. The problem is exacerbated by the fact that the measurements on the two beams are usually made at different times, often separated by a few minutes. If multiple beams (east, west, north, and south) are used to estimate the mean wind, then a trade-off occurs – more beams may in principle produce higher reliability, but the use of measurements at different regions of space introduces extra uncertainty. No matter how the averaging is done, the full horizontal vector wind speed relevant to the region of space in which the turbulence is being determined, while measurable, is not known exactly, and associated errors are a few ms^{-1}. The impossibility of truly measuring the exact mean wind in the region of the scatterers adds an unerasable uncertainty to the estimate of the beam-broadened spectral width. (The problem could in principle be reduced by using say the spaced antenna method to measure the winds, since both components are measured simultaneously, but this introduces other problems: for example, the spaced antenna

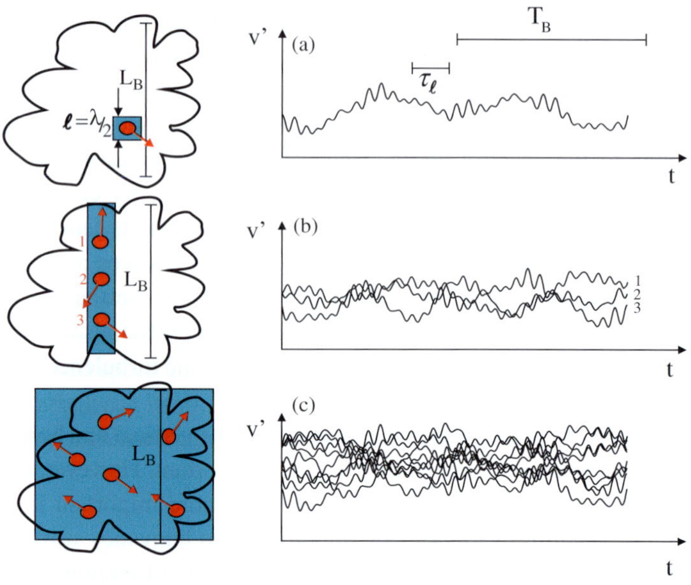

Figure 7.12 The effect of radar volume on measured $\overline{v'^2}$ values. (a) The motion of a single scatterer. In any time interval τ_ℓ, only a limited amount of velocity fluctuation occurs, but in the larger time interval T_B, the full possible range of velocity fluctuations occurs, as larger scales become more effective. In (b) and (c), the radar volume is assumed to be larger (shown by the shaded regions to the left), and many scatterers contribute to the final signal received, each with different velocity, so a full range of Doppler velocities is experienced in a very short time.

4.5° half-width); data were collected for 12 min in order to emphasize the effect (normally data-lengths are of the order of 2 or 3 min at MF, and tens of seconds at VHF). In normal VHF experiments, where the sampling times are less (typically a few tens of seconds), the effect may not be so dramatic.

Thus measurements of turbulent energy dissipation rates are best made using a vertical beam when doing mesospheric work. Ironically, for tropospheric and stratospheric work, where the horizontal fluctuating motions associated with gravity waves are much weaker, off-vertical beams give a better indication of turbulence strengths, provided that the tilt is not too severe (say less than 12 degrees), because specular reflectors can dominate the signal recorded with vertical beams in that case.

The contribution due to turbulence can be envisaged as follows and is illustrated in Figure 7.12. Backscatter occurs predominantly from scatterers with scales of the order of the radar half-wavelength, but these scatterers are carried around by the larger scales. The mean-square fluctuating velocity measured by the radar is then the integrated effect from scales of the order of the radar half-wavelength out to scales comparable with the radar volume (e.g., *Hocking*, 1983a; *Sato and Woodman*, 1982b).

For radars with pulse lengths and beam-widths comparable to the buoyancy scale of turbulence, scales even beyond the buoyancy scale may contribute to the mean-square fluctuating velocity, although fortunately with reduced contributions. Let us say that the

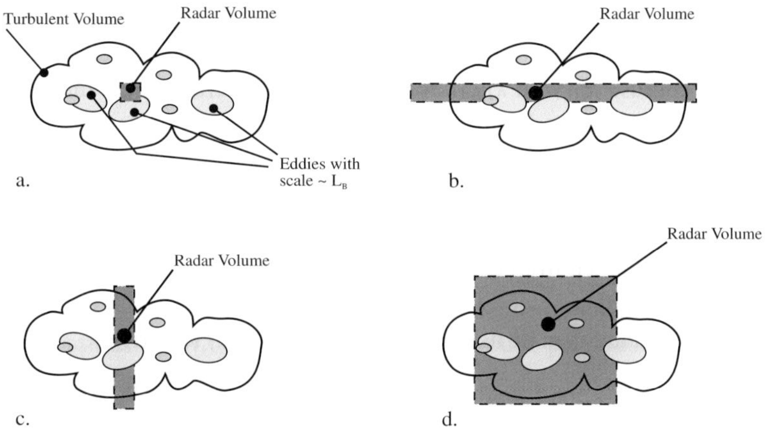

Figure 7.13 Possible radar volumes relative to the scale of a typical patch of turbulence.

measured mean-square fluctuating velocity is due to a fraction $\mathcal{F}$ from scales within the inertial range of turbulence, and the remaining contribution comes from scales within the buoyancy range. The exact value of $\mathcal{F}$ depends on the radar configuration, sampling time, etc., and for the present we will not concern ourselves with its evaluation. Figure 7.13 shows different possible relationships between the scales associated with the radar volume and the scale of the turbulence. Each of these scenarios give different fractions $\mathcal{F}$. We will quantify these effects shortly in terms of $\mathcal{F}$.

We begin by writing (following *Hocking*, 1983a; *Hocking*, 1986) that the velocity variance observed with the radar is

$$\overline{v^2} = \int \Theta_{tt}(k_t)dk_t, \tag{7.39}$$

where $\Theta_{tt}(k_t)$ is the transverse one-dimensional spectrum function (e.g., *Batchelor*, 1953) for the direction radial from the radar (also see Chapter 10). *Hocking* (1988) used the longitudinal spectrum, but since we view the turbulence with the radar from a direction perpendicular to the mean flow, it makes better sense to use the transverse spectrum. The integration is performed over all scales which can affect the radar measurements, which for VHF radars means scales out to the radar pulse length or the buoyancy scale of turbulence, whichever is larger. For a radar pulse length of say, 600 m, this means that scales well into the buoyancy range will be effective, since the thicknesses of these layers are often well below 600 m (e.g., *Barat*, 1982; *Crane*, 1980b) and the inertial range/buoyancy range transition scale is usually several times less than the layer thickness (e.g. *Barat*, 1982). If it is assumed that a fraction $\mathcal{F}$ of the measured velocity variance resides in the inertial range and the rest in the buoyancy range, we may write that the measured velocity variance $\overline{v^2}$ obeys the relation

$$\int_{-k_B}^{-k_{LB}} \Theta_{tt}(k_t)dk_t + \int_{k_{LB}}^{k_B} \Theta_{tt}(k_t)dk_t = \mathcal{F}\,\overline{v^2}, \tag{7.40}$$

half-power of about $0.44 \times 2\pi/\ell_r$. This different situation means that for $\ell_r \ll L_B$, the following relation applies (e.g., *Labitt*, 1979; *Bohne*, 1982, (Appendix C)):

$$\varepsilon \simeq 1.3 \left(\overline{v^2}\right)^{3/2} / L_r. \tag{7.51}$$

The constant (1.3) has changed considerably compared to that in (7.44) and (7.45). The reason for this change is that Equation (7.51), which uses a constant of 1.3, assumes that there is no buoyancy scale, and assumes that the $k^{-\frac{5}{3}}$ law applies over all scales; thus Fourier scales of small wavenumber, although only a small contribution to the pulse, make a large contribution to the integral in (7.40).

We now return to consolidating these two approaches. While they seem to make different assumptions, *Hocking* (1996a) was able to develop a model that covered both cases. Figure 7.14 shows the radar/atmospheric conditions that are required for each of these formulas to be valid. Most of the time, Equation (7.48) is valid, but for high resolution radars using short data samples, Equation (7.51) is more appropriate, as indicated by the portion of Figure 7.14(a) labelled Labitt. In the rest of Figure 7.14(a), corrections that need to be made to Equation (7.48) according to *Hocking* (1996a) are shown, assuming that gravity waves do not contribute substantially to the oscillations. The corrections are typically 5 or 10%. If gravity waves are a significant contributor to the oscillations and satisfy the expected gravity wave spectral behavior (discussed in more detail in Chapter 10), then the corrections to Equations (7.44) and (7.48) are shown in Figure 7.14(b), where we assume that

$$\varepsilon = 2.24 \frac{\left(\overline{v^2}\right)^{3/2}}{L_B} \frac{1}{c_f} = 0.37 \overline{v^2} \omega_B \left[\frac{1}{c_f}\right]^{2/3}, \tag{7.52}$$

where c_f is a near-constant parameter which changes weakly as a function of pulse-length, beam-width, buoyancy scale, and relative gravity wave contribution. Typical values are shown in Figure 7.14. It should also be noted that even if $\ell_r \ll L_B$, but

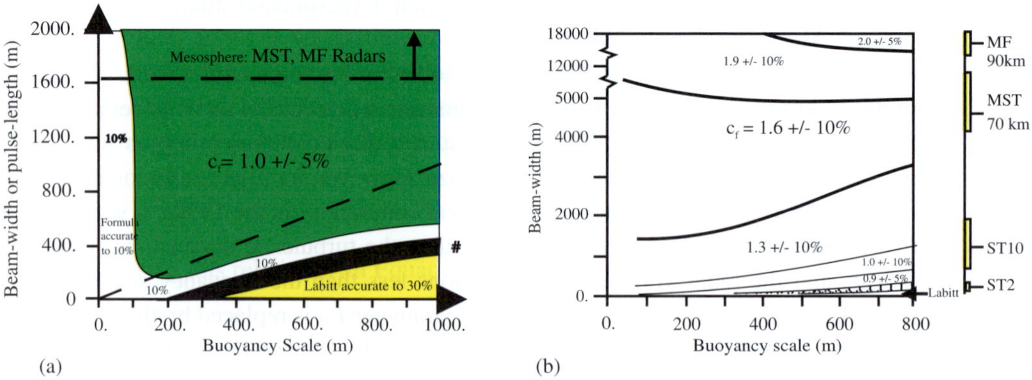

Figure 7.14 (a) The combinations of radar scale and turbulence scale that require application of the Labitt formalism and the buoyancy-scale formalism. (b) Correction factors to account for the fact the gravity-wave fluctuating motions are integrated into the turbulence measurements.

data lengths of 40 seconds to a minute or so are used in forming the power spectra, Equations (7.44) and (7.48) are still better estimators of ε; see Figure 7.12. It turns out that even in many meteorological applications, Equation (7.51) should not be used, and Equation (7.48), or the improved Equations (7.52), are the most apt.

The relations (7.42–7.52) (whichever is applicable) may be used to determine the turbulent energy dissipation rate if one knows the contribution to the spectrum from turbulent fluctuations. One warning is needed, however. Ideally, the Brunt–Väisälä frequency should be measured locally at the same time, but this is often not possible, and it is not unusual to use climatological values for ω_B. This should be done with caution, and the full impact of this approach should be an area for future study.

If the value of c_f is taken to be 1.6, as suggested for a typical value in Figure 7.14(b), then (7.52) becomes

$$\varepsilon = 1.40\frac{\left(\overline{v^2}\right)^{3/2}}{L_B} = 0.27\overline{v^2}\omega_B \qquad \text{with reservation.} \tag{7.53}$$

Interestingly, this agrees fairly well with the result produced by dimensional analysis by *Roper* (2000); *Roper and Brosnahan* (2005). However, the term "with reservation" hints that the formula may be incomplete – an issue that we will now address.

These results discussed above are all theoretical, or even derived by dimensional analysis. Proper experimental tests were missing for many years. However, finally in 2014, some useful experimental tests between radar and aircraft became available in the published literature. *Dehghan et al.* (2014) peformed a 3-way test between instrumented aircraft, radar, and commercial aircraft flying into and out of Detroit. The measurements were made with a 40.68 MHz radar located near Harrow, Canada (42.039N, 82.892W) which is within 43 km of the Detroit airport.

The final result of these tests showed that for lower tropospheric measurements, the appropriate equation was

$$\varepsilon = c_\varepsilon \overline{v^2}\omega_B = c_{\varepsilon T}\overline{v^2}/T_B, \tag{7.54}$$

where T_B is the Brunt–Väisälä period, $c_\varepsilon = 0.6 \pm 0.2$ and $c_{\varepsilon T} = 3.8 \pm 1.8$.

Ironically, this is closer to the values derived by *Hocking* (1983a) ($c_\varepsilon = 0.45$ and $c_{\varepsilon T} = 2.8$), rather than the value of 0.27 proposed in (7.53), and as also used in *Dehghan and Hocking* (2011); *Hocking* (2011). The experimental value is also close to the value used by *Nastrom* (1997) who used $c_\varepsilon = 0.49$ and $c_{\varepsilon T} = 3.1$.

However, it needs to be emphasized that this is more of a coincidence than a proof that the earlier derivations were valid, since the newer derivations include more physical processes, including a better treatment of the relative roles of gravity waves and turbulence.

It is also noteworthy that the aicraft used in the studies of *Dehghan et al.* (2014) only flew at altitudes of 1–3 km, and since gravity waves are relatively weak at low altitudes, this estimate of c_ε may not be valid at higher altitudes. Larger contributions from gravity waves will reduce c_ε.

So while a value of $c_\varepsilon = 0.6 \pm 0.2$ may be relevant for the lowest altitudes, these measurements are not inconsistent with a value in the range 0.45 to 0.49, as proposed

Hence

$$\tau \sim \frac{v^2}{\varepsilon} \sim \left(\frac{\ell}{\tau}\right)^2 . \frac{1}{\varepsilon}, \tag{7.58}$$

so that

$$\tau \sim \ell^{\frac{2}{3}} \varepsilon^{-\frac{1}{3}}. \tag{7.59}$$

Assuming that the growth and decay of eddies produces an autocorrelation function with a half-peak-half-width $\tau_{1/2}$, we may approximate $\tau_{1/2}$ to τ in the above expression. If the autocorrelation function is taken to be Gaussian, then its Fourier transform is also Gaussian, with a half- power half-width of $0.22/\tau$, and we will denote this as f_{dc} , where "dc" stands for decorrelation. Thus

$$f_{dc} \simeq \frac{0.22}{\tau} \sim 0.22\ell^{-\frac{2}{3}} \varepsilon^{\frac{1}{3}}, \tag{7.60}$$

where ℓ can be taken to be of the order $\lambda/2$.

Contrast this to the contribution due to fluctuating motions, which contribute to scales of the order of the buoyancy scale, L_B. In this case, we have already seen in Equation (7.44) that if we take $\mathcal{F}$ as about 0.5, then

$$\varepsilon \simeq 2.2 \frac{v_{RMS}^3}{L_B}. \tag{7.61}$$

Then the half-power half-width of the spectrum due to the fluctuating motion of the scatterers is given by

$$f_{\frac{1}{2}\text{fluct}(m)} \simeq 0.9 \left(\frac{2}{\lambda}\right) \varepsilon^{\frac{1}{3}} L_B^{\frac{1}{3}}. \tag{7.62}$$

Hence the ratio of spectral half-widths due to the eddy motions and the decorrelation time of the eddies is

$$\frac{f_{\frac{1}{2}\text{fluct}(m)}}{f_{dc}} \sim 4 \left[\frac{L_B}{\lambda/2}\right]^{\frac{1}{3}}. \tag{7.63}$$

Physically, this arises because the spectral width associated with the scatterer movement is related to the buoyancy scale L_B (since we have seen that this width is due to the integrated effect of all scales up to L_B), whilst the decorrelation time depends only on the scale of the scatterers.

For a typical case with $\lambda/2$ equal to 3 m, and L_B equal to say 200 m, the ratio is about 16. Since the total spectral width due to these two components combined is equal to the square root of the sum of the squares, the correction due to the decorrelation time in this case would be only a fraction of a percent. Thus, provided the buoyancy scale is greater than the Bragg backscatter scale by a few times, the decorrelation time of the eddies is only a minor correction to the spectral width and can usually be ignored.

Nevertheless, there are times when the dissipation time of the eddies becomes a major factor in determining the spectral width. An example of this occurs in association with polar mesosphere summer echoes, as discussed by *Hocking and Röttger* (1997).

It was mentioned earlier that information about the level of turbulence also exists in the correlation functions and can be obtained from the full correlation analysis technique using spaced antennas. Indeed, one of the output parameters of the analysis is a parameter which is usually denoted as $T_{\frac{1}{2}}$. It represents the correlation function half-width that would be measured with a radar which moved along the ground with the velocity of the mean wind in the scattering region. Spectral beam-broadening has been removed from this parameter, although the effects of wind-shear have not. Thus the parameter $0.22/T_{\frac{1}{2}}$ can be used in place of $f_{\frac{1}{2}\mathrm{fluct}(m)}$ in all the discussions above; the main potential problem is that there may be increased contributions from buoyancy waves if the polar diagram of the system is wide.

Provided the effects of gravity waves can be adequately separated, or even shown to be relatively unimportant, the procedure described above allows radars to be used to extract estimates of atmospheric turbulence.

It is also possible (in principle) to infer the turbulent diffusion coefficient for a turbulent layer through the relation

$$K_M = c_2\varepsilon/\omega_B^2, \tag{7.64}$$

(e.g., *Lilly et al.*, 1974; *Weinstock*, 1978a, b). The constant c_2 is quoted in the literature to have a variety of values, ranging from about 0.25 to 1.25. In earlier years, the most commonly accepted value seems to have been 0.8 (*Weinstock*, 1978a), but more recently there seems a general consensus of a value around 0.25 to 0.5 (*Fukao et al.*, 1994; *Nastrom and Eaton*, 1997a; *Roper and Brosnahan*, 2005). Ideally it is also necessary to know the Brunt–Väisälä frequency averaged over the turbulent layer, but unfortunately it is not always possible to find this. Some authors use climatological values, but it is better to use radiosonde determinations where possible.

A word of warning must be given here, because Equation (7.64) assumes that the turbulence is homogeneous throughout the atmosphere. In fact, turbulence is often spatially intermittent, with regions of laminar flow separating regions of turbulence. This fact alters the process of deriving diffusion coefficients greatly (see *Hocking*, 1999a). We will examine this concept again later in the chapter.

The preceding discussions have concentrated solely on the so-called "spectral-width" method of turbulence determination. There are other methods, and one of these is based on measurements of absolute backscattered powers. We shall therefore now turn our attention to the information that can be deduced from measurements of backscattered powers.

7.4 Power measurements

One simple but effective method for deducing information about the scatterers is to record the backscattered power. In many experiments, powers are compared in a relative way; for example, power variations as a function of time and height are usually studied in most experiments. Even this simple process can give useful results, but it is even more effective if the radar can be calibrated in an absolute sense. Calibration requires some careful work by the user, but if it is done, it is then possible to convert the measured powers to effective reflection coefficients, backscatter cross-sections, or perhaps estimates of the turbulence intensity. (The parameter actually calculated depends largely on the scattering mechanism, and we will consider ways of determining this shortly.) Such calibration not only allows better comparisons to be made world-wide, but also allows better comparison with theory. It has already been discussed in Chapter 5. We will now consider applications of such calibrations.

In order to show how this calibration can be applied, it is a useful exercise to look in more detail at the mathematical formulation of the scattering process. We will begin by considering the simplest case, namely that of reflection from stratified steps.

7.4.1 Modeling the reflection and scattering processes

Horizontally stratified structure

In Chapters 3 and 4, we touched on the issue of the convolutive nature of scattering from the atmosphere. In Chapter 3, we considered the situation largely qualitatively, although Figure 3.28 was fairly detailed. In Chapter 4, we derived the expression for the interaction between a pulse and the scattering/reflecting atmosphere, and showed the exact nature of the convolution. In the sections following, we will demonstrate some extra subtleties about some special issues, and in particular the nature of efficiency as a function of scale.

Consider first, and for simplicity, a horizontally-stratified atmosphere which has fluctuations in the refractive index in the vertical direction but none horizontally. In fact, we will see later that this is not such an unreasonable model and has some real applicability in the atmosphere. A pulse of the form $g_1(t - z/c)\cos[\omega_c(t - z/c)]$ is transmitted, where $f_c = \omega_c/(2\pi)$ is the carrier frequency. At $z = 0$, this is a pulse which varies in time as $g_1(t)\cos(\omega_c t)$. This can be written as $g(\xi) = g_z(\xi)\cos(k_{\mathrm{eff}}\xi)$ where $k_{\mathrm{eff}} = 2\omega_c/c = 4\pi/\lambda$ (λ being the radar wavelength) and $\xi = ct/2$ is a length coordinate which closely matches the height of the scatterers (e.g., *Hocking and Röttger*, 1983). If the refractive index fluctuations are described by $n(z)$, then the radiowave reflection coefficient profile is given approximately by $r(z) = \frac{1}{2}(\mathrm{d}n/\mathrm{d}z)$ (e.g., Section 3.11.1, and *Hocking and Vincent*, 1982a). The reflected complex amplitude as a function of height is then given by

$$a(z) = \left\{ \frac{r(z)}{z} \right\} \otimes g(z), \tag{7.65}$$

where $\otimes$ stands for convolution, as shown in Equation (4.21) and as noted in several references (e.g., *Hocking and Röttger* 1983, 1997, and references therein). (This expression is accurate for VHF scatter, although if absorption is high or the pulse is significantly dispersive, more careful approaches are necessary, such those given by *Hocking and Vincent* (1982b), or even full-wave equations (e.g., *Budden*, 1965)). For uniformity with earlier discussions, we will persist with writing the reflection coefficient profiles as $r(z)$, (despite the possible conflict with the more common use of r as range) and will consider only the one-dimensional case with height as z.

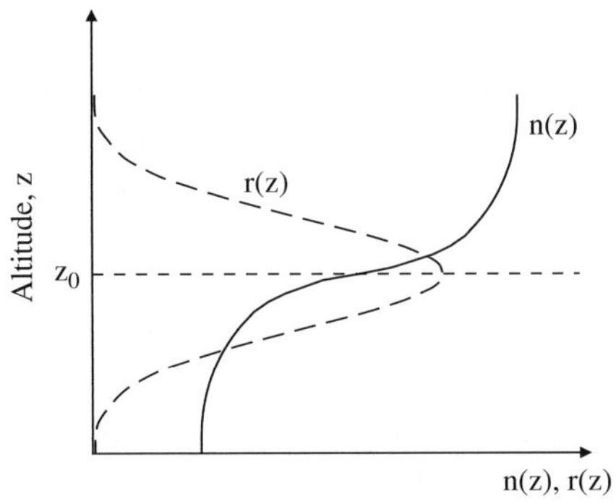

Figure 7.16 A smoothed "step" in refractive index as a function of height (solid line). The broken line shows the gradient of the step as a function of height. The gradient has a Gaussian profile.

To begin, it is of interest to examine what happens when reflection occurs from a single step of some finite thickness. The easiest step to deal with is one of the type with

$$r(z) \propto \mathrm{e}^{-\frac{(z-z_0)^2}{d^2}}, \qquad (7.66)$$

as shown in Figure (7.16). In this case, the refractive index profile is a step of finite thickness. It is continuous, continuously differentiable, contains no sharp changes, and is easily dealt with mathematically. Note that although d is a measure of the step depth, it is probably not the best measure of this depth. A better measure of the step depth might be the distance between the two points where the reflection coefficient falls to one half of its maximum value, or

$$d_\Xi = 2\sqrt{\ln 2}\, d. \qquad (7.67)$$

The choice of the variable d_Ξ is consistent with Figure 3.28.

The convolution can be done numerically, but it is instructive to examine the process using a slightly different approach. From Fourier theory (e.g., *Bracewell*, 1978)), the convolution can be evaluated by Fourier transforming each of $g(z)$ and $r(z)$, multiplying the respective Fourier transforms $G(k)$ and $R(k)$, and then re-Fourier transforming the product. This process is considered in some detail by *Hocking and Röttger* (1997) and was presented in this book in Figure 3.28 in Chapter 3. We simply note here that the Fourier transform $G(k)$ of $g(z)$ exists in two narrow bands centered at $k_c = \pm 4\pi/\lambda$, while $R(k)$ is a function of Gaussian form centered at $k_c = 0$. The narrower the step (smaller d_Ξ), the wider the function $R(k)$, as illustrated in Figure 3.28. Hence the product of the Fourier transforms is larger in amplitude, and has a larger area under the function, when the step is narrower. In fact, the peak amplitude of the product is

$$A_p(k_c) \propto \mathrm{e}^{-\frac{\pi^2 d_\Xi{}^2}{\ln 2\lambda^2}}, \qquad (7.68)$$

(*Hocking and Röttger* (1997), equation (16)). Clearly, once d_Ξ exeeds λ, the backscattered power is very small. In fact, even if $d = \lambda/4$ ($d_\Xi = 0.42\lambda$), the reflected amplitude is 0.08 times that for a step of zero width (i.e., a sharp discontinuity). The power will therefore be reduced by 22 dB compared to a single step. Many authors have taken this to infer that only steps much less than about a quarter wavelength in thickness will ever be seen by radar, and this may well be true for, say, MF radars. However, with coherent integration and the greater sensitivity of modern radars, particularly VHF radars, it is not so easy to adopt this argument; VHF radars can often see such steps even if reduced in efficiency by 20 dB, and they are indeed capable of detecting layers with a depth d of about $\lambda/4$. However, it is true that beyond this depth, the efficiency falls off remarkably quickly; for example, if $d = \lambda/2$ ($d_\Xi = 0.84\lambda$), the power is reduced by 80 dB, and even VHF radars would not normally detect such a step. *Hocking* (1987a) has discussed this point and suggests that some "specular reflectors" seen by VHF radars have typical depths with $2d$ = 3–4 m, or d_Ξ between 2.5 and 3 meters; in other words, the steps are right on the edge of the region of detectability.

It needs to be emphasized that we have taken care to avoid using a "step" which includes any discontinuities in any of its derivatives. If such a discontinuity (in either the function or any of its derivatives) is included, it will indeed produce strong backscatter, but viscosity and diffusion will ensure that in the real world such "Heaviside" steps cannot exist. Our representation is a physically realistic one.

Another useful model is that of "Fresnel scatter," a model known for many years in D-region MF studies, (e.g., *Austin et al.*, 1969; *Gregory and Vincent*, 1970; *Manson et al.*, 1969). It was given renewed popularity by *Gage et al.* (1981) in respect to VHF studies. In this model, horizontal stratification is again assumed, but $n(z)$ is assumed to vary randomly, so the atmosphere can be thought of as a series of thin slabs sitting atop each other, each with slightly different refractive indices. Despite some initial controversy, it is relatively easy to show that in this case the backscattered power is proportional to the pulse length (*Hocking and Röttger*, 1983). If one includes the decrease in reflected power as a function of height z, then one finds that the power received by a radar takes the form

$$P_{Rx} = \frac{\alpha^2 P_{Tx} A_e^2}{4\lambda^2 z^2} \left[F_{\text{spec}}(\lambda)\overline{M_n}\right]^2 (\Delta z), \tag{7.69}$$

where P_{Rx} is the received power, α is the array efficiency, P_{Tx} is the peak transmitted power, A_e is the array effective area, λ is the radar wavelength, z is the height of reflection, $\overline{M}$ is the mean generalized refractive index gradient, and $F_{\text{spec}}(\lambda)$ is a "calibration constant" which must be determined empirically for each radar. The term Δz represents the radar pulse-length. In the case that $\overline{M_n}$ varies substantially within one pulse-length, this formula needs some modification, as described by *Hocking and Röttger* (1983). Later developments of this model have been discussed by *Gage et al.* (1985) and *Green and Gage* (1985).

Three dimensional structures

The next extension to these models is to allow the scattering medium to have non-constant structure in the horizontal direction as well. An example might be fully developed isotropic turbulence, in which the refractive index has random fluctuations caused by the turbulent velocity field. In this case, the theory for relating the backscattered signal to the turbulence intensity has been fairly well developed (*Hocking*, 1985; *Ottersten*, 1969a, b; *Van Zandt et al.*, 1978). The backscattered power depends not only on the intensity of the turbulence, but also on the mean refractive index gradient in which the turbulence exists. In the mesosphere, the latter term is largely determined by the electron density gradient, in the stratosphere by the temperature gradient, and in the troposphere by the temperature and humidity gradients. Expressions for these potential refractive index gradients are given below. Expressions for evaluation of the turbulence intensity from measurements of the absolute backscattered power will be left until after the following section on calibration of a radar.

In the un-ionized atmosphere, and for centimeter and meter radio waves, the potential refractive index gradient is given by (*Tatarski* (1961), with update in the constants from *Larsen and Röttger* (1982)):

$$M_n = \frac{-77.6 \times 10^{-6}}{T^2} p \left[1 + \frac{15500}{T} q_{wp}\right] \left[\frac{dT}{dz} + \Gamma_a - \frac{7800}{1 + \frac{15500}{T} q_{wp}} \cdot \frac{dq_{wp}}{dz}\right]. \quad (7.70)$$

This equation was given as (3.288), but is repeated here for ease of access. The various variables were defined in that equation.

In the ionosphere, the relevant potential refractive index gradient is given by *Hocking* (1985) as

$$M_e = \frac{\partial n}{\partial N} \left[N \frac{\omega_B^2}{g} - \frac{\mathrm{d}N}{\mathrm{d}z} + \frac{N}{\rho} \frac{\mathrm{d}\rho}{\mathrm{d}z} \right], \quad (7.71)$$

where N is the electron density, n is the refractive index, and ρ is the neutral density of the atmosphere. The Brunt–Väisälä angular frequency is represented by ω_B^2, and g represents the acceleration due to gravity (also see Appendix A, Section A.6). Note that (7.71) can also be written as

$$M_e = \frac{\partial n}{\partial N} \left[\frac{N}{\gamma P} \frac{\mathrm{d}P}{\mathrm{d}z} - \frac{\mathrm{d}N}{d\mathrm{d}z} \right], \quad (7.72)$$

(e.g., *Thrane and Grandal*, 1981) where γ is the ratio of specific heats at constant pressure and constant volume.

For a VHF radar,

$$\frac{\partial n}{\partial N} = \frac{1}{2} \pi^{-1} r_e \lambda^2, \quad (7.73)$$

where r_e is the classical electron radius. At HF and MF, the relation between N and n becomes more complex (see Equations (3.125) and (3.128) in Chapter 3).

Let us now turn our attention to the subject of calibrating the radar, so we may then see how to use the above expressions to determine turbulence intensities.

7.4.2 Converting received powers to backscatter cross-sections

As discussed at the beginning of this section, in order to calculate parameters like backscatter cross-sections of the scatterers or the reflection coefficients of the reflectors,

7.4.3 Determination of turbulence intensities from measurements of received power

Once it can be ascertained that turbulence is the main cause of the radiowave scatterers, it is possible to convert the received powers to parameters which describe the turbulence. One key parameter is the "(potential) refractive index structure constant," usually denoted as C_n^2. Formulas for its determination were presented in Chapter 5. Here, we briefly repeat these expressions.

If the turbulence obeys the classical Kolmogoroff inertial spectrum, then the spectrum of refractive index fluctuations is given by (*Tatarski*, 1961, 1971)

$$\Phi_n(k_x, k_y, k_z) = 0.033 C_n^2 \mid \underline{k} \mid^{-\frac{11}{3}}, \tag{7.78}$$

where a normalization has been chosen such that $\int\int\int_{-\infty}^{\infty} \Phi(\underline{k}) d\underline{k} =< n^2 >$. Thus C_n^2 is a parameter which indicates the level of refractive index fluctuation. C_n^2 was derived in the equations leading up to Chapter 5, Equation (5.138), viz.

$$C_n^2 \approx 66.4 \frac{P_{Rx} r^2 \lambda^{\frac{1}{3}}}{P_{Tx} A_{Rx} L_{Tx} L_{Rx} \ell_{rr}}. \tag{7.79}$$

Where P_{Rx} is the received power, r is the range to the target, λ is the radar wavelength, P_{Tx} is the transmitted power, A_{Rx} is the actual receiver area, L_{Tx} is a number less than 1 and greater than zero which takes account of losses in transmission power, L_{Rx} is an efficiency term for reception, and ℓ_{rr} is the pulse resolution. Note that in contrast to (5.138), the term α no longer exists as it has been absorbed into L_{Rx}.

Appropriate relations can also be easily derived for the case in which the transmitter and receiver are separate systems, as discussed in Chapter 5.

C_n^2 is a useful parameter, but an even more useful one is, of course, the kinetic turbulent energy dissipation rate, ε. It is possible to relate C_n^2 to ε in the following way.

Starting from *Tatarski* (1961) (p44, equation 3.19), we have

$$C_n^2 = a^2 N' \varepsilon^{-\frac{1}{3}}, \tag{7.80}$$

where N' is a parameter defined by

$$N' = K_n M_n^2 \tag{7.81}$$

for a stratified environment. The constant a^2 has been measured to be about 2.8. Using the definition of the turbulent Prandtl number $P_K = K_M / K_T$ (where K_T is the diffusion coefficent of heat), defining $\alpha' = P_K^{-1}$, taking $K_n = K_T$ for now, and using the relation (7.64), viz.

$$K_M = c_2 \varepsilon / \omega_B^2, \tag{7.82}$$

we see that

$$\varepsilon = \left[\frac{C_n^2 \omega_B^2}{a^2 \alpha' c_2 M_n^2} \right]^{\frac{3}{2}}. \tag{7.83}$$

A similar expression was derived by *Van Zandt et al.* (1978) and noted by *Hocking* (1985), although a slightly different proof was used in the second case, with the result that

$$\varepsilon = \left[\frac{C_n^2 \omega_B^2}{a^2 \alpha' M_n^2 \left[\frac{R_{i(c)}}{b^{\frac{2}{3}}} \right]} \right]^{\frac{3}{2}}, \tag{7.84}$$

where $R_{i(c)}$ is the critical Richardson number at which turbulence should develop, and b is yet another constant relating the kinetic energy dissipation rate to the mean windshear. In fact, *Van Zandt et al.* (1978) and *Hocking* (1985) took $b = 1.0$, so b did not appear explicitly in their expressions; with hindsight, this assumption was not wise. The first expression (7.83) is derived in a more fundamental way, and requires fewer assumptions than the second (7.84), and it is better to use the former.

Hocking (1992) further developed this formula, to more properly consider the true values for a and b, and in fact showed that there was an additional Richardson number dependency. The final expression was

$$\varepsilon = \left[\gamma \frac{C_n^2 \omega_B^2}{M_n^2} \right]^{\frac{3}{2}}, \tag{7.85}$$

where

$$\gamma = \frac{3}{22} \frac{\mid 1 - R_i \mid}{\mid R_i \mid}. \tag{7.86}$$

Other expressions have been assumed for γ and are discussed in some detail by *Hocking and Mu* (1997). Some of these alternative expressions recognize that the turbulent Prandtl number may differ from unity and some are developed in terms of the flux Richardson number instead of the gradient Richardson number. The correct choice is still debatable. Further discussion can be found in Chapter 11.

An extra complication arises if the turbulence does not fill the radar volume, and indeed this often appears to be the case. It appears that in the stratosphere and mesosphere, turbulence occurs in relatively thin layers with thicknesses ranging from a few tens of meters to perhaps a kilometer or so, but generally of the order of 100 m. At any one instant, only a small fraction of the radar volume contains turbulence, and this should be taken into account when calculating ε. In other words, the radar measurement of C_n^2 is actually too small by a factor F_t, where F_t is the fraction of the radar volume that is filled with turbulence at any one time. Thus one normally calculates

$$C_n^2(\text{turb}) = C_n^2(\text{radar})/F_t, \tag{7.87}$$

where C_n^2(radar) is the value determined from the radar measurements. *Van Zandt et al.* (1978, 1981) have developed models for the variation of F_t as a function of atmospheric conditions, enabling estimates of ε to be made. Furthermore, one is often interested in the mean value of ε averaged over the radar volume, so Van Zandt *et al.* suggested calculating the quantity

$$\overline{\varepsilon} = F_t \varepsilon_{\text{turb}}. \tag{7.88}$$

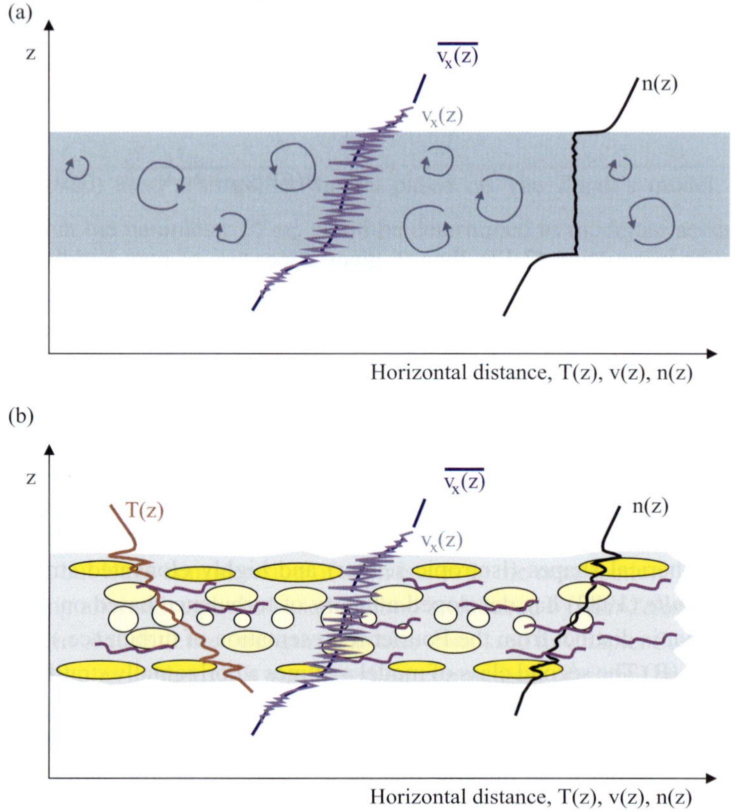

Figure 7.17 Idealized views of two models for turbulence in the atmosphere. In (a), the partially closed circles represent the rotation of eddies, and the top and bottom of the shaded region represent a relatively sharp transition between the turbulent region and the non-turbulent air. Assumed velocity fluctuations are shown by the irregular fluctuations to the left, and the mean profile of the velocity is shown by the heavier line. The proposed refractive index variation is shown by the solid line on the right. In (b), it is proposed that the turbulent layer does not cease so abruptly in the vertical direction and progresses from isotropic turbulence near the center of the layer to greater and greater degrees of anisotropy as the edge of the layer is approached. The eddies are represented by ellipses that are more oblate towards the edges. Mean and small-scale temperature, velocity, and refractive index variations are shown by the various solid lines.

Peltier et al., 1978). For the purposes of this book, however, these models are simply noted as a type of extension to the simple models proposed above.

Another model which may give a physical basis to model B is the proposal that the specular reflectors might be damped gravity waves (e.g., *Hocking*, 1987a; *Van Zandt and Vincent*, 1983, and references therein) or even viscosity waves, the latter being capable of existing at very short wavelengths. This latter model was first introduced by *Hooke and Jones* (1986) with regard to formation of turbulent layers with thicknesses of a few tens of meters in the boundary layer. It was then extended to describe specularly-reflecting layers at even smaller vertical scales by *Hocking et al.* (1991), *Hocking* (1996b) and *Hocking* (2003a). Note that criticisms of the existence of these

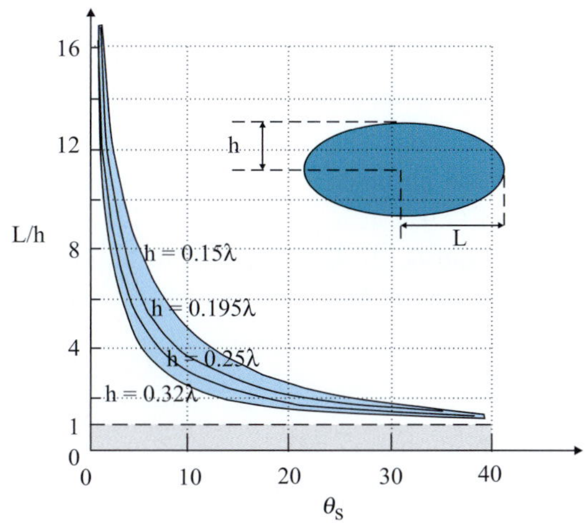

Figure 7.18 Relation between θ_s and the length-to-depth ratios of idealized scattering eddies.

waves on mathemetical grounds by *Fritts and Alexander* (2003) are unquestionably in error, since wave-solutions of the diffusion equation are encountered in many areas, including surface propagation of electromagnetic waves along conducting media and heat waves in geophysics. There is no *mathematical* restriction to the existence of these waves.

Having now established that both models have some physical basis, let us concentrate on the simpler models, since these form an excellent basis for later discussion of any of the more complex models.

With regard to model A, it should be noted that θ_s gives a direct measure of the length-to-depth ratio of the scatterers. Figure 7.18, from *Hocking* (1987a), shows this relationship.

What techniques, then, can be used to determine the nature of these scatterers?

7.5.1 Experimental techniques to determine the nature of the scatterers

The following section describes a variety of techniques which may be used to determine information about the nature of the scatterers, and some of the results obtained so far are discussed. The list is not, however, exhaustive.

Methods utilizing different beam configurations

One of the simplest methods to investigate the aspect sensitivity of the scatterers is to simply point a narrow beam vertically and then at several off-vertical angles. The variation in power P as a function of beam tilt angle θ is then related to θ_s. It was shown earlier (Equation (7.32)) that

$$P(\theta_T) \propto e^{-\left[\frac{(\sin\theta_{\rm eff}-\sin\theta_T)^2}{\sin^2\theta_0}+\frac{\sin^2\theta_{\rm eff}}{\sin^2\theta_s}\right]}, \tag{7.92}$$

where θ_{eff} is defined by Equation (7.18), θ_T is the beam tilt direction from the vertical, and the polar diagram of the radar beam is assumed to be of the form $\exp\{-(\sin^2\theta)/(\sin^2\theta_0)\}$ (see Equation (7.21)).

A typical experiment which might be performed is to compare the powers received with a vertical and an off-vertical beam, and use them to deduce θ_s. Utilizing Equations (7.18) and (7.92), it is possible to derive the following simple relation between $P(\theta_T)/P(0)$, θ_T and θ_0. If R_θ is defined to be $\ln\{P(0)/P(\theta_T\}$ (or $R_\theta = 0.23026 R_{\text{dB}}$, where R_{dB} is the ratio of $P(0)/P(\theta_T)$ expressed in dBs), then

$$\theta_s^2 \simeq \frac{\theta_T^2}{R_\theta} - \theta_0^2. \tag{7.93}$$

This is a simplification of (7.33).

Typical variations of $P(\theta)$ show an approximately Gaussian fall-off out to about 5 to 10°, and then an approximately constant value beyond this point, indicating possibly isotropic turbulence, with more anisotropic scatterers either embedded or nearby (e.g., *Doviak and Zrnić*, 1984; *Hocking et al.*, 1990). Typical values of θ_s are often in excess of 8° in the troposphere, whilst in the stratosphere at VHF, values can be as small as 3 to 4°. The tropopause can sometimes be a region of highly anisotropic scatter. Figure 7.19, from *Hocking et al.* (1986), summarizes some measurements made with the SOUSY radar in Germany (after correction for an error in the original paper). Note the tendency for the scatterers to become anisotropic as the lower stratosphere is approached from below, and then to become slightly more isotropic in the high stratosphere.

Figure 7.20 shows a contour plot illustrating high anisotropy of scatterers at or slightly below the tropopause, and then a rapid development of turbulence immediately above

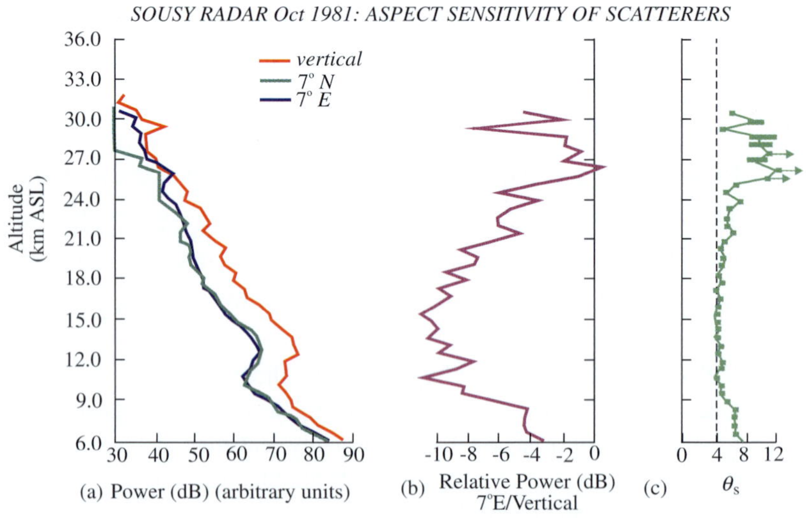

Figure 7.19 Plots from the SOUSY radar: (a) Mean power as a function of height; (b) Ratio of off-vertical to vertical beam powers as a function of height; and (c) Corresponding θ_s values.

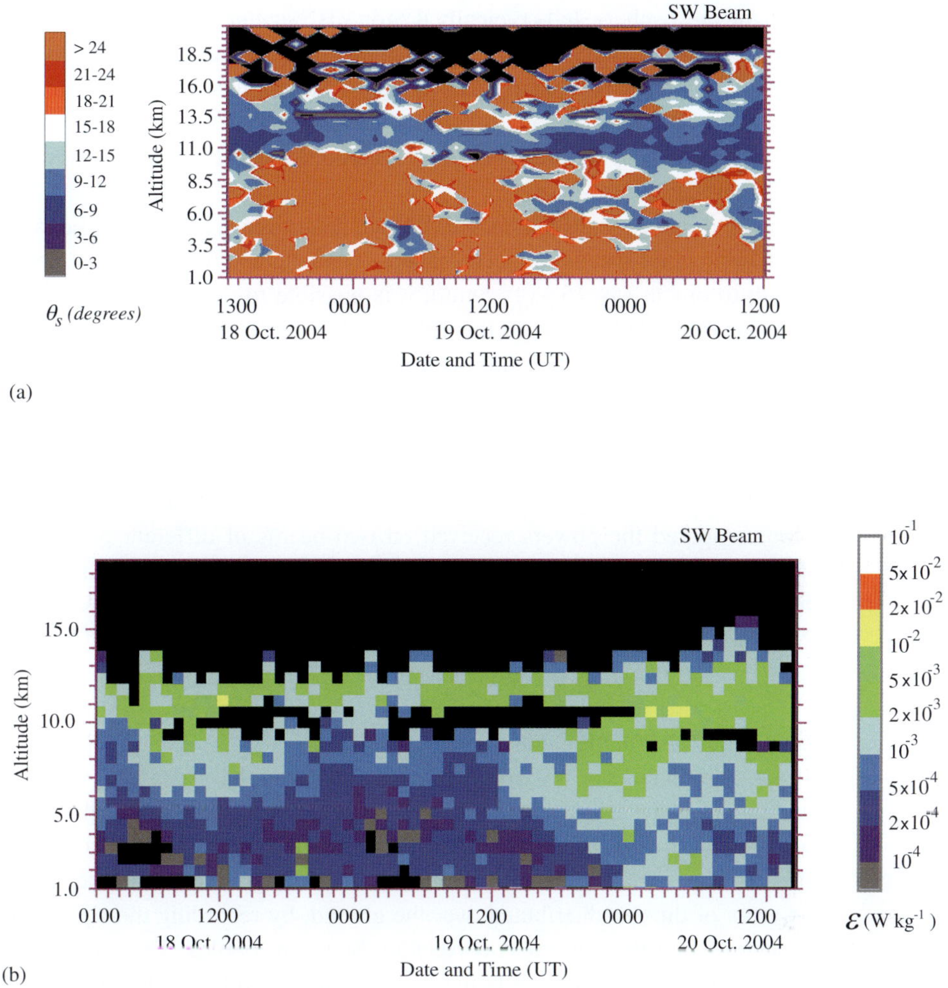

Figure 7.20 (a) θ_S as a function of height and time for the Alwin VHF radar in Northern Norway. A layer of significant anisotropy is evident at about 10–12 km altitude. (b) Turbulence kinetic energy dissipation rates as a function of height and time, covering a similar time frame to the graph shown in (a). Note a layer of "missing data" at 10–12 km altitude, which corresponds to highly specular echoes, and then a rapid increase to strong turbulence just above this region. These data were provided by Dr. Werner Singer of the Institute for Atmospheric Physics in Germany.

this. Development of an upper layer of turbulence like this is consistent with the model proposed theoretically by *Van Zandt and Fritts* (1989), for situations where gravity waves pass from a region of lower stability into a region of high stability.

We emphasize here, however, that the tropopause is not always this specular in nature. Sometimes it can be a turbulent region, and specularity can be weak or non-existent. We will discuss ways to identify the tropopause later in this chapter and again in the chapter on meteorological phenomena.

In the mesosphere, θ_s is typically 4 ° for VHF scatter below 75 km, although on occasion isotropic scatter is also seen. Above 80 km, VHF measurements give θ_s to be about 6 to 8 °. At MF, θ_s is typically 2 to 5 ° below 80 km, increasing to about 8 to 15 ° above 80 km (e.g., *Lindner*, 1975a, b; *Vincent and Belrose*, 1978). *Reid* (1990) has summarized the various mesospheric measurements, and one of his figures was shown as Figure 2.8 in Chapter 2.

An alternative means that may be used to determine θ_s is to utilize Equation (7.18). By comparing wind speeds deduced using the DBS method for a beam pointed at say 5 ° off-zenith to one at say 15 ° off zenith, it is possible to deduce θ_s from (7.18), assuming that the value deduced with the 15 ° beam is the true wind speed. *Hocking* (2001b) has used an approach like this with a radar at Resolute Bay, but he used radiosonde data as a reference for the winds. An alternative is to use spaced antenna methods to determine the true wind speed. Then comparisons with the DBS measurements may allow determination of θ_s.

Another interesting determination of θ_s was made by *Vincent and Belrose* (1978), who compared the powers received on two beams of different polar diagram widths and then used the resultant ratios of powers to determine θ_s. The method yielded results consistent with determinations made by other techniques discussed in this section.

Spatial correlation methods

If one illuminates the sky from a transmitting array that has a very wide polar diagram and monitors the electric field received at the ground, then the variation of electric field as a function of position is simply the diffraction pattern of the scattering irregularities. The spatial autocorrelation function over the ground can be determined by using an array of dipoles distributed over the ground, by recording the signal on each dipole separately and then by cross-correlating between dipoles. The spatial autocorrelation function so produced is simply the Fourier transform of the effective polar diagram (i.e. the combined polar diagrams of the radar beam and the scatterers). If the e^{-1} width of the effective polar diagram is θ_{sb}, then the spatial lag at which the amplitude of the complex autocorrelation function falls to 0.5 is approximately $15.2/\theta_{sb}$ radar wavelengths, where θ_{sb} is expressed in degrees (e.g., *Briggs*, 1992). (This corrects an earlier estimate of $12.0/\theta_{sb}$, which arose from an incorrect derivation in *Briggs and Vincent* (1973), and then was propagated through several papers in the literature (e.g., *Hocking*, 1987a; *Hocking et al.*, 1989).)

Thus a useful technique for determination of the polar diagram of backscatter is to produce the spatial autocorrelation function in the manner described and then Fourier transform it. Such a technique has been utilized by *Lindner* (1975a, b) in order to study the aspect sensitivity of mesospheric scatterers at an MF frequency of 1.98 MHz. For example, Lindner found typical values for θ_s of about 2 to 5 ° below 80 km, and 10 to 15 ° above. *Lesicar et al.* (1994); *Lesicar and Hocking* (1992) have presented similar analyses. These results are consistent with other observations using beam-swinging techniques (e.g., *Hocking*, 1979).

Spectral methods

It was noted earlier in regard to discussions about extraction of turbulence from spectra that in many cases the main contribution to the spectral width was spectral-broadening due to the finite width of the polar diagram of the radar beam. At the time, this was a nuisance, but now it can be turned to good effect. The effective polar diagram is the product between the polar diagram of the radar and the backscatter polar diagram of the scatterers. As seen in Equation (7.28), if θ_{sb} is the e^{-1} half-width of the effective polar diagram (i.e. the product of the backscatter polar diagram and the radar beam polar diagram), then adapting (7.28), we produce

$$\frac{1}{\sin^2 \theta_{sb}} = \frac{1}{\sin^2 \theta_0} + \frac{1}{\sin^2 \theta_s}. \tag{7.94}$$

But from equation (7.34) the beam-broadening of the spectral width is

$$f_{\frac{1}{2}b} = \frac{2}{\lambda}(1.0) \mid v_{\text{hor}} \mid \theta_{\frac{1}{2}}. \tag{7.95}$$

The total spectral half-power half-width is given approximately by

$$f^2_{\frac{1}{2}\text{expt}} = f^2_{\frac{1}{2}b} + f^2_{\frac{1}{2}\text{fluct}}, \tag{7.96}$$

if we ignore the contribution due to wind-shear. The contributions due to wind-shear are generally less in magnitude than the beam-broadening contribution, and although caution is advised in ignoring this term, it is often acceptable, at least for the experiment proposed below.

Then we can apply our experimentally measured spectral widths to place upper limits on θ_s. That is, if we calculate

$$\theta'_{\frac{1}{2}} = \frac{\lambda}{2} \frac{f_{\frac{1}{2}\text{expt}}}{\mid v_{\text{hor}} \mid}, \tag{7.97}$$

this value is a useful upper limit to $\theta_{\frac{1}{2}\text{eff}}$, the half-power half-width of the combined polar diagram of the scatterers and the radar beam. In the case that it can be shown that $f_{\frac{1}{2}b} \gg f_{\frac{1}{2}\text{fluct}}$, as often happens, then $\theta'_{\frac{1}{2}}$ is a good estimate of $\theta_{\frac{1}{2}\text{eff}}$. Then $\theta_{sb} = \theta_{\frac{1}{2}\text{eff}}/\sqrt{\ln 2}$, and Equation (7.94) can be used to deduce upper limits to θ_s. In the special case that a relatively wide beam is used, so that $\theta_0 \gg \theta s$, $\theta_{sb} \approx \theta_s$.

The above principles have been used (e.g., *Hocking et al.*, 1986; *Hocking*, 1987a; *Hocking*, 1987b) to make estimates of backscatter polar diagram half-widths. The method of using fading times as a crude indicator of "specularity," as done, for example, by *Rastogi and Röttger* (1982) may also be considered as a primitive special case of this method, although that procedure does not really pay proper consideration to the role of the mean wind in determining the fading time through beam-broadening. *Woodman and Chu* (1989) have used similar techniques. However, rather than just using the spectral width and assuming Gaussian polar diagrams, as done here, they have used the whole spectrum and the one-to-one correspondence between the polar diagram of backscatter and the spectrum to determine additional detail about the actual shape of the polar diagram of backscatter and so the irregularities themselves. However, their theory was derived using a narrow-beam assumption, and then was applied to a wide-beam situation, which is not entirely valid. An improved discussion of a similar model, which more properly represents a wide beam, has been presented by *Briggs* (1992).

A procedure like this is very useful if there are several types of scatterers in the beam. For example, if scatterers and reflectors described by models A and B both exist in the same radar volume, the spectrum will not be Gaussian, but will comprise two portions: a narrow central component corresponding to the specular reflectors, and a wider component corresponding to the model A scatterers. As it turned out, *Woodman and Chu* (1989) saw no evidence of model B reflectors, but this is likely to be because their spectra were averaged over 45 min, whilst specular reflectors, if they exist, are likely to be relatively short-lived.

Indeed, evidence for the coexistence of the two types of scatterers in the same region of space has been given by *Hocking* (1987a) and is illustrated in Figure 7.21. The data are presented because they show yet another useful means of determining information about the scatterers, as well as making the point that both specular reflectors and turbulent scatterers do seem to coexist.

These data were obtained using a hybrid of the beam-swinging and spectral approaches. Two beams were used, one vertical and one off-vertical. A strong signal of very narrow width was seen with the vertical beam, but nothing else, whereas on the off-vertical beam two separate contributions to the spectra were seen; first, a broader component corresponding to isotropic backscatter received through the main lobe of the beam, and secondly, the same narrow spectrum as seen with the vertical beam. Clearly,

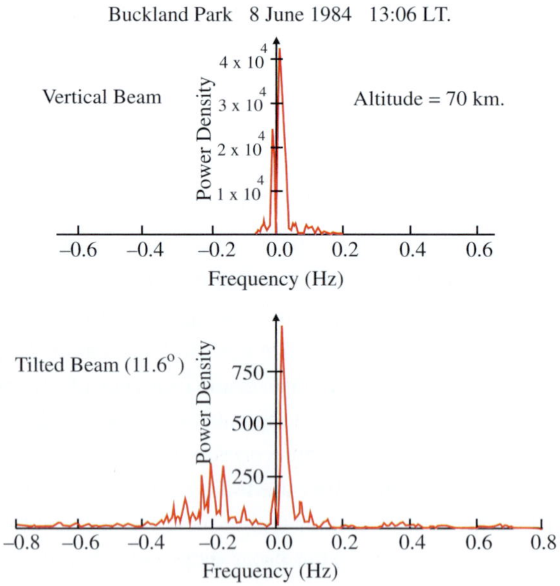

Figure 7.21 Demonstration of the near-simultaneous existence of both specular reflectors and turbulence (see text for details).

the second component was due to leakage from overhead. Comparison of the powers in the specular component observed with the narrow beam and the more isotropic component show that the specular component is some 70 times stronger. The model discussed in Figure 7.17(b) may apply in some cases, but certainly does not here, as it is unlikely that the anisotropic scatterers at the layer edges would be so much stronger than their counterparts in the center of the layer. Thus, this figure does indeed suggest the coexistence of both models, whilst at the same time demonstrating yet another useful technique to determine the aspect-sensitivity and nature of the scatterers.

Amplitude distributions

The preceding techniques have been designed to make measurements of θ_s and are particularly powerful if model A is valid. However, there is a useful method which allows the validity of model B to be tested, and which has been used with varying degrees of success in recent years. This is the use of amplitude distributions (e.g., *Hocking*, 1987b; *Kuo et al.*, 1987; *Rastogi and Holt*, 1981; *Röttger*, 1980a; *Sheen et al.*, 1985; *Vincent and Belrose*, 1978; *Von Biel*, 1971, 1981, among others).

There are many variations of this technique, but only the simplest will be discussed here in order to illustrate the method. If scatter is due to an ensemble of roughly similar scatterers, as might occur in a turbulent patch, then the amplitudes of the resultant distribution will have a so-called "Rayleigh distribution" (*Rayleigh*, 1894). (These distributions were also studied in a different context in Chapter 3, Sub-section 3.8.5.) If, however, there is also a much stronger single specular scatterer in addition to these weaker scatterers, the distribution changes to a so-called "Rice distribution" (*Rice*, 1944, 1945). Figure 7.22 shows how these distributions change as the specular component is made larger. Each curve is parameterized by a parameter called the "Rice parameter"

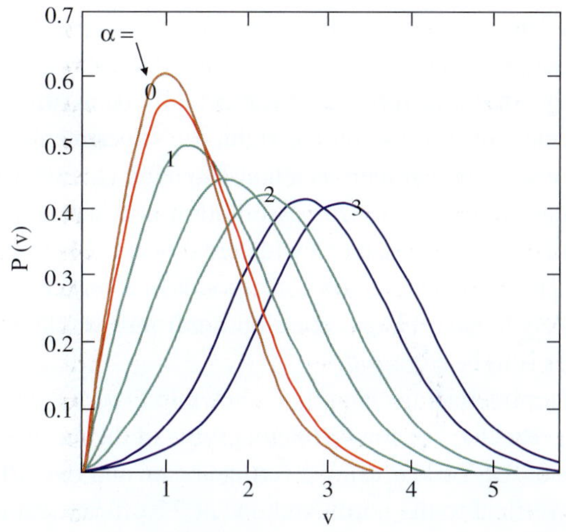

Figure 7.22 The Rice distributions for different ratios of specular to random powers.

α, which is a measure of the strength of the specular component divided by the RMS "random" component. For a Rayleigh distribution, this parameter is zero.

In principle, by making histograms of the amplitudes of the received signal and comparing them to the above curves, it is possible to determine if there is a single dominant scatterer within the radar beam. More complex variations on this process exist, including looking at the phase distributions (e.g., *Röttger*, 1980a) and using more complex distributions such as the Nakagami-M distribution (e.g., *Sheen et al.*, 1985; *Kuo et al.*, 1987). The latter generalization is particularly useful if the specular component has undulations on it and causes focusing and de-focusing of the reflected radiation.

Unfortunately, as with almost all techniques, complications exist. For example, if there is more than one specular reflector in the radar volume, then the amplitude distribution changes, and if there are more than about four, the distribution begins to look almost Rayleigh-like again. Furthermore, if one uses relatively short data sets (less than about 10 min of data), statistical effects can cause a set of scatterers that should produce a Rayleigh distribution to produce a Rice distribution, which wrongly suggests the existence of a specular component. On the other hand, geophysical variability precludes the use of very long data sets. For example, if the scattering medium were purely turbulent, we would expect a Rice parameter of zero, but the strength of the turbulence would vary significantly on time scales of 10 to 30 minutes as it grows and dies. Therefore, although data lengths of less than 10 or 5 minutes should be used, these will be statistically unreliable. To properly utilize the so-called Rice parameter, one must look at the distributions of the Rice parameter itself; the calculation of several non-zero Rice parameters is not in itself evidence for a non-Rayleigh distribution. The correct interpretation of the Rice parameter is discussed by *Hocking* (1987b), and Figure 7.23(a) shows the distribution of the Rice parameters deduced using Monte-Carlo computer simulations from a purely Rayleigh process for different data lengths. The data length is expressed as a non-dimensionalized unit by dividing by the correlation time $\tau_{1/2}$, which is the half-value half-width of the autocorrelation function of the signal. Figure 7.23(b) shows a comparison of this theoretical form with experimental measurements taken with an MF radar. In the first case (left-hand side), the experimental and theoretical values look very similar in form, so it can actually be assumed that these data were Rayleigh in character. In the case on the right, the experimental data stretch to larger values of α, demonstrating that there is a non-Rayleigh character to the data, probably indicating a specular contribution. This assumption is consistent with the beam types, since the left-hand data correspond to a wide radar beam, which would be dominated by turbulent scatter. The right-hand graph corresponds to a vertically directed narrow beam, which is more likely to have a larger contribution from specular reflectors, and so a non-Rayleigh character is to be expected.

Another interesting example is shown in Figure 7.24, which was taken from *Hocking* (1987b). This figure shows a height profile of the mean Rice parameter ($<\alpha>$) measured with the SOUSY radar using a vertical beam and two off-vertical beams, one directed at 7° off-vertical to the north, and one at 7° off-vertical to the east. Note the increase in $< \alpha >$ just above the tropopause when observing with the vertical beam, indicating the presence of a few dominant reflectors within the radar volume in the stratosphere. Note

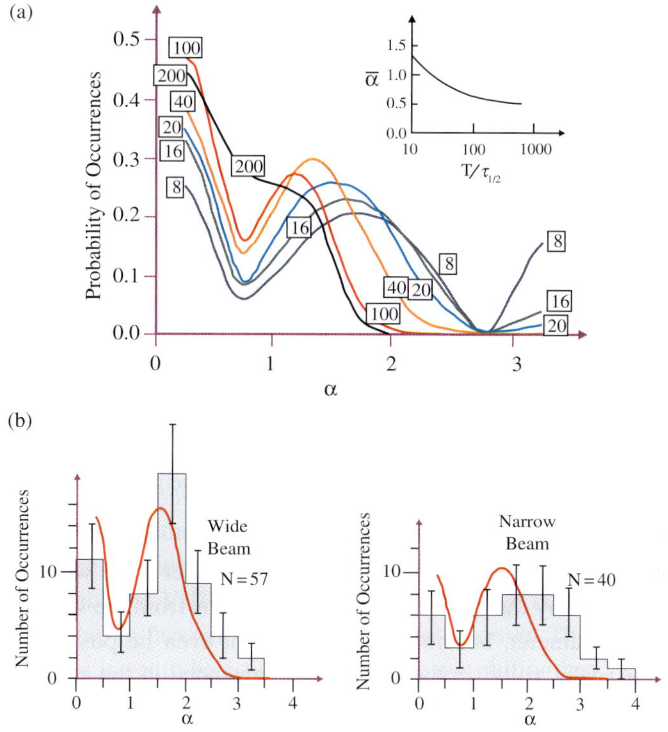

Figure 7.23 (a) Theoretical distributions of the Rice parameter that would be expected when data sets that are relatively short are used. Data lengths (expressed as a multiple of the fading time) are shown in the boxes. The inset shows how the "mean" Rice parameter changes as a function of data length. Note it is not zero until the data set becomes very very long. (b) Distribution of Rice parameters using short time series for two different beams used to observe a scattering layer observed at Buckland Park, Australia, with an MF radar (*Hocking*, 1987b). Solid curves represent the expected "$\alpha = 0$" distribution. See text for details.

also that there is still a non-Rayleigh character to the scattering process on the north beam, but on the east beam the mean Rice parameter is fairly constant with height and consistent with a Rayleigh process.

Several possible interpretations of these results are possible – the data alone do not tell the whole story, but help eliminate various options and strengthen the case for others. However, the process should involve specular reflectors in the mix. It is possible that there truly were specular reflectors, but these were tilted back and forth by waves and other events, possibly slowly tilting the reflectors more severely in the north-south direction, allowing the tilts to be large enough that they could even be seen by an off-vertical beam at times. The tilts would not need to be as large as 7° – even tilts as small as 3–4° from horizontal could allow useful reflections to be seen coming in through the edges of the beam. This could relate to the existence of gravity waves with wave-fronts that are preferentially aligned in the east-west direction, either due to local generation of the waves or fortuitous external generation of the waves at some distance to the north or south. Yet another model might be of even larger-scale partially tilted specular reflectors that orientate with normal vectors in the north-south vertical plane,

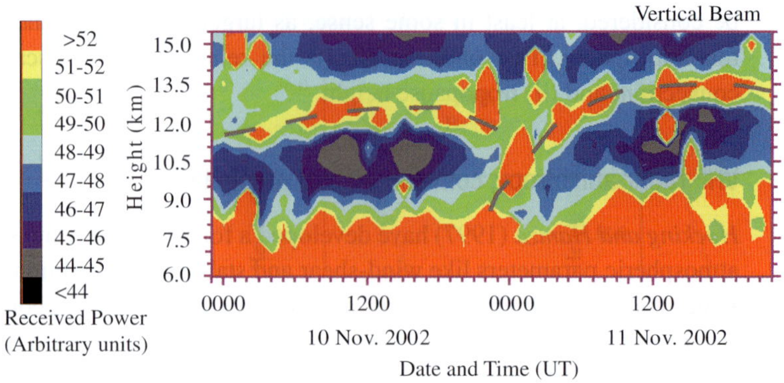

Figure 7.25 Tropopause motions during a period of complex movement. The tropopause is highlighted by the broken brown line. This might be a tropospheric fold or a high-level occluded front.

to higher aspect sensitivity or to increased turbulence depends on location and circumstances. However, the use of enhanced power on the vertical beam as a way to locate the tropopause does seem generally applicable. *Gage and Green* (1982b) have formalized a procedure to determine tropopause heights from radar data, though the method is in need of more testing.

7.7 Less easily determined target parameters

The discussion so far has concentrated on parameters that can be inferred fairly directly from the radar measurements. There are, however, other parameters which can be deduced with a little extra work. For example, *Vincent and Reid* (1983) showed how, by using two off-vertical beams, measurements of the gravity-wave and turbulent momentum fluxes could be calculated. The momentum flux is not actually a target parameter and so has not been discussed here greatly. Nevertheless, it is a parameter which affects the targets and knowledge about it is most desirable. Some further discussions about momentum fluxes will be presented in Chapter 11.

Another example is the Brunt–Väisälä frequency. Normally this is very difficult to measure by radar methods, but if the mean winds are light, then spectral analysis of the time series of *vertical* velocity measurements can be used to measure it. That is, the spectrum shows a cutoff at the Brunt–Väisälä frequency (with little organized wave activity at periods less than the Brunt–Väisälä period), and knowledge of the BV frequency in turn allows determination of the temperature gradient as a function of height (e.g., *Röttger*, 1980b, 1986). The temperature profile can then be found by integrating the temperature gradient up from the ground. This method has been implemented by *Revathy et al.* (1996) with good results, but it does appear limited to conditions of very quiet winds. This is necessary in order to avoid Doppler shifting effects associated with the gravity waves which blur out the relatively sharp edge of the gravity-wave spectrum. Some examples of the application of this method will be shown in Chapter 12.

It is also possible to determine the temperature at the regions of scatter by using the so-called "radio acoustic sounding system," or RASS. This method uses the radar to track the speed of artificially generated sound pulses, thereby allowing the atmospheric temperature to be found as a function of height. There are many publications associated with RASS, and it is not appropriate, nor is there sufficient space, to consider it in detail in this chapter. Interested readers are referred to *Tsuda et al.* (1994), and references therein, for more extended discussions. Some discussions may also be found in Chapter 10 of this book.

The atmospheric diffusion coefficient is also a parameter that can be determined to some extent with MST radars. This was already discussed in connection with Equation (7.64), where we considered a simple relation between K_M and ε. In truth, however, this can only be used to determine the diffusion coefficient within a turbulent layer (e.g., *Fukao et al.*, 1994). If mean diffusion coefficients over much larger scales are required, a different approach must be used. One approach is to use procedures as proposed by *Dewan* (1981) and *Woodman and Rastogi* (1984), who have shown how careful measurements of the temporal and spatial distribution of the occurrence of thin turbulent layers can be used to infer the mean turbulent diffusion coefficient in the stratosphere, as distinct from the diffusion coefficient within a turbulent layer. However, at larger scales, other diffusive processes may come into play, like Stokes diffusion (e.g., *Hocking and Walterscheid*, 1993; *Walterscheid and Hocking*, 1991). The various types of diffusive processes are discussed by *Hocking* (1999a). Essentially, diffusion is a more complex process than has been commonly believed, and in fact the modes of diffusion are scale dependent. Further discussion appears in Chapter 11.

High resolution studies can also be used to infer something about the nature of the scatterers; for example, *Röttger and Schmidt* (1979) used a resolution of 30 m to observe cat's-eye structures in the stratosphere, confirming that at least some of the observed turbulent layers are due to dynamical instability. *Reid et al.* (1987) have observed similar features in the mesosphere. Other studies which allow information about the nature of the scatterers to be obtained include, for example, those by *Klostermeyer and Rüster* (1980, 1981) and *Yamamoto et al.* (1987, 1988b); in these studies, relations between power bursts and buoyancy-wave oscillations were investigated.

Meteors represent another type of radar target that can be scrutinized by MST radars. The echoes are actually produced by reflection from an ionized plasma trail produced as a meteor particle enters the atmosphere, and the signals produced can be very strong. Meteor echoes are discussed separately in Chapter 10.

Another powerful method for determining the environment in which the scatterers exist is the so-called differential absorption experiment, in which powers received for backscatter from the ionospheric D-region using different polarizations are compared in order to determine the electron density as a function of height. A more detailed discussion of this method will be given in Chapter 10.

Of course, by using long time series of velocities, one can determine other characteristics of the scattering region, like the buoyancy-wave spectra, tidal amplitudes, planetary wave amplitudes, and a host of dynamical quantities. In a broad sense, one might like to think of these as "target parameters" of a sort, but they are beyond the scope of the current chapter, and some are further discussed in Chapter 11.

8 Digital processing of Doppler radar signals

The goal of atmospheric radar signal processing is to efficiently extract physically significant parameters from the radar signal with minimal bias and error. The signal received from a typical radar is *analog*, which is conceptually best recorded by paper charts and other continuous recording devices. Practical analysis of radar signals is performed on computers, however, since these allow a vast array of processing techniques and imaging methods. Hence it is normal practice to digitize the recorded signals, as already alluded to in Chapter 4. The digitization stage has typically taken place after conversion to baseband, i.e., $I(t)$ and $Q(t)$. However, recent advances in DSP integrated circuit technology have allowed the digitization stage to be moved to the IF level, and even the RF level, in the radar system. Radars that make extensive use of digital technology are often referred to as *digital radar*. Many advantages exist such as perfect gain balance between the in-phase and quadrature channels and better filter characteristics. Although an important and emerging technology, this chapter will assume that digitization is accomplished at the baseband level. The interested reader is referred to several texts on the topic of digital radar/radio (e.g., *Pace*, 2000).

This chapter will concentrate on signal-processing techniques. It will include spectral analysis, and there will be some overlap especially with Chapter 7. Although we have assumed that the reader is familiar with Fourier analysis and spectra, we will reintroduce some of these concepts here, in preparation for the more detailed discussions to follow. There may also be some modest overlap with other earlier chapters, but again mainly to set the scene for later discussions.

In Chapter 4, we demonstrated that the demodulated radar signal returned from a point scatterer at range r with radial velocity v_r can, in the simplest case, be considered to have the following form:

$$\underline{y}(t) = \underline{\tilde{A}}\mathrm{e}^{-\mathrm{i}2k_0 r_0}\mathrm{e}^{-\mathrm{i}2k_0 v_r t}\underline{p}\left(t - 2\frac{r(t)}{c}\right).$$

Here we assume the particle is at range r_0 at time $t = 0$, and that the time of recording is short enough that the scatterer radial velocity can be considered constant over the recording interval, so that $r(t) = r_0 + v_r t$. This may not always be a realistic assumption if the recording interval becomes tens of seconds or more, but this formula still helps us in dealing with the mathematics of the scatter, at least conceptually.

Many of the topics discussed in this chapter can be generally applied to digital signal processing problems. Nevertheless, it is important to emphasize the particular signal which we will be processing. Although simplified, atmospheric scatter can be thought of as originating from a large (possibly even infinite) number of point scatterers. Given this simplification and assuming an additive noise process $\underline{n}(t)$, the baseband radar signal from atmospheric volume scattering will be represented by the following equation:

$$\underline{y}(t) = \lim_{N_s \to \infty} \sum_{\ell=1}^{N_s} \underline{\tilde{A}}_\ell e^{-i2k_0 r_{0\ell}} e^{-i2k_0 v_{r\ell} t} \underline{p}\left(t - 2\frac{r_\ell(t)}{c}\right) + \underline{n}(t), \tag{8.1}$$

where N_s is the number of scatterers within the resolution volume, $r_{0\ell}$ is the position of the ℓth particle at time 0, and $v_{r\ell}$ is the radial velocity of the ℓth particle. The amplitude $\tilde{A}_\ell$ incorporates the scatterer effects and the weighting of the two-way antenna pattern. Subsequent sections will make use of (8.1) for the investigation of frequency content and estimation of atmospheric parameters.

Examples of time-series data from the EISCAT radar and other radars were provided in Chapter 4. The EISCAT data were obtained from mesospheric echoes and it was suggested that their characteristics could be representative of the atmospheric conditions during the experiment. One way of determining their characteristics is to simply observe the temporal behavior of the time-series data. The amplitude is related to the strength of echo (which depends on the degree of permittivity fluctuations at the Bragg scale, see Chapter 3), while the frequency of oscillation in the time-series is proportional to the radial velocity. For a vertically pointing radar, of course, the radial velocity is an estimate of the vertical velocity. While time-series data can be extremely useful, and are often more computationally efficient to process, their frequency-domain counterpart can often provide easier interpretation and have numerous advantages. Sample time-series data (left panel) and the so-called *Doppler spectra* of the data (right panel) are provided in Figure 8.1. These graphs are representative of data obtained by most MST radars.

The Doppler spectra were estimated using the *periodogram* algorithm, discussed more extensively later in this chapter. This is based on Fourier analysis and can often provide more easily interpreted information about the frequency content of the time-series data. As seen in Chapter 7, Equation (7.10), the first moment of the Doppler spectrum provides an estimate of the mean radial velocity. For time-series data, one would be required to count zero-crossing, or otherwise estimate the frequency, in order to determine the radial velocity. For the lowest gate in the example, the time-series data resemble noise with very little coherent signal content. This is reflected in the periodogram which, in contrast to the other gates, shows no dominant peak.

Atmospheric radar signal processing can be conducted in the time and/or frequency domains, with advantages and disadvantages for each. This chapter will focus on the fundamental theory behind these analysis techniques and will discuss how one can extract physical parameters in an efficient and statistically significant manner.

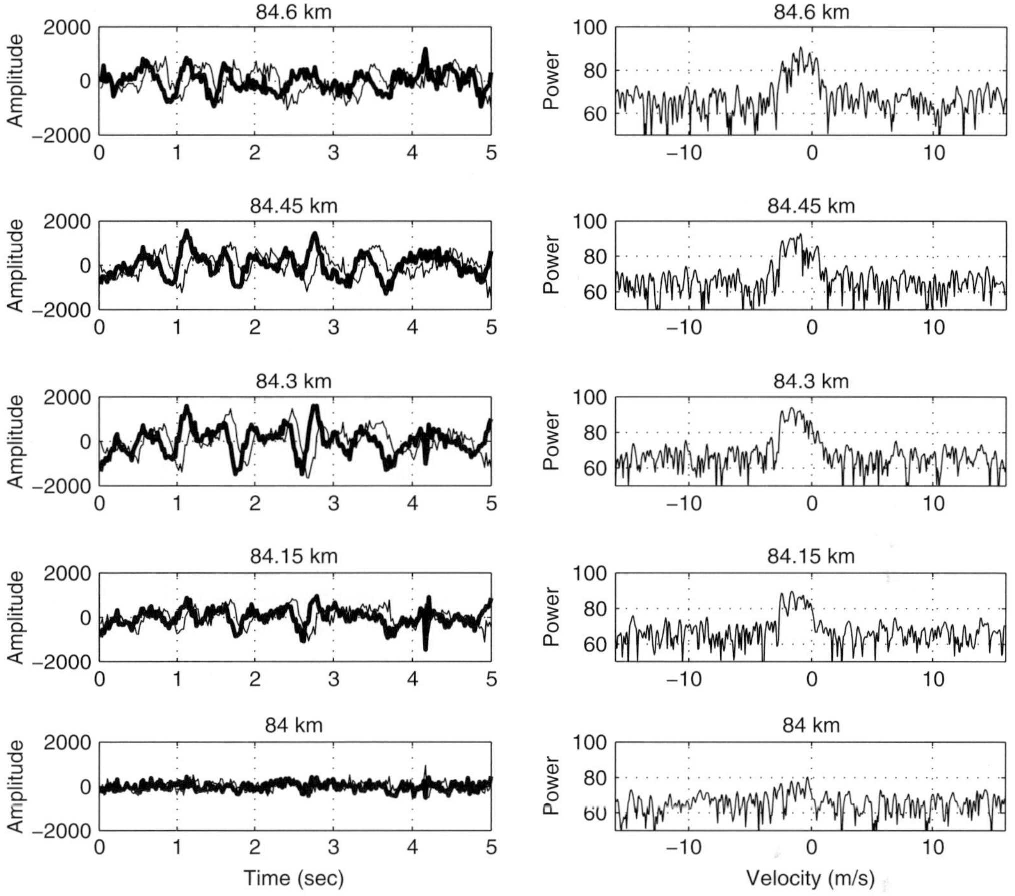

Figure 8.1 Example of time-series data (left panel) and corresponding Doppler spectra (right panel) for mesospheric data. The data were obtained using the EISCAT radar during the summer.

8.1 Analog-to-digital conversion

We have previously discussed the need to pulse-modulate the transmitted radar signal in order to track the slowing varying phase resulting from atmospheric scatter. Given this inherent sampling process, a discrete-time signal, or time series, can be generated for a particular gate by simply grouping samples from a sequence of pulses. As emphasized already, the sampling interval is denoted by T_s and it is this parameter, along with the number of coherent integrations, which dictates the aliasing velocity. In order to be processed, the discrete-time samples must be converted to a digital format by a procedure called *quantization*.

Quantization converts the amplitude of the discrete-time samples to a finite number of voltage levels. The number of levels is determined by the length of the binary word

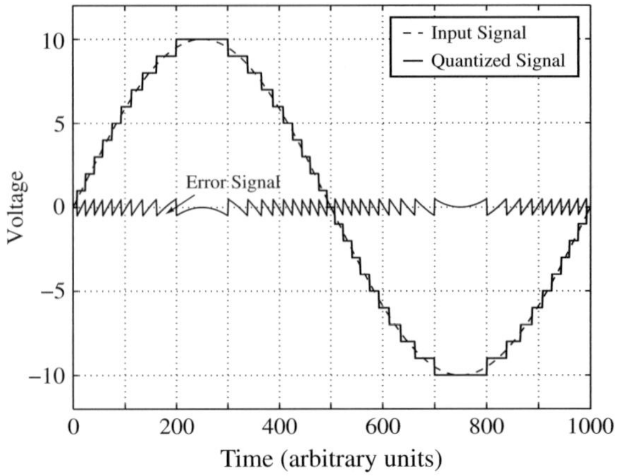

Figure 8.2 Effects of severe quantization on a sampled sinusoidal signal. The error signal is the difference of the original signal and the quantizer output and is bounded by $\pm\Delta/2$.

used to represent the levels. For example, a 12-bit quantizer produces $2^{12} = 4096$ levels. An example of a coarsely quantized sinusoid is shown in Figure 8.2.

The curve fluctuating about zero and denoted as the *Error Signal* is the difference between the original signal and the quantized signal. Given a voltage range of $\pm y_{\max}$ and L quantization levels, the so-called step size is given by

$$\Delta = \frac{2y_{\max}}{L}.$$

Assuming a uniform distribution of error values over the range $\pm\Delta/2$, the variance of the quantization error can be shown to have the following form (as also presented in Chapter 5, Sub-section 5.2.11):

$$\sigma_Q^2 = \frac{\Delta^2}{12}. \tag{8.2}$$

It should be emphasized that the effects of quantization cannot normally be reversed. Therefore, one should show caution in the choice of word size and maximum voltage level. Obviously, the quantization error given in (8.2) could be reduced by increasing the number of levels L. This requires an increase in the word length and more data storage. An additional problem, particular to radar signals, can arise in relation to the input voltage levels. Given the large dynamic range of atmospheric radar signals (70–90 dB), a rather weak signal will only use a small fraction of the range $\pm y_{\max}$. As a result, the relative quantization error will increase for this case. If $y_{\max}$ were decreased in order to mitigate this effect, the possibility of signal clipping could occur for the case of strong returns. Because of this problem, some radars employ automatic gain control (AGC) circuits which attenuate strong signals and amplify weak signals before quantization. Most MST radars do not use AGC, however, and it is typical to adjust the gain according to the average expected signal levels.

Remember in a radar that there are two time-series involved. The first occurs during a single pulse, where the time-series represents the backscattered signal from increasingly more distant scatterers, and typically has a duration of the order of a millisecond. Although it is a time-series, it reflects the different signals from different ranges, and so is considered a function of range. The second type of time-sequence comprises data sampled over multiple pulses, with a different sequence for each sampled range, and has lengths of typically seconds and tens of seconds. Here we are mainly interested in the second time-series, though the theory we will present can also be applied to the range-data. After the sampling and quantization process, the atmospheric radar signal will consist of a finite set of digital data points, which are samples of $y(t)$, with a separate time series for each range gate. In the MST case, when using systems that beat the received signal down to baseband, each receiver produces a pair of in-phase and quadrature signals, which are considered as a single complex time-series, denoted $\underline{y}(t)$. For notational clarity, the sampled radar signal will be denoted by $\underline{y}[n]$ where

$$\underline{y}[n] = \underline{y}(nT_s) \qquad n = 0, \pm 1, \pm 2, \ldots \tag{8.3}$$

In some cases, we will only allow positive n. Some important points to remember about the discrete-time Doppler radar signal $\underline{y}[n]$ are:

- The signal is a discrete-time, complex random process.
- Frequency content, or spectral shape, is not known a priori.
- Data length is limited by temporal resolution requirements.
- The signal can be contaminated by noise, clutter and/or interference.

A major goal will be to estimate the frequency content of $\underline{y}[n]$ from a finite record of Doppler radar data. Since $\underline{y}[n]$ is often a random process (especially for turbulent scatter), the frequency content is generally obtained from the so-called *power spectral density* (PSD). Before discussing the problem of spectral estimation, several important topics related to time-series analysis and basic digital processing will be reviewed.

8.2 Time-domain processing

As already noted, the scattered signal received from the atmosphere can be considered to be a sum of Doppler shifted frequencies, each component due (in a simplistic view) to the radial motion of a single scatterer. Hence each scatterer can be loosely associated with a particular complex exponential (in-phase and quadrature) or sinusoidal oscillation in time. Therefore it is natural to transform the signal using Fourier analysis, and use the resulting spectrum for interpretation, since Fourier analysis is especially well suited to sinusoidal components.

However, this picture oversimplifies the situation. In reality, any scatterer has a finite lifetime, so that its associated Doppler shifted sinusoidal oscillation has a finite lifetime as well. The scatterer may also accelerate during the time interval. Consequently the scattered signal is associated with a frequency-*band* of sinusoidal oscillations. Other complicating features such as rotational motion and variations in backscatter reflectivity can also mean that there is not always a one-to-one correspondence between scattered signals and individual Fourier components. But even then, it is still generally sensible to apply Fourier methods.

But there are cases where this idea breaks down. One extreme example is that of lightning, in which the received signal is not due to scatter, but to bursts of radio-frequency interference generated by the lightning itself. Such events can be very short-lived, lasting less than half a second. Another example is a meteor echo. Meteors entering the atmosphere produce ionized plasma trails which can reflect incident radiowaves, and when these signals return to the receiver they can produce increases in received power. But such trails last typically only a few tenths of a second, so again have very short durations, just like lightning. Digital sampling errors can also occur at times (generally rarely), and these can also produce short lived but sudden changes in apparent signal power.

Figure 8.3 shows an example with both spikes and meteor signals. The corresponding spectrum is shown in part (c). The short-lived meteor corresponds to a broad spectrum, and is most easily identified in the time domain. Indeed generally a signal is easiest to identify if it is concentrated into a short interval – so short bursts are easiest to see in the time domain, and have broad spectra, while long-lasting pure sinusoidal oscillations are easiest to see in the Fourier domain, where they appear as narrow spikes. The lightning spikes in Figure 8.3 have Fourier components that are either DC offsets or small amplitude sinusoidal oscillations in the frequency domain, and are hard to identify at all in the spectra. Impulsive and intermittent interference is also often best examined in the time domain.

Another case in which time-domain analysis is most useful is in the case of slow drifts in the in-phase and quadrature components. Sometimes this can occur due to ground echoes, and sometimes temperature drifts in the transmitter can cause mild frequency shifting in the transmitter pulse, for example. Figure 4.21, upper part, shows an example where the in-phase and quadrature components contain slow drifts in time. A polynomial fit has been applied, shown as the white line in that figure. Such non-sinusoidal drifts are most easily identified and (if undesirable) removed in the time-domain.

Other examples can be found in which time-series analysis is preferable to Fourier analysis. The astute researcher will recognize that both time-series and spectral analysis are available, and will select the most appropriate tool. Even if spectral processing is the preferred mode of analysis, it is still often a good idea to look at the time series from time to time, since the very different perspective seen in the temporal and frequency domains can give complementary views of the signal. This is especially true if there are single large dominant peaks, representing dominant spectral components. Viewing in the temporal domain can tell the observer whether the signal associated with this peak persisted over the whole time series, or only a part of it, or tell the observer about other features not easily seen in the spectral domain.

The rest of this chapter will largely focus on spectral processing, but as noted, the temporal domain should not be ignored, and can often give useful additional information. Wavelet analysis is a further extension of this concept, though we will not have sufficient space to discuss it in this book.

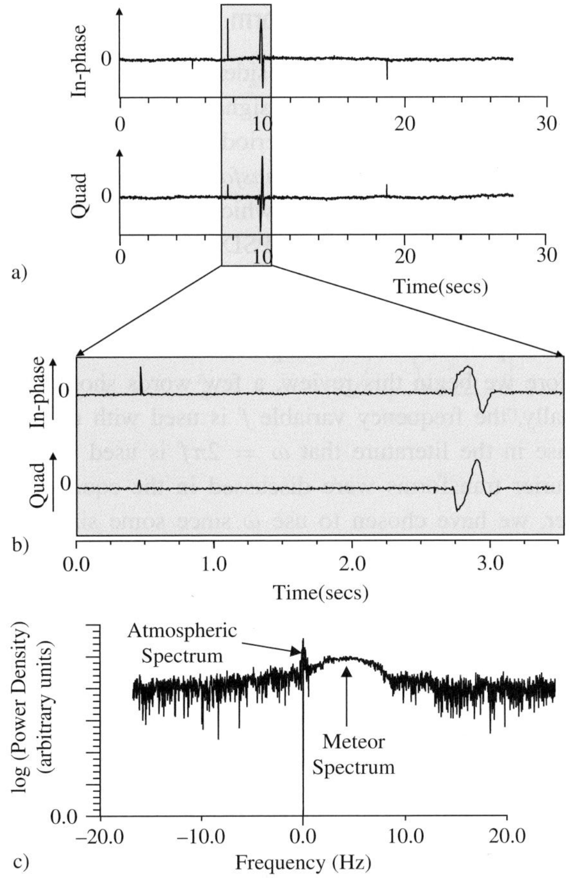

Figure 8.3 (a) Example of a data record containing spikes (possibly due to lightning, or maybe a digitization error) and a meteor. Both are clearly evident in the temporal domain. (b) An expanded view of (a), as indicated by the arrows. (c) The spectrum of the time series. The broad spectral peak corresponds to the meteor, and the lightning spikes essentially produce a DC offset in the spectrum.

8.3 Brief review of Fourier analysis

Given that the Doppler radar signal is often pulse-modulated, and can involve a variety of wave-forms, it is important to have a thorough understanding of discrete-time Fourier analysis. The frequency content (Doppler radial velocity) will be estimated using the discrete Fourier transform (DFT) or its computationally efficient counterpart, the fast Fourier transform (FFT) (*Cooley and Tukey*, 1965). The DFT is derived from the continuous-time Fourier transform and has many of the same properties. Therefore, a short review of continuous-time Fourier analysis will be provided before presenting a summary of discrete-time theory. Many basic texts on signal and system analysis exist providing additional review material (e.g., *Oppenheim et al.*, 1983).

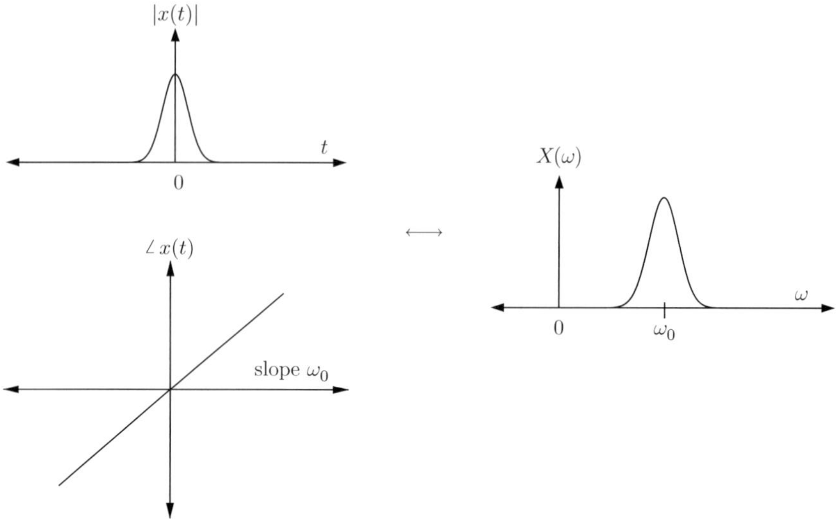

Figure 8.5 Same as Figure 8.4, except for multiplication by a complex sinusoid. This time we show the full complex function on the left, representing it as a magnitude and a phase. A shift in the frequency domain causes a linear phase variation in the time domain. As shown, the slope of the phase variation is equal to the frequency shift.

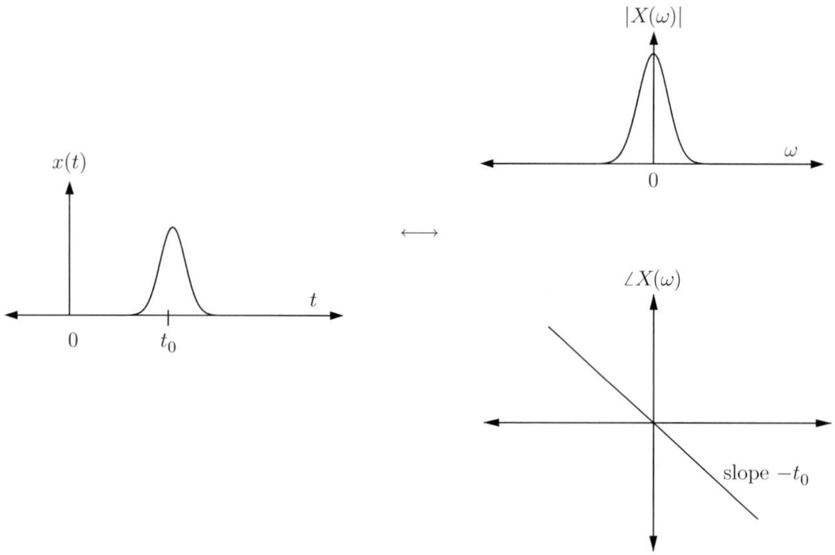

Figure 8.6 A time delay causes a linear variation in the phase of the Fourier transform. The magnitude of the Fourier transform is not affected.

In a subsequent chapter, the spaced antenna method of estimating the horizontal wind will be introduced. This method depends on the time shifting property, since the cross-correlation function between signals from spaced receivers will possess time delays related to the horizontal wind and the relative locations of the receiver antennas.

Linearity

The integral form of the Fourier transform (8.4) can be used to show that the transform is linear and homogeneous, so that

$$\underline{a}_1\underline{x}_1(t) + \underline{a}_2\underline{x}_2(t) \leftrightarrow \underline{a}_1\underline{X}_1(\omega) + \underline{a}_2\underline{X}_2(\omega). \tag{8.10}$$

Many of the analytical signal models dealt with for Doppler radar signal analysis involve either a summation or integral (*e.g.*, 8.1). As such, linearity of the Fourier transform is imperative.

Convolution

Many basic texts on signal and system analysis typically begin with a thorough coverage of the convolution operator, given its importance for the calculation of the output of a linear system. In fact, we have already used convolution for our discussion of the matched filter, and applied it to radar backscatter theory in Chapter 4, Equation (4.20) and the associated section. We have also used convolutions in regard to various discussions pertaining to Figure 3.28. It can be shown that convolution in the time (frequency) domain results in a multiplication in the frequency (time) domain i.e.,

$$\underline{x}(t) \otimes \underline{y}(t) \leftrightarrow \underline{X}(\omega)\underline{Y}(\omega), \tag{8.11}$$

$$\underline{x}(t)\underline{y}(t) \leftrightarrow \frac{1}{2\pi}\underline{X}(\omega) \otimes \underline{Y}(\omega), \tag{8.12}$$

where we have used "$\otimes$" to indicate convolution (some texts use the symbol $*$). This property is important for the understanding of many of the concepts to be covered in later sections of this chapter. For example, the effect of a finite-length data set on the estimation of Doppler spectral width is dependent on this property.

Duallty

Given the similarity between the Fourier and inverse Fourier transform integrals, the *duality* property should not come as a surprise. For the Fourier transform pair $\underline{x}(t) \leftrightarrow \underline{X}(\omega)$, the following relationship holds:

$$\underline{X}(t) \leftrightarrow 2\pi\underline{x}(-\omega). \tag{8.13}$$

In essence, the duality property states that if a functional form exists in the time domain and results in a particular frequency domain signal, the opposite relationship holds.

For example, let us define the pulse function $p_\tau(t)$ as a boxcar function of the form

$$p_\tau(t) = \begin{cases} 1 & -\tau/2 \le t \le \tau/2 \\ 0 & \text{otherwise.} \end{cases} \tag{8.14}$$

The Fourier transform is given by the so-called sinc function.

$$\mathcal{F}[p_\tau(t)] = \tau \text{ sinc}\left(\frac{\tau\omega}{2\pi}\right), \tag{8.15}$$

where for a variable ξ, $\text{sinc}(\xi) = \sin(\pi\xi)/(\pi\xi)$.

Now suppose that the *frequency response* (rather than the pulse) takes a boxcar form, with formula given by

$$\underline{H}_B(\omega) = \begin{cases} 1 & -B/2 \le \omega \le B/2 \\ 0 & \text{otherwise.} \end{cases} \tag{8.16}$$

This is a low pass filter with sharp edges, allowing frequencies with magnitude less than $\frac{B}{2}$ to pass. Then, as expected from duality, the inverse Fourier transform of the low-pass filter (LPF) frequency response, called the *impulse response*, is given by a time-domain sinc function, viz.

$$\mathcal{F}^{-1}[\underline{H}_B(\omega)] = \frac{B}{2\pi}\text{sinc}\left(\frac{Bt}{2\pi}\right). \tag{8.17}$$

It should be noted that an impulse response of such an ideal LPF is non-causal (i.e., a sinc function must extend from $-\infty$ to ∞, and it is not possible to generate a function that began at $-\infty$), and therefore not realizable. In reality, a less-than-ideal LPF must be implemented (in particular, the phase will vary across the filter), with resulting spectral leakage and time delay.

From this example, the reader can see that duality simply implies the reciprocity of the time and frequency domains, at least from a theoretical perspective.

Parseval's theorem

As mentioned in the beginning of this chapter, the Doppler radar signal is a random process and the Fourier transform of any time-series that is determined in any measurement is only one realization out of many possibilities. However, all such data-sets are derived from the same population and have common statistical properties.

Therefore, it is necessary to study the statistical nature of the Doppler radar signal in order to understand the frequency content. This is accomplished by estimating the PSD, which is the Fourier transform of the autocorrelation function. These topics will be covered in subsequent sections but are related to *Parseval's theorem*, which states that the energy of a signal can be calculated in either the time or frequency domain. In other words,

$$\int_{-\infty}^{\infty} |\underline{x}(t)|^2 \, dt = \frac{1}{2\pi}\int_{-\infty}^{\infty} |\underline{X}(\omega)|^2 \, d\omega. \tag{8.18}$$

An interesting interpretation of Parseval's theorem is that $|\underline{X}(\omega)|^2$ can be thought of as a distribution of energy over frequency. Later, we will make use of this important property in order to calculate the spectral moments of the Doppler spectrum.

8.3.2 Discrete-time Fourier transform

In the previous section, the continuous-time Fourier transform was reviewed. Because the signal associated with atmospheric scatter is often buried in noise, yet the coherence time can be quite long, it is often useful (or even necessary) to transmit a sequence of pulses and monitor the signal over several hundreds of pulses. Therefore, the signal resulting from MST radar observations is inherently sampled.

The sampled points are therefore not continuous, but occur at discrete points spaced by some interval T_s (an equivalent procedure of course exists for spatially discrete data as well). There are two ways to treat this problem. The most elegant is to consider the resulting time-series as a truly continuous data-set, multiplied by a regular series of delta-functions at spacing T_s. The properties of a regular series of delta-functions under Fourier transformation are well known, and so with this slight modification, all of the preceding theory concerning continuous Fourier transforms may be applied.

However, because this book is designed also as something of a handbook, we will treat discrete digitization as a specific subset of Fourier transforms, and consider the changes required in dealing with digital sampling explicitly.

Let us assume that a continuous-time sinusoid with frequency f is sampled every T_s s, resulting in the following discrete-time sinusoid:

$$\cos(2\pi ft)|_{t=nT_s} = \cos(2\pi fT_s n),$$

where n is an integer. Assuming that T_s is chosen to satisfy the Nyquist sampling theorem ($1/T_s = f_s > 2f$), the normalized discrete-time frequency of the sampled cosinusoid, defined by

$$\Omega = 2\pi f/f_s, \tag{8.19}$$

is bounded by $\pm\pi$ and has "units" of rad/sample. An uppercase omega is used for discrete-time frequency in order to differentiate it from the continuous-time case.

For any data sequence which is absolutely summable, the discrete-time Fourier transform (DTFT) exists and is defined as follows:

$$\underline{X}(\Omega) = \sum_{n=-\infty}^{\infty} \underline{x}[n]\mathrm{e}^{-\mathrm{i}\Omega n}. \tag{8.20}$$

It should be emphasized that $\underline{x}[n]$ is discrete in time but the DTFT is continuous in Ω. Therefore, the inverse DTFT has a similar form to the continuous-time inverse Fourier transform with the exception of the limits of integration, viz.,

$$\underline{x}[n] = \frac{1}{2\pi}\int_{-\pi}^{\pi} \underline{X}(\Omega)\mathrm{e}^{\mathrm{i}\Omega n}\,\mathrm{d}\Omega. \tag{8.21}$$

As before, the DTFT pair will use the following notation:

$$\underline{x}[n] \leftrightarrow \underline{X}(\Omega). \tag{8.22}$$

Periodicity of DTFT

Many of the properties of the DTFT are similar to those of the Fourier transform. An interesting exception is the periodicity of the DTFT, see Figure 8.7. The following derivation shows that the DTFT repeats every 2π in discrete-time frequency:

$$\begin{aligned}\underline{X}(\Omega + 2\pi) &= \sum_{n=-\infty}^{\infty} \underline{x}[n]\mathrm{e}^{-\mathrm{i}(\Omega+2\pi)n}\\ &= \sum_{n=-\infty}^{\infty} \underline{x}[n]\mathrm{e}^{-\mathrm{i}\Omega n}\\ &= \underline{X}(\Omega).\end{aligned}$$

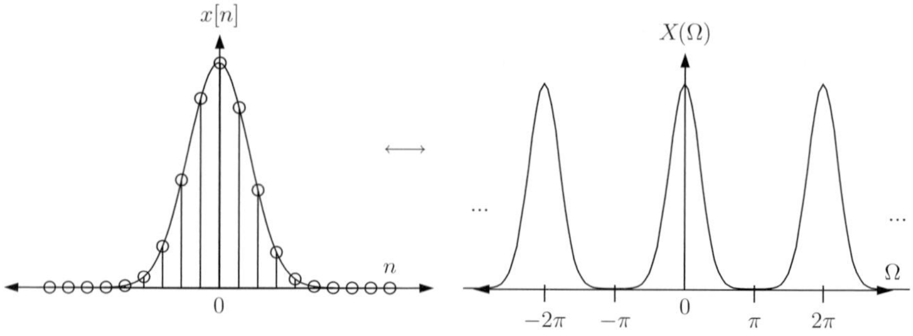

Figure 8.7 Depiction of the effect of time sampling on the Fourier transform. The Fourier transform is 2π periodic due to time sampling. As a result, plots of $\underline{X}(\Omega)$ usually only include values between $\pm\pi$.

Therefore, it is not necessary to plot $\underline{X}(\Omega)$ over a range larger than $\pm\pi$. Of course, this property results from the sampling process of $\underline{x}(t)$.

This same repetition can be seen by considering the discretely-sampled time series as a product of a "true" continuous time-series and an infinite series of delta-functions at spacing T_s. The Fourier transform is the convolution of the respective Fourier transforms of the original time-series and of the series of delta functions. The latter Fourier transform is another series of regularly spaced delta-functions in frequency space at frequency steps of $1/T_s$, so the final Fourier transform is simply the Fourier transform of the continuous time-series repeated at steps of $1/T_s$ (assuming that the Fourier transform of the continuous time-series is constrained to be non-zero only between $\pm 1/(2T_s)$, and that all Fourier amplitudes are zero outside this range).

The concepts just discussed can be utilized to develop a more pragmatic form of Equation (8.20). First, there are only N points available, so the limits can be reset to cover only $n = 0$ to $n = N - 1$. All other points are taken as zero, since we do not know them. We may think of this time series as an infinite (unknown) one multiplied by a window which equals unity where the data exist, and zero elsewhere.

Second, we recognize that the only useful frequencies are between normalized values of $-\pi$ and $+\pi$. If it is assumed that this frequency domain is divided into L equally spaced steps, then we may write

$$\underline{X}(\Omega) = \sum_{n=0}^{N-1} \underline{x}[n]\mathrm{e}^{-\mathrm{i}\Omega n}, \tag{8.23}$$

where Ω assumes values of $-\pi + \frac{2\pi\ell}{L}$, with ℓ assuming values between 0 and $L - 1$. Writing this out explicitly gives a sequence of values

$$\underline{X}_\ell = \sum_{n=0}^{N-1} \underline{x}[n]\mathrm{e}^{-\mathrm{i}(-\pi + \frac{2\pi\ell}{L})}. \tag{8.24}$$

Parseval's theorem

For the discrete-time case, the time-domain integral in Parseval's theorem is replaced by a summation. Further, the discrete-time frequency variable Ω is limited over $\pm\pi$, and we can write

$$\sum_{n=-\infty}^{\infty} |\underline{x}[n]|^2 = \frac{1}{2\pi}\int_{-\pi}^{\pi} |\underline{X}(\Omega)|^2 \, \mathrm{d}\Omega. \tag{8.25}$$

Of course, the general concept that energy can be calculated in either the time or frequency domains is intact.

This equation can be written more pragmatically for a time series of N points as

$$\sum_{n=0}^{N-1} |\underline{x}[n]|^2 = \frac{1}{2\pi}\int_{-\pi}^{\pi} |\underline{X}(\Omega)|^2 \, \mathrm{d}\Omega. \tag{8.26}$$

8.3.3 Discrete Fourier transform (fast Fourier transform)

As stated earlier, the DTFT is a continuous function over the frequency variable Ω. Although quite useful for theoretical analysis, the DTFT is not used to process actual Doppler radar data.

In practice, a special choice of L in Equation (8.24) is made – namely, $L = N$. This choice happens to be the optimum one for retaining all the information about the spectrum but allowing storage of the smallest number of points. In this case, Equation (8.24) can be modified. We do two things: first, we take $L = N$, and secondly, rather than covering normalized frequencies from $-\pi$ to π, we consider the range 0 to 2π.

Thus by uniformly sampling Ω with N samples, as shown in Figure 8.8, the so-called discrete Fourier transform (DFT) is obtained.

$$\begin{aligned} \underline{X}_k &= \underline{X}(\Omega)|_{\Omega=2\pi k/N} \qquad k = 0, 1, 2, \ldots, N-1 \\ &= \sum_{n=0}^{N-1} \underline{x}[n]\mathrm{e}^{-\mathrm{i}2\pi kn/N}. \end{aligned} \tag{8.27}$$

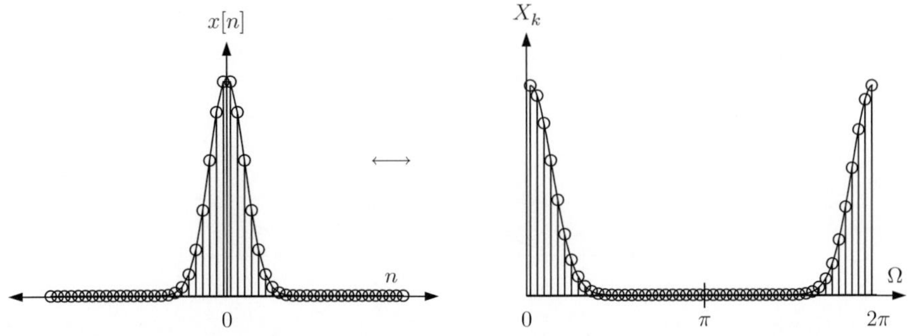

Figure 8.8 Conversion from DTFT to DFT. The DFT is simply a frequency sampled version of the DTFT. One should be careful in the interpretation, however, since it is typical for the sampled normalized frequencies Ω to be taken between 0 and 2π.

The samples of Ω are actually taken over the range $2\pi[0, \frac{1}{N}, \cdots, \frac{N-1}{N}]$. Given the periodicity of the DTFT, this sampling range does cover the necessary range of Ω although some care must be taken in the interpretation of the frequency samples. Because the time series and the Fourier transform have the same number of points, this construction turns out to be extremely useful for computer studies of the frequency content of discrete data. Furthermore, the finite summation in (8.27) guarantees the existence of the DFT.

Since the frequency variable of the DFT is only defined at a finite number of values, the inverse DFT can be calculated using the following summation equation:

$$\underline{x}[n] = \frac{1}{N}\sum_{k=0}^{N-1} \underline{X}_k e^{i2\pi kn/N} \qquad n = 0, 1, 2, \ldots, N-1. \tag{8.28}$$

An interesting example of the computation of the DFT is provided by a sampled version of the pulse function. Recall that the Fourier transform of the continuous-time pulse function is a sinc function, see (8.15). For computational simplicity, let us define a four-point pulse function as $p_4 = [1\ 1\ 1\ 1]$, which are four samples within the pulse. Taking a four-point DFT results in the Fourier transform being $P_k = [4\ 0\ 0\ 0]$, which can easily be verified from (8.27). It is interesting that the resulting DFT does not have the expected shape of a sinc function. The reason for this unexpected result lies in the sampling of the frequency variable Ω.

The sinc function that was expected from the four-point pulse is shown in Figure 8.9 as a solid line. The four samples of the DFT are shown as circles. Note that three of the four frequency samples have values of zero. Therefore, the resulting DFT $P_k = [4\ 0\ 0\ 0]$ is correct but it is the interpretation of the values where one must be careful. An obvious question is why only use N samples for the DFT? In fact, more frequency samples can easily be taken by a process called *zero padding*. By concatenating a finite number of zeros to the data, one can effectively sample frequency on a finer scale. For example, the resulting DFT values are plotted in Figure 8.10 for a total number of frequency samples of 8, 32, and 64. Note the number of actual data points is still set to four. It is only the number of concatenated zeros which is changed.

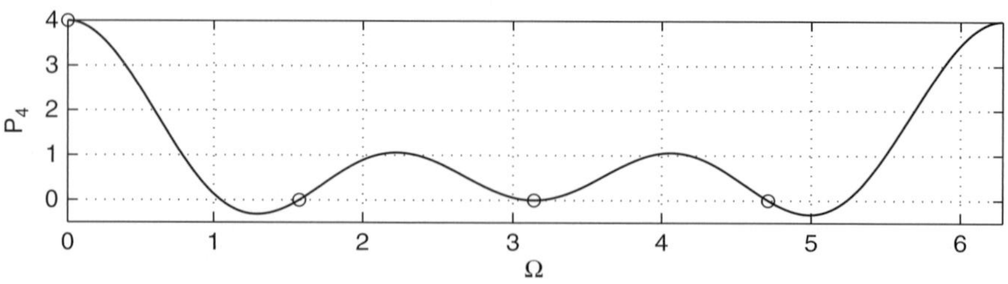

Figure 8.9 DFT of four-point pulse function. Note that three of the four frequency samples, shown as circles, lie close to the nulls of the sinc function. However, they are sufficient to fully define the spectrum.

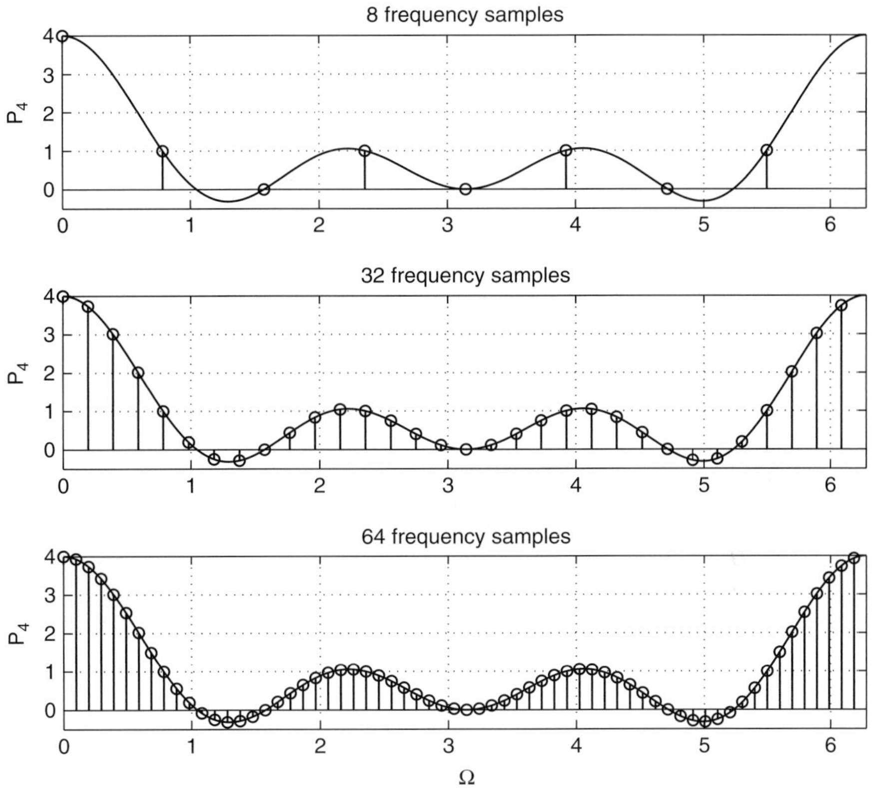

Figure 8.10 Example of the effect of zero padding on the DFT. By the concatenation of zeros to the actual data, the DFT automatically results in a finer frequency sampling. It should be emphasized, however, that the resolving capability of the DFT has not changed and is still dictated by the length of the non-zero data.

As more Ω samples are taken, the DFT more closely resembles the sinc form expected from the continuous-time Fourier transform. Of course, the number of frequency samples is limited due to computer memory. In addition, the computational burden increases with number of samples irrespective of the fact that the additional data are zero. This computational problem directly leads to the true usefulness of the DFT and the introduction of the fast Fourier transform (FFT) (*Cooley and Tukey*, 1965).

The FFT is simply a more computationally efficient method of calculating the DFT. From (8.27) and (8.28), it is observed that the number of complex multiplications for the calculation of either the DFT or inverse DFT is on the order of N^2. By clever manipulation of the summations, a rather dramatic computational saving can be obtained.

We will begin by defining the so-called *twiddle factor*,

$$\underline{W}_N = \mathrm{e}^{-\mathrm{i}2\pi/N}, \tag{8.29}$$

which results in the following equivalent reformulation of the DFT as

$$\underline{X}_k = \sum_{n=0}^{N-1} \underline{x}[n] \left(\underline{W}_N\right)^{(kn)} \qquad k = 0, 1, 2, \ldots, N-1. \tag{8.30}$$

We will assume that N is a power of two for reasons that will soon become apparent. By dividing $\underline{x}[n]$ up into odd and even values of n, we define the following two new discrete-time signals (this process of reducing the number of points in a sample, or reducing the sampling rate, is referred to as "decimation" in signal-processing jargon):

$$\underline{x}^{(e)}[n] = \underline{x}[2n]$$
$$\underline{x}^{(o)}[n] = \underline{x}[2n+1],$$

for $n = 0, 1, 2, \ldots, \frac{N}{2} - 1$. The corresponding $\frac{N}{2}$-point DFTs for these new sequences are given by

$$\underline{X}_k^{(e)} = \sum_{n=0}^{\frac{N}{2}-1} \underline{x}^{(e)}[n]\underline{W}_{N/2}^{kn} \qquad k = 0, 1, 2, \ldots, \frac{N}{2} - 1$$
$$\underline{X}_k^{(o)} = \sum_{n=0}^{\frac{N}{2}-1} \underline{x}^{(o)}[n]\underline{W}_{N/2}^{kn} \qquad k = 0, 1, 2, \ldots, \frac{N}{2} - 1.$$

It can be verified that the original DFT of $\underline{x}[n]$ can be obtained from $\underline{X}_k^{(e)}$ and $\underline{X}_k^{(o)}$ using the following relationships:

$$\underline{X}_k = \underline{X}_k^{(e)} + \underline{W}_N^k \underline{X}_k^{(o)} \qquad k = 0, 1, 2, \ldots, \frac{N}{2} - 1$$
$$\underline{X}_{\frac{N}{2}+k} = \underline{X}_k^{(e)} - \underline{W}_N^k \underline{X}_k^{(o)} \qquad k = 0, 1, 2, \ldots, \frac{N}{2} - 1.$$

The calculation of $\underline{X}_k^{(e)}$ and $\underline{X}_k^{(o)}$ each require $(N/2)^2$ multiplications. Further, combining these two functions into $\underline{X}_k$ requires an additional $N/2$ multiplications by the twiddle factor. Therefore, the total number of multiplications is $N^2/2 + N/2$. Recall that the original DFT required N^2 multiplications. As a result, the computational saving is $N^2/2 - N/2$.

The next step in the derivation of the FFT is discovered by noticing that the computations of $\underline{X}_k^{(e)}$ and $\underline{X}_k^{(o)}$ are actually $\frac{N}{2}$-point DFTs in themselves. Thus, these calculations can also be decimated in time as was accomplished with the original DFT. It should now be apparent that this decimation process could be continued until we are left with the trivial case of a two-point DFT, when the twiddle factor would become –1. Given N is a power of two, it can be shown that this so-called *decimation-in-time* FFT algorithm requires multiplications on the order of $N \log_2(N)/2$ resulting in the following significant computational savings:

$$\text{FFT computational savings} = N^2 - N\log_2(N)/2. \tag{8.31}$$

For a typical data length of 512 points, for example, the DFT would require 262 144 multiplies while the FFT would need only 2304! Obviously, the FFT is used in all MST

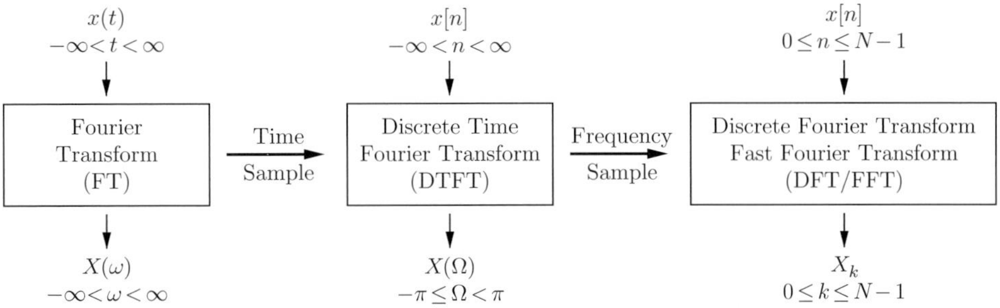

Figure 8.11 Summary of Fourier transforms.

radar applications requiring the computer calculation of the frequency content of the signal.

Figure 8.11 is provided to summarize the relationships between the various forms of the Fourier transform and their time-domain counterparts.

The continuous-time Fourier transform is useful for general signal/system analysis. For theoretical analysis of discrete signals, the discrete-time Fourier transform is used. Finally, computational problems, such as the analysis of actual MST radar signals, employ the discrete Fourier transform, usually in its computationally efficient form, the FFT.

8.4 Digital filtering concepts

In this section, fundamental issues related to digital filter design will be reviewed (*Oppenheim and Schafer*, 1975). The theory presented is not only important for general filtering issues related to time series analysis but is also imperative for a complete understanding of Doppler spectral estimation, which will be covered in subsequent sections.

8.4.1 *z*-transform and frequency response

The *z*-transform is often used in the analysis of discrete-time signals and systems. In many ways, it can be thought of as the discrete-time equivalent of the Laplace transform. Given a discrete-time sequence $\underline{x}[n]$, the *z*-transform and inverse *z*-transform are defined by the following equations:

$$\underline{X}(\underline{z}) = \sum_{n=-\infty}^{\infty} \underline{x}[n]\underline{z}^{-n}, \tag{8.32}$$

$$\underline{x}[n] = \frac{1}{j2\pi} \oint_{cc} \underline{X}(\underline{z})\underline{z}^{k-1}\, d\underline{z}, \tag{8.33}$$

where $\underline{z}$ is a complex variable. The contour integral in the inverse *z*-transform is usually avoided by the use of a table of *z*-transform pairs and properties available in numerous

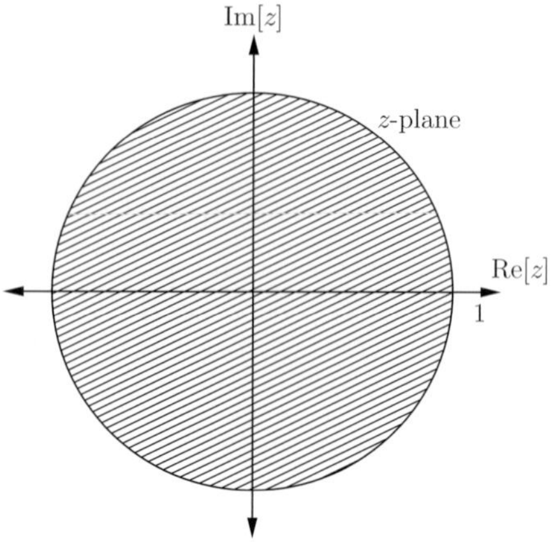

Figure 8.12 The complex $\underline{z}$-plane. The unit disk is emphasized by the hatched area.

texts (e.g., *Kamen and Heck*, 2000). Existence of the z-transform in (8.32) is dictated by the convergence of the infinite summation. Typically, convergence is obtained only over a certain region of the complex variable $\underline{z}$, which is aptly called the *region of convergence* (ROC).

The DTFT (8.20) is a special case of the z-transform. By the addition of a control term to the exponent of the DTFT, it is possible to calculate the z-transforms of many signals whose DTFTs do not exist. As a result, the z-transform is used most often in discrete-time system analysis. If the frequency content of a discrete-time signal is desired, the DTFT can be obtained from the z-transform using the following simple relationship:

$$\underline{X}(\Omega) = \underline{X}(\underline{z})\big|_{\underline{z}=\mathrm{e}^{\mathrm{i}\Omega}} . \tag{8.34}$$

In order to use this transformation, however, the ROC of the z-transform must contain the unit circle on the $\underline{z}$-plane which is a graphical representation of all possible values of the complex variable $\underline{z}$.

The $\underline{z}$-plane is shown in Figure 8.12 with the unit circle emphasized, which is defined where $\underline{z}$ has unit magnitude. The importance of the unit circle can be understood by studying the substitution of $\underline{z} = \mathrm{e}^{\mathrm{i}\Omega}$ (note unit magnitude) into Equation (8.34), which results in the DTFT of $\underline{x}[n]$, thereby providing its frequency content.

As we know from continuous-time system theory, the frequency response of a filter is obtained from the Fourier transform of the impulse response. The same holds true for discrete time where the frequency response of a discrete-time filter is given by

$$\underline{\mathcal{H}}(\Omega) = \underline{H}(\underline{z})\big|_{\underline{z}=\mathrm{e}^{\mathrm{i}\Omega}} = \sum_{n=-\infty}^{\infty} \underline{h}[n]\underline{z}^{-n}, \tag{8.35}$$

where $\underline{h}[n]$ is the impulse response of the filter.

A simple example of the impulse response of a discrete-time LPF is given by

$$h[n] = 0.5^n u[n],$$

where $u[n]$ is a discrete unit step function defined by

$$u[n)] = \begin{cases} 1 & n \geq 0 \\ 0 & \text{otherwise.} \end{cases} \tag{8.36}$$

We would like to examine the spectral characteristics of the filter. This is accomplished by first taking the z-transform of $\underline{h}[n]$ to obtain the transfer function $\underline{H}(\underline{z})$. Hence

$$\begin{aligned} \underline{H}(\underline{z}) &= \sum_{n=-\infty}^{\infty} 0.5^n u[n] \underline{z}^{-n} \\ &= \sum_{n=0}^{\infty} 0.5^n \underline{z}^{-n} \\ &= \sum_{n=0}^{\infty} \left(\frac{0.5}{\underline{z}} \right)^n \\ &= \frac{1}{1 - 0.5\underline{z}^{-1}} \\ &= \frac{\underline{z}}{\underline{z} - 0.5}. \end{aligned}$$

The closed-form solution of the infinite summation holds for $|\underline{z}| > 0.5$, which defines the ROC. As observed, the example filter has a *zero* at $\underline{z} = 0 + 0\mathrm{i}$ and a *pole* at $\underline{z} = 0.5 + 0\mathrm{i}$. For a filter to be stable, it can be shown that all poles must lie within the unit circle, which is the case for this example. The frequency response of the filter is obtained by substitution, viz.,

$$\begin{aligned} \underline{\mathcal{H}}(\Omega) &= \left. \frac{\underline{z}}{\underline{z} - 0.5} \right|_{\underline{z} = \mathrm{e}^{\mathrm{i}\Omega}} \\ &= \frac{\mathrm{e}^{\mathrm{i}\Omega}}{\mathrm{e}^{\mathrm{i}\Omega} - 0.5}. \end{aligned}$$

A plot of the magnitude and phase of $\underline{\mathcal{H}}(\Omega)$ is provided in Figure 8.13.

As previously stated, this example corresponds to a LPF, given the higher gain for small values of Ω. In general, it is the location of the poles and zeros which dictates the type and characteristics of the filter. For example, if an additional pole was added at the original position ($\underline{z} = 0.5$), the roll-off of the filter would be steeper. If the pole location was moved from $\underline{z} = 0.5$ to $\underline{z} = -0.5$ on the $\underline{z}$-plane, the resulting filter would have a high-pass characteristic. This simple example has shown how discrete-time filter characteristics can be studied using the z-transform.

8.4.2 Digital filter design

It is often convenient to describe a digital filter by a difference equation. Defining the filter input and output by $\underline{x}[n]$ and $\underline{y}[n]$, respectively, the following difference equation provides a general filter model:

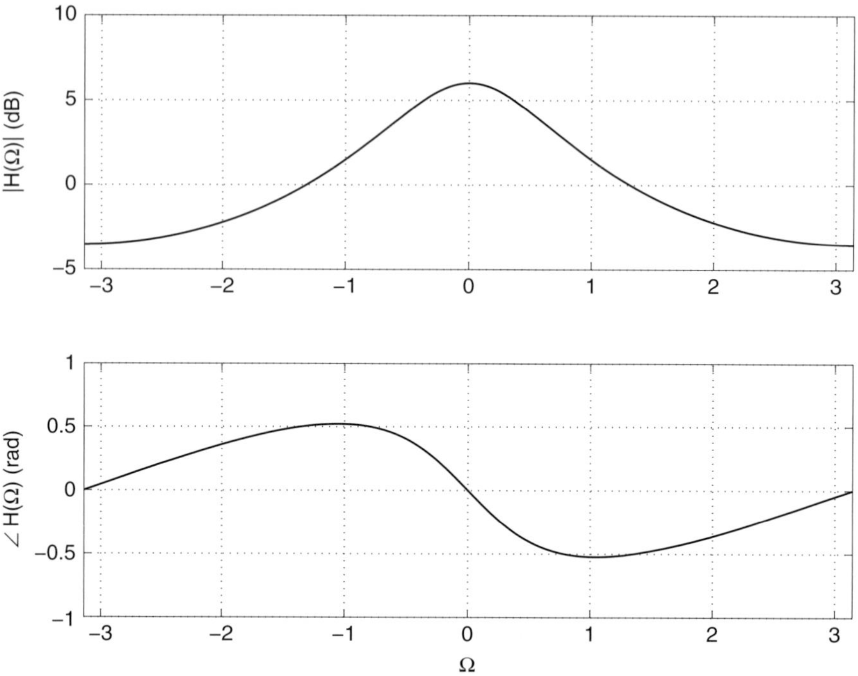

Figure 8.13 Frequency response of the example filter with impulse response $h[n] = 0.5^n u[n]$, denoted $\underline{\mathcal{H}}(\Omega)$ in the text. The magnitude in dB and the phase in radians are shown in the top and bottom panels, respectively.

$$\underline{y}[n] + \sum_{k=1}^{p} \underline{a}_k \underline{y}[n-k] = \sum_{k=0}^{q} \underline{b}_k \underline{x}[n-k], \qquad (8.37)$$

where $\underline{a}_k$ and $\underline{b}_k$ are possible complex filter coefficients for the infinite impulse response (IIR) and finite impulse response (FIR) parts of the filter, respectively. The filter order is denoted by p (IIR) and q (FIR). As the names imply, the impulse response of an IIR filter has an infinite length. The FIR filter impulse response is of finite length, which has computational advantages since a simple correlation process can be used to implement the filter. FIR filters are most often used in the matched filtering process of an MST radar. Using the delay property of the z-transform, the following transfer function of the digital filter can be obtained:

$$\underline{H}(\underline{z}) = \frac{\underline{Y}(\underline{z})}{\underline{X}(\underline{z})} = \frac{1 + \underline{b}_1 \underline{z}^{-1} + \underline{b}_2 \underline{z}^{-2} + \cdots + \underline{b}_q \underline{z}^{-p}}{1 + \underline{a}_1 \underline{z}^{-1} + \underline{a}_2 \underline{z}^{-2} + \cdots + \underline{a}_p \underline{z}^{-p}}, \qquad (8.38)$$

where $\underline{b}_0$ has been assumed to be equal to unity.

A simple example of a digital IIR ($p = 1$) LPF was given in the previous section. Nevertheless, an obvious question is how to choose the filter coefficients in (8.38). Many algorithms exist to estimate the filter parameters based on various forms of optimality

(*Oppenheim and Schafer*, 1975). One useful approach is to make use of the extensive literature on the design of analog filters and use a mapping from continuous to discrete time. This is the concept behind the *bilinear transformation* given below:

$$\underline{s} = \frac{2}{T_s}\frac{\underline{z}-1}{\underline{z}+1}. \tag{8.39}$$

This maps the continuous-time Laplace transform variable $\underline{s}$ to the discrete variable $\underline{z}$. Given a continuous-time transfer function, it is possible to convert the filter to discrete-time using the bilinear transformation. Care must be taken, however, in the application of this transformation since it represents a nonlinear mapping from continuous frequency to discrete frequency. As such, filter specifications given in continuous time may not hold for discrete time. This problem is well known and can be eliminated through a process called *prewarping* (*Kamen and Heck*, 2000).

Fortunately, several commercial software packages exist for the design of digital filters based on both analog design principles and direct digital implementations. Important filters include Butterworth, Chebyshev, and elliptic, for example, with each having its own advantages and disadvantages. It is important to recognize that these are examples of IIR filters. As mentioned earlier, FIR filters have some implementation advantages and are often used in MST radar applications.

An illustrative example of an FIR filter application is now provided. In this example, a discrete-time signal is generated by sampling the summation of three equal-amplitude sinusoids with frequencies of 5, 25, and 30 Hz. We will assume for simplicity that the sinusoids are purely real, described by

$$x[n] = \sin(2\pi 5n/100) + \sin(2\pi 25n/100) + \sin(2\pi 30n/100),$$

where the sampling rate was chosen to be 100 Hz. It is desired to filter this signal with an FIR filter in order to retrieve the 5 Hz signal. Obviously, an LPF is necessary and we will choose the cutoff frequency of the filter to be 15 Hz with an order of 64. For FIR filters, the order is equivalent to the length of the impulse response. Using a commercially available filter design package within the programming language *Matlab*, the filter shown in Figure 8.14 was obtained using the window method (*Oppenheim and Schafer*, 1975). Of course, the use of filter design packages, such as those provided in *Matlab*, eliminates the need for a deeper understanding of z-transforms, design theory, etc. However, this background theory is useful for studying more advanced concepts that arise in more advanced research areas.

The impulse response has the expected sinc shape given the rectangular shape of the desired LPF frequency response. The magnitude of the frequency response is shown in the bottom panel, from which it is apparent that the 25 and 30 Hz signals should be significantly attenuated in the output of the filter.

The previously mentioned input signal $x[n]$, made up of three sinusoids, is shown in the top panel of Figure 8.15. While we took x to be real, it could have been complex, so we now allow this. By convolving $\underline{x}[n]$ with the impulse response $\underline{h}[n]$ of the LPF, the output signal $\underline{y}[n] = \underline{x}[n] \otimes \underline{h}[n]$ was obtained (bottom panel).

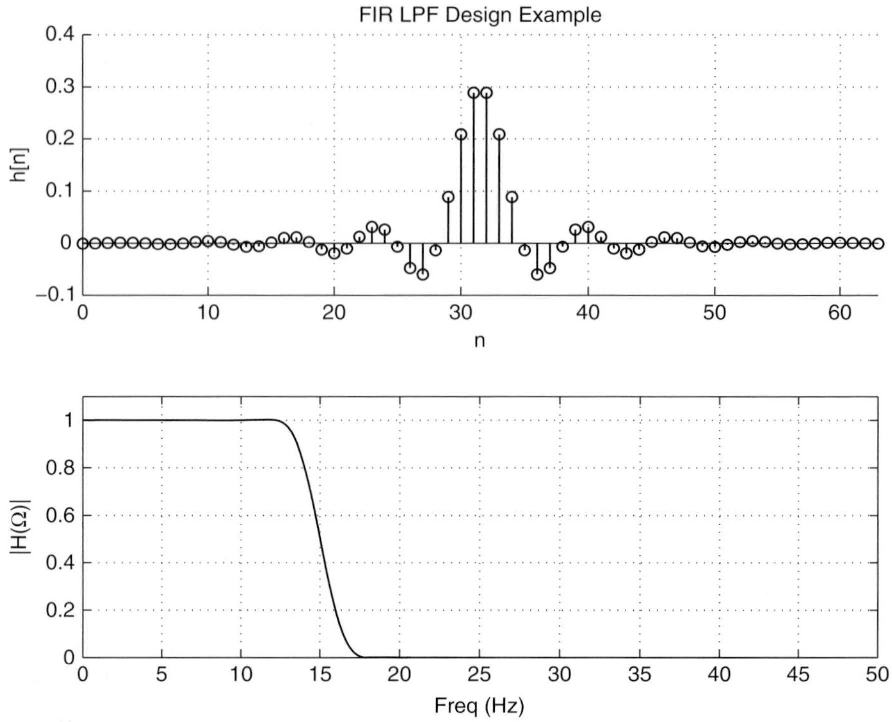

Figure 8.14 Example of FIR LPF with cutoff frequency of 15 Hz and order of 64. The top and bottom panels show the impulse response and magnitude of the frequency response, respectively. Note that the gain of the filter has dropped to 0.5 at the cutoff frequency.

From the results, it is fairly obvious that the LPF performed conceptually as desired, retaining the 5 Hz signal while suppressing the others. However, a word of warning must be sounded here. The impulse response should never be non-zero before the time that the impulse was applied; the response can only be non-zero after that time, by causality. This is resolved by recognizing that the assumed filter function has zero phase across all frequencies, whereas a real filter will have a frequency-dependent phase variation which forces the response to be zero at $t < 0$. However, for purposes of description of the nature of the frequency response, this slightly unrealistic model is adequate.

An interesting artifact is the near-zero level in $\underline{y}[n]$ at the beginning of the signal. This effect is caused by the initialization of the length-64 FIR filter at the beginning of the convolution process. Of course, this is well understood but does induce an inherent time delay in the filtered signal. It is interesting to note the analogy of this FIR filtering effect with pulse compression, described in Chapter 4. Compression of the transmitted pulse is equivalent to the convolution (filtering) process where the impulse response is replaced with the compression code. The delay seen here in the output of the filter is equivalent to the minimum height limitation seen in pulse compression applications.

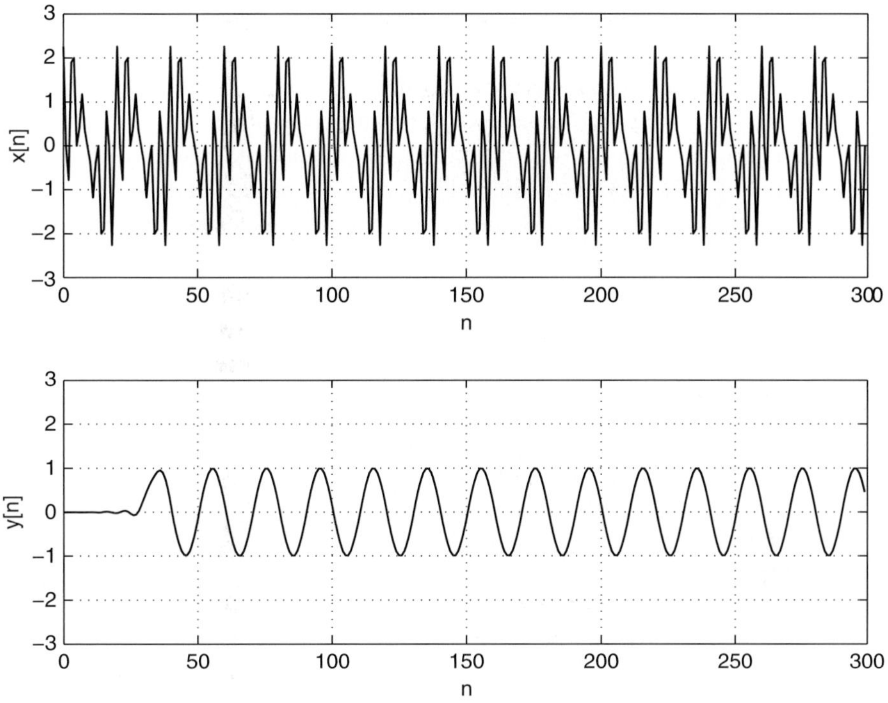

Figure 8.15 Example of a discrete-time signal made up of a summation of three sinusoids of frequencies 5, 25, and 30 Hz (top panel). Output sequence (bottom panel) of an order-64 LPF with a cutoff frequency of 15 Hz.

8.5 Review of random processes

In this section, a brief summary of the fundamental theory of random processes will be provided. A complete coverage would include probability, random variables, and statistics before attempting to study the topic of random processes. For our purposes, however, it will be assumed that the reader has a fundamental grasp of these topics. If this is not the case, numerous excellent texts on the subject provide sufficient detail for the interested reader (e.g., *Papoulis*, 1965; *Kay*, 1987).

Emphasis will be placed on the second-order statistics of the Doppler radar signals, since these are related to backscattered power and can typically be considered stationary over the dwell time of normal radar operations. Therefore, it will be assumed that the Doppler radar signal $\underline{y}[n]$ is a discrete random process and is wide-sense stationary (WSS), or second-order stationary. A discrete WSS random process adheres to two characteristics. The first is that the mean of the random process does not vary with time, viz.

$$\underline{\mu}_y = E\left\{\underline{y}[n]\right\} = \frac{1}{N}\sum_{n=0}^{N-1}\underline{y}[n]. \tag{8.40}$$

Here, we have used E to indicate the mean, or *expectation* value, of the time series. We have also taken the first point to be $\underline{y}[0]$, rather than $\underline{y}[1]$, i.e., we have used a summation convention starting at zero rather than unity, and we will continue with this convention for the rest of the chapter. Readers who prefer to start at $\underline{y}[1]$ will need to adjust the equations to suit. It is a little unusual to include the dummy of summation (n) inside the $E\{\underline{y}[n]\}$ expression, since it is not really needed here, but we include it simply to emphasize that $\underline{y}$ is a sequence. Of course n is needed in the second, more explicit, expression.

The second characteristic is that the autocovariance function (ACF) depends only on the lag k. As such, the ACF is given by the following equation:

$$\underline{r}_{yy}[k] = E\left\{\underline{y}[n]\underline{y}^*[n-k]\right\} = \frac{1}{N}\sum_{n=0}^{N-1} \underline{y}(n)\underline{y}^*(n-k), \tag{8.41}$$

where k is an integer and the $*$ represents the complex conjugate. Note that in the above equation, we need to take cases where $n-k$ is less than zero to have a value $\underline{y}(n-k) = 0$.

Note that if the autocovariance function is normalized to unity at zero lag, it is called the *autocorrelation* function by statisticians, although engineers use a slightly different definition, as discussed in Section 4.8.3 in Chapter 4.

Equivalently, (8.41) may be written as

$$\underline{r}_{yy}[k] = E\left\{\underline{y}^*[n]\underline{y}[n+k]\right\} = \frac{1}{N}\sum_{n=0}^{N-1} \underline{y}^*(n)\underline{y}(n+k). \tag{8.42}$$

This may readily be proven by substituting $n' = n - k$ in Equation (8.41) and then changing the dummy of summation n' back to n after the process is complete. In the latter case, we take cases where $n + k \geq N$ to have zero value.

The form of the equation shown in (8.42) is actually the more common definition of the ACF, since it leads more naturally to the correct definition of the cross-covariance function (see later). Nevertheless, we will still often use Equation (8.41) for the ACF in this chapter.

In future portions of this text, we will sometimes write the summation in (8.41) as

$$\sum_{n=0}^{N-1} \underline{y}(n)\underline{y}^*(n-k) = \sum_{n=0}^{N-1} \underline{y}^*(n-k)\underline{y}(n) = \mathbf{y_k}^{\dagger}\mathbf{y_0}, \tag{8.43}$$

where $\mathbf{y_0}$ is a *column matrix* given by

$$\mathbf{y}_0 = \begin{bmatrix} \underline{y}(0) \\ \underline{y}(1) \\ \vdots \\ \underline{y}(N-1) \end{bmatrix}, \tag{8.44}$$

(with $\underline{y}(i)$ being zero if $i < 0$), and where $\mathbf{y_k}^{\dagger}$ is a *row matrix* described by

$$\mathbf{y_k}^{\dagger} = \left[\underline{y}^*(-k), \underline{y}^*(-k+1), \ldots, \underline{y}^*(N-1-k)\right]. \tag{8.45}$$

Note that the bold-face **y** represents a matrix (possibly with complex elements), while a non-bold-face represents a scalar, and of course an underlined scalar indicates a complex scalar. Row matrices will generally have the symbol †, while column matrices do not have a † symbol. In this notation, a vector is often represented as a column matrix.

The operator † is called the *Hermitian operator*. The Hermitian operator acts as follows: for any matrix, the rows become the columns, and the columns become the rows. This process is called *transposing* the matrix: the first row is mapped to the first column, the second row becomes the second column, etc., and conversely. At the same time, all the elements of the matrix are converted to their complex conjugate. This process produces the so-called *Hermitian adjoint*.

Mathematically, we can write

$$\mathbf{y}^{\dagger} = \left(\mathbf{y}^{T}\right)^{*}, \tag{8.46}$$

where the superscript T means transpose, and the superscript $*$ means complex conjugate.

If the initial matrix is a row matrix, the Hermitian adjoint is a column matrix, and conversely. For scalars, the Hermitian is simply the complex conjugate. However, we will generally use a * for scalar complex conjugates, and † for matrix adjoints. If a matrix is a square matrix, and it happens that the matrix and its adjoint are equal, the matrix is referred to as *Hermitian*. We will employ this matrix and vector representation several times within this chapter.

Finally, we note that in Equations (8.41) and (8.42) we have divided by N. It actually makes sense to divide by $N - k$, since there are only $N - k$ overlapping values from the original time-series. This is commonly done in calculating the autocovariance function, e.g.,

$$\underline{r}_{yy}[k] = E\left\{\underline{y}[n]\underline{y}^{*}[n-k]\right\} = \frac{1}{N-k}\sum_{n=0}^{N-1}\underline{y}(n)\underline{y}^{*}(n-k). \tag{8.47}$$

This point will be discussed further later in this chapter.

It should be noted that the ACF is similar to the *autocovariance sequence* (ACS) for discrete random processes if the mean of the process is zero.

At first, the WSS assumption may not seem to be reasonable for Doppler radar signals, given the ever-changing nature of the atmosphere. However, the assumption is valid for the short dwell times over which many of our techniques are applied. Recall that a finite data set is one of our limitations of the radar signal and the dwell time is typically on the order of seconds to tens of seconds.

For many spectral estimation techniques, the ACF is constructed in the form of the covariance matrix. Assuming a zero-mean process, the $m \times m$ covariance matrix is given by the following equation:

$$\mathbf{R}_{y} = \begin{bmatrix} \underline{r}_{yy}[0] & \underline{r}_{yy}^{*}[1] & \cdots & \underline{r}_{yy}^{*}[m-1] \\ \underline{r}_{yy}[1] & \underline{r}_{yy}[0] & \cdots & \underline{r}_{yy}^{*}[m-2] \\ \vdots & \vdots & \ddots & \vdots \\ \underline{r}_{yy}[m-1] & \underline{r}_{yy}[m-2] & \cdots & \underline{r}_{yy}[0] \end{bmatrix}. \tag{8.48}$$

non-parametric methods. This emphasis is justified since atmospheric radar signals seldom follow any prescribed model. Furthermore, the atmosphere is a dynamic fluid and the backscattered signals are ever-changing and their Doppler frequency content is not known a priori.

For the Doppler radar spectral estimation problem, our goal is to estimate $S(\Omega)$ for $\Omega \in [-\pi, \pi)$ from a *finite-length* data set $\{\underline{y}[0], \ldots, \underline{y}[N-1]\}$. This is an ill-posed problem since we are attempting to estimate $S(\Omega)$ for *all* values (infinite) of $\Omega \in [-\pi, \pi)$ but only have N data points. Solutions to this classic estimation problem can be categorized into two approaches. The first approach uses finite-dimensional models to describe the underlying random process. As mentioned before, however, atmospheric radar signals do not lend themselves to such techniques. Moreover, the order of the model must be determined, further complicating the spectral estimation problem. The second approach is based on the concept of *resolution reduction*, which essentially means estimation of $S(\Omega)$ for only a finite set of Ω values. By approaching the problem in this manner, we are reducing the resolution of the spectral estimate and are implicitly assuming that the PSD is constant over the bandwidth of each frequency bin.

Techniques based on *resolution reduction* are also called *filter-bank methods* and will be the focus of our analyses of atmospheric radar data. The first technique described – *the periodogram* – is the most often used in the MST radar community due to its computational efficiency.

8.6.1 Periodogram and correlogram

The most commonly used spectral estimation method is the so-called *periodogram*. By using definition (8.52), after subtracting out the mean (expectation) value, and assuming a finite data set, the following spectral estimator can be obtained:

$$\lim_{N\to\infty} E\left\{\frac{1}{N}\left|\sum_{n=0}^{N-1}\underline{y}[n]e^{-i\Omega n}\right|^2\right\} \quad\Longrightarrow\quad \hat{S}_P(\Omega) = \frac{1}{N}\left|\sum_{n=0}^{N-1}\underline{y}[n]e^{-i\Omega n}\right|^2, \tag{8.53}$$

where the symbol ˆ represents an estimator.

Using definition (8.50) and truncating the ACF to the maximum lag of $N-1$, the so-called *correlogram* can be obtained, viz.,

$$\sum_{k=-\infty}^{\infty} \underline{r}_{yy}[k]e^{-i\Omega k} \quad\Longrightarrow\quad \hat{S}_C(\Omega) = \sum_{k=-(N-1)}^{N-1} \hat{\underline{r}}_{yy}[k]e^{-i\Omega k}. \tag{8.54}$$

Note that in (8.50), the true ACF was used. Of course, the true ACF is not known and must be estimated. It is typical to use the biased estimator of the ACF given by the following equation:

$$\begin{aligned} \hat{\underline{r}}_{yy}[k] &= \frac{1}{N}\sum_{n=k}^{N-1}\underline{y}[n]\underline{y}^*[n-k] \qquad 0 \le k \le N-1 \\ \hat{\underline{r}}_{yy}[-k] &= \hat{\underline{r}}^*_{yy}[k]. \end{aligned} \tag{8.55}$$

Note a few changes in this formula compared to Equations (8.41), (8.42), and (8.47). First, the limits seem different. However, the expression is the same – in the above equations, we allowed a full coverage of values of n, but set parameters that were outside the index range 0 to $N-1$ to zero. Here, we simply do not perform the calculations involving these artificially constructed zeros. In addition, we do not bother with k values less than zero, since we know that the ideal ACF should obey the second equation in (8.55), so we avoid the extra computational load of evaluating them, and simply map them from the $k \geq 0$ case. In addition, we have divided by N out the front, but in Equation (8.47) it was suggested it may be better to divide by $N-k$. We have retained division by N here because it simplifies the conceptual mapping between the autocovariance and spectrum, but in a practical sense it does make sense to do the division. However, it is also wise in a practical sense to limit the number of lags k – typically it is not advisable to use lags in excess of 20% of the total data-length.

The biased ACF estimator produces a positive semidefinite sequence, which implies that negative values of the PSD estimates cannot occur. Of course, this is an extremely valuable characteristic and is the reason why the biased ACF estimator is most often used.

Several comments related to $\hat{S}_P(\Omega)$ and $\hat{S}_C(\Omega)$ are now presented:

- It can easily be shown that $\hat{S}_P(\Omega) = \hat{S}_C(\Omega)$. As a result and because of computational efficiency, it is typical in MST radar applications to use $\hat{S}_P(\Omega)$.
- $\hat{S}_P(\Omega)$ and $\hat{S}_C(\Omega)$ are statistically unreliable estimates. It will be shown that these estimates of the PSD have a variance, or uncertainty, which does not decrease with increasing data length. Methods to mitigate this limitation will be presented.
- As emphasized previously, the DTFT uses a frequency variable Ω which is continuous. For implementation, the discrete Fourier transform (DFT), or fast Fourier transform (FFT), is used which essentially samples Ω over the $-\pi$ to π domain. Values of Ω_k are given by

$$\begin{aligned}\Omega_k &= \frac{2\pi}{N}k \quad k = 0, \ldots, N-1 \\ &= \left\{0, \frac{2\pi}{N}, 2\frac{2\pi}{N}, \ldots, (N-1)\frac{2\pi}{N}\right\}.\end{aligned}$$

  Therefore, the periodogram spectral estimator can be written in the following form:

$$\hat{S}_P(\Omega_k) = \hat{S}_P\left(\frac{2\pi}{N}k\right) = \frac{1}{N}\left|\sum_{n=0}^{N-1} \underline{y}[n]\mathrm{e}^{-\mathrm{i}\frac{2\pi}{N}kn}\right|^2,$$

  where the subscript k represents the sampled frequency index.
- When N is not a power of 2, or when more spectral samples are desired, the data may be *zero padded* to a higher power of 2 to take advantage of the FFT algorithm, e.g.,

$$\underbrace{\{\underline{y}[0], \underline{y}[1], \ldots, \underline{y}[N-1], 0, 0, \ldots, 0\}}_{\text{length } L\ (>N)}$$

$$\Omega_k = \frac{2\pi}{L}k \qquad 0 \le k \le L-1.$$

However, algorithms also exist that allow dealing directly with data-sets which are not factors of 2, but can be factorized in other ways. For example, if there are 154 points in the sequence, this can be factored as $2 \times 7 \times \times 11$, and this factorization can be employed to perform an FFT-like algorithm which still has efficient computational ability. This is referred to as a mixed-radix method. An example is presented by *Singleton* (1969).

Returning to the topic of zero-padding, we note that since the actual length of the data-set is the same, no increase in resolution is possible by zero padding. The zero-padding process is a type of interpolation. An example of zero-padded periodogram PSD estimates is shown in Figure 8.16 for three different amounts of padding. The spectral estimates have been plotted with circles to emphasize the interpolation effect.

Zero-padding is a nice interpolation procedure in that it introduces no new frequencies. If a time series is Fourier transformed, then the Fourier transform is zero-padded, and transformed back to the original time series, a new time series appears with additional points filled in between the previous existing points. The new data-set

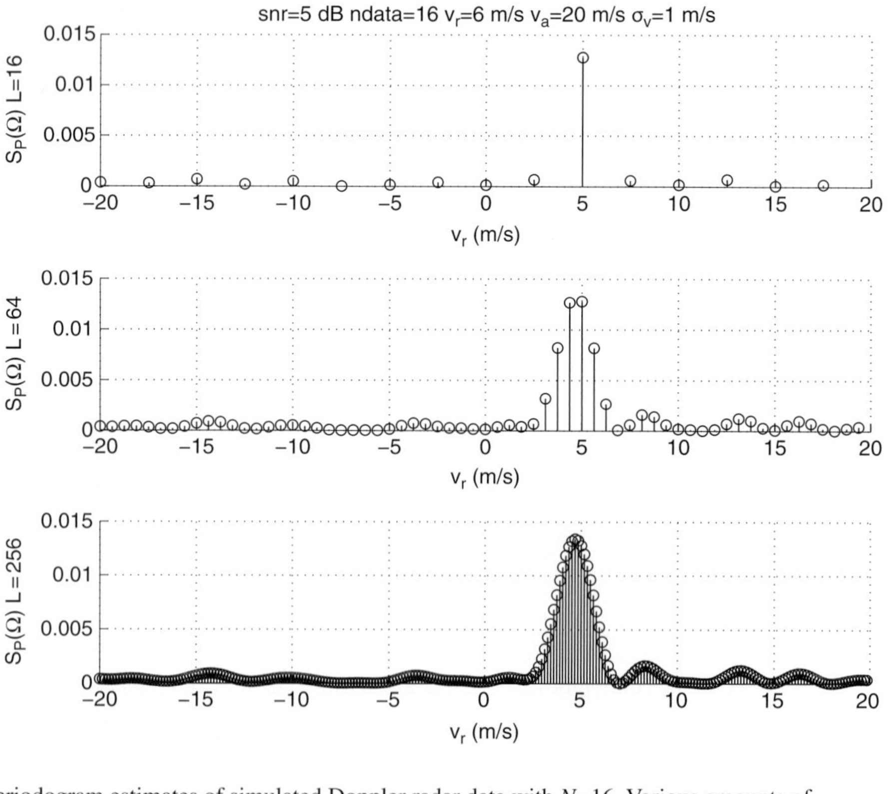

Figure 8.16 Periodogram estimates of simulated Doppler radar data with N=16. Various amounts of zero-padding were implemented.

can be assured of being a true interpolation without introduction of *any* new Fourier components.

- The DFT, which is used for the periodogram, is plotted versus samples of Ω over $\pm\pi$, which are in units of rad/sample. In order to convert to more useful units (ms^{-1}) for MST radar applications, the following conversion is used, as illustrated in Figure 8.16:

$$v_r = -\frac{\lambda}{4\pi T_s}\Omega. \tag{8.56}$$

Error analysis of the periodogram

Any estimator must be analyzed in order to determine the statistical significance of the results. The mean-squared error (MSE) is often used to judge the *quality* of an estimator. This stems from the fact that the MSE is given by the sum of two terms related to bias and variance of the estimator. For the periodogram, the MSE would be given by the following equation:

$$\mathrm{MSE} = \mathrm{var}\left\{\hat{S}_P(\Omega)\right\} + \left|\mathrm{bias}\left\{\hat{S}_P(\Omega)\right\}\right|^2. \tag{8.57}$$

For the case of the periodogram, we will now study the MSE by investigating the bias and variance terms separately.

The *bias* of any estimator is calculated by first taking the expected value of the estimator. An unbiased estimator (a desirable quality) would have an expected value equal to the actual value of the parameter being estimated. Given the equivalence of the periodogram and correlogram, the bias for both methods can be calculated by using the form of the correlogram alone. We use

$$\begin{aligned} E\{\hat{S}_P(\Omega)\} &= E\left\{\sum_{k=-(N-1)}^{N-1} \hat{\underline{r}}_{yy}[k]\mathrm{e}^{-\mathrm{i}\Omega k}\right\} \\ &= \sum_{k=-(N-1)}^{N-1} E\left\{\hat{\underline{r}}_{yy}[k]\right\}\mathrm{e}^{-\mathrm{i}\Omega k}. \end{aligned}$$

The expected value of the biased ACF estimate is given by the following equation:

$$E\left\{\hat{\underline{r}}_{yy}[k]\right\} = \left(1 - \frac{|k|}{N}\right)\underline{r}_{yy}[k].$$

Therefore, the expected value of the periodogram has the following form:

$$E\left\{\hat{S}_P(\Omega)\right\} = \sum_{k=-\infty}^{\infty} w[k]\underline{r}_{yy}[k]\mathrm{e}^{-\mathrm{i}\Omega k}, \tag{8.58}$$

where

$$w[k] = \begin{cases} 1 - \frac{|k|}{N} & , \quad k = 0, \pm 1, \ldots \\ 0 & , \quad \text{else} \end{cases} \tag{8.59}$$

is a so-called *window function* in the lag-k domain. This specific form of the window function is called the *Bartlett* window, or *triangular* window.

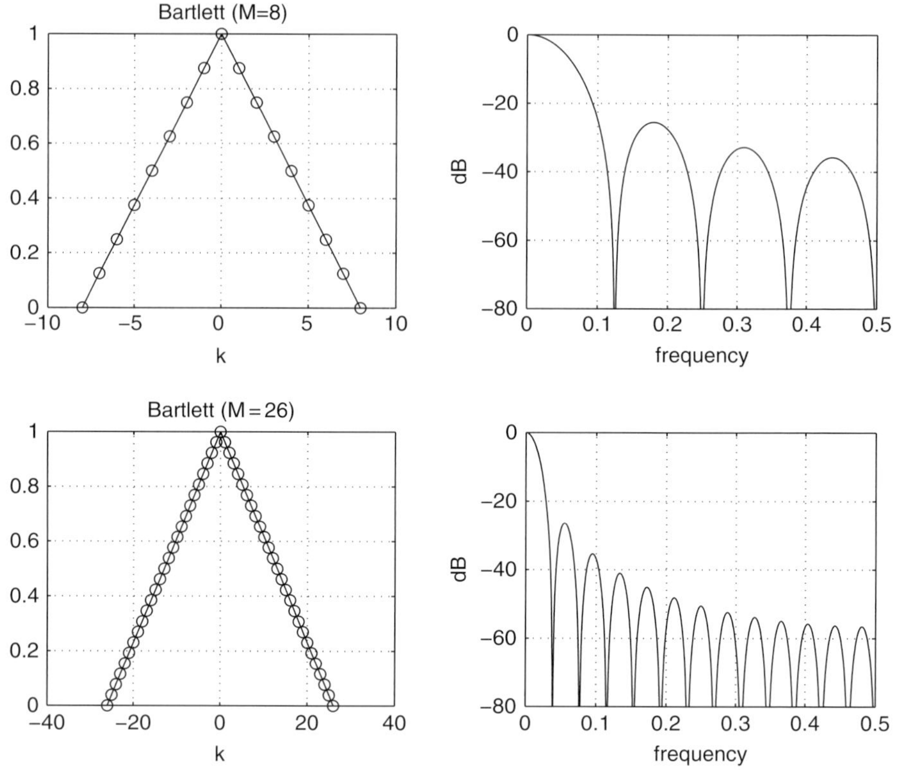

Figure 8.22 Time and frequency domain Bartlett window for M equal to 8 and 26.

Examples of the Bartlett and Hanning lag windows are provided in Figures 8.22 and 8.23, respectively. The figures show both the time and frequency domain characteristics of the windows for two different values of M.

As stated previously, a smoother transition from the peak to the zero of the lag window generally results in lower side-lobes and a wider main lobe.

Given that the periodogram is used in practice more often than the correlogram, due to the efficiency of the FFT algorithm, it is typical to directly window the time-series data (temporal window) rather than the ACF estimates, as is represented in Equation (8.54). It can be shown that the lag window is equal to the autocorrelation function of the temporal window. Therefore, if the temporal window is designed with a smooth transition from the peak to the zero, the resulting lag window would have a similar characteristic. This relationship should be considered when designing the temporal window function used with the periodogram algorithm.

8.6.3 Averaged periodogram method – Bartlett method

Using the Blackman–Tukey method, the variance of the spectral estimate was reduced by windowing the ACF estimates before calculation of the Fourier transform. As a result,

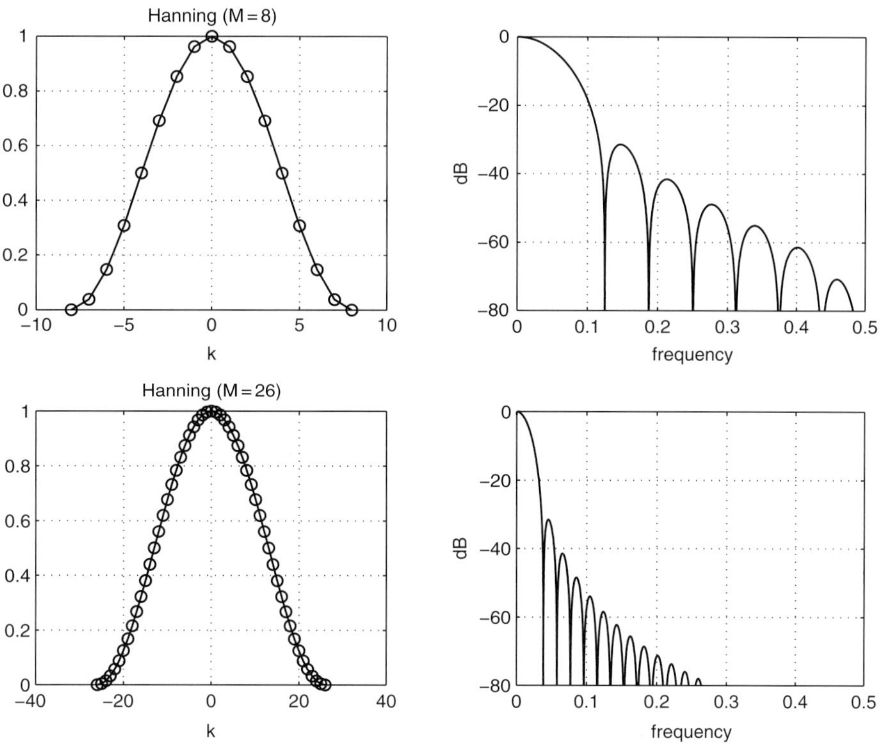

Figure 8.23 Time and frequency domain Hanning window for M equal to 8 and 26.

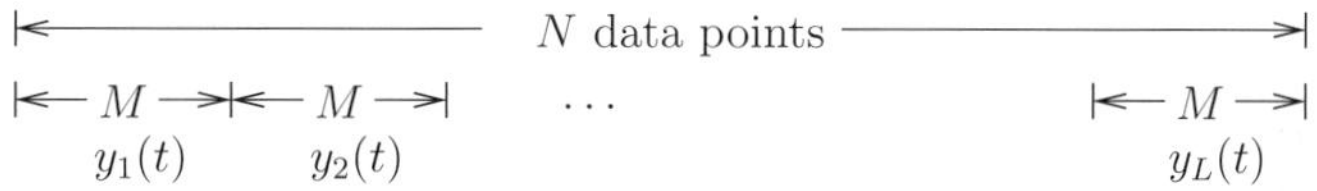

Figure 8.24 The method by which data sets are segmented for the Bartlett method (averaged periodogram).

the effect of the inferior estimates of the ACF for large lag was reduced. Of course, it should be apparent that this procedure trades resolution for reduced variance. Another, more obvious, way to reduce variance would be to simply average independent periodogram estimates of the PSD. Of course, if the data length were kept constant for each of these periodogram estimates, the temporal resolution would be reduced. In this case, one would need to consider stationarity in the underlying process.

Another approach to variance reduction via averaging could be to segment the original data set into a number of smaller-length data sets and calculate the periodogram of each. This concept is illustrated in Figure 8.24. Of course, the shorter data-sets would result in reduced frequency resolution but the averaging process would, in turn, produce a reduction in variance of the spectral estimate. The technique is often used in atmospheric radar applications and is appropriately called the *average periodogram method*, or the *Bartlett method* (*Bartlett*, 1950).

The last few special cases dealt with multiplication in the time domain and convolution in the frequency domain, with the intent of smoothing the spectra. Next, we turn to a different strategy. Here, we will not try to smooth the data, but to find better ways to suppress the effects of unwanted spectral lines. This requires us to remember that after signal is passed through a filter, the resultant spectrum is a *product* between the "true" spectrum and the receiver spectrum in the *frequency domain*, which corresponds to a *convolution* in the *time* domain.

8.6.5 Capon method

As stated previously, the periodogram and its variants can be thought of as filter-bank methods where the filter design process is a result of the Fourier transform. The periodogram filter has an FIR form with filter coefficients derived from the Fourier transform vector with finite length. As such, significant, adverse side lobe effects were evident and it was necessary to develop mitigation schemes, such as windowing and spectral averaging. It is interesting to note that many of the problems with the periodogram could be avoided if the input data were considered when designing the filter. For example, if it were known that significant ground clutter (zero Doppler frequency) existed in the signal, the filter coefficients used for the filter-bank spectral estimation could be designed to produce a null at frequencies near zero. If this process of adjusting the filter coefficients to the data could be achieved in general, a so-called *adaptive* algorithm would result. Just such an algorithm was originally developed for seismic applications by *Capon* (1969) and is appropriately termed the *Capon* method but is also referred to as the *minimum variance* method.

As a reminder of the process of filtering, consider Figure 8.26. In this figure, an input waveform is split and passed through a number (in this case ten) of different filters. For each filter, the resultant spectrum is the complex product of the original spectrum (real and imaginary components) and the filter function (also complex). However, the output is still a time series, denoted $\underline{\tilde{y}}_k$, where k varies from 1 to the total number of filters. The user then finds the variance of the resultant time series. This equals the area under the combined spectrum, by Parseval's theorem (Equation 8.18).

This filter bank could easily be built in a laboratory, and indeed represents a realistic instrument that might have been built in the days before fast computers.

Examination of Figure 8.26 shows that the output of filter 1 will be dominated by noise, while filter 2 will be dominated by spectral spike A. Spectral spike B will dominantly affect filter 3, and spike C will best be seen through filter 4. Spectral spike D will appear mainly in filter 6, but spike E is split between filters 8 and 9. This can be remedied by choosing finer frequency resolution, and it may also be possible to choose filters with narrower widths. However, we do not have complete freedom in choosing these filter shapes, and allowed choices are dependent on data length and hardware filter selection.

But perhaps the main point to notice can be shown with filter 3 and spectral line B. It may be seen that spectral line A passes right through the first side-lobe of the filter – an

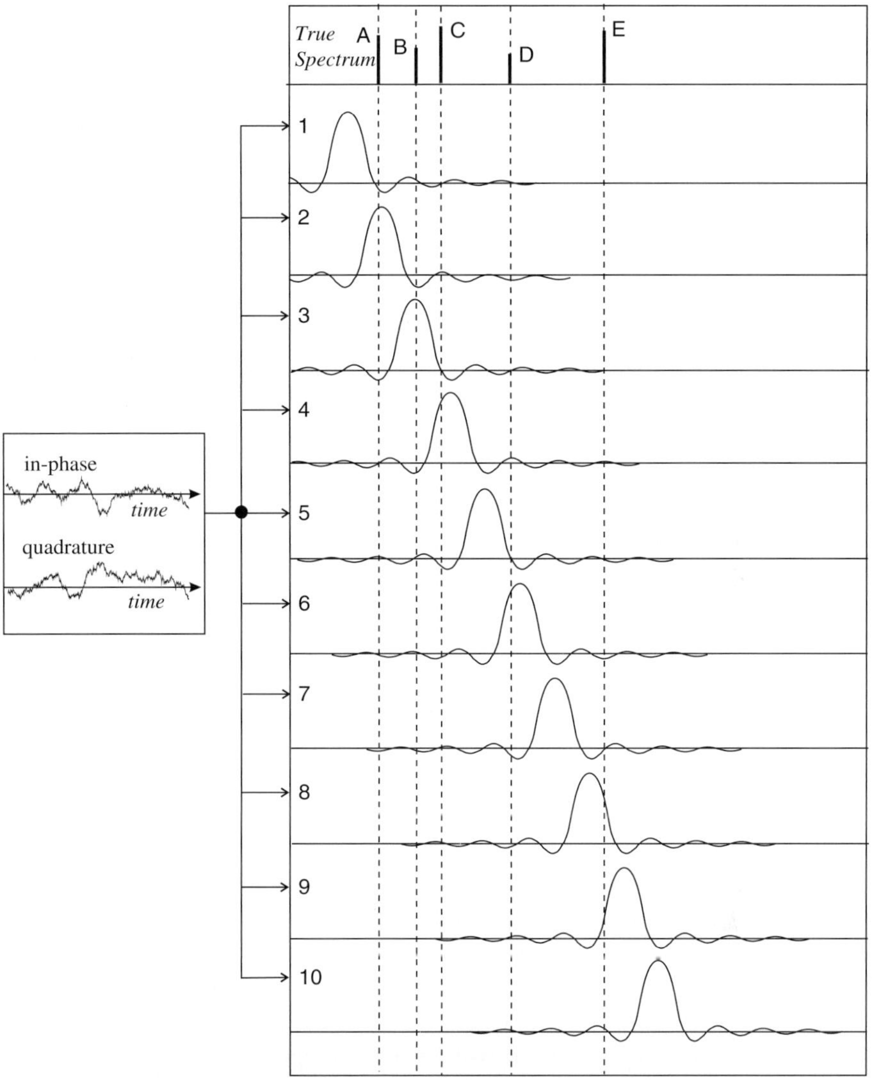

Figure 8.26 Application of a filter bank. The signal on the left is split into a number of identical waveforms, and each is passed through a different filter. Only the real part of the filter function is shown – an imaginary part also exists. The same is true for the "true" spectrum. In this case, all filters are identical, but are slid to the right by an equal frequency step from filter to filter. See text for more details.

undesirable effect. So spectral line A will contaminate the output of filter 3, assuming the filter is designed to emphasize spectral line B.

Our purpose is as follows: to design a bank of filters that may *vary* in shape for each different frequency offset, and designed in such a way that *all* of the dominant spectral lines except the one of interest coincide with points where the filter function is zero.

This will reduce contamination in each filter, allowing emphasis of (optimally) only one spectral line in the filter output.

Note that some texts discuss Capon's method as a way to eliminate *unwanted* spectral lines. This oversimplifies the situation, and is only one special case – the intent is in fact to eliminate the effect of *all* spectral lines except the one for which the filter is currently being designed. The idea is to sweep through the spectrum, locate the dominant spectral lines, and for each such line design a filter that maximizes throughput of that line but suppresses the impact of all the others.

Mathematically, we would like to design an FIR filter (essentially a bandpass filter, as shown in Figure 8.26), with an impulse response which is represented by a sequence of numbers $\{\underline{h}_0, \underline{h}_1, \ldots, \underline{h}_m\}$, and this sequence is *data-dependent*. We will *choose* a suitable value for the integer m, which will remain fixed throughout the development. We cover the interval from 0 to m, so there are in fact $m+1$ filters. The value $m+1$ represents the length (number of points) of the temporal function which will be convolved with the original time series.

In order to simplify the mathematics, we now define a vector **h** as follows:

$$\mathbf{h} = \begin{bmatrix} \underline{h}_0^* \\ \underline{h}_1^* \\ \vdots \\ \underline{h}_m^* \end{bmatrix}. \tag{8.65}$$

Then

$$\mathbf{h}^\dagger = \left[\underline{h}_0, \underline{h}_1, \cdots, \underline{h}_m\right]. \tag{8.66}$$

Note that the elements of **h** are $\underline{h}_j^*$, rather than $\underline{h}_j$. This is in contrast to Equation (8.44), where the elements are not asterisked. However, this is still an acceptable convention, as long as $\mathbf{h}^\dagger$ contains the complex conjugates of **h** as its elements. By using this approach we introduce better symmetry into the ensuing equations.

The actual elements of the convolving function that we seek are still $\underline{h}_0, \underline{h}_1, \ldots, \underline{h}_m$. **h** and $\mathbf{h}^\dagger$ are simply used to make the equations more concise.

The Fourier transform will be the appropriate filter function. Because **h** varies in functional form, depending on the selected frequency, we will denote each **h** with a subscript to associate it with the selected normalized frequency.

Basics of the method

The design process will now be summarized. We emphasize that the following discussion is not required if the user simply wants to obtain the desired answer, and indeed many of the functions we develop in this section need never be determined by the user. If only a final answer is required, skip immediately to Equation (8.75). The next few pages are designed to assist the reader who desires an understanding about the principle of the method (which we highly recommend, since better understanding is a valuable tool for better interpretation).

We consider different filters for each selected frequency, with the selected frequency being denote by Ω_s. We will eventually consider the calculations for a variety of choices of Ω_s (generally chosen by stepping the value from 0 through to 2π in step sizes chosen by the user), and we will refer to the selected frequency at each step as the "frequency of optimum response" (for the filter, not necessarily for the data), or FOOR.

We now consider one particular choice of Ω_s. The output time series of the filter designed for frequency Ω_s, denoted by $\underline{\tilde{y}}_{\Omega_s}[n]$, is given by a convolution, as shown below:

$$\underline{\tilde{y}}_{\Omega_s}[n] = \sum_{k=0}^{m} \underline{h}_{\Omega_s,k}\underline{y}[n-k] = \mathbf{h}_{\Omega_s}^{\dagger} \begin{bmatrix} \underline{y}[n] \\ \underline{y}[n-1] \\ \vdots \\ \underline{y}[n-m] \end{bmatrix}.$$

Note that the convolution does not involve any complex conjugates of $\{\underline{h}\}$ or $\{\underline{y}\}$, in contrast to the cross-covariance, which does (e.g. Equation (8.49)). The time series $\underline{\tilde{y}}_{\Omega_s}[n]$ will differ for each frequency Ω_s considered.

As discussed in regard to Figure 8.26, the output we are interested in is simply the total power produced by the filter – which equals both the area under the combined spectrum *and* the variance of the signal $\underline{\tilde{y}}_{\Omega_s}$. The two are equal by Parseval's theorem. We will consider the signal variance.

Then the corresponding output power for a selected normalized frequency Ω_s is given, after subtraction of the mean, by the following equation:

$$E\{|\underline{\tilde{y}}_{\Omega_s}[n]|^2\} = E\left\{ \mathbf{h}_{\Omega_s}^{\dagger} \begin{bmatrix} \underline{y}[n] \\ \underline{y}[n-1] \\ \vdots \\ \underline{y}[n-m] \end{bmatrix} \begin{bmatrix} \underline{y}^*[n] & \underline{y}^*[n-1] & \cdots & \underline{y}^*[n-m] \end{bmatrix} \mathbf{h}_{\Omega_s} \right\}. \tag{8.67}$$

Note that in this equation, the resultant is an *expectation, or mean value*, and it is formed by averaging over all N points in the time series, just as we did in Equations (8.41) to (8.48). In other words, the term inside the curly brackets must be found for all n, from $n = 0$ to $n = N - 1$, and the terms need to be then added and divided by N. In cases where the index μ of the $\underline{y}$ element is less than zero (e.g. $n < m$, so $\mu = n - m < 0$), the value of $\underline{y}(\mu)$ is set to zero.

Equation (8.67) can therefore be reduced to

$$E\{|\underline{\tilde{y}}_{\Omega_s}[n]|^2\} = \mathbf{h}_{\Omega_s}^{\dagger} \mathbf{R}_y \mathbf{h}_{\Omega_s}, \tag{8.68}$$

where $\mathbf{R}_y$ is the autocovariance matrix of actual data (see Equation (8.48)).

As we saw in Section 8.4.1, the frequency response (from the transfer function), of the bandpass filter is obtained by taking the DTFT of the impulse response. Adapting Equation (8.24), with L becoming m, and $\underline{x}$ changing to $\underline{h}_{\Omega_s}$, and using $\underline{H}_{\Omega_s}$ in place of $\underline{X}$, and evaluating the expansion *only* for the special case that $\Omega = \Omega_s$, gives

$$\underline{H}_{\Omega_s}(\Omega_s) = \sum_{k=0}^{m-1} \underline{h}_{\Omega_s}[k] \mathrm{e}^{-\mathrm{i}\Omega_s n}. \tag{8.69}$$

By writing this in matrix/vector notation, the following form of the frequency response is produced:

$$\underline{H}_{\Omega_s}(\Omega_s) = \mathbf{h}_{\Omega_s}^{\dagger} \begin{bmatrix} 1 \\ \mathrm{e}^{-\mathrm{i}\Omega_s} \\ \vdots \\ \mathrm{e}^{-\mathrm{i}m\Omega_s} \end{bmatrix} = \mathbf{h}_{\Omega_s}^{\dagger} \mathbf{a}_{\Omega_s}(\Omega_s), \tag{8.70}$$

where by definition we have that

$$\mathbf{a}_{\Omega_s}(\Omega) = \begin{bmatrix} 1 \\ \mathrm{e}^{-\mathrm{i}\Omega} \\ \vdots \\ \mathrm{e}^{-\mathrm{i}m\Omega} \end{bmatrix}. \tag{8.71}$$

Note that $\underline{H}_{\Omega_s}(\Omega_s)$ is a scalar, since it is only evaluated at $\Omega = \Omega_s$. Nevertheless, a generalized function $\underline{\mathcal{H}}_{\Omega_s}(\boldsymbol{\Omega})$ can be found, which represents the full range of frequencies available. We use the symbol $\underline{\mathcal{H}}$ to emphasize that it is a function of a vector, rather than a function of a scalar, although numerically its value at $\Omega = \Omega_s$ is just $\underline{H}_{\Omega_s}(\Omega_s)$. To calculate this function, simply apply the expression for $\underline{H}_{\Omega_s}$ given in Equations (8.69), or equivalently (8.71), for each element of the vector $\boldsymbol{\Omega}$ given by

$$\boldsymbol{\Omega}(\Delta\Omega, L) = \begin{bmatrix} 1 \\ \mathrm{e}^{-\mathrm{i}\Delta\Omega} \\ \vdots \\ \mathrm{e}^{-\mathrm{i}\ell\Delta\Omega} \\ \vdots \\ \mathrm{e}^{-\mathrm{i}(L-1)\Delta\Omega} \end{bmatrix}, \tag{8.72}$$

where $\Delta\Omega$ is the selected step size between successive frequencies (usually chosen to be very small, to give optimal visual resolution), and for all ℓ from 0 to $L-1$, where $L = \frac{2\pi}{\Delta\Omega}$. $\Delta\Omega$ should also be chosen so that Ω_s is included in the sequence, or at least lies very close to an element of the sequence.

Although we never explicitly use this form in our calculations, it is often useful to look at the shape of $\underline{\mathcal{H}}_{\Omega_s}$ as a function of the generalized frequency, and examples will be given later.

We now return to discussion of how to optimally design our filter. One design criterion will be to minimize the output power of the filter. By doing so, we will be able to design a filter with minimal gain outside the passband of the filter. However, we must also constrain the frequency response to have unity gain at the center frequency or the first criterion would force the filter coefficients to be zero! This constrained design criterion can be stated in the following minimization problem:

$$\min_{\mathbf{h}} \mathbf{h}_{\Omega_s}^{\dagger} \mathbf{R}_y \mathbf{h}_{\Omega_s} \qquad \text{subject to} \qquad \mathbf{h}_{\Omega_s}^{\dagger} \left[\mathbf{a}_{\Omega_s}(\Omega_s)\right] = 1. \tag{8.73}$$

The obvious advantage for such an adaptive design concept is that out-of-band interference will automatically be minimized, since the best way to achieve such a minimization is to place the frequencies with largest power in regions where the filter response is zero.

The solution to (8.73) can be obtained using standard constrained optimization approaches (*Luenberger*, 1984) with Lagrange multipliers (*Palmer et al.*, 1998). By minimizing output power with the constraint of unity gain at the center frequency of the filter, the following FIR filter coefficients are derived:

$$\mathbf{h}_{\Omega_s} = \frac{\mathbf{R}_y^{-1}\left[\mathbf{a}_{\Omega_s}(\Omega_s)\right]}{\left[\mathbf{a}_{\Omega_s}{}^{\dagger}(\Omega_s)\right]\mathbf{R}_y^{-1}\left[\mathbf{a}_{\Omega_s}(\Omega_s)\right]}. \tag{8.74}$$

Note that the filter impulse response depends on the data through $\mathbf{R}_y^{-1}$. Therefore, this spectral estimation method is *data-dependent*, or *adaptive*, unlike the periodogram.

When considering computational complexity, it is also important to note that the inverse of the covariance matrix need only be computed once for all FOOR. Frequency selectivity is achieved through changing Ω_s in (8.74) for the frequency band of interest, and is normally swept over the entire band of frequencies. The output power is used as a spectral estimate, which by definition is the power of the random process over all frequencies after weighting by the applied filter.

An example of the frequency response of the Capon bandpass filter is shown in Figure 8.27. Here, we have plotted the function $\mathcal{H}_{\Omega_s}$ as a function of the elements of Ω at very close steps, giving rise to a function that appears continuous. For this example, input data were simulated with five complex sinusoids with normalized frequencies of ± 0.2, ± 0.3, and 0.09, which are shown as the unity amplitude pulses. In this case, no noise is added, so as to best illustrate the Capon filter's interference rejection capabilities.

In this example, the Capon filter can be thought of as a bandpass filter with FOOR at 0.0 and -0.1 normalized frequency for the top and second panels, respectively. Although the frequency response looks similar to what one might expect from the periodogram (peak at the center frequency and side-lobes outside the main lobe), an important difference is how the nulls in the response are adaptively "placed" at frequencies where undesired signals (or at least undesired frequencies for this particular choice of FOOR) occur. As expected by the unity constraint in (8.73), the gain at the FOOR of the filter is always unity. Since the design criterion of the Capon filter attempts to minimize output power, the resulting frequency response automatically places nulls at the strongest signals outside the region of the FOOR. In both the upper and second panels, for example, note the obvious nulls at ± 0.2, ± 0.3, and 0.09. By attenuating the interfering signals, higher resolution and reduced bias (i.e., "spectral leakage") can in principle be obtained over that of the periodogram.

However, note that when the FOOR coincides with a large spectral peak in the original data, the filter actually achieves a **maximum** *(or close to it) at this point, rather than a null*, as seen in Figure 8.27(c).

The situation is seen even more clearly in Figure 8.28,which shows a contour plot of successive filter response curves $\mathcal{H}_{\Omega_s}$ for different choices of FOOR, Ω_s. The normalized values of Ω_s are plotted on the ordinate. The depth of grey coloring represents the

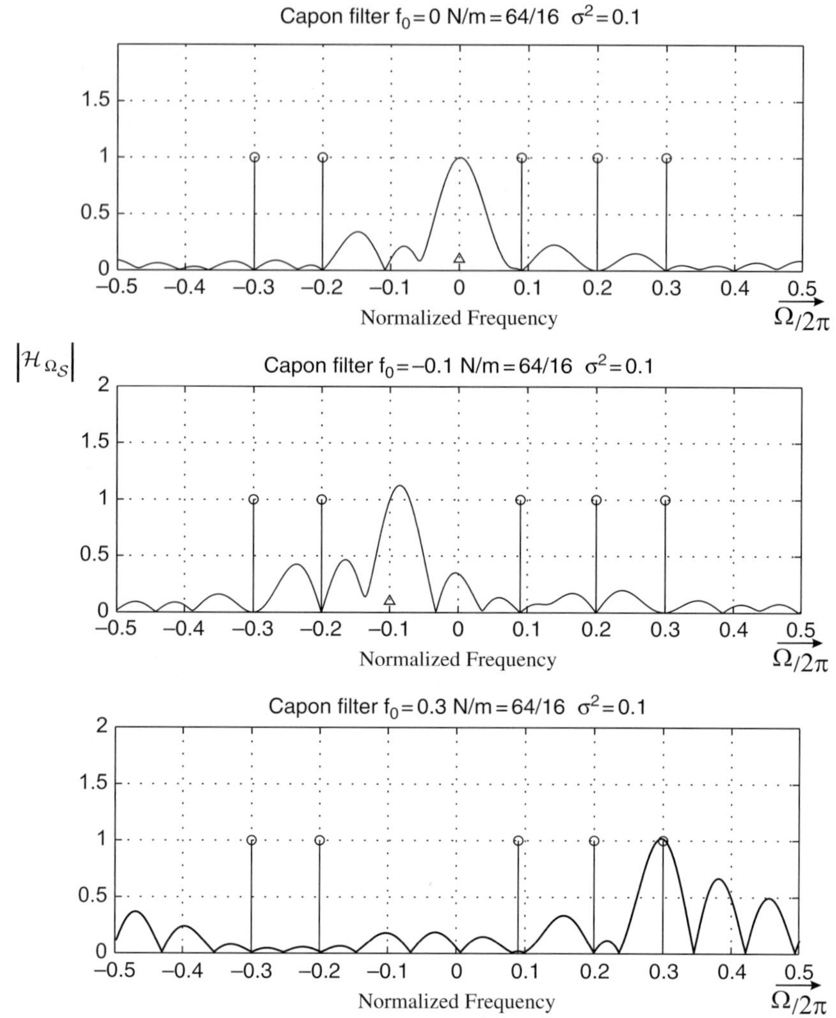

Figure 8.27 Frequency response $\mathcal{H}_{\Omega_s}$ of the Capon filter for a simulation in which the original time-series was formed using five complex sinusoids with normalized frequencies of ± 0.2, ± 0.3, and 0.09. The FOOR (frequencies of optimum response) of the filter for the top and middle panels are 0.0 and -0.1 normalized frequency, respectively (where we have further normalized the Ω by dividing by 2π). For the lowest panel, the FOOR is 0.3. The thin vertical lines with small circles on top show the lines of interest (these being used to create the original time series), and are *not* part of the filter. Note that the upper two graphs have nulls at all of the spectral lines used to create the original signal, since the chosen FOOR are *not* one of the five spectral lines used to create the time series. The lowest graph also has nulls at most, but not all, of the spectral lines – at the FOOR it is a maximum, since this frequency was one of the lines used to create the original time series, and this filter is designed to accept this frequency component.

response $\mathcal{H}_{\Omega_s}$, (with light shading representing the strongest response and dark shading the weakest) and the normalized frequencies Ω used to determine $\mathcal{H}_{\Omega_s}$ are plotted on the abscissa. The values of $\underline{H}_{\Omega_s}(\Omega_s)$ (Equation (8.70)) correspond to the points where $\Omega = \Omega_s$, i.e. the bright line sweeping diagonally from bottom right to upper left. Note

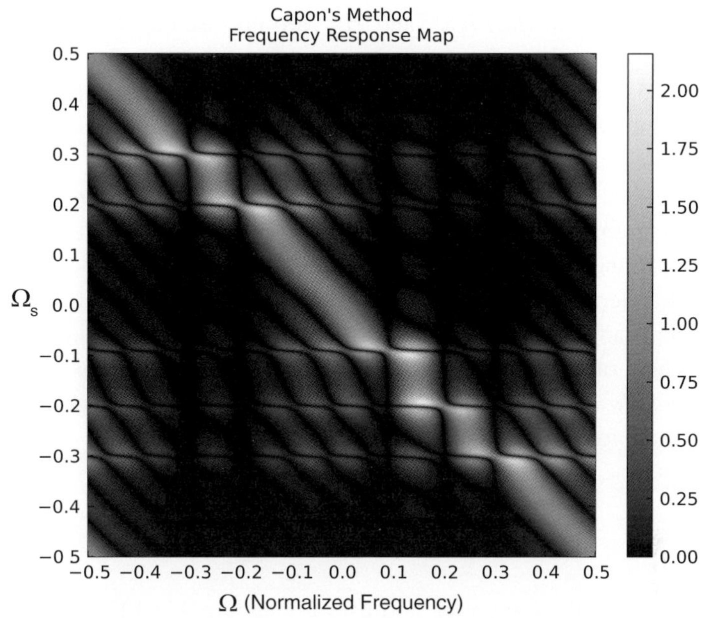

Figure 8.28 A contour map generalizing the three graphs shown in Figure 8.27. A succession of filters like those in Figure 8.27 were calculated, for different FOOR (Ω_S), with each response function ($\mathcal{H}_{\Omega_s}$) being plotted as a function of the normalized frequency over all frequencies. The FOOR is shown on the ordinate, and the frequencies swept for each spectrum are shown on the abscissa. (This map and the third graph in Figure 8.27 were produced by Marcial Garbanzo-Salas.)

that at the dominant frequencies, the values of $\underline{\mathcal{H}}_{\Omega_s}(\Omega)$ are generally deep minima, which show as vertical lines sweeping up the graph. *However*, when the value of Ω on the ordinate approaches Ω_s, the minima very quickly turn to maxima, permitting the passage of this line through this filter. Once Ω exceeds Ω_s, the value of $\underline{\mathcal{H}}_{\Omega_s}$ again returns to a deep null.

Capon – the final solution

For estimation of the Doppler spectrum, we are not necessarily interested in the filter impulse response or its frequency response. More important is obtaining the power spectral density of the output of the filter. This is achieved by simply calculating the power in the signal $\tilde{y}_{\underline{\Omega}_s}[n]$, as already indicated in our preamble to this section, where we commented that "the user then finds the variance of the resultant time series. This equals the area under the combined spectrum, by Parseval's theorem (Equation 8.18)."

We therefore simply find

$$\begin{aligned} E\{|\underline{\tilde{y}}[n]|^2\} &= \mathbf{h}^{\dagger}_{\Omega_s}\mathbf{R}_y\mathbf{h}_{\Omega_s} \\ &= \frac{1}{\mathbf{a}^{\dagger}_{\Omega_s}(\Omega)\mathbf{R}_y^{-1}\mathbf{a}_{\Omega_s}(\Omega)}. \end{aligned}$$

Notice that $\{h_i\}$, $\underline{H}_{\Omega_s}$, and $\underline{\mathcal{H}}_{\Omega_s}$ are not used specifically in the final calculation – their roles were simply to determine the optimum $\underline{\tilde{y}}$ arrays, which finally led us to (8.75). It is interesting that in the end, we only need the autocovariance matrix and the arrays $\mathbf{a}_{\Omega_s}$, while all of the data-dependent character of the filters is passed through the autocovariance matrix.

When determined as a function of Ω_s, this equation provides the *shape* of the Doppler spectrum but the units are not correct for a power spectral density. In order to correct this minor problem, the power output of the filter must be divided by the bandwidth of the filter, which in this case is a bandpass filter. Several methods exist for estimating the bandwidth of an adaptive filter but we will use the simplest, which is that given a length of the filter $(m + 1)$, the bandwidth can be approximated by $1/(m + 1)$ (*Stoica and Moses*, 2005). Therefore, the Capon estimate of the power spectral density is given by the following equation:

$$\hat{S}_{\mathrm{CM}}(\Omega) = \frac{m+1}{\mathbf{a}^{\dagger}_{\Omega_s}(\Omega_s)\hat{\mathbf{R}}_y^{-1}\mathbf{a}_{\Omega_s}(\Omega_s)}. \tag{8.75}$$

Although the Capon method uses a length $m + 1$ FIR filter, it can achieve better resolution than would be expected. In fact, it has been empirically shown that the Capon method can outperform the periodogram in terms of resolution in some circumstances, especially in regard to identification of dominant peaks in data-sets with only a few strong peaks. The method has been used in several other atmospheric radar applications and it has been shown that with due care it can be robust and provide excellent resolution and interference rejection capabilities (*Palmer et al.*, 1999; *Yu et al.*, 2000; *Palmer et al.*, 2001; *Chilson et al.*, 2001b). Again using simulated atmospheric radar data, the Capon method and periodogram were computed for ten realizations, and are shown is the upper and lower panels of Figure 8.29, respectively.

Although the resolution enhancement provided by the Capon method is difficult to observe in this example with a single Gaussian peak, the reduced variance is apparent. In practice with interference, ground clutter, etc., the interference rejection capabilities of adaptive methods, such as the Capon method, are more evident.

However, the method is not robust under all circumstances, as shown clearly by *Garbanzo-Salas and Hocking* (2015). It does have weaknesses when dealing with small spectral signals, which it tends to eliminate; it functions something like the human eye, picking out and amplifying the dominant Fourier components (and it does this very well). The claim that it does a better job at suppressing the side-lobes is also misleading – it suppresses all smaller peaks in favor of larger ones. It does very badly when there are more large spectral peaks than there are degrees of freedom in the filter. It also does poorly when estimating the width of Gaussian spectra, and also reproduces spectra with initially box-car spectra quite poorly (*Li and Stoica*, 1996). The method should not be considered as fully robust, but as one of many alternatives to standard Fourier analysis, each with their own positive and negative features, and each requiring some degree of care in application. The reader is referred to some of the books and references mentioned elsewhere herein, as well as *Marple* (1987), for some of these alternatives (Prony

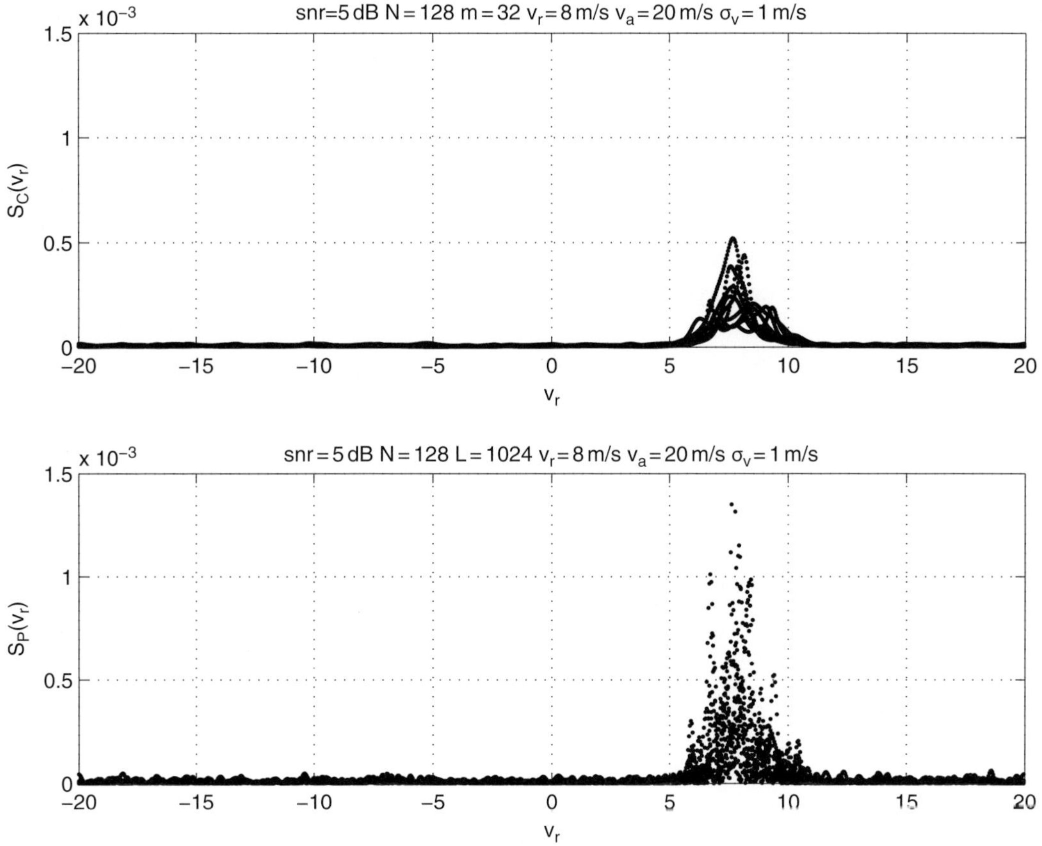

Figure 8.29 A comparison of spectral estimation results using the Capon and periodogram methods, respectively. The reduced variance is noted in the Capon method.

method, maximum entropy method (MEM), minimum variance, multiple signal classification (MUSIC) etc., each of which has its own special capability). Algorithms for spectrally analyzing data-sets containing non-equally-spaced data are also worth knowing about (e.g., *Ferraz-Mello*, 1981; *Lomb*, 1976; *Scargle*, 1982). Wavelet analysis (e.g., *Lehmann and Teschke*, 2001) and other procedures like running (sliding) spectra and the S-transform, which analyse data in windows that slide through a larger data-set, have also been popular in recent years (e.g., *Stockwell et al.*, 1996), but again, this goes beyond the intent of this chapter.

8.7 The atmospheric Doppler spectrum

In the previous section, we introduced the concept of spectral estimation as a tool for understanding the frequency content of Doppler radar signals. We would now like to connect this frequency content to the characteristics of the media under investigation, i.e., the atmosphere. This section will involve some repetition of some of the concepts discussed in Chapter 7, but with slightly different perspectives.

In a similar sense to Equation (8.1), the coherently detected signal from an infinite number of scattering points can be written as follows. We assume that data are collected over a time interval $[0, T]$, and that during this time, the radial velocities of the scatterers are all broadly constant. As discussed earlier, this assumption is not necessarily true, especially in the presence of turbulence, but even if it breaks down, the idea behind the proposal helps us to at least understand the basis of some of our equations. Under these assumptions, we may write

$$\underline{y}(t) = \lim_{N_s \to \infty} \sum_{\ell=1}^{N_s} \underline{\tilde{A}}_\ell \mathrm{e}^{-\mathrm{i}2k_0 r_{0\ell}} \mathrm{e}^{-\mathrm{i}2k_0 v_{r\ell} t}, \tag{8.76}$$

where $r_{0\ell}$ gives the distance to the scatterer at time zero, and $v_{r\ell}$ is the radial velocity of the ℓth scatterer during the recording interval (assuming no acceleration). Each target has an independent radial velocity, echo power, and range. Normally, scatterers close to one another will have correlated characteristics due to the natural continuity of the atmosphere. For this qualitative discussion, however, we will assume this form of the signal.

Given the linearity of the Fourier transform and assuming a deterministic signal, the continuous-time Fourier transform of $\underline{y}(t)$ can be shown to have the form

$$\underline{Y}(\omega) = \mathcal{F}\left[\underline{y}(t)\right] = 2\pi \lim_{N_s \to \infty} \sum_{\ell=1}^{N_s} \underline{\tilde{A}}_\ell \, \delta(\omega - \omega_{d\ell}) \, \mathrm{e}^{-\mathrm{i}2k_0 r_{0\ell}}, \tag{8.77}$$

where $\omega_{d\ell}$ is the Doppler frequency for scatterer ℓ. The Dirac delta function is denoted by $\delta(\,)$. A depiction of $|\underline{Y}(\omega)|$ is provided in Figure 8.30, where the height of individual delta functions is meant to illustrate variations in signal strength.

Due the *Doppler sorting* process of the Fourier transform, it is evident why $\underline{Y}(\omega)$ holds the frequency content of $\underline{y}(t)$. If the power spectrum were calculated from $|\underline{Y}(\omega)|^2$, the resulting function would be proportional to the so-called Doppler spectrum, which

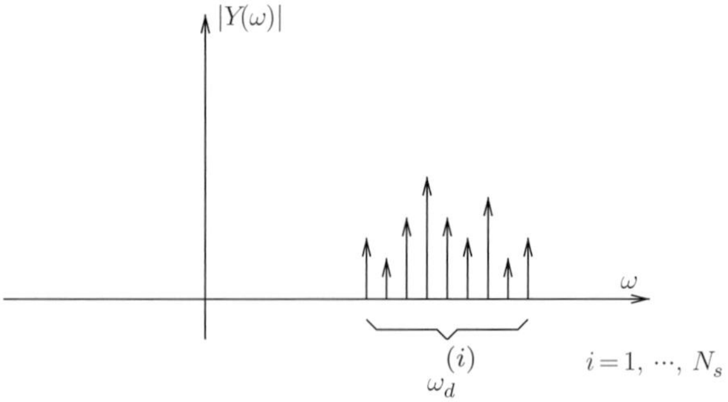

Figure 8.30 Magnitude of the Fourier transform of a radar signal produced from N_s scattering points, with the Doppler-shifted frequency offset being given by $\omega_d^{(i)}$.

represents a *power-weighted* distribution of Doppler frequencies within the resolution volume of the radar. Of course, Doppler frequency is proportional to radial velocity so the Doppler spectrum holds the velocity content, as well. The mean of this distribution would provide information on the average flow within the volume. The spread in the frequency distribution would give a measure of the variability of velocities, which is related to several factors, including turbulence, wind-shear, etc. (Once again we emphasize that this description is over-simplified, since scatterers may often show changes in radial velocity over the course of the sampling period, and in truth an individual scatterer will be associated with multiple spectral lines, but the concept as described is useful for producing a general picture of the relation between scatterers and spectra.)

In order to properly interpret the characteristics of the Doppler spectrum, it is important to understand its expected functionality. As just mentioned, one of the main quantities which dictate the shape of the Doppler spectrum is the distribution of the wind field. A constant mean flow in the wind field is represented by the vector $\mathbf{u} = \begin{bmatrix} u & v & w \end{bmatrix}$. By convention, $+u$ is defined as the zonal component and is directed $W \rightarrow E$ (*westerly* or *eastward*). The positive meridional component is in a *southerly* or *northward* direction $S \rightarrow N$ and is denoted by $+v$. The vector element $+w$ is the vertical component and is directed away (up) from the radar. Using this coordinate convention, we can define the zenith and azimuth angles as θ and ϕ, respectively, as shown in Figure 8.31. It is common to denote the velocity vector as

$$\vec{u} = u\hat{i} + v\hat{j} + w\hat{k}, \tag{8.78}$$

where $\hat{i}$, $\hat{j}$, and $\hat{k}$ are unit vectors along the x, y, and z directions respectively. Note the convention of using $\vec{u}$ for the *total* wind vector *and also* the zonal component of flow – this is common in working with atmospheric flow: u is *not* the magnitude of $\vec{u}$.

The radial velocity would then be given by the following equation:

$$v_r = \vec{u} \cdot \left(\frac{x\hat{i} + y\hat{j} + z\hat{k}}{\sqrt{x^2 + y^2 + z^2}} \right) = \vec{u} \cdot \left[\sin\theta \sin\phi\hat{i} + v \sin\theta \cos\phi\hat{j} + w \cos\theta\hat{k} \right], \tag{8.79}$$

where the vector $\left(\frac{x\hat{i} + y\hat{j} + z\hat{k}}{\sqrt{x^2+y^2+z^2}} \right)$ is the unit vector in the direction of the beam. It should be noted that the elements of $\begin{bmatrix} \sin\theta \sin\phi & \sin\theta \cos\phi & \cos\theta \end{bmatrix}$ are often called the *direction cosines*.

After performing the dot product, the following equation can be obtained for the radial velocity:

$$v_r = u \sin\theta \sin\phi + v \sin\theta \cos\phi + w \cos\theta. \tag{8.80}$$

Note that this applies only for the coordinate system shown in Figure 8.31: the equation will differ if ϕ is taken as the angle anticlockwise from east.

Several factors affect the distribution of radial velocities within the resolution volume of the radar, some of which are due to the atmosphere and others due to the radar. The important effects are now outlined.

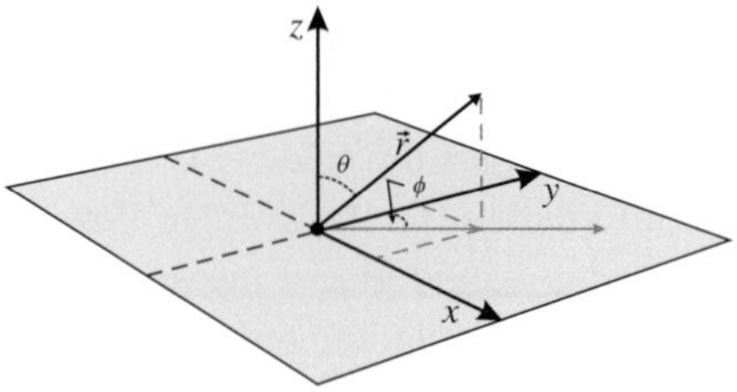

Figure 8.31 Defined coordinate system for azimuth and zenith angles. Note that the azimuthal direction is defined as clockwise from the y-axis i.e., clockwise from north. In other parts of the text, ϕ is defined as *anticlockwise* from the x-axis i.e., clockwise from east. Both conventions are common, and the reader should watch for this.

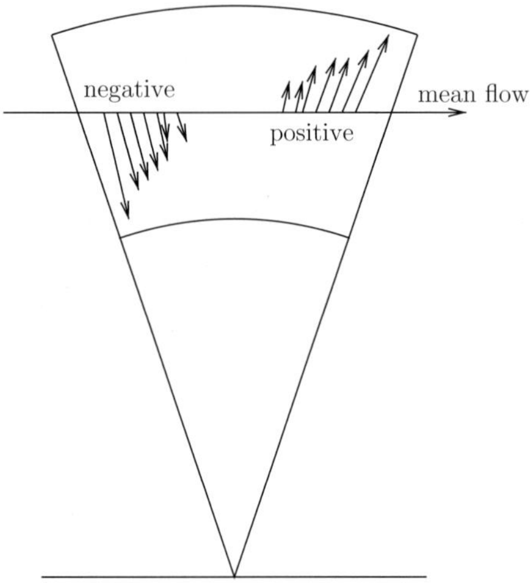

Figure 8.32 Beam broadening effect for finite beam-width and a vertically pointing radar.

Beam broadening: Due to the finite beam-width of the radar, radial velocities sampled by $\underline{y}(t)$ will have an inherent variation across the beam. A vertically pointing beam is illustrated in Figure 8.32, where a constant mean flow is shown to produce a systematic variation of v_r across the beam. This process was discussed in greater detail in Chapter 7. Indeed, all of the topics listed below were treated extensively in Chapter 7, and are included here simply as a summary.

This effect is termed *beam broadening* and is unavoidable given that extremely narrow beam-widths are possible only through large aperture antennas, which can be unrealistic at MST radar wavelengths.

Wind-shear: Geophysical departures from a constant mean flow can affect the Doppler spectrum. Gradients can be caused by divergence, vorticity, or other deformations in the wind field.

Turbulence: Isotropic turbulence will produce spatially correlated three-dimensional fluctuations in the wind field. Anisotropic turbulence can have a similar effect but has a restricted flow in the vertical direction.

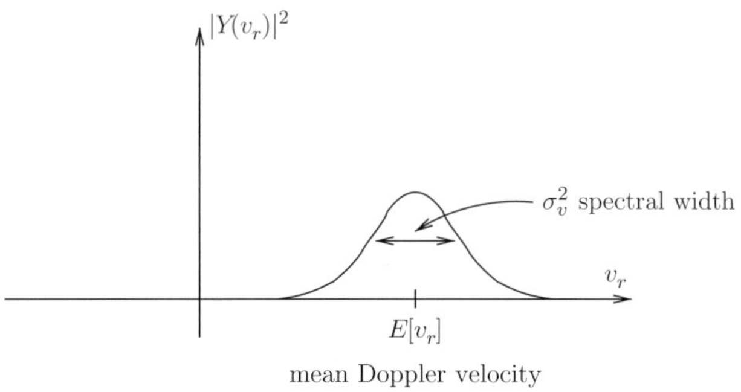

Figure 8.33 Gaussian shape of the mean Doppler spectrum. In reality an individual spectrum is quite spikey and non-Gaussian, but after sufficient averaging it approaches this shape. Note that the reference to the spectral width and σ_v^2 is meant to be qualitative – in reality the double-arrow shown is not σ_v^2 but would be closer to $2\sigma_v$.

Precipitation: Precipitation particles of various sizes and type have different fall speeds. Shorter wavelength atmospheric radars are particularly sensitive to precipitation echoes. Variations in fall speed obviously cause a spread of radial velocities. It is interesting to note that VHF radars are sensitive to both precipitation and clear-air turbulence and can be used to estimate drop-size distribution, which is important for rainfall rate estimation, for example.

Since the distribution of radial velocity is affected by many independent sources (hence allowing the central limit theorem to be invoked), and given that the Doppler spectrum is a power-weighted distribution of radial velocities within the resolution volume, and that the beam profile across the beam is broadly Gaussian, and that the power profile along the beam (radially) is broadly Gaussian (after consideration of the receiver filter response), it is widely assumed that the Doppler spectrum has a Gaussian shape on average. Accordingly, the Gaussian Doppler spectrum can be described by three parameters: total area, first moment (mean), and second moment about the mean, and is shown in Figure 8.33.

The second moment about the mean is typically denoted by σ_v^2 and will contain contributions from each of the sources stated above. Therefore, the use of σ_v^2 for estimation of turbulent intensity, which is important for atmospheric studies, must be carried out carefully (e.g., *Hocking*, 1983a; *Dehghan and Hocking*, 2011).

8.8 Estimation of spectral moments

In Section 8.7, an argument was made that the Doppler spectrum is a power-weighted distribution of velocities within the resolution volume of the radar. Since the source of velocity variability was due to many independent sources, an experimentally validated assumption was made that the Doppler spectrum possesses a Gaussian functionality. However, no specific spectral model was put forward. We would now like to formalize the spectral model and derive systematic methods by which the spectral moments can be obtained.

The power spectral density, or Doppler spectrum, of an atmospheric signal will be assumed to have the following Gaussian form:

$$S(\Omega) = \frac{\sqrt{2\pi}P_S}{\sigma_\Omega} \exp\left[-\frac{(\Omega - \Omega_d)^2}{2\sigma_\Omega^2}\right] + P_N, \tag{8.81}$$

where the peak of $S(\Omega)$ is at Ω_d and the spectral width is dictated by σ_Ω. The additive white noise is represented by the constant noise level P_N. As stated previously, the average power in the signal can be calculated by integration of the $S(\Omega)$ over the domain of $\pm\pi$. Then

$$\begin{aligned}\frac{1}{2\pi}\int_{-\pi}^{\pi} S(\Omega)\mathrm{d}\Omega &= \frac{1}{2\pi}\int_{-\pi}^{\pi} \frac{\sqrt{2\pi}P_S}{\sigma_\Omega} \exp\left[-\frac{(\Omega - \Omega_d)^2}{2\sigma_\Omega^2}\right]\mathrm{d}\Omega + \frac{1}{2\pi}\int_{-\pi}^{\pi} P_N \mathrm{d}\Omega \\ &= P_S + P_N.\end{aligned}$$

This simple calculation shows that by integrating the Gaussian spectral model, we obtain the *signal* power P_S plus the *noise* power P_N. A graphical depiction of the noise and signal power is given in Figure 8.34.

From the inverse DTFT relationship between the PSD and ACF shown in Equation (8.51), it should be evident that identical information exists in either the time or frequency domains. For example, the average power can also be calculated from the ACF at lag zero, and the relation between the powers measured by the ACF and the PSD is

$$\begin{aligned}\underline{r}_{yy}[k]\Big|_{k=0} &= \frac{1}{2\pi}\int_{-\pi}^{\pi} S(\Omega)\mathrm{e}^{\mathrm{i}\Omega k}\mathrm{d}\Omega\Big|_{k=0} \\ &= P_S + P_N.\end{aligned}$$

Due to the relationship between the ACF and the PSD, algorithms to estimate the spectral moments can be developed in either the time or frequency domains. Therefore, the form of the ACF, assuming a Gaussian spectral model in Equation (8.81), is needed.

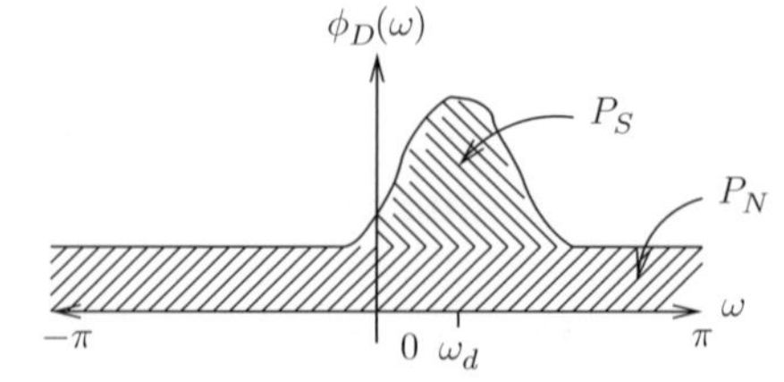

Figure 8.34 Average power and noise for the Gaussian spectral model. Recall also the importance of the *detectability*, which is not shown here, but represents the height of the spectral peak above the noise divided by the standard deviation per unit frequency of the noise (see Figure 4.22).

The following outlines the derivation of the inverse DTFT of the Gaussian spectral model (8.81). Begin with

$$\underline{r}_{yy}[k] = \frac{1}{2\pi}\int_{-\pi}^{\pi} S(\Omega)\mathrm{e}^{\mathrm{i}\Omega k}\mathrm{d}\Omega$$
$$= \frac{1}{2\pi}\int_{-\pi}^{\pi} \frac{\sqrt{2\pi}P_S}{\sigma_\Omega}\exp\left[-\frac{(\Omega-\Omega_d)^2}{2\sigma_\Omega^2}\right]\mathrm{e}^{\mathrm{i}\Omega k}\mathrm{d}\Omega \;+\; \frac{1}{2\pi}\int_{-\pi}^{\pi} P_N\mathrm{e}^{\mathrm{i}\Omega k}\mathrm{d}\Omega.$$

The solution of this inverse transform requires the following inverse DTFT pair:

$$\mathrm{e}^{-\frac{\sigma_\Omega^2 k^2}{2}} \longleftrightarrow \frac{\sqrt{2\pi}}{\sigma_\Omega}\mathrm{e}^{-\Omega^2/2\sigma_\Omega^2} \qquad \Omega \in [-\pi,\pi].$$

Using this, and the frequency shifting property of the DTFT, the ACF for the assumed Gaussian spectral shape can be derived to have the following form:

$$\underline{r}_{yy}[k] = P_S\mathrm{e}^{-\frac{\sigma_\Omega^2 k^2}{2}}\mathrm{e}^{\mathrm{i}k\Omega_d} \;+\; P_N\delta[k], \tag{8.82}$$

where $\delta[k]$ is the Kronecker delta function, which has a value of unity at $k = 0$ and zero elsewhere (resulting in a spike in the function at zero lag). The magnitude of $\underline{r}_{yy}[k]$ possesses a Gaussian form but its width is inversely related to the spectral width σ_Ω. The phase term $\mathrm{e}^{\mathrm{i}k\Omega_d}$ in the ACF is due to the Doppler shift.

Important methods of estimating the spectral moments with both (8.81) and (8.82) will now be presented. Subsequently, the statistical characteristics of the estimators will be investigated using numerical simulation.

8.8.1 Time domain estimators (autocovariance method)

Moment estimators presented in this section will be derived from the model ACF $r_{yy}[k]$ given in (8.82). This time domain method is often referred to as the *autocovariance method*, with a special name of *pulse pair processor* used for the radial velocity estimator, since it will be seen that a minimum of two pulses are needed to estimate the radial velocity.

As seen in the example from the previous section, the total power in the signal (P_T) can be estimated directly from the ACF by substituting $k = 0$ into Equation (8.82), giving

$$P_T = \underline{r}_{yy}[0] = P_S + P_N. \tag{8.83}$$

Note that the total power includes the noise power, which has no atmospheric significance. However, it is often the relative power as a function of space and time that is important in MST radar applications. While determination of absolute power is preferable, for many studies the relative power (signal-to-noise ratio) is often deemed sufficient.

From (8.82), it is readily observed that the Doppler frequency is embedded in the phase of $r_{yy}[k]$. Therefore, by taking the phase of the ACF at $k = 1$, for convenience, the noise term is eliminated resulting in a phase value of Ω_d. Using the relationship between

Doppler frequency and radial velocity of $\Omega_d = 2\pi f_d = 2\pi(-\frac{2}{\lambda}v_r) = -\frac{4\pi}{\lambda}v_r$, the so-called *pulse pair processing (PPP)* estimator of the radial velocity can be obtained:

$$v_r = -\frac{\lambda}{4\pi T_s}\angle \underline{r}_{yy}[1]. \tag{8.84}$$

Note that T_s has now been added as a multiplicative factor on k (1 in this case) in order to convert the temporal units to seconds resulting in velocity units for the estimator of v_r.

Although we have assumed a Gaussian form in the derivation of (8.84), it can easily be shown, and has been known from the earliest days, that this assumption is overly restrictive (e.g., *Woodman and Guillen*, 1974; *Doviak and Zrnić*, 1993, among many other references). In fact, the only assumption needed is that the PSD is an even function about Ω_d, which is reasonable under many atmospheric situations. This fact can be seen by manipulating the ACF–PSD relationship in the following manner:

$$\begin{aligned}
\underline{r}_{yy}[k] &= \frac{1}{2\pi}\int_{-\pi}^{\pi} S(\Omega)e^{i\Omega k}d\Omega \\
&= e^{ik\Omega_d}\frac{1}{2\pi}\int_{-\pi}^{\pi} S(\Omega)e^{ik(\Omega-\Omega_d)}d\Omega \\
&= e^{ik\Omega_d}\frac{1}{2\pi}\int_{-\pi}^{\pi} S(\Omega)\left[\cos k(\Omega-\Omega_d) + i\sin k(\Omega-\Omega_d)\right]d\Omega \\
&= e^{ik\Omega_d}\underbrace{\frac{1}{2\pi}\int_{-\pi}^{\pi} S(\Omega)\cos k(\Omega-\Omega_d)d\Omega}_{\text{real}}.
\end{aligned}$$

The final reduction in the above derivation can only be made if the integral of $S(\Omega)\sin k(\Omega-\Omega_d)$ is zero, which is true if $S(\Omega)$ is even about Ω_d. Therefore, we have shown that the phase of the ACF is equal to $k\Omega_d$, which is the desired result. Of course, perfectly symmetric spectra are seldom the case, resulting in a small bias, a point of which the reader should be aware. (As a point of note, irrespective of whether S is symmetric about its peak or not, $|\underline{r}_{yy}|$ is *always* symmetric about $k = 0$ when S is real. This is because $\int_{-\pi}^{\pi} S(\Omega)\cos(\Omega k)d\Omega = \int_{-\pi}^{\pi} S(\Omega)\cos(\Omega(-k))d\Omega$ and $\int_{-\pi}^{\pi} S(\Omega)\sin(\Omega k)d\Omega = -\int_{-\pi}^{\pi} S(\Omega)\sin(\Omega(-k))d\Omega$, so the real part of the Fourier transform is always symmetric, and the imaginary part is always anti-symmetric if S is real. Hence the modulus is symmetric. However, in regard to measurement of the radial velocity of the scatterers, if S is not symmetric about Ω_d, this asymmetry contributes to the rate of change of phase as a function of k in the autocovariance function, so the rate of change of phase contains not only the effect of the radial velocity but also additional terms related to the asymmetry of the spectrum.)

As in the case of the radial velocity estimator, we will study the form of the ACF (8.82). By doing so, we observe that the second moment about the mean σ_Ω is contained in the magnitude of $\underline{r}_{yy}[k]$. Therefore, a reasonable approach would be to first take the magnitude of the ACF, which would eliminate the phase term not needed. By taking the ratio of two different lags of the ACF, say $\underline{r}_{yy}[1]$ and $\underline{r}_{yy}[2]$, the unknown signal power (P_S) cancels and results in the following equation:

$$\left|\frac{\underline{r}_{yy}[1]}{\underline{r}_{yy}[2]}\right| = \exp\left[\frac{3}{2}\sigma_\Omega^2 T_s^2\right].$$

Note that T_s has again been included for unit conversion purposes.

Solving for the spectral width and converting to velocity units, the following equation can be obtained:

$$\sigma_v = \frac{2v_a}{\pi\sqrt{6}}\sqrt{\ln\left|\frac{\underline{r}_{yy}[1]}{\underline{r}_{yy}[2]}\right|}. \tag{8.85}$$

Any two lags could be used to estimate the spectral width. However, lag 1 and 2 provide the most reliable estimates and avoid the use of lag 0, where the noise power would complicate the estimator.

The previous time-domain estimators for the radial velocity and spectral width are in terms of the actual ACF. For implementation, $\underline{r}_{yy}[k]$ will be estimated using the biased ACF estimator, viz.,

$$\hat{\underline{r}}_{yy}[k] = \frac{1}{N}\sum_{n=k+1}^{N} \underline{y}[n]\underline{y}^*[n-k], \qquad 0 \le k \le N-1.$$

The *biased* ACF estimator is used in order to guarantee a positive semi-definite PSD estimator. In practice, this is justified since the variance of $\hat{r}_{yy}[k]$ increases for large k and should, therefore, be de-emphasized in any spectral estimator. Of course, if only lag-1 and lag-2 ACF values are used, as in the above estimators, this subtle point would be unimportant.

8.8.2 Frequency domain estimators

We will now derive the frequency domain counterparts of the moment estimators provided in the previous section. These methods will make use of estimates of the Doppler spectrum. It should be emphasized that any of the spectral estimation techniques introduced in Section 8.6 could be used.

As shown previously, the echo power can be estimated from either the time or frequency domains. From the Doppler spectrum, the total power is given by

$$P_T = \frac{1}{2\pi}\int_{-\pi}^{\pi} S(\Omega)\mathrm{d}\Omega = P_S + P_N. \tag{8.86}$$

The radial velocity and spectrum width can be thought of as the first moment and the second moment about the mean, respectively, of the Doppler spectrum. Therefore, it is convenient to make an analogy to the moments of a probability density function (pdf) in the field of statistics. The total area of a pdf is unity and the analogy can be made by normalizing the Doppler spectrum by its total area, viz.,

$$S_N(\Omega) = \frac{S(\Omega)}{\frac{1}{2\pi}\int_{-\pi}^{\pi} S(\Omega)\mathrm{d}\Omega}.$$

Obviously, the total area of $S_N(\Omega)$ is unity, as in the case of a pdf. This normalized Doppler spectrum will now be used to estimate the radial velocity and spectrum width.

On a more practical note, actual implementation of spectral estimation methods involves sampling $S(\Omega)$ at a finite number of frequencies. These samples are denoted by Ω_k as shown here:

$$S(\Omega_k) = S\left(\frac{2\pi k}{N}\right), \qquad k = 0, 1, \ldots, N-1$$

where N is the number of frequency samples. In this case, the normalized spectrum is obtained by the following manipulation:

$$S_N(\Omega_k) = \frac{S(\Omega_k)}{\sum_{k=0}^{N-1} S(\Omega_k)}. \tag{8.87}$$

From (8.87), the first moment, or Doppler radial velocity, can be estimated using standard moment calculations from statistics. Assuming that zero Doppler frequency corresponds to an index of $N/2$,then

$$v_r = \left[\sum_{k=0}^{N-1}(k - \frac{N}{2})S_N(\Omega_k)\right]\left(\frac{2v_a}{N}\right), \tag{8.88}$$

where v_a represents the radial velocity corresponding to the Nyquist frequency. The summation generates the first moment in index units of k. Shifting k by $N/2$ creates an appropriate range of values for the first moment of k, assuming the zero Doppler frequency corresponds to an index of $N/2$, which is often the case for most FFT algorithms. If the user maps the last half of the spectrum to the negative frequency range, there is no need to subtract $\frac{N}{2}$. The multiplicative term simply converts the output to units of velocity.

Note that an alternative and improved relation was also presented in Equation (7.9), and the use of spectral fitting has also been considered (Chapter 7). This will be briefly discussed again shortly.

Several problems can arise from using the simple estimator given in Equation (8.88). First, the noise level term P_N in the Doppler spectrum can bias the v_r estimate toward zero. Solutions include noise level estimation/removal and thresholding the spectral amplitudes, for example. An important method of noise level estimation has been developed by *Hildebrand and Sekhon* (1974). *Yamamoto et al.* (1988b) and *Hocking* (1997a) resolved this problem by using functional fitting to the spectrum, generally using a Gaussian fit plus a DC offset, where the DC offset represents noise. This has become more commonplace as computer systems have become faster.

Another problem exists when the actual v_r approaches the aliasing velocity v_a. Since the Doppler spectrum has a finite width, the spectrum can wrap around v_a causing a predictable bimodal shape. If this occurs, the estimated radial velocity will be severely biased. This problem can be solved if the radial velocity does not actually get larger than v_a but just approaches. The following estimator of v_r uses the modulo function to unwrap the Doppler spectrum before calculation of the radial velocity (*Doviak and Zrnić*, 1993):

$$v_r = \left[\sum_{k=k_m-\frac{N}{2}}^{k_m+\frac{N}{2}-1} \left(k - \frac{N}{2}\right) S_N(\Omega_{k'}) \right] \left(\frac{2v_a}{N}\right), \tag{8.89}$$

where $k' = \mathrm{mod}_N(k)$ is the modulo-N value of k, which is necessary when the Doppler spectrum is partially aliased. The index k_m is a preliminary estimate of the v_r location, where k_m is typically obtained from the maximum in the Doppler spectrum.

An example of the results obtained using (8.89) is provided in Figure 8.35.

Given the actual radial velocity of 3 ms^{-1}, the low-SNR case exhibited a significant bias toward zero. Noise level estimation could mitigate this bias. The bias caused by aliasing is illustrated in Figure 8.36 where the radial velocity was set to 3 ms^{-1}and 19 ms^{-1}with an aliasing velocity of 20 ms^{-1}.

For v_r equal to 3 ms^{-1}, the estimate is close to the actual but for v_r equal to 19 ms^{-1}the results are extremely poor. Obviously, this is due to the wrapping of the spectrum around v_a. Although not shown, Equation (8.89) virtually eliminates this bias.

A much better way to resolve this problem is to reduce the amount of coherent integration. Then aliasing frequencies can be hundreds of meters per second or even more

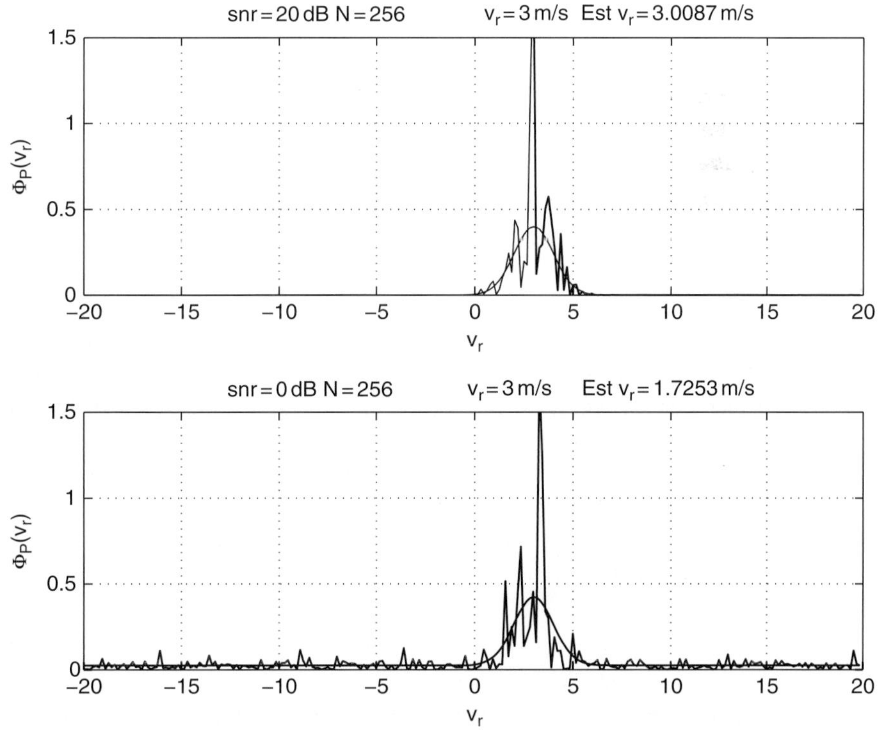

Figure 8.35 Radial velocity estimation example with $v_r = 3$ ms^{-1}and SNR values of 20 dB and 0 dB.

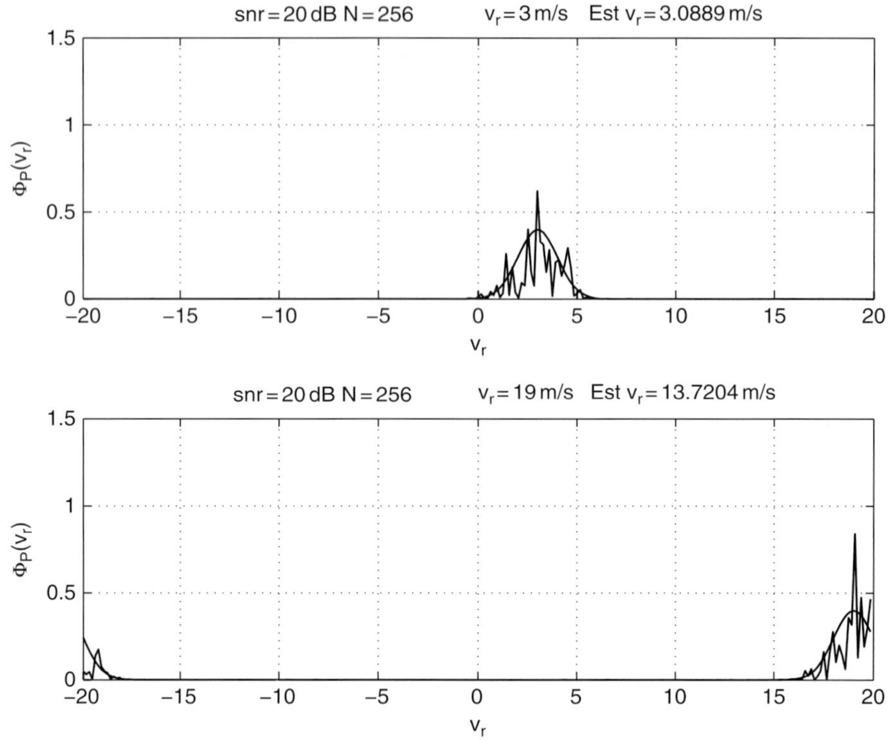

Figure 8.36 Radial velocity estimation example for two different radial velocities. The example illustrates the aliasing bias.

(*Hocking*, 1997a), so the spectra are never aliased, at least for realistic atmospheric circumstances. This requires that the user employ spectral fitting, since in that case the most important parameter is the *detectability*, rather than the signal-to-noise ratio, and the detectability is independent of the amount of coherent integration, as already discussed in Chapters 4 and 7.

Using the concepts developed for the frequency domain v_r estimators, the derivation of the spectral width estimator is a simple extension. Given the normalized Doppler spectrum, the second moment about the mean is given by the following equation with the appropriate conversion to velocity units:

$$\sigma_v^2 = \left[\sum_{k=k_m-\frac{N}{2}}^{k_m+\frac{N}{2}-1} (k - k_m)^2 S_N(\Omega_{k'}) \right] \left(\frac{2v_a}{N} \right)^2 . \tag{8.90}$$

As before, k_m is a rough estimate of the first moment in k units. Of course, the radial velocity estimate, presented in the previous section, could be used to refine the spectral width estimate by providing a more robust estimate of k_m.

Several important studies of the statistical characteristics of the presented spectral moment estimators have been undertaken. In particular, *Zrnić* (1979) provides a comprehensive study of both time and frequency-domain methods and covers several practical problems regarding the scanning weather radar case. As already discussed, *Yamamoto et al.* (1988a) conducted a study focused on MST radar applications with an emphasis on frequency-domain methods, and *Hocking* (1997a) advocated fitting methods with limited coherent integration as an optimal solution. *Yamamoto et al.* (1988a) statistically compared moment-based techniques (similar to the procedures presented here) to least squares fitting methods. It was shown that fitting methods can produce superior results under low-SNR environments.

9 Multiple-receiver and multiple-frequency radar techniques

9.1 Introduction

As we have already discussed, there are many competing factors that must be taken into account in order to optimally investigate the atmosphere through radar observations. One of the more notable examples is the Doppler dilemma. Obviously one would like to select an inter-pulse period (IPP) corresponding to a sufficiently large Nyquist velocity interval. Here *sufficiently large* means a velocity range that encompasses most of the anticipated radial velocities to be observed. The range of Nyquist velocities is extended by decreasing the IPP. However, decreasing the IPP also reduces the maximum unambiguous range that can be measured. Ideally one would like to maintain a large Nyquist velocity (short IPP) and large maximum unambiguous range (long IPP) – hence the dilemma. Another example involves the disparity between the desire to improve range resolution and improve radar sensitivity. Range resolution can be improved by decreasing the radar pulse width; however, this means a decrease in the amount of power that illuminates a scatterer and corresponding decrease in detectability. That is, the desire to increase the detectability of atmospheric signals by transmitting longer radar pulses is at odds with the need to improve range resolution.

In many cases, techniques have been developed that allow us to work around the compromises that arise in designing radar experiments. For example, pulse compression (discussed in Chapter 4) is used to improve range resolution without compromising the signal-to-noise ratio (SNR) (*Schmidt et al.*, 1979). By and large, such techniques are known to introduce corresponding undesirable side effects. For the case of pulse compression, either the existence of some level of range side-lobes, or a decrease in temporal resolution, are a by-product of complementary codes.

In this chapter, we discuss how the use of multiple-receiver and multiple-frequency techniques can be used in atmospheric remote sensing as a means of improving angular and range resolution respectively. Before proceeding, we should clarify one point of nomenclature. The term *multiple-receiver* will be used throughout this chapter to describe a radar system that is capable of receiving and recording atmospheric signals on two or more spatially separated antennas or groups of antennas. The myriad names associated with interferometric techniques were discussed in Chapter 2, Section 2.15.6: here, we will discuss in detail just a subset of these, but the points discussed will cover to some extent all the techniques. In the introduction to Chapter 5, we had a discussion about the meanings of the term multistatic and bistatic. Cases where the receivers were relatively close, and cases where they are far away, were distinguished. For the applications to be considered here, the spatial separation between the receiving antennas is assumed to be relatively small. Likewise, *multiple-frequency* refers to a radar system capable of transmitting and receiving atmospheric signals from two or more discrete frequencies. Here again, the frequency separations are considered to be relatively small. The meaning of *relatively small* will become clear within the context of the forthcoming discussion.

Many multiple-receiver techniques used for atmospheric research were developed for application in the ionosphere (e.g., *Mitra*, 1949; *Briggs et al.*, 1950; *Woodman*, 1971; *Farley et al.*, 1981), where the need for enhanced radar resolution is particularly acute. At ionospheric altitudes, the beam-width of the probing radar becomes an important consideration. For example, for a radar having a beam-width of 3°, the horizontal resolution at an altitude of 80 km is 4 km. Additionally, for wind estimation methods such as VAD (velocity azimuth display) and DBS (Doppler beam swinging), the separation between sampling volumes at a height of 80 km can easily become 15 to 20 km. Over such large distances, the underlying assumption of having a uniform or linearly varying wind field across space becomes difficult to fulfill.

Correspondingly, in order to improve the sensitivity of the radar when probing the atmosphere at such large ranges, one must typically transmit long pulses of radiowaves. Although pulse compression techniques are commonly used for such applications, one can also rely on multiple-frequency techniques to improve resolution. Such techniques are particularly germane in the case of thin "scattering layers," resulting from strong vertical gradients in temperature, humidity, or electron densities (see various discussions about specular reflectors in earlier chapters, for example). These scattering layers can occur individually or collectively and may not be resolvable using conventional radar resolution. As discussed near the end of Chapter 2, and as we shall discuss below, these techniques fall broadly into parametric techniques (frequency domain interferometry) or atmospheric imaging techniques (range imaging).

Although spaced antenna and multiple-frequency atmospheric radar techniques were first developed for measurements in the upper mesosphere and lower thermosphere, the need for high-resolution atmospheric radar measurements is not limited to these regions. In the atmospheric boundary layer, variability in the thermodynamic and kinematic fields can exist on spatial scales smaller than the region sampled by a wind profiler operating in DBS mode. As in the case of ionospheric observations, non-uniformity in wind and reflectivity fields across the different beams can also be problematic in such cases. Figure 9.1 gives a schematic of some of the complicated air-flows that might be expected in the boundary layer close to a radar. One can mitigate the effect by averaging over "sufficiently long" time intervals, but the requisite amount of averaging is not known a priori. Furthermore, the averaging process limits our ability to observe rapidly evolving phenomena.

Moreover, there are numerous examples of structures in the lower atmosphere that manifest themselves within narrow, well-confined height intervals. There have been many observations of persistent layers in the atmosphere, attributed to both turbulence and sharp gradients in temperature and humidity (*Dalaudier et al.*, 1994; *Muschinski*

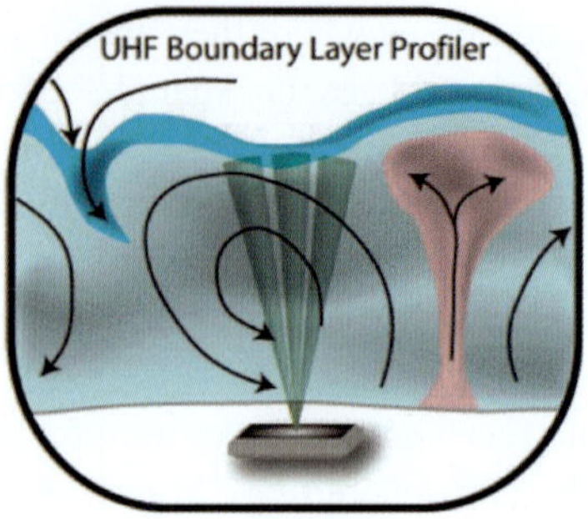

Figure 9.1 Potential complex thermodynamic and kinematic structures within the convective atmospheric boundary layer along with a UHF wind profiler operating in a DBS mode.

and Wode, 1998; *Luce et al.*, 2002; *Balsley et al.*, 2003). In addition to being intriguing from a dynamical point of view, such layers are interesting for their impact on radar backscatter (*Luce et al.*, 1995; *Muschinski*, 1997; *Muschinski and Wode*, 1998; *Chilson et al.*, 2001b). Finally, there are numerous reports pertaining to the relevance and observation of gravity waves in stably stratified flows in or near the atmospheric boundary layer (e.g. *Gossard et al.*, 1970; *Gossard*, 1990; *Eaton et al.*, 1995). Often, the features of these waves such as amplitude, period, and thickness of the embedded layers cannot be satisfactorily resolved with conventional boundary-layer profilers that provide height resolutions of 60–100 m (*Angevine et al.*, 1993).

In Figure 9.2, we show one example of the types of atmospheric structures that can be revealed by profiling radars in the boundary layer. The image shows data collected with a frequency-modulated continuous-wave (FMCW) radar. FMCW radars were discussed in Chapter 5, Section 5.1.1, and are capable of retrieving profiles of the backscattered power from the atmosphere with range resolution of down to one meter (e.g., *Richter*, 1969; *Eaton et al.*, 1995). However, FMCW radars (commonly operated at a 10 cm wavelength) are hampered by limitations such as restricted altitude coverage (due to the wavelength), sensitivity to biological scatterers, and difficulties in measuring Doppler velocities with good range resolution, which preclude their use for operational measurements.

The FMCW radar used to collect the data shown in Figure 9.2 swept across a narrow frequency band centered at 3 GHz (or a wavelength of 10 cm). One can see an example of Kelvin–Helmholtz billows in the height range 1000–1700 m in the time interval 14:00 to 15:30h GMT. Below that (at about 500 m and below), the dynamic evolution of the top of the boundary layer can be seen. We should note that the local time of the observations was 08:00–10:00. The speckles that appear throughout the data result from backscatter from migrating miller moths (*Acronicta leporina*). Contamination of radar data by birds and insects is a common problem at some frequencies, especially in the K, X, C, S, L, and UHF bands.

For the following discussion, we will make a distinction between the terms *spaced antenna* or *correlation* techniques, *radar interferometry*, and *radar imaging*. Although these approaches are closely related, there are distinct and important differences. *Spaced*

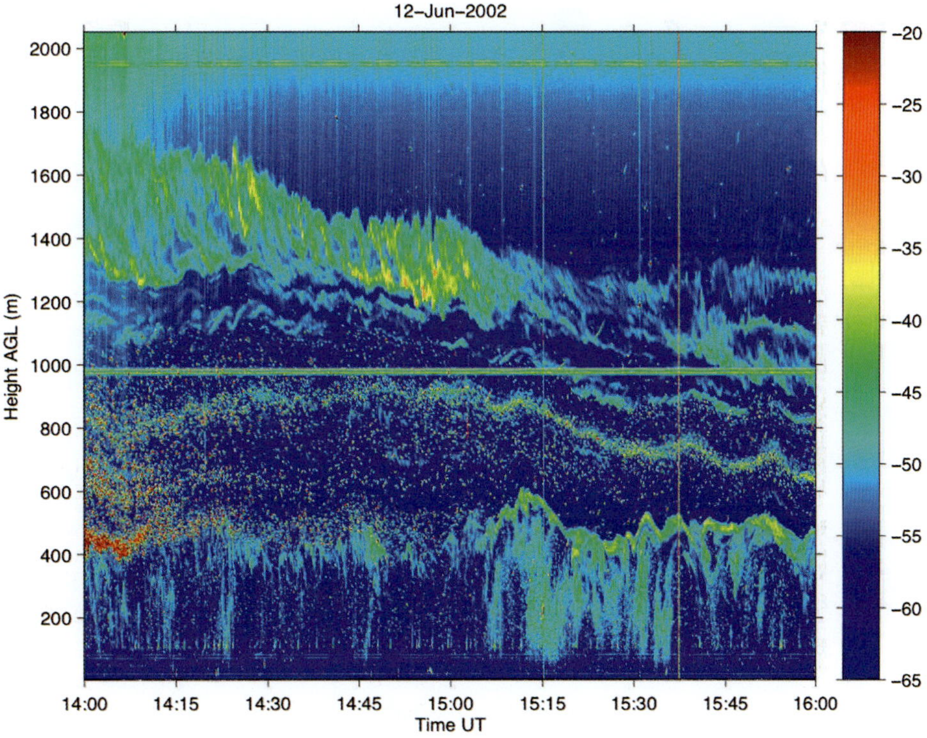

Figure 9.2 Time height intensity plot of the backscattered power from an S-band (3 GHz) frequency-modulated continuous-wave (FMCW) radar. The radar was sampling the atmosphere directly overhead and is capable of vertical resolution of about 2 m. The power is expressed logarithmically with an arbitrary reference point. The radar was operated near Boulder, Colorado.

antenna correlation techniques are mainly used to measure wind speeds, and do not look at the actual scattering structure in the sky in too much detail. The term *radar interferometry* implies the utilization of signal processing methods to locate and track the motions of a particular scatterer or a collection of scatterers. In contrast, the term *radar imaging* is being used here to describe a technique of reconstructing atmospheric reflectivity fields and their velocities within a nominal resolution volume of the probing radar through inversion algorithms, thereby increasing spatial resolution. Each method takes advantage of radar signals collected using multiple antennas and/or frequencies, and they are described in greater detail below.

To illustrate the underlying concepts, we begin by considering the simple case of scatter from a single point as detected by two receiving antennas. The two antennas are separated by a distance d; the wavelength of the radiowave being transmitted and received is λ; therefore, the wavenumber is $k = 2\pi/\lambda$. The location of the scatterer is displaced from zenith by an angle θ as measured along the baseline of the two antennas. The situation is depicted on the left-hand side of Figure 9.3.

The phase difference between the two received radar signals will be simply a function of the angle θ. Here we will use the plane wave approximation. That is, although a

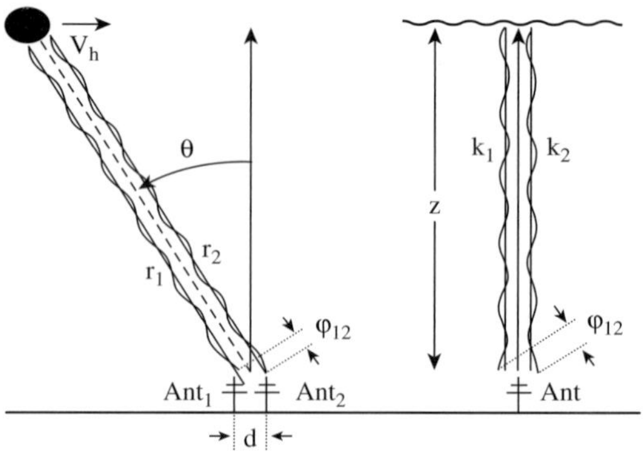

Figure 9.3 The principles of multiple-receiver and multiple-frequency interferometry. On the left, radiowaves having a single frequency are scattered from a single point and received by two spatially separated antennas. The phase difference $\varphi_{12} = kd \sin\theta$. On the right is shown the case in which radiowaves at two distinct frequencies are scattered from an elongated surface or layer parallel to the ground and received by a monostatic radar operating at the two frequencies. Here the phase difference is given by $\varphi_{12} = 2(k_2 - k_1)z$. Note that in the left-hand case, the two lines adjacent to the symbol φ_{12} could be considered as parts of the wavefronts, but in the right-hand figure, this is not true – the wavefronts will be parallel to the ground, and the lines merely point to the parts of the wave near the antenna, and have no other significance.

spherical wave is scattered by the hard scatterer, if the distance between the scatterer and antennas is much larger than a Fresnel radius ($\sqrt{r\lambda}$), where r is the range to the scatterer, then the received signals can be treated as plane waves. The phase difference is then given by

$$\varphi_{12} = \varphi_1 - \varphi_2 = k\,d\,\sin\theta. \tag{9.1}$$

Therefore, by sampling the radar signals received by the two antennas and measuring the phase difference, one can directly obtain the location of the scatterer. This is the simplest case of radar interferometry, or more specifically, SDI (spatial domain interferometry). It is dependent on the assumption that the scattering object *really is* a point scatterer, as discussed in Chapter 2, Section 2.15.3 – if this assumption cannot be made, the resolution limits of normal Fourier theory apply.

In the next section, we demonstrate how additional receive antennas can be used to locate the scatterer in three dimensions and how the scatterer's velocity can be obtained.

In the previous example, we considered a case of interferometry using multiple receive antennas. It is also possible to conduct interferometric measurements using multiple frequencies. Consider the case of scatter from an elongated surface or layer that is oriented parallel to the ground and located at a height z. In this example, a monostatic radar operating at two distinct frequencies ω_1 and ω_2 will be used. The corresponding wavenumbers are given by $k_1 = \omega_1/c$ and $k_2 = \omega_2/c$. Again it is assumed that plane waves are detected by the antenna. If the transmitted radiowaves for both frequencies have the same initial phase, then the phase difference between the received signals is given by

$$\varphi_{12} = \varphi_1 - \varphi_2 = 2(k_1 - k_2)z. \tag{9.2}$$

It is clear that the height of the surface or layer above the radar can be obtained from the value of φ_{12}, assuming that the phase can be measured. A warning is needed here – to determine the height exactly, the full phase difference needs to be known, even if it exceeds 2π. In some cases, the phase difference can be many times 2π. But in practice, the phase difference is only measured to modulo(2π), so only the fractional portion of the phase difference is known. Then extra effort is needed to determine the full phase rotation, and it is not always possible to do this.

9.2 Mathematical framework to describe the radar signal

In the last section, we briefly showed how a radar operating in a multiple-receiver mode could be used to locate the position of a point scatterer. The treatment of the problem was primarily intended to illustrate the concepts of calculating the angle of arrival. For the discussions that follow, we require a more rigorous mathematical framework.

We begin by introducing some of the notation that will be used here to generalize the standard single-receiver, single-frequency radar equation to that for multiple receivers and frequencies. This is not meant to be an exhaustive treatment. More general treatments of radar scatter from the clear atmosphere (not necessarily for the case of multiple-receiver or multiple-frequency) can be found in *Liu and Yeh* (1980); *Doviak and Zrnić* (1984); *Liu and Pan* (1993), and *Yu and Palmer* (2001), among others.

We first consider the case for point scatterers and then generalize the treatment to that for atmospheric scatter. The equations to be presented have intentionally been kept simple for the sake of illustrating the techniques to be described. We will consider both multiple-receiver and multiple-frequency scenarios together in the equations.

9.2.1 Scatter from a single scatterer

Imagine that we transmit a propagating radiowave from an antenna located at the origin of some coordinate system. A Cartesian coordinate system will be used, and a vector $\vec{r}$ will be written as $x\hat{i} + y\hat{j} + z\hat{k}$, with $\hat{i}, \hat{j}$ and $\hat{k}$ being unit Cartesian vectors in the x, y, and z directions respectively. We will denote this vector using the shorthand $\left[x\ y\ z\right]$.

Now consider a scatterer located at $\vec{r}$. The radiowave scatters off an isolated object (point scatterer) located at the position $\vec{r}$ that is moving with velocity $\vec{v} = \dot{\vec{r}}$. The angular position of $\vec{r}$ can be conveniently expressed using cartesian coordinates or direction cosines such that

$$\begin{aligned} \hat{a} &= [x\ y\ z]/r \\ &= [\theta_x\ \theta_y\ \theta_z], \end{aligned} \tag{9.3}$$

where $\hat{a}$ is the unit vector directed towards the point scatterer. Then $\vec{r} = \hat{a}r$. The direction cosines are defined here as $\theta_x \equiv \sin\theta \sin\phi$, $\theta_y \equiv \sin\theta \cos\phi$, and $\theta_z \equiv \cos\theta$. A diagram illustrating the geometry discussed above is presented in Figure 9.4. Note that in our case we have defined ϕ as clockwise from the y axis, since in some of our analysis y will be northward. Then ϕ is the azimuthal angle clockwise from true north.

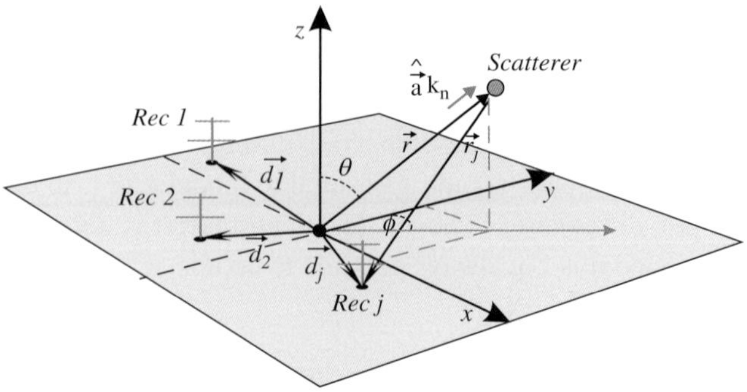

Figure 9.4 The coordinate system used for describing scatter from a point scatterer in a radar interferometry mode.

We assume that the scatterer is sufficiently far removed from the antenna that the backscattered signal can be treated as plane waves. As we have seen in Chapter 4, Equation (4.8), we can write that for the case of a monostatic radar operating at a frequency ω_0, the plane waves can be written as

$$\underline{E}_R(t) = \tilde{A}e^{i\{\omega_0 t-(2k_0 r+2k_0 v_r t)+\psi\}}, \tag{9.4}$$

where $k_0 = \omega_0/c$. $\tilde{A}$ is real here, and has an associated term ψ in the exponent of e which makes the combination $\tilde{A}e^{i\psi}$ complex. So this term plays the same role as the complex constant $\underline{\tilde{A}}$ in Equation (4.8). As in that earlier equation, we again recognize that $\tilde{A}$ is also a function of the scatterer's location and velocity. That is, $\tilde{A} = \tilde{A}(\hat{\vec{a}}, r, \omega)$. In our work in this chapter, we will denote

$$\omega \equiv -2k_0 v_r, \tag{9.5}$$

and we use the "positive away from the radar" convention. Note this definition is different to that used in Chapter 3, where ω was the *total* Doppler-shifted angular frequency. Here, ω is just the Doppler shift, and the total Doppler-shifted frequency is $\omega_0 - \omega$, so that a radial velocity *towards* the radar leads to an increase in frequency – see the discussion following Equation (4.7). This change in definition of ω is adopted to accommodate the approach often used in some of the key texts in the field of interferometry.

For simplicity, we will simply use $\tilde{A}$ when discussing scatter from a point scatterer. However, it will be recognized that $\tilde{A}$ does depend on angle, range, and so forth when discussing scatter from distributed or multiple scatterers. Additionally, we are not showing the pulse shape here, also for simplicity (see Chapter 4). Recall that for a monostatic radar, the transmit and receive antennas are co-located. Now if the backscattered signal is received at a displaced antenna j located at $\vec{d_j} = \left[x_j\ y_j\ z_j\right]$, then (9.4) must be re-written as

$$\underline{E}_R(\vec{d_j}, t) = \tilde{A}e^{i\{\omega_0 t-(k_0 r+k_0 r_j+2k_0 v_r t)+\psi\}}, \tag{9.6}$$

where $\vec{r_j} = \vec{d_j} - \vec{r}$ is the position vector from the scatterer to the receive antenna j. The transmit antenna is still taken to be at the origin. This formula recognizes that the wave travels to the scatterer along a path of length $r = |\vec{r}|$ and back to the jth receiver along a path of length $r_j = |\vec{r}_j|$, so the total path is the sum of the two.

We can simplify (9.6) by expressing r_j in terms of r. First, using the law of cosines, we can write

$$r_j = r\left(1 + \frac{d_j^2}{r^2} - 2\frac{\vec{r}\cdot\vec{d}_j}{r^2}\right)^{1/2}. \tag{9.7}$$

By neglecting the second term on the right-hand side of (9.7), since $d_j \ll r$, and using a binomial expansion for the remaining terms, we find that

$$r_j \simeq r - \hat{a}\cdot\vec{d}_j. \tag{9.8}$$

Since we are dealing with a single point scatterer, and we assume that $\vec{r}$ and $\vec{r}_j$ are very nearly anti-parallel, the radar scatter is zero for all values of the wave vector $\vec{k}$ except when $\vec{k} = \hat{a}k_0$. Finally, substituting (9.8) into (9.6) we obtain

$$\underline{E}_R(\vec{d}_j, t) = \tilde{A}e^{i\{(\omega_0+\omega)t-(2k_0r-k_0\hat{a}\cdot\vec{d}_j)+\psi\}}. \tag{9.9}$$

The incident electromagnetic wave on the receive antenna or group of antennas, after being sampled using a coherent detector, can be expressed as a complex output voltage (see Chapter 4). Using the above results, we write this expression as

$$\underline{s}(\vec{d}_j, t) = \underline{\tilde{A}}e^{i(\omega t-2k_0r+k_0\hat{a}\cdot\vec{d}_j)}, \tag{9.10}$$

where the phase term ψ has been incorporated into the new complex constant $\underline{\tilde{A}}$. The complex radar voltage given by (9.10) can be further generalized to include multiple frequencies. The frequencies are given by ω_n, and we write the wavenumbers for the frequencies as $\vec{k}_n = \hat{a}k_n$, where $k_n = \omega_n/c$. This results in

$$\underline{s}(\vec{d}_j, k_n, t) = \underline{\tilde{A}}e^{i(\omega_nt-2k_nr+k_n\hat{a}\cdot\vec{d}_j)}. \tag{9.11}$$

The term ω_nt in the exponent of e deserves some consideration. Recall that $\omega_n = -2k_nv_r$. In general, even if there are multiple frequencies, they are relatively close in value. For example, for a system with frequencies close to 50 MHz, two typical transmitted frequencies used in an interferometry experiment might be $\omega_{01} = 2\pi\times 49.6$ MHz and $\omega_{02} = 2\pi \times 50.4$ MHz. A typical value of v_r might be 5 ms^{-1}. Then evaluating $2k_nv_r$, with $k_n = \frac{\omega_n}{c}$ gives 10.388 radians and 10.558 radians for the lower and upper carrier frequencies. The difference is 0.17 radians, and the offset of each from the midpoint is 0.085 radians, or about 5 degrees of phase. This is quite small, yet the numbers used are quite typical. Generally government licensing constraints limit the range of frequencies available, so the range of frequencies that might be used in MST applications are contained within a relatively narrow band ($\Delta f_{\text{max}} = 1 - 5$ MHz) centered at some nominal operating frequency for the particular radar. At higher frequencies, the spread of frequencies relative to the central frequency can be even smaller.

Since interferometry using multiple receivers can get quite complicated mathematically, it is prudent to make simplifying approximations wherever possible. Such a simplification is possible here. Rather then use k_n in the calculations of ω_n, it is common practice to use the wavenumber of the center frequency of all the transmitted

of these concepts have already been presented in Chapter 8 for the case of random processes observed as a function on time. Here we will generalize the discussion to include spatial and frequency components.

We begin by considering a situation in which radar signals are received by two antennas located at $\vec{d}_i$ and $\vec{d}_j$ with respect to the transmit antenna. Furthermore, the radar is operated using two alternating frequencies ω_n and ω_m. These signals, which we express as $\underline{s}(\vec{d}_i, k_n, t)$, are treated statistically as random processes. We now make two simplifying but reasonable assumptions regarding $\underline{s}(\vec{d}_i, k_n, t)$: first, we assume that $E\left\{s(\vec{d}_i, k_n, t)\right\} = 0$, and secondly, $\underline{s}(\vec{d}_i, k_n, t)$ is wide-sense stationary, or WSS (the concept of WSS was discussed in Chapter 8). We should note that for our applications, the assumptions of WSS need only apply over short time, distance, and frequency scales.

Throughout the rest of this chapter, we will employ two functions which represent statistical processes slightly differently. One will be the set of auto- and cross-covariance functions, the other will be the set of auto- and cross-correlation functions.

First, we repeat Equation 8.49 from Chapter 8 as follows: assuming two WSS random processes $\underline{x}[n]$ and $\underline{y}[n]$, the cross-covariance function is defined by (e.g., *Marple*, 1987, page 116) (with R replacing r for this chapter):

$$\underline{R}_{xy}[k] = E\left\{\underline{x}[n+k]\underline{y}^*[n]\right\}, \tag{9.17}$$

assuming that $\underline{x}$ and $\underline{y}$ have had any mean values removed.

Here, for ease of reference, we now repeat (and expand on) the comments following Equation (8.49). Engineers regard Equation (8.49), without removal of the mean, as the formal definition of the cross-correlation function (CCF). They regard the *cross-covariance* as the same function, but calculated *after* removal of the means from x and y, as given by (9.17). Statisticians use the same definition as engineers for the cross-covariance function, but a *quite different one* for the cross-correlation function, which they consider to be the autocovariance function, but divided by the square-root of the product of the variances of $x[n]$ and $y[n]$, σ_x^2 and σ_y^2. This has the result that if x and y are equivalent, the auto- and cross-correlation functions are both normalized to unity at zero lag. Further discussion can be found in *Marple* (1987), pages 115–116, among other references. This form of the CCF can also be obtained by dividing $\underline{x}$ and $\underline{y}$ (after removal of their respective means) throughout by the square roots of their respective standard deviations, so that each has a standard deviation of 1.

As shown following (8.49), (9.17) can also be written as

$$\underline{R}_{xy}[k] = E\left\{\underline{x}[n]\underline{y}^*[n-k]\right\}. \tag{9.18}$$

As also discussed following (8.49), this is the expression for x cross-correlated with y – for y cross-correlated with x, x and y change roles: the real parts of the two functions are mirror images of each other about the ordinate axis, and the imaginary parts are inverted mirror images of each other.

In this chapter, we will employ the cross-covariance function $\underline{R}$ as defined by (9.18), and will on occasion use the *statistician's definition* of the cross/auto-correlation function. We have adopted this strategy because the section on full correlation analysis, which employs correlation functions to determine wind speeds, utilizes this form of the correlation function.

Then we now define the general **space-frequency-time cross-covariance function**, or cross-covariance sequence, as

$$\underline{R}(\vec{\delta}_{ij}, \Delta k_{nm}, \tau) = E\left\{\underline{s}(\vec{d}_i, k_n, t)\underline{s}^*(\vec{d}_i - \vec{\delta}_{ij}, k_n - \Delta k_{nm}, t - \tau)\right\}, \tag{9.19}$$

where $\vec{\delta}_{ij} = \vec{d}_i - \vec{d}_j$ and $\Delta k_{nm} = k_n - k_m$ are the vector separations in antenna locations and wavenumber respectively. As before, * is the complex conjugate operator and $E\{\ \}$ is the expected-value operator.

As discussed above, some confusion exists in the literature regarding nomenclature and symbols associated with covariance and correlation functions. At the risk of laboring the point, the terminology used in the present discussion is listed below. In all cases the term covariance sequence may be used interchangeably with covariance function. We will primarily refer to covariance functions as our un-normalized form of functional comparisons, and the term "correlation" will be preserved (at least in this chapter) for a normalized version of the covariance. However, again we emphasize that the engineering definition of the correlation function is quite different to the statistician's form.

Space-frequency-time covariance function: This is the most general expression of the covariance function and has already been defined as Equation (9.19).

Space-time covariance function: In certain applications, we are only concerned with the multiple-receiver component of our mathematical formulation. In this case we can limit our discussion to space-time covariance function, which is given by

$$\underline{R}(\vec{\delta}_{ij}, \tau) = E\left\{\underline{s}(\vec{d}_i, t)\underline{s}^*(\vec{d}_i - \vec{\delta}_{ij}, t - \tau)\right\}. \tag{9.20}$$

Frequency-time covariance function: When dealing with the case of pure multiple-frequency measurements, it suffices to consider the frequency-time covariance function given by

$$\underline{R}(\Delta k_{nm}, \tau) = E\left\{\underline{s}(k_n, t)\underline{s}^*(k_n - \Delta k_{nm}, t - \tau)\right\}. \tag{9.21}$$

Autocovariance function: Some authors use the term autocorrelation or autocovariance function (ACF) to discuss the self-covariances between signals. It is the special case that $i = j$ in $\vec{\delta}_{ij}$ and/or $n = m$ in Δk_{nm}.

Cross-covariance function: The space-frequency-time covariance function for the case of $\vec{\delta}_{ij} \neq 0$ and/or $\Delta k_{nm} \neq 0$ is a cross-covariance function. This can be considered as the complement of the ACF.

We should at this point mention that different terminology is used in the literature for correlation/covariance functions. The word "correlation" is often used where we have used "covariance." In some texts, Equation (9.20) is referred to as the space-time autocorrelation function (ACF), the cross-correlation function, or the spatiotemporal correlation function. We will simply refer to Equation (9.20) or expressions similar to it as covariance functions, and normalized forms as correlation functions.

Given the information available in (9.12) and (9.16), what additional information can be extracted about the scatterer or scatterers? We have already seen that the phase differences between two received signals provide the angular location and range of a point scatterer in the case of multiple receivers and multiple frequencies respectively. As we will discover, these equations are useful for obtaining additional information as well. One method of extracting this information is through the use of correlation or covariance functions.

Let us consider the case of backscatter from a single hard scatterer observed with a multiple-receiver radar system using only a single radar frequency for now. If we substitute (9.10) into (9.19), then we find

$$\underline{R}(\vec{\delta}_{ij},\,\tau) = \left|\underline{\tilde{A}}\right|^2 \mathrm{e}^{\mathrm{i}(\omega\tau + k_0\hat{\vec{a}}\cdot\vec{\delta}_{ij})}. \tag{9.22}$$

Equation (9.22) expresses the covariance function for multiple-receiver observations of a single hard scatterer (i.e. one that does not change form in time), and we have ignored the effect of the radar pulse. For these reasons, $\underline{R}$ has constant amplitude over all time, which will not be true when we introduce soft scatterers for which their scattering characteristics may evolve with time. If signals from three non-collinear receive antennas are available then (9.22) can in principle be used to determine the position and velocity of the scatterer. To demonstrate this, we now consider a case for which there are three receive antennas located in three-dimensional space at $(0, 0, 0)$, $(\xi_0,\ 0,\ 0)$, and $(0,\ \eta_0,\ 0)$. An illustration is given in Figure 9.6. The transmit antenna coincides with antenna 1. For this situation, the values of $\vec{\delta}_{ij}$ are given as $\vec{\delta}_{21} = [\xi_0\ 0\ 0]$, $\vec{\delta}_{31} = [0\ \eta_0\ 0]$, and $\vec{\delta}_{23} = [\xi_0\ -\eta_0\ \ 0]$. Using these values, we next examine the phase angles given by the covariance function (9.22) when $\tau = 0$. The phase angles are simply given by

$$\angle\underline{R}(\vec{\delta}_{ij}, 0) = k_0\hat{\vec{a}}\cdot\vec{\delta}_{ij}. \tag{9.23}$$

Therefore the phase angles for two particular antenna pairs, $\angle\underline{R}(\delta_{21}, 0) = k_0\xi_0\theta_x$ and $\angle\underline{R}(\delta_{31}, 0) = k_0\eta_0\theta_y$ can be used to uniquely locate the angular position of the scatterer provided $-\pi < k_0\xi_0\theta_x \leq \pi$ and $-\pi < k_0\eta_0\theta_y \leq \pi$. For a pulsed Doppler radar,

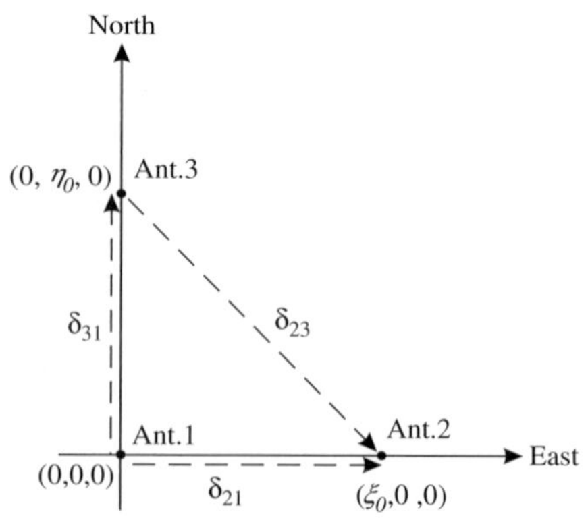

Figure 9.6 Possible antenna locations for a spaced antenna experiment.

the range of the scatterer is given in the conventional way by gating the samples of the transmit pulse as discussed in Chapter 4. The radial velocity of the scatterer can easily be found by setting $\vec{\delta}_{ij}$ to zero in (9.22) to produce

$$\angle \underline{R}(0, \tau) = \omega\tau, \tag{9.24}$$

where again $\omega \equiv -2k_0 v_r$. This can be done separately for the coherently detected signals at each of the antennas. Note that although we typically have a limited number of lags in space in our samples, we will often have many lags in time.

Next let us consider the covariance function for a *mono-static* but *dual-frequency* radar application for the case of a single scatterer. The wavenumbers of the two frequencies with subscripts n and m are given by k_n and k_m, respectively. Again we express wavenumber k_m relative to wavenumber k_n such that $k_m = k_n - \Delta k_{nm}$. Note that although $\vec{k}_m$ and $\vec{k}_n$ are really vectors, they will be parallel for any given scatterer so we can treat them as scalars. Now the covariance function takes the form

$$\underline{R}(\Delta k_{nm}, \tau) = E\left\{\underline{s}(k_n, t)\underline{s}^*(k_n - \Delta k_{nm}, t - \tau)\right\}. \tag{9.25}$$

This time we substitute (9.13) into the covariance function to get

$$\underline{R}(\Delta k_{nm}, \tau) = \left|\tilde{A}\right|^2 \mathrm{e}^{\mathrm{i}(\omega\tau - 2\Delta k_{nm} r)}. \tag{9.26}$$

As before, we examine the phase angle of the covariance function for the case of $\tau = 0$. This gives

$$\angle \underline{R}(\Delta k_{nm}, 0) = -2\Delta k_{nm} r, \tag{9.27}$$

which can be used to locate the range of the scatterer *within* a given range gate. As was true for the multiple-receiver case, the radial velocity of the scatterer is simply obtained by examining the covariance functions given by (9.26) for $\tau = 0$. In order to assure that no range ambiguities arise, one should select the frequency spacing to match that of the resolution of the conventional range gate. Recall that $\Delta r = c\Delta t/2$, where c is the speed of light and Δt is the pulse width. We would like the values of (9.27) to range from 0 to 2π along the line-of-sight dimension of a particular sampling volume. We can achieve this by requiring that $2\Delta k \Delta r = 2\pi$ or alternatively $\Delta f = 1/\Delta t$. For example, if the range resolution is $\Delta r = 300$ m, then the frequency spread should be 500 kHz. Note that the phase information given in (9.23) and (9.27) reflects the results that were obtained using the simplistic arguments presented in the beginning of the chapter.

Having explored the significance of the various covariance functions within the framework of multiple-receiver and multiple-frequency radar measurements, we now turn again to the case of scatter from distributed or multiple scatterers. As before, the space-frequency-time covariance function is given by (9.19). Instead of (9.12) we now substitute (9.16) into (9.19) to obtain the expression

$$\underline{R}(\vec{\delta}_{ij}, \Delta k_{nm}, \tau) = E\left\{\underline{s}(\vec{d}_i, k_n, t)\underline{s}^*(\vec{d}_i - \vec{\delta}_{ij}, k_n - \Delta k_{nm}, t - \tau)\right\}$$
$$= E\Bigg\{ \iiiint\!\!\iint \underline{W}(\hat{\vec{a}}, r)\underline{\tilde{A}}(\hat{\vec{a}}, r, \omega) e^{i[\omega t - 2k_n r + k_n \hat{\vec{a}}\cdot\vec{d}_i)]}$$
$$\times \underline{W}^*(\hat{\vec{a}}', r')\underline{\tilde{A}}^*(\hat{\vec{a}}', r', \omega') e^{-i[\omega'(t-\tau) - 2(k_n - \Delta k_{nm})r' + (k_n - \Delta k_{nm})\hat{\vec{a}}'\cdot(\vec{d}_i - \vec{\delta}_{ij})]}$$
$$\times \mathrm{d}\hat{\vec{a}}\,\mathrm{d}r\,\mathrm{d}\omega\,\mathrm{d}\hat{\vec{a}}'\,\mathrm{d}r'\,\mathrm{d}\omega' \Bigg\}. \tag{9.28}$$

This expression is generally intractable until we make some simplifying but reasonable assumptions. First, consider the amplitude of the radar signal as being an independent zero-mean random variable. That is, we assume that

$$E\left\{\underline{\tilde{A}}(\hat{\vec{a}}, r, \omega)\right\} = 0. \tag{9.29}$$

We further assume that

$$E\left\{\underline{\tilde{A}}(\hat{\vec{a}}, r, \omega)\underline{\tilde{A}}^*(\hat{\vec{a}}', r', \omega')\right\} = E\left\{\left|\underline{\tilde{A}}(\hat{\vec{a}}, r, \omega)\right|^2\right\} \delta(\hat{\vec{a}} - \hat{\vec{a}}')\,\delta(r - r')\,\delta(\omega - \omega'), \tag{9.30}$$

where $\delta(\ )$ is the Dirac delta function (not to be confused with $\vec{\delta}_{ij}$; the former has no subscripts). Using (9.29) and (9.30), the unwieldy expression given in (9.28) can be simplified to

$$\underline{R}(\vec{\delta}_{ij}, \Delta k_{nm}, \tau) = \iiint E\left\{\left|\underline{\tilde{A}}(\hat{\vec{a}}, r, \omega)\right|^2\right\} |\underline{W}|^2(\hat{\vec{a}}, r)$$
$$\times e^{i(\omega\tau - 2\Delta k_{nm} r + k_n \hat{\vec{a}}\cdot\vec{\delta}_{ij} + \Delta k_{nm}\hat{\vec{a}}\cdot\vec{d}_i + \Delta k_{nm}\hat{\vec{a}}\cdot\vec{\delta}_{ij})}\mathrm{d}\hat{\vec{a}}\,\mathrm{d}r\,\mathrm{d}\omega$$
$$= \iiint b(\hat{\vec{a}}, r, \omega)|W|^2(\hat{\vec{a}}, r) e^{i(\omega\tau + \varphi)}\mathrm{d}\hat{\vec{a}}\,\mathrm{d}r\,\mathrm{d}\omega, \tag{9.31}$$

where

$$\varphi = \varphi(\vec{\delta}_{ij}, \Delta k_{nm}) = -2\Delta k_{nm} r + k_n \hat{\vec{a}}\cdot\vec{\delta}_{ij} + \Delta k_{nm}\hat{\vec{a}}\cdot\vec{d}_i + \Delta k_{nm}\hat{\vec{a}}\cdot\vec{\delta}_{ij}, \tag{9.32}$$

and $b(\hat{\vec{a}}, r, \omega)$ is a new quantity known as the *brightness spectrum*, given as

$$b(\hat{\vec{a}}, r, \omega) = E\left\{\left|\tilde{A}(\hat{\vec{a}}, r, \omega)\right|^2\right\}. \tag{9.33}$$

We write $|W|^2$ simply as W^2, and recognize that both b and W^2 are real functions.

The term "brightness" has its origins in radio astronomy. In fact, there are considerable similarities between the signal processing tools to be discussed in this chapter and those found in radio astronomy. The brightness spectrum expresses the power density of the backscattered signal over all space and for all Doppler frequencies. Therefore, backscattered power for a particular receive antenna j and frequency n observed using a pulsed Doppler radar system is given simply as

$$P_0 = E\left\{\left|\underline{s}(\vec{d}_j, k_n, t)\right|^2\right\} \tag{9.34}$$
$$= \iiint b(\hat{\vec{a}}, r, \omega) W^2(\hat{\vec{a}}, r)\,\mathrm{d}\hat{\vec{a}}\,\mathrm{d}r\,\mathrm{d}\omega.$$

In this case, a proper definition of P_0 requires us to use the un-normalized form of $\underline{s}(\vec{d}_j, k_n, t)$. As we discuss later in the chapter, it is not possible to directly obtain the brightness spectrum through radar measurements. The best that can be achieved is an estimate of the brightness using various inversion techniques. This is the essence of radar imaging.

9.3 Spaced antenna methods

We have established a framework which can be used to discuss multiple-receiver and multiple-frequency radar measurements. Now we can begin introducing in more detail some of the techniques that call upon this framework. One of the oldest of these techniques is known in broad terms as the spaced antenna (SA) technique. It was originally devised as a means of measuring winds in the upper atmosphere during the 1940s (e.g., *Mitra*, 1949; *Briggs et al.*, 1950). In the SA technique, the atmosphere aloft is normally illuminated by a single vertically oriented radar beam. The radiowaves interact with the irregularities in the atmospheric refractive index, which leads to Bragg or Fresnel scatter. The scattering layer acts like a reflectional diffracting screen, so that the backscattered portions of the radiowaves form a diffraction pattern on the ground, which experiences both spatial and temporal variability in accordance with the motions of the atmosphere. If the bulk motion of the atmospheric medium in which the refractive index variations are embedded is non-zero, then the projected diffraction pattern on the ground will likewise be found to propagate. The vector describing the bulk motion of the diffraction pattern projected onto the ground is called the trace velocity v_T. An illustration of this is given in Figure 9.7.

Historically, work in this field has dealt mainly with correlation functions, which are normalized as per our comments earlier in this chapter, i.e. workers used the "statistician's definition" of the correlation functions. There was no need to even know the absolute values of the signal, since the main parameters were spatial and temporal lags, which could be found from the normalized autocovariance (correlation) functions.

9.3.1 Fundamental concepts

In the SA technique, the time-varying electric field amplitudes associated with the backscattered diffraction pattern are monitored using two or more spatially separated receiving antennas. From the resulting radar signals, autocorrelations and

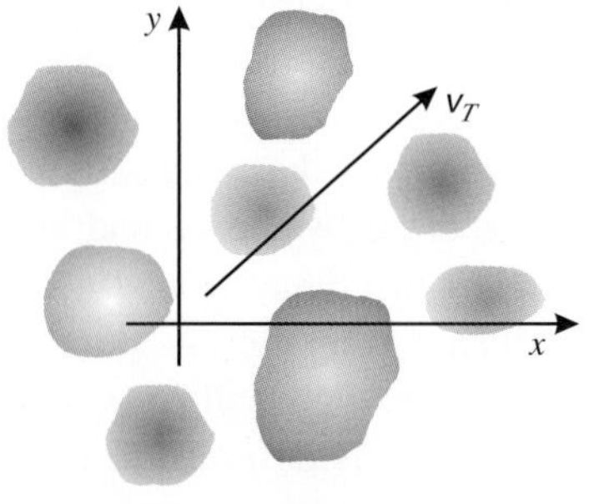

Figure 9.7 The diffraction patterns formed on the ground resulting from atmospheric radiowave scatter and the vector describing its bulk motion.

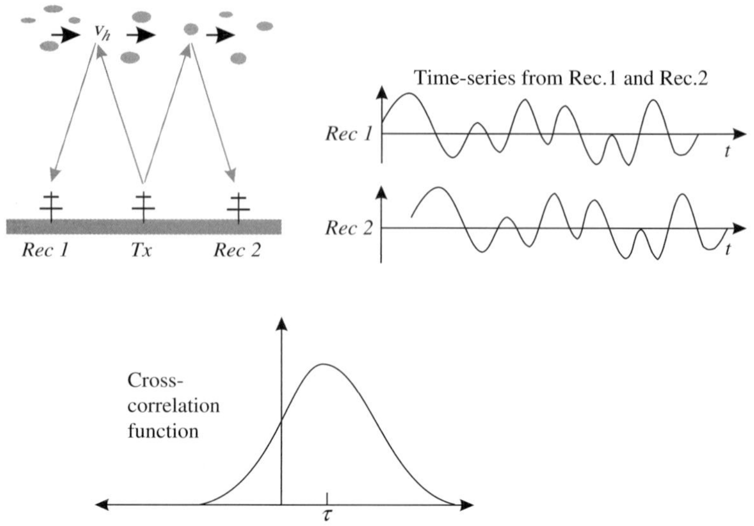

Figure 9.8 Showing how a simple spaced antenna configuration can be used to monitor the wind speed aloft using correlation functions. See text for details.

cross-correlations are computed and used to determine the underlying motions of the scattering field responsible for the backscattered waves.

To illustrate this, we begin by imagining the simple case of an electromagnetic wave transmitted from a single antenna and backscattered by a collection of atmospheric scatterers. This discussion follows that presented in *Briggs* (1980). The returned signal is measured at two spatially separated receiving antennas: $Rx\,1$ located at $(-d,\ 0,\ 0)$ and $Rx\,2$ located at $(d,\ 0,\ 0)$. The transmit antenna is located at the origin of this coordinate system, as shown in Figure 9.8. The structures that are responsible for the backscattered signals are assumed to be advected by a background wind, which flows parallel to the baseline of the two receive antennas. We further imagine that the scatterer's motion can be completely described by the Taylor frozen flow hypothesis. That is,

$$n(\vec{r},\ t) = n(\vec{r} - \vec{v}t,\ 0), \tag{9.35}$$

where $n(\vec{r},\ t)$ is the refractive index of the scatterers at location $\vec{r}$ and time t and $\vec{v}$ is their velocity. This is equivalent to stating that the scattering field observed at some initial time t is the same as the scattering field observed at a later time $t + \tau$, which has been displaced by a distance $\vec{v}\tau$. (Note that this assumption will be relaxed when we deal with the most general form of the theory.)

Under these assumptions, the radar signal received by the first antenna $Rx\,1$ will be similar to that received by the second antenna $Rx\,2$, with the exception that the signal from $Rx\,1$ *leads* the signal from $Rx\,2$. We can study the temporal relationship between the two signals using a cross-correlation analysis as shown in Figure 9.8. In this example, the two signals are shifted in time by an amount τ. Therefore, the diffraction pattern on the ground is observed to move with a phase velocity (trace velocity) given by $v_T = 2d/\tau$.

We can continue with this analysis by asking the question: How does the trace velocity relate to the actual velocity of the scatterers? That is, by observing the phase velocity

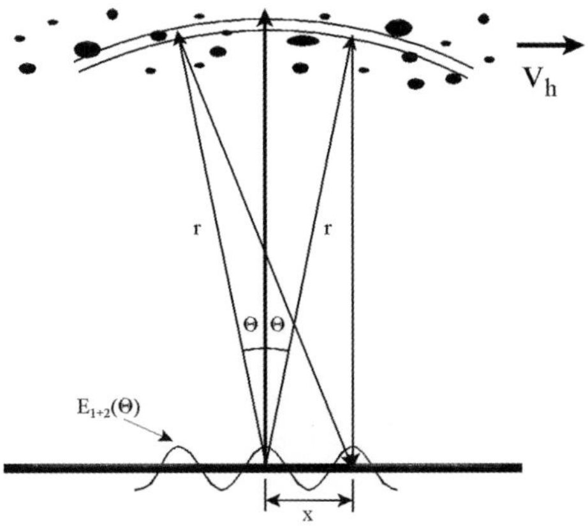

Figure 9.9 The diffraction pattern measured at the surface results from a superposition of radiowave scatter from the ensemble of "scatterers" aloft.

of the moving interference pattern observed on the ground, can we arrive at the velocity of the scatterers aloft? Following the treatment presented in (*Briggs*, 1980), we again consider a collection of uniformly distributed scatterers that are being advected with a mean wind having a horizontal velocity given by v_h and the direction of the wind is parallel to the baseline of the two antennas. This time, we will examine the backscattered signal not only at the locations of the receive antennas but along the entire extent of the antenna baseline. We will take this as the x-axis (see Figure 9.9). For the sake of illustration, we again consider a two dimensional scattering model and the simple case in which the scatterers experience no random motions; that is, they are frozen in.

The electric field projected onto the ground in the two-dimensional scattering model results from contributions from scatters coming from all zenith angles. We begin by considering only those contributions coming from angles $+\theta$. The electric fields resulting from scatter coming from angles $-\theta$ and θ are both given by

$$\underline{E}_R(\theta, t) = \underline{\tilde{A}} e^{i\{(\omega_0+\omega)t-(2k_0r-k_0\hat{\vec{a}}\cdot\vec{x})+\psi\}}, \tag{9.36}$$

where $\vec{x}$ gives the location along the antenna baseline being considered. Making the substitutions that $\omega = -2k_0v_r$, $v_r = v_h \sin\theta$ and $k_0\hat{\vec{a}} \cdot \vec{x} = k_0 x \sin\theta$ produces

$$\underline{E}_1 = \underline{E}_R(-\theta, t) = \underline{\tilde{A}} e^{i\{(\omega_0+2k_0v_h \sin\theta)t-(2k_0r+k_0x\sin\theta)+\psi_1\}} \tag{9.37}$$

and

$$\underline{E}_2 = \underline{E}_R(\theta, t) = \underline{\tilde{A}} e^{i\{(\omega_0-2k_0v_h \sin\theta)t-(2k_0r-k_0x\sin\theta)+\psi_2\}}. \tag{9.38}$$

For this simple example, we have assumed that the strengths of the backscattered signals from $-\theta$ and θ are equal. Combining the two signals we get

$$\underline{E}_{1+2} = 2\underline{\tilde{A}} e^{i(\omega_0 t-2k_0r+\psi_1+\psi_2)} \cos(2k_0v_h \sin\theta t - k_0x\sin\theta + \psi_1 - \psi_2). \tag{9.39}$$

The amplitude of the interacting waves is given by

$$2\underline{\tilde{A}} \cos(2k_0v_h \sin\theta t - k_0x\sin\theta + \psi_1 - \psi_2), \tag{9.40}$$

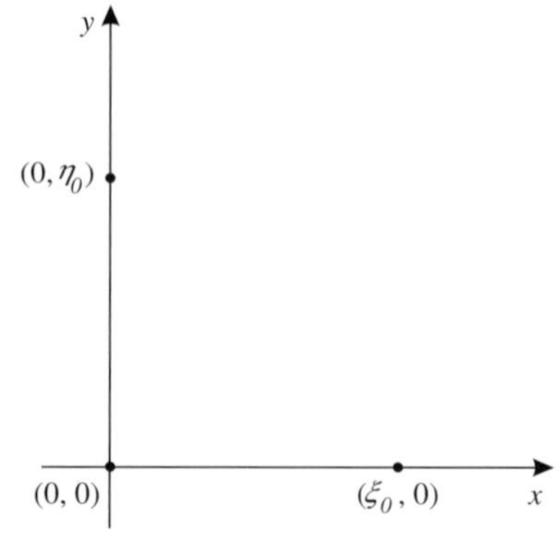

Figure 9.10 Possible antenna configuration for use in spaced antenna measurements.

and moves along the ground with a phase speed of

$$v_\phi = v_T = \frac{2k_0 v_h \sin\theta}{k_0 \sin\theta} = 2v_h. \tag{9.41}$$

This gives the classical result that the trace velocity of the diffraction pattern projected onto the ground is twice the actual velocity of motion of the particles. Of course, in reality we would need to consider the summation of signals from a range of angles.

How can the horizontal wind vector be determined if it is not aligned with one of the antenna baselines? In answering this question, we again look to *Briggs* (1984). Begin by considering the configuration of antenna elements shown in Figure 9.10, where in contrast to Figure 9.6, we have given the coordinates in only two dimensions and assumed that the antennas are on flat, level ground. Using this configuration, one can sample the diffraction pattern resulting from the backscattered signals at three locations. Note that this triangular arrangement of three *non-collinear* receive antennas constitutes a minimum configuration required in order to estimate the horizontal wind vector. The spacing between the antenna elements is assumed to be small compared to the individual features within the diffraction pattern projected onto the ground.

Imagine a localized maximum in the diffraction pattern as it passes across the three receive antennas depicted in Figure 9.10. Such a scenario is shown in Figure 9.11. In general, each antenna will perceive a different cross-section of the "hill" in the diffraction pattern, shown by the three broken lines. As discussed in *Briggs* (1984), if the feature in the diffraction pattern is large compared to the antenna spacing, then to a good approximation, a straight line that is perpendicular to the direction of propagation will connect the maxima in these three cross-sections. This is shown as the solid line in Figure 9.11. The time required for the line of maximum to pass from the antenna at the origin to one at $(\xi_0, 0)$ can be written as

$$t_x = \frac{\xi_0 \sin\phi}{v_T}, \tag{9.42}$$

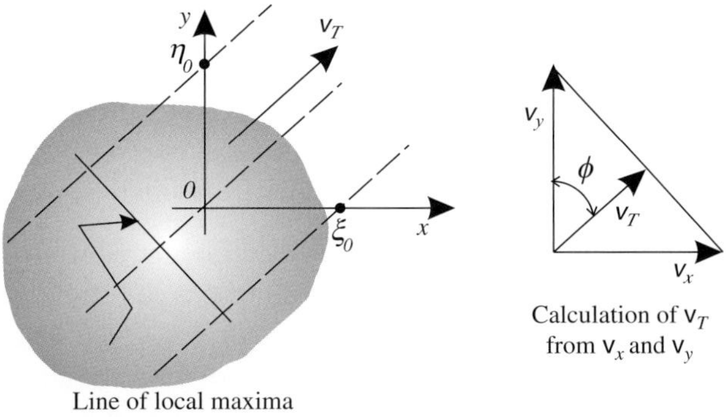

Figure 9.11 A diffraction pattern projected onto the ground as it passes over the receiving antennas located at (0, 0), $(\xi_0, 0)$, and $(0, \eta_0)$. Each receiver "sees" the portion of the "hill" indicated by the broken lines.

where ϕ is as shown in Figure 9.11. Likewise, the time required for the line of maximum to pass from the origin to $(0, \eta_0, 0)$ is given by

$$t_y = \frac{\eta_0 \cos\phi}{v_T}. \tag{9.43}$$

If the values of t_x and t_y are measured (for example through the use of cross-correlation functions), then v_T and ϕ can easily be found as

$$v_T = v_x \sin\phi = v_y \cos\phi \tag{9.44}$$

and

$$\phi = \tan^{-1}\frac{v_y}{v_x}, \tag{9.45}$$

where $v_x = \xi_0/t_x$ and $v_x = \eta_0/t_y$. Note that Equation (9.44) also means that

$$\frac{1}{v_x^2} + \frac{1}{v_y^2} = \frac{\sin^2\phi}{v_T^2} + \frac{\cos^2\phi}{v_T^2} = \frac{1}{v_T^2}. \tag{9.46}$$

Note that v_x, v_y are *not* components of v_T, and indeed both exceed v_T.

In general, localized features (hills) in a diffraction pattern can take on complex forms. Therefore, we do not always expect the line of maximum to be perpendicular to the trace velocity. However, if a sufficient number of radar samples is collected, then many hills will be represented in the resulting data set. If the shape of these localized features is approximately circular on average, then the mean values of t_x and t_y will still produce the correct value for v_T and ϕ. More complicated cases are discussed below.

9.3.2 Full correlation analysis (FCA)

In the idealized treatments presented above, several factors have not been taken into consideration. They can affect the calculated correlation functions in such a way as to impact the estimation of the wind field. In particular, one must consider the following:

1. The diffraction pattern that is projected onto the ground can be systematically elongated in a direction not necessarily either perpendicular or parallel to the mean flow of the wind.
2. The diffraction patterns in reality can, and most likely will, evolve over time. In other words, we cannot assume the Taylor frozen-in hypothesis.

Full correlation analysis (FCA) was therefore introduced, this being a technique which uses the auto- and cross-correlation functions to estimate the wind velocity in such a way that it allows for these effects.(e.g., *Briggs*, 1984; *Briggs et al.*, 1950).

Earlier we showed that the general form of the correlation function could be expressed as Equation (9.28). Furthermore, the average return power is given by Equation (9.34). So the normalized correlation function is simply Equation (9.28) divided by Equation (9.34). We now, following the historical evolution of this subject, construct a correlation function based on Equation 9.17, but normalized to unity at zero lag (i.e., the statistician's version of the correlation function), namely

$$\underline{\rho}(\xi, \eta, \tau) = \frac{E\{\underline{s}(x+\xi, y+\eta, t+\tau)\underline{s}^*(x, y, t)\}}{E\{|\underline{s}(x, y, t)|^2\}}. \tag{9.47}$$

Note that instead of defining the antenna spacings collectively using $\vec{\delta}_{ij}$, we have here defined them separately along the directions of the x- and y-axes using η and ξ respectively. Here η and ξ are spatial lags projected onto the x- and y-axes, respectively.

Using Equation (9.47), we can now discuss certain properties of the correlation functions more easily and with a bit more transparency. For the case of wide sense stationarity, the correlation functions must satisfy certain properties by definition. Three of these properties, which are relevant for the present discussion, are listed below.

1. **Property 1:** The autocorrelation is conjugate symmetric in ξ, η, and τ:

$$\underline{\rho}\,(-\xi, -\eta, -\tau) = \underline{\rho}^*\,(\xi, \eta, \tau), \tag{9.48}$$

and therefore

$$\left[|\underline{\rho}\,(-\xi, -\eta, -\tau)\right| = \left[|\underline{\rho}\,(\xi, \eta, \tau)\right|. \tag{9.49}$$

2. **Property 2:** The mean-squared value of the random process is equal to the autocorrelation evaluated at zero lag:

$$\underline{\rho}\,(0, 0, 0) = 1. \tag{9.50}$$

3. **Property 3:** The autocorrelation is upper bounded by its value at zero lag:

$$\underline{\rho}\,(0, 0, 0) \geq \left[|\underline{\rho}\,(\xi, \eta, \tau)\right|. \tag{9.51}$$

As a consequence of Equation (9.48), the correlation must be reflection invariant. That is, $\left|\underline{\rho}\,(\xi, \eta, \tau)\right|$ can only be a function of the products of ξ, η, and τ, specifically ξ^2, η^2, τ^2, $\xi\tau$, $\eta\tau$ and $\xi\eta$. Therefore, we can construct a simple function that is consistent with these requirements via

$$\left|\underline{\rho}\,(\xi, \eta, \tau)\right| = f\left(A\xi^2 + B\eta^2 + C\tau^2 + 2F\xi\tau + 2G\eta\tau + 2H\xi\eta\right), \tag{9.52}$$

where the coefficients have been chosen here to be consistent with those given in *Briggs* (1984). We note that $f(\cdot)$ in Equation (9.52) is an arbitrary *real* function; however, from Equations (9.50) and (9.51), we know that $f(0, 0, 0) = 1$ and $|f(\xi, \eta, \tau)| \leq |f(0, 0, 0)|$ for all values of ξ, η, and τ. By knowing the parameters A through H we can describe the shape of the ellipsoidal autocorrelation functions. Next we show how to solve for these coefficients.

The set of equations given by $|\underline{\rho}(\xi, \eta, \tau)| =$ constant describes a family of concentric ellipsoids in (ξ, η, τ)-space, which are similar in shape. We now wish to consider a coordinate system in which the ellipsoids are aligned along the τ-axis. This is convenient when examining how $\underline{\rho}(\xi, \eta, \tau)$ evolves for non-zero values of τ. That is, we wish to explore the temporal evolution of the ellipsoidal shapes in a Lagrangian reference frame. This is achieved through the coordinate transformations $\xi \rightarrow \xi - V_x\tau$ and $\eta \rightarrow \eta - V_y\tau$, where V_x and V_y are the trace velocities along the x- and y-axis respectively. This choice can be understood by asking the question: Where will $\rho(\xi, \eta, \tau)$ be most correlated for non-zero values of τ? This means that the moving correlation function maintains its form as it moves, so its value at $[(\xi - V_x\tau), (\eta - V_y\tau)]$ in the shifted frame is the same as described by (9.52). In the new coordinate system (see Figure 9.12), we can write

$$\left|\underline{\rho}(\xi, \eta, \tau)\right| = f\left(A(\xi - V_x\tau)^2 + B\left(\eta - V_y\tau\right)^2 + K\tau^2 + 2H(\xi - V_x\tau)\left(\eta - V_y\tau\right)\right). \tag{9.53}$$

We then equate common terms in Equations (9.52) and (9.53). For example, consider equating the terms involving $\xi\tau$. In (9.52), this is just $2F\xi\tau$. In (9.53) this is

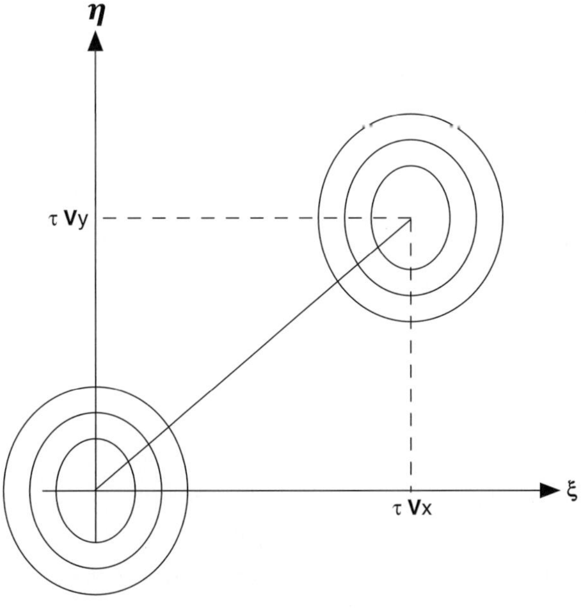

Figure 9.12 Translation of an atmospheric diffraction pattern along the ground and how the trace velocity of the pattern can be calculated.

$-2AV_x\xi\tau - 2H\xi V_y\tau$. These two terms may be equated to give $AV_x + HV_y = -F$. Other similar equations can be developed for $\eta\tau$, etc. As a result we produce

$$\begin{aligned} AV_x + HV_y &= -F \\ HV_x + BV_y &= -G \\ K = C - AV_x^2 - BV_y^2 &- 2HV_xV_y. \end{aligned} \tag{9.54}$$

Consequently, we can determine the values of V_x and V_y, provided we can arrive at estimates for the values of the coefficients.

Using Equations (9.52) and (9.53) we can devise a framework that allows us to experimentally determine the various coefficients for the different representations of the correlation functions. Consider an arrangement of three antennas placed on the ground forming a triangle as shown in Figure 9.10. The autocorrelation taken at lags $\xi = \xi_0$ and $\eta = 0$ can be written as

$$\left|\underline{\rho}\left(\xi_0,\ 0,\ \tau\right)\right| = f\left(A\xi_0^2 + C\tau^2 + 2F\xi_0\tau\right). \tag{9.55}$$

This expression attains a maximum value at $\tau = \tau_x^{'}$. We can find the value of τ for which the expression attains a maximum by setting the derivative of $f\,(\ )$ with respect to τ at $\tau = \tau_x^{'}$ equal to zero, viz.,

$$\left.\frac{\delta f\left(\xi_0,\ 0,\ \tau\right)}{\delta\tau}\right|_{\tau=\tau_x^{'}} = f_\tau\left(A\xi_0^2 + C\tau_x^{'2} + 2F\xi_0\tau_x^{'}\right)\left(2C\tau_x^{'} + 2F\xi_0\right) = 0, \tag{9.56}$$

where $f_\tau\,(\ )$ is the derivative of $f\,(\ )$ with respect to τ. This implies that

$$\tau_x^{'} = -\frac{F}{C}\xi_0. \tag{9.57}$$

Likewise, we can examine autocorrelation at lags $\xi = 0$ and $\eta = \eta_0$ through the expression

$$\left|\underline{\rho}\left(0,\ \eta_0,\ \tau\right)\right| = f\left(B\eta_0^2 + C\tau^2 + 2G\eta_0\tau\right) \tag{9.58}$$

and

$$\left.\frac{\delta f\left(0,\ \eta_0,\ \tau\right)}{\delta\tau}\right|_{\tau=\tau_y^{'}} = f_\tau\left(B\eta_0^2 + C\tau_y^{'2} + 2G\eta_0\tau_y^{'}\right)\left(2C\tau_y^{'} + 2G\eta_0\right) = 0, \tag{9.59}$$

which implies

$$\tau_y^{'} = -\frac{G}{C}\eta_0. \tag{9.60}$$

Note that ξ_0 and η_0 are known and $\tau_x^{'}$ and $\tau_y^{'}$ can be estimated from observations. So we are able to experimentally determine F/C and G/C.

We should note at this point that estimates of the zonal and meridional wind can be obtained simply by calculating $\xi_0/\tau_x^{'}$ and $\eta_0/\tau_y^{'}$, respectively. These are the so-called "apparent" velocities. If the observed turbulent structures are truly frozen and do not evolve over time, then these values correspond to the actual velocity of the atmosphere.

However, if the atmospheric structures do evolve with time, then $\tau_x^{'}$ and $\tau_y^{'}$ contain contributions from both the horizontal displacement and the temporal evolution of the atmospheric structures, and so the cross-spectra will be reduced in width compared to a non-evolving case. Consequently, the apparent velocity will always be greater than or equal to the "true" velocity. The method outlined in the following paragraphs provides the true velocity. As discussed in *Briggs* (1984), the assumptions made in arriving at the solution for the true velocity in the full-correlation analysis may not always be met, so one may wish to calculate the apparent velocity along with the true velocity as a quality check.

If we now assume that the auto- and cross-correlations have the same functional forms, we can develop a method of estimating the remaining parameters describing the shape of the ellipsoidal correlation function. The autocorrelation function calculated at each of the three antenna locations should be the approximately the same. We can express the average of these autocorrelation functions as $\underline{\rho}(0, 0, \tau)$. We then define a temporal lag τ_x such that

$$\left|\underline{\rho}\,(\xi_0,\ 0,\ 0)\right| = \left|\underline{\rho}\,(0,\ 0,\ \tau_x)\right|. \tag{9.61}$$

This is illustrated in Figure 9.13. The equality holds when

$$A\xi_0 = C\tau_x^2 \tag{9.62}$$

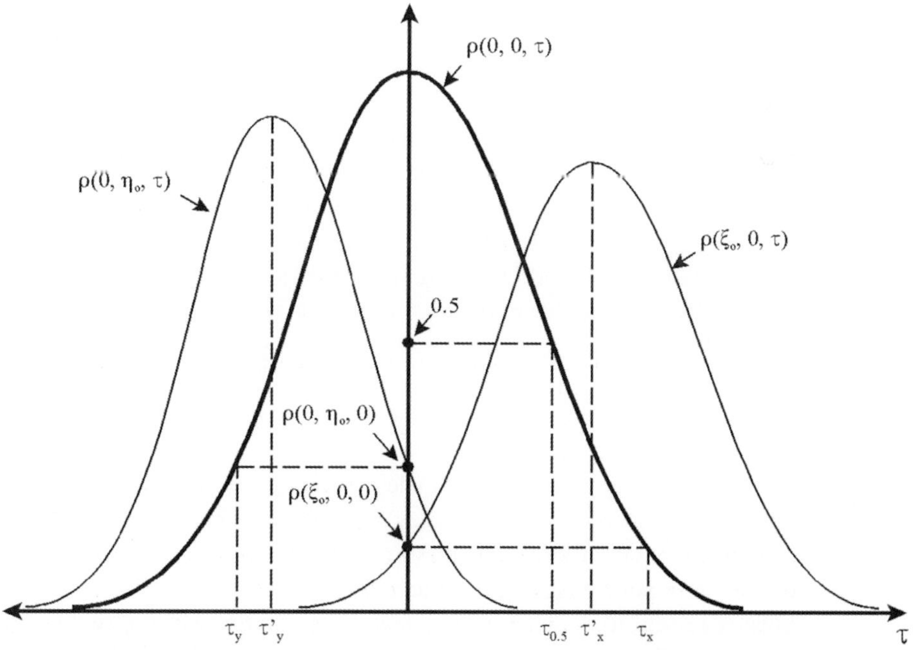

Figure 9.13 Representations of the correlation function along with some of the parameters derived from them, which are used in FCA calculations.

or

$$\tau_x^2 = \frac{A}{C}\xi_0. \tag{9.63}$$

This provides a means of evaluating the ratio B/C since τ_x^2 can be estimated from observations. Similarly, we can define τ_y such that

$$\left|\underline{\rho}\,(0,\ \eta_0,\ 0)\right| = \left|\underline{\rho}\left(0,\ 0,\ \tau_y\right)\right|, \tag{9.64}$$

which leads to

$$B\eta_0 = C\tau_y^2 \tag{9.65}$$

or

$$\tau_y^2 = \frac{B}{C}\eta_0. \tag{9.66}$$

Finally, we equate the cross-correlation corresponding to points $\left(\xi_0,\ 0\right)$ and $(0,\ \eta_0)$ with the mean autocorrelation to define a point τ_{xy} given by the equation

$$\left|\underline{\rho}\,(\xi_0,\ \eta_0,\ 0)\right| = \left|\underline{\rho}\left(0,\ 0,\ \tau_{xy}\right)\right|, \tag{9.67}$$

which results in

$$A\xi_0^2 + B\eta_0^2 + 2H\xi_0\eta_0 = C\tau_{xy}^2 \tag{9.68}$$

or

$$\tau_{xy}^2 = 2\xi_0\eta_0\frac{H}{C} + \xi_0^2\frac{A}{C} + \eta_0^2\frac{B}{C}. \tag{9.69}$$

The values of A/C and B/C have already been solved for, so this equation can be used to estimate H/C.

As we shall show, knowing the ratios of A/C, B/C, F/C, G/C, and H/C is sufficient to calculate the horizontal drift of the diffraction pattern (and correspondingly the horizontal wind aloft). However, we need to find C in order to describe the ellipsoidal shape of the correlation function. We can calculate this from the averaged autocorrelation function $|\underline{\rho}(0, 0, \tau)|$ by finding the temporal lag $\tau_{0.5}$ at which the autocorrelation function has a magnitude of 0.5. In other words we find C by solving

$$|\underline{\rho}(0, 0, \tau_{0.5})| = |\underline{\rho}(C\tau_{0.5}^2)| = 0.5. \tag{9.70}$$

Solving for C requires that we assume some reasonable analytic form of the autocorrelation function, such as a Gaussian.

Having the coefficients needed to describe the correlation function allows us to examine various properties of the atmospheric flow which produced the observed diffraction patterns. Here we only consider how to extract the speed and direction of the trace velocity. We have already shown in Equation (9.54) how the trace velocities V_x and V_y are linearly related to the coefficients A, B, F, G, and H. Having two equations and two unknowns allows us to determine these trace velocities. Note that these equations can also be solved using the coefficient ratios that we found earlier by dividing through by C. As before the magnitude and direction can be found using Equations (9.44) and (9.45).

We can generalize the following discussion such that it can be applied to an arbitrary placement of receive antennas. As before, it will be necessary to have at least three antenna locations (not positioned along a common line) to estimate the two-dimensional horizontal wind vector. For a given pair of antennas i, j, let the separation between the two along the x- and y-axes be ξ_{ij} and η_{ij}, respectively. Then the equation for the autocorrelation function, which is similar to Equation (9.52), becomes

$$\left|\underline{\rho}\left(\xi_{ij}, \eta_{ij}, \tau_{ij}\right)\right| = f\left(A\xi_{ij}^2 + B\eta_{ij}^2 + C\tau_{ij}^2 + 2F\xi_{ij}\tau_{ij} + 2G\eta_{ij}\tau_{ij} + 2H\xi_{ij}\eta_{ij}\right). \quad (9.71)$$

In a similar manner to our earlier treatment, we wish to find the temporal lag at which the autocorrelation attains a maximum. Whereas the lags that we found through Equations (9.56) and (9.59) corresponded to the x- and y-axes, respectively, here we are finding the temporal lag along an arbitrary baseline between the receiving antennas i and j. The temporal lag can be found through the equation

$$\left.\frac{\delta f\left(\xi_{ij}, \eta_{ij}, \tau_{ij}\right)}{\delta \tau_{ij}}\right|_{\tau=\tau_{ij}'} = f_\tau\left(\tau = \tau_{ij}'\right)\left(2C\tau_{ij}' + 2F\xi_{ij} + 2G\eta_{ij}\right) = 0, \quad (9.72)$$

where

$$f_\tau\left(\tau = \tau_{ij}'\right) = f_\tau\left(A\xi_{ij}^2 + B\eta_{ij}^2 + C\tau_{ij}'^2 + 2F\xi_{ij}\tau_{ij}' + 2G\eta_{ij}\tau_{ij}' + 2H\xi_{ij}\eta_{ij}\right) \quad (9.73)$$

is the derivative of Equation (9.71) with respect to τ. A general solution to Equation (9.72) requires that the value of τ_{ij}' be given by the equation

$$\tau_{ij}' = -\frac{F}{C}\xi_{ij} - \frac{G}{C}\eta_{ij}. \quad (9.74)$$

We let $\rho\left(0, 0, \tau_{ij}\right)$ be the average autocorrelation function calculated for each antenna. We can define a temporal lag τ_{ij} such that

$$\left|\underline{\rho}\left(\xi_{ij}, \eta_{ij}, 0\right)\right| = \left|\underline{\rho}\left(0, 0, \tau_{ij}\right)\right|. \quad (9.75)$$

This is demonstrated in Figure 9.14. Then from Equation (9.52) we can write

$$\tau_{ij}^2 = \frac{A}{C}\xi_{ij}^2 + \frac{B}{C}\eta_{ij}^2 + 2\frac{H}{C}\xi_{ij}\eta_{ij}. \quad (9.76)$$

We now have a system of equations that we can use to solve for the coefficients, or, at a minimum, the coefficients normalized by C. We can then use these values to calculate the horizontal wind speed and direction. The *true* wind speed is one half of V_T, as given by Equations (9.44) and (9.45), but using V_x, V_y, and V_T in place of v_x, v_y, and v_T. The *apparent* velocity is found by also using these two equations, but using v_x and v_y as shown in Equations (9.42) and (9.43), where the time lags between receivers i and j are found simply as the time lag at the peak of the appropriate cross-correlation function.

The full correlation analysis that we have presented provides insight into the spaced-antenna approach, but it should be noted that other methods have been advanced over the years. These include treatments by *Doviak et al.* (1996), *Holloway et al.* (1997), and *Zhang et al.* (2003). A review of various spaced antenna techniques for wind estimation can be found in *Doviak et al.* (2004). Generally speaking, the most common techniques

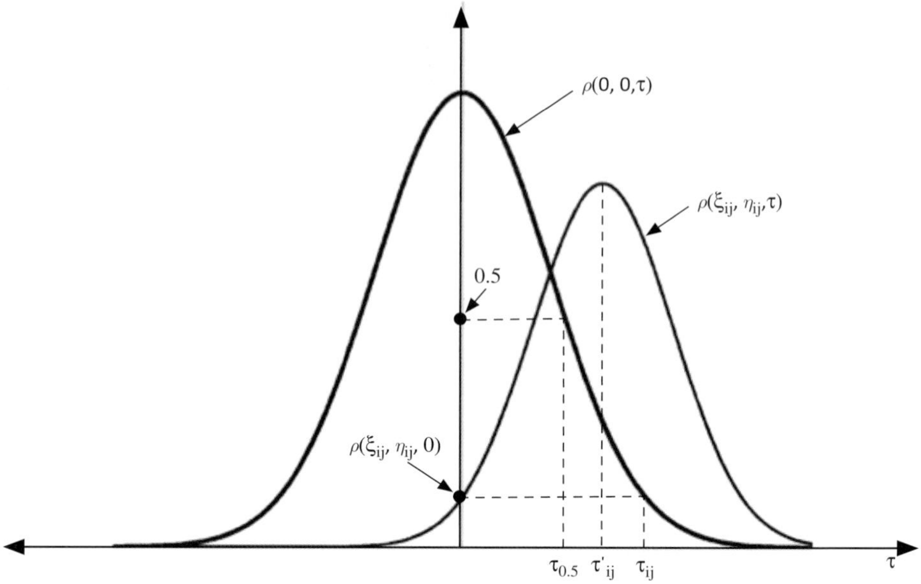

Figure 9.14 Representation similar to that in Figure 9.13 except that here we consider arbitrary placement of two antennas *i* and *j*.

are the Briggs FCA method, the intersection method, and the slope-at-zero-lag method. Among these, there are sub-variants based on a Gaussian model approach and a direct finite difference method. *Praskovsky and Praskovskaya* (2003) presented an alternative theory based not on correlation functions but rather on structure functions.

Regardless of the details, the spaced-antenna method has found application in many areas, including medium frequency (MF) measurements of winds in the lower mesosphere (e.g., see Chapter 2, Section 2.5), and in many tropospheric applications.

9.4 Interferometry

As mentioned above, we will refer to interferometry as a means of extracting information about atmospheric scatterers through analysis of correlation and/or covariance functions, or even spectra. The methods described below have their origins in works by *Pfister* (1971) and *Woodman* (1971). Before proceeding, it may be instructive to consider the fundamental principles used within the Doppler beam swinging, spaced antenna, and radar interferometry frameworks as a means of estimating the magnitude and direction of horizontal winds. Here we will consider scatter from localized gradients in the refractive index (Bragg scatter) as the source of the returned radiowave signal. A schematic representation of these techniques is provided in Figure 9.15. As discussed in earlier chapters, using the DBS technique, the atmosphere aloft is sampled using three or more non-collinear radar beams, which are directed at or near zenith. From the resulting radial velocity data, the wind field can be retrieved through some form of a matrix inversion technique.

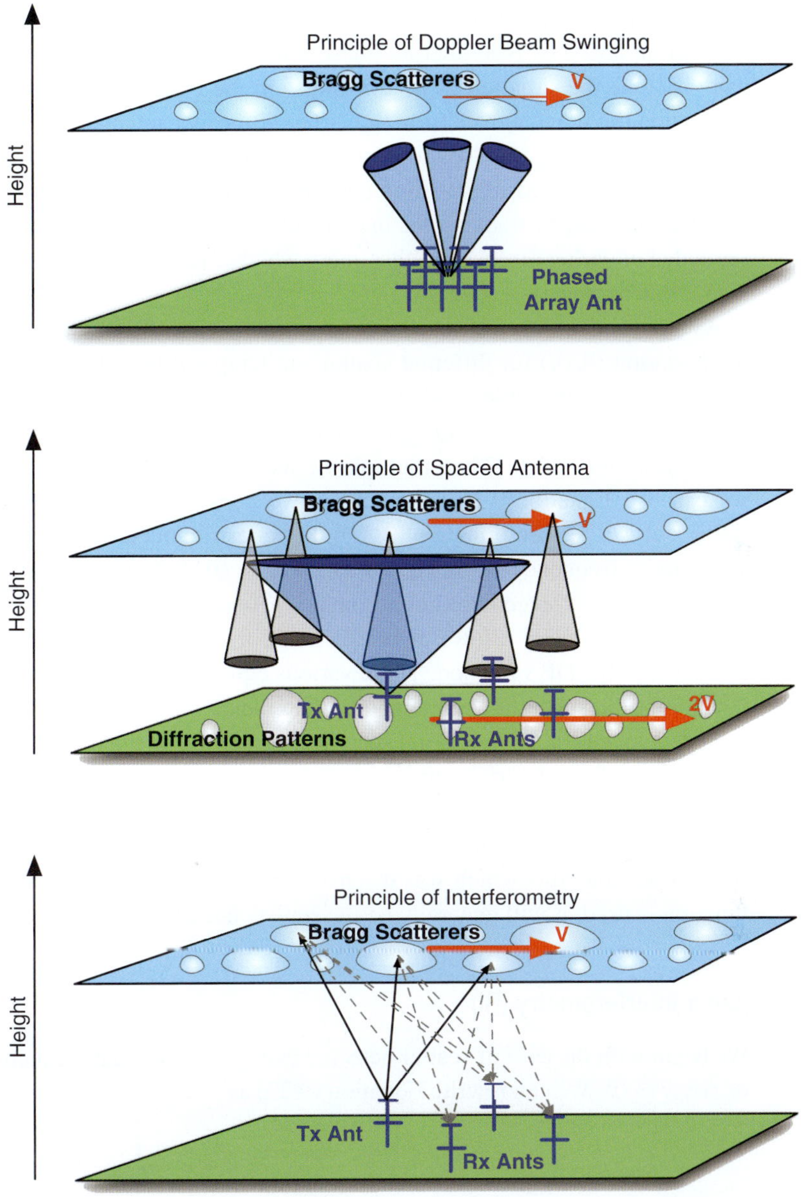

Figure 9.15 Three techniques used to estimate the wind speed using clear-air radar: Doppler beam swinging, spaced antenna, and interferometry. In the examples provided, each relies on coherent Bragg scatterers to produce the returned signal.

As we have seen from our discussion of the spaced antenna technique, the atmosphere is illuminated using a single beam, and diffraction pattern projected onto the ground is observed using three or more antennas or antenna groups (subarrays). Here one can imagine "blobs" of Bragg scatterers, which absorb the incoming radiowaves and re-emit them partly towards the surface (see Figure 9.15). The wind aloft is obtained by observing the speed at which the diffraction patterns move along the surface using correlation analysis. Note that no phase information is needed. For interferometry, we can visualize the situation as if "hot spots" in the Bragg scattering field are tracked in time using the input signals from the different receive antennas.

Broadly speaking, the locations of the hot spots are obtained by measuring the phase offsets in the received signals from the different antennas. The wind speed and direction are then obtained by tracking the movement of the hot spots. Nevertheless, the reader is reminded of earlier discussions that query whether such hot spots exist (e.g., see Section 2.15.7 in Chapter 2).

As an introduction, we can revisit the case of scatter from a point source. We examine Equation (9.19) for different spatial and temporal lags. For the case of backscatter observed with a multiple-receiver, as expressed in Equation (9.22), we showed that for $\tau = 0$, the phase angle of the covariance function is simply given by $\angle \underline{R} = k_0 \hat{\vec{a}} \cdot \vec{\delta}_{ij}$ (see Equation (9.23)). When multiple frequencies and a single antenna are used, then we can use Equation(9.26). Here again we consider the case when $\tau = 0$ to find $\angle R = -2\,\Delta k_{nm}\, r$ (see Equation (9.27)). If we restrict ourselves to $(\vec{d}_i = \vec{d}_j = 0)$ using a single radar frequency ($k_n = k_m = k$ or $\Delta k_{nm} = 0$), (i.e., a monostatic radar), the phase angle of the covariance function becomes $\angle \underline{R} = \omega\tau = -2\,k_0\, v_r\, \tau$ (see Equation (9.24)). This equation simply allows us to determine the radial velocity of the scatterer, as is normal with the DBS method. These various equations then allow us to find the angular position of the scatterer and the appropriate radial velocity. Whereas this treatment has been developed for a single point scatterer, we show that the formalism can be extended to hot spots of volumetric scatter.

Here, we also wish to make the point that some of the subsequent analysis will be done using covariance functions $\underline{R}$, to avoid the potential conflict in definitions regarding correlation functions which was discussed earlier in the chapter. When the correlation function is used, it will generally be the statistician's version.

9.4.1 Radar interferometry (RI)

We begin with the case of scatter detected using a single radar frequency ($k_n = k_m = k_0$ or $\Delta k_{nm} = 0$). We can rewrite Equation (9.23) as

$$\angle \underline{R} = k_0 \left[\theta_x \hat{\vec{i}} \cdot \vec{\delta}_{ij} + \theta_y \hat{\vec{j}} \cdot \vec{\delta}_{ij} + \theta_z \hat{\vec{k}} \cdot \vec{\delta}_{ij} \right], \tag{9.77}$$

where $\hat{\vec{i}}, \hat{\vec{j}}, \hat{\vec{k}}$ are unit vectors in the Cartesian coordinate system and θ_x, θ_y, and θ_z are the direction cosines for the antenna pair (ij), viz.,

$$\begin{aligned} \theta_x &= \sin\theta \sin\phi \\ \theta_y &= \sin\theta \cos\phi \\ \theta_z &= \cos\theta. \end{aligned} \tag{9.78}$$

These definitions are consistent with Figure 9.4.

Therefore, we can solve for the direction cosines if we have values of φ_{ij} from three or more receiving antennas (three of which must be non-collinear). Within this framework, we can also solve for v_r by finding the phase at zero lag of the correlation function through Equation (9.24). Assume that there are (at least) three independent scatterers, which are detected by the radar and which are moving with velocity $\vec{v}$ during a time interval such that $\vec{v}$ does not change significantly while the observations are being made. Under these conditions we can write

$$\begin{bmatrix} \theta_x^{(1)} & \theta_y^{(1)} & \theta_z^{(1)} \\ \theta_x^{(2)} & \theta_y^{(2)} & \theta_z^{(2)} \\ \theta_x^{(3)} & \theta_y^{(3)} & \theta_z^{(3)} \end{bmatrix} \begin{bmatrix} u \\ v \\ w \end{bmatrix} = \begin{bmatrix} v_r^{(1)} \\ v_r^{(2)} \\ v_r^{(3)} \end{bmatrix}, \tag{9.79}$$

where the mean wind within the scattering region is $\vec{u} = u\,\hat{\vec{i}} + v\,\hat{\vec{j}} + w\,\hat{\vec{k}}$, and where the superscripts in parentheses represent the measurements corresponding to the different scatterers. That is, $\theta_x^{(i)}$, $\theta_y^{(i)}$, and $\theta_z^{(i)}$ are the direction cosines for scatterer i, while $v_r^{(i)}$ is its radial velocity. This equation looks very similar to the DBS technique except that the various "beam directions" are not obtained by physically steering the beam. They are obtained by simply having multiple scatterers in the beam and locating them by the phase differences between spatially separated receivers. This approach is analogous to imaging Doppler interferometry (IDI). For an overview of IDI see *Brosnahan and Adams* (1993).

When considering the case of scatter from atmospheric fields, we will be interested in mean values for the velocity and position, which may be confined to certain hot spots in the reflectivity field. We achieve this by rewriting Equation (9.31) as

$$\underline{R}(\vec{\delta}_{ij}, \tau) = \mathrm{e}^{\mathrm{i}(\bar{\omega}\tau+\bar{\varphi})} \iiint b\left(\hat{\vec{a}}, r, \omega\right) W^2\left(\hat{\vec{a}}, r\right) \mathrm{e}^{\mathrm{i}[(\omega-\bar{\omega})\tau+(\varphi-\bar{\varphi}))} \mathrm{d}\Omega\, \mathrm{d}r\, \mathrm{d}\omega, \tag{9.80}$$

where

$$\bar{\omega} = \iiint \omega\, b\left(\hat{\vec{a}}, r, \omega\right) W^2\left(\hat{\vec{a}}, r\right) \mathrm{d}\Omega\, \mathrm{d}r\, \mathrm{d}\omega \tag{9.81}$$

and

$$\bar{\varphi} = \iiint \varphi\, b\left(\hat{\vec{a}}, r, \omega\right) W^2\left(\hat{\vec{a}}, r\right) \mathrm{d}\Omega\, \mathrm{d}r\, \mathrm{d}\omega, \tag{9.82}$$

with $\mathrm{d}\Omega$ being a differential unit of solid angle. This formulation includes several assumptions about the atmospheric scatter, which we will not elaborate on here in the interest of brevity. However, this equation implies that we can use Equations (9.24) and (9.23) to find the mean position and mean velocity of the atmospheric field in a similar fashion to the approach given for the hard scatterer.

We now consider an example of how the concepts just presented can be applied to actual radar observations by considering measurements collected using the MU radar in Japan (*Palmer et al.*, 1995a, b). The MU radar was configured to enable transmission of a vertical beam and reception of backscattered signals from the atmosphere using three separate groupings (subarrays) of antennas. A Doppler sorting technique is then used to create "slices" within the sampled volume, representing regions of constant measured radial velocity (see the upper panel in Figure 9.16). It should be noted that the wind field is assumed to be uniform across the sampled volume. In the diagram, the slices of constant radial velocity are aligned perpendicular to the wind field vector, corresponding to the case when the wind vector is oriented parallel to $\vec{\delta}_{ij}$. In the general case, the

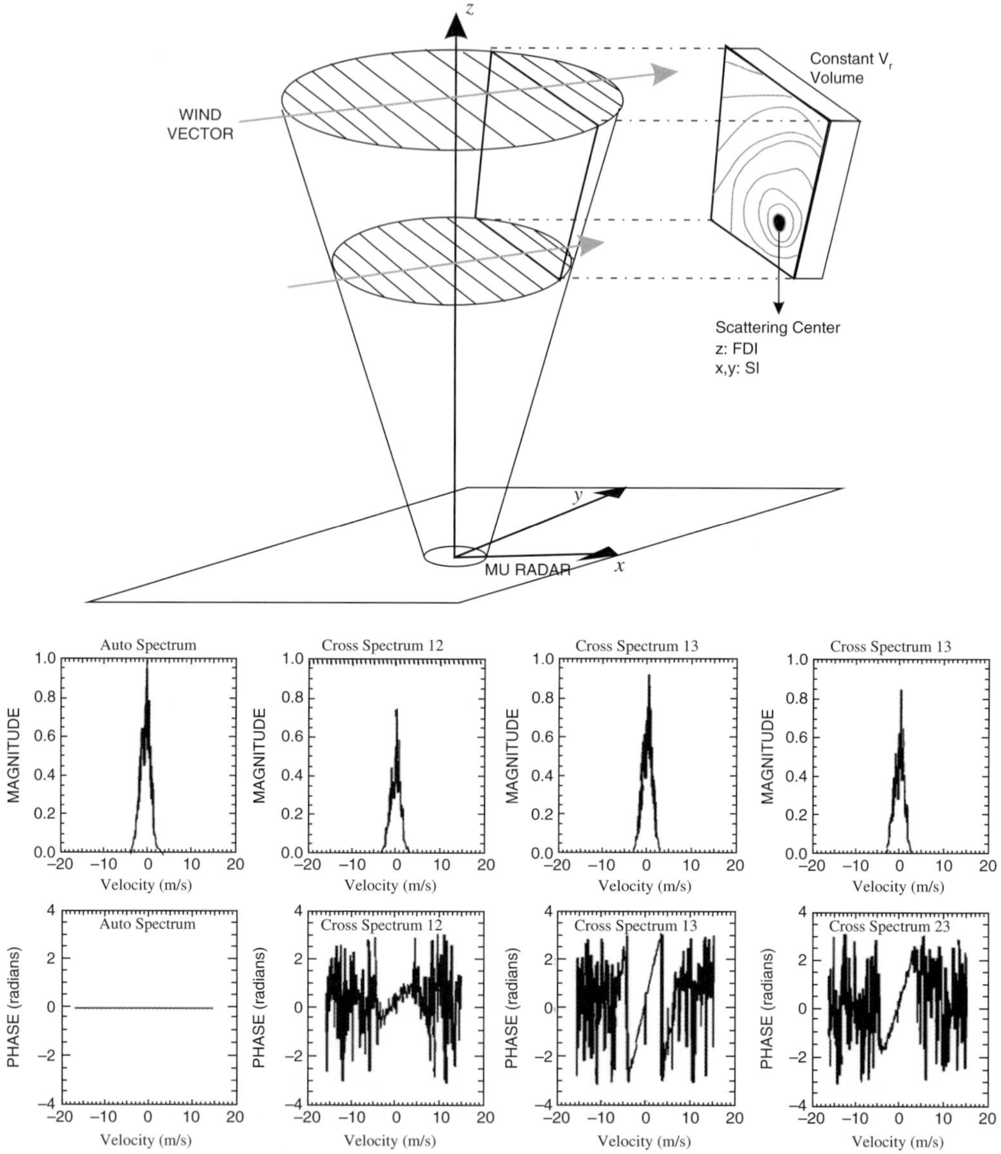

Figure 9.16 The upper panel provides an illustration of how Doppler sorting can be used to retrieve the wind speed and direction using the IDI technique (*Palmer et al.*, 1995a). The auto and cross spectra shown were collected using the MU Radar. The antenna was configured to allow reception on three separate subarrays (*Palmer et al.*, 1995b). (Upper panel reprinted with permission from John Wiley and Sons.)

orientation of the slices (determined by $\vec{\delta}_{ij}$) is not aligned with the wind, which is why we need at least three non-collinear antennas. Received signals for the three antenna subarrays were used to compute auto and cross spectra, from which the wind speed and direction can be calculated. Examples of the amplitude and phase of these spectra are provided in Figure 9.16. Note that the phase of the cross spectra varies linearly with velocity (Doppler sorting) across the velocities that have significantly large amplitudes. The slope of the phase variation as a function of velocity depends on the orientation of the wind direction relative to the direction of $\vec{\delta}_{ij}$.

9.4.2 Frequency domain interferometry (FDI)

Now we consider the case of a monostatic radar operating at two separate (but closely spaced) frequencies. The antenna is located at the origin of the Cartesian coordinate system that we presented earlier. Under these conditions Equation (9.31) becomes

$$\underline{R}(\Delta k_{nm}, \tau) = \iiint b\left(\hat{\vec{a}}, r, \omega\right) W^2\left(\hat{\vec{a}}, r\right) \mathrm{e}^{\mathrm{i}(\omega\tau+\varphi)} \mathrm{d}\hat{\vec{a}}\, \mathrm{d}r\, \mathrm{d}\omega, \tag{9.83}$$

(b being given by Equation (9.33)), with

$$\varphi = -2\,\Delta k_{nm}\, r\,. \tag{9.84}$$

The (normalized) correlation function is then given by

$$\underline{\rho}(\Delta k_{nm}, \tau) = \underline{R}(\Delta k_{nm}, \tau)/P_0, \tag{9.85}$$

where P_0 is defined in Equation (9.34). For the sake of simplicity, we will find the cross-correlation function (statistician's version) between the two time-series recorded using the two wavenumbers k_1 and k_2 at zero lag. Furthermore, we will let $W^2\left(\hat{\vec{a}}, r\right) = 1$. We will do this for wavenumbers k_1 and k_2. This gives us

$$\underline{\rho}(\Delta k_{12}) = \frac{1}{P_0} \iint b\left(\hat{\vec{a}}, r\right) e^{-\mathrm{i}2\,\Delta k_{12}\, r} \mathrm{d}\Omega\, \mathrm{d}r. \tag{9.86}$$

The correlation at zero lag calculated from two streams of time-series data at a chosen (fixed) range, $\underline{s}(k_n, t)$ and $\underline{s}(k_m, t)$ (where successive values of t refer to samples at this fixed height on successive pulses), is given as

$$\underline{\rho}(\Delta k_{12}) = \frac{\left\langle \underline{s}(k_1, t)\left[\underline{s}(k_2, t)\right]^* \right\rangle}{\sqrt{\left\langle \left|\underline{s}(k_1, t)\right|^2\right\rangle \left\langle \left|\underline{s}(k_2, t)\right|^2\right\rangle}}. \tag{9.87}$$

With this framework, we can develop the mathematical formulation used for frequency domain interferometry (FDI) as discussed in (*Kudeki and Stitt*, 1987). Along those lines, we introduce a parametric model to describe the shape of a horizontally oriented scattering layer in the reflectivity field. This model assumes a Gaussian distribution in height having a center at z_l and with a standard deviation of σ_l:

$$\underline{\rho}(\Delta k_{12}) = \frac{1}{\sqrt{2\pi\sigma_l^2}} \int \mathrm{e}^{-(z-z_l)^2/(2\sigma_l^2)} \mathrm{e}^{-\mathrm{i}2\,\Delta k_{12}\, z}\, \mathrm{d}z. \tag{9.88}$$

It should be noted for clarity that this formula describing a Gaussian variation in reflectivity is quite different to the formulas used in relation to the scatterers shown in Figure 7.18 in Chapter 7. In the earlier case, the individual scatterers were on the scale of a half a wavelength in size, and represented true scattering entities. In the above case, the Gaussian variation is on a much larger scale, and is just a crude parameterization of how the strength of scatter is assumed to vary across the depth of the layer. It does not consider at all the shapes and sizes of the actual scattering entities, or indeed anything at all about what actually *causes* the radio-scatter. It is simply assumed that the layer scatters most strongly in the center and more weakly at the edges, and that the variation from the middle to the edges falls off in a Gaussian manner.

As an aside, we should note that this Gaussian assumption is actually not a very good approximation in many cases. For example, Figure 7.17(b) shows that in turbulent layers, localized variability in the refractive index is greatest at the edges of the layer, and is often weakest in the middle, where the medium is adiabatically mixed. Nevertheless, we will continue with the assumption embodied by (9.88), since historically this was one of the earlier assumptions. Further generalizations can be left to the reader to develop, once the basic principles are understood.

Our next task is to evaluate (9.88). For simplicty, we will assume that the pulses transmitted at the two different frequencies both have the same phase at $z = 0$, which we will take to be zero. Looking at the form of the equation it is clear it has the same functional character as F_{c1} in Equation (3.208) in Chapter 3, so we can evaluate the expression by simply finding the Fourier transform of $f(z) = \mathrm{e}^{-(z-z_l)^2/(2\sigma_l^2)}$, and replacing k with $2\Delta k_{12}$. Since f is just a Gaussian with a shift of z_l we simply need to find the Fourier transform of $\mathrm{e}^{-(z)^2/(2\sigma_l^2)}$ and multiply by $\mathrm{e}^{-\mathrm{i}(2\Delta k_{12})z_l}$ (by the Fourier transform shift theorem). The result is therefore

$$\underline{\rho} = \mathrm{e}^{-2(\Delta k_{12})^2\sigma_l^2}\mathrm{e}^{-\mathrm{i}2\Delta k_{12}z_l}. \tag{9.89}$$

(If the reader is unfamiliar with the proof discussed above, an alternative is as follows: first use the change of variables $z' = z - z_l$, and recognize that the integration is between $-\infty$ and ∞, to write $\int \mathrm{e}^{-(z-z_l)^2/(2\sigma_l^2)}\mathrm{e}^{-\mathrm{i}2\,\Delta k_{12}\,z}\,\mathrm{d}z$ as $\int \mathrm{e}^{-(z')^2/(2\sigma_l^2)}\mathrm{e}^{-\mathrm{i}2\,\Delta k_{12}\,(z'+z_l)}\,\mathrm{d}z'$. This may be written as $\mathrm{e}^{-\mathrm{i}2\Delta k_{12}z_l}\int \mathrm{e}^{\{-[(z')^2+\mathrm{i}2\,\Delta k_{12}\,z'2\sigma_l^2)]/(2\sigma_l^2)\}}\,\mathrm{d}z$. Notice this has already produced the term $\mathrm{e}^{-\mathrm{i}2\Delta k_{12}z_l}$ out the front. The rest of the derivation involves modifying the exponent in the integrand to $\frac{-\left[\left(z'+\mathrm{i}(2\sigma_l^2)\Delta k_{12}\right)^2+4\sigma_l^4(\Delta k_{12})^2\right]}{2\sigma_l^2}$. The term $\mathrm{e}^{-2\sigma_l^2(\Delta k_{12})^2}$ comes out to the front of the integral, and with the substitution $z' + \mathrm{i}(2\sigma_l^2)\Delta k_{12} = z''$, the integral clearly integrates to $\sqrt{2\pi}\,\sigma_l$, leaving (9.89).)

Now we can use (9.89) to relate the properties of the scattering layer to the magnitude and phase of $\underline{\rho}$ calculated using Equation (9.87), namely

$$\left|\underline{\rho}\right| = \mathrm{e}^{-2(\Delta k_{12})^2\sigma_l^2} \tag{9.90}$$

and

$$\angle\underline{\rho} = -2\Delta k_{12}\,z_l. \tag{9.91}$$

These are the same expressions as those originally given by *Kudeki and Stitt* (1987). An additive phase correction is needed if the waves k_1 and k_2 have different phases at $z = 0$.

As an example of how FDI works, consider a narrow scattering layer located within a radar range gate with a resolution of 300 m. The position of the layer is 100 m above

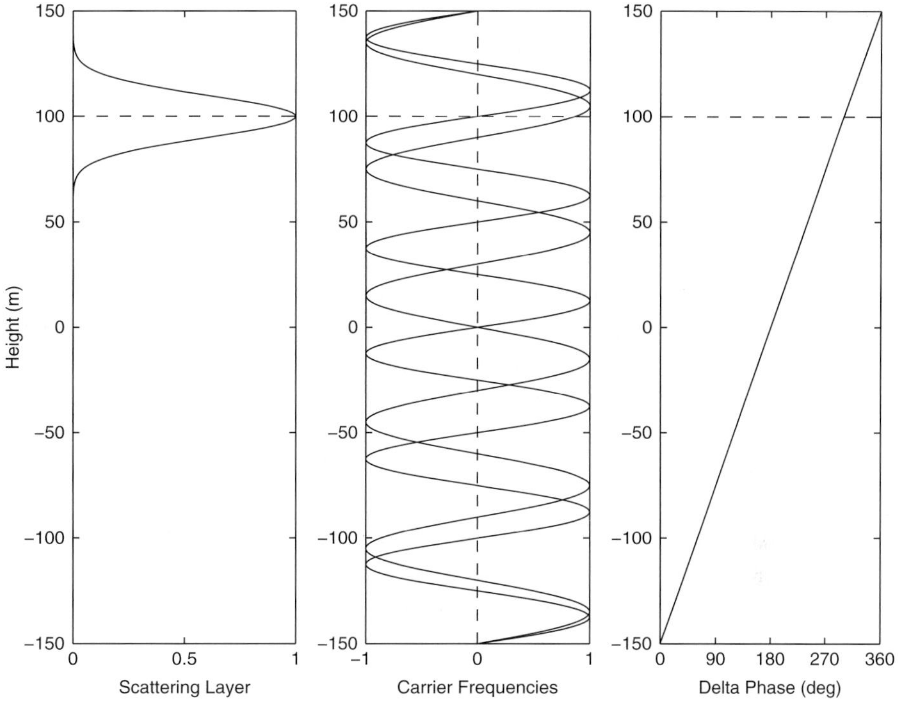

Figure 9.17 How frequency domain interferometry can be used to locate the position of a scattering layer to range accuracy better than one pulse-length (range-gate). The range resolution of the radar in this case is 300 m, and the phase differences between the received signal at different frequencies allow accuracy to 20 m or so.

the center of the range gate as seen in Figure 9.17. Two radar frequencies are used when operating the radar. In the example shown, the phase difference Δ_{12} would be about 270°, which can in turn be mapped into a corresponding location *within* the range gate using Equation (9.91). The intensity of the scatter associated with the layer can be estimated by evaluating the magnitude of the correlation function as shown in Equation (9.90). Note that this technique does not allow us to estimate the characteristics of two scattering layers within a range gate.

The derivation given above, although accurate, has been simplified for the sake of illustration. A more general derivation is given in *Franke* (1990). Additional information regarding uncertainty estimates can be found in the appendix of *Franke et al.* (1992). There have also been several investigations into issues such as the dependence of FDI estimates on the choice of the scattering layer model (*Chu and Chen*, 1995), limited horizontal extent and advection (*Luce et al.*, 2000a, b), and tilts of the scattering layer or radar beam (*Luce et al.*, 2000b).

9.5 Imaging

Techniques involving interferometry rely on differential phase information to help locate regions of enhanced reflectivity. Angular position is achieved through multiple receivers, and range information is found from multiple frequencies. The imaging technique involves a somewhat different approach. Phase and amplitude information are still used, but the goal is to construct a robust approximation of what the underlying reflectivity field must have been in order to have produced the observed signals at the different antennas or frequencies. Some form of inversion process is used to carry out this transformation.

It is not necessary to have a-priori information about the structure of the field. Typically the final product is presented in the form of a brightness function. In the discussion to follow, we will refer to the overall method of estimating the brightness associated with the reflectivity field as coherent radar imaging (CRI, also referred to as "angular imaging" (AIM)). The process of mapping the angular components of the field through multiple receivers will be referred to as angular imaging (AIM), while the term "range imaging" (RIM) will be used for multiple-frequency procedures which probe the range dependence of the field. Only the basics will be covered; however, additional information can be found in such texts as *Woodman* (1997), *Palmer et al.* (1998), *Hysell* (1996), *Chau and Woodman* (2001), *Yu and Palmer* (2001), *Palmer et al.* (1999), *Luce et al.* (2001a), *Yu and Brown* (2004) and *Röttger* (2013). We note that the formulation of the problem for AIM and RIM is similar.

For our discussion of imaging we will focus on a covariance-based treatment of the inversion problem. Equivalent representations using spectra are also often employed (*Woodman*, 1997; *Palmer et al.*, 1998; *Hysell*, 1996; *Chau and Woodman*, 2001; *Yu and Palmer*, 2001). We begin by defining a visibility function, which is based on Equation (9.31) evaluated at $\tau = 0$, or

$$\underline{V}_I\left(\vec{\delta}_{ij}, \Delta k_{nm}; \omega\right) = \iint b\left(\hat{\vec{a}}, r; \omega\right) W^2\left(\hat{\vec{a}}, r\right) \mathrm{e}^{\mathrm{i}\varphi} \mathrm{d}\hat{\vec{a}}\, \mathrm{d}r, \tag{9.92}$$

where b and W^2 are real functions (as discussed earlier) and

$$\varphi = -2\Delta k_{nm} r + k_n \hat{\vec{a}} \cdot \vec{\delta}_{ij} + \Delta k_{nm} \hat{\vec{a}} \cdot \vec{d}_i + \Delta k_{nm} \hat{\vec{a}} \cdot \vec{\delta}_{ij}. \tag{9.93}$$

It should be noted that Equation (9.92) is solved independently for every possible value of ω. This can be useful when exploring how atmospheric structures evolve as a function of Doppler frequency. The atmospheric brightness can now be found by taking the inverse Fourier transform of the visibility function. The problem is that we have only a limited number of estimates of the visibility function with which to perform our inversion calculation. This is discussed in more detail below.

9.5.1 Multiple-receiver imaging

For the discussion of multiple-receiver imaging, we will consider only the single-frequency case ($k_n = k_m = k_0$). The visibility function can be simplified as

$$\underline{V}_I\left(\vec{\delta}_{ij}; \omega\right) = \int b_a\left(\hat{\vec{a}}; \omega\right) W_a^2\left(\hat{\vec{a}}\right) \mathrm{e}^{\mathrm{i}\, k_n \hat{\vec{a}} \cdot \vec{\delta}_{ij}} \mathrm{d}\hat{\vec{a}}, \tag{9.94}$$

where

$$b_a\left(\hat{\vec{a}};\, \omega\right) = \int b\left(\vec{a}, r;\, \omega\right) W_r^2\left(r\right) \mathrm{d}r \tag{9.95}$$

is the *angular brightness*.

Before we proceed, it will be instructive to examine conceptually how AIM is used to estimate the angular brightness of the atmosphere. When using phased array antennas to electronically steer a radar beam, (for example when operating a radar in a DBS mode), phase shifters are used to physically adjust the phases across the antenna elements to achieve a desired beam orientation, as discussed in Chapter 5. In this application, only a single data stream is recorded for each beam orientation. With imaging, however, a "wide" beam is transmitted and multiple-receiver channels (multiple data streams) are used to record the backscattered signals, with each separate data stream corresponding to recordings from a distinct and spatially separated antenna or antenna subarray. Digital beam steering is then achieved by appropriately adjusting the phase (and sometimes the amplitude) of the different data streams. Data from each receiver are then combined to form a synthesized antenna beam directed along a prescribed orientation.

A visual depiction of this procedure for beam forming is shown in Figure 9.18. Typically a set of desired beam orientations, which span the extent of the transmitted beam, will be determined before the processing begins. Ideally one would like to select a large number of narrow beams (giving better angular resolution of the reflectivity field), but the number of possible independent beam-direction options is ultimately driven by the number and spacing of the receiving antenna array. The crux of all angular imaging techniques is to devise a method to decide what phase and amplitude offsets are needed

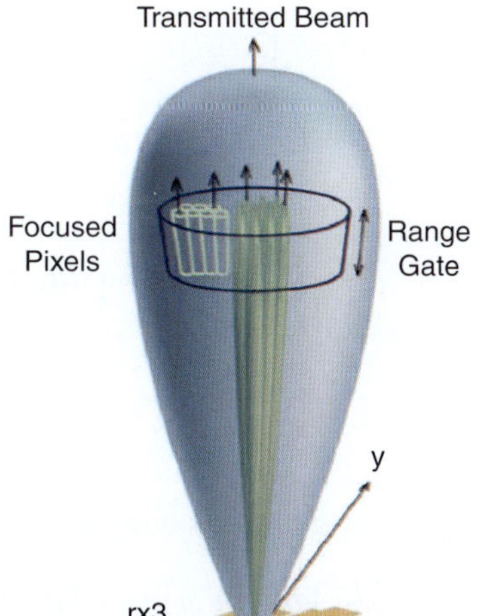

Figure 9.18 How synthesized beams are formed in the AIM technique as a means of probing the angular structure of the reflectivity field. The image is reproduced from (*Palmer et al.*, 2005). (Reprinted with permission from the American Meteorological Society.)

for each data stream to achieve the desired beam orientations. That is, we wish to find an appropriate weighting vector to be used while summing the signals.

We begin by examining the relationship between angular brightness $b_a\left(\hat{\vec{a}};\, \omega\right)$ and the visibility $\underline{V}_I\left(\vec{\delta}_{ij};\, \omega\right)$. Consider the complex radar time-series of signals received from M spatially separated and non-collinear antennas. These signals are represented as $\underline{s}(\vec{d}_i,\, t)$ using a form of Equation (9.16), where $i = 0, \ldots, M-1$. We wish to create a composite time-series data stream composed of the weighted sums of the original data. This is represented by

$$\underline{\widetilde{s}}(t) = \left[\underline{w}\right]^{\dagger}\left[\underline{s}\right](t) = \sum_{i=0}^{M-1} \underline{w}_i^{*}\,\underline{s}(\vec{d}_i,\, t), \tag{9.96}$$

where $\underline{w}_i$ is used to represent components of the weight matrix $\left[\underline{w}\right]$ and the dagger (†) denotes the Hermitian (complex conjugate) operator. Note that $\left[\underline{w}\right]$ is a column matrix, so $\left[\underline{w}\right]^{\dagger}$ is a row matrix. The column matrix $\left[\underline{w}\right]$ can also be thought of as a vector, since in matrix theory it is normal to represent a vector as a column matrix. However, note that $\underline{s}(\vec{d}_i,\, t)$ and $\underline{\widetilde{s}}(t)$ are shown as continuous functions. In some cases these might be discretized, and the time considered at L points with steps $\ell = 0$ to $L-1$; if that is needed, the matrix consideration makes it easy to extend the equations since one need simply extend $\left[\underline{s}\right]$ to a matrix with M rows and L columns.

At this point, the choice of $\left[\underline{w}\right]$ is completely open. Later we will seek to determine the elements of $\left[\underline{w}\right]$ that are needed to optimally retrieve estimates of the brightness distribution as a function of angle.

The brightness distribution can be found by first calculating the autocovariance function from the composite time-series data $\underline{\widetilde{s}}(t)$ through the equation

$$\begin{aligned}
\underline{\widetilde{R}}(\tau) &= E\left\{\underline{\widetilde{s}}(t)\underline{\widetilde{s}}^{*}(t-\tau)\right\} \\
&= E\left\{\sum_{i=0}^{M-1} \underline{w}_i^{*}\,\underline{s}(\vec{d}_i,\, t)\left(\sum_{j=0}^{M-1} \underline{w}_j^{*}\,\underline{s}(\vec{d}_j,\, t-\tau)\right)^{*}\right\} \\
&= \sum_{i=0}^{M-1}\sum_{j=0}^{M-1} \underline{w}_i^{*}\, E\left\{\underline{s}(\vec{d}_i,\, t)\underline{s}^{*}(\vec{d}_j,\, t-\tau)\right\}\underline{w}_j \\
&= \sum_{i=0}^{M-1}\sum_{j=0}^{M-1} \underline{w}_i^{*}\,\underline{R}(\vec{\delta}_{ij},\, \tau)\underline{w}_j.
\end{aligned} \tag{9.97}$$

The weighted angular brightness is now simply the Fourier transform of $\underline{\widetilde{R}}(\tau)$. That is

$$\begin{aligned}
W_a^2\left(\hat{\vec{a}}\right) b_a\left(\hat{\vec{a}};\, \omega\right) &= \mathcal{F}\left\{\underline{\widetilde{R}}(\tau)\right\} \\
&= \sum_{i=0}^{M-1}\sum_{j=0}^{M-1} \underline{w}_i^{*}\, \mathcal{F}\left\{\underline{R}(\vec{\delta}_{ij},\, \tau)\right\}\, \underline{w}_j \\
&= \sum_{i=0}^{M-1}\sum_{j=0}^{M-1} \underline{w}_i^{*}\, \underline{V}_I\left(\vec{\delta}_{ij};\omega\right)\, \underline{w}_j,
\end{aligned} \tag{9.98}$$

where $F\{\ \}$ denotes the Fourier transform. The elements of the so-called visibility matrix are simply the spectra given by

$$\underline{V}_I\left(\vec{\delta}_{ij};\,\omega\right) = E\left\{F\left\{\underline{s}(\vec{d}_i,\,t)\right\}\left[F\left\{\underline{s}(\vec{d}_j,\,t)\right\}\right]^*\right\}$$
$$= E\left\{\underline{S}(\vec{d}_i;\,\omega)\underline{S}^*(\vec{d}_j;\,\omega)\right\}, \tag{9.99}$$

where $\underline{S}(\vec{d}_j;\,\omega)$ is the Fourier transform calculated from $\underline{s}(\vec{d}_i,\,t)$. The reader should note that the term $\underline{S}(\vec{d}_i;\,\omega)\underline{S}^*(\vec{d}_j;\,\omega)$ is the *cross-spectrum* of the signals between the two receivers. It is in general complex. Now the question becomes one of estimating the values of the weights contained within $[\underline{w}]$. That is, for a prescribed beam direction, how should one select the components of $[\underline{w}]$?

9.5.2 Estimation of the weighting vector

The task of estimating the weight vector is essentially an inversion problem. We wish to focus the sensitivity of the radar in a specific direction. This is accomplished by adjusting the phase and/or amplitude of the time-series data from the radar in such a way as to produce constructive interference at the desired location in space. Although there are many algorithms to accomplish this, we will consider only two of them in the present discussion. These are the so-called Fourier and Capon methods. The treatment of additional techniques can be found, for example, in *Palmer et al.* (1998), *Luce et al.* (2001a), *Chau and Woodman* (2001), *H'elal et al.* (2001), *Yu and Palmer* (2001), *Smaïni et al.* (2002).

In the Fourier method, the weighting vector is chosen by simply adjusting the phases of the radar signals. This is similar to what is done, for example, in post-statistics steering. The weighting vector is written as

$$[\underline{w}]_\omega = \begin{bmatrix} e^{-i\,k_0\hat{\vec{a}}\cdot\vec{d}_1} \\ e^{-i\,k_0\hat{\vec{a}}\cdot\vec{d}_2} \\ \vdots \\ e^{-i\,k_0\hat{\vec{a}}\cdot\vec{d}_M} \end{bmatrix}. \tag{9.100}$$

Having an estimate of the weighting vector, the brightness estimate is then obtained by substituting Equation (9.100) into Equation (9.98).

The Capon method of obtaining the weighting vector utilizes adaptive signal processing. That is, the atmospheric signal is used when estimating $[\underline{w}]$. A constrained optimization is used as described in *Palmer et al.* (1998). The solution of the optimization can also be seen in Equation (8.74) in Chapter 8. Modifying the variables to match our current problem leads to

$$[\underline{w}]_c = \frac{[\underline{V}_I]^{-1}[\underline{e}]}{[\underline{e}]^\dagger[\underline{V}_I]^{-1}[\underline{e}]}. \tag{9.101}$$

where $[\underline{e}]$ contains the same elements as those in Equation (9.100). We have simply used $[\underline{V}_I]$ to represent the visibility matrix $\left[\underline{V}_I\left(\vec{\delta}_{ij};\,\omega\right)\right]$. Again, the brightness estimate is obtained by substituting Equation (9.101) into Equation (9.98). Next we present two examples of how AIM has been used to investigate complex atmospheric structures.

Motivated by the desire to observe small-scale temporal and spatial features within the atmospheric boundary layer, a research team at the University of Massachusetts developed the turbulent eddy profiler (TEP) (*Mead et al.*, 1998). The radar operates at UHF and can accommodate up to 64 different complex data streams. An experiment was conducted in Massachusetts in 2003 using TEP with the antenna arranged into a "Big Y" configuration (*Palmer et al.*, 2005). During the experiment sporadic rain showers were observed with the radar. The precipitation descended through several Bragg scattering layers. Using the recorded data and AIM signal processing, *Palmer et al.* (1998) were able to successfully resolve features of both the precipitation and the Bragg scattering layers at a spatial scale of about 50 m × 50 m × 30 m. Moreover, since the radial velocity of the scatter associated with the precipitation was distinctly different to that for the clear-air turbulence, it was possible to discriminate between the two contributions through Doppler sorting. That is, the authors were able to apply the imaging technique to the signals from the precipitation and the clear-air scatter independently. An example of the precipitation falling through the Bragg scattering layer is shown in Figure 9.19.

We next consider observations collected with the VHF middle atmosphere Alomar radar system, referred to as MAARSY (*Latteck et al.*, 2012). This system was discussed briefly in Chapter 6. The antenna array of this MST radar consists of 433 three-element Yagi elements, which cover an area of roughly 6300 m^2. The array of antenna elements can be grouped to form various subarrays that provide data streams, which can be separately recorded on 16 different receive channels. This allows for angular image processing. One particular antenna configuration is shown in Figure 9.20. The radar has been used to investigate polar mesosphere summer echoes (PMSE). These echoes occur in polar regions during summer at heights near that of the mesopause and are

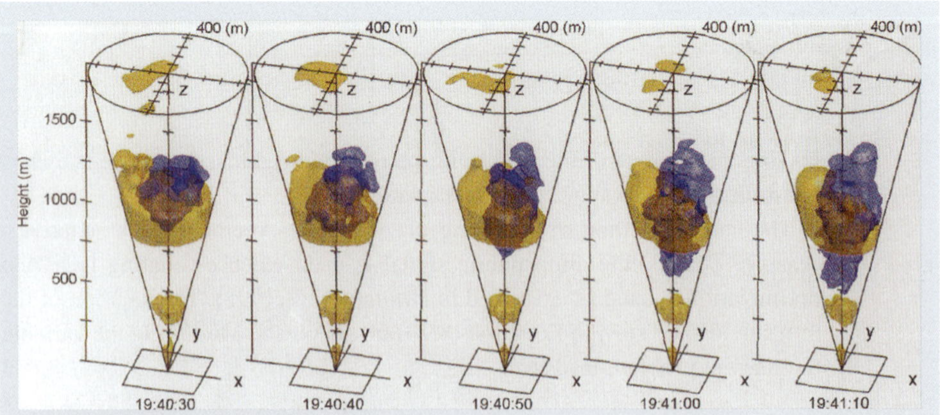

Figure 9.19 Observations of precipitation (blue fields) and clear-air scatter (beige fields) observed using the UHF turbulent eddy profiler. Signal processing was achieved using AIM. Image reproduced from *Palmer et al.* (2005). (Reprinted with permission from the American Meteorological Society.)

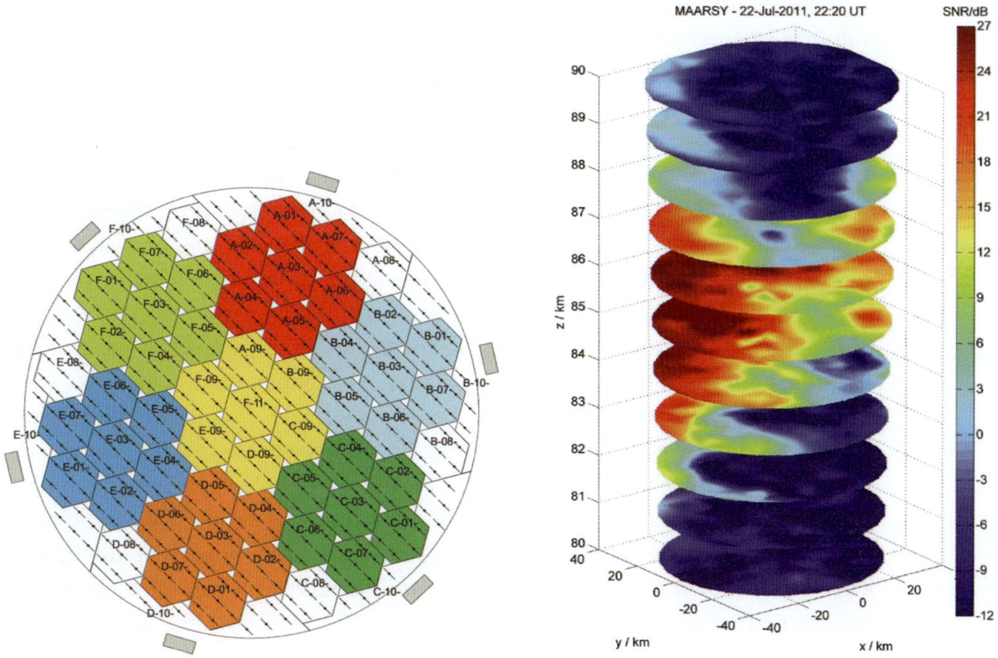

Figure 9.20 The antenna array of MAARSY arranged in one of many possible subarray configurations taken from (*Renkwitz et al.*, 2013). Also shown is an example of data collected corresponding to polar mesosphere summer echoes, which was processed using AIM. The image is from *Latteck et al.* (2012). (Reproduced with permission from John Wiley and Sons.)

attributed to ionized ice particles. They were discussed briefly in Chapter 2, and will be further discussed in Chapter 10. One example of data from a PMSE layer is provided in Figure 9.20. The different slices correspond to the sampled range gates during the observations. Each slice displays varying degrees of structure on spatial scales smaller than the beam-width of the radar, each slice having been determined using angular imaging. Being able to probe PMSE at fine spatial scales allows us to better understand the nature of the scatter responsible for the phenomenon and to investigate the underlying dynamic structure of the atmospheric field.

9.5.3 Multiple-frequency imaging

Now we consider the case of multiple-frequency imaging using a monostatic radar. This application was first put forward by *Palmer et al.* (1999) and then independently by *Luce et al.* (2001a). We will refer to this technique as range imaging (RIM). It will become clear that RIM is analogous to AIM except that the former utilizes frequency diversity (also called frequency agility) whereas the latter relies on spatially separated antennas. The atmosphere can readily support fine-scale (in the vertical extent) persistent structures, which can be difficult to resolve using conventional pulsed Doppler radar systems. Although pulse compression techniques can be used to improve range resolution (see Chapter 4), range imaging offers several advantages over pulse compression techniques.

To illustrate the principles of range imaging, we consider radiowave scatter caused by sharp vertical gradients in the temperature and humidity fields commonly associated with the entrainment zone at the top of the convective boundary layer (CBL). As shown in the top two panels of Figure 9.21, the potential temperature is expected to be approximately constant in height within the mixed layer of the CBL and then sharply increase within the capping entrainment zone. Correspondingly, the specific humidity is also uniform in height through the mixed layer and then exhibits a sharp decrease at the height of the entrainment zone. The actual values of these two parameters are not important and have not been shown in the figure. Rather, we are interested in their rates of change with height, i.e., $\Delta\Theta/\Delta z$ and $\Delta q/\Delta z$, because these are associated with an increase in reflectivity. This was seen in Equation (3.288), which emphasized the importance of the potential refractive index gradient. In the idealized example depicted in Figure 9.21, the entrainment zone and corresponding peak in reflectivity occur at a height of 1200 m. Here we are ignoring contributions to the reflectivity from other sources such as turbulence. Also shown in the plot of reflectivity is a representation of observed values of reflectivity (red connected circles) using a vertically pointing radar with a range resolution of 200 m. Note that the observed reflectivity is reduced due to the effects of volume averaging and the range weighting function.

The principles of RIM are illustrated in the lower panels of Figure 9.21, where we focus on a single range gate which contains the layer of enhanced reflectivity. In this case, the center of the range gate is located at a height of 1150 m. In the RIM technique, the aim is to enhance the sensitivity of the radar in specified height bins located within the nominal range gate, which are sometimes referred to as subsets. In the example shown, four different carrier frequencies are used to probe the atmosphere. In general, these radiowave signals will have arbitrary phases as shown in the "STD" case (blue lines). However, using RIM, the phases can be modified through signal processing to create phase coherency at the height of the different subsets (red lines). In the example shown, the center of the subgate corresponds to the height of the peak in reflectivity (height of the entrainment zone). Although we used scatter from the top of the boundary layer for this example, scattering layers can and do occur at many heights within the atmosphere, as shown in Figure 9.2.

We can now explore RIM within a mathematical framework. The visibility function for this case can be simplified as

$$\underline{V}_I\,(\Delta k_{nm};\,\omega) = \int b_r(r;\,\omega) W_r^2(r) \mathrm{e}^{-\mathrm{i}2\Delta k_{nm} r} \mathrm{d}r, \tag{9.102}$$

where

$$b_r(r;\,\omega) = \int b(\hat{\vec{a}},\,r;\,\omega)\, W_a^2(\hat{\vec{a}})\, \mathrm{d}\hat{\vec{a}} \tag{9.103}$$

is the *range brightness*. Note the similarities with angular imaging. Analogous to our approach for angular imaging, we consider the complex radar time-series signals

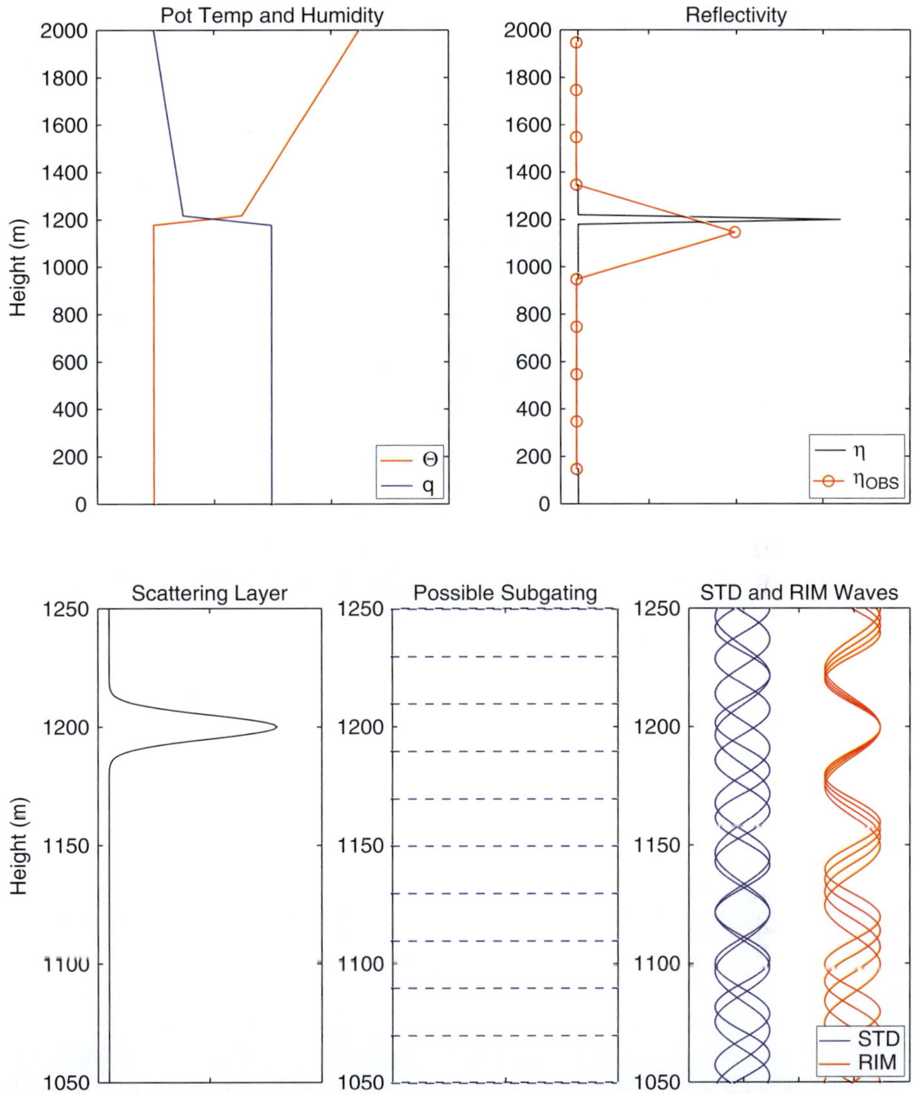

Figure 9.21 The working principles of RIM for the case of enhanced backscatter from a region of the atmosphere corresponding to the top of the convective boundary layer. Using the RIM technique, multiple frequencies are used (here four) to sample the atmosphere. The phases of the received signals are adjusted after the fact in signal processing to achieve coherency at a particular desired height. See text for further discussion.

received using N different frequencies. The signals are represented as $\underline{s}(k_n, t)$ using a variant of Equation (9.16), where $n = 0, \ldots, N-1$. We create a composite time-series data stream composed of the weighted sums of the original data:

$$\underline{\widetilde{s}}(t) = \left[\underline{w}\right]^{\dagger} \left[\underline{s}\right](t) = \sum_{n=0}^{N-1} \underline{w}_n^* \underline{s}(k_n, t). \tag{9.104}$$

The brightness distribution can be found by calculating the autocovariance function for $\underline{\widetilde{s}}(t)$:

$$\begin{aligned}\underline{\widetilde{R}}(\tau) &= E\left\{\underline{\widetilde{s}}(t)\underline{\widetilde{s}}^*(t-\tau)\right\}\\ &= E\left\{\sum_{n=0}^{N-1} \underline{w}_n^* \underline{s}(k_n, t)\left(\sum_{m=0}^{N-1} \underline{w}_m^* \underline{s}(k_m, t-\tau)\right)^*\right\}\\ &= \sum_{n=0}^{N-1}\sum_{j=0}^{N-1} \underline{w}_n^* E\left\{\underline{s}(k_n, t)\underline{s}^*(k_m, t-\tau)\right\}\underline{w}_m\\ &= \sum_{n=0}^{N-1}\sum_{m=0}^{N-1} \underline{w}_n^* \underline{R}(\Delta k_{nm}, \tau)\underline{w}_m. \end{aligned} \tag{9.105}$$

The weighted angular brightness is then simply the Fourier transform of $\underline{\widetilde{R}}(\tau)$:

$$\begin{aligned}W_r^2(r)\, b_r(r; \omega) &= F\left\{\underline{\widetilde{R}}(\tau)\right\}\\ &= \sum_{n=0}^{N-1}\sum_{m=0}^{N-1} \underline{w}_n^* F\left\{\underline{R}(\Delta k_{nm}, \tau)\right\}\underline{w}_m\\ &= \sum_{n=0}^{N-1}\sum_{m=0}^{N-1} \underline{w}_n^* \underline{V}_I\left(\Delta k_{nm}; \omega\right)\underline{w}_m. \end{aligned} \tag{9.106}$$

The elements of the visibility matrix are the cross-spectra given by

$$\underline{V}_I(\Delta k_{nm}; \omega) = E\left\{\underline{S}(k_n; \omega)\underline{S}^*(k_m; \omega)\right\}. \tag{9.107}$$

In the Fourier treatment of the problem, we define a steering vector as given by

$$\left[\underline{w}_\omega\right] = \begin{bmatrix} \mathrm{e}^{-\mathrm{i}\,2k_1 r} \\ \mathrm{e}^{-\mathrm{i}\,2k_2 r} \\ \vdots \\ \mathrm{e}^{-\mathrm{i}\,2k_N r} \end{bmatrix}. \tag{9.108}$$

The Capon weighting vector is calculated using Equation (9.101), where $\left[\underline{e}\right]$ is given by the components shown in Equation (9.108).

Range imaging (RIM) has been implemented on various radar systems and demonstrated in several field experiments. Some examples of these applications made at VHF include *Luce et al.* (2001a), *Chilson et al.* (2001b), *Palmer et al.* (2001), *Muschinski et al.* (2001), *Smaïni et al.* (2002). The technique has also been implemented at UHF by *Chilson et al.* (2003).

We now present data from a boundary layer experiment conducted in Colorado at the Boulder Atmospheric Observatory in 2002. For the experiment, a 915 MHz boundary-layer radar (Väisälä) was operated near an S-Band (center frequency of 2.9 GHz) FMCW wind profiler (Army Research Laboratory). The boundary-layer radar was modified to allow operation at four frequencies (914.0, 914.33, 915.33, and 916.0 MHz). The nominal range resolution of the boundary-layer radar was 210 m, whereas the resolution

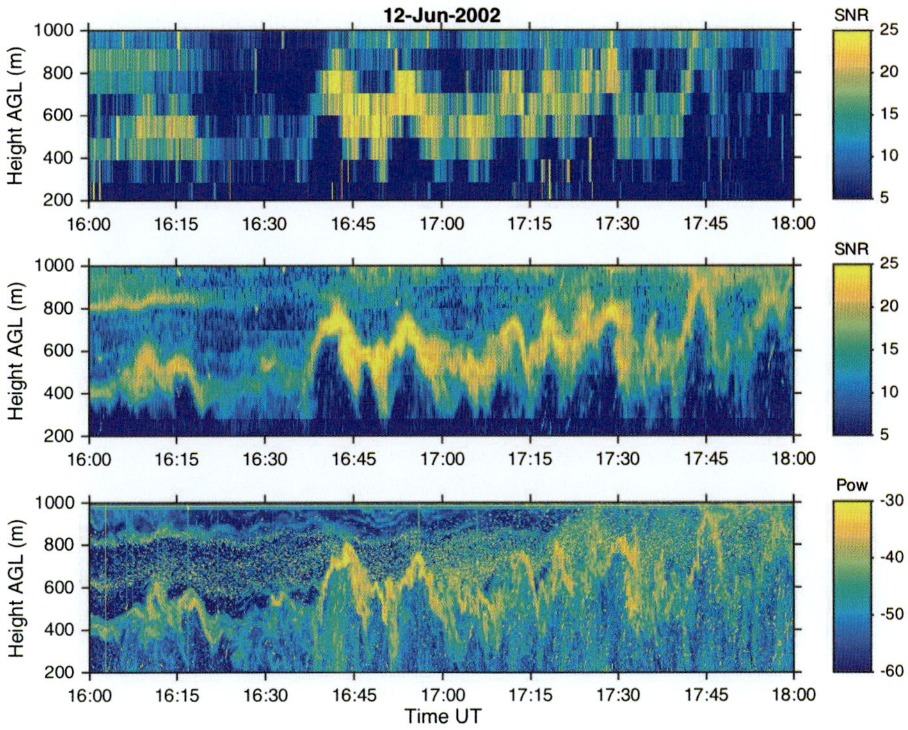

Figure 9.22 Radar observations of the atmospheric boundary layer collected using UHF pulsed Doppler and S-band FMCW wind profilers. Uppermost panel: conventional signal processing was performed. The same data were used to create the image shown in the middle panel, except RIM processing was used. The bottom panel shows observations from an FMCW that was located near the pulsed Doppler system. The power is expressed logarithmically with an arbitrary reference point. The range resolution of the FMCW radar was about 2 m.

of the FMCW radar was 2 m. The radar systems were separated by 100 m. Shown in Figure 9.22 is an example of data collected during a 2 hour period. The upper two panels show results from the boundary-layer radar. In the uppermost plot, conventional radar signal processing was performed on the recorded data at the four frequencies separately and these results were averaged. In the middle plot, the same data streams were used, but a Capon-based RIM processing procedure was used. It is obvious that the RIM processing reveals more detail, but do the details reflect actual structures in the atmosphere or are they artifacts of the processing? Independent observations from the FMCW wind profiler show very similar features to those obtained using RIM. The sub-gate spacing used for the RIM processing was 20 m whereas the resolution of the FMCW radar was 2 m. Results from this experiment provide evidence that the RIM signal processing procedure was indeed able to capture finer spatial details in the atmospheric reflectivity structure than was possible using conventional signal processing, although the warnings of *Garbanzo-Salas and Hocking* (2015) (and also discussed in Chapter 8) should be heeded. In this case, the true thicknesses of the layers (as seen with the FMCW data) are generally less than the resolution even of the Capon method, so the natural tendency of the Capon method to artifically thin the widths of the peaks works to the apparent advantage of the interpretation. A more interesting test would have been to see if layers that were truly say 30 m deep were properly reproduced as having a depth of 30 m, or whether the method would artificially contract them to 20 m. Blind application of such methods is to be discouraged: nevertheless, the method does have clear advantages when used wisely.

10 Extended and miscellaneous applications of atmospheric radars

10.1 Introduction

In this chapter, we discuss various extended, and in some cases unusual, applications of MST radar. These may be special cases of general MST techniques, or specific applications of the technique applied to special cases, or even quite unusual applications which are a substantial deviation from "normal" MST standard practices. If such a topic fits well in another chapter, it may appear there – if it is somewhat of an exception, or has a sightly unusual methodology, or is not really an operational technique, it may appear here. Polar mesosphere summer echoes are an example of an "extended" application. While the techniques used to study these unusual echoes are really the same as for other MST studies, the unusual physics associated with the scatterers that produce these echoes makes them of particular interest. Lightning study is an example of a slightly "miscellaneous" application, in that the techniques are a little unusual (high PRFs, and the events are very short lived). Meteor study is an example of a slightly non-standard application that has grown into a substantial field all of its own. Differential absorption is a technique developed early in the days of radar in the 1960s and 1970s which has had a rebirth in the last decades, and deserves a brief mention here. Precipitation study with MST radars is a relatively mature field, but is still a secondary application, so is also included here.

Each of these fields has a significant role in its own right, but extended discussion of them would simply take up too much space, and would spread the intended application of this book beyond its original goals. Hence the topics are summarized briefly in this chapter – maybe too briefly for some, but we have tried to give sufficient references that interested readers may expand their knowledge through these references.

This book is intended to concentrate on experimental and analysis techniques, and the underlying processes (both geophysical and technical) that guide the experiments and their design. Examples of the latter include the basic theory of turbulence, and the theory behind gravity waves (see the next chapter, and also some small discussion in Chapter 2). We do not discuss the many thousands of papers that give specific results, such as details of solar-cycle variations of atmospheric behavior at particular sites, or global details about tidal phases, except where they guide us to better development of new techniques. So, despite the importance of such measurements, these types of papers are not discussed in this book, nor in this chapter.

10.2 PMSE and PMWE

One topic of considerable interest is PMSE – polar mesosphere summer echoes. These are strong scattering layers seen at VHF and even UHF that occur in the altitude region 80 to 95 km, and were introduced in Chapter 2, Section 2.14. Figure 10.1 shows a typical PMSE layer.

The interesting thing about these layers is that they should not be visible at all with radar, at least not if they are due to traditional neutral atmospheric turbulence. As discussed in Section 2.14, in turbulence theory, perturbations are highly damped if the wavelengths of their 3-D spatial Fourier components are significantly smaller than the so-called turbulence inner scale, given by

$$\ell_0 \approx \kappa_\ell \eta_K \simeq \kappa_\ell \left(\nu^3/\varepsilon\right)^{\frac{1}{4}}, \tag{10.1}$$

where $\kappa_\ell \approx 7$–14 and where η_K is called the Kolmogoroff microscale. Graphs of ℓ_0 were presented by *Hocking* (1985), and the key figure has also been reproduced in Chapter 11, Figure 11.25; its specific details will be discussed further in that chapter. For our purposes, it is most important to note that the Bragg scale for a 50 MHz radiowave (about 3 m) coincides with the inner scale at an altitude of about 70 km, and so scales beyond 80 km altitude should be strongly damped due to the effects of molecular viscosity overpowering the inertial effects within the turbulence.

The occurrence of these layers signalled that the cause of the scatter involved some new physics, beyond that of traditional turbulence. In the following paragraphs, we will

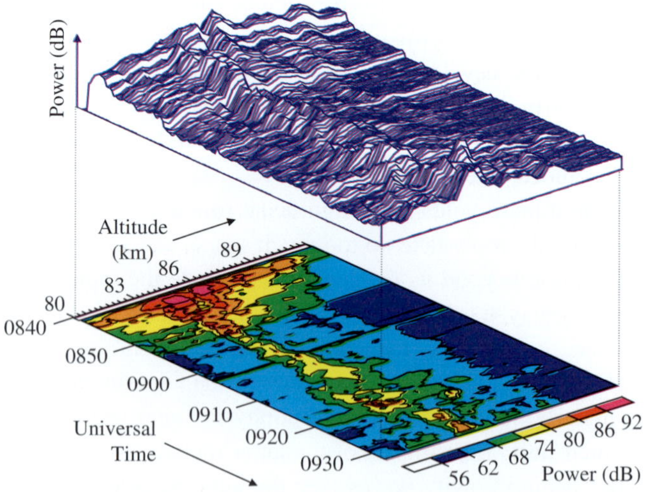

Figure 10.1 A contour height–time–intensity (HTI) plot of a polar mesosphere summer echo, overlaid with a three-dimensional representation of the HTI. The event was recorded on 30 June 1998, with the Eiscat (European incoherent scatter) 224 MHz radar. Temporal resolution was 10 s and vertical resolution was 300 m. From *Hocking and Röttger* (1997). (Reprinted with permission from John Wiley and Sons.)

first discuss some of the characteristics of these models and then consider proposed theories for their existence.

The behavior of the echoes was not that unusual, exhibiting many features seen by more classical scatterers. Their strong tendency to layering was a bit unusual, but not extremely so. Temporal variability occurred on time-scales of hundreds of minutes down to a few tens of seconds (e.g., *Röttger et al.*, 1990b; *Röttger*, 1994; *Czechowsky and Rüster*, 1997). Significant vertical spatial variability was evident, with variations occurring on the scale of a few hundreds of meters (e.g., *Röttger et al.*, 1990a). Spectra produced from the raw data were variable in form, sometimes narrow, sometimes wide. Generally the echo powers and spectral widths did not show particularly strong correlation (e.g., *Röttger et al.*, 1990b; *Czechowsky and Rüster*, 1997). Echoes with narrow spectra were often also aspect-sensitive, which should be of no surprise (e.g., *Czechowsky et al.*, 1988; *Huaman and Balsley*, 1998), though the very fact that aspect-sensitivity existed was of interest, suggesting that elongated horizontal structures existed at times (compare with Chapter 7, Section 7.5). The echoes were quite dominant when seen using VHF radars. UHF radar echoes were rarer and had poorer signal-to-noise. However, when visible, they allowed excellent vertical resolution studies due to the wider bandwidths associated with UHF radars (e.g., *Röttger and La Hoz*, 1990; *Hoppe et al.*, 1994; *Hocking and Röttger*, 1997).

PMSE have also been associated with special high-level mesospheric clouds that occur at typically 82–83 km altitude, slightly below the PMSE. These are called noctilucent clouds (NLC), and can be seen under twilight conditions with a naked-eye by ground-based observers in the northern summer seasons at 50 °N–65 °N latitude. While it was tempting to believe the phenomena should be related, correlations have not been consistent. Occasionally both PMSE and NLC have occurred in the same volume (e.g., *Nussbaumer et al.*, 1996; *Lübken et al.*, 2004), but spatial coincidence has not always been evident (e.g., *Taylor et al.*, 1989). *Bremer et al.* (2003) performed a longer-term study between the two phenomena, and indicated the existence of a significant level of long-term correlation, at least within the European sector.

NLC activity seems to have increased in the last few decades (*Thomas et al.*, 1989), so these clouds (and potentially PMSE) have been proposed as potentially related to global warming in the troposphere. It has been argued that the fraction of methane in the troposphere may have risen (*Pearman and Fraser*, 1988) since the time of the industrial revolution. Methane released in the troposphere is believed to reach the stratosphere, where it is converted (through a sequence of reactions) into water vapor. From there, it is possible that it may work its way upwards to the upper mesosphere and lower thermosphere (e.g., *Thomas*, 1991; *Olivero and Thomas*, 2001). Ice-crystal growth then occurs due to the enhanced water vapor concentrations. (Because the pressure is below that of the triple point of water, it cannot exist as a liquid at these heights.) These same ice aerosols could also be responsible for the scatterers that cause PMSE, as will be seen shortly.

Earlier reviews of PMSE were presented by *Cho and Röttger* (1997) and *Rapp and Lübken* (2004). The introduction presented in *Swarnalingam et al.* (2009a) also provides a compact overview.

10.2.1 Geographical distribution

One of the most significant clues about the cause of these strange PMSEs is their geographical distribution, and their relative strengths at different locations. Their location near the summer pole, in the coldest place in the Earth's atmosphere (see Chapter 1, Figures 1.20 and 1.23) clearly indicated a temperature dependence, but what else might be involved? Perhaps meteor dust? Or maybe water-vapor content? Or maybe some more complex chemistry? The geographical distribution might give some clues in this regard. This investigation had an interesting side-effect, in that it required that the various radars which studied PMSE be properly calibrated, so that PMSE strengths could be compared in an absolute sense. The procedures for doing such calibrations were outlined in Chapter 5, Section 5.6.3.

As noted already, PMSE detections were first made in Alaska (*Ecklund and Balsley*, 1981), and then in Europe (*Kelley et al.*, 1987; *Czechowsky et al.*, 1979). Sample graphs were shown in Chapter 2, Figures 2.16 and 2.24. Since all these detections were quite strong, it was initially assumed that PMSE would be ubiquitous throughout the arctic, since the entire arctic summer mesopause should be cold, as described in Chapter 1.

However, as it turned out, the distribution was far from uniform. A VHF (51.5 MHz) radar installed at Resolute Bay (75 °N) just prior to the 21st century (*Hocking et al.*, 2001b) showed weak scatter, even after allowance for its low peak power (10 kW) (*Huaman et al.*, 2001). It seemed therefore possible that there was significant variability in PMSE strengths even within the polar regions. Initially *Rapp and Lübken* (2004) proposed that the Resolute Bay radar was improperly calibrated. Therefore staff from the Institute of Atmospheric Physics in Germany visited the radar with the operators of the radar (including W. Hocking) and additional calibrations were undertaken. The radar was carefully calibrated by various methods, including against galactic noise and a noise source (*Swarnalingam et al.*, 2009a). The efficiency of the radar was also calculated (see additional discussions in Chapter 5 of this book). Previous calibration constants were confirmed as accurate.

Furthermore, PMSE were also measured with a newer VHF radar at Eureka (80 °N) on Ellesmere Island in far northern Canada. That radar was established in 2007. It was concluded that PMSE at both Resolute Bay and Eureka are quite a bit weaker than their European and Alaskan counterparts, although Resolute Bay is the weakest. Both sites are deep inside the auroral oval, and in fact are close to the north magnetic pole. They are therefore less susceptible to solar proton and electron precipitation from the auroral oval. Thus it seems likely that electron and proton precipitation may be an extra factor in enhancing the PMSE backscatter cross-sections, in addition to cold temperatures. With regard to latitudinal spread, PMSE have been observed at latitudes as low as 52 °N (e.g., *Reid et al.*, 1989; *Zecha et al.*, 2003), although 60 ° is a more common lower limit.

A map of the locations of polar sites in the period 2010–2015 is shown in Figure 10.2. The new PANSY radar at Syowa has only just been installed (see Chapter 6) and other key sites for PMSE study include radars at the EISCAT sites, Kuhlungsborn, Yellowknife, Resolute Bay, Eureka, Poker Flat, and Svalbard in the northern polar regions, and Rothera, Wasa, Syowa, Mawson, Davis, and McMurdo in the southern hemisphere.

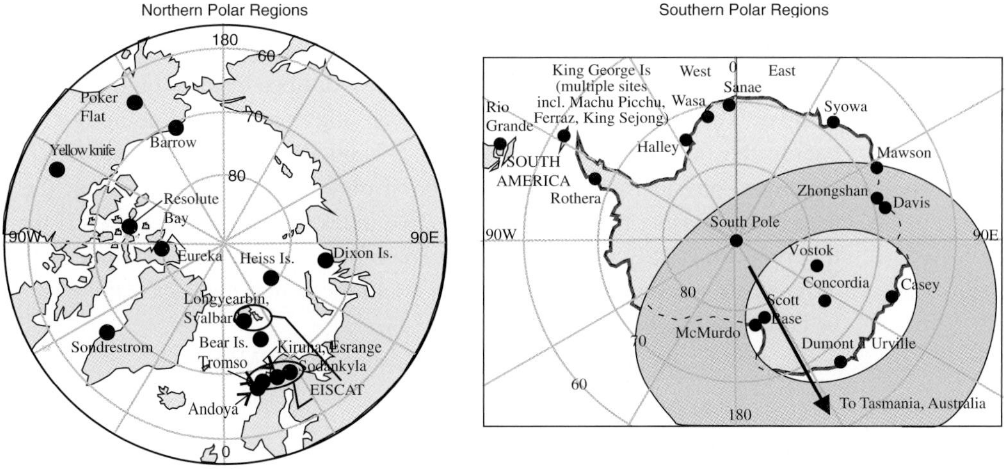

Figure 10.2 Location of polar MST/ST radars circa 2010–2012. These sites are key to understanding the geographical distribution of PMSE. The auroral oval is shown for the southern hemisphere. From *Hocking* (2011). (Reprinted with permission from Elsevier.)

With regard to the south polar regions, PMSE were late in being found, though there were several searches for them. Ron Woodman even built a small radar on a ship and searched for PMSE on a trip to the Antarctic regions. PMSE were finally observed for the first time in the southern hemisphere in the summer of 1994 by a 50 MHz radar located at 62 °S (*Woodman et al.*, 1999). Subsequently they were seen at other southern sites (*Morris et al.*, 2006, 2007; *Kirkwood et al.*, 2007 and references therein). On average, PMSE seem somewhat stronger in the northern hemisphere, but both follow similar seasonal variations. The availability of multiple radars working at different frequencies also allowed multi frequency studies.

The suite of radars which were able to study PMSE was clearly growing by 2005, but even more radars were then brought into the mix, this time including smaller radars with wider beams. These radars were more mobile than the larger ones. A specially built wide beam meteor radar was modified for PMSE studies at Resolute Bay, comprising only 16 transmitter antennas, and SKiYMET meteor radars at Yellowknife (62 °N) and Andenes (69 °N) were also adapted for PMSE investigations (*Swarnalingam et al.*, 2009b). Antarctic sites (e.g., *Latteck et al.*, 2008; *Kirkwood et al.*, 2007) were also used, and significant variability within polar regions, and even between poles, was confirmed.

The possible impact of the auroral oval was discussed above. On a related note, electron precipitation can be important. Such events comprise the mass injection at high latitudes of electrons with energies from 10 keV to several hundred keV. This can happen during both day and night. Consequently, electron densities in both the D- and E-region increase. This could lead to associated increases in electron gradients, which, by Equation (7.71), increases the potential refractive index gradient and hence the backscattered signal. This seems to be partly verified by a few studies relating ionization increases to PMSE strength/activity (*Bremer et al.*, 2000; *Morris et al.*, 2005; *Barabash et al.*,

2002a). However, this is not the full story. Extremely large increases in precipitation can have the opposite effect and make PMSE actually disappear (*Morris et al.*, 2005; *Rapp et al.*, 2002; *Barabash et al.*, 2002b). Such a disappearance would not be a complete surprise. The electron density gradient is an important term in defining the strength of backscatter (again see Equation (7.71)). It is possible that in the cases of strongest precipitation, physical processes might be at play that actually smear out the gradients, thereby decreasing the potential refractive index gradient. Hence even if the scatterers that produce PMSE are present (be it some sort of turbulence or other dynamical process), weak electron density gradients could mean that the backscattered signal would be correspondingly weak.

We now ask the question: What could cause these PMSE?

10.2.2 Reasons for PMSE

The key parameter to consider in explaining PMSE is the molecular diffusion coefficient, ν. We have already seen it appear in Equation (10.1). It is important to recognize that ν is the molecular diffusion coefficient for the neutrals, whereas we are actually interested in the diffusion of electrons. So we should more properly use as our cut-off parameter η_K the quantity $\eta_e = \left(\frac{D_e^3}{\varepsilon}\right)^{1/4}$, where D_e is the diffusion coefficient of the electrons. The use of ν arose only because we assumed that the electrons were being dragged around by the neutral particles, due to collisions, but if the electrons have quasi-independent motions, then this assumption may not be valid.

However, it could be argued that the scatterers are not due to turbulence, but might perhaps be some sort of quasi-specular reflectors, like a (possibly undulating) step in electron density. Such a step was drawn in Figure 7.16 in Chapter 7. If this has a step thickness d_Ξ, the step will be destroyed by diffusion in a time scale of $t_d \simeq d_\Xi^2/D_e$. If we assume that the electrons are dominated by neutral motions, then we can assume $D_e \approx \nu$. At 85 km altitude, ν is typically 3–5 m^2s^{-1} (as verification, the reader can check Figure 11.33 in Chapter 11, which will be discussed in detail later). Hence for a 3 m step, $t_d \simeq 9/3$ to $9/5$, or typically 2–3 s. This is too short for a sustained step.

So no matter how we look at it – either requiring the Kolmogoroff microscale to be smaller for turbulence studies, or for the step to be longer-lived for specular reflection, we are forced to consider the posssibility that the rate of electron diffusion is much slower than that of the neutrals. This forces us to consider the relative effects of electron and neutral diffusion. To do this, we look at parameter S_c defined as

$$S_c = \frac{\nu}{D_e}, \tag{10.2}$$

where ν is the kinematic viscosity coefficient of the neutral gas. This quantity is called the "Schmidt number." Our proposal is that this number should be substantially greater than unity, indicating that the electron diffusivity is considerably smaller than the kinematic viscosity of the neutral gas.

We now need a physical explanation as to why S_c should be large. The focus should not be on the electrons, but rather on the ions. For if the ions diffuse slowly, the electrons will follow closely. Initial ideas were based around the possible existence of large cluster ions, perhaps made of large collectives of water molecules. However, theories show that the diffusion coefficient in such cases depends on the reduced mass of a typical neutral particle and the ion or ion cluster, which is often of the order of the mass of the neutral particle. This is insufficient to explain the effect. But an alternative model is one of large, *highly ionized* particles, perhaps with charge – positive or negative – of hundreds of times the electronic charges (i.e. they have lost or gained a large number of electrons and are no longer neutral). Then these particles are surrounded by, and drag along with them, large numbers of free electrons. These entities are referred to as "dressed aerosols." The idea was introduced by *Cho et al.* (1992). Such particles would have a small diffusion coefficient as a result of their sluggish response to collisions within the gas.

Hill (1978) had developed a multipolar diffusion theory which could be easily adapted to the dressed aerosol hypothesis, and *Cho et al.* (1992) applied it in the mesopause environment. They assumed an environment of electrons, positive ions, and charged ice particles. These authors found that if the plasma charge balance is dominated by *negatively* charged ice particles, then ice particles and electrons will maintain anti-correlated fluctuations due to Coulomb repulsion, which would lead to the required slow diffusion. This concept was further supported during subsequent rocket probe experiments, when charged aerosols were indeed found in the mesopause region (*Havnes et al.*, 1996).

Figure 10.3 shows the effects of varying the Schmidt number on radar reflectivity, and demonstrates how large Schmidt numbers strongly enhance the radar scatter over and above the more "normal" scatter.

This theory, and its dependence on ice-particles, also provides a link to noctilucent clouds. Ice particles may form as small aerosols at altitudes of 85 km or so, where they play a role in creating PMSE, and then fall, growing by accumulating more water vapor as they fall. Eventually they become large enough to be seen as NLC at slightly lower altitudes.

Extensive experiments of various types have been undertaken to see if these large Schmidt numbers can be verified experimentally. Investigations of backscatter cross-sections at multiple radar frequencies (e.g., *Rapp et al.*, 2008), as well as in-situ measurements (e.g., *Lübken et al.*, 1994, 1998; *Havnes et al.*, 1996) have been common for this purpose. Other studies have included examination of radar decay times (*Hocking and Röttger*, 1997). It seems S_c may be as high as 400 or more at times.

The concept of dressed aerosols is not the only mechanism proposed for explaining the slow electron diffusion, but most models do rely on the existence of cold summer mesopause temperatures, and a high Schmidt number is usually also a requirement.

As an example of an alternative mechanism, *Havnes et al.* (1992) has proposed a theory based on dust-hole scatter. The idea is dependent on the proposed existence of air vortices (whirls) with scale sizes of several tens of meters. Plasma dust particles (perhaps left over from meteors) are then assumed to fall through these whirls. It is proposed that the dust cannot penetrate within an inner forbidden zone of the vortex, the dimensions of which are defined by a balance involving neutral drag, gravity, and perhaps centripetal forces. As a result, a gradient in dust concentration will occur at

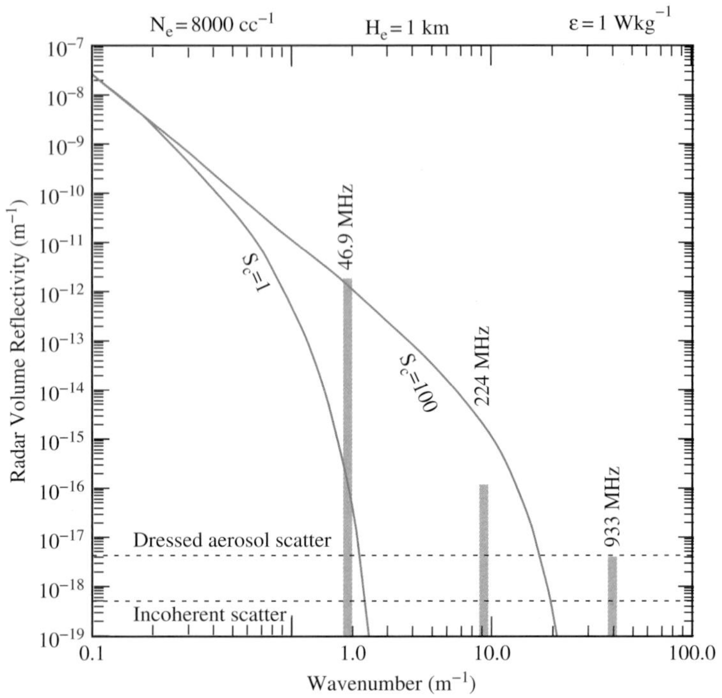

Figure 10.3 Comparison of the radar backscatter efficiency as a function of Bragg wavenumber (radians per meter) for different Schmidt numbers (from *Cho and Kelley*, 1993). For cases of large Schmidt number (sluggish ion and electron movement relative to the neutrals) the spectrum has an extended "tail." (Reprinted with permission from John Wiley and Sons.)

the boundary of the forbidden zone. A corresponding gradient in the electron and ion densities will result if the dust is charged. If sufficiently sharp, these density gradients will scatter radar waves.

La Hoz (1992) was also interested in the role of charged dust particles. For a plasma containing charged dust particles with multiple elemental charges, it was proposed that in order to maintain charge neutrality, clouds of opposite charges will be formed around each dust particle. If these clouds are mainly comprised of electrons, LaHoz has proposed that they will produce radiowave scatter of sufficient strength to explain PMSE.

Currently, however, it is fair to say that the dressed aeorosol theory is the lead contender for explaining PMSE.

While these theories all relate to explanation of anomalously slow electron diffusion, another aspect that needs to be considered is the mode of formation of the scattering entities. The first contender is generally turbulence, but not all PMSE are consistent with turbulent eddies. Scatter can be strongly anisotropic, indicative of specular reflections, as discussed elsewhere in this book (e.g. Chapters 2 and 7). On occasion, PMSE occurs without co-existing turbulence, so this possibility is a real one.

Hocking et al. (1991) and *Hocking* (2003a) had earlier proposed a model for viscosity wave production in the atmosphere to explain specular reflections, assuming that the viscosity waves might be generated during reflection of gravity waves. In order to explain PMSE, they adapted the model to consider reflection of infrasound at steps of temperature or wind-shear, which would in turn generate viscosity waves of a scale suitable to reflect VHF radiowaves. They determined that radiowaves could be reflected from these viscosity waves with strength sufficient to eplain PMSE.

In the winter of 2004, *Kirkwood et al.* (2006) observed highly aspect-sensitive echoes, which moved horizontally with the speed of sound, possibly indicating the existence of these viscosity waves. This is not to say that all PMSE should be due to such waves, but simply that the proposal is a suitable model for explaining cases when specular reflection is indeed evident. (These authors referred to the viscosity waves as damped ion-acoustic waves, but they are essentially the same.)

For the case of isotropic scatter with associated slow fading, the concept of fossil turbulence has been invoked. Normally this is not considered valid for the atmosphere: it exists under the ocean surface, where the rate of diffusion of salt is much slower than the rates of diffusion of momentum, so that after the velocity fluctuations have died out, there remain clumps of incompletely mixed salty water. It is not normally considered to apply to air, since the rates of diffusion of constituents and the rates of diffusion of momentum (velocity) are similar (Prandtl number of the order of unity). However, in cases of large Schmidt number, it is possible that within turbulence, the neutral perturbations may re-coalesce and smooth out after turbulence is complete, but the electron/ice/ion perturbations may persist longer as a form of "fossil plasma." Such a proposal was presented by, for example, *Cho et al.* (1996) and *Rapp and Lübken* (2003). This was the way in which they explained PMSE echoes in the absence of active turbulence. However, *La Hoz et al.* (2006) suggest that the estimated electron diffusion times used in the work of *Rapp and Lübken* (2003) are overestimates.

Figure 10.4 shows an example which appears to demonstrate initial specular scatter, followed by a breakdown process, and then showing ensuing strong turbulence. The work was originally presented by *Pan and Röttger* (1996), and was also discussed in *Cho and Röttger* (1997). The data were produced using a radar with spatial interferometry capabilities; the upper two graphs show classical HTI (height-time-intensity) plots and running power spectra, while the bottom one shows cross-spectra (also called coherence spectra) taken between signals recorded on two displaced receiver antennas. Lack of coherence between signals on the two spaced antennas is taken as an indicator of turbulent scatter, while good coherence is taken as evidence of specular reflections.

The need for further research on the nature of PMSE exists.

10.2.3 Other mesospheric echoes

Other polar echoes have attracted some attention. One example is studies of PMSE at medium and high frequencies, as reported by, for example, *Karashtin et al.* (1997), *Jones et al.* (2004) and *Jarvis et al.* (2005). However, for radars working at 2–3 MHz, the Bragg scales are 50–75 m, and so no special charged ice particles are needed to

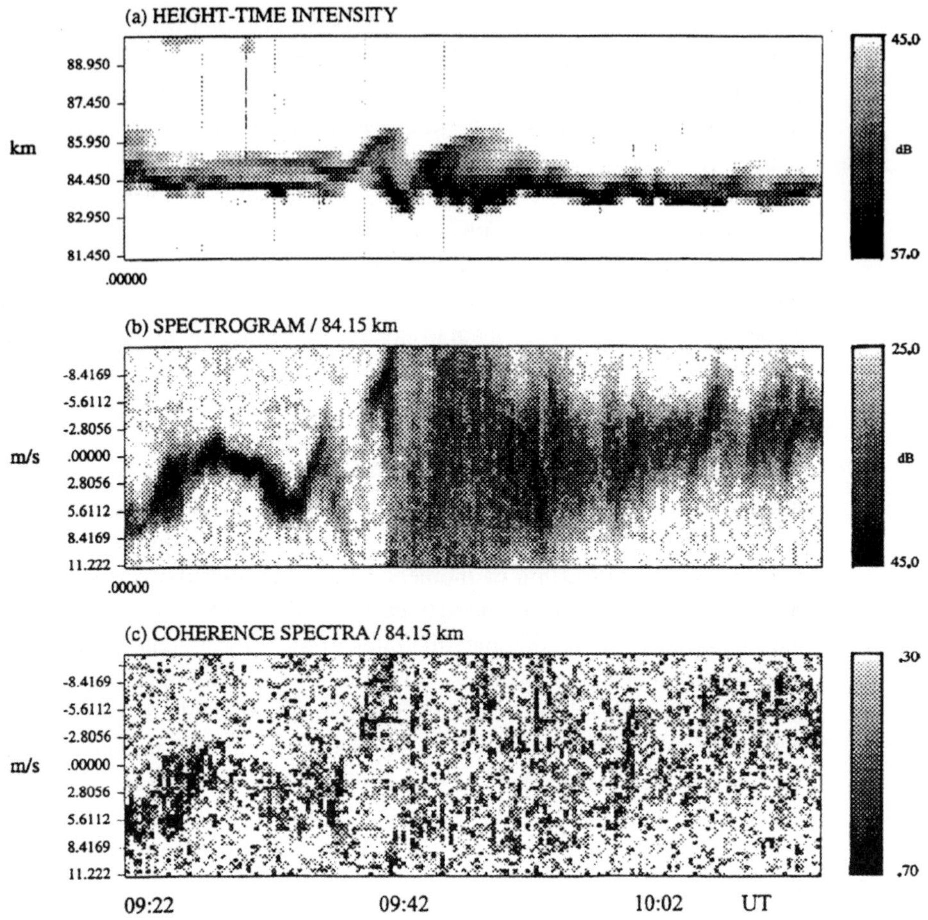

Figure 10.4 (a) Height-time-intensity plots; (b) Running spectra (spectrogram); (c) Coherence spectra produced between two spatially distinct receiving antennas, for one particular PMSE event. From *Pan and Röttger* (1996) and *Cho and Röttger* (1997). (Reprinted with permission from John Wiley and Sons.)

explain the scatter – Bragg scales with these wavelengths are already inside the neutral inertial range. Hence there is nothing special about such detections, and they should be visible even if $S_c = 1$. Indeed *Jones et al.* (2004) have noted a completely different annual variation for MF scatter compared to PMSE, and MF scatterers should not really be considered in the same category as the special cases discussed in the last section. Even for frequencies of 28–30 MHz (e.g., *Jarvis et al.*, 2005), the Bragg scale is 5 m, which is only a little less than the neutral turbulence inner scale at 85 km altitude, so anomalously low diffusion rates are not really needed to explain echoes at these frequencies.

Polar mesosphere winter echoes (PMWE) are also another polar phenomenon of some interest. Of course mesospheric echoes at 75 km have been well known since the earliest

VHF studies (see Figure 2.16), but since the discovery of PMSE, these lower altitude winter-time scatterers have also been given their own special new name. In principle, they have no need for a non-unity Schmidt number, and can be normal turbulence. They often appear in connection with solar proton events, and have been discussed by *Kirkwood et al.* (2006). Despite the fact that their existence was not entirely a mystery, they were still considered in some sense to be anomalous due to purported unusual characteristics. In particular, interferometric techniques were used to infer that the most intense echoes moved at horizontal speeds close to the speed of sound, and were quite specular. Initial suggestions that PMWE were unusual began in 2002 and finally led to a suggestion (*Kirkwood et al.*, 2006) that some were not due to turbulence, but rather due to localised, transient viscosity-wave-like (*Hocking*, 2003a) disturbances. These authors referred to them as ion-acoustic waves, by analogy with similar waves in ionospheric plasmas, and considered their cause to be related to partial reflection of infrasonic waves, following proposals from *Hocking* (2003a). PMWE have been less frequently observed than PMSE.

The proposal that viscosity waves produce some PMWE (and also PMSE) does not preclude the production of significant scatter of turbulence, and it does seem that the majority of the echoes (especially those of weak to modest strength) are, in the main, due to atmospheric turbulence. Studies during proton events in January 2005 which employed simultaneous application of multiple instruments, including rocket-based in-situ ones (*Lübken et al.*, 2006, 2007) concluded that the PMWE on that occasion were due to ordinary neutral turbulence. A theoretical model based on turbulence theory, using simultaneously measured kinetic turbulent energy dissipation rates and electron densities, was used to estimate backscatter strengths, and these were consistent with measured reflectivities. This contrasts to the studies presented by *Belova et al.* (2005). Summaries on non-turbulent origins of PMWE can be found in *Kirkwood* (2007), with other reports by *Belova et al.* (2005) and *Belova et al.* (2008). Mean characteristics of mesospheric winter echoes at high and mid latitudes observed from 2001 to 2005 have been summarized by *Zeller et al.* (2006), who especially emphasized turbulent scatter.

On a somewhat different note, another class of echoes that has generated some interest has been that of artificially induced mesospheric (ionospheric D-region) scatterers. These are man-made scattering structures induced by ionospheric heating from large radiowave heater facilities. Transient and decay effects associated with the ionospheric D-region can be studied by turning the heaters on and off, which in turn produce irregularities that can be probed and studied using MST radars. Such studies take place at all latitudes, though due to the distribution of heater facilities, a large portion of such studies have taken place in polar regions (e.g., *Thide et al.*, 1983; *Frolov et al.*, 1999; *Kagan et al.*, 2000; *Chilson et al.*, 2000; *Belova et al.*, 2008; *Hysell*, 2008).

Finally, it should be noted that mesospheric echoes at non-polar latitudes have been perhaps overshadowed by their polar-latitude equivalents, but mid-latitude and tropical mesospheric echoes also have their own unique characteristics which still require explanation. Anisotropic scatter and specular reflections can occur from these regions at MF, HF, and VHF (e.g., *Bremer et al.*, 2006), and need to be further studied.

10.3 Meteor studies

10.3.1 Introduction and radar design

A brief historical overview of the area of meteor physics was given in Chapter 2, Section 2.6, including the early days of the field, the "drop-out" in activity during the 1970s and 1980s, and the substantial rebirth in the 1990s, to the extent that meteor physics is now a large field in its own right.

In this section, we will concentrate less on the history and more on recent techniques. Meteor studies involve several different types of radar, from large powerful ones that can see head-echoes to smaller ones comprising only a few antennas which rely on specular reflection from the ionized trails. Since our primary focus is atmospheric studies using meteor trails, and since the primary instruments used for that purpose are based on relatively low-cost interferometric radars, it is these that will be the main focus of our discussions here. Many such radars are either SKiYMET radars developed by Mardoc Inc. and Genesis Software Pty Ltd. (*Hocking et al.*, 2001a), or clones of them, so these radars will be used as a reference point.

One of the special design aspects which helped make these radars more effective was the antenna layout of the interferometer. The main one employed a special five-receiver spaced antenna pattern (*Hocking*, 1997a; *Hocking et al.*, 1997; *Jones et al.*, 1998; *Rhodes et al.*, 1994). The general layout of the five-receiver array is shown in Figure 10.5. The most common arrangement of antennas, which appears in the form of a cross, is shown, but other options exist. For example, if antenna A_4 is moved to A_4', and A_3 is moved to A_3', the array will still function properly.

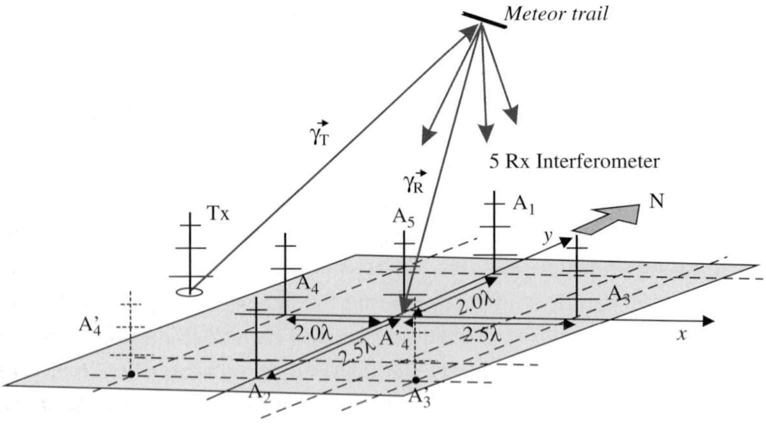

Figure 10.5 Possible antenna positions for a five-receiver interferometric array. The primary array is drawn as solid lines, with antennas being labelled A_1 to A_5. The transmitter array can be placed anywhere, but generally within a few wavelengths of the receivers. The path from the transmitter to the meteor trail and back to the receivers is shown schematically – in reality, the reflected paths to the five receivers are almost parallel. Other possibilities for the receiver positions exist; see the text for details.

A suitable array can be built with three receiver antennas, but the antennas need to be less than a half-wavelength apart, in order to remove angular ambiguities in the detection angle. However, such close spacing introduces serious impedance-coupling issues and so destroys the ability to produce reliable phases. The five-receiver system largely eliminates angular ambiguities (except in regard to noise effects) but at the same time keeps the antennas far enough apart that mutual coupling is not an issue.

As discussed above, moving A_4 to A_4' and A_3 to A_3', thereby producing a T-shape, also functions properly and removes angular ambiguities. Indeed shifting *only one* of A_3 or A_4 *still* allows the interferometer to function. As long as the displacement along an east-west line between the eastern antenna relative to the center, and the western antenna relative to the center, is a half-wavelength, and similarly for the north-south pair, the array will work. The displacements do not even need to be in the same line – for example if A_4 is moved to A_4', the east-west antennas (A_3, A_5, and A_4) are no longer in a straight line, yet the removal of ambiguities still works. Other pairings of spacings can apply (e.g., 3.5 and 3.0 wavelength-pairs), but phase differences need to be detemined to higher precision for larger antenna spacings (to avoid 2π ambiguities), which can be more sensitive to the signal-to-noise ratio. Designs using other than five receiver antennas have been used. For example, an effective four-element interferometer with good angular discrimination was described by *Poole* (2004). *Hocking* (1997a), *Hocking and Thayaparan* (1997) and *Hocking et al.* (2001b) also used an earlier four-antenna version, but with diminished angular resolution. In general, however, the five-receiver systems seems to give optimum cost-to-benefit ratio.

10.3.2 Winds and temperatures

In regard to MST studies, the main parameters deduced by meteor radar are winds, temperatures, and momentum fluxes. In the earlier 1990s, measurement of winds was the main focus, and the techniques used followed the description given in Chapter 9, Section 9.4.1 of this book. One of the earliest verifications of the new design was a series of good comparisons of meteor radar winds with other techniques, including lidar methods (e.g., *Franke et al.*, 2005) and ionosonde comparisons (e.g., *Jones et al.*, 2003). The meteor method has become something of a standard for wind measurements.

In the late 1990s and into the early 2000s, there were two new developments of note. One was development of ways to measure mesopause temperatures with meteor radar, the other was the implementation of procedures to determine momentum fluxes. We will discuss the temperature developments in this sub-section, and momentum fluxes in the one that follows.

Temperature measurements begin with the recognition that the rate of expansion of underdense meteor trails depends on the ambipolar diffusion coefficient, and that the ambipolar diffusion coefficient in turn depends on temperature and pressure. The relation is (e.g., *Chilson et al.*, 1996)

$$D_a = K_* \frac{T^2}{p}. \tag{10.3}$$

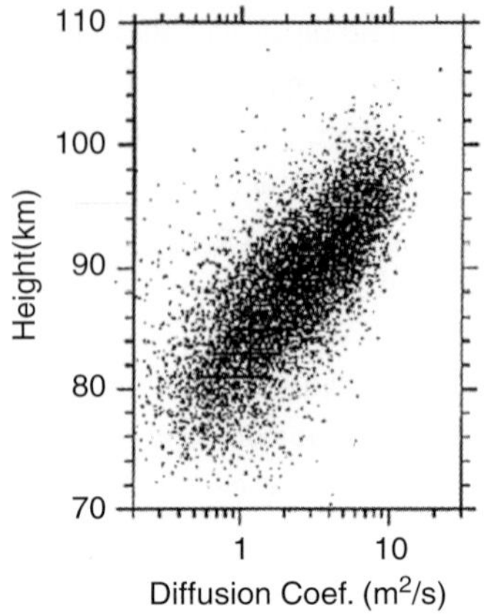

Figure 10.6 Scatterplot of ambipolar diffusion coefficient versus height in the meteor region. Data were determined using meteor measurements detected by the MU radar (from *Tsutsumi et al.* (1994)). (Reprinted with permission from John Wiley and Sons.)

The quantity K_* has a value determined from theoretical studies and laboratory experiments, and is usually taken to be a constant. The ambipolar diffusion coefficient can be found from the rate of exponential decay of the amplitude of meteor trails as observed by radar (e.g., *Hocking et al.*, 1997). If one extracts ambipolar diffusion coefficients for many meteors, and plots the values as a function of height, a graph like Figure 10.6 results. The graph of the log of the diffusion coefficient versus height is fairly linear, and lends itself to the possibility of using it to determine the temperature in the region if K_* and p are known, a concept that was turned into a viable technique by *Hocking et al.* (1997).

However, one problem with the method was the need to know the pressure, and these turned out to be unreliably documented. Therefore *Hocking* (1999b) developed a method which avoids the need for a-priori pressure measurements. The method relies instead on measurement of the slope of the log of the inverse decay time versus height, bypassing the need for pressure data. In place of pressure data, it does need an approximate global model of the temperature gradient around the mesopause (*Hocking et al.*, 2004), but this can be found more reliably than the pressure. In addition, the new method did not need an estimate for K_*, removing another level of uncertainty from the equations. The method has shown good comparisons with optical techniques (*Hocking et al.*, 2004). For those who choose to use the first method, which requires p and K_*, *Younger et al.* (2008) has made numerical studies of the dependence of K_* on wavelength.

Due to the scatter evident in Figure 10.6, large numbers of meteors (preferably thousands, and at least hundreds) should be used in any detemination of temperature, and normally the value determined is an average around the mesopause region. The scatter occurs due to a variety of reasons, not least of which is the limited height resolution; if the vertical resolution is say 2 km, the variations in decay times over a 2 km height interval due simply to pressure changes in that zone can be important (*Hocking*, 2004a). Generally scatter is more severe at lower temperatures when the scale height is less. Beam-width is also important – the scatter is less for radars with narrower beams. However, despite all these issues, the method has proven to be a reliable technique for temperature determinations, with good agreement with OH and other optical techniques, as seen for example in *Singer et al.* (2004a) and *Hall et al.* (2004), among many others.

Height-dependent temperatures can be determined when sufficient meteors exist. At least some thousands of meteors are required for any one determination, as reliable fitting is crucial. The method combines the techniques from *Hocking et al.* (1997) and *Hocking* (1999b) to obtain both a pressure and a temperature at the mean height of the meteors. Then the pressure is used to determine a scale-height. It is then assumed that the pressure varies smoothly with altitude. Finally a height profile of temperature is produced by using the measured height profile of the ambipolar diffusion coefficient, combined with the pressures. The method has been introduced and applied by both *Hocking et al.* (2007b) and *Kumar* (2007).

The temperature-gradient procedure can also be used to determine tides in the temperature, but special compensation is needed in the analysis because the tide causes the temperature gradient to change during the course of the day. The details are presented by *Hocking and Hocking* (2002), where numerical tidal models were compared to sample data. Analysis of tides in the northern polar regions using similar methods was reported by *Singer et al.* (2003).

10.3.3 Momentum fluxes

In Chapter 1, Section 1.3.4, the concept of momentum fluxes was introduced. It has been known for many years that knowledge about momentum fluxes is important in defining the mean flow of the atmosphere. In 1983, the so-called "dual beam" radar experiment was introduced for mesospheric studies by (*Vincent and Reid*, 1983), and it became a useful tool for radar measurements of momentum flux throughout the mesosphere, troposphere, and lower stratosphere. However, it generally required large antenna arrays with narrow steerable beams. Because of the need for large antennas, the number of sites which could make such measurements was limited, especially for mesospheric studies (e.g., *Murphy and Vincent*, 1993). However, a few discrete studies were made: *Dutta et al.* (2007) studied optimization of the pointing angle for momentum flux measurements, and *Kudeki and Franke* (1998) discussed measurements and the statistical reliability of the method with the Jicamarca radar. Measurements in the lower atmosphere troposphere and stratosphere have been presented by *Worthington and Thomas* (1996). Other radar-based mesospheric measurements were presented by *Fritts et al.* (2006); *Janches et al.* (2006); *Fritts and Janches* (2008). *Sasi and Deepa* (2001) presented observations with the Gadanki radar in India. *Thorsen et al.* (1997) developed a method which could be used with spaced antenna MF radars, but it does not seem to have been widely applied. Some other techniques, based on optical methods, were also developed and used (e.g., *Espy et al.*, 2006).

However, despite these investigations, measurements were only possible at select sites. It was within this background that *Hocking* (2005) introduced a method for measuring momentum fluxes with interferometric meteor radars. This method enabled the user to not only determine momentum fluxes, but all of the parameters $\overline{u'^2}$, $\overline{v'^2}$, $\overline{w'^2}$, $\overline{u'w'}$, $\overline{v'w'}$ and $\overline{u'v'}$. Of course results were restricted to the meteor region (typically 75 to 95 km altitude), but at least having momentum fluxes world-wide in this regime was a big step forward for middle atmosphere momentum studies.

The method made use of measurements of locations and radial velocities of a large number of meteors recorded in some pre-specified time-frame – typically 1 or 2 hours. This information was then collated in a 6×6 matrix with elements which depended on various sums related only to the geometries of the trails. Specific details were presented in *Hocking* (2005), appendix A. This matrix then needed to be inverted and multiplied by a column matrix which could be determined from measurements related to the trail positions and radial velocities. This process then produces the six main parameters discussed above. The method is quite general, and the dual-beam method discussed above is a subset of this method.

The procedure can be applied with a small radar like the SKiYMET design discussed in Figure 10.5: a large array with narrow beams is not required to implement it. However, it does need to be remembered that these smaller meteor radars detect a large fraction of their meteors at 50–60 ° off-zenith, which means trails at opposite azimuthal directions can be separated by distances of up to 300 km horizontally. It may be that statistical stationarity cannot be assumed over such large distances. Other issues, such as the azimuthal distribution of meteors in any typical time interval, also come into play. The statistics of the method has been tested by *Vincent et al.* (2010), and the conclusions are that the method works provided sufficient numbers of meteors are used (typically 30 or more per time-height bin), in agreement with the earlier conclusions of *Hocking* (2005). Further tests comparing radars in different locations and with different basic designs have also been undertaken by *Fritts et al.* (2010) and *Fritts et al.* (2012).

Another important point relates to the ways in which the final analysis is carried out. While meteors need to be grouped into hourly or two-hourly bins, this can be done in various ways. If the user is interested for example in diurnal variations averaged over a month, one could lump data into say hourly bins, one for each of the 720 hours of a 30 day month, and then average the various fluxes binned into times of day. Alternatively, it would be possible to bin all data from a specified hour for all days of the month, and apply the analysis to superposed data collected in this way. This was the procedure adopted by *Hocking* (2005). The two methods do not always give the same results, and comparisons between them can give a further measure of the reliability of the data. *Andrioli et al.* (2010) has introduced yet another interesting idea. In this scheme, the meteor data are used to determine hourly or two-hourly mean winds, and then these are fitted by suitable tidal and longer-period waves to determine a smoothly behaving long-term wind variation, properly embodying tides and longer period oscillations. Then the user goes back to the raw data, and for each meteor, removes the component of the radial velocity associated with the long-period variations. Momentum flux parameters are then determined on the residuals. The authors claim that this technique produces superior results.

Measurements using the method are ongoing, and to date appear to be producing good results (e.g., *Placke et al.*, 2015).

10.3.4 Additional miscellaneous meteor-related studies

Now we choose to look at some advances with these radars in other areas, even including studies of astronomical parameters (as distinct from atmospheric ones).

Determination of the entrance speeds of meteors into the atmosphere is one such area. With the relatively fast sampling rates used with the SKiYMET-type systems (> 2000 Hz in many cases), the signal can be sampled fast enough that the diffraction pattern at the front of the meteor trail can be recorded as it sweeps past the antennas. From this information, the entrance speeds of meteors (typically tens of kms^{-1}) can be obtained (e.g., *Cervera et al.*, 1997; *Hocking*, 2000).

The radars can also be used to determine meteor radiants. *Singer et al.* (2000) showed an example of studies of a meteor shower – in this particular case the Leonids 1999 storm – while *Jones and Jones* (2006) presented an example of the determination of the direction of the radiant of a shower, though there are many other publications in this field.

Simple studies of diurnal and seasonal variations in numbers of meteors detected per hour (meteor fluxes) have also been undertaken by many authors. For example, *Singer et al.* (2004b) studied the diurnal and annual cycle of non-shower meteor fluxes in the polar regions. There is a strong daily cycle at the equator, with lowest counts in mid-afternoon and largest count rates in the early morning (local time). The annual variation at these latitudes is quite modest. In contrast, at the poles it is the annual cycle which shows the greatest minimum-to-maximum variation, with maxima in summer and minima in winter. The diurnal variation at the poles is relatively flat.

While most interferometric meteor radars generally work in the upper HF and VHF bands (typically 20 to 55 MHz), some researchers have used medium frequencies as low as 2 MHz (e.g., *Tsutsumi et al.*, 1999). Use of these lower frequencies allows meteors to be detected to higher altitudes, even up to well above 100 km.

Other possibilities for studies of meteors include investigations of head echoes with MST radar (e.g., *Janches et al.*, 2003), and long duration meteor trails (e.g., *Bourdillon et al.*, 2005). Some SKiYMET (and similar) radars have been upgraded to include outlier stations which can be used for orbit determinations (e.g., *Baggaley et al.*, 1994; *Jones et al.*, 2005; *Brown et al.*, 2008; *Younger et al.*, 2009; *Janches et al.*, 2013).

Networks of radars can give additional information; one such network is a pole-to-pole SKiYMET network called *Axonmet* (see *Hocking et al.*, 2010). *Elford and Taylor* (1997) used Faraday rotation from meteor echoes to obtain electron densities. *Elford* (2001) has discussed other possible applications relating to meteor radars, and the applications seem to keep growing.

For further reading regarding meteor-astronomy, see *Hawkes et al.* (2005), for example.

10.4 Tropospheric temperature measurements and RASS

Use of profilers to measure temperature in the troposphere over the radar has been discussed at several points in this book to date, notably Sections 2.16 and 7.7. These discussions considered methods involving atmospheric stability studies, gravity-wave studies and RASS. Of the three, RASS is the most reliable, but is limited in applicability due to its tendency to be very acoustically noisy. However, some useful measurements have been made.

We will not dwell further on this topic, except to remind the reader that the radars do have some capability for temperature measurement. As a reminder of the possibilities, we present Figure 10.7 from *Adachi et al.* (1993).

We will not add anything futher on this topic here.

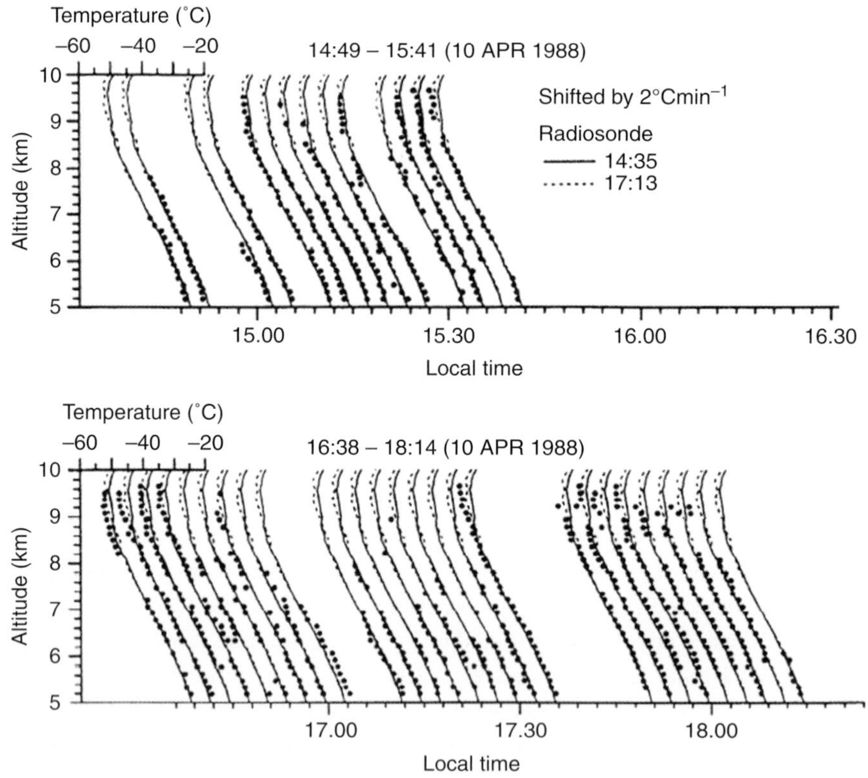

Figure 10.7 Temperature profiles in the troposphere produced using the radio acoustic sounding system (RASS) technique (small dots) compared to radiosonde measurements (lines). The good accuracy and excellent temporal resolution of the RASS method are both clearly apparent. The data were collected using the MU (middle and upper atmosphere) VHF radar, which is able to employ fast beam-steering to allow measurements up to 10 km in height and higher (from *Adachi et al.*, 1993). (Reprinted with permission from John Wiley and Sons.)

10.5 Water in the troposphere and stratosphere

Water in the troposphere occurs as either ice, liquid, or gas (vapor). When considered as a gas, we refer to its content in terms of the humidity of the atmosphere. The humidity affects the backscatter cross-section of the incident waves and we will discuss this shortly. When in the form of ice or water, it occurs as discrete entities, specifically as drops, ice-crystals, or snowflakes (dendrites). Collectively, these are called hydrometeors. Ice particles are hard to see directly by radar, since the dielectric constant (relative permittivity) of ice is about 1.3, and so quite close to that of air, allowing little contrast between the air and the ice. Water has a dielectric constant closer to 80, and so is more capable of producing measurable backscattered radiation. In general water droplets produce greater backscatter (relative to the neutral air) at higher frequencies. In this section, we will discuss the effect of water in the atmosphere in two parts: one will be the effect of scatter from droplets, and the other will be the impact of humidity.

10.5.1 Precipitation measurements with ST radar

In Chapter 2, Section 2.17, a brief history was presented of the early days of precipitation studies using VHF radar, concentrating on the 1980s to the early 2000s. We will not repeat this here, except to say that occasionally the area is revisited, though it is still not entirely mainstream. *Yamamoto et al.* (2009) has compared Mie Lidar measurements of rainfall terminal velocities relative to the air motion with radar data. *Williams and Gage* (2009) presented an evaluation of the errors in the drop-size distribution method. We will not dwell further on this topic, but will now turn to a related topic of more recent vintage.

10.5.2 Measuring humidity with ST radar

The ability to *remotely* measure humidity is not available to many instruments, but would be of significant value to meteorology. Probably the best instrument to date is the microwave radiometer, but even they have still not demonstrated ideal capabilities. Normally they need to be supported by other instruments (e.g., *Bianco et al.*, 2005). Radiosondes are still generally considered the standard reference. However, radiosondes also show failures when the air temperature gets too low. So attempts have been made to use profilers to determine humidity as a function of height, based on pioneering work by *Tsuda et al.* (1997b) and *Tsuda et al.* (2001).

The backscattered signal received by a radar depends in part on the potential refractive index gradient, (as seen in any of Equations (3.288), (3.295), (7.70), (7.85), (7.87), and/or (7.88)), which in turn depends on the humidity gradient, so the backscattered signal received by windprofiler radars depends in part on humidity gradients. Indeed, the humidity term is often the *major* contributor to the potential refractive index gradient in the lower troposphere (e.g., *Hocking and Mu* (1997)). If the potential refractive index can be determined as a function of height, and if, in addition, the temperature profile is known (either by RASS or from radiosondes), then, in principle, it is possible to determine humidity gradients as a function of height and then integrate the humidity gradients from the ground upward (or from the tropopause downward), producing an overall humidity profile.

However, determination of the refractive index gradient is not trivial. The procedure is first to find the potential refractive index gradient M_n by re-arranging Equation (7.89) to the form

$$M_n = \pm \sqrt{\frac{\gamma \overline{C_n^2} \omega_B^2}{F_t^{1/3}} \varepsilon^{-1/3}}. \tag{10.4}$$

Determinaton of $|M_n|$ therefore requires C_n^2 (from calibrated radar measurements), the Brunt–Väisälä frequency (from RASS or radiosondes), the fraction of the radar volume filled with turbulence, F_t (discussed shortly), and the turbulent kinetic energy dissipation rate (also discussed shortly).

It may then be seen from Equation (3.288) that if M_n is known, the humidity gradient may be found as long as the temperature profile is known. This eventually leads to a vertical profile of the specific humidity, q_{wp}.

Complications are as follows. First, the radar needs to be accurately calibrated. Secondly, the signal strength received by the radar depends on the gradient squared, so the sign of M_n cannot be found from the radar signal. However, to create a humidity profile from M_n requires the actual value of $M_n(z)$, including the sign. Therefore the humidity profile cannot be determined without additional knowledge – namely the sign of M_n at each height step (see 10.4). Thirdly, additional parameters like the turbulent energy dissipation rate and fraction of active turblence in the atmosphere are needed. These items are discussed below.

The calibration of a radar and extraction of C_n^2 from radar signal strengths have been discussed at some length in Chapter 5 (e.g., equations (5.135), (5.136), (5.137) and/or (5.138)), and also a little in Chapter 7. So here, we accept that measurement of C_n^2 is possible, using established methods already discussed earlier in this book. The key issue becomes determination of the sign of the humidity gradient.

One of the more thorough methods for humidity-profile determination presented to date is that of *Furumoto et al.* (2003). Due to the aforementioned complications, the intention in that work was not to obtain totally independent profiles. Rather, the procedure represented something of a "half-way" house: the intent was simply to develop a method that fills in missing humidity data between reference profiles determined every few hours, where the reference profiles might be made, for example, by 12-hourly radiosonde measurements. We will refer to the reference profiles as "anchor points." The procedure was designed to track the behavior of the humidity profiles between the anchor points. While not as ideal as producing completely independent measurements of humidity, knowing the behavior between anchor points could still be a valuable capability in short-term precipitation forecasting. We will follow this particular reference as a guide to the various issues that need to be dealt with.

Furumoto et al. (2003) employed a profiler radar and a RASS system, each running on a continuous basis, to produce real-time capability. Radiosondes were launched from time to time to provide anchor points, typically every 3–12 hours, depending on local weather variability. Precipitable water vapor measurements by GPS satellite were also desirable as a reference. The radar data could then be used to fill in the humidity behavior for the period of time between radiosonde launches. The radar-derived humidity profiles can be considered to some extent as a sophisticated interpolation scheme in height and time.

Here we discuss how the various parameters are found. First RASS, together with occasional radiosonde data, can be used to calculate the Brunt–Väisälä frequency ω_B^2. The turbulence strength ε can be estimated from the spectral width of the Doppler spectrum, after compensation for beam-broadening, as described in Chapter 7. The turbulent kinetic energy dissipation rate ε has uncertainties of the order of a factor of 2 or 3, but because it appears in Equation (10.4) as $\varepsilon^{-1/3}$, this uncertainty is reduced in impact.

The fraction of the radar volume filled with turbulence, F_t must be determined, and this needs to be done using a climatological model (e.g., *Warnock and Van Zandt*, 1985; *Cohn*, 1995). *Furumoto et al.* (2003) used the model of *Warnock and Van Zandt* (1985).

Once these parameters, and a suitable humidity profile, have been determined at an anchor point, the subsequent processing assumes that all parameters change relatively slowly and smoothly with time. In order to constrain the profiles of $q_{wp}(z)$, measurements of precipitable water vapor (PWV) may be determined from continuous GPS data. The sign of dq_{wp}/dz is allowed to change, but only in a smooth and continuous manner, guided by the GPS measurements of PWV. Once a new anchor point is created, the process starts again.

At times, GPS data are not available. In such cases, an assumed correlation between M_n and $-\omega_B^2$ (*Tsuda et al.* (2001)) is invoked, which is based on an adiabatic assumption. The correlation is only valid in relatively stable conditions, and cannot be applied under conditions such as passage of rain clouds, which is unfortunately the time that humidity data could be the most useful for forecasting.

Other developments in regard to humidity studies have continued, including co-located measurements at VHF and UHF or L-bands (e.g., *Mohan et al.*, 2001; *Furumoto et al.*, 2005, 2006, 2007; *Imura et al.*, 2007). *Furumoto and Tsuda* (2001) have performed additional studies of the effects of humidity on turbulence echo power. *Bianco et al.* (2005) have looked at joint applications of windprofilers and microwave radiometers to produce humidity profiles. At the current time, no fully satisfactory radar method has been developed, though attempts continue.

10.6 Other specialized meteorological topics

A variety of specialized studies and techniques have been developed in regard to meteorological studies. We will leave extended discussions of these issues to our chapter on meteorology. Nevertheless, we will briefly identify them here.

One example is identification of the tropopause, and tracking of its motions; this is a major capability of windprofiler and MST radars. The techniques for doing this have already been mentioned in Chapters 2 and 7. Applications of this ability are multiple. Knowledge of the height of the tropopause can be useful information in applying inversion methods for satellite measurements. Tracking the tropopause height can also be of value for studies of the motion of frontal systems, as seen in Chapter 7. These measurements can also be useful for studies of stratosphere–troposphere exchange (STE). Jumps in the height of the tropopause are also associated with ozone exchange between the stratosphere and the troposphere (e.g., *Hocking et al.*, 2007a). Due to the meteorological nature of these capabilities, we will concentrate on these in more detail in Chapter 12.

Other special topics of meteorological interest include measurements of precipitation by radar (already discussed), specialized development of boundary-layer VHF radars, calibration of radars using precipitation, and the relation between convection, windshear and turbulence anisotropy.

One meteorological area that is less well developed, however, but which has unique requirements and potential, is studies of lightning by MST radar. The primary emphasis of our discussion on lightning will be on measurement and detection methods, so to some extent it fits better in this technique-orientated chapter. However, we will nonetheless begin the section with a short review of the mechanisms of lightning generation.

10.7 Lightning detection with windprofiler radars

In highly developed clouds, particularly ones developed by strong convection, complex temperature and humidity differences exist, which create hydrometeors (water droplets and ice crystals). These vigorous motions can lead to charged ice and water particles, which can lead to layers of excess negative and positive charge in the clouds. These in turn lead to strong electric fields, and eventually strong discharges of ionized particles in the air, called lightning. MST radars may be used to study lightning events.

Here, we first present an overview of the basics of electric fields and lightning production in the atmosphere, and then extend the discussion to considerations about how MST radars have been used to study lightning.

10.7.1 The mechanics of lightning

It is easy to recognize that the strong shears present in dynamically vigorous clouds may lead to breaking and tearing apart of water droplets and ice crystals, and that the resultant fragments might not be electrically neutral, as one fragment may take away a slightly higher percentage of electrons and the other will be left with a correspondingly lower percentage. But in order for electric fields to develop, there must be a *systematic* separation of positive and negative charge.

One such proposed charging process is called the graupel-ice mechanism. Supercooled water – namely water that stays in its liquid form even below its freezing point because of the lack of any nucleus around which a crystal structure can form – is a common feature in the atmosphere. When snow in the atmosphere encounters supercooled water, ice crystals form instantly on the outside of the snow, leading to an ice-snow hybrid call graupel. Collisions between graupel and ice particles then become a major source of charge generation, since, for reasons not fully understood, negative charge is transferred preferentially to the larger graupel particles during these encounters. The graupel falls faster than the positively charged ice particles because of the balance between gravity and frictional drag – the graupel is in effect more aerodynamic. This results in a negative charge region in the lower part of the storm and a positive charge region above it. Other processes exist, but this is one mechanism that is relatively well understood.

This leads to a model for the cloud charge structure during a thunderstorm called the *tripole model*. In this model, a main upper layer of positive charge develops at the cloud top, while a main negative charge results at mid-levels in the cloud, both being due to the ice-graupel interaction. These two layers are usually assumed to be equal and opposite in charge. A smaller region of positive charge at the base of the cloud arises due to space charge redistribution around the thundercloud and due to the corona near the ground. The lower positive charge is thought to be very small in magnitude and not present in all thunderclouds. These charges in the cloud reside on hydrometeors, which are various liquid or frozen water particles in the atmosphere. Such a model has been used for many years in simulations of cloud physics.

The same model also explains the so called "fair-weather" electric field. The clouds in thunderstorms act like giant batteries, with the positive terminal at the top. At the same time the cloud induces a negative charge in the ground (particularly due to the lower, weaker positively-charged layer in the tripole model), so the ground acts as the negative battery terminal. The complex electrical interactions within the cloud, and in the cloud–ground system, provide the internal charge supply of the battery. This combination of events drives a large scale flow of positively and negatively charged ions, with the positive ions flowing upward to the ionosphere, then spreading laterally, and being drawn down to the negative ground. Any free negative ions, or electrons, take the reverse route. The result is a vertical electric field that acts downward in regions far from the thunderstorm, which has a field strength of about 200–400 V/m. Stronger and more complex fields develop within the cloud. Typical electric fields within the cloud, and between the cloud and the ground, can be as strong as 5000–10 000 V/m. The continued presence of at least a few thunderstorms at any time somewhere on the planet keeps the fair-weather electric current flowing and the global fair-weather electric field intact. Further discussion can be found in texts like *Uman* (1971), among others.

More recent studies have shown that other complex electric charge distributions can develop that have even greater structure than the tripole model (e.g., *Stolzenburg et al.*, 1998), with four and even six separate layers of charge within the clouds.

Other mechanisms that have also been discussed for the generation of charge include splintering of hailstones carrying charges of different density, and induction charging, whereby moving charged particles could lead to several charge centers in the cloud. All these scenarios produce a growing complex charge re-distribution in the convective cloud system.

If the electric potential created by these charge separations exceeds a critical limit (several hundred thousand volts per meter), a so-called dart leader establishes an ionized channel from the ground, or from another part of the cloud system, to the positive region aloft. Along this ionization channel, a very intense spark then often develops. This is the main lightning stroke, which propagates from the cloud and discharges the elevated electrical field. The stroke behaves like a highly conducting thin wire, or a system of multi-furcated wires, which carry an intense current building up in milliseconds. Due to its very large current, it creates very localized, extremely high temperatures of some tens of thousands Kelvin. The highly conducting and rapidly developing stroke acts as a generator of a broad spectrum of electro-magnetic waves, called sferics. The stroke also acts as a highly conductive scatterer or reflector of electro-magnetic (radar) waves. The extremely elevated density change due to the very hot lightning channel causes a shock wave heard as thunder, and potentially generates acoustic or infra-sonic waves. Sprites can also be produced, which are related to discharge to the ionosphere, which occur above thundercloud systems, but discussion of these is outside the scope of this book.

The production of sferics, and the existence of highly conducting channels of ionized air, lead to two quite different ways to study lightning by radio methods.

10.7.2 VHF radar and radio observations of lightning

We now turn to a discussion of how radar and radio tehniques can be used to study these lightning events.

Of course, a radar cannot observe accumulation and distribution of electrical charges, or measure electrical potentials and their development in the clouds. It can follow the development of air motions in the cloud and it can also detect scatter from hydrometeors, as discussed elsewhere in this book, and these parameters are part of the early precursors of lightning discharges. But the location of the lightning channels is another potential capability of radar and radio tehniques.

Two primary methods exist – one using the sferics produced by the lightning, and the other employing the high scattering efficiency of the lightning plasma. The second requires a transmitter, and is well suited to MST techniques. The first requires only a network of receivers and GPS timing. We will discuss the first method first.

Each new lightning strike has a somewhat unique signature, often involving multiple short, impulsive events. The lightning strokes produce sferics which are strongest in the 60–66 MHz band, but still have good power down to a few MHz and up to 100 MHz and more. The so-called *lightning mapping array* or LMA (*Thomas et al.*, 2004), receives signal on widely distributed antennas (many tens and even hundreds of km apart), and each lightning strike has a unique signature. However, if multiple strikes from different sources occur at similar times, a single antenna will receive all these signals at once. Several antennas may receive the same signals, but the timing may differ on different antennas, depending on time of travel. This makes untangling the information difficult. The LMA method does pattern searches on the signals received by all the antennas, looking for common patterns on each receiver. Once such patterns are detected, the temporal displacements between similar signal-patterns on different receivers are used to perform triangulation and hence locate the source of the sferics. The process is computer-intensive, and uses lightning amplitudes only. Currently, the LMA is the primary commercial instrument for lightning location over scales of hundreds of km. Future developments include interferometric applications, which had not been possible until the recent development of digitizer cards capable of GHz sampling.

The second method is more applicable to MST-type radars. The first attempts at this method appear to be with the Chung-Li radar (*Röttger et al.*, 1995). Subsequent papers include *Petitdidier and Laroche* (2005) and *Beres et al.* (2010). In this method, a radar pulse is transmitted with (optimally) a broad transmitter polar diagram, and three or more receiving antennas are used for reception, permitting interferometry. The receiver antennas are relatively closely spaced, in contrast to the LMA – in this case, spacings are only a few radar wavelengths or less. The signal received after reflection from the lightning channel is then recorded on all receivers. The major complicating factor is that the reflected signal is received at the same time as the sferics are generated, so the

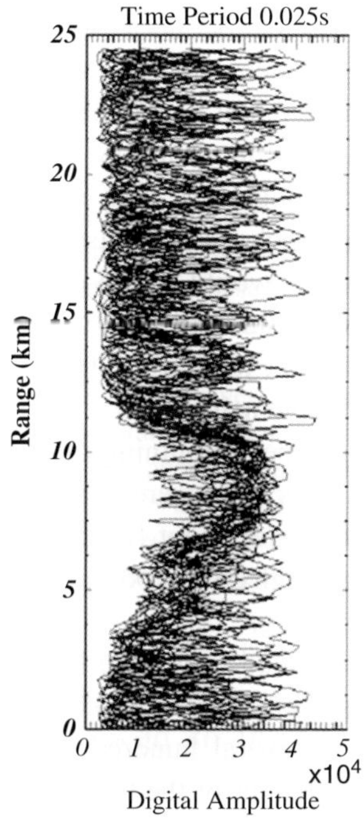

Figure 10.8 Height profiles of successive samples (overlaid) during a lightning stroke, using a wide-band filter with simultaneous radar transmission of a 2 km long radar pulse. The sferics are apparent as the signal with the rapid variations as a function of height, while the reflected pulse is apparent as a smoother variation embedded in the sferics at 6 to 11 km range (adapted from *Beres et al.*, 2010).

it was assumed that most of the observed strong lightning echoes were not received through the main beam of the radar antenna.

10.7.4 VHF radar interferometer observations of lightning

Interferometric measurements of some lightning strokes will now be discussed. Here we will concentrate on data recorded with the Chung-Li VHF radar in Taiwan. The general instrumental set-up of the measurements was equivalent to that described by *Röttger et al.* (1995). Two of the main antenna modules of 2×64 Yagi antennas were used for transmission at about 2 x 35 kW peak power. Three single four-element Yagi antennas on a short baseline were used for reception. These were set up in a triangle with vertex separations of half a radar wavelength (2.88 m) in order to avoid angular ambiguities due to multiple interferometer lobes, which would occur if the main transmitting antenna modules were also used for reception. The mutual coupling between these receiving Yagis was measured at −25 dB, which was regarded to be sufficient for at least initial measurements, though it is at the limits of acceptability, since the re-radiated signal from one antenna to a second has 6% of the amplitude of the signal received on the second, leading to potential phase errors of 3–4 degrees.

A special phase and amplitude calibration was designed, which allowed the user to simultaneously feed the same signal, distributed via properly calibrated couplers, to the three Yagi antennas. The data were taken by using three complex channels sampled simultaneously, using 300 m range resolution. A short coherent integration time allowed the actual time resolution to be quite good at 4.8 ms. The phase calibration and the data acquisition were both performed with the same system setup as that used for lightning echo recording. This permitted a more precise phase and amplitude calibration of the interferometer.

The basic principle of spatial domain interferometry was explained in Chapter 9. The setup is shown schematically in Figure 10.11. Using three receiving antennas

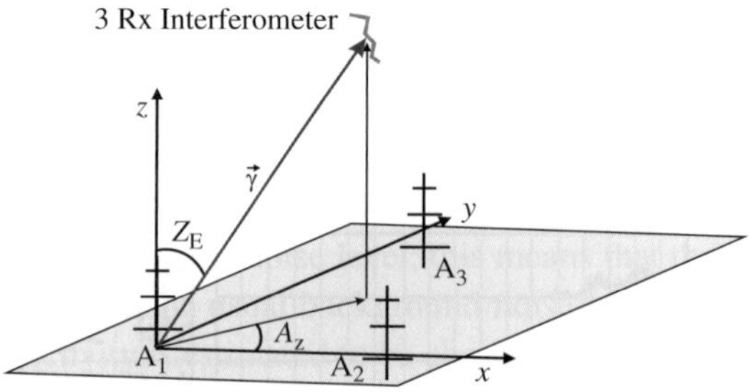

Figure 10.11 Configuration of receiving antennas for three-receiver interferometric measurements of lightning echoes. The coordinate x points to the east direction, y to the north direction, and z to the vertical. The vector γ points to the radar target, namely the location of the scattering/reflection point on the lightning stroke. The receiving antennas A_1, A_2, and A_3 are on the ground ($z = 0$), and are assumed to be in a horizontal plane. The transmitting antenna points vertically into the zenith direction, and is (for simplicity) assumed to be co-located with the receivers. Its beam width should illuminate the radar targets aloft.

A_1, A_2, and A_3 located in the x–y–z coordinate system shown in Figure 10.11, phase differences $\Delta\varphi_{12}$ and $\Delta\varphi_{13}$ were measured between antenna-pairs 1–2 and 1–3 respectively. The zenith angle Z_E and the azimuth angle A_Z are then given by the following equations:

$$A_Z = \arctan\left[\frac{(\Delta\varphi_{13}x_2 - \Delta\varphi_{12}x_3)}{(\Delta\varphi_{12}y_3 - \Delta\varphi_{13}y_2)}\right], \tag{10.5}$$

and

$$Z_E = \arcsin\left[\frac{\lambda}{2\pi}\frac{(\Delta\varphi_{12}x_3 - \Delta\varphi_{13}x_2)}{(x_3y_2 - x_2y_3)\sin A_Z}\right]. \tag{10.6}$$

Note that more advanced interferometric array distributions may be used, such as a five-receiver system like that used for meteor studies, and as shown in Figure 10.5 (e.g., *Beres et al.*, 2010).

Sample plots determined from interferometric analysis are shown in Figures 10.12 and 10.13, and details are discussed in the captions.

The polar plots are most easily interpreted if it is assumed that fixed range corresponds to fixed height, so it can be assumed that the lightning strokes are close to overhead, but this may not be a valid assumption. In Figure 10.13, the center point is the location of the radar, the maximum circle has a radius of 10 km. The small circles around the center point represent the area illuminated at the given range by the main beam of the transmitting antenna. Each single data point is the position determined at time steps of 4.8 ms. The lower diagram in Figure 10.13 shows the corresponding amplitude variation as function of time. Only echoes exceeding the amplitude limit given by the

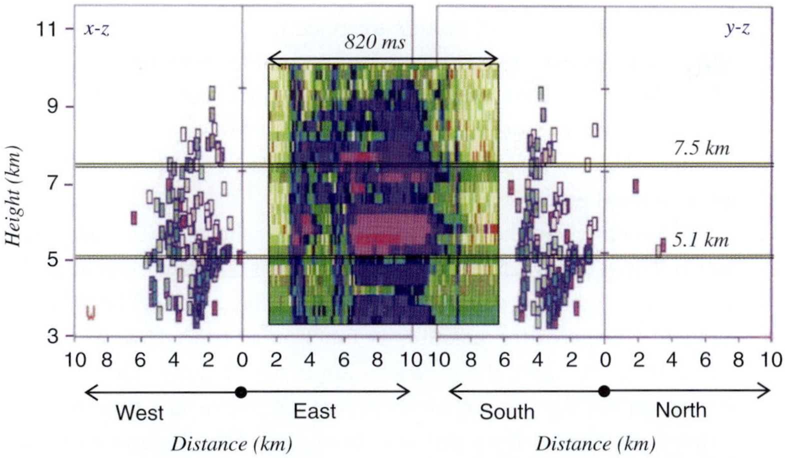

Figure 10.12 Interferometric measurements of lightning with the Chung-Li radar. Plots of the echo locations in the west–east (x–z) and north–south (y–z) planes are shown on the left and right respectively, while the central image shows the range–time–intensity plot, using a similar color scheme to that in Figure 10.9. The display covers a time of 820 ms. Ranges of 5.1 and 7.5 km are indicated (which are used in Figure 10.13). The data were from a thunderstorm on 13 August 1995, and were recorded at 16:13:08 local time.

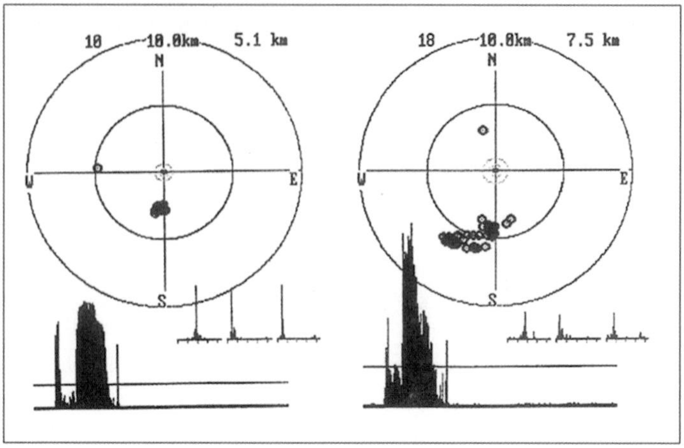

Figure 10.13 Lightning occurrences in a 2-D polar form, for ranges of 5.1 and 7.5 km. The largest circles in each polar plot represent a radius of 10 km. These plots match the corresponding heights indicated in Figure 10.12. The upper panels show the polar plots, while the graphs at the bottom show overlaid plots of amplitude vs range for the period, for one receiver. The other small insets show the temporal evolution of the phase as a function of time on each of the three receivers, for the ranges specified on the corresponding polar plots. The small circles in the middle correspond to radar beam-widths, and are discussed in greater detail in the body of the text.

straight line are plotted in the polar plot. The three graphs right below the polar plot show the echo phase distributions (0–360 °) of the echoes in each of the three receiver channels.

Several "branches" of scatter are evident in the polar plots of Figure 10.13, where the different branches show as clusters of accumulated locations. The temporal development also tells us that the scattering point frequently moves deterministically along a certain track. This indicates that the scattering point follows a certain structure on the lightning channel. The increase of the noise level due to sferics causes some scatter in the positions of individual points within the clusters.

It is apparent in the polar plots that the echo positions are not in the main antenna beam, but at about 2.5 km to 7.5 km south of the radar. It is also observed that the position of the lightning echo at 7.5 km range has a hook-like shape which results from a gradual change of the position as function of time. However, events can also be seen in the data set where the positions jump significantly. The explanation is that the echo region on the lightning channel is moving during the time the echo persists, and that echoes can also be from different branches of the lightning channel. This gives rise to the interpretation that the lightning scattering process on 52 MHz is likely to be highly aspect sensitive, which in turn lets us assume that the scattering is from overdense ionization in the lightning channel. Signal comes from one specular point until such time as the channel is no longer specular (it either dies or re-orientates), and then another channel at another location may form or re-orientate so that it becomes specularly reflecting, giving the appearance of a jump.

These studies are earlier ones from the 1990s, and more detailed studies will no doubt be possible with improved technology in the future. Nevertheless, we have described here some of the important technical set-up details which may help define future studies.

10.8 Studies above the mesosphere – plasma and ionospheric processes

While the focus of MST radar research is clearly on the mesospheric regions and below, the region above the mesospause can still be usefully studied with these radars. In this region the air is less dense than in the mesosphere, electron and ion densities are markedly greater, and molecular viscosity increases to large values – sufficient to highly damp turbulence above 110 km altitude. The lower thermosphere and upper mesosphere have much in common, both being affected by the same waves (planetary and gravity waves) propagating from below. Turbulent processes at around 85 to 95 km, and gravity wave events, can couple the regions (e.g., *Hocking*, 1996c), so that the combined region is often referred to as the MLTS (mesosphere lower thermosphere) region, or simply the MLT. The electron densities in the D-region, which is part of the mesosphere, are also sufficient to allow a mixture of (weak) plasma processes and neutral dynamics to combine in unique ways: PMSE are but one example. Of course, to some extent the ionspheric radar at Jicamarca in Peru was the birthplace of MST radar, so the link between these fields of research is not surprising. Many of the larger MST radars, like the MU radar, PANSY and MAARSY (all discussed in Chapter 6) were even designed and built to allow some level of ionospheric research. The same is true for the EAR (equatorial atmospheric radar) in Indonesia (e.g., *Fukao et al.*, 2003). The feedback between researchers in the two fields is often substantial, and in 2014 the first joint iMST1 workshop, which included reseachers from both the ionospheric (i) and MST areas, was held in Sao Paulo, Brazil. Some researchers work quite comfortably in both regions, and the two communities frequently collaborate.

In this section, then, we will touch on some areas where MST research has useful impact on ionospheric studies, and conversely. The review will not include all available research, but will highlight a few cases of significant interest.

Imaging is one area where the research areas coalesce. Such techniques are useful in ionospheric studies of diverse phenomena like the so-called "150 km echoes" (discussed shortly), equatorial spread-F, range-spread meteor trails, quasi-periodic (QP) echoes (e.g., *Yamamoto et al.*, 1991) and the equatorial electrojet, among others. The overlap with MST studies was noted several times in Chapter 9.

Not surprisingly, following the discussions in Chapter 9, two types of imaging techniques have been used. The first is direct beam-steering, the second interferometry. Beam-steering studies have been reported by, among others, *Kelly and Heinselman* (2009), and interferometric studies have been reported by *Hysell and Chau* (2006) and others. Each of these references contains multiple links and references to other related studies.

Examples of beam-steering systems include the PFISR (Poker Flat incoherent scatter radar in Alaska) and the RISR (Resolute Bay incoherent scatter radar in Northern Canada), each of which is an example of an AMISR (advanced modular incoherent scatter radar), as discussed by *Kelly and Heinselman* (2009) and *Semeter et al.* (2009), among many others. These are multi-million dollar steerable active phased-array radars with over 1000 separate antennas. Transmitters are small units of typically 1–2 kW peak power, each of which is attached to a separate antenna, much like the PANSY design discussed in Chapter 6. Each unit also has its own transmit-receive switches and front-end receivers, and all can be phased independently.

Both meteors (e.g., *Sparks et al.*, 2009) and PMSE (e.g., *Nicolls et al.*, 2009 and references therein) have been studied using beam-steering with PFISR. *Hysell and Chau* (2006) developed interesting new ideas to study the ionosphere using interferometry with only a small number of antennas. Such a situation does not lend itself to unique determinations of the scattering field, due to limited degrees of freedom, so these authors turned instead to statistical inversion theory, endeavoring to utilize more fully all available information in the data inversion. The imaging algorithm is based on an implementation of the MaxEnt method developed for radio astronomy; more details can be found in *Hysell and Chau* (2006).

Some large MST radars, especially the MU radar, have even been used for E- and F-region studies. We should note that the scattering irregularities in the E- and F-region result from plasma instabilities and magnetic field-aligned irregularities. This is in contrast to the irregularities in the MST region, which are generally consequences of the structure of the neutral atmosphere. (For more experimental discussions on ionospheric processes, see *Röttger*, 2014). Partly as a result of these different generation mechanisms, E- and F-region irregularities usually have large spectral width (small coherence) and high mean velocities. Thus, in contrast to MST radar applications, complementary pulse-codes cannot be applied. Codes used for ionospheric studies are either single pulse, multi-pulse or Barker codes. Antennas and other system configuration are basically comparable to MST radars, except that often the beam-directions have to be perpendicular to the Earth's magnetic field, since some of the most interesting ionospheric irregularities are field-aligned.

In some cases, the upper MST region and the lower ionosphere can be controlled by similar processes, such as gravity waves and other propagating phenomena arriving from the troposphere. Often experimental and theoretical concepts developed in ionospheric research find application in MST studies, and vice versa.

Examples of F-region studies include both equatorial and mid-latitude investigations (e.g., *Woodman*, 2009; *Fukao et al.*, 2003; *dePaula and Hysell*, 2004). With regard to the E-region, a variety of phenomena are of interest, including sporadic E, blanketing E_S (ESB), the equatorial electrojet (EEJ), kilometer scale waves, counter electrojet (CEJ), and day/night-time plasma irregularities (e.g., *Patra et al.*, 2002, 2009; *Chau and Kudeki*, 2006a, and references therein). E-region studies using SuperDarn can also have overlap with MST studies (e.g., *St-Maurice et al.*, 2007). For more detailed discussions on these special plasma instabilities we refer the reader to *Kelley* (1989); *Fukao and Hamazu* (2014).

10.8.1 150 km echoes

Unusual echoes which have coherent characteristics, but which come from unexpectedly high altitudes of typically 150 km, also exist in the upper ionosphere. These are referred to as "150 km echoes." At the time of writing of this book, these echoes were unexplained, though we will give some new insights in this section. Sample publications regarding these phenomena include *Chau et al.* (2009); *Chau* (2004); *Chau and Kudeki* (2006a); *Choudhary et al.* (2004); *Tsunoda and Ecklund* (2007, 2008); *Nicolls et al.* (2009); *Patra and Rao* (2007); *Patra et al.* (2008), and references therein. Typical range–time–intensity plots are shown in Figure 10.14.

There are two main features that need to be explained with regard to these echoes. The first is determination of the reason that scatterers with scales of 3 m exist at all at these heights; they are clearly not due to classical neutral atmospheric turbulence, since the Kolmogoroff microscale at these heights is of the order of hundreds of meters, so any 3 m scales would be deep within the viscous range, and heavily damped. Most likely the scatterers are generated by plasma processes.

The second curious aspect of the scatterers is the necklace shape, and the strings of regions of high and low scatter. This is in fact quite easily explained, and was first presented by *Hocking* (2014). The structure is proposed to be due to highly Doppler-shifted

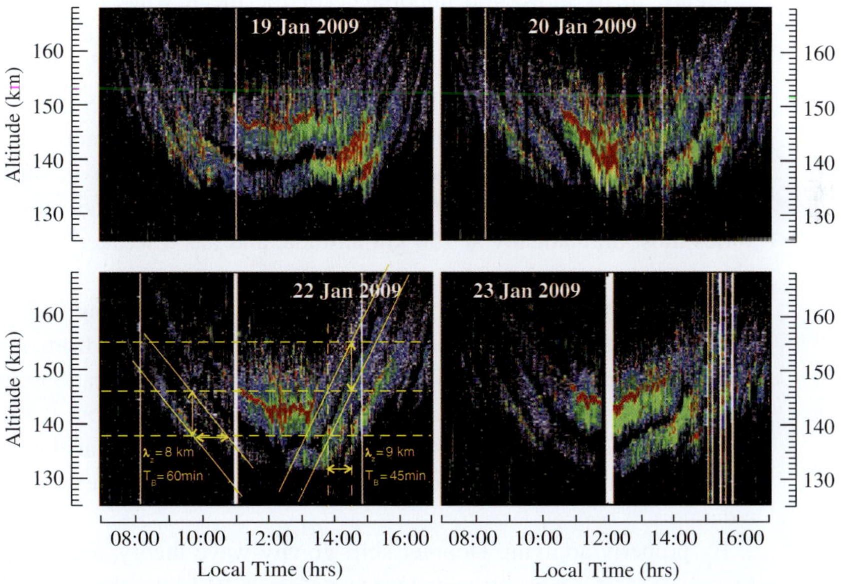

Figure 10.14 Range–time–intensity plots of so-called 150 km echoes recorded at Jicamarca, Peru on the dates indicated. Since these echoes look to some extent like a necklace, they are sometimes referred to as "necklace echoes." A series of yellow lines are shown in the bottom left-hand plot; these are used to describe characteristics about the echoes. Working on the assumption that these lines are gravity-wave wavefronts, appropriate ground-based periods and vertical wavelengths are shown. These are discussed further in the main text (adapted from *Chau and Kudeki*, 2013).

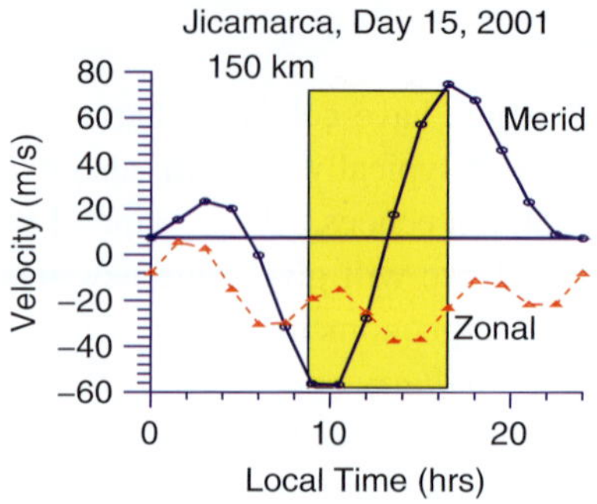

Figure 10.15 Winds determined for 150 km altitude over Jicamarca in January 2001. The HWM07 model from *Drob et al.*, 2008, was used for these calculations.

gravity waves, combined with a diurnally varying mean wind (a tide) and a filtering effect due to viscosity at these upper altitudes.

Figure 10.15 shows the "normal (undisturbed) winds" over the Jicamarca site in January 2001, as determined from *Drob et al.*, 2008, Version HWM071308E_DWM07B104i. Note in particular the rapid wind change in the meridional component from early morning to later afternoon, with a change from −60 to 60 m/s. This occurs around the same time that the necklace structure starts and finishes.

Figure 10.16 shows the importance of the mean wind in radar measurements of the structure. Assuming that the lines of enhanced scatter correspond to some line of enhanced instability parallel to the wavefronts of the wave, and that the gravity wave moves in a generally northward or southward direction, then the direction of the north-south wind defines the upward or downward motion of the fronts relative to the radar, largely irrespective of the speed of the wave, since the mean wind speed significantly exceeds the horizontal phase speed of the wave. The mean wind varies most dramatically at 150 km altitude, and has a weaker variability at the lower heights.

Since the radar echoes move downward in the morning, when the winds are most southward, then the wavefronts must slope upward from north to south, as shown in Figure 10.16(b). The movement of the echo locations moves upward in the afternoon, as the winds change to strongly northward, but the orientation of the wavefronts stays the same as in the morning. Assuming that the wave is generated below 150 km altitude, this indicates that the waves are generated to the north of the radar.

By properly applying Doppler shift gravity-wave theory, and using the winds from Figure 10.15, it is possible to deduce from Figure 10.14 the characteristics of the relevant gravity wave, such as the phase speed of the wave and its intrinsic period, where "intrinsic" refers to the characteristics seen if moving with the mean wind. However, exact (non-relativistic) theory must be used, since the wind speeds far exceed the horizontal phase speed of the wave. In other words, the expression

$$(c_\phi + v_{\mathrm{hor}})T_{\mathrm{intr}} = \lambda_{\mathrm{grd}} = c_{\mathrm{grd}}T_{\mathrm{grd}} \tag{10.7}$$

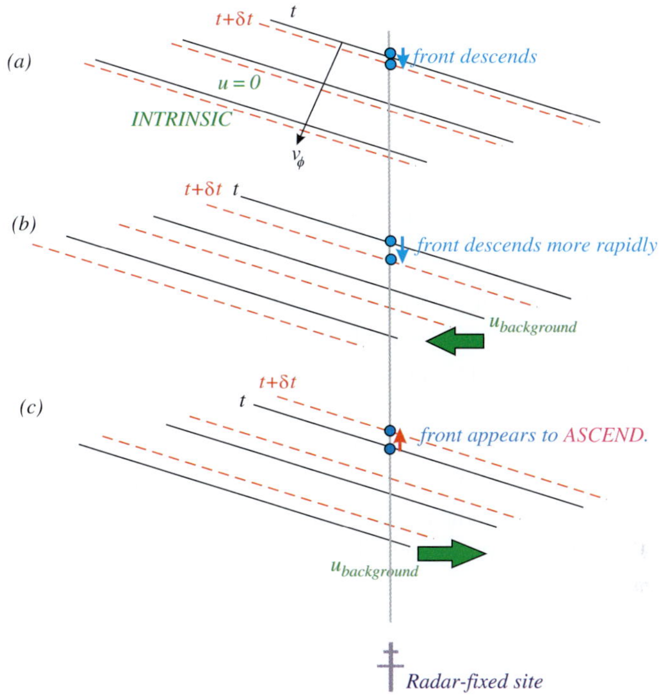

Figure 10.16 Wavefronts of a gravity wave with a phase velocity moving down and to the left, corresponding to a group velocity moving up and to the left. The front is shown at successive times t and $t + \delta t$ for situations of (a) no mean wind, (b) a strong mean wind to the left, dragging the wavefronts more rapidly to the left (enhancing the rate of downward motion seen by the radar) and (c) a strong mean wind to the right, actually overpowering and reversing the apparent vertical motion of the wavefront as seen by the radar.

must be used, where c_ϕ is the intrinsic horizontal phase speed, v_{hor} is the background wind-speed, T_{intr} is the intrinsc wave period, λ_{grd} is the ground-based wavelength, c_{grd} is the horizontal wave velocity as measured from the ground, and T_{grd} is the wave-period as measured from the ground.

In addition, the standard dispersion relations for the intrinsic wavelengths, phase-speeds and frequencies of a gravity wave may be used, as well as the fact that the vertical and horizontal wavelengths will be the same in both the ground-based and intrinsic reference frames. More details about these relationships can be found in Chapter 11.

One question that remains is the issue of the spacing between successive fronts. It seems similar on all days, with a vertical wavelength of around 8–10 km (see the lower left-hand graph in Figure 10.14), and a ground-based period of around an hour or so (60 mins in the morning and 45 mins in the afternoon in the figure discussed). Why are the structures so similar? This in fact turns out to be a property of the value of the viscosity at these heights.

In order to understand this, we need to look at the theory of diffusion. Often this quantity is ignored in gravity-wave theory, but at altitudes of 150 km, it is a major damping mechanism. If we consider a perturbation, the lifetime of that entity can be calculated in terms of its size using similitude analysis. The perturbation expands due to diffusion, and eventually smears out. In the case of a wave, the time taken for the perturbation maxima to expand into the perturbation minima can be found. In the latter case, this damps the wave amplitude, and in the extreme case, destroys it. However, in the intermediate case, when the wave exists as a damped oscillation, the diffusion actually changes the wave speed. For example, in calculating the speed of sound in air, adiabatic processes are assumed, and the result $c_s = \sqrt{\gamma \frac{P}{\rho}}$ is produced, where P is the background pressure and ρ the mean air density. Here, γ is the ratio of the specific heat at constant pressure divided by the specific heat at constant volume. Once diffusion dominates, the adiabatic assumption cannot be applied, and the speed of sound approaches $\sqrt{\frac{P}{\rho}}$. In the paragraphs that follow, the effect of diffusion on the gravity wave periods will be discussed.

Consider the standard diffusion equation, and let us examine first the case of a perturbation in a highly non-linear environment. Then we may assume that the perturbations are comparable to the mean quantities, so the standard diffusion equation for the density ξ of a selected constituent, viz.

$$\frac{\partial \xi}{\partial t} = \nu \nabla^2 \xi, \tag{10.8}$$

becomes, via similitude analysis,

$$\frac{\xi}{t} \sim \nu \frac{\xi}{x^2}, \tag{10.9}$$

(where t is a “typical” time scale and x is a “typical” spatial scale), leading to

$$t \sim \frac{x^2}{\nu}. \tag{10.10}$$

For a gravity wave, the wavelength is the perpendicular distance between two successive wavefronts, and this is very similar to the vertical wavelength λ_z. We can take x to be one quarter of λ_z, since the maxima and minima diffuse into each other and we can regard the destruction of the wave to occur when the maxima and minima “meet in the middle.” We replace t with the wave period, giving

$$T \sim \frac{1}{16} \frac{\lambda_z^2}{\nu}. \tag{10.11}$$

An alternative approach may be adopted. If high levels of nonlinearity are *not* assumed, but rather it is assumed the perturbations are due to waves of the form $\underline{\xi} = \underline{\xi_0}\{\exp \mathrm{i}(kx + \underline{m}z - \omega t\}$, (where $\underline{m}$ may be complex), then substitution into (10.8) gives

$$\mathrm{i}\omega\xi \approx \nu(\underline{m}^2 + k^2)\xi. \tag{10.12}$$

Note we have assumed a 2-D spatial system, with the x-axis aligned along the direction of wave propagation. Taking $|\underline{m}| \gg k$ (i.e. the vertical wavelength λ_z is much less than the horizontal one, as is generally true for gravity waves), this leads to

$$\underline{m} = \sqrt{\mathrm{i}\frac{\omega}{\nu}}, \tag{10.13}$$

which has solutions

$$\underline{m}_{1,2} = \pm\left[\frac{1}{\sqrt{2}}\sqrt{\frac{\omega}{\nu}} + \frac{\mathrm{i}}{\sqrt{2}}\sqrt{\frac{\omega}{\nu}}\right]. \tag{10.14}$$

This analysis is very similar to that presented in *Hocking et al.* (1991), although the purposes differ. The second part is the most important for our purposes, since it shows the wave damps with a $\frac{1}{e}$ scale of

$$\zeta_z = \frac{2\pi}{\mathrm{Im}\{\underline{m}_{1,2}\}} = 2\pi\sqrt{2}\sqrt{\frac{\nu}{\omega}}, \tag{10.15}$$

where ζ_z is also the vertical wavelength of the wave, λ_z. Then for a period T, we may write

$$T = \frac{1}{4\pi}\frac{\lambda_z^2}{\nu}. \tag{10.16}$$

This is very similar to equation (10.11), except here the constant is $\frac{1}{4\pi}$ and in the earlier equation it was $\frac{1}{16}$.

We need to remember that the amplitude of the wave falls by $1/\mathrm{e}$ over approximately one wavelength.

However, it needs to be remembered that as a gravity wave propagates upward, its amplitude may grow as the density decreases, with the amplitude increasing according to

$$A \propto \mathrm{e}^{z/(2H)}, \tag{10.17}$$

where H is the scale height which may be taken to be of the order of 10 km, depending on temperature and molecular species. Hence although the gravity waves with periods of the order of T given by (10.16) are attenuated as they propagate upward with a $1/e$ height of λ_z, they also continue to grow in amplitude due to the density decrease with increasing height. While the amplitude decrease will dominate for wavelengths less than about 20 km, the waves will be non-zero for longer than equation (10.16) might indicate. So for a parameter that defines significant suppression of the wave amplitude, we need to use a large vertical wavelength. If we consider say 3 times λ_z, then the suppression is e^3 times, which will represent significant suppression, notwithstanding any amplitude increase due to scale-height effects. So using this as our criterion, we write that the approximate cutoff period is

$$T_c = \frac{1}{16}\frac{(3\lambda_z)^2}{\nu} = \frac{1}{2}\frac{\lambda_z^2}{\nu}. \tag{10.18}$$

The new constant is $\frac{1}{2}$; since this is an approximation, we round it to unity. So we finally write the expression defining the transition to heavily damped waves as

$$T_c = \frac{\lambda_z^2}{\nu}. \tag{10.19}$$

Taking the formula

$$\nu \approx (3.8 \pm 0.4) \times 10^{-7}\frac{T_a^{0.69}}{\rho} \tag{10.20}$$

as the expression for the kinematic viscosity as a function of height, where $\mathcal{T}_a$ is the atmospheric temperature and ρ is the atmospheric density, (from *Rees*, 1989), we may then deduce a profile of allowed maximum wave periods as a function of vertical wavelength and height. Data concerning mean atmospheric densities can be found at the MSIS/NASA web-site. Longer period waves than those so-deduced will be heavily damped.

Hence allowed wave periods lie between the Brunt–Väisälä period (typically 10 to 20 min in the 100–160 km region) at the low end, and the periods defined above at the upper end. For a wave of vertical wavelength of 5 km, the viscosity-defined upper period is around 10 min, so waves with a vertical wavelength of 5 km and less are essentially non-existent. Waves with a vertical wavelength of 10 km are constrained to periods between the Brunt–Väisälä period and about 1–2 hours.

Therefore at heights of 140–160 km, the joint filtering effects of the BV period (due to the properties of gravity waves) and viscosity, act to allow waves with periods of typically 0.2 to 1–2 hours, and vertical wavelengths of around 10 km, to dominate. Longer period waves must have correspondingly longer vertical wavelengths – well over 10 km. These periods ($\sim$ 1–2 hours) and wavelengths ($\sim$ 10 km) match well with the observed values shown in Figure 10.14. Note that there is no need that there be any special properties associated with the source region – only that it launches a broad spectrum of waves propagating southward from a region north of the radar. The waves launched from this region can in fact be quite isotropic – it only matters that the region is north of the radar. Filtering takes care of the rest of the mechanism for wave-selection.

We will not extend this discussion further – sufficient information has been presented to show that the necklace pattern is quite predictable by this theory. The key to understanding the necklace shape is the Doppler shifting due to the large tides, as shown in Figure 10.15, and the key to understanding the limited ranges of wavelengths and timescales is the filtering effects due to the limits at short periods due to the Brunt–Väisälä frequency, and limits to the period at the upper end due to viscous damping. Other patterns will emerge if the tidal motions do not produce large daily changes from strong positive to strong northward winds over the course of the day.

10.8.2 Other ionospheric research

While the D-region is dominated by neutral particles, and ionization is relatively weak compared to the E-region and higher, some powerful incoherent scatter radars can make direct use of the ionized portion of the region. Of course even for VHF radars the potential refractive index gradient depends on the electron density (Equation 7.71, Chapter 7), so electrons are even important for VHF scatter, but use of the incoherent scatter is a different technique. In this context, E- and D-region incoherent spectra can sometimes be measured with large radars like the Arecibo radar (e.g., *Mathews*, 1984b), and this has recently been achieved by the Jicamarca radar as well (e.g., *Chau and Kudeki*, 2006b).

Another experiment of some interest is so-called "heater experiments." In these, the ionosphere is heated by high-power MF and HF waves, and then observed by other radars and optical means; these were discussed briefly in Section 10.2.3.

The "differential absorption experiment" (DAE) is another useful ionospheric method to probe the D-region and mesosphere. The technique allows measurements of electron densities and collision frequencies. In recent years, this experiment has been re-introduced on occasion (e.g., *Singer et al.*, 2008; *Holdsworth et al.*, 2002; *Vuthaluru et al.*, 2002) and we will shortly give a brief review of this procedure.

The behavior of localized ionized plasmas in the upper D-region is also of some interest, since such processes are an interesting hybrid of electromagnetic forces and neutral drag effects. One such type of localized ion/electron plasma is a meteor trail. As a probe into this area, *Hocking* (2004b) has studied the spatial variation of the rates of diffusion of meteor trails. The expectation was that the diffusion rates might be larger at particular angles in the sky, depending on the strength and alignment of the magnetic field lines, especially at heights above typically 92–94 km. Then again, it was possible that drag forces due to the neutrals might damp such effects, and nothing might be seen at all. But neither situation occurred. The decay rates did indeed show zones of the sky above 92–94 km altitude where the typical decay rates were higher than those at other zenithal-azimuthal coordinates. However, these positions were *not* locked but varied diurnally. The zones of maxima typically followed an elliptical path overhead, rotating 360 degrees in azimuth during the course of a day. This was a surprise, and the cause is still unexplained. Tidally controlled electric fields in the lower thermosphere have been proposed as one possible cause.

10.9 D-region scatter and the differential absorption experiment

One interesting technique that was developed quite early in the studies of the upper atmosphere was the DAE.

The earliest serious observations of D-region radar echoes were performed by *Gardner and Pawsey* (1953). They used an intensity-modulated electron beam to expose a trace on a moving photographic film to record the data. The history of DAE was discussed briefly in Chapter 2.

Once these observations had been made, it was natural to see if it was possible to use this information to determine the electron density in the D-region as a function of height and time. Gardner and Pawsey used their original studies to introduce a technique which permitted them to measure *electron densities* in the 60–85 km altitude region, and this was the DAE.

10.9.1 DAE (the differential absorption experiment)

The DAE technique is based on the birefringent nature of the ionosphere. It draws on the fact that there are two characteristic waves, O and X, with different absorption and reflection properties, as discussed in Chapter 3. Many subsequent studies were performed using this technique, and a review with a large number of references to these later works can be found in *Manson and Meek* (1984).

Here, we will briefly outline the method.

A linear wave propagating along the z-axis through a vacuum has an electric field which varies in the manner

$$\underline{E}(z,t) = \underline{E}_o e^{i(k\,z-\omega t)}. \tag{10.21}$$

However, if the medium is not a vacuum, but has a spatially varying refractive index n, then we must write

$$\underline{E}(z,t) = \underline{E}_o e^{i(\int_0^z \underline{k} dz' - \omega t)}, \tag{10.22}$$

where $\vec{k}$ is the wavenumber, with scalar value $\underline{k}$, and $\underline{k}$ may be complex. If such a wave moves from a transmitter at the ground to a reflection level at height z, then we can write its electric field at this level as

$$\underline{E}(z,t) = \frac{1}{z}\underline{E}_o e^{-i\omega t} e^{i\int_0^z \omega n_R/c\, dz'} e^{i\int_0^z \omega (in_I)/c\, dz'}, \tag{10.23}$$

where $\underline{n} = n_R + in_I$. Note that the $\frac{1}{z}$ term accounts for the fact that the radiation is transmitted from a point transmitter.

The key term is the last one, because it describes the attenuation of the signal due to ionospheric absorption. It is the ratio of these absorption terms which we will be using to determine the electron density.

If the signal is then reflected or scattered back to the ground, then the total amplitude attenuation due to ionospheric absorption *alone* is $e^{-2(\omega/c)\int_0^z n_I\, dz'}$ where the extra factor of 2 is introduced because the radiation must pass the same trajectory twice. Note the n_I must be positive for absorption to occur. We will write $n_I = \chi_a$, and then the absorption term is

$$e^{-2(\omega/c)\int_0^z \chi_a\, dz'}. \tag{10.24}$$

The actual intensity received at the ground also depends on the type of scattering process (reflection or volume scatter), and the reflection coefficient (for reflection) or backscatter cross-section (for volume scatter). It also depends on the depth of the fluctuations in electron density at the height of scatter. Gardner and Pawsey developed their theory in terms of specular partial reflection. We will let the ratio of the reflected amplitude to the incident amplitude at the height of reflection be denoted as $R_{X,O}$.

In the DAE, the experimenter transmits alternately X mode and O mode radiowaves, and then looks at the ratio of the amplitudes received at the ground. The effects of the range dependence cancel out in forming this ratio, and we are left with the following expression for the ratio of the amplitudes received at the ground as

$$\frac{A_X}{A_O} = \frac{R_X e^{-\int_0^z 2\kappa_X dz'}}{R_O e^{-\int_0^z 2\kappa_O dz'}}, \tag{10.25}$$

where $\kappa_O = \frac{\omega}{c}\chi_{aO}$ and $\kappa_X = \frac{\omega}{c}\chi_{aX}$. (Note that we are using a different notation than Gardner and Pawsey, with the roles of R and A interchanged.)

We now need to make some observations about $R_{X,O}$ and χ_{aX} / χ_{aO}.

First, the ratio R_X/R_O is equal to $(\Delta n_X)/(\Delta n_O)$, the ratio of the (relatively small) changes in refractive index at the reflection or scattering level. Provided that the refractive index is moderately close to 1, it turns out that this ratio is dependent much more strongly on the mean collision frequency than on the refractive index, and we can generally assume that in the D-region the ratio is independent of the electron density N. Thus at any height we can make a pretty good estimate of this ratio.

Then we may take logs and re-write the previous equation as

$$\int_0^z 2(\kappa_X - \kappa_O)dz' = \ln\left(\frac{R_X}{R_O}\right) - \ln\left(\frac{A_X}{A_O}\right), \tag{10.26}$$

and finally we take the derivative with respect to z to obtain

$$2(\kappa_X - \kappa_O) = \frac{\mathrm{d}}{\mathrm{d}z}\ln\left(\frac{R_X}{R_O}\right) - \frac{\mathrm{d}}{\mathrm{d}z}\ln\left(\frac{A_X}{A_O}\right). \tag{10.27}$$

Now it turns out that for small N and collision frequencies typical of those in the D-region, the quantity $(\kappa_X - \kappa_O)$ is proportional to the electron density N. The constant of proportionality is dependent on the collision frequency ν_{ec}.

Thus if we have a suitable program for determining the refractive indices (preferably using the full Sen–Wyller formulation e.g. see Equation (3.128) in Chapter 3, or the modified Appleton–Hartree equation that precedes it), then we can:

(i) Determine R_X/R_O at any height (assuming this to be largely independent of N); and
(ii) Use our measured values of $\frac{\mathrm{d}}{\mathrm{d}z}\ln(\frac{A_X}{A_O})$ to determine
(iii) $(\kappa_X - \kappa_O)$.
(iv) Finally, we again use our Sen–Wyller computer program to convert $(\kappa_X - \kappa_O)$ to a value for N.

Note that we need to have good estimates of $\nu_{ec}(z)$ in order to apply this method; one expression often used is

$$\nu_{ec}(z) = (6.4 \pm 0.4) \times 10^7 \mathrm{ps}^{-1}, \tag{10.28}$$

where p is the pressure in millibars. (e.g., see *Manson and Meek*, 1984).

In any real application, it is the user's responsibility to produce plots of the terms R_X/R_O and $(\kappa_X - \kappa_O)$ as a function of height z, for different values of N. The height dependence comes about principally through the collision frequency term ν_{ec}, so this should be fairly well known. As a rule one finds that these curves lie on top of each other for different values of N – or at least up to values of typically 4×10^3 cm^{-3}, and up to heights of about 85 km. This therefore gives the user an idea of the sort of ranges of height and electron density over which the technique is valid.

Although amplitude ratios are the most commonly used, it is also possible to use the rate of change of phase difference between the modes as a function of height to deduce the electron density. The principles are similar to the ones above, and the following expression is obtained:

$$N(z) = \left[\frac{\mathrm{d}}{\mathrm{d}z}(\varphi_{RX} - \varphi_{RO}) - \frac{\mathrm{d}}{\mathrm{d}z}(\varphi_X - \varphi_O)\right] \Big/ \left[\frac{2\omega}{c}(\beta_X - \beta_O)\right], \tag{10.29}$$

where $n_{RX} = \beta_X N$ and $n_{RO} = \beta_O N$, $n_{(RX,RO)}$ being the real parts of the refractive index. The phase changes due to the reflection process (φ_{RX} and φ_{RO}) need to be found theoretically, though are usually identical or have a difference of π^c.

We will not pursue this further here.

In disturbed conditions, such as electron precipitation events, $N(z)$ can exceed the limits of 4×10^3 cm^{-3} discussed above, and the method begins to break down.

In early work, the X and O modes which were transmitted were generally purely circularly polarized radiation – usually achieved by transmitting on perpendicularly orientated dipoles fed with a phase difference of $\pm\pi/2$ radians. Circular polarizations can only strictly be used to describe these modes at the magnetic poles; at all other latitudes the X and O modes are elliptical (and linear at the equator). Whilst it turns out to be a reasonable approximation to use circular polarizations even at mid-latitudes, it is better if the true elliptically polarized waves are transmitted. More recent techniques do this, and also record the full details of the received ellipse, so that X and O mode contributions can be more readily resolved. Such devices are called *polarimeters*. These more complex methods are discussed in *Von Biel* (1977); *Meek and Manson* (1981); *Manson and Meek* (1984). An alternative procedure to produce the polarization ellipse is to measure the so-called Stokes parameters. This involves measurement of the full received intensity, two linear components at 90° to each other, and one mode of circular polarization. Using these, a complete picture of the polarization of the incoming waves can be constructed, and the characteristic modes can be developed from there. A more complete discussion can be found in *Hecht and Zajac* (1974), Section 8.12.1, pages 266–271. Key parameters that are measured are denoted as S_0, S_1, S_2, and S_3. These are also sometimes denoted as I, Q, U, and V. Use of Stokes parameters is common in radio astronomy, but has not been generally adopted in MST work.

To date, we have effectively assumed that the mechanism which produces the backscattered radiation is a form of specular reflection. But in reality this is not always true, and volume (turbulent?) scatter may also be important, especially above 80 km altitude. Some authors developed a modified form of the DAE assuming volume scatter (again see Manson and Meek for details), but there was always a certain amount of controversy about these modified equations.

In addition, the DAE works best when it is known that the radiation is backscattered (or reflected) from directly overhead. If volume scatter is important, and wide beams are used, the presence of obliques can severely contaminate the signal.

For the above reasons, it therefore is of great importance if the nature of the scattering/reflection mechanism is properly understood. This has been discussed in many sections of this book, but the issue is not fully resolved.

The DAE was a very popular experiment in the 1960s and 1970s, but fell from favor for some decades. It has recently been rejuvenated, (e.g., *Singer et al.*, 2011). Examples of recent results can be found therein, and a selection of graphs from that reference are shown in Figure 10.17, including comparisons with other techniques, examples of daily variability, and special events (a proton event is shown). It should be noted that these data were taken with a 3 MHz radar with a relatively narrow beam, which helps resolve issues in distinguishing between specular and turbulent scatter. Use of a narrow

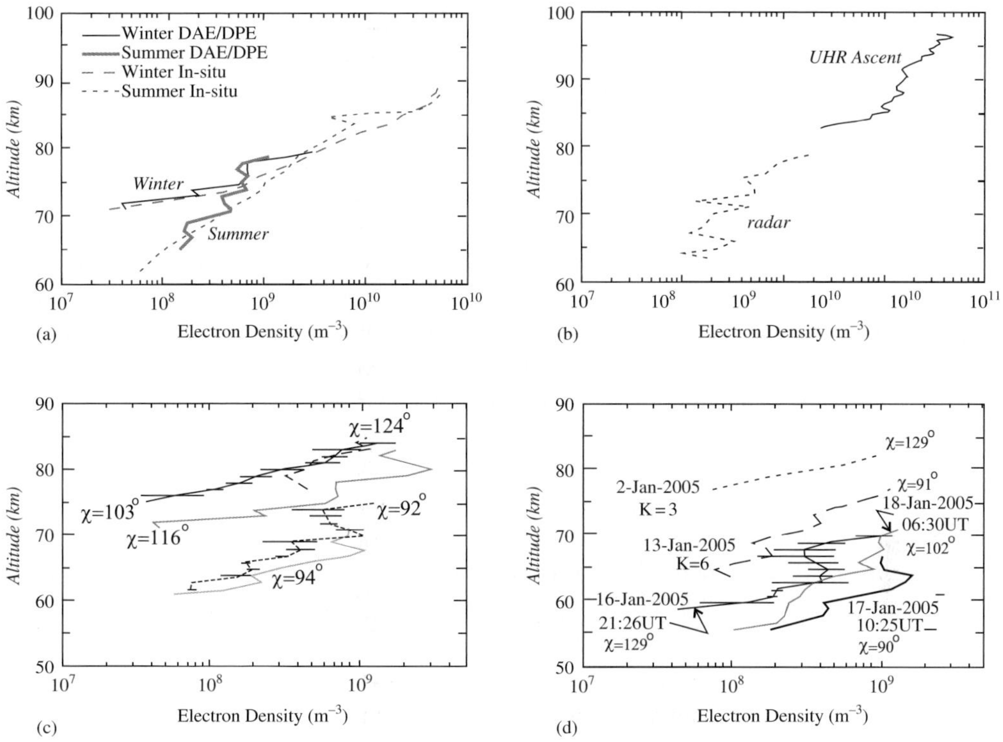

Figure 10.17 A variety of measurements of electron density compared to other techniques. See the text for details. All graphs were taken from *Singer et al.* (2011). (Reprinted with permission from Elsevier.)

beam improves the capabilities of the radar over wide-beam radars. The radar is located at Saura on the island of Andoya in northern Norway, which is located close to the Andenes rocket range.

The graphs of Figure 10.17 involve a considerable amount of description, and so in order to keep the captions manageable, the figures are described below.

Figure 10.17 (a) shows the mean electron density profiles from DAE/DPE measurements (full lines) and electron density profiles using in-situ radiowave propagation measurements (broken lines) at Andenes during comparable ionospheric conditions in summer and winter. Summer-time solar zenith angles were around 58 °, and winter-time angles were around 120 ° (i.e. the sun was below the horizon). Values of 10.7 cm solar flux were around 100 in summer and 90 in winter. The radiowave propagation experiments involved reception of a HF signal transmitted from the ground by a receiver on the rocket, and monitoring the rate of rotation of the plane of polarization of the wave. This rate of rotation relates to the rate of change of electron density as the rocket flies through the D-region. Faraday rotation was discussed in Chapter 3. Several frequencies were used, varying between 1 and 8 MHz.

Figure 10.17(b) shows electron density profiles from radar observations and an in-situ impedance probe. The impedance probe measured the UHR (upper hybrid resonance) frequency of the ionospheric plasma. The UHR technique is least reliable below 90 km due to the need for collision-frequency corrections below 90 km. The DAE/DPE method works best when electron densities are less than about 5×10^9 m^{-3}. Although the datasets do not overlap, a degree of continuity is obvious. Rocket data were recorded during a special Japanese campaign at Andenes (denoted DELTA) on 13 Dec 2004, at 00:34 GMT. The radar data were recorded between 01:02 and 01:08 GMT on the same day. The solar zenith angle at the time was 130° (night-time).

Figure 10.17(c) illustrates the diurnal variation of electron density in winter during quiet solar and geomagnetic activity at Saura, northern Norway. The data were taken on 4 December 2004, at the following solar zenith angles and times (solar zenith angle is the angle of the sun from overhead, and a value in excess of 90° means the sun has set). Hence these data refer to post-sunset conditions. All times are GMT. $\chi = 92°$, time = 11:02 to 11:11. $\chi = 94°$, time = 12:02 to 12:29. $\chi = 103°$, time = 15:02 to 15:11. $\chi = 116°$, time = 17:02 to 17:41. $\chi = 124°$, time = 19:02 to 19:17. A general decrease in electron density is evident as the sun sets lower. Representative error bars are shown on the $\chi = 92°$ profile.

Figure 10.17(d) demonstrates the usefulness of DAE measurements during a proton event, showing the increase in electron density during the event. A solar proton event occurred centered on 17 January, 2005. Profiles are shown for the days prior to, during, and after the event. For reference, values expected during conditions of mildly disturbed conditions (K = 3 and 6) are also shown. Representative error bars are shown on the 16 January profile (from *Singer et al.* (2011)).

10.9.2 Passive radar

One interesting development – which seems to have stalled of late – is the idea to build a radar without a transmitter, since the transmitter is one of the most expensive components of a radar. The concept also avoids the need for obtaining radio licences. The principle is based around using other transmitters as the source of the radiowaves, especially commercial radio-stations. The idea is to record the ground-wave of the transmitter, and also record signal coming from the atmosphere/ionosphere, and to deconvolve the latter against the former, thereby determining signal components due to atmospheric scatter. The objective is very computationally demanding, since the ground-wave and the signal from the air and ionosphere are intermingled, and the deconvolution problem is very difficult. Interferometry is required if any sort of height-resolution is to be obtained. The method has been discussed in more detail by *Sahr and Lind* (1997), and we will not elaborate further in this text.

10.10 Astronomical applications

MST radars have also been used in astronomical applications. In the main, these are applications that use astronomical features for processes like calibration, rather than for astronomical studies per se. The main exception to this rule is in regard to meteor studies, where meteor orbits and mass-flux studies have been made, but these have been discussed earlier in this chapter, and so will not be reconsidered here. Key examples of astronomical applications for calibration purposes include calibration of the MU radar using the moon (see Chapter 5, and *Sato et al.* (1989)), as well as calibration of the polar diagram shape using galactic radio sources (e.g., *Carey-Smith et al.*, 2003). A more recent, and equally interesting application, is the use of signals fron astronomical sources to measure phase offsets of antenna elements within large phased arrays (e.g., *Chau et al.*, 2014). However, these types of applications are beyond the primary topics of this book, and we leave the interested reader to pursue them through the references given.

10.11 Final comments

Many other topics have been presented in the literature on more abstract and diverse applications of MST radar, but the linkages with the mesosphere, stratosphere, and troposphere are weaker, and for this reason we have left them out of this review.

11 Gravity waves and turbulence

11.1 Introduction

Wind motions in the atmosphere can cover a wide range of temporal and spatial scales. They may include variations on annual, seasonal, monthly, daily, hourly, and minute scales, down to scales of seconds. Spatially, motions may cover global scales down to synoptic (continental-sized), meso- (city-sized) and microscales (e.g., *Ahrens*, 1999, Figure 10.1). Windprofiler radars can study all of these scales. However, we cannot possibly cover all of them in this chapter. Larger scale motions (including planetary waves and tides) can be studied well with satellites and in-situ instruments carried by balloons and rockets, as well as numerical computer models. While profilers can also contribute here, it is at the smaller scales that windprofilers really make their best contributions. We will therefore concentrate in this chapter on synoptic, mesoscale and microscale motions, with strongest emphasis on the last two. The primary focus will be on height regions where MST radars have made a significant contribution, restricting discussion to the troposphere, lower stratosphere (below 25–30 km), and the upper mesosphere and lower thermosphere (60 to 100 km altitude). Other height regimes will be discussed primarily in their relation to these regions.

In meteorology, atmospheric mesoscales motions refer to spatial scales between a few kilometers and one or two hundred kilometers, and temporal scales of the order of minutes to a few hours. In the troposphere, mesoscale events include thunderstorms, tornadoes, and various types of local circulations like land and sea breezes and valley breezes. Typical synoptic scale events include hurricanes, high and low pressure systems, and frontal systems.

Some or all of these events may be quite familiar to many readers. In fact, these events are only really dominant in the lowest few kilometers of the atmosphere. MST atmospheric radars can be used to investigate these phenomena, and this has been done in the past (e.g. *Strauch et al.* (1984); *Gage et al.* (1991a); *Webster and Lukas* (1992); *Teshiba et al.* (2001) (and references therein); *Röttger and Larsen* (1990); *Hooper and Pavelin* (2003), among others). Since this book has a special chapter on meteorology, these events will not be pursued here in any detail.

When MST radars are used for studies to heights of ten kilometers and more, and even into the upper atmosphere, a different class of mesoscale/synoptic scale motion becomes apparent. This motion is often well organized and can propagate over large distances. The motions observed belong to the class of *gravity waves*, which have characteristics

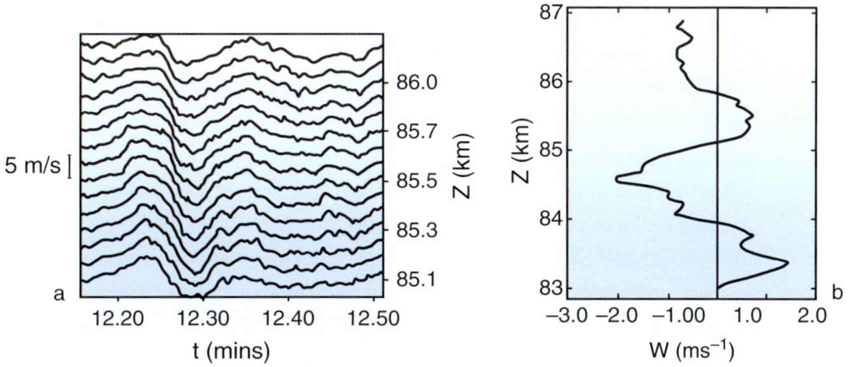

Figure 11.1 Example of vertical velocity fluctuations w' as a function of (a) time t, and (b) height z, in the mesosphere (from *Czechowsky et al.*, 1989).

uniquely different from the other events of similar scales discussed above. The first part of this chapter will be devoted to these gravity waves, since windprofiler radars have made an enormous impact on our understanding of these phenomena. We will not discuss all aspects of gravity waves, since this would consume too much of this book. Rather, after some theoretical introduction to the nature of gravity waves, we will discuss aspects to which radars can make useful contributions. For more detailed discussions, the reader will often be referred to the paper by *Fritts and Alexander* (2003), which contains a higher level discussion of many gravity wave aspects.

Gravity waves are atmospheric waves which propagate in the free air. Individual waves cause velocity, pressure, and density fluctuations of a sinusoidal nature, and these waves carry momentum flux and energy with them as they propagate. The waves are observed as quasi-sinusoidal oscillations in wind velocity and temperature as a function of both time and height; an example of the vertical velocity component of the wind is shown in Figure 11.1.

Gravity waves are generated in a multitude of ways. The simplest types to visualize are the waves generated by flow of air over a corrugated boundary. The air begins to oscillate over the corrugations and in so doing generates propagating waves. The waves have the curious property that the phase velocity and group velocity are almost perpendicular to each other, and when the phase group velocity is upward, the phase velocity is downward. If such a wave loses no energy as it propagates, then the exponential decrease of atmospheric density with increasing height results in an exponential increase of amplitude with increasing height. Thus waves which start at ground level with velocity amplitudes of a few centimeters per second attain amplitudes of several tens of meters per second at altitudes of 70 km and above. In reality, the waves do in fact lose some energy and momentum as they rise in height, the most dramatic losses occurring in the regions above 70 km altitude.

Gravity waves are also called buoyancy waves, and sometimes *internal buoyancy waves* or *internal gravity waves*, to emphasize the fact that they are not surface waves. It is generally believed that these waves are the main cause of middle atmosphere wind and temperature fluctuations with periods of a few hours and less and even down to a few minutes. There is a school of thought which suggests that at least some of these motions are due to two-dimensional turbulence. This is a minority view, however, at least for upper atmospheric motions. The issue is more contentious in the lower atmosphere (e.g., *Lilly*, 1983, 1989).

Turbulence studies represent another area where radars can and have been used productively. This refers to random and quasi-random fluid motions which are highly non-linear, and are usually the end product of a series of successively more and more non-linear events. Both two- and three-dimensional turbulence exist, but our primary focus will be on the three-dimensional form, which occurs around scales less than a few km. Two-dimensional turbulence (if appplicable) occurs on larger scales, typically 10 km to several thousand km. The three-dimensional form of turbulent motions show a cascade from large scales to smaller ones, and at the smallest scales, where internal velocity shears are the greatest, the end result is dissipation of kinetic energy of motion to molecular kinetic energy, and hence to heat. Two-dimensional turbulence reverse cascades, from smaller scales of a few kilometers out to scales of hundreds and even thosands of kilometers. In the upper troposphere and above, the major process for production of three-dimensional turbulence is breakdown of wave motions, and especially includes breakdown of gravity waves, which is partly why these topics are included in the same chapter. Turbulence has already been discussed to some extent in earlier chapters, especially Chapter 3, where the interplay between radar backscatter and the turbulence spectrum was discussed. Discussion of turbulence in this chapter will therefore be somewhat abbreviated.

The next sections begin with a brief account of the importance of gravity waves. Following this, a simple descriptive model is shown which demonstrates how a gravity wave is generated. Following this section, the fluid equations will be examined in more detail to see (briefly) how the waves arise. Then the implications and practical ramifications of these waves in the atmosphere will be examined. A discussion about turbulence, including its production and impact, will follow that.

11.2 Gravity waves

11.2.1 The importance of gravity waves

In the 1960s and 70s, many scientists regarded gravity waves as simply idle curiosities. Many people concurred that they had no real impact on atmospheric motions at any sort of important scale. This attitude has now changed.

Gravity waves carry momentum flux and energy between different points in the atmosphere. If a gravity wave is generated at a source region (for example, a mountain) and dissipates somewhere else, this amounts to a transfer of energy and momentum from the first point to the second. When energy and momentum are deposited in the dissipation region, they can alter the mean flow. Meteorologists have realized in the last decades that computer models have not achieved their full potential of predicting mean winds, or making good forecasts, partly because they had not been including gravity wave generation and dissipation in their models. A considerable amount of effort is being turned towards proper parameterization of gravity waves in meteorological models. We will discuss parameterization briefly later in this chapter.

In the upper regions of the atmosphere, especially the stratosphere and mesosphere, gravity waves have huge effects. For example, in the mesosphere, it has been found that by including gravity waves in computer models, the directions of the mean winds have in some cases been reversed relative to the expected wind directions deduced without inclusion of gravity waves. This phenomenon occurs because the waves deposit momentum and energy as they dissipate; the momentum and energy are deposited in the atmosphere and affect both the mean winds and the temperature distribution. In particular, as discussed in Chapter 1, Section 1.3.4, the momentum deposited by the waves acts to alter the mean eastward (zonal) winds, in some cases even reversing them compared to the direction expected on the basis of radiative theory alone. This changed zonal wind then drives a north-south (meridional) wind circulation, and this in turn causes upwelling at some latitudes and down-welling at others. As a result, the temperatures in the upper atmosphere are also substantially altered relative to their radiative situation (see Chapter 1, Figure 1.23). One dramatic result of note is the phenomenon of the polar summer mesopause becoming much colder than even the winter polar mesopause. More detail about the way in which these waves modify the mean wind circulation and thence the latitudinal temperature distribution can be found in *Holton* (1982) and *Holton* (1983), and will be elaborated upon later in this chapter. Computer models that take gravity waves into account produce wind and temperature results that agree better with observations than do the older predictions.

Gravity waves are ubiquitous, and demonstrate a surprising degree of consistency in wave activity across different latitudes, time, and geographical location. This consistency refers to both total spectral power and the form of the wave spectra. From summer to winter, the variation in amplitude is less than a factor of 3, and the variations as a function of latitude are also relatively small. It has been proposed that the waves can be described by a common "universal" spectrum of almost invariant shape and amplitude, and the term universal spectrum is often encountered in the literature (*Van Zandt*, 1982). This apparent universality will be considered herein.

Thus gravity waves are major transporters of momentum and energy in the atmosphere. They represent a significant portion of the small-scale atmospheric variability. For all these reasons, and more, they are important phenomena for study.

11.2.2 A simple description of the generation of gravity waves

In order to see how gravity waves arise, first consider waves on the surface of the ocean. In that case, an interface exists between a region of high density and a region of low density, as illustrated in Figure 11.2. When the sea water is displaced vertically, an excess of mass above the mean level of the water results, so that the surface begins to fall. The water surface then overshoots its equilibrium position and suffers a restoring force due to the water around it. This results in the downward motion of the surface

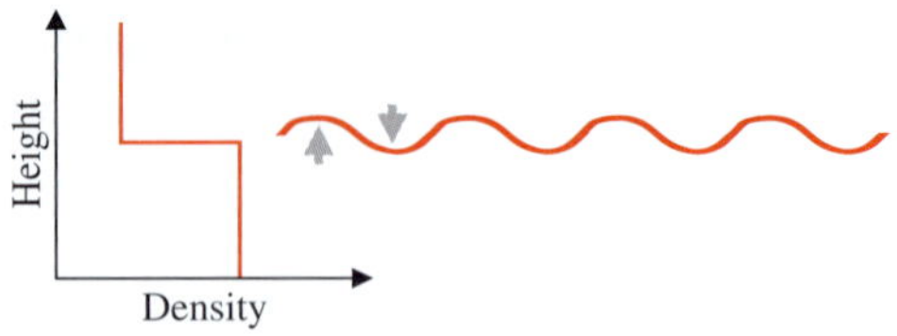

Figure 11.2 Density as a function of height for the special case of the interface between the sea and the air.

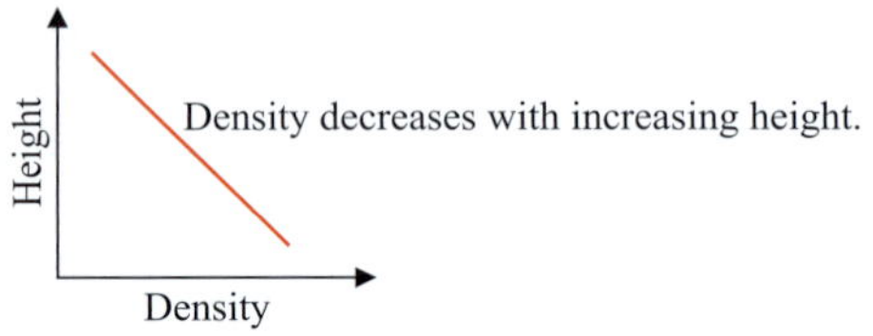

Figure 11.3 Approximation of the variation of density with height in the atmosphere.

slowing, and eventually stopping, and then returning towards the equilibrium position. As it approaches the equilibrium, however, it is travelling at non-zero speed. It passes through, to be displaced once again above the equilibrium level. Hence an oscillation is established, which constitutes the motion within the wave. This motion propagates along the surface.

The major factor in the above mechanism is that the water is more dense than the air. However, a discontinuity in density is not necessary in order to have this effect. Even if the density decreases monotonically with height, as shown in Figure 11.3, a similar process can eventuate for a displaced parcel of air, leading to regular oscillations in three dimensions in the air. Waves of this type exist throughout the atmosphere. They are actually called *internal gravity* (or *internal buoyancy*) waves, to distinguish them from waves which occur at a surface interface.

These waves can be generated in a variety of ways – flow over mountains is one very common production method. We will give an example of such a generation mechanism shortly.

The waves grow larger in amplitude as they propagate vertically. This is because they conserve energy per unit volume and as they travel higher in the atmosphere, the neutral density decreases. However, the kinetic energy per unit volume is just (1/2 × density × velocity squared), so that if this is to be maintained as the density decreases, then the velocity of oscillation of the wave must increase. Hence the waves get larger in amplitude as they move upwards.

In the troposphere, the velocity amplitudes of these waves are a few centimeters per second; in the mesosphere, they can have amplitudes of several meters per second. Their effect becomes very pronounced in the mesosphere and they impact on many areas of mesospheric dynamics. They can be important at all levels, however.

It was mentioned above that flow over mountains is one way by which gravity waves can be generated. An easy way to physically understand the characteristics of gravity

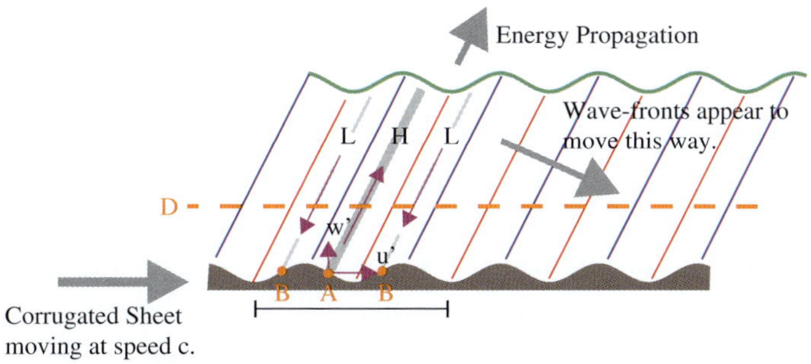

Figure 11.4 Movement of a corrugation through a body of air (adapted from *Dunkerton*, 1981).

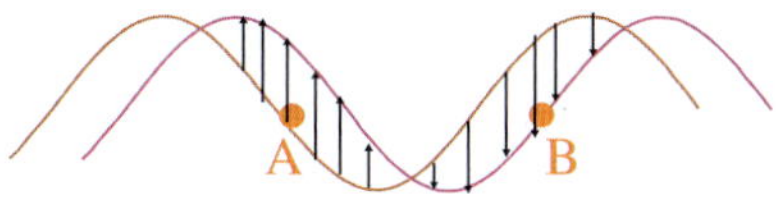

Figure 11.5 Position of corrugation at two successive time steps.

waves is to use a similar configuration to a mountain, Specifically, consider a regularly repeating series of mountains. Visualize a corrugated sheet moving through a fluid. This is shown in Figure 11.4.

As the corrugation moves through the air, the air is forced to oscillate, and a wave propagates away. The way in which the wave is created can be seen by looking at the motions induced on particles of air adjacent to the corrugation. Concentrate on the region shown by the horizontal bar at the base of the diagram. Remember that the things described for this section repeat themselves over and over as one follows along the corrugations.

To begin our understanding of the above figure, first look at Figure 11.5. Consider that the red curved line represents the corrugations at one instant in time, and the purple line represents the corrugations a short time later. Then the surface of the corrugation will appear to have moved upward (and forward) in the region shown by the upward arrows, while in the region shown by the downward arrows, the surface appears to have fallen.

Now let us return to Figure 11.4. At the point A, the corrugated surface will appear to be moving upward with increasing time, forcing the air in this region up and forward. Thus the air particles will achieve a velocity component w' which is upward, and a forward component u'. This will "push" the air along the line of the purple arrow shown sloping up and to the right. As a result, the air will be somewhat compressed, increasing the pressure along this line (indicated by the grey sloping region), so this whole grey area will be a region of *high* pressure. Conversely, at the regions indicated by B, the corrugations will appear to be falling away, so the air in this region will also fall. Hence the w' and u' air-velocity components here will be downward and to the left. At the same time, the air will spread out along the grey broken lines, resulting in a lowering of pressure. Thus the broken grey lines will be regions of *low* pressure.

The vertical displacements will follow the corrugations. Parcels of air that have maximum displacement will be those which move along the *blue* lines shown sloping upwards and to the right. These parcels will cool as they rise, by adiabatic cooling, so these blue lines will also be regions of the wave where it is coldest (which is why they have been drawn in blue). Parcels of air which lie on the *red* lines will be displaced *downwards*, since they line up with *low points* in the corrugations. Because they are displaced downwards, they will be *heated* by adiabatic processes, and thus parcels along this line will be the warmest of all (which is why these lines have been drawn in red).

Hence, by using simple logic, it has been possible to describe how the displacement, velocity, pressure, and temperature vary within a buoyancy wave. Furthermore, it is expected that the whole structure will move along with the corrugations. To an outside observer, the wave-fronts (i.e. the red, grey, and blue sloping lines) will appear to move in the direction of the grey arrow sloping down and to the right (labelled *Wave-fronts appear to move this way*). This arrow therefore indicates the apparent propagation direction of the wave. The rate at which these fronts appear to move is called the *phase speed* of the wave.

At the same time, remember that the air at the point A was forced up and to the right. Thus this is the direction in which energy propagates. Therefore the waves carry energy up into the atmosphere as indicated by the short arrow at the top of the diagram. This arrow is labelled *Energy Propagation*. (This case is different to the case of other waves, like ocean waves or sound waves, where the energy direction and the phase direction are usually the same.)

Note that all of this has been described from the frame of reference of the air and it is seen that a "propagating wave" is produced. If the situation is viewed from the frame of reference of the *corrugations*, however, the whole system would appear *stationary*. This is why lee-waves over mountains appear to be stationary – a ground-based observer is essentially observing them from the frame of reference of the corrugations. However, relative to the *air* (which is moving), the lee-waves are in fact propagating phenomena.

Figure 11.6 shows how the horizontal and vertical wavelengths of a gravity wave are defined. The blue lines represent phase fronts of temperature minima, and the red lines represent phase fronts of temperature maxima. The *horizontal wavelength* is defined by the horizontal distance between identical wave-fronts, while the *vertical wavelength* is defined by the vertical distance between identical wavefronts. Note that the distances between the temperature minima have been used in this diagram, and the same values

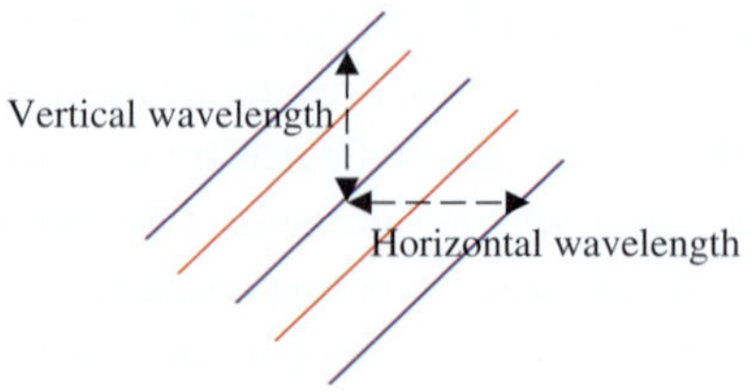

Figure 11.6 Definition of horizontal and vertical wavelengths.

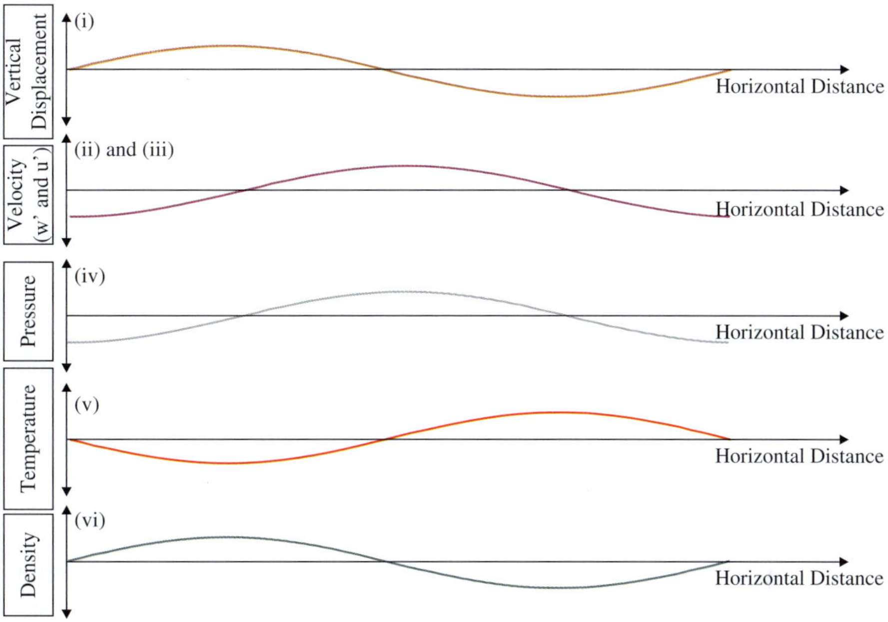

Figure 11.7 Relation between perturbations in various parameters shown in Figure 11.4 as a function of horizontal distance, specifically along the line labelled D. This was for the case that the corrugations pushed into the air. In the case of a gravity wave, it is more likely that the air would blow against static corrugations, such as a mountain range. So if the reader would like to see how the perturbations look in the direction of the wave propagation, the figures above need to be read from right to left, and the directions of the horizontal wind need to be reversed. Hence in the second figure, the horizontal and vertical winds will be 180° out of phase. Actual phases are also shown in Equations (11.18) to (11.22), as will be seen shortly.

would have been found if the distances between the maxima had been used. The diagram is *not* to scale. As a general rule, the horizontal wavelengths are much larger than the vertical ones. Typical vertical wavelengths are in the range of a few hundred meters out to 10 or 20 km, while typical horizontal wavelengths can reach hundreds and thousands of kilometers.

Now consider taking a slice through one oscillation of the wave at the level D shown in Figure 11.4. Figure 11.7 shows temporal variations for the parameters of: (i) vertical displacement, (ii) vertical velocity, (iii) horizontal velocity, (iv) pressure, and (v) temperature. We also show the variation of (vi) density of the displaced parcel of air relative to the mean background temperature at the same height. The form followed can be simply understood from the knowledge that the air is most dense when the air is coldest, so that the density will be least when the temperature is highest, and conversely. Note that these graphs show only the *form* of the oscillation – they do *not* say anything about the magnitudes.

We now need to consider an extra point, and that relates to the *stability* of the atmosphere. Figure 11.8 shows the motion of a parcel of air moving under the influence

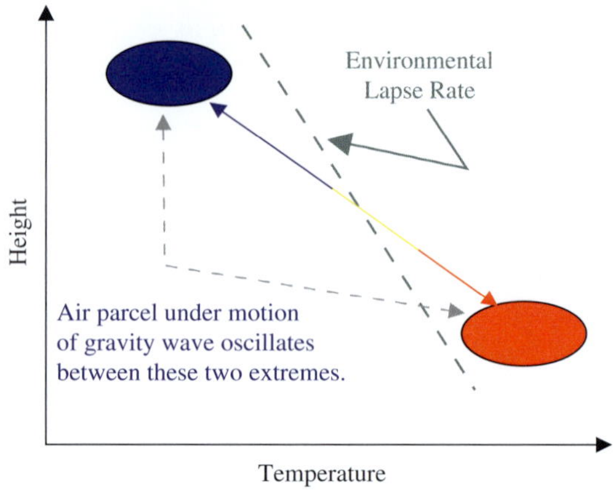

Figure 11.8 Demonstration of oscillation of a parcel of air in the atmosphere.

of a buoyancy wave, plotted in terms of height and temperature coordinates. The parcel moves adiabatically, which means it heats and cools according to the adiabatic lapse rate. At the top of its oscillation, it is coolest, and also has its lowest density, due to adiabatic expansion as it rises. However, the thing that matters most is not its density as a function of position in its oscillation, but its density relative to its surrounding environment. On the same graph, the *environmental lapse rate* has been drawn. Note that at the top of its oscillation, the parcel of air is cooler than its surroundings. This is consistent with a *stable* atmospheric situation. If the environmental lapse rate is *unstable*, gravity waves *cannot* exist, since a parcel of air which is displaced vertically will continue to rise, rather than oscillate. At the bottom of its motion, the parcel of air is warmer (and less dense) than its surroundings, so it is forced to rise again.

Hence, the existence of gravity waves in the atmosphere requires stable conditions. Furthermore, if a gravity wave *is* generated in a stable region, and propagates upwards, it can encounter an unstable region at a greater height. If this happens, the wave cannot propagate through this region, and will reflect from it, or deposit some of its energy. In a similar vein, if a gravity wave encounters a region where its horizontal phase speed equals the speed of the background wind, it will also not be able to propagate above that height. This latter situation is a bit like the gravity wave's version of a sonic boom – the wave grows to very large amplitude because it is moving along with the air, feeding the oscillation in phase, and eventually gets so big that it destroys itself. In the process, it dumps energy and momentum into the atmosphere. Processes of wave dissipation (like the ones described above) are very important when we come to discuss energy and momentum transport in the atmosphere. These matters will be further discussed later in this chapter.

We mentioned above that generation and propagation of gravity waves require stable atmospheric conditions. With suitable mathematics, it is possible to make even more precise statements. For example, if the environmental lapse rate is known, it is possible to determine a special period of oscillation called the *Brunt–Väisälä period* (or sometimes (rarely) the Väisälä–Brunt period. Numerically, this period can be found from Equation (1.58), viz.

$$\tau_B = \frac{2\pi}{\sqrt{\frac{g}{T}\left[\Gamma_a - \Gamma_e\right]}}, \tag{11.1}$$

where Γ_a is the adiabatic lapse rate and Γ_e is the environmental lapse rate. This period is special because it corresponds to the period which a parcel of air would have if it were displaced in the air and then allowed to oscillate freely. It is impossible for a gravity wave to have an intrinsic period (meaning the period measured if the wave was observed from the frame of reference of the air in which the wave is embedded) that is less than the Brunt–Väisälä period. Typically, the Brunt–Väisälä period in the troposphere is about 10 minutes. In the stratosphere, it is about 5 minutes. It is possible in some situations that Γ_e is greater than Γ_a (i.e. that the temperature falls off as a function of height more rapidly than the adiabatic lapse rate); in this case, such parcel oscillation is not possible.

As with any wave, the horizontal wavelength, the intrinsic period of the wave and the intrinsic horizontal phase speed (i.e., the phase speed measured by an observer moving with the mean wind) c_{int} are related through the usual relation $c_{\text{int}} = \lambda/T$, where λ is the horizontal wavelength and T is the intrinsic wave period.

However, gravity waves are different to many other waves in that they have special laws which dictate how their direction of propagation relates to their period and wavelength. We can define the direction of propagation using Figure 11.6, which shows some sample phase fronts of the wave. Recall that the blue lines represent phase fronts of temperature minima, and the red lines represent phase fronts of temperature maxima.

The ratio between the vertical wavelength and the horizontal wavelength uniquely defines the orientation and propagation direction of the wave. For waves with periods much larger than the Brunt–Väisälä period (say greater than 1 hour) and less than a few hours, the following relation approximately applies:

$$\lambda_x/\lambda_z = T/\tau_B, \tag{11.2}$$

where λ_x is the horizontal wavelength, λ_z is the vertical wavelength, and T is the wave period. Note that $\lambda_x > \lambda$ and $\lambda_z > \lambda$, as seen in Figure 11.6. This relation changes for very long and very short periods: a more exact relation will be given shortly.

It is also possible, using mathematical relations, to determine relationships between the magnitudes of the temperature oscillations, velocity oscillations, displacements, and density and pressure amplitudes. These relations are called "polarization relations." We will derive them shortly.

In the above discussions, air flow relative to the ground was considered as a major source of gravity waves. However, gravity waves may also be generated by other processes. These include atmospheric convection (e.g., see *Alexander and Rosenlof*, 1996), and the violent turbulent motions at the tops of thunderstorms. Other sources include shear excitation and geostrophic adjustment, and wave–wave interactions (see *Fritts and Alexander*, 2003, for more details). They can also be generated by frontal systems, and even during solar eclipses (e.g., *Ball*, 1979). Not all of these sources will be discussed just yet, but we will note that gravity waves are extremely prolific in the atmosphere and exist at all times and all places. They are as common as waves on the surface of the ocean, although they cannot always be discerned visually. They can, however, frequently be detected with radar.

11.2.3 The fluid dynamical equations of motion

We have now given a descriptive view of gravity wave generation and propagation. At this point, a more mathematical perspective will follow.

We begin with the equations of fluid motion. The equations considered are the standard fluid-dynamical equations, viz.

$$\frac{D\vec{u}}{Dt} + 2\vec{\Omega} \times \vec{u} - \vec{g} + \frac{1}{\rho}\vec{\nabla}p - \nu\nabla^2\vec{u} = 0, \tag{11.3}$$

$$\frac{D\rho}{Dt} = \frac{1}{c_s^2}\frac{Dp}{Dt}, \tag{11.4}$$

$$\frac{D\rho}{Dt} + \rho\vec{\nabla}\cdot\vec{u} = 0, \tag{11.5}$$

$$\frac{D\Theta}{Dt} = \frac{\kappa}{\rho}\nabla^2\Theta, \tag{11.6}$$

where D/Dt represents differentiation following the motion (also called the *advective derivative*). The first equation is three-dimensional, and is the Navier–Stokes equation. It is essentially Newton's second law for a fluid parcel. The second equation is a combination of the first law of thermodynamics, Newton's second law and the continuity equation, the third is the continuity equation, and the fourth represents Fick's law for heat transport. The total velocity is $\vec{u}$, the density is ρ, $\vec{\Omega}$ is the Earth's angular velocity (with magnitude Ω), $\times$ means cross product, $\vec{g}$ is the acceleration due to gravity = $[0, 0, -g]$, p is the pressure, c_s^2 is the speed of sound squared, $\vec{\nabla}$ represents the gradient differential operator and "$\cdot$" means the dot product. Θ represents potential temperature, κ the heat diffusion coefficient, and ν is the kinematic viscosity coefficient (which is of course just the molecular viscosity coefficient divided by the atmospheric density). Note the inclusion of the Coriolis pseudo-force, as discussed following Equation (1.7) in Chapter 1, which arises due to our decision to view the flow from a non-inertial frame of reference fixed to the surface of the rotating Earth.

Of course the basic sets of equations can be expressed in different ways. Our choice is somewhat different to the choice made by *Hines* (1960), for example, although both express the same physics. Our choice is designed to more clearly express the internal gravity-wave branch of the equation solutions. There are in fact two separate sets of solutions, namely an acoustic branch, which contains the commonly occurring *sound waves*, and a second branch – the internal gravity wave branch – which is the subject of interest here. Acoustic waves can exist at angular frequencies *higher than* $\omega_a = \frac{\gamma g}{2c_s}$, which is referred to as the acoustic cutoff (γ being the usual ratio of specific heats). Buoyancy (gravity) waves exist at angular frequencies *less than* the Väisälä–Brunt frequency, ω_B (Equation (1.52), discussed in Chapter 1). Typically the acoustic cutoff period is in the range of 4 to 5 mins, while ω_B is often greater than 5 mins, so it is possible that a "forbidden zone" may exist between the two regimes. Acoustic waves are largely longitudinal (compressional) and are evident most strongly in the pressure and density fluctuations, while gravity waves have both a transverse and a longitudinal component, but with the transverse component being dominant.

Both acoustic waves and gravity waves play roles in atmospheric dynamics, but from the point of view of energetics and forcing, it is the internal gravity wave branch that dominates. Infrasound, with periods of seconds to one or two minutes, does appear to have importance on occasions, as seen in Chapter 10. But with periods typically from microseconds to a few tens of seconds, acoustic waves are also less amenable to radar studies than their gravity-wave cousins. In addition, the reader should be well familiarised with their characteristics, since they are commonly a topic in first or second-year university physics courses. Therefore, due to the fact that they are energetically of less importance, the fact that they are less amenable to gravity wave studies, and the fact that they should already be well understood by the reader, we will concentrate on gravity waves for the rest of this chapter.

In the following sections, it will be briefly shown how Equations (11.3) to (11.6) are modified for dealing with gravity waves, and the important parameters required in any useful study will be defined.

11.2.4 The approximations of the equations of motion for gravity wave studies

The preceding equations describe the motions for all fluids. However, the equations are highly nonlinear, mainly due to the term $\vec{u} \cdot \vec{\nabla}$ in $\frac{\mathrm{D}}{\mathrm{D}t}$. Hence for tractable solutions, it is often useful to simplify them somewhat to study particular types of flow. Simplification takes various forms, depending on the nature of the problem. For wave solutions which are not too excessive in amplitude, it is normal to assume solutions proportional to $\mathrm{e}^{\mathrm{i}(\vec{k}\cdot\vec{x}-\omega t+\phi)}$. Examples of such solutions include buoyancy (gravity) waves, tides, planetary waves and vortical modes (e.g., *Houghton*, 1977; *Gossard and Hooke*, 1975; *Hines*, 1960; *Dong and Yeh*, 1988; *Yeh and Dong*, 1989). All such oscillations carry momentum and energy from source regions (very often in the troposphere) to the stratosphere and mesosphere, and so are important for the dynamics of the upper regions. Gravity waves (as mentioned already) are particularly important: as discussed in Chapter 1, and as will be discussed further in this chapter, the momentum flux deposited in the mesosphere by these waves is enough to significantly modify the mean flow state at mesospheric heights (*Matsuno*, 1981; *Holton*, 1982, 1983).

In the case of gravity waves, one assumes solutions of the form $\Psi = \Psi_0 \mathrm{e}^{\mathrm{i}(\vec{k}\cdot\vec{x}-\omega t+\phi)}$ (where Ψ can be any one of the velocity components, the pressure, the density, or temperature, and may be complex) and then substitutes these assumed forms into the equations of motion, retaining only first-order terms. ω refers to the ground-based angular frequency of the wave, and $\vec{k}$ is the wavenumber vector (= $[k, l, m]$), while the position $\vec{x}$ is written as (x, y, z) (Cartesian coordinates).

Here, we adopt a complex-number notation, following the type-I strategy discussed in Chapter 3, Section 3.3.1, in which we represent the variables as complex numbers, even though they are not, because this represents a convenient and easy way to deal with phase variations. As per usual with this convention, all calculations are done using complex numbers, but at the completion, real parts are taken (either explicitly or implicitly).

Within these equations, we will assume that the angular frequency ω is purely real, and likewise we will assume no horizontal amplitude change, so we take k as real. However, the wave may grow or decay as it rises up in height, so we represent this by allowing m to be complex, written $\underline{m}$.

The effects of viscosity will be ignored, and a mean wind of zero will be assumed (for now) for simplicity. It will also be assumed that the wave propagates in the x–z plane.

In addition, only first-order perturbation terms will be retained. The fact that we are considering only these term will be indicated by placing a small "hat" over the variable, viz,

$$\text{a first order perturbation in } \underline{\Psi} \text{is represented by } \hat{\underline{\Psi}}. \tag{11.7}$$

Note this is a little different to Ψ', which would represent all perturbative terms, rather than just the linear ones. Then the following non-hydrostatic equations result:

$$-\mathrm{i}\omega\hat{\underline{u}} - f_c\hat{\underline{v}} = -\mathrm{i}k\hat{\underline{\psi}}, \tag{11.8}$$

$$-\mathrm{i}\omega\hat{\underline{v}} + f_c\hat{\underline{u}} = 0, \tag{11.9}$$

$$-\mathrm{i}\omega\hat{\underline{w}} + \hat{\underline{r}}g = -(\mathrm{i}\underline{m}\hat{\underline{\psi}} - \hat{\underline{\psi}}/\overline{H}), \tag{11.10}$$

$$-\mathrm{i}\omega\hat{\underline{r}} - \hat{\underline{w}}\,\omega_B^2/g = -\mathrm{i}\frac{\omega\hat{\underline{\psi}}}{\overline{c_s^2}}, \tag{11.11}$$

$$-\mathrm{i}\omega\hat{\underline{r}} + \mathrm{i}k\hat{\underline{u}} - \hat{\underline{w}}/\overline{H} + \mathrm{i}\underline{m}\hat{\underline{w}} = 0. \tag{11.12}$$

Here $\overline{c_s^2}$ is the mean squared speed of sound at the height of the wave, and f_c is the Coriolis parameter, equal to $2\Omega\sin\theta$, where θ is the latitude. (Note that since Ω is expressed as *radians per second*, so f_c is also in radians per second. This can be confusing since we normally associate the symbol f with cycles per second (Hz), but this is not the case here.)

The $\vec{\underline{k}}$ vector is given by $[k, 0, \underline{m}]$, where it is assumed that the wave propagates in the $(x - z)$ plane for simplicity, with z being vertical. The parameter $\underline{m}$ is the vertical wavenumber, which is complex, equal to $m_R + \mathrm{i}m_I$, where $m_I = -1/(2\overline{H})$, and describes an exponential increase in amplitude with increasing height. While we wrote $\vec{k} = [k, 0, \underline{m}]$, for purposes of propagation direction determinations, we regard the k vector as real, equal to $[k, 0, m_R]$. $\overline{H}$ is the scale height and is given by (to a reasonable approximation)

$$\overline{H} = \frac{K_B\overline{T}}{mg}, \tag{11.13}$$

where K_B is Boltzmann's constant and m is the mean molecular mass. The complex velocity perturbation is $\hat{\underline{\vec{u}}} = (\hat{\underline{u}}, \hat{\underline{v}}, \hat{\underline{w}})$, and a velocity may be considered a perturbation so long as it is substantially less than c_s. The term $\hat{\underline{\psi}} = \hat{p}/\overline{p}$. The quantity $\hat{\underline{r}}$ equals $\hat{\rho}/\overline{\rho}$, and ω_B is the Brunt–Väisälä frequency, which satisfies

$$\omega_B^2 = \frac{g}{\overline{T}}\frac{\partial \overline{T}}{\partial z} + \epsilon\frac{g}{\overline{H}} = \frac{g}{\overline{\Theta}}\frac{d\overline{\Theta}}{dz}, \tag{11.14}$$

where $\epsilon = R/c_p$, R being the gas constant for air and c_p being the specific heat of air at constant pressure. The symbol Θ represents potential temperature. These equations can also be found in *Walterscheid and Hocking* (1991), although in that paper a Rayleigh drag parameterization was also included. The first three equations are the momentum equations. The next equation is a form of the first law of thermodynamics, and the last is the continuity equation.

When these equations are solved, one finds that various waves can exist, but they must satisfy particular relations between the wave frequency and wavenumbers; the so-called "dispersion relation." In particular, gravity waves are confined to have intrinsic frequencies in the range between the Brunt–Väisälä frequency (the natural frequency of oscillation of a displaced air parcel in the atmosphere, typically in the range 5 to 10 minutes below 100 km altitude) and the "inertial frequency," the lower frequency limit.

The inertial angular frequency, which we will denote as ω_i, equals the Coriolis parameter, f_c (defined above). *Houghton* (1977), page 99, problem 8.1, shows that the inertial period is $T_i = \frac{2\pi}{f_c}$, or $f_i = \frac{f_c}{2\pi}$, giving $\omega_i = f_c$.

The relations between wave-velocity amplitudes and the temperature, density and pressure amplitudes are also defined uniquely by these equations. These relations are called "polarization relations," and will be presented shortly. There are also different levels of approximation depending on whether particular terms are ignored or retained in the linearized approximations of the equations of motions. The dispersion relations and the polarization relations differ slightly depending on the form of the equations used. Examples include the Boussinesq approximation (the simplest), the hydrostatic approximation, the anelastic approximation, the non-hydrostatic and fully compressible solutions. The Boussinesq aproximation considers the case that all density perturbation terms that are not multiplied by the acceleration due to gravity g are ignored. The anelastic case refers to the case that $\frac{\partial \rho}{\partial t}$ (which is embedded in $\frac{\mathrm{D}\rho}{\mathrm{D}t}$ in (11.4)) is ignored. We will not discuss these differences here – such discussions can be found in any reasonable book on fluid dynamics. Rather, we will begin with the most general linearized solution (e.g. *Walterscheid and Hocking*, 1991), and then develop simplifications which are of value for experimentalists.

The most general dispersion relation for these equations takes the form

$$m_R^2 = \frac{\omega_B^2 - \omega^2}{\omega^2 - \omega_i^2}k^2 - \frac{1}{4H^2} + \frac{\omega^2}{c_s^2}, \tag{11.15}$$

(also see *Gossard and Hooke*, 1975). *Walterscheid and Hocking* (1991) actually developed an even more complex version which incorporated a Rayleigh drag term (i.e., a drag force proportional to the instantaneous horizontal velocity). The reader may refer to that article for higher precision, but our intent here is to make things simpler, not more complex, with the idea of developing expressions that can easily be used in simple mechanistic thought experiments. A simpler expression for the dispersion relation, the Boussinesq approximation, is

$$m_R^2 = \frac{\omega_B^2 - \omega^2}{\omega^2 - \omega_i^2} k^2. \tag{11.16}$$

For a wide range of frequencies, somewhat larger than ω_i and somewhat smaller than ω_B, this dispersion relation can be reasonably well approximated by

$$\frac{m_R}{k} = \frac{\omega_B}{\omega}. \tag{11.17}$$

When dealing with thought experiments, or making estimates, Equation (11.15) can be unnecessarily complicated. Under such circumstances, Equation (11.17) can be a suitable substitute for the dispersion relation. Of course this should not be used in a computer program, where the full relation should be employed. The relation is quite reasonable for periods of more than about three times the Brunt–Väisälä period.

We will now present the polarization relations. These are not all exact, but represent a useful set of equations for understanding gravity wave effects. As an example, Equation (11.18) comes from (11.12) with the density term and the term involving the scale height ignored. For exact polarization and dispersion relations, see *Walterscheid and Hocking* (1991).

In the following equations the mean wind will be allowed to be non-zero, equal to $[\overline{u}, 0, 0]$. Equation (11.20) requires a little explanation, which will be presented following the list of equations:

$$\underline{\hat{w}} = -\frac{k}{m_R}\underline{\hat{u}} = \frac{\omega}{(\overline{u} - c_\phi) m_R}\underline{\hat{u}}, \tag{11.18}$$

$$\underline{\hat{v}} = -\mathrm{i}\frac{\omega_i}{\omega}\underline{\hat{u}}, \tag{11.19}$$

$$\underline{\hat{\Theta}} = -\frac{\mathrm{d}\overline{\Theta}}{\mathrm{d}z}\underline{\hat{\xi}} = \frac{-\mathrm{i}}{\omega}\frac{\mathrm{d}\overline{\Theta}}{\mathrm{d}z}\underline{\hat{w}} = \frac{-\mathrm{i}}{\omega}\frac{\omega_B^2}{g}\overline{\Theta}\underline{\hat{w}} = \frac{-\mathrm{i}}{m_R(\overline{u} - c_\phi)} \cdot \frac{\mathrm{d}\overline{\Theta}}{\mathrm{d}z}\underline{\hat{u}}, \tag{11.20}$$

where $\underline{\hat{\xi}}$ is the vertical displacement, and

$$\frac{\underline{\hat{\Theta}}}{\overline{\Theta}} = -\frac{\underline{\hat{\rho}}}{\overline{\rho}}, \tag{11.21}$$

$$\underline{\hat{p}} = \underline{\hat{u}}\overline{\rho}(c_\phi - \overline{u}). \tag{11.22}$$

In these equations, $c_\phi - \overline{u} = \omega/k$ is the so-called "intrinsic phase speed" of the wave, or the wave phase speed with respect to the mean wind at the height of the wave. Equations (11.18) and (11.19) are exact for the Boussinesq approximation, whilst (11.21) and (11.22) are really only valid for $\omega_i \ll \omega \ll \omega_B$, although they do in fact apply quite well over much of the range of observed frequencies.

As noted above, Equation (11.20) requires some explanation. The parameter $\underline{\hat{\Theta}}$ is *not* the variation of potential temperature that the parcel experiences as it moves up and down, i.e., it is *not* the deviation from the equilibrium position. After all, a displaced parcel of air moves adiabatically, and so it maintains a *fixed* potential temperature.

Its deviation from equilibrium must by definition be zero. The quantity $\underline{\hat{\Theta}}$ *actually represents the difference between its own potential temperature and that of its immediate environment*. This is crucial to recognize.

This explains the first equality in (11.20). The potential temperature of the parcel at a displacement $\underline{\hat{\xi}}$, relative to the value at the equilibrium position, is always zero. The potential temperature of the environment is simply $\frac{d\overline{\Theta}}{dz}\underline{\hat{\xi}}$. The difference between the potential temperature of the parcel and that of the environment is then just $(0 - \frac{d\overline{\Theta}}{dz}\underline{\hat{\xi}})$, as presented in the equation.

The second equality in this equation then comes from recognizing that $\underline{\hat{\xi}} = \hat{\xi}_0 e^{i\{kx+\underline{m}z-\omega t\}}$, and the vertical velocity is just its time derivative, or $-i\omega\underline{\hat{\xi}}$. Hence $\underline{\hat{\xi}} = i\frac{\underline{\hat{w}}}{\omega}$. Subsequent expressions in (11.20) are simply obtained by suitable substitution of earlier equations.

The most important dynamical and thermodynamic parameters used in gravity wave studies are the velocity fluctuations, and the temperature, density, and pressure fluctuations. The most important wave diagnostics are direction and speed of propagation, wave periods (or frequencies), and vertical, horizontal, and total wavelengths. Various additional "derived" quantities are of value too. These include momentum fluxes; that is, terms like $\overline{\rho\underline{\hat{u}}\underline{\hat{w}}}$, where the overbar may be either a space or time average. This particular term refers to the vertical flux of eastward momentum, and also the eastward flux of vertical momentum. Terms like $\overline{\underline{\hat{p}}\underline{\hat{w}}}$ refer to the vertical flux of energy.

As noted, momentum fluxes can also be used to determine body forces in the atmosphere. Such forces act on and change the mean flow. As an important and representative example, $\frac{d\overline{\rho\underline{\hat{u}}\underline{\hat{w}}}}{dt}$ gives the drag force per unit volume. The drag force per unit mass is obtained by dividing by $\overline{\rho}$. We will discuss these forces in more detail later in this chapter.

11.2.5 Saturation theory and the "universal spectrum"

In the previous section, many of the various parameters that are used to describe gravity waves were discussed. One other class of wave-parameters has been left out of the discussion until now, and this is the class relating to the so-called "universal spectrum." It is now appropriate to discuss these. We will discuss them separately because studies of this spectral form have tended to dominate the literature on gravity wave studies in recent years.

The concept of gravity waves as a major contributor to motions in the upper atmosphere was first introduced by *Hines* (1960), and that paper has led to many publications describing wave motions in the atmosphere. Initially most studies concentrated on individual "sightings" of single quasi-monochromatic waves, and these have been discussed in various reviews (e.g., *Rastogi*, 1981; *Fukao et al.*, 1979; *Vincent and Ball*, 1977, to list but a few).

A change from the "monochromatic" perspective came in 1982 when *Van Zandt* (1982) proposed, after a series of spectral studies, that the distribution of wave spectral densities was largely invariant as a function of latitude, longitude, time of year, and

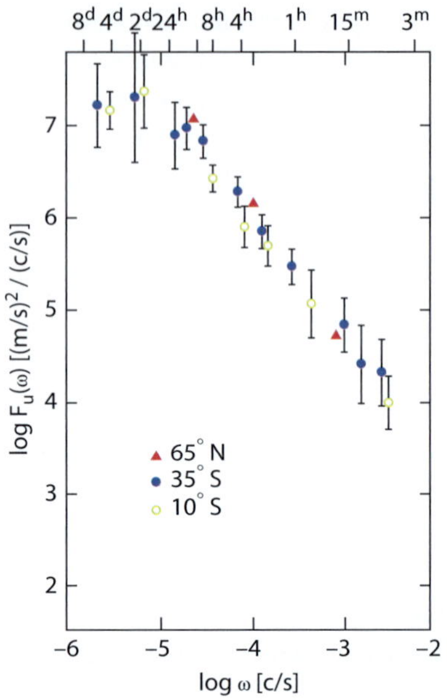

Figure 11.9 Spectra of horizontal fluctuations as a function of frequency for an altitude of around 85 km. The data are for 65 °N (triangles), 35 °S (filled circles) and 10 °S (open circles). From *Van Zandt* (1985a). (Reprinted with permission from John Wiley and Sons.)

underlying geophysical terrain. This author looked at the so-called universal spectrum of oceanic underwater gravity waves originally proposed for internal waves (*Garrett and Munk*, 1972, 1975), and developed a similar model for atmospheric waves. Figure 11.9 shows spectra of wave energy densities plotted as a function of frequency for wave motions at 85 km altitude, using data recorded at 35 °S, 10 °S and 65 °N. These spectra are not normalized in any way, yet clearly the spectra are similar in power and form. A similar universality was found when spatial spectra were plotted as a function of wavenumber. The forms of the spectra proposed by *Van Zandt* (1982) are:

$$E(k_\lambda, \omega) = E_0 \frac{c^*}{[1 + k_\lambda/k_*]^p} \cdot \frac{2.0}{\pi} 2\pi\omega_i \frac{\omega^{-t}}{k_*}, \tag{11.23}$$

and

$$E(m_\lambda, \omega) = E_0 \frac{c^*}{[1 + m_\lambda/m_*]^q} \cdot \frac{2.0}{\pi} 2\pi\omega_i \frac{\omega^{-t}}{m_*}. \tag{11.24}$$

Here again, ω_i is the inertial (angular) frequency. In this case k_λ and m_λ are taken as inverse wavelengths $(1/\lambda_x, 1/\lambda_z)$, although this definition is somewhat unconventional; normally $2\pi/\lambda_x$ and $2\pi/\lambda_z$ might be expected. Also note that m_λ is only the real part of the complex vertical wavenumber in this case. The term c^* is an empirical constant.

It can be shown that p must equal q if the spectrum is separable in k and ω. In practice, however, it turns out that separability is not exactly valid, but nevertheless, it is not too bad an approximation. After studying a small set of experimental data, *Van Zandt* (1982) proposed that $p = 2.4$, $t = 5/3$, $q = 2.4$, and $c^* = 1.4$ if m_λ and k_λ are in units of cycles per meter. *Van Zandt* (1982) also took m_* to be about 10^{-4} cycles per meter, corresponding to a vertical wavelength of about 10 km. More recent results, based on further data, suggest that t is around 1.5–2, while p and q may be about 2–3. It is not entirely practical to define these constants too precisely since, as we will see later, the universality turns out to be less universal than originally envisaged. These variances will be discussed later, and indeed the relevance of the universality is still an open question. Nevertheless, for now we will continue with the assumption of its accuracy simply in order to complete the theoretical development.

We now return to the principles of the theory. The quantities m_* and k_* represent typical values which separate the "low" wavenumber part of the spectrum, where E is almost independent of m_λ or k_λ, from the "high" wavenumber part of the spectrum, where the spectrum is almost of the form m^{-q} or k^{-p}. It should be emphasized that a value for $m_* = 10^{-4}$ cycles per meter applies only in the troposphere and changes as a function of altitude. *Sidi et al.* (1988) have suggested refinements and improvements to the above model.

Spectra of wave motions became the focus of many investigations (e.g., *Van Zandt*, 1985a, b; *Scheffler and Liu*, 1985; *Smith et al.*, 1985; *Vincent*, 1984; *Nastrom et al.*, 1987; *Fritts and Vincent*, 1987; *Meek et al.*, 1985; *Dewan et al.*, 1984). To summarize briefly, it has been found that whilst there is some degree of universality, there is still some departure from this, though generally by less than a factor of 3 or 4 at any one height. The GASP experiments (*Nastrom et al.*, 1987) found wave activity in the upper troposphere to be 2–3 times greater over mountainous terrain than over flat land or the sea. Studies of wave activity as a function of season in the mesosphere have also shown variations of a factor of 2 or 3, and these will be reported later. Occasionally exceptional sites have been found to exist with anomalously enhanced wave activity, the Antarctic Peninsula being one case (*Hertzog et al.*, 2008). This site is special because it is essentially the only land at 60 to 65 ° latitude in the southern hemisphere, so upper tropospheric and stratospheric winds, being largely free of drag, can reach very high speeds as they circulate the earth there. On encountering the peninsula, the wave flow over the mountains can cause waves to be launched in large numbers.

The question also arises as to how many waves can be found in the atmosphere at any one time. Many spectra are determined using averages over at least a day or more, but *Sica and Russell* (1999a, b) have attempted to determine the number of waves present at any one moment. They found typically only a few dominant modes in the mesosphere. This result needs to be studied further, because the prevailing view is that there are a large number present at any one time. *Fritts and Alexander* (2003) (pages 3–19) have discussed this issue in greater detail, but complete resolution of the question has not yet been achieved.

The spectral density of the velocity spectra tends to grow with increasing height, at least at lower altitudes up to about 70 km. This is expected as the waves rise into regions of decreasing density but conserve total kinetic energy. Above that the amplitudes tend to become more constant, suggesting energy loss as they rise. In other words the waves seem to "saturate" in some way. This saturation effect happens first at the higher frequencies at the lowest heights, and then happens at lower and lower frequencies as the waves rise in altitude. The "rollover" point where the wave energies change from a power-law behavior at high frequencies to a flatter form at lower frequencies therefore changes with height.

Description of the wave spectrum is one thing, but explaining why it has the shape that it does is another matter entirely. We cannot possibly do justice to all the available models here, but we will briefly refer to several of the more common ones. We will try not to make any statements regarding which model will be the most important.

Perhaps the earliest model to explain the universality and the saturated nature of the spectrum was one that predicted that as the waves grow, they become unstable, generate small amounts of turbulence, lose energy and thereby reduce their amplitude to become stable once again. This process repeats at various altitudes, with the actual altitudes at which breaking occurs dependent on the way in which the waves add up. This is to say that, at any one altitude, the resultant velocity and temperature profiles are due to the sum of all contributing waves. At some altitudes, they will add in such a way as to give an unstable Richardson number (depending on the phases of the waves). At this altitude, turbulence will occur, while at other nearby altitudes, the arrangement of phases may not result in instability. This process is variously called "convective adjustment" (*Fritts and Rastogi*, 1985), or in some cases "shedding." The "saturation theory" has been proposed and reported in several papers (*Dewan and Good*, 1986; *Smith et al.*, 1987; *Weinstock*, 1984a; *Tsuda et al.*, 1989; *Fritts and Chou*, 1987; *Gardner et al.*, 1993a; *Gardner*, 1994). Further extensive developments of these theories also include the work of *Hines* (1991a, b, c, 1993, 1996). *Desaubies and Smith* (1982) have studied processes of this type in the oceans and examined the statistics and expectation of regions of instability arising. These patches of turbulence occur in much the same way as whitecaps appear on the surface of the ocean. The same concept has been considered for the troposphere by *Fairall et al.* (1991), and for the middle atmosphere by *Hocking* (1991) and *Hines* (1991a), *Hines* (1991b), and *Hines* (1991c). *Sica and Thorsley* (1996) have demonstrated the existence of these atmospheric whitecaps using lidar observations. However, the convective adjustment model is only one of several used to explain the structure of the spectrum. We will discuss others shortly.

This leads to the rollover wavenumbers (m_* and k_*) changing with height, as discussed a few paragraphs back. At wavenumbers much higher than this value (short wavelengths), the spectral shape is of the form m^{-q}, whilst at much longer vertical wavelengths, the spectral shape is nearer to flat. The rollover vertical wavelengths are smallest at the lowest altitudes. This behavior is shown in Figure 11.10 from *Smith et al.* (1987).

A further addition to the model is introduction of a dependence on the atmospheric stability, so that the spectral form of the vertical wavenumber in the high wavenumber limit takes the form

$$P(m) \propto \omega_B^2/m^3, \tag{11.25}$$

where again m is just the real part of the vertical wavenumber. This equation therefore includes a new dependence – namely a dependence on the Brunt–Väisälä frequency ω_B. Other variations of the universal theory include modifications to take account of

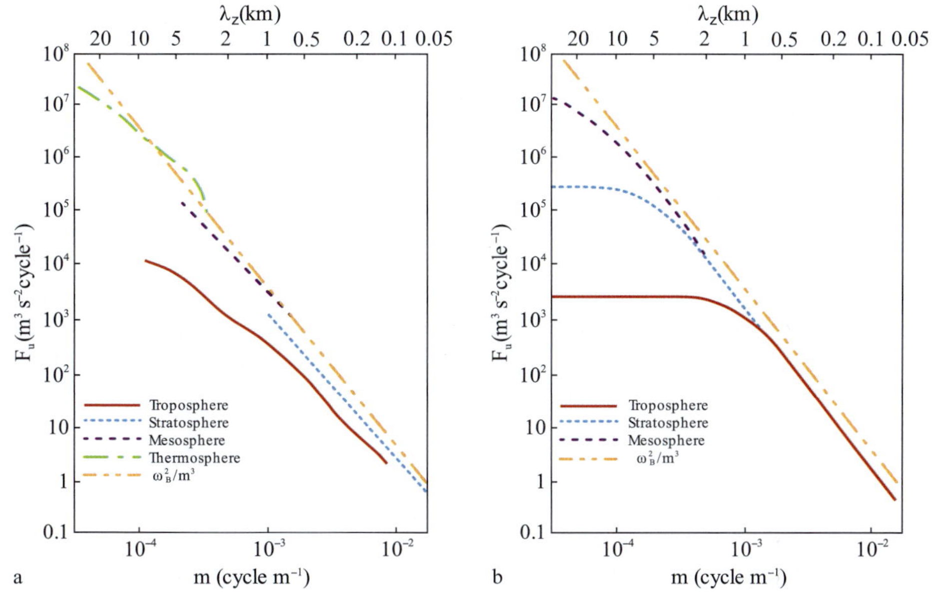

Figure 11.10 (a) Measured atmospheric spectra F_u of horizontal gravity wave fluctuations as a function of vertical wavenumber for different height regimes. The various line-types are indicated in the figure. The dash-double-dot line shows the line $F_u \propto \omega_B^2/m^3$, where ω_B is the Brunt-Väisälä frequency. (b) Model spectra proposed by *Smith et al.* (1987) for the troposphere, stratosphere, and mesosphere. Note how the "roll-off" point (which marks the transition between the m^{-3} part of the spectrum and the flatter part) varies as a function of height (from *Smith et al.*, 1987). (Reprinted with permission from the American Meteorological Society.)

variations in wave amplitude as a function of height due to variations in static stability (*Van Zandt and Fritts*, 1989) and Doppler shifting due to the mean wind (*Van Zandt and Fritts*, 1987).

While earlier studies concentrated on the total energies and associated horizontal wind fluctuations, the vertical wind fluctuations have also been studied in some detail. The spectrum of these fluctuations shows a fairly flat spectrum, but in light wind conditions there is a peak in spectral density just before the Brunt–Väisälä frequency, and then a sharp cutoff at frequencies higher than the Brunt–Väisälä frequency. Figure 11.11 shows an example. Other observations at multiple sites have been presented by, for example, *Ecklund et al.* (1985).

Despite a reasonable (but perhaps not definitive) degree of certainty about the general spectral form of gravity wave densities in the atmosphere, the exact reasons for the development of this universal spectrum are complex and as yet our understanding is incomplete. Dewan's models were based purely on similitude analysis. Other studies have invoked the principle of convective adjustment to explain the spectrum (discussed above), while others have invoked diffusive processes and wave nonlinearities. *Gardner et al.* (1993a, b); *Gardner* (1994, 1996, 1998) have extended the theory to consider anisotropic wave fields, and to consider non-separable spectral wave forms.

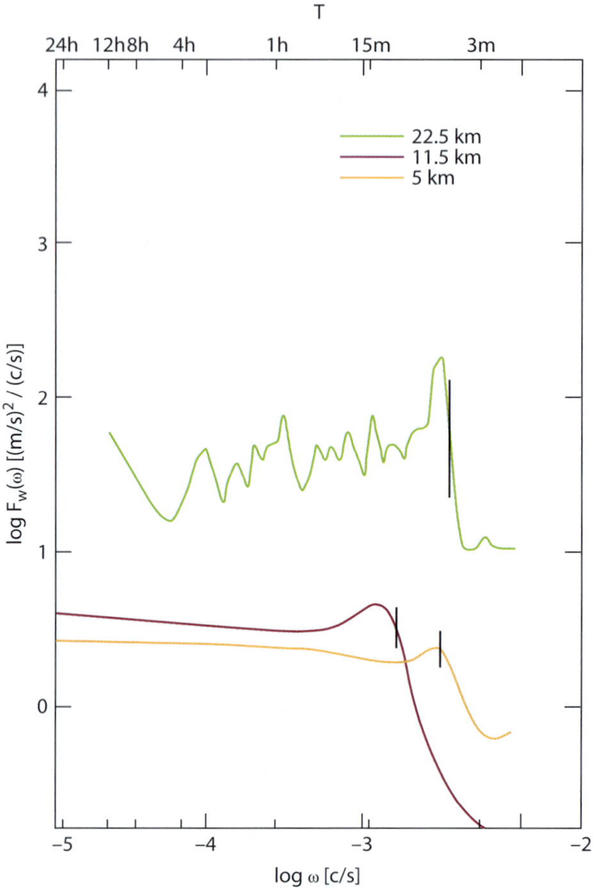

Figure 11.11 Some typical frequency spectra of vertical velocity fluctuations under quiet conditions, emphasizing a spectral peak close to the Brunt–Väisälä frequency. The curves show the basic shape; absolute values are only meaningful for the lower two curves. The upper curve refers to an altitude of 22.5 km, the middle one to about 11.5 km, and the lowest to about 5 km altitude (from *Van Zandt*, 1985a). (Reprinted with permission from John Wiley and Sons.)

An alternative model for energy and momentum deposition is that the energy and momentum are deposited in "catastrophic events," in which waves break and completely destroy themselves (e.g., *Andreassen et al.*, 1994; *Fritts et al.*, 1996b; *Werne and Fritts*, 1999). These events may also generate secondary waves, and in this case nonlinear wave–wave interactions can be important. Wave–wave interactions may play an important role (e.g., *Muller et al.*, 1986). Still others have invoked diffusive gravity wave models (e.g., *Weinstock*, 1976, 1982, 1984a, 1990; *Gardner*, 1994, 1996). *Hines* (2001), adapting the work of *Allen and Joseph* (1989), has suggested that the reason for the m^{-3} law relates to the fact that a ground-based observer views the waves from an Eulerian perspective, whereas a more natural reference frame is a Lagrangian one. Further developments in this area were presented by *Chunchuzov* (2002). However, a more recent paper by *Klaassen* (2010) shows that this so-called "Doppler spread theory" has a serious mathematical problem in that it generates Jacobians-of-transformation which depart significantly from unity. Theory dictates that such departures would be unphysical, and this publication seems to have substantially damped support for the model.

We will not attempt to identify the most successful model here. The interested reader is referred to *Fritts and Alexander* (2003), Section 6, for one set of views of the current relative importance of these models, but it should be noted that this is still only one of several current perspectives.

It should once again be emphasized that the above discussions are all based on a theory of quasi-universality and assume that the irregular motions are due to a spectrum of buoyancy waves. It was noted earlier that a few workers believe that the irregular motions are due to two-dimensional turbulence, but it does appear that at least in the stratosphere and mesosphere the wave theory is quite compelling. For motions in the troposphere, the argument is still much more alive.

With regard to radar studies, there are several areas in which windprofilers can provide useful contributions to help determine the accuracy of these various models. The first is in the successful identification of wave sources. The second is to help determine the relative importance of shedding versus catastrophic collapse of gravity waves.

In the next section, ways in which gravity waves can be measured in the atmosphere will be considered, with concentration on radar and windprofiler techniques.

11.2.6 Measurement techniques for gravity waves

Gravity waves can be identified and measured in a variety of ways. Their impact can be seen in clouds, often showing as regular bands of clouds across the sky. For quantitative scientific work, multiple methods are available, based on both in-situ and remote sensing techniques. In-situ techniques include balloons, aircraft, and rockets, and many types of parameters can be measured, including pressure, density, and wind velocity. With regard to remote sensing, the waves can be sensed because of chemical reactions that produce optical emissions (especially in the mesosphere), or via visual effects like those in clouds, or via undulations produced in scattering characteristics of radiowaves, or through direct measurement of density perturbations via lidar. Optical imaging methods are very good for measuring horizontal wavelengths. Many hundreds, if not thousands, of publications present results of such measurements and it would be impossible to discuss them in any detail here. A few representative ones are listed below. With regard to rocket measurements, techniques may involve smoke or TMA (tri-methyl aluminum) trail releases (e.g., *Dewan et al.*, 1984), falling spheres released at altitude (e.g., *Jones and Peterson*, 1968; *Eckermann and Vincent*, 1989) or release of chaff (e.g., *Widdel*, 1987) to name a few. *Dieminger et al.* (1996) gives a slightly more extensive, but not overwhelming, discussion of rocket techniques. Balloons carry many types of instrumentation, usually including temperature sensors and anemometers, and can measure up to 30 or 40 km altitude into the stratosphere. Lidar experiments usually infer density and temperature, and the fluctuations in these parameters can then be used to study gravity wave motions (e.g., *Chanin and Hauchecorne*, 1981; *Shibata et al.*, 1986; *Hauchecorne et al.*, 1987). As far as radar experiments go, most techniques are based on measurement of wind velocities, either by the spaced antenna method (e.g., *Hocking et al.*, 1989) or interferometric methods (see Chapter 9 in this book), or using tilted beams and Doppler measurements as discussed in Chapter 7 and elsewhere herein. We will concentrate here mainly on radar techniques.

However, radar methods are not as obvious as they might seem. It is true that if it is possible to measure the instantaneous velocity vector with good resolution and high accuracy, then determination of winds is straightforward. But with the Doppler method, only one component of the wind can be measured in any one beam. When looking at long term average winds, it is possible to assume that the vertical wind is zero, and it is possible to vectorially add two wind components in two orthogonal beams to produce the mean vector wind. With gravity wave studies, this is not always valid. It may be satisfactory for waves with periods of many hours, but for short-term oscillations with periods of tens of minutes or horizontal wavelengths of a few km, such additions cannot be employed, since each beam will look at a different part of the wave, and those parts may have significant phase differences. In addition, it is not valid to ignore the vertical components of the wind. It is possible to use spaced antenna measurements to obtain all three components of the wind simultaneously, but these can have their own complications, such as wide radar beams. Design of a good experiment needs to consider these effects.

As discussed earlier, one important parameter in wave studies is the momentum flux. In principle, this requires simultaneous measurements of one horizontal component of the wind, plus the vertical velocity, since terms like $\overline{\underline{\hat{u}}\underline{\hat{w}}}$ are needed. However, there is a better way to do this than by measuring $\underline{\hat{u}}$ and $\underline{\hat{w}}$ independently. This technique, developed originally at around the same time by *Lhermitte* (1983) and *Vincent and Reid* (1983), used pairs of oppositely directed tilted beams to make direct measurements of the vertical flux of horizontal momentum. This method is often referred to as the "dual beam" method of momentum flux measurement. Measurement of momentum flux in this way has become almost a field in its own right. The procedure was introduced descriptively in Chapter 1, Section 1.3.4.

In this method, two beams are pointed off-vertically in, say, the eastward and westward directions, at zenith angle θ. The radial velocity variance is measured for each beam, and then the difference is proportional to $\overline{\underline{u}'\underline{w}'}$. (Note that we now use $\underline{u}'$ instead of $\underline{\hat{u}}$, etc., since in an experiment the complete perturbation velocity is measured, rather than a first order linearized term.) Specifically,

$$\overline{\underline{u}'\underline{w}'} = \frac{\overline{v_{r1}^2} - \overline{v_{r2}^2}}{2 \sin 2\theta}, \tag{11.26}$$

where $\overline{v_{r1}^2}$ is the mean square velocity fluctuation in the eastward beam and $\overline{v_{r2}^2}$ is the mean square fluctuation in the westward beam. Similar application for a north and south pair of beams produces estimates of $\overline{\underline{v}'\underline{w}'}$. As discussed in Chapter 10, this method was eventually generalized to interferometric techniques and used in meteor–wind studies (Section 10.3.3). The same principles can be used with any reliable interferometer.

For measurements of gravity wave horizontal wavelengths, the best methods are optical ones, since they involve essentially imaging the sky using appropriately selected filters. Measurements of wave periods and wavelengths by radar are more difficult, mainly because of the much poorer resolution and because often there are only a few available beams. Determination of the vertical wavelengths is fairly straightforward, since it only requires profiles of the wind speed as a function of height, as seen in Figure 11.1. However, determination of horizontal wavelengths, phase velocities, and intrinsic frequencies is more difficult. The best available methods are multi-station observations and (to a limited degree) radar beam-swinging methods (e.g., *Vincent and Reid*, 1983). Nevertheless, some useful applications have been made. However, these data are still useful supplements to other methods, as all methods have limitations. Even optical methods can be limited by their own instrumental and geophysical effects, including the fact that optical methods may sometimes preferentially select certain wave modes, especially ducted modes (e.g., *Isler et al.*, 1997; *Gardner and Taylor*, 1998), and many optical emissions are confined to limited altitudes.

11.2.7 Overview of some important gravity wave parameters

Radars are well suited to study gravity wave spectra, because they can record long sequences of radial velocities which may then be Fourier transformed. Radars are the best instruments for long-term data sets of many days and even months. Optical techniques usually require night-time, low-moonlight conditions, so data-sets are choppy and intermittent, with missing data during daylight hours. However, for gravity-wave spectra, analysis is usually performed on single beams, and it needs to be remembered that the measured parameter, which is usually the radial velocity, is actually a mixture of horizontal and vertical velocity components.

Radars are therefore well suited to determination of gravity-wave spectra. They have been especially useful for examination of the universality of the spectra, and for examining any seasonal variations. In this section, we will look at the implications of some such studies.

Some of the following results are somewhat tentative, but they do seem to produce a self-consistent picture. Following ealier comments, it should be recognized that there is no such thing as a "perfectly universal spectrum," since there are undeniably seasonal and geographical variations. Nevertheless, in any one height range in the stratosphere or mesosphere over a large part of the globe and the whole year, the variations in total wave RMS amplitude seem to be less than a factor of 3 or so, which is considered a small variation. Workers in the field do talk of universal spectra, but recognize the implicit approximations.

Typical frequency spectra have already been discussed and illustrated in Figure 11.9. Spectra recorded over periods of only a few hours generally do not show universal structure, and can be very variable. However, once temporal periods of many days are used, all tend to show similar shapes. Periods generally cover the range from about 5 minutes up to the inertial period (generally 10–20 hours for most latitudes), and integrated power under the spectra also tend to vary by less than a factor of 3 or 4. The spectral shape is generally a power law, of the type $P(f) \propto f^{-t}$, with t in the range between about 1.3 and 2. Such values are not inconsistent with the values proposed following Equations (11.23) and (11.24).

It should be noted that the estimates of typical scales outlined above were often made from studies of seemingly monochromatic waves, and although this is generally an allowable procedure, one must be careful. The dangers of studying monochromatic events in the presence of a spectrum of waves have been highlighted by *Eckermann and Hocking* (1989).

Recall from Equation (11.25) that according to that theory, the power spectral density is proportional to m^{-3}, and that vertical wavelengths range from 100 or 200 m up to typically tens of km but most of the energy is in wavelengths larger than 1–2 km (Figure 11.10). A power law can also be used to describe horizontal wave spectra, being typically proportional to k^{-2} (*Fritts et al.*, 1989), with horizontal wavelengths ranging from thousands of kilometers down to 5 km or so (*Reid*, 1986). *Fritts et al.* (1989) suggested that a typical horizontal scale is about 300 to 500 km. It should be emphasized that a power of –3 for the vertical wavenumber spectra is not consistent with a power law of –2 for the horizontal spectra unless the spectra are not separable in ω, k, and m, as was seen following Equations (11.23) and (11.24) (e.g., *Gardner et al.*, 1993a, b). Separability requires that $p = q$ in (11.23) and (11.24). Nevertheless, it should be noted that this k^{-2} power law is based on only a few observations made by *Fritts et al.* (1989) and there was large variability between the few spectra measured by these authors. Whether the spectra are separable is still debatable. Of course they do not need to be, but it makes some forms of analysis easier if they are.

The relative contribution of downgoing and upgoing waves is another important parameter. It is generally felt that the troposphere is the major source of gravity waves that reach the mesosphere. This belief is supported by studies of so-called "rotary" spectra by *Vincent* (1984) and *Eckermann and Vincent* (1989), who showed that upward propagating waves were responsible for at least 65% of the measured wave energy, and it is likely that this is a lower limit. To understand this statement, recall that the wave motions in the horizontal plane will be elliptical, with the ratio of semi-major to semi-minor axis being dependent on the ratio of the wave frequency relative to the inertial frequency (see Equation (11.19)). If decomposed into circular modes, as done by *Eckermann and Vincent* (1989), then upward propagating elliptical waves will have both an upward and a downward component, so the downward propagating component will be artificially amplified. (Of course this argument can also be used in reverse to claim that upward moving waves are artificially amplified too, so care is needed; the argument is not supposed to be foolproof but is simply offered as a possibility that the percentage of upgoing waves might be underestimated.) In general, most researchers accept this as evidence that a high percentage of sources of the waves is in the troposphere, and this in itself is an important result. Sources will be discussed in greater detail later in this chapter.

The next item in this list seems somewhat out of place, but is needed in order to consider some subsequent items. This is a discussion about the ways in which waves break. Breaking of waves leads to turbulence and eventual deposition of momentum and energy into the atmosphere. Our purpose here is to show some simple formulas which apply to wave-breaking, and especially to highlight the ways in which these equations relate to the phase velocity c_ϕ of the wave.

Waves can break by various mechanisms, but here we concentrate only on two of them – critical level interactions and convective instability. More extensive summaries of other wave-breaking mechanisms will come in a separate subsection later.

Critical level interactions occur when the *intrinsic phase speed* approaches zero. The process acts like a sonic boom, when the speed of an aircraft approaches the speed of sound. In the aircraft case, waves that are generated add in phase with the pre-existing waves, resulting in large amplitudes that lead to a shock-front. In the case of gravity waves, a particle at say the crest of a wave front keeps being fed with more perturbations by the same phase of the wave. Furthermore, under such circumstances, the vertical component of the group velocity approaches zero, and the upward rising wave begins to pile up its energy on itself, resulting in very large amplitudes which then cause the waves to break. Such interactions are called critical level interactions and are an important means by which turbulence is generated in the upper atmosphere. Since this happens when the intrinsic phase speed approaches zero, it requires $\overline{u} = c_\phi$. Clearly knowledge of the phase speed c_ϕ is important in this case.

In the case of convective breakdown, there is also a dependence on intrinsic phase speed, as we will now show. Convective breakdown requires that, at some phase of the wave, the *total* local potential temperature gradient approaches zero and then becomes unstable. The critical condition is that

$$\frac{\partial \hat{\Theta}}{\partial z} + \frac{\mathrm{d}\overline{\Theta}}{\mathrm{d}z} = 0. \tag{11.27}$$

But from Equation (11.20),

$$\frac{\partial \hat{\Theta}}{\partial z} = \mathrm{i}m\hat{\Theta} = \mathrm{i}m\frac{-\mathrm{i}}{m_R(\overline{u} - c_\phi)} \cdot \frac{\mathrm{d}\overline{\Theta}}{\mathrm{d}z}\underline{\hat{u}} = \frac{\hat{u}}{\overline{u} - c_\phi)}\frac{\mathrm{d}\overline{\Theta}}{\mathrm{d}z}, \tag{11.28}$$

where we have taken $m = m_R$ and assumed that the dissipative (imaginary) part of the wavenumber is small. Substitution in (11.27) gives

$$(\frac{\underline{\hat{u}}}{\overline{u} - c_\phi)} + 1)\frac{\mathrm{d}\overline{\Theta}}{\mathrm{d}z} = 0. \tag{11.29}$$

Hence the critical condition for instability is

$$\underline{\hat{u}} = c_\phi - \overline{u}. \tag{11.30}$$

The wave will therefore break if the wave amplitude exceeds the intrinsic phase speed of the wave. Breaking can happen anywhere that this is satisfied, but the least extreme case will be when $\underline{\hat{u}}$ equals the peak velocity, so we consider the condition for convective breakdown to be when the peak horizontal velocity amplitude exceeds the intrinsic phase velocity of the wave, viz.,

$$\hat{u}_0 = c_\phi - \overline{u}, \tag{11.31}$$

where u_0 is the magnitude of the peak wave amplitude. Hence once again we see that the phase speed of the wave is an important parameter to consider when we deal with wave-breaking. This term will arise again shortly when we discuss the effect of waves on the large-scale circulation of the atmosphere.

The preceding discussion now leads us to the area of wave velocities – both phase and group. We need to know about both the velocity directions and their magnitudes. If the phase velocity of a wave is known, then given the wave frequency and wavelengths, the group velocity can be deduced, so we will concentrate only on phase velocities for now.

Directional information gives information about several aspects of the waves. First, it can give information about wave sources, and secondly it relates to filtering and gravity-wave refraction in the atmosphere. These topics are of sufficient importance that they have been left to separate subsections, which will be presented shortly.

With regard to horizontal phase velocities, knowledge is somewhat limited. They are known to be in the range between 0 and 100 ms^{-1}, but otherwise the distribution of phase velocities is poorly documented. Distinction needs to be made between the phase speed as seen from the ground, and the intrinsic phase speed, the latter being the speed relative to the mean wind. Knowledge of this parameter is especially important in regard to wave-breaking, as seen above.

11.2.8 Seasonal and latitudinal variations

Although extensive studies of seasonal, latitudinal, and longitudinal variations of gravity wave activity are still far from complete, the number of works with such an agenda has grown in recent years. Studies of gravity wave activity in the mesosphere, stratosphere and troposphere as a function of time and latitude have included *Hirota* (1984), *Meek et al.* (1985), *Eckermann and Vincent* (1989), *Eckermann et al.* (1995), *Fritts and Vincent* (1987), *Manson et al.* (1999), *Eckermann and Preusse* (1999), *Preusse et al.* (1999), *Preusse et al.* (2000), *Tsuda et al.* (2000), *Manson et al.* (2002), *Gavrilov et al.* (2002), *Preusse et al.* (2002), to name a few. The paper by *Manson et al.* (1999) especially examined the latitudinal and seasonal variation of wave energy.

One of the few global studies of gravity waves using rocket data was presented by *Hirota* (1985). In this study, wind and temperature data were recorded by meteorological rockets from a variety of stations situated around the globe. The availabilty of simultaneous measurements of both winds and temperatures made this study unique, although of course it was limited by the fact that each profile was a snapshot in time, so continuity of data was limited. The fact that the data also cover the height interval above 30 km is also special, since the region from about 25 to 65 km altitude is a region where radars tend to be blind. The large amount of data made statistical studies viable. A summary of the results is shown in Figure 11.12. RMS values of the second derivative of wind and

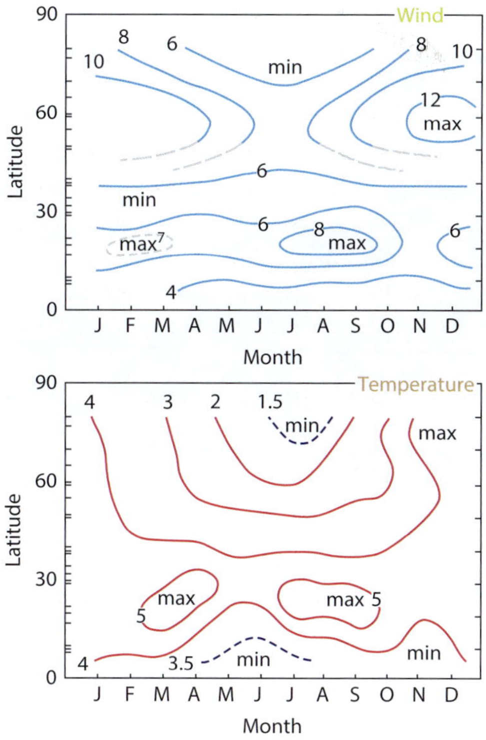

Figure 11.12 Latitude–time cross-sections of wind (upper graph) and temperature (lower graph) activity. The actual numbers refer to RMS values of the second derivative. Ticks on the left axis denote the latitudes of the rocket stations used to produce this plot. (From *Hirota*, 1985).

temperature as a function of height are presented, measured in the altitude range 30–70 km. Because of the nature of rocket profiles, wave periods could not be distinguished, and so these RMS values collectively include waves of all periods.

The data show an annual oscillation in wave intensity at high latitudes (max. in winter). This transitions to a semi-annual oscillation in the tropics and at low latitudes, with a maximum at the equinoxes. Wind speed and temperature second derivatives are generally in the range 6–10 ms^{-1} km^{-2} and 2–5 K km^{-2}. Typical vertical wavelengths are of the order of 5–10 km, giving the vertical wavenumber m as approximately 1 rad km^{-1}. Therefore an RMS value of the second derivative of approximately 6 ms^{-1} km^{-2} means RMS amplitudes of $\simeq$ 6 ms^{-1} and temperature fluctuations of around 2–5 K. RMS wave amplitudes from rocket data in the height region 30 to 60 km altitude have also been presented by *Eckermann and Vincent* (1989), who found horizontal RMS velocity amplitudes of around 5 ms^{-1} at 30 km altitude, rising to around 9 ms^{-1} at 55 km altitude. Typical RMS temperature fluctuations were around 3–6 K, so these values are consistent with those of Hirota.

Figure 11.13 shows stratospheric observations of global gravity wave activity determined by GPS satellite observations (*Tsuda et al.*, 2000). In this case, latitudinal–longitudinal variability is emphasized rather than seasonal/latitudinal variability. Note

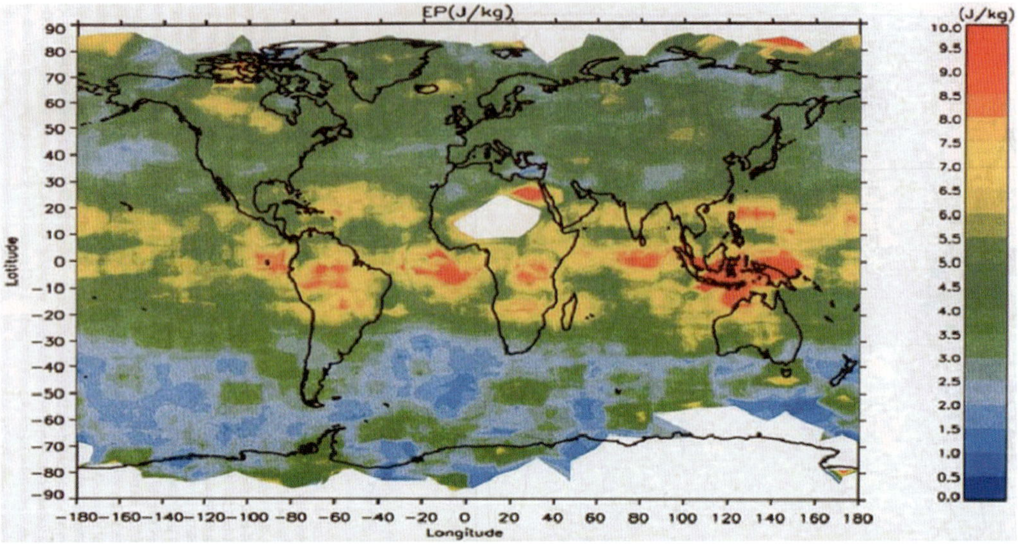

Figure 11.13 GPS-based measurements of gravity wave activity on a global scale (from *Tsuda et al.* (2000)). (Reprinted with permission from John Wiley and Sons.)

the tendency for the wave activity to be highest over large mountains and over equatorial regions of strong convection.

Radar methods have been used by both *Meek et al.* (1985) and *Vincent and Fritts* (1987) to monitor mesospheric wind fluctuations They produced wind profiles as a function of time and height, and spectrally analyzed the time series and divided the spectra into frequency bands. Figure 11.14 shows a sample from *Vincent and Fritts* (1987) of the mean square amplitudes in the period range 1 to 8 hours, plotted as a function of height and time of year. The RMS values of the wind fluctuations are typically 10 to 20 ms^{-1} at 60 to 100 km altitude. A semi-annual oscillation as a function of season is evident in Figure 11.14 below 80 km, and this becomes more like an annual variation at higher altitudes. This semi-annual oscillation below 80 km is shifted in phase by 180° relative to that found by *Hirota* (1984); in this case, maxima occur in summer and winter, whilst Hirota found the maxima to occur at the equinoxes. However, the data presented by *Hirota* (1984) were mainly for stratospheric heights. In addition there was no frequency filtering applied by *Hirota* (1984). Thus we cannot tell whether there was a genuine change in seasonal characteristics at around 70 km altitude, or whether the apparent differences between the two sets of data occur because the data presented by Hirota were in fact dominated by near-inertial frequencies, which would make any comparisons of limited value.

Further studies of this type, by radar and other techniques, are still very much needed.

11.2.9 Refraction, turning levels, and wave ducting

In the previous section, we discussed various parameters concerning gravity waves, and also discussed some of the ways that they break down. The latter topic related to

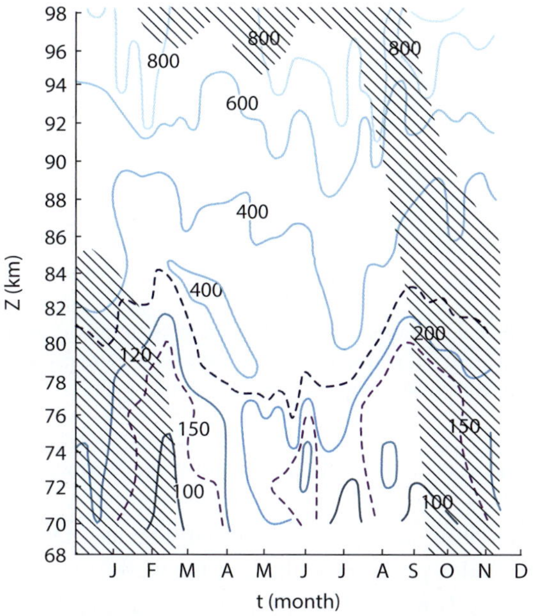

Figure 11.14 Height–time cross-section of mean square east-west amplitudes of gravity waves with periods in the range 1 to 8 hrs, observed at Adelaide, Australia. The altitude is denoted by z, and t is the time (months). Shaded regions indicate times and altitudes at which the mean wind was easterly (westward). (From *Vincent and Fritts*, 1987.) (Reprinted with permission from the American Meteorological Society.)

the interaction of the waves with the mean-state atmosphere. But the interaction with the atmosphere is not confined to wave-breaking. There are other forms of interaction, which we will now discuss.

Gravity waves do not always propagate in straight lines. They can suffer refraction and reflection. They can become trapped in wave-guides and can suffer a variety of fates. They can also break and deposit energy and momentum. In this section, some of these effects will be examined, with concentration on non-breaking effects.

The issue of refraction has especially been considered by *Eckermann* (1992) and *Marks and Eckermann* (1995), who recognized that refraction could occur not only as a function of height (which is well known), but also as a function of latitude. Thus studies of the distribution of wave motions in the upper regions of the middle atmosphere might not properly reflect the distribution at the source regions. Large-scale focusing of waves can also exist. *Ball* (1979) studied wave propagation for waves produced by an eclipse, and here again horizontal ray-path bending and focusing were evident. Determinations of preferred propagation directions for gravity waves must recognize such effects.

Refraction as a function of height is perhaps an even more important consideration. Figure 11.15 shows the typical trajectory of a gravity wave packet propagating through the atmosphere, and bending in the ray path is clearly evident. If the wave

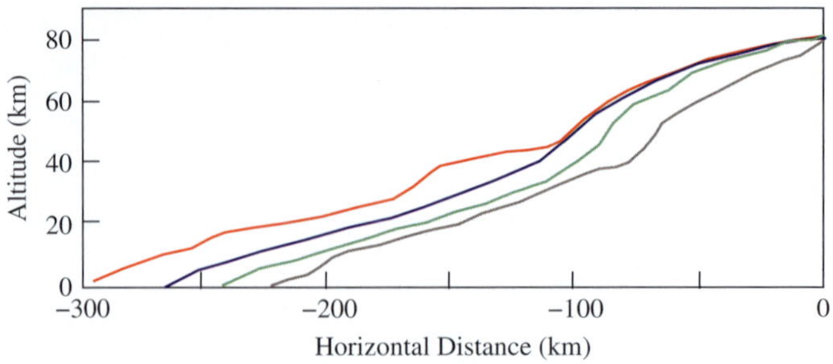

Figure 11.15 Ray-tracing calculations in 2-D for different assumptions about the dispersion relations and assumed mean wind and temperature profiles for a gravity wave. In this case the wave has been reverse-ray-traced, starting at a point at $x = 0$, $z = 80$.

approaches a critical level (where the wave frequency approaches zero, or at least the inertial frequency), the vertical group velocity becomes zero and no propagation occurs. In such circumstances, the wave usually breaks. On the other hand, if the variation in refractive index produces a tilting of the wavefronts towards vertical, the frequency can approach the Brunt–Väisälä frequency. In this case, the wave packet will reflect backwards (a "turning level"). If the vertical temperature and wind structures are suitable, the wave may be reflected from an upper level and then be re-reflected at a slightly lower displaced height, leading to reflection back and forth between the lower and upper levels of a "wave duct." Gravity waves may travel for hundreds of kilometers in such a trapped mode, and the wave is said to be "ducted." Such ducted waves have vertically aligned wavefronts and so are often quite easily visible with optical instruments. This can lead to a bias in gravity wave statistics, since waves seen from the ground are often ducted waves because of the ease with which they can be detected. Mis-identification of ducted waves as freely propagating waves can bias ground-based estimates of gravity wave momentum fluxes (*Fritts and Alexander*, 2003). Observations of ducted waves have been presented by *Isler et al.* (1997), *Walterscheid et al.* (1999), and *Taylor et al.* (1995a). Generally, ducting is more likely with waves of shorter horizontal wavelength (*Fritts and Alexander*, 2003).

Another phenomenon, which is often related to ducting, is the existence of "solitons," or "bores," which are a type of ducted gravity wave comprising only a few dominant oscillations, and with a rapid onset spatially. They contain an abrupt starting edge and often have ripples trailing in their wake (e.g., *Taylor et al.*, 1995b; *Swenson and Espy*, 1995; *Dewan and Picard*, 1998; *Smith et al.*, 2003, among others). These have been detected, and theoretical models for their existence have been developed (e.g., *Dewan and Picard*, 1998, 2001). They are not vertically propagating gravity waves, and their impact on the overall energetics of the atmosphere is unknown. However, when they do exist, they seem to generate a considerable impact on their local environment. An example from *Smith et al.* (2003) is shown in Figure 11.16. They also occur in the troposphere, the Gulf of Carpentaria *Morning Glory* in Australia being one common example (*Smith*, 1988). The paper by *Smith et al.* (2003) gives a good overview and a useful set of related references about this phenomenon.

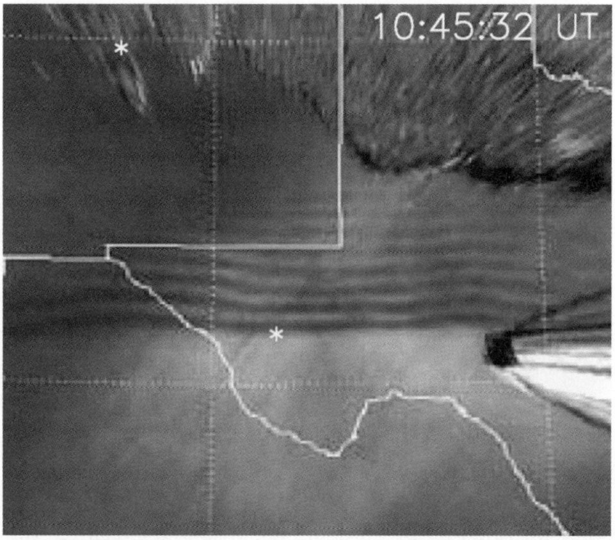

Figure 11.16 An atmospheric bore at around 80–90 km altitude (from *Smith et al.*, 2003). (Reprinted with permission from John Wiley and Sons.)

11.2.10 Sources of gravity waves

Despite a clear recognition of the importance of gravity waves in helping to define the general atmospheric circulation, and their importance for the transport of momentum and energy in the middle atmosphere, as well as their role in the generation of turbulence and mixing, the origin of the waves is still an issue in need of further investigation. There seems to be evidence that they are often generated in the troposphere (e.g. *Eckermann and Vincent*, 1989), but this is not always so. Events have been seen where mesospheric gravity waves were generated in situ by sprite activity above thunderstorms, for example.

Typical tropospheric sources are considered to be convection, thunderstorms, frontal systems, wind-shears, and flow over orography, among others. Convection and frontal systems especially have been theoretically studied by *Alexander and Rosenlof* (1996), and experimental studies have been performed. Geostrophic adjustment (see *Fritts and Alexander*, 2003, pages 3–11) is also another potential source. Some such studies have identified frontal systems as sources of gravity waves (*Belu and Hocking*, 2000). *Fritts and Nastrom* (1992) suggest that typically 30–40% of waves are generated by orographic effects, 20–25% by the jet stream, about 20% by frontal effects, about 15% by convective processes, and the rest by other events. Despite some literature about sources, this still remains a largely open field for study.

The mesosphere is an ideal place to study gravity waves because of their relatively large magnitudes at these heights. Correlations have been found between observations of gravity waves and possible sources like tropospheric thunderstorms. These include some limited satellite studies, but many of these have been limited in scope because they used limb-viewing geometry (e.g., *Dewan et al.*, 1998; *McLandress et al.*, 2000). An extensive study of gravity wave activity as a function of position around the globe was presented by *Tsuda et al.* (2000) using GPS data. These were stratospheric observations.

One significant tool that can be utilized in studies of wave sources is reverse ray tracing. If it is possible to measure the horizontal and vertical wavelengths of some monochromatic waves at one height, determine the frequency, and also be able to determine background conditions from wind radars (at least down to typically 60–70 km altitude), there will be occasions when the observed waves can be tracked back to their source. Ray-tracing models with varying degrees of capability have been developed by *Marks and Eckermann* (1995), *Zhong et al.* (1995), and *Belu* (1998).

The exact path followed by a real wave depends on the details of the background temperature and wind profiles (Figure 11.15). The path deduced in a numerical model can also depend on the type of model (Boussinesq, hydrostatic, anelastic, etc.) that is assumed. *Belu* (1998) has studied the different possible paths as a function of variability of the mean profiles about the seasonal mean, and finds that for some frequencies and wavelengths and seasons, the path varies only a little from one realization to the next, whereas for other cases, the path is extremely sensitive. Figure 11.17 from (from *Belu*, 1998; *Belu and Hocking*, 2000) shows how the source position of a gravity wave can vary for one particular season as a function of wavelength and period. Figure 11.17 was created by performing a large number of ray-tracing simulations, like those shown in Figure 11.15, for an assumed wave observed at 90 km altitude and then reverse ray tracing back to the ground to find the source. The quantity σ shows the standard deviation of the deduced source over many realizations of the mean wind and mean temperature profiles. Clearly, certain combinations of wavelength and period have greater uncertainty in position than others. Knowledge of this type of information is important in order to be able to determine the accuracy of ray-tracing methods.

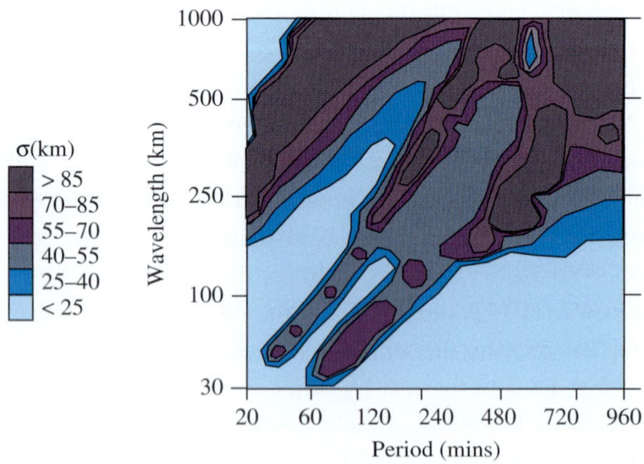

Figure 11.17 Typical gravity wave source position uncertainty as a function of period and horizontal wavelength for July, 40 °N. This graph refers only to waves with downward and eastward phase velocity, and allows ±10% fluctuations in CIRA temperatures and mean winds. (From *Belu and Hocking*, 2000).

Further discussion about these issues is beyond the scope of this chapter, but the interested reader is referred to *Fritts and Alexander* (2003) and references therein for more extended discussions about wave sources. Some additional discussion about meteorological sources can also be found in the meteorology chapter of this book.

11.2.11 Directions of propagation

Radars are capable of determining directions of gravity wave propagation, and there is a great need for knowledge of this parameter. Usually these determinations are made utilizing the known polarization relations for gravity waves or, on occasion, the so-called Stokes' parameters. In cases of clearly defined monochromatic waves, hodograph analyses can be employed, in which the velocity vector is plotted for different heights or times on a two-dimensional graph. The tips of the vectors often trace out an ellipse, and then the polarization relations may be used to determine directions (e.g., *Shimizu and Tsuda*, 1997). Knowledge about the preferred directions of gravity waves helps in understanding both wave sources and wave filtering, although separating out the two effects can be difficult. From the point of view of filtering, there is an expectation that waves will be seen to be propagating preferentially against the mean wind at the upper heights, due to the fact that waves that were propagating parallel to the wind will be lost through critical level interactions. This occurs because as the wave propagates up into the mesosphere, the phase speed of the wave approaches the mean wind speed, and when the two are equal, the wave grows in amplitude and breaks. Waves propagating against the flow do not suffer this fate; the process will be described in more detail shortly. Searches for clear signatures in directions of propagation, however, are often inconclusive. *Maekawa et al.* (1987) and *Fan et al.* (1991) suggested that critical level filtering does play a large role in defining preferred directions, while *Vincent and Fritts* (1987), and *Nakamura et al.* (1993) noticed a tendency for preferred meridional propagation in winter. Preferred propagation towards the north-east and south-west was detected by *Nakamura et al.* (1998) in summer. Results also vary depending on altitude (stratosphere vs. mesosphere) and location. *Vincent and Stubbs* (1977), *Vincent and Fritts* (1987), and *Hocking* (1983b) found a preference for winter-time north-south propagation in the altitude region 80–100 km over Adelaide, Australia. These results have already been discussed earlier in this chapter in the context of upward and downward energy propagation. A more detailed study by *Eckermann and Vincent* (1989) in the region 30–60 km altitude showed an east-west preference in summer and a north-south preference in winter in southern Australia. As discussed, part of the reason for this north-south preference in winter may be related to filtering of gravity waves due to critical level interactions with the mean flow (see shortly). However, *Eckermann and Vincent* (1989) suggest that the distribution of sources may also be important in establishing these preferred orientations. *Ebel et al.* (1987) have studied the effects on gravity waves of filtering by the mean wind. Figure 11.18 shows one example of directional anisotropy from *Eckermann and Vincent* (1989), showing average anisotropy for the height region 33 to 58 km. As a rule, anisotropy was greatest at the greater heights. Other papers that have tried to examine the preferred sense of direction in the middle atmosphere include *Vincent and*

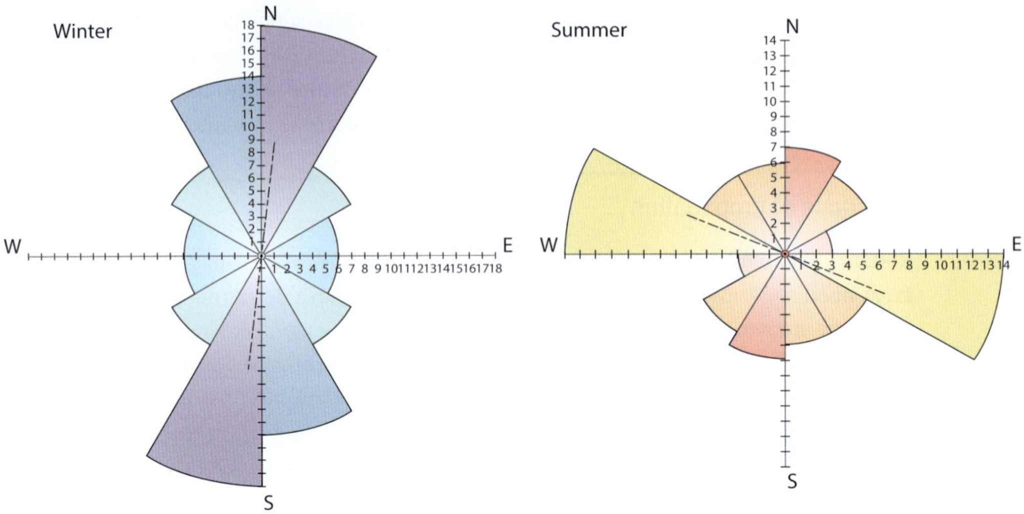

Figure 11.18 Angular histogram of directions of propagation for gravity waves deduced for the stratosphere by rockets (*Eckermann and Vincent*, 1989.) (Reprinted with permission from Springer .)

Alexander (2000) and *Manson and Meek* (1988). *Manson et al.* (1999) applied a special analysis involving wind-velocity ovals and reported mesospheric data recorded with MF radar for a variety of stations.

11.2.12 Breakdown, convective adjustment (shedding), and catastrophic collapse

Gravity waves break by a variety of processes, some of which have already been discussed in and around Equation (11.31). These earlier discussions were in regard to critical levels and convective instability. Indeed, while not mentioned at the time, the two processes are related: it was mentioned that the critical level mode piles up its energy and so grows in amplitude, but of course the condition $\hat{u_0} > c_\phi - \overline{u}$ is easily satisfied as the latter term approaches zero as well. It turns out that the condition for convective instability is one of the more easily satisfied ones, so this is a major cause of turbulence in the atmosphere.

In terms of general breakdown mechanisms, it is best to think in terms of the Richardson number (see the end of Chapter 1). Waves grow as they increase in altitude and it is possible for the wave to produce internal Richardson numbers below 0.25 and thence break. A variety of mechanisms can permit this breakdown. One simple mechanism is convective instability, as already discussed, in which the wave has a Richardson number that is negative, meaning that the temperature gradient is unstable. For a wave, this occurs when the wave amplitude exceeds the intrinsic wave velocity $c_\phi - \overline{u}$, as discussed in Equation (11.31).

Shear instability (often in the presence of an already existing mean shear) (e.g., *Fritts and Yuan* (1989); *Reid et al.* (1987)) is another possible breakdown mode. However, while in the case of convective instability the motions become rotational and then chaotic quite quickly, in the case of wind-shear induced turbulence, it is common for the atmosphere to develop a short-wavelength localized wave that then coils up and eventually breaks. The well-known Kelvin–Helmholtz instability is one such scenario.

Indeed, speaking of the production of waves, one must be careful in assuming that gravity waves always break into turbulence. For example, *Klostermeyer* (1989, 1991) has proposed that waves may also lose energy by nonlinearly converting to waves of other frequencies and wavenumbers, and even "cascade" down to waves of very short vertical wavelength (parametric instabilities). This impacts the rate at which energy may be deposited in the atmosphere – while the motion is still in the form of waves, energy transport is reversible and cannot be considered as injecting heat into the atmosphere.

Other breakdown scenarios in addition to critical levels, convective overturning and Kelvin–Helmholtz instabilities include Rayleigh–Taylor, Holmboe, vertical-pair, slant-wise (*Hines*, 1988), parametric sub-harmonic (*Klostermeyer*, 1991), Beaumont, resonant, and oblique instabilities. Klaassen and colleagues (*Sonmor and Klaassen* (1997); *Klaassen* (2003); *Yau et al.* (2004)) have expended considerable effort in determining the conditions required for different types of breakdown to occur. These have been succinctly summarized in *Klaassen* (2003) and references therein. These analyses were performed using Floquet analysis and revealed a more complex picture of wave breakdown than that obtained by simpler Richardson-number arguments.

A major issue associated with gravity wave breakdown is the determination of whether waves break by catastrophic collapse, or by shedding. Shedding is also called "convective adjustment" (*Fritts and Rastogi*, 1985) and has already been discussed in Section (11.2.5), where the causes of the so-called universal spectrum were discussed. We will not discuss it again here, except to remind the reader that the concept assumes that turbulence occurs in small layers whenever the Richardson number drops below 0.25. This results in small amounts of turbulence being created, which absorb enough energy out of the gravity waves to cause the wave amplitudes to decrease towards stable values, whereupon the turbulence dies out. Hence waves do not break catastrophically in this scheme, but simply shed sufficient energy that they can maintain a stable spectrum. These patches of turbulence occur in much the same way as whitecaps appear on the surface of the ocean. The same concept has been considered for the troposphere by *Fairall et al.* (1991), and for the middle atmosphere by *Hocking* (1991) and *Hines* (1991a, b, c). Early advocates of such models (each with their own variation) included *Dewan and Good* (1986); *Smith et al.* (1987); *Weinstock* (1984a); *Tsuda et al.* (1989); *Fritts and Chou* (1987); *Gardner et al.* (1993a); *Gardner* (1994). Further extensive developments of these theories also include the work of *Hines* (1991a, b, c, 1993, 1996), although the latter ones were later (correctly) called into question by *Klaassen* (2010).

Recent modeling studies have suggested, however, that waves do not always shed their energy in short concise packets, but often break down catastrophically and deposit all of their energy locally in a relatively small region of space (e.g., *Fritts and Alexander*, 2003, Section 6.2.1, and discussions therein), like ocean waves breaking on a beach. The exact balance between shedding and catastrophic collapse is still unknown and is an area to which radar studies can contribute in future years. An associated issue is that of determination of the number of waves present in the atmosphere at any one time, as already discussed (e.g., *Sica and Russell*, 1999a, b). A small number of waves would lend themselves to catastrophic collapse ideas, whereas shedding might be more likely to occur if the atmosphere contains a large number of waves at any one time.

11.2.13 Momentum fluxes, drag forces, and energy fluxes

Momentum fluxes have a very dramatic effect on the mean wind, especially in the mesosphere. To begin, consider a typical mean-wind profile which is produced under conditions of radiative equilibrium (i.e., conditions in which the heating of the Earth and the Coriolis forces associated with rotation are in balance). In such a profile, the eastward mean wind increases monotonically with increasing altitude up to about 100 km altitude. To see this, we look at Figure 11.19. Concentrate only on parts (a) and (b) for now.

In a situation of a non-rotating Earth, with no clouds and no form of drag, the summer pole is warmer than the winter pole, being exposed to sunlight while the winter pole is in darkness. A temperature gradient therefore exists between the poles, as shown in Figure 11.19(a). This attempts to drive a north-south flow, which would in principle happen for just long enough that sufficient air would build up at the winter pole to produce a reverse gradient to counter the initial flow (other inertial-driven flows may develop, but we will ignore these for now). But now let the Earth be rotating. The same summer to winter flow develops as for the non-rotating Earth, but now we have a Coriolis force, at least as viewed from a frame of reference on the Earth (see the discussion following Equation (1.7) in Chapter 1 for a discussion of the "reality" of the Coriolis force). The Coriolis force re-orientates this into an east-west flow, as shown in Figure 11.19(b). In turn, this east-west flow produces a north-south Coriolis acceleration of its own, which acts with equal magnitude and in the opposite direction to the temperature gradient (yellow lines in Figure 11.19). As a result, no north-south flow occurs and a purely zonal flow remains. The wind speeds are larger at the higher altitudes largely due to the decreasing density with increasing height, so that the speeds associated with a specified momentum flux are larger.

Now ignore parts (c) and (d) of this figure for now, and turn to Figure 11.20. The mean wind profile just described is presented as the line labelled $\overline{u}_{\text{initial}}$. Suppose that gravity waves with phase velocities of all orientations propagate from the ground upward. Our primary interest will be those that propagate in the east-west plane. Two representative wave packets are shown in Figure 11.20. We assume that the phase velocities vary between say +60 ms^{-1} (eastward) and –60 ms^{-1} (westward). Eastward propagating waves eventually encounter an altitude where their phase speed equals that of the mean wind, so they will break by critical-level interactions. Thus they are prevented from propagating above this level, and also impart their momentum to the mean flow at this point. The higher we go into the atmosphere, the lower is the ratio of eastward propagating waves to westward propagating ones, since many of the eastward-propagating ones have been "filtered" out lower down. Above about 70 km, these westward moving waves now begin to break, but not necessarily by critical level interactions. Breaking could, for example, occur due to convective instability (Equation (11.31)), since as the waves grow with height (due to the decreasing density

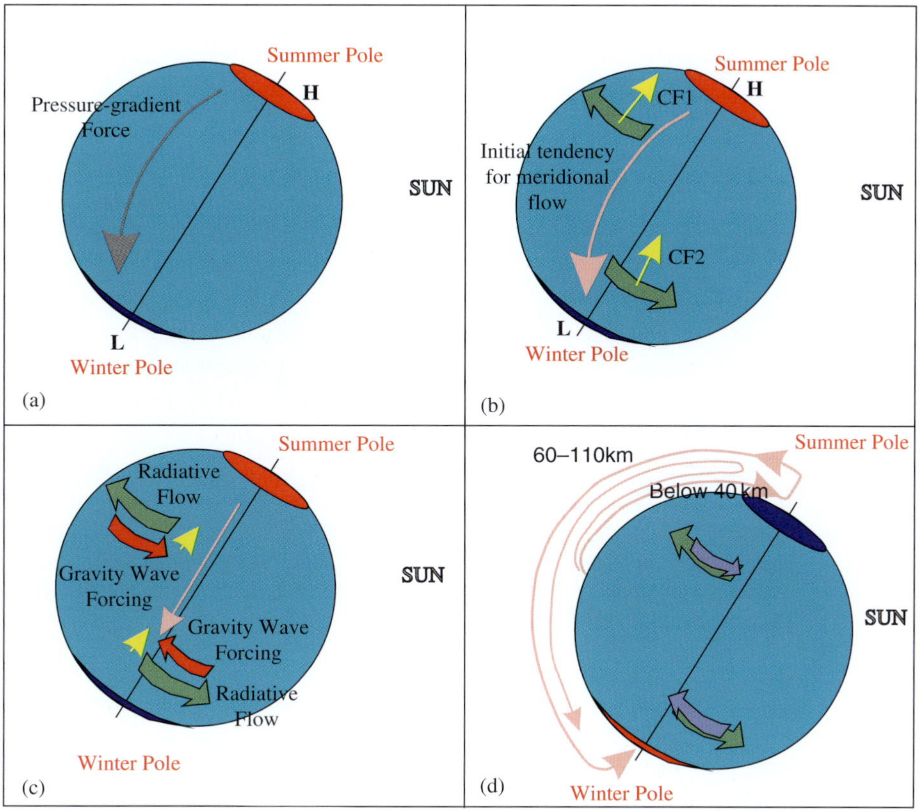

Figure 11.19 The way in which filtering-defined anisotropic propagation of gravity waves can alter the mean flow of the middle atmosphere. (a) The pressure gradient force due to heating of the summer pole. (b) The green thick arrows show the induced zonal wind flows due to the effect of the Coriolis force applied to the pressure gradient force. The yellow lines CF1 and CF2 show the Coriolis forces produced by the wind flow. If no other forcing acts, these yellow lines exactly balance the initial pressure gradient force, resulting in zero pole-to-pole flow, and a purely zonal flow. (c) Gravity wave forcing is shown in red, and reduces the zonal flow. As a consequence, the Coriolis forces due to the zonal winds (CF1 and CF2 in (b)) are reduced to the smaller yellow vectors seen in this figure. (d) The initial pressure gradient force is now stronger than the yellow vectors in (c), and so drives a flow above 60 km from the summer to the winter pole. This is shown by the pink arrows. Note that continuity must be achieved, which is done by drawing the upper flow into the lower Hadley circulation. The lower level circulation in the summer hemisphere is drawn, but the winter low-level circulation is more complex and is not included. (See Chapter 1, Figure 1.20 for more details about this lower level flow.) The green arrows represent the wind flow at approximately 50 to 85 km, and the purple arrows represent the winds which may exist at altitudes above 85 km, where the gravity wave forcing may be so strong that the winds are reversed in direction compared to those underneath.

at higher altitudes) their velocity amplitude increases and Equation (11.31) becomes satisfied. This imparts westward momentum to the mean wind. As a result, this slows and even reverses the mean wind above 70 km, as shown by the "final" profile in Figure 11.20. The drag force term which describes the acceleration on the mean wind is

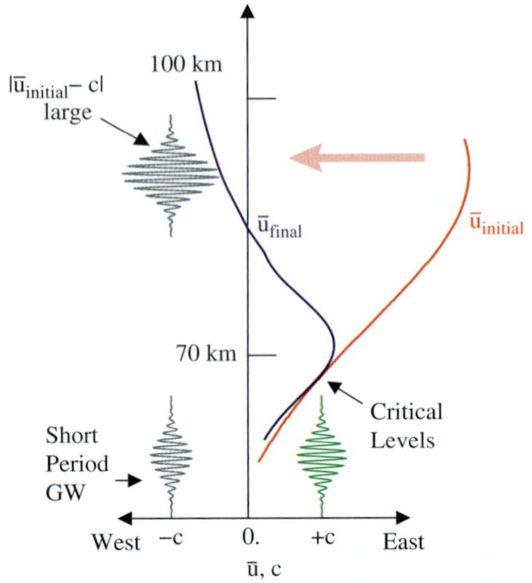

Figure 11.20 Showing how an initially isotropic gravity distribution at the ground leads to a mean flow weakening and then a reversal in the mesosphere. (Adapted and substantially modifed from *Thayaparan et al.*, 1995.) See text for details.

$$F_d = -\frac{1}{\overline{\rho}}\frac{\mathrm{d}}{\mathrm{d}z}\overline{\rho}\overline{\underline{u}'\underline{w}'}\,, \quad \text{or approximately} - \frac{\mathrm{d}}{\mathrm{d}z}\overline{\underline{u}'\underline{w}'}\,, \tag{11.32}$$

where the last term arises if we assume that the density varies relatively slowly with height relative to the term $\overline{\underline{u}'\underline{w}'}$. Note that F_d is a force per unit mass (acceleration).

This drag force also determines the mean north-south wind, and even the vertical winds and temperature distribution, as will now be explained. To see this, return to Figure 11.19, especially parts (c) and (d). The dominant westward propagating gravity waves reduce the strength of the mean zonal winds above 70 km altitude (shown by the red arrows in Figure 11.19(c)). Now the north-south Coriolis force is also reduced, since $\overline{u}$ is reduced, and so this Coriolis force due to the mean flow no longer balances the temperature gradient force. As a result a net north-south wind results at the upper heights (greater than say 50–60 km and up to 90 km and more). Equilibrium occurs when

$$\frac{\mathrm{d}}{\mathrm{d}z}\overline{\underline{u}'\underline{w}'} = f_c\overline{v}, \tag{11.33}$$

where $\overline{v}$ is the mean north-south wind and f_c is the Coriolis parameter $= 2\Omega\sin\theta$, θ being the latitude and Ω the rotation rate of the Earth ($= 7.27 \times 10^{-5}\mathrm{rad\,s^{-1}}$).

The gravity-wave forcing at the upper levels is often strong enough to reverse the flow from the radiative equilibrium case at those levels. In any event, regardless of whether the zonal flow is reversed or just weakened, the non-zero meridional flow now forces air to accumulate at the winter pole and be drawn out of the summer pole. This induces vertical movement at the poles, with rising air at the summer pole, and falling air at the winter pole (Figure 11.19(d)). (A similar diagram was shown in Chapter 1, Figures 1.19 and 1.20, but with less explanation.) Rising air produces adiabatic cooling, and falling air produces adiabatic warming, so the winter pole is warmed, and the summer pole is cooled. The result is that the summer mesopause is a very cold region of the Earth's atmosphere, with temperatures as low as 120 K.

The full summer to winter pole-to-pole flow occurs at altitudes above about 60 km, and a partial effect is seen as low as 40 km. Of course with the build-up of air expected at the winter pole due to this subsidence, it is necessary to have a source of supply for air in the summer hemisphere, and the falling air on the winter hemisphere must eventually leave the winter pole. The process by which this occurs is complex: air is supplied via flow upward and out of the equator, moving towards the summer pole, where it then moves up even further, and then flows towards the winter pole. At the winter pole, sinking occurs all the way down to the lower stratosphere, where the air is absorbed into the low-level polar circulation. This is illustrated in Chapter 1, Figure 1.20. Since the density of the upper altitude air is very low, it has little influence on the lower-level circulation and is drawn into that flow without any real impact.

The resultant temperature distributions with and without radiative equilibrium are shown in Figure 11.21.

To see how these events play out in practice, consider a 10 ms^{-1} north-south wind at 45° latitude ($f_c\bar{v} \simeq 10^{-3}\,\mathrm{ms}^{-2}$). This requires a value for $\frac{\mathrm{d}}{\mathrm{d}z}\overline{\hat{u}\hat{w}}$ of around 1 $\mathrm{m^2s^{-2}km^{-1}}$, or about 80 $\mathrm{ms^{-1}day^{-1}}$. Measurements of momentum fluxes by radar are crucial in order to determine the validity and details of this momentum flux balance scenario. The few investigations made so far, using Equation (11.26), seem to

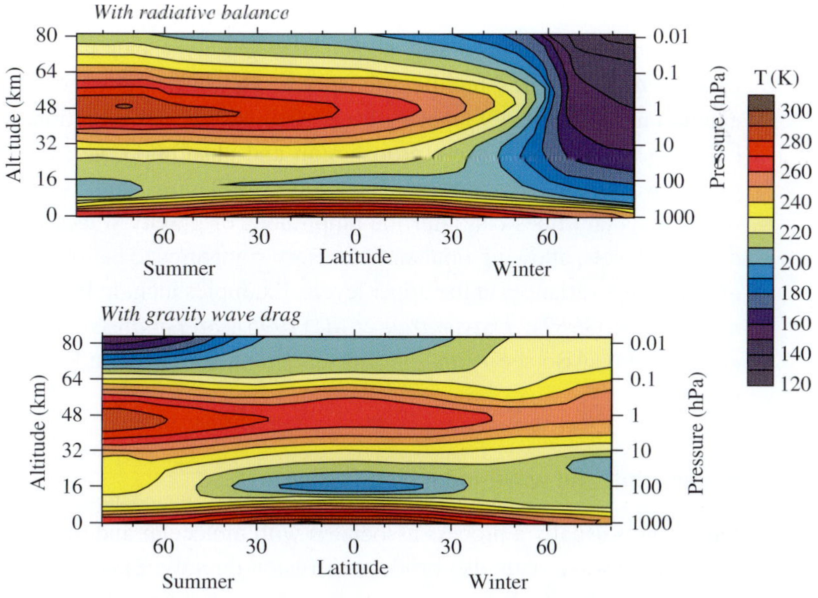

Figure 11.21 Temperature distributions as a function of latitude and height for cases of (a) radiative equilibrium, and (b) gravity wave impact. These graphs are an extended version of Figure 1.23 shown in Chapter 1.

measure values of this order (e.g. *Vincent and Reid*, 1983; *Fritts and Vincent*, 1987; *Reid and Vincent*, 1987), but there have only been occasional measurements. More are needed in order to confirm that the gravity wave drag is responsible for the observed meridional winds. More extensive discussions can be found in *Fritts and Alexander* (2003).

In the discussions above we have concentrated on $\overline{u'w'}$ and its height derivatives. But it is worth recalling that the north-south main winds can also be partly balanced by other momentum forcing such as terms like $\frac{d}{dy}(\overline{u'v'})$. More studies of all these different momentum fluxes and body forces are badly needed before the dynamics of the atmosphere can be better understood. An appreciation of the role of gravity waves, and better measurements of the parameters outlined above, are crucial if this goal is to be attained. *Becker* (2012) provides useful extra reading regarding these concepts.

The other topic of this subsection is energy fluxes. These have not been studied in much detail at all, and generally momentum fluxes have taken priority. Nevertheless, some measurements have been made. *Vincent* (1987) reported estimates of energy fluxes between 7×10^{-4} W m^{-2} for waves in the period range 50–60 mins, and $\simeq 10^{-2}$W m^{-2} when integrated over all wave periods, at altitudes of 80–90 km. *Czechowsky et al.* (1989) have made measurements of the upward and downward fluxes of gravity wave energy at the top of the troposphere in Germany and found values in the range 0.08 to 0.1 W m^{-2}. The upward flux exceeded the downward flux by about 0.014 W m^{-2}. Again, more measurements are needed.

11.2.14 Mean flow interactions

In the previous section, it was shown how gravity waves may impact the mean circulation of the Earth's middle atmosphere. However, the effect of gravity waves can extend far beyond this. Planetary wave oscillations, and even tidal oscillations, can all be affected by gravity wave propagation and absorption. Critical level absorption of gravity waves can force new planetary wave oscillations, and conversely planetary wave oscillations and tides can alter the amplitudes of gravity waves that penetrate them at different levels, allowing tidal and planetary signatures to be imprinted on the measured gravity wave variances at the upper levels. Examples include *Walterscheid* (1981); *Fritts and Vincent* (1987); *Thayaparan et al.* (1995); *McLandress and Ward* (1994); *Zhong et al.* (1995). Also see *Fritts and Alexander* (2003), Section 8.2, for a more extensive review.

11.2.15 Stokes' drift and wave-induced diffusion

Diffusion is usually a process associated with molecular and turbulent processes, but in fact gravity waves can also produce diffusion directly. This comes about via a process called Stokes' drift. When a gravity wave completes one full period of oscillation, its three components of velocity each return to the values that they had one period previously. One might expect that a particle operating under the influence of this wave should complete an elliptical path and also return to its original position as well. In truth, the particle returns to a position displaced slightly from its original location (see *Coy et al.* (1986) and *Hall et al.* (1992), and references therein, for a discussion of this process). This small displacement is called Stokes' drift. Normally its impact is considered to be of little consequence, but on occasion it can be important. *Coy et al.* (1986) and *Hall et al.* (1992) considered its impact in producing abnormally large vertical velocities in association with polar mesosphere summer echoes. *Walterscheid and Hocking* (1991) and *Hocking and Walterscheid* (1993) have examined the impact when the effects of combining the Stokes' drifts of an ensemble of waves are considered. The result is that a particle subject to Stokes' drift from a spectrum of linear gravity waves that are harmonically related in period executes a random walk, even after integral numbers of cycles of the lowest frequency wave (even though in principle the particle should have returned to its original position at this time). Particle displacements over periods of many hours can be many kilometers, and the process looks to all intents like diffusion. In the case of nonlinear waves, the process is amplified. *Weinstock* (1982) has also examined the possible diffusive effect of an ensemble of waves.

Hence a spectrum of gravity waves can lead to atmospheric mixing of constituents even without production of turbulence. In the case of purely linear waves, there can be no net diffusion of momentum, as discussed by *Hocking and Walterscheid* (1993), so the diffusion is somewhat misleading. Once the waves become even slightly nonlinear, however, the process can be important. In addition, it appears that this Stokes' diffusion, when acting on a pre-existing constituent gradient, can produce genuine constituent diffusion, even for the case of linear waves. We will see shortly that turbulence in the middle atmosphere often occurs in thin layers, and so the processes by which diffusion occurs over scales of several kilometers vertically still require clarification. Stokes' diffusion, and other gravity wave related diffusive processes, are models that may be important when considering large scale diffusion in the atmosphere (*Hocking*, 1999a).

11.2.16 Local gravity wave effects

We have concentrated in the preceding sections on global effects of gravity waves at all levels of the atmosphere. There are also some quite unique and distinctive effects which occur at a local level. One example of such an effect is the Chinook winds.

The Chinook winds are strong breezes which blow down from the mountains onto the plains on the eastern side of the Rockies in Western Canada. These winds can often be periodic. In other words, the winds can be quite strong, and then die down to low values, and then increase again to large values, and so forth, with times between the "quiet" periods being typically 10 minutes. This periodicity, when it occurs, is due to gravity waves. Similar periodicities can often be found in other mountain down-slope winds.

The periodicities which arise are associated with gravity waves which are produced by air flow over the mountains. These waves lead, of course, to lee-waves and lenticular clouds. If they become intense enough, they may produce Kelvin–Helmholtz secondary instabilities generated at the shear interface between low-level wind jets and upper-level decelerated winds, and these instabilities can play a key role in the generation of downslope wind pulsations. This process has been discussed in more detail by *Peltier and Scinocca* (1990).

Other examples of local gravity wave effects exist. One key equation in this regard is the Taylor–Goldstein equation. This equation results when it is assumed that waves are generated with sinusoidal horizontal variation and sinusoidal temporal variation, but for which the vertical structure of temperature and mean winds is locally quite complex. For example, the vertical fluctuating velocity may be of the form

$$\underline{\hat{w}}(x, z, t) = \underline{W}(z) e^{i(kx-\omega t)}, \tag{11.34}$$

where $\underline{W}(z)$ might be non-sinusoidal. This is different to the previous discussions about gravity waves in the sense that in the earlier cases it was assumed that the vertical structure was quite simple; now the vertical structures are allowed to be quite general. These structures can lead to wave trapping, wave ducting, and other phenomena. The structure of the "cavity" can also define the growth of some gravity waves and suppress others. For this case, the relevant equation is the Taylor–Goldstein equation. One simple form of this equation is

$$\frac{d^2 W}{dz^2} + \left[\frac{\omega_B^2}{(\overline{u} - c_\phi)^2} - \frac{1}{(\overline{u} - c_\phi)} d^2\overline{u}/dz^2 - k^2 \right] W = 0, \tag{11.35}$$

although more complex versions exist (e.g., *Merrill*, 1977; *Merrill and Grant*, 1979; *Gossard and Hooke*, 1975). *Merrill* (1977) applied the Taylor–Goldstein equation to a cavity to demonstrate the growth of one wave mode out of geophysical noise fluctuations at the expense of other modes. In general, solution of these equations can be very important in determining the behavior of gravity waves in regions of complex vertical structure.

There are many other local effects associated with gravity waves, but there is insufficient space to address them all.

11.2.17 Gravity wave parameterization for meteorological models

Computer models need to incorporate and recognize the effect of small-scale forces like turbulence and waves on the outcomes of their simulations. In the past, some form of dissipation like eddy diffusion or Rayleigh drag was adequate. These "drag" mechanisms can, in the most extreme case, pull the mean winds back to zero speed. Often, however, the experimental wind fields were found to be reversed relative to the radiative equilibrium situation. Only waves, and especially gravity waves, can produce flow reversal relative to the radiative balance situation. With the recognition in the 1980s of the importance of gravity waves for defining large-scale circulation, gravity waves became an important ingredient of circulation models, both in the middle atmosphere and in the troposphere. However, it is impossible, even with modern computers, to include a full gravity wave spectrum in such models. Therefore, it is necessary to find simple ways to parameterize the effects of these waves, so that their main impacts can be incorporated without a need for excessive increase in computer resources.

Many schemes have been developed to do this (e.g., *Kim et al.*, 2003). It is important that such algorithms properly represent the energy content of the waves, the rate of energy loss as a function of height, and the momentum flux and body forces associated with the waves. The first such model was due to *Lindzen* (1981), who developed an equivalence between wave forcing and eddy diffusion. Others have been developed and are under test. They are important for determination of middle atmosphere circulation and for tropospheric numerical weather forecasting. Different types of waves need to be incorporated – orographic, convectively forced, jet-stream generated, frontal system waves, and so forth. Thus a good parameterization also requires knowledge of wave sources as functions of time, season, latitude, and longitude. The intermittency of these sources is also important to know (*Alexander*, 1996).

We will not delve too deeply into these models here, since at present they are more the realm of computer modelers, but radar scientists need to know at least of the importance of this development. However, we will note that in recent years computer models have become sufficiently powerful that they can even generate realistic gravity waves within them. Further discussion can be found in *Fritts and Alexander* (2003), section 7. Other useful references include *Holton* (1982); *McLandress* (1998); *Medvedev and Klaassen* (2000); *Warner and McIntyre* (2001); *Schmidt et al.* (2006); *Becker* (2012); *Sato et al.* (1999); *Hamilton* (2006); *Alexander et al.* (2010); *Miyoshi et al.* (2014).

11.3 Turbulence in the upper atmosphere

Once gravity waves break, they generally produce turbulent motions. Understanding turbulence is important not only for the energetics of the atmosphere, but also because it is a primary cause of radar scatter.

We have already briefly discussed turbulence in various aspects earlier in this book (especially in Chapters 3 and 7), chiefly in regard to its effect in causing radar scatter, and the methods needed to determine turbulence parameters from radar measurements. This required knowledge of the spectral characteristics of the turbulence and the various important scales of the turbulence (buoyancy scale, Kolmogoroff scale, inner scale, etc.). We will briefly re-examine these features in this chapter, but we will also present an overview of the general characteristics of turbulence in the atmosphere, and especially look at its impact on atmospheric dynamics and energetics.

11.3.1 Turbulence structure above the boundary layer

Turbulence in the upper atmosphere is different in several ways to turbulence at ground level. Whereas at the ground frictional and drag forces both drive and define boundary layer turbulence, in the upper atmosphere, and certainly in the upper troposphere, stratosphere, and mesosphere, turbulence is driven largely by wave events which break and interact with each other. Turbulence affects its environment in at least two main ways: it may heat the fluid in which it exists, and it causes diffusion of momentum, heat, particles, and atmospheric constituents. Turbulence occurs on a wide range of scales, but in this book, most discussion will be concentrated on small-scale turbulence, that is, scales less than about 5 km in size, where turbulence is at least quasi-isotropic and can truly be called three-dimensional turbulence.

A variety of parameters are important in describing turbulence. From the perspective of radar scatter, knowledge about the internal structure, and the spectral distributions, is a necessity. From the point of view of larger scale dynamics, there are other important parameters. The rate at which turbulence causes heating of its environment, ε, and the rates at which momentum (K_M) and heat (K_T) diffuse are some of the most important. In theory, the rates of diffusion of momentum and heat differ, but in practise they are often taken to be similar.

The quantity ε refers to kinetic energy dissipation, but the "kinetic" term is often implied and left unstated. Furthermore, ε has a second meaning – apart from being the rate of atmospheric heating, it is also the rate at which energy flows between different scales, as we will see soon. But energy transport does not have to be only in the form of kinetic energy. Due to the existence of temperature structure, and the associated buoyancy, there is another form of energy storage in a turbulent region – potential energy. This can also be dissipated, and in some cases can actually be negative. A proper treatment of turbulence recognizes these two forms of ε, denoted as ε_K (for kinetic) and ε_P (for potential). When no subscript is used, ε_K is inferred.

Many measurements made in the mesosphere assume that turbulence obeys Kolmogoroff inertial range theory (*Kolmogoroff*, 1941a, b; *Kolmogoroff*, 1991a, b; *Tatarski* 1961), and this applies between a minimum scale called the "inner scale," ℓ_0, and a buoyancy (outer) scale, called the buoyancy scale, L_B.

Various levels of sophistication exist for theories connected with turbulence, but Figure 11.22 gives a general overview. It is envisaged that turbulence is rotational, and that some mechanism produces turbulence at large scales. These rotational motions produce "eddies" that drive rotations at yet smaller scales, which drive rotations at yet smaller scales, and so forth, until the right-hand region is reached. (The concept of an eddy is to some extent an idealization, and true motions are more complex (see Section 2.15.7 in Chapter 2, and also *Hocking and Hamza* (1997).) The quantity k refers to Fourier components of the turbulent region. In the region on the right, the scales are smallest, but the second derivative of the velocity fluctuations is largest, so that $\nu \frac{d^2u}{dz^2}$, the viscous dissipation of momentum, is greatest. There are three main regimes, although some authors introduce others. The first is the buoyancy regime, where the eddies are large enough that gravity still plays a role in defining them. Typical scales in this region can be of the order of kilometers to tens of meters, depending on the energy dissipation rate. This regime has been described in some detail by *Weinstock* (1978a) who actually considered it as two separate regions. He also noted that this region not only cascades energy to smaller scales, but can radiate gravity waves away – especially in the lower-wave-number part of the regime. Energy density as a function of k varies proportionally to $k^{-5/3}$. At scales smaller than the buoyancy range, the *inertial range* exists. In this regime, buoyancy efffects are much weaker than in the buoyancy range, and there is little energy loss. So energy essentially just cascades through the region.

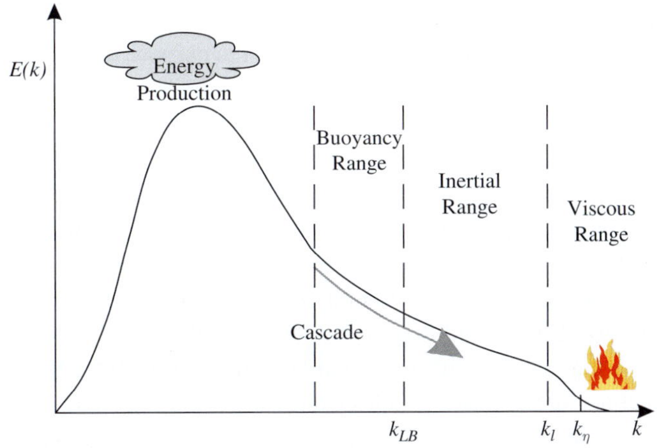

Figure 11.22 The energy spectrum of a typical turbulent region of the atmosphere as a function of wavenumber. Largest scales (Fourier wavelengths) are on the left, while the smallest scales (largest wavenumbers) are on the right. Energy cascades from the production region on the left to the viscous region (or viscous range) on the right, where energy is finally deposited as heat. Key wavenumbers are marked. See text for details.

The eddies are considered to be statistically isotropic, although recent work suggests this is an oversimplficiation (*Hocking and Hamza*, 1997). Dimensional analysis (*Kolmogoroff*, 1941a, b) can be used to show that the energy density varies as $E(k) = \alpha\varepsilon^{2/3}k^{-5/3}$, where $\alpha = 1.53$ (also see *Tatarski*, 1961, 1971, and Appendix A in this book), ε being the rate at which energy cascades through the scales, and also the rate at which energy is finally deposited as heat at the smaller scales. On the far right, small-scale wind-shears become large, and terms like $\nu\frac{d^2u}{dz^2}$, (where ν is the kinematic molecular viscosity), become dominant. Hence the eddies are heavily damped, and the energy density drops away rapidly as k increases. The exact details of the way in which the energy drops away vary, depending on the theoretical model, but it is either a rapidly decaying power law as a function of k, with exponent of the order of -9 to -7, or an exponential decay, or even a combination of power laws and exponential decay. Some of the different options are discussed by *Tchen* (1954); *Lübken et al.* (1987); *Driscoll and Kennedy* (1985); *Lübken* (2014). *Tchen* (1954) actually proposed an additional subrange between the inertial and viscous ranges, referred to as the "Tchen range" but that seems to have dropped out of favor. Occasionally reference to other subranges like the viscous-convective subrange can be encountered. The most popular model at presents however, seems to be that of *Driscoll and Kennedy* (1985). *Lübken* (2014) shows a comparison between some different models. For most of the purposes of this book, the differences are not too significant, unless we are dealing with scatter from scales within the viscous range, in which case the form of the viscous part of the spectrum becomes important (*Lübken*, 2014).

In Figure 11.22, various scales are indicated along the abscissa (k_{LB}, k_l, and k_η). These need a little explanation. There is an implicit conflict here. By definition, the inverse of k

is $\lambda_k/(2\pi)$, where λ_k is the wavelength of the corresponding spatial Fourier oscillation. However, in deriving quantities like k_η, the purpose is to find representative scales which somehow define the transition between regimes (e.g., a border between the inertial and viscous regions). Such calculations do not involved visualization of a Fourier scale, but an entity like an eddy. So this represents a conflict between visualizations of the type of scale envisaged. In the case of the scale η_K, which is supposed to represent an eddy that has equal aspects of the inertial range and the viscous range (we will see the proof later), the scale deduced is actually something of the order of 5 to 10 times smaller than the wavelength at the break-point of the spectrum. Another example can be taken from *Hocking* (1987a), and *Briggs and Vincent* (1973). In this case, they looked at an ellipsoid scattering "eddy" with Gaussian cross-section of the form $n \propto n_0 \exp\{z^2/h^2\}$, where h defines the $1/e$ half-width of the eddy. This eddy has a full width at half-maximum amplitude of $2\sqrt{\ln\, 2}h$. These authors showed that maximum radar scatter occurred from the eddy if $h = 0.195\lambda$ where λ is the wavelength of the probing radiowave. Then the "width" of the eddy is $2\sqrt{\ln\, 2}\; 0.195 \times 2 \times \lambda_B$, where $\lambda_B = \lambda/2$ is the Bragg backscatter scale. So the eddy width is about 0.65 times λ_B. Hence the dominant reflective Fourier scale is 0.65 times the Bragg scale. So the eddy width is almost one half of the dominant Fourier scale. Once again, there is a conflict in definition between the Fourier scales and the eddy scales.

Because of these apparent inconsistencies, workers in turbulence theory take an interesting approach. If a critical number like η_K is based on a dimensional derivation, the corresponding wavenumber scale is not denoted as $2\pi/\eta_K$, but rather simply as $1/\eta_K$. They regard such scaling constants as being of the right order to within a factor of 2π, but do not care too much if it does not exactly match the original intent exactly. Scales formed in this way become a *definition* which helps them in normalizing their equations, and they then leave it to experiment or other analyses to determine any relevant scaling constants. This can lead to confusion between experimentalists and theorists – for example, a theoretician might consider k_η as $1/\eta_K$, whereas an experimentalist might regard k_η as $2\pi/\eta_K$. As seen above, neither is correct in any absolute sense of the word, since comparing eddies and oscillations is somewhat meaningless. Readers should be aware of this potential confusion as they flip between experimental and theoretical papers. In the end, experimental measurements are needed to define a more accurate transition, but sometimes the original theoretical parameter is still maintained. This is the case here, where the scale η_K is maintained, and a new parameter ℓ_0 is the more experimentally accurate transition scale between the inertial and viscous range. In fact it gets even more confusing – the Kolmogoroff microscale is defined by looking at diffusion of constituents, but in dealing with turbulence, we discuss many parameters – velocity fluctuations, neutral density variations, temperaure variations, electron density variations, and so forth. The relation between the inner scale and the Kolmogoroff microscale is different, depending on the parameter under study (e.g., it is different for temperature and wind perturbations by a factor of about 50%).

For our purposes, the key scales are η_K, ℓ_0, and L_B. While the original intent of η_K was to find a transition scale between the inertial and viscous ranges, it did not produce this, but in fact produced a scale well into the viscous region (as seen in Figure 11.22), necessitating *another* scale to define the true transition. This scale is ℓ_0, which is of the order of 5 to 10 times η_K (which can be shown by using structure-function theory (e.g., *Tatarski*, 1961)). These points all need to be borne in mind in the following discussions.

Once the basic premises of turbulence are established, the mathematical theory of turbulence is quite rigorous. But some problems do arise in the basic assumptions. Is it valid to assume that no energy loss to heat take place in the inertial range? Is it valid to assume that the situation is homogeneous? How important is intermittency? Is it valid to assume that the nature of the turbulence is independent of the mechanism of generation? There is an element of assumption within the underlying principles, which leads to issues of interpretation and ultimately uncertainties in the measurement of the parameters. Is the "energy dissipation rate" the optimum parameter to measure? Some authors distinguish between kinetic energy dissipation rates and potential energy dissipation rates, for example. Is it too simplistic to consider that diffusion can be represented by a single parameter K? In addition, measurement of many of the parameters is difficult, especially the kinetic energy dissipation rate ε and the eddy diffusion rate K. It is often considered to be good accuracy if these can even be measured to within a factor of 2 or 3.

Given the accuracy with which measurements have been possible, and even the possibility of uncertainty in interpretation, it is often only worth using approximate relations between the various parameters. Such relations will be discussed in due course. In this text, we shall adopt the theory discussed above as a basis, and will make measurements of things like ε and K. Typical inner scales and buoyancy scales as a function of altitude have been presented by *Hocking* (1985) and will also be shown later. Without these types of parameterizations, we can do nothing constructive at all! But all the time, the warnings sounded above should be borne in mind. For now we note that typical values of L_B are of the order of kilometers down to a few tens of meters, depending on turbulence strength, and typical values of ℓ_0 and η_K are of the order of millimeters and centimeters in the lowest atmosphere and rise to values of meters in the mesosphere. The values L_B and ℓ_0 become similar around 100–110 km altitude.

This "approximate" approach applies not just to the turbulence parameters that we measure, but also to the form of the spectrum assumed. We accept the basic hypothesis that the turbulence obeys the rules of the Kolmogoroff theory of inertial range turbulence. This is often questioned as a valid assumption, and no doubt it becomes less true as one approaches the upper levels around 100 km altitude, where the buoyancy scales and inner scales approach each other. Nevertheless, the little experimental data available suggest that the turbulence at least tries to tend to a Kolmogoroff spectral shape (e.g., *Zimmerman et al.*, 1971; *Booker and Cohen*, 1956; *Blix et al.*, 1990; *Lübken*, 1997), at least in conditions of weak to moderate wind-shear. For stronger wind-shears, other theories (e.g., *Tchen*, 1954) have occasionally been invoked. Theoretical studies such as those by *Hill and Clifford* (1978) and *Driscoll and Kennedy* (1985) also show that there is something like an inertial range of turbulence with the classical Kolmogoroff shape, although interesting departures occur near the scales at which viscous energy dissipation occurs. In addition to questions about the appropriateness of assuming a Kolmogoroff spectrum, it is also noteworthy that the upper atmosphere is an especially difficult region to study. It is, for example, too low for in-situ satellite measurements, yet too high for aircraft. Even measurements of ε and K must be made by somewhat indirect means and are therefore difficult. Given the tendency for the atmosphere to at least try to approach an "inertial" spectrum, and given that an experimental bias will be followed in these pages, it will be assumed in this chapter that the Kolmogoroff theory may be approximately applied.

Turbulence, by its nature, is very variable in intensity, but there are some general features about its altitudinal distribution that can be commented on. Clearly it can be quite intense in the Earth's boundary layer and troposphere, especially during storm conditions. However, the assumption that turbulence is weak above the boundary layer (as is often assumed) is far from true. Indeed, some of the largest energy dissipation rates per unit mass anywhere in the atmosphere occur above 60 km altitude, where values of ε can easily exceed the values seen in a modest storm at the ground. In the boundary layer, turbulence is often caused by orographic effects. Above the tropopause, the main sources of turbulence are almost certainly gravity waves and (to a lesser extent) tides and even planetary waves. These generate turbulence by processes such as nonlinear breaking, shear instabilities, convective overturning and critical-level interactions (*Lindzen*, 1981; *Teitelbaum and Sidi*, 1976; *Sidi and Teitelbaum*, 1978; *Hodges*, 1967; *Jones and Houghton*, 1971). Measurements of turbulence by rocket techniques (e.g., *Blamont and Barat*, 1967) have shown that turbulence often appears in horizontal laminas of thicknesses of a few kilometers, interspersed with non-turbulent regions, and it appears that turbulence is both spatially and temporally intermittent. Turbulence appears to occur in patches; *Anandarao et al.* (1978), *Teitelbaum* (1966), and *Zimmerman and Murphy* (1977) have presented data to suggest that turbulence occurs between 30% and 80% of the time, with the lower percentage occurring at lower heights. Radar studies, and particularly high resolution VHF studies, have also confirmed the intermittent and layered nature of turbulence at these higher altitudes (*Czechowsky et al.*, 1979; *Röttger et al.*, 1979; *Woodman et al.*, 1980; *Woodman*, 1980; *Sato et al.*, 1985; *Sato and Woodman*, 1982b). With regard to the upper troposphere and stratosphere, balloon studies have shown layer thicknesses that vary from 2 or 3 meters up to 1000 m (*Wilson et al.* 2011), with the overall distribution broadly following a power law, and with the thinnest layers being almost 10^4 times more common than the 1 km thick layers.

Generally, turbulence is important to an upper altitude of somewhere between 90 and 110 km (the exact limit varies with time within this range), whereupon the atmospheric viscosity becomes so large that it quickly damps any tendency for turbulence to form. This transition region is called the turbopause. We will discuss this "cap" to the turbulent regime in more detail later.

When considering turbulent flows in the atmosphere, the mean state motion is used. This is an adaptation of Equation (11.3), where the x-component of the velocity, u, is replaced by $\overline{u} + u'$, $\overline{u}$ being the mean speed, and u' being the fluctuating x component. Similar substitutions for the y and z components are made. The exact details of this transformation can be found in *Houghton* (1977).

Note that in the following pages, we will have a slight change in notation. In the previous section on gravity waves, we used the convention $\underline{\hat{\rho}}$ to represent density perturbations, for example. However, that convention really meant that this was the solution to the linearized form of the equations. In the following pages we will revert to the use of a prime to indicate perturbations from the mean e.g. ρ', and this perturbation term will represent all motions after subtraction from the mean. We will also remove the underline, and essentially take all components as real. Complex theory can be used to solve the equations at times, but we will not delve too much in any more theory, and so the simple non-complex representation will be adequate.

In the upper atmosphere, then, the resultant equation for the x-component of velocity is (e.g., *Dieminger et al.*, 1996, Section 1.2.2):

$$\frac{\partial \overline{u}}{\partial t} + (\overline{u}, \overline{v}, \overline{w}) \cdot \vec{\nabla} \overline{u} = \frac{1}{\rho_0} \frac{\partial \overline{p}}{\partial x} - \left[\frac{\partial}{\partial x} \overline{(u')^2} + \frac{\partial}{\partial y} \overline{(u'v')} + \frac{\partial}{\partial z} \overline{(u'w')} + \nu \nabla^2 (\overline{u} + u') \right], \tag{11.36}$$

where $\overline{u}$ is the mean x-component (eastward) of velocity, u' represents the fluctuating x-component of velocity, $\overline{v}$ is the mean y-component (northward) of velocity, $\overline{w}$ is the mean vertical component of velocity, and v' and w' are appropriate fluctuating components. In addition, ρ_0 is the mean density, $\overline{p}$ is the mean pressure, and ν is the kinematic viscosity coefficient. Similar equations exist for the y and z components. Here, the Coriolis force has been ignored (compared with Equation (11.3)), and also the force due to gravity.

This equation looks very much like the standard Navier–Stokes equation for a fluid (Equation 11.3), except that the total velocity vector $\underline{\vec{u}}$ has been replaced by the mean velocity vector $\overline{\underline{\vec{u}}}$, and additional terms like $d/dz\left(\rho\overline{u'w'}\right)$ now exist. As seen in Chapter 1, Equation (1.13), terms like $\rho\overline{u'w'}$ are examples of "Reynolds' stresses."

It should be noted that the term $\nu\nabla^2(u + u')$ has been left in. This is not normal practice, since it is usually assumed that the viscosity is small. Indeed, this is true below about 95–100 km altitude, but it is important to note that it may not be negligible, especially at heights where the kinematic viscosity exceeds the turbulent diffusion coefficient. *Fujiwara et al.* (2004) ignored this term and interpreted energy loss above 100 km over the EISCAT radar as due to turbulence dissipation, whereas in fact turbulence hardly occurs at all at those heights due to the high kinematic viscosity. Rather, the wind-shear gradients at the upper heights, combined with the large kinematic viscosity, enable the energy of fluid motion to pass to heat by direct molecular transport with no need for turbulence at all. Even turbulence ultimately acts through this term; the main role of turbulence is simply to produce wind-shears that have sufficient gradients at the smallest scales that this term can be important. The energy deposited as heat per unit mass and per unit time is

$$\varepsilon = 2 \int_0^{\infty} \nu k^2 E(k) dk, \tag{11.37}$$

where $E(k)$ is the total energy spectrum (e.g., *Hocking*, 1999a, and references therein). In regions of high kinematic viscosity, heat deposition can be achieved without the need for turbulence, since the combination of shears and high viscosity allows rapid frictional heating. Note also that if $E(k) \propto k^{-5/3}$ in the inertial range, then there will be contributions to heating even from higher wavenumbers in the inertial range, since

$E(k)k^2 \propto k^{1/3}$. The heating becomes much stronger in the viscous range, of course, where $E(k)$ falls off more sharply so the k^2 terms are more dominant.

The term $\rho\overline{u'w'}$, a component of the Reynolds' stress, represents the vertical flux of horizontal momentum. These terms have already been seen in regard to gravity wave transport, although in those cases the terms were referred to as "momentum flux" terms.

As noted, the above equation looks very much like Equation (11.3) except that a new set of terms involving the Reynolds' stresses has been introduced.

Since the random motions due to molecular collisions lead to diffusion, it might be expected that the random motions due to turbulence should do likewise. So by analogy with diffusion, we introduce a term that looks like $\nu\nabla^2\underline{u}$. We write this term as $K_{zz}\nabla^2\underline{\overline{u}}$ and use it to replace the term $\overline{\frac{\partial}{\partial z}(u'w')}$. Similar terms like $K_{xy}\nabla^2\underline{\overline{u}}$ might be used to replace the term $\overline{\frac{\partial}{\partial y}(u'v')}$, etc. In molecular flow, the kinematic viscosity is defined by the relation $f = -\rho\nu d/dz(\overline{u'w'})$ where f is the drag force per unit area. In the case of flow with fluctuating motions, the Reynolds' stress acts like the viscous drag, and either by noting the similarity between the Reynolds' stress and the viscous drag, or by comparing (11.3) and (11.36) (with ν in (11.3) replaced by the turbulent diffusivity K_{zz}), it can be seen that the momentum diffusivity K_{zz} is defined through the relation

$$\rho\overline{u'w'} = -\rho K_{zz}\frac{d\overline{u}}{dz}. \tag{11.38}$$

At the molecular level, and with the simplest theory, the particle diffusion coefficient and the kinematic viscosity are identical, (e.g., *Tabor*, 1969): after all, the molecular viscosity is just the momentum diffusion coefficient, and the momentum of a particle is just its mass multiplied by its velocity, so the rate of diffusion of momentum *per unit mass* (kinematic viscosity) and rate of diffusion of particles are in essence the same thing. Subtle differences will arise in more complex theory, but for the sake of developing our new turbulent diffusion coefficient term, the assumption is appropriate.

We therefore make the same assumption at the turbulent level, viz. the turbulent diffusion coefficient K_{zz} (also denoted K_M, where the M stands for momentum) is also called the (vertical) turbulent viscosity.

At this point, we need to raise a concern. The development has been done purely as an analogy. But the case of molecular collisions is a true random walk, while in turbulence, the existence of eddies means that it is not. As a particle moves around, driven by the eddy motions, there is a scaling effect that makes the dispersion increase quite dramatically as time increases, up to a limit associated with the buoyancy scale. The details of this process will be discussed later – for now, we simply highlight that the analogy with molecular diffusion may have its limits.

Returning to our new hypothesized turbulent diffusion coefficient, we note that even in this simple theory, there is an asymmetry in the rate of diffusion as a function of the direction being considered, so additional diffusion coefficients are needed (e.g. K_{xz}, K_{yz}, etc.). However, these differences are usually more apparent at larger scales. (Some very preliminary estimates of large-scale horizontal diffusion coefficients have been made by *Ebel* (1980).) In most of this section, it is the effects of turbulence at scales less than about 5 km (small scale) that interest us. At such scales, the rate of diffusion is approximately independent of direction; the rates of diffusion in the vertical and horizontal are at least similar to within a factor of 2 or 3, so concentration will mostly be on K_{zz}. We also note here that the values of the vertical diffusion coefficient vary as a function of scale, because the processes of diffusion are scale dependent (e.g., *Hocking*, 1999a, Table 1).

The above development of K_M has been based on mixing of the turbulent velocities, or equivalently the momentum mixing. Since molecular momentum diffusion is also referred to as kinematic viscosity, K_M is also referred to as turbulent viscosity. A very similar approach can be taken with regard to temperature fluctuations, in which case the molecular heat diffusion coefficient κ_T is replaced by the turbulent heat diffusion coefficient, K_T. Because heat is carried by the molecules in the form of their speed, the rate of diffusion of atmospheric neutral density is also controlled by K_T.

In the case of molecular flow, the ratio of viscosity divided by thermal diffusivity, ν/κ_T, is given a special name, the Prandtl number, denoted here as P_{pr}. For air it is about 0.7. In turbulence, the quantity $P_K = K_M/K_T$ is defined as the turbulent Prandtl number. Often it is assumed (without justification) that the turbulent Prandtl number is also about 0.7. Experiments do not always support this. For example, *Justus* (1967) has made measurements with rockets which suggest that P_K may have a numerical value of about 2 or 3. Recent model calculations have indeed indicated that the diffusivities of momentum (*Garcia and Solomon*, 1985) and heat (*Strobel et al.*, 1987; *Strobel*, 1989) seem to be quite different, and (*Fritts and Dunkerton*, 1985) have offered physical reasons why P_K might be quite large when considered over long time scales and large spatial scales. However, these arguments are restricted to long time scales, so the sit uation within turbulent patches is somewhat uncertain. Given this uncertainty, K_T and K_M are often treated as a similar parameter, usually denoted by K, despite the fact that $P_K \neq 1$. This is not entirely unreasonable if we recognize the difficulty of measuring P_K. Even some attempts at measurement of P_K have large errors, so there is a tendency to err on the side of caution. The need to consider K_M and K_T as separate may become more acute in the future.

Another important parameter used in many turbulence theories is the Reynolds' number. In laminar flow, this is often taken as a typical scale (for example the diameter of a pipe through which the fluid flows), multiplied by a typical velocity and divided by ν. In atmospheric flows, a typical scale is the buoyancy scale, L_B, and a typical velocity associated with this can be written as v_L. The buoyancy scale is a scale roughly associated with a transition region between inertial range turbulence and the buoyancy range. Thus

$$R_e = \frac{L_B v_L}{\nu}. \tag{11.39}$$

This can also be written more usefully as

$$R_e = \frac{K_M}{\nu}. \tag{11.40}$$

Here we return to a warning sounded earlier about the dangers of creating too much of an analogy between molecular diffusion and turbulent diffusion. Sometimes this is even taken to further extremes than we have applied; it is sometimes noted that the molecular diffusion coefficient is proportional to the mean free path between collisions for a molecule, multiplied by the molecular speed. Sometimes a "diffusion coefficient" is likewise inferred in the turbulence case by multiplying a typical turbulent speed by a typical "scale." This works in a broad sense, but one must be very careful. *Bradshaw* (1975) sounds warnings against such crass comparisons. Indeed, one can recognize that there is not a simple correspondence between the molecular and turbulent diffusion cases in the following manner. If a cloud of gas is released in air and expands by molecular diffusion, then the mean square radius of this cloud expands according to a law of the type $r^2 = 2\nu t$, where t represents time since the moment of release. However, this is not true for turbulent diffusion, for in that case the cloud expands according to a law of the type $r^2 \propto t^3$, at least out to values of r comparable with the size of the largest eddy. This occurs because, as the cloud expands, larger scale eddies become more important in the diffusion process (*Batchelor*, 1977). Thus, whilst many developments of "turbulent parameters" have their basis in comparisons with molecular diffusion processes, one must be very wary about this procedure.

It has been emphasized that in this book we will concentrate mostly on small-scale turbulence, since that is most amenable to radar studies. Nevertheless, a few words about the large scale distribution of turbulence are appropriate at this time.

Turbulence in the mesospheric and stratospheric regions is not homogeneous, and tends to occur in layers separated by regions of laminar flow, as shown in Figure 11.23. This occurs because gravity waves tend to break in horizontal layers, and stay linear in other regions. Indeed, this has already been discussed to some extent when the modes of gravity wave breakdown were discussed earlier (also see *Hines*, 1991a, b, c, 1993, 1996; *Fairall et al.*, 1991; *Hocking*, 1991; *Sica and Thorsley*, 1996). The existence of such layering alters the modes of turbulence diffusion, since the processes by which diffusing molecules pass from one turbulent layer to the next must be considered. Models of this process have been presented by *Dewan* (1981) and *Woodman and Rastogi* (1984), and summarized also by *Hocking* (1991, 1999a). These authors have presented a process involving stochastic and intermittent creation and destruction of turbulent layers.

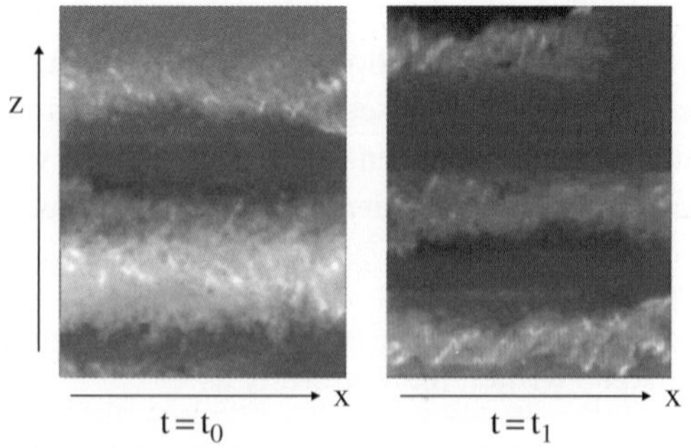

Figure 11.23 Artist's impression of layers of turbulence in the middle atmosphere at two different times. The layers are separated by laminar regions, and white areas indicate stronger turbulence. Layer thicknesses may vary between tens of meters even to kilometers.

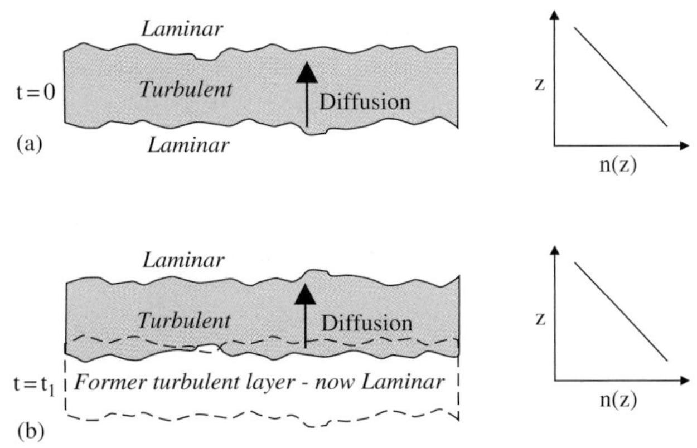

Figure 11.24 Diffusion due to layered turbulence.

Figure 11.24 shows the type of process. In (a) ($t = 0$), traditional counter-gradient diffusion can occur across the turbulent layer, but once the top of the layer is reached, any diffusion is very slow and occurs only by molecular diffusion. At a later time, purely by chance, another turbulent layer appears above the old one (where the old one has died out) and diffusion may now occur across this new layer, allowing particles to progress further along the counter-gradient direction. The overall rate of diffusion depends not only on the strengths of turbulence but also on the rates of creation and the lifetimes of these turbulent layers. Further developments of this model have been reported by *Vanneste* (2004).

Additionally, *Walterscheid and Hocking* (1991) and *Hocking and Walterscheid* (1993) have demonstrated the possible importance of Stokes' diffusion as a possible mechanism for diffusion of constituents in the atmosphere. This was discussed in Section 11.2.15 of this chapter. All these processes need to be recognized, in addition to the homogeneous diffusive processes present within a patch of turbulence. The diffusion rates are in fact scale dependent, since different mechanisms dominate at different scales.

11.3.2 The key scales of turbulence

In regard to Figure 11.22 we discussed some important scales. In this section, the intent is to show the origins of these scales. We shall do this using similitude analysis.

To begin, the Kolmogoroff microscale will be derived. Consider an eddy, which we will consider to be circular for simplicity, which is rotating at angular speed ω. Its kinetic energy will be $\frac{1}{2}I\omega^2$, where I is the moment of inertia. For simplicity, treat it as a spinning sphere. A sphere of uniform density and radius r has moment of inertia $I = \frac{2}{5}mr^2$, where r is the radius. For the approximate purposes of this derivation, consider that $\frac{2}{5}$ is roughly unity, so we can consider the kinetic energy as mv^2 where m is the mass of the eddy and v is a typical speed of a parcel within the eddy (located at a distance somewhat less than the radius of the eddy from its center). Assuming that the eddy has a scale that puts it in the inertial range, the only energy loss is transfer of energy to smaller scales.

While the idea of the inertial range is that energy will be gained from larger eddies as it is simultaneously dissipated to smaller eddies, we will consider just the loss processes. If the eddy dissipates in time τ, the mean kinetic energy dissipation per unit mass and per unit time is $mv^2/m/\tau$, or $\varepsilon \simeq v^2/\tau$. Parcels within the eddy will both rotate and move radially, but the typical speed v will be of the order of the circumference at the radius of the typical parcel divided by the time for one revolution τ_0, say. Ignoring factors of 2π, we write $\tau_0 \simeq r/v$. Since $\varepsilon \simeq v^2/\tau$, we may use $v \simeq (\varepsilon\tau)^{1/2}$ to write

$$\tau_0 \simeq r/(\varepsilon\tau)^{1/2}. \tag{11.41}$$

In the above derivation, we have assumed the scale of the eddy is in the inertial range. Now we consider an eddy in the viscous range, and ultimately we will consider that the eddy has a little of the character of both an inertial eddy and a viscous one. So for the viscous case, we consider that the eddy does not simply rotate – diffusion takes place within it. If we imagine the eddy rotating clockwise, then at the top a parcel within the eddy moves to the right, and at the bottom to the left. If each parcel spreads and diffuses, some of the momentum of each parcel will approach the center, and when air volumes from the upper and lower parts of the eddy "collide," they will have opposite momenta and annihilate. (It is by this viscous momentum transport that the energy of the eddy heats the atmosphere.) By this argument the parcel lifetime will be of the order of $\tau \simeq r^2/\nu$, where ν is the kinematic molecular viscosity. Now, as discussed, we consider an eddy which is close to the transition region between the inertial and viscous ranges, so both of the above equations may apply simultaneously. We will denote this intermediate scale as $r = \eta_K$. We can then assume that the rotational lifetime of the eddy will be comparable to its diffusive lifetime. Then we may use $\tau_0 \simeq \tau$ in (11.41), which may be solved to give $\tau \simeq [\eta_K/(\varepsilon^{1/2})]^{2/3}$. But from the diffusion equation, τ is also given by $\tau \simeq \eta_K^2/\nu$. Equating the two expressions for τ gives

$$\eta_K \simeq \left(\nu^3/\varepsilon\right)^{\frac{1}{4}}, \tag{11.42}$$

which is the Kolmogoroff microscale. As noted earlier, this scale turns out not to be at the boundary of the inertial and viscous ranges, but is still a useful "scaling" quantity. As mentioned earlier, the scale turns out to be deep within the viscous region, so it does not match our original intent, but still serves its purpose of being at least somewhere in the vicinity of the transition region. Scales of this size and smaller should be heavily dissipated, and so not seen by radar. Hence radars with Bragg scales of this order should not see turbulence. As noted, the true transition scale is about 7 times larger than η_K, as derived by *Tatarski* (1961), for example. Figure 11.25 shows a graph of the inner scale as a function of height in the atmosphere, for realistic ranges of energy dissipation rates. At the ground the inner scale is of the order of a few millimeters to a few centimeters, while at 85 km altitude, the typical inner scale is 10 m or more. At even greater heights, the scale becomes larger, and ultimately this growth in the smallest scale leads to the development of the turbopause, which we will discuss shortly.

Now we are in a position to derive an expression for the buoyancy scale. In this case, we actually lock in a time scale. The period of a parcel of air allowed to oscillate freely

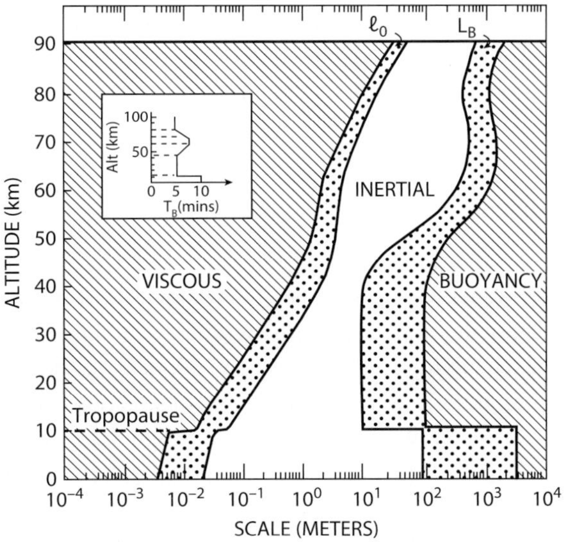

Figure 11.25 Inner and buoyancy (outer) scales, shown by the dotted areas (buoyancy scale to the right, inner scale to the left), for turbulence in the atmosphere as a function of height. Realistic kinetic energy dissipation rates were used. (From *Hocking*, 1985. Details of assumptions made are outlined in that paper.) (Reprinted with permission from John Wiley and Sons.)

in the air is the Brunt–Väisälä period, so we surmise that within the buoyancy range, eddies will have periods of rotation comparable to this. Then given that the dimensions of ε are m^2s^{-3}, we propose that we can write

$$\varepsilon \sim L_B^2 \tau_B^{-3} \sim L_B^2 \omega_B^3, \tag{11.43}$$

where we ignore terms of order 2π. Rearranging gives

$$L_B \sim \varepsilon^{\frac{1}{2}} \omega_B^{-\frac{3}{2}}. \tag{11.44}$$

In oceanography, the quantity $L_O = \varepsilon^{\frac{1}{2}} \omega_B^{-\frac{3}{2}}$ is called the Ozmidov scale. In atmospheric studies, we rely on the more detailed derivation of *Weinstock* (1978a) to write that

$$L_B \approx \frac{2\pi}{0.62} \varepsilon^{\frac{1}{2}} \omega_B^{-\frac{3}{2}} \approx 10 \varepsilon^{\frac{1}{2}} \omega_B^{-\frac{3}{2}}. \tag{11.45}$$

The equation has been experimentally verified, at least in a few cases, by *Barat* (1982) using stratospheric balloon studies.

Finally, we turn to the eddy viscosity. As described, we can really only consider turbulent diffusion to "look" like molecular diffusion at scales greater than L_B, so these large eddies somewhat play the role of molecules in the atomic case. Then the diffusion coefficient is of the order of $L_B \times v_L$, where v_L is a typical velocity of the eddy. We take the typical velocity as L_B/τ_B, so we produce

$$K \sim L_B \left(\frac{L_B}{\tau_B} \right) \sim L_B^2 \omega_B \sim \varepsilon \omega_B^{-2}, \tag{11.46}$$

where again we have ignored factors of 2π. The constant of proportionality is found largely from experiment, but can also be found from more careful mathematical treatments. Most determinations give a value for the scaling constant of between 0.25 and 1.0. We will discuss this constant later.

11.3.3 The turbopause

The eddy viscosity K_M depends on ε and ω_B. Measurements show that values of ε generally increase with increasing height, as gravity waves grow more rapidly with amplitude and break more often, and more vigorously, at higher altitudes. But the increase in K is not huge, changing from typically 1 to 10 m^2s^{-1} at 60 km altitude to around 100 m^2s^{-1} at 100 km altitude. The kinematic viscosity ν, on the other hand, increases exponentially with increasing height in the atmosphere, changing from 0.1 m^2s^{-1} at 60 km altitude to 100 m^2s^{-1} at 100 km. Somewhere around 100 to 110 km (depending on local energy deposition rates and local dynamics), the two diffusion coefficients become equal. In other words, the Reynolds' number is approximately 1. This height is called the turbopause.

At this altitude, the buoyancy scale and the inner scale also become almost coincident (e.g., see *Hocking*, 1987a). Hence inertial range turbulence cannot exist at all near and above the turbopause. The scales at which turbulence generation could occur are comparable to those at which viscous forces are important, and any mechanism which attempts to induce turbulence is very rapidly damped.

Short-term changes in the height of the turbopause mainly arise because of variations in K, particularly through its dependence on ε, which can be very intermittent and variable. As seen in Equation (10.20), ν depends on density and temperature, which tend to vary more slowly than the strength of turbulence. Since $\nu \simeq K$ at the turbopause, $\varepsilon \simeq \nu\omega_B^2$ here. Larger values of ε allow the turbopause to exist at larger values of ν, pushing the turbopause height up.

The turbopause shows quite clearly with rocket vapor trail measurements, which are releases of luminescent gaseous compounds from rockets as they fly upward. They remain suspended in the air, and are then distorted by local wind motions and turbulence. By photographing from the ground, images of turbulent motions can be captured. The trails appear turbulent up to the turbopause, and then quite suddenly become laminar above that height. The reason for the rapid change lies largely in the exponential increase of ν with increasing height. The technique shows various stages of turbulence development. When a vapor trail forms, it first diffuses by molecular processes until a time t_η, after which the trail begins to show the distortive effects of turbulence. The kinematic viscosity increases exponentially with height, and near the turbopause this transition time typically increases from less than 1 minute to greater then 2 minutes in less than about 5 km of altitude. Thus the trail appears laminar for considerable time at the greater heights. This, coupled with the higher damping which turbulence experiences due to the larger kinematic viscosity, results in the appearance of a rapid transition to laminar flow in the vapor trails. (There are some observers who feel that this is insufficient to explain the rapidity of the change in trail structure with height, and that extra physical processes are at play, but this is still unresolved.) Weak turbulence can at times be seen up to altitudes as high as ~130 km (*Rees et al.*, 1972), but this is rare. The turbopause does truly represent a level above which turbulence plays only a minor role.

The turbopause shows significant fluctuation in height, both in seasonal and day-to-day variation. *Danilov* (1984) has collated a large set of measurements of this height, largely using data acquired from rocket-based experiments. Large scatter and only a weak trend as a function of season were apparent.

11.3.4 Turbulence structure functions and spectra

In order to make useful measurements of turbulence in the middle and lower atmosphere, it is necessary to understand the mathematical description of the turbulence. This is usually done with spectral and structure function descriptions. Many of these have already been discussed in earlier chapters, since they were needed in order to interpret the radar signals (Chapters 3 and Chapter 7, for example). A good summary of these functions can also be found in Appendix A of this book.

However, the relevant spectra were presented in a somewhat ad hoc manner, being introduced as needed without real proof. Here, we discuss these equations again, in a slightly more methodical manner. We will not prove all the expressions, but will try to at least present them in a more logical sequence.

When considering turbulence functions, it is necessary to consider the nature of the measurements being undertaken. For example, the full three-dimensional nature of the turbulence can be considered, and a three-dimensional spectrum can be obtained. Alternatively, the measurement might be made by a sensor moving in a line through the fluid – in this case, a different spectrum will result. It is also necessary to consider the parameter being studied. If it is a scalar, like say the concentration of an inert gas, the spectral fluctuations will differ compared to the fluctuations in refractive index. In the latter case, the intensity of fluctuation is affected by adiabatic expansion and compression of the parcels of air as they move up and down. If it is velocity fluctuations that are of interest, then three separate components are involved, and each of these can be considered separately, or they can be considered collectively to give a total energy spectrum. We will not consider all of these here, but just recap the main ones. For more information, see the earlier chapters, and also Appendix A.

The first function to consider is the structure function. This is the key function that *Tatarski* (1961) used to introduce the basics of turbulent structure. The spectra are then derived from the structure functions. For practical applications, the structure function is often measured by a probe moving in a line through the fluid. For velocity fluctuations, it is common to distinguish between motions parallel to and motions transverse to the trajectory of the sensor. The *parallel* structure function is given by

$$D_{\parallel} = \overline{\mid u_{\parallel}(\vec{x}+\vec{r}) - u_{\parallel}(\vec{x}) \mid^2}, \tag{11.47}$$

where $u_{\parallel}$ is the *component* of the fluid flow vector at each point in the direction parallel to $\vec{r}$ (i.e., in the direction parallel to the line joining $\vec{x}$ and $\vec{x}+\vec{r}$).

A *transverse structure function* may also be defined as

$$D_\perp = \overline{|\, u_\perp(\vec{x}+\vec{r}) - u_\perp(\vec{x})\,|^2}, \tag{11.48}$$

where $u_\perp$ is the *component* of the fluid flow vector at each point in the direction perpendicular to $\vec{r}$ (i.e., in the direction parallel to the line joining $\vec{x}$ and $\vec{x}+\vec{r}$).

There is also a *total* structure function

$$D_{\rm tot} = \overline{|\, \vec{u}(\vec{x}+\vec{r}) - \vec{u}(\vec{x})\,|^2}. \tag{11.49}$$

Note that this involves a *vector difference*.

All of these structure functions are important.

Tatarski (1961) uses the structure function approach to derive the Kolmogoroff form of both the structure function and the spectrum within the inertial range. First, the hypothesis is made that the turbulence is isotropic, so is independent of direction. Therefore, any of the structure functions listed above should depend only on the *magnitude* of $\vec{r}$. Then it is assumed that within the inertial range, neither viscous effects, nor effects associated with the generation of the turbulence (wind-shear, temperature gradients, buoyancy, etc.), play a role in defining the structure function. Hence the structure function should depend only on the rate that energy passes through the eddies from large to small scale, viz., ε. It is then assumed that the radial vector and ε relate to the structure function as a product of power laws, viz.,

$$D = D_r(r) = \alpha_D \varepsilon^a r^b. \tag{11.50}$$

Assume α_D is a dimensionless constant, and recognize that the dimensions of D are $[L]^2[T]^{-2}$, the dimensions of ε are $[L]^2[T]^{-3}$ and the dimensions of r are $[L]$. Consistency of dimensions on each side requires that

$$2 = 2a + b$$
$$-2 = -3a\,,$$

which gives $a = \frac{2}{3}$ and $b = \frac{2}{3}$.

Once the structure function is known, the covariance function can be found through the standard relation $D_{ij} = \overline{u^2}\left[1 - \rho_{ij}\right]$ (where ρ is the normalized covariance function), and then the spectrum can be found as a Fourier transform of ρ.

Some texts do the derivation using dimensional analysis on the spectrum, but as will be seen, there are many types of spectra, so these proofs lead to confusion about which form is being used. The structure-function approach leads to no such ambiguities.

Having developed values for a and b, we may now write that for Kolmogoroff inertial-range turbulence, the *parallel* structure function has the form

$$D_\parallel = C_v^2 r^{2/3}, \tag{11.51}$$

where $C_v^2 = C\varepsilon^{2/3}$, C being a universal constant.

Similar expressions exist for the other structure functions. For example, the transverse structure function is given by

$$D_\perp = \frac{4}{3} C_v^2 r^{2/3}. \tag{11.52}$$

Note especially the factor $\frac{4}{3}$. The reasons for this are subtle, but arise from the fact that when one forms a longitudinal structure function, part of the turbulent motions are "lost" to the mean, so subtract out to zero. The "total" structure function in the inertial range is

$$D_{\rm tot} = \frac{11}{3} C_v^2 r^{2/3}. \tag{11.53}$$

Careful experiments in the boundary layer give a value for C of about 2.0.

Similar structure functions can be calculated for scalar quantities like concentrations and temperature. For temperature fluctuations

$$D_T = C_T^2 r^{2/3} \tag{11.54}$$

in the inertial range, and for a general scalar variable ξ,

$$D_\xi = C_\xi^2 r^{2/3}. \tag{11.55}$$

Although turbulence is fundamentally a rotational motion, it is common to also represent the fluctuations in terms of spectra. Some of the major spectra are briefly discussed below. These are simply derived as mathematical mappings from the structure functions (e.g., *Tatarski*, 1961, 1971). The reader should pay particular attention to the fact that there are many different forms of spectra.

For scalars, the full three-dimensional function for Kolmogoroff inertial range turbulence is

$$\Phi_\zeta(\vec{k}) = 0.033 C_\zeta |\vec{k}|^{-11/3}. \tag{11.56}$$

This function has been chosen to be normalized so that when integrated over all possible wavenumbers it has a value of unity (e.g. see Equation (3.291), Chapter 3). Note it is *not* a "$-5/3$" power law. A monostatic radar looking into a turbulence patch sees only one of these Fourier components, with wavefronts aligned perpendicular to the view direction, and with wavelength equal to one half of the radar wavelength (the so called Bragg scale discussed elsewhere in this text, especially in Chapter 3).

With regard to velocity fluctuations, a similar spectrum exists which describes the kinetic energy per unit wavenumber volume in (k_x, k_y, k_z) space. This function is

$$\mathcal{F}(\vec{k}) = A\varepsilon^{2/3} k^{-11/3}, \tag{11.57}$$

where $k = |\vec{k}|$ is the length of the vector $\vec{k}$, and $A = \frac{11\Gamma(\frac{8}{3})\sin(\frac{\pi}{3})}{24\pi^2} C \simeq 0.061C$ (*Tatarski*, 1971). For homogeneous isotropic turbulence, this function is isotropic. Pictorially, one can visualize this as a solid sphere in (k_x, k_y, k_z)-space which has highest density at the center and decreasing density as $|\vec{k}|$ increases, where $\mathcal{F}$ represents the density. Spectra can also be independently defined for the three velocity coordinates u, v, and w. Note that both the scalar and velocity spectra have a functional dependence of the form $k^{-11/3}$.

Because these functions are isotropic, they are often integrated over a shell of radius k to give a new function. For example, in the case of the velocity spectrum, such an integration gives

$$E(k) = 4\pi k^2 \mathcal{F} = \alpha\varepsilon^{2/3} k^{-5/3}, \tag{11.58}$$

where $\alpha = 4\pi A = \frac{11\Gamma(\frac{8}{3})\sin(\frac{\pi}{3})}{6\pi} C = 0.76655C$ (e.g., see *Tatarski* (1971); *Batchelor* (1953)).

If $C = 2.0$ is used, then

$$E(k) = 1.53\varepsilon^{2/3}k^{-5/3}.$$

Different authors use different values for the constant 1.53. Any values between 1.35 and 1.53 are common. Note, however, that if one adjusts this constant, then the constant C also needs adjustment. $C = 2.0$ is often used because it has been measured with good accuracy at least in the lower atmosphere (e.g., *Caughey et al.*, 1978).

For the scalar case, a similar integration leads to the function

$$E_\xi(k) = 0.132\pi C_\xi^2 k^{-5/3} = 0.415C_\xi^2 k^{-5/3}. \tag{11.59}$$

The other key spectral form is that found when a measuring device moves linearly through the turbulent region. In the case of the scalar quantities, a spectrum

$$S_\xi(k_x) = \int_{-\infty}^{\infty}\int_{-\infty}^{\infty} \Phi_\zeta(\vec{k})dk_y dk_z = \int_{-\infty}^{\infty}\int_{-\infty}^{\infty} 0.033C_\xi|\vec{k}|^{-11/3}dk_y dk_z \tag{11.60}$$

is produced. Evaluating gives

$$S_\xi(k_x) = 0.125C_\xi^2 k_x^{-5/3}, \qquad -\infty < k_x < \infty\,. \tag{11.61}$$

We have assumed that the probe is moving in the x direction, but the same formalism applies for any direction.

Because of the obvious symmetry, experimentalists often "fold" their negative spectral densities over onto their positive ones. Then the following functions result:

$$S_\xi(k_x) = 0.250C_\xi^2 k_x^{-5/3}, \qquad 0 < k_x < \infty\,. \tag{11.62}$$

Similar equations may be developed for the velocity spectra. A summary can be found in Appendix A and *Hocking* (1999a).

Note that Equation (11.62) and related equations have $k^{-5/3}$ laws, and so does (11.59). However, these equations are conceptually different; (11.59) represents an integration over a shell of radius k in three-dimensional k-space, whilst (11.62) represents a spectrum determined by a probe moving in a straight line through the turbulence. It is a common mistake for novice researchers to confuse the two spectra when they speak of the $k^{-5/3}$ law, which can lead to the propagation of considerable confusion. It is important to conceptually distinguish these spectra. This was also the reason that we used the structure-function approach to develop the Kolmogoroff inertial-range laws.

Finally, the relation between radar backscatter and the strength of turbulence should be considered. The quantity C_n^2 is the refractive index structure constant and represents the degree of turbulent variability of refractive index within the turbulent patch. It is *not* always proportional to the rate of turbulent kinetic energy dissipation. For example, intense turbulence in the presence of an adiabatic lapse rate in temperature, in dry air, and without free electrons present, produces a value of C_n^2 of zero (e.g., *Hocking and Mu*, 1997).

In Equation (3.291) in Chapter 3, we used Equation (11.56) to derive that the radio backscatter reflectivities σ_s and η_s relate to refractive index structure constant C_n^2 through

$$\sigma = 0.00655\pi^{4/3}C_n^2\lambda^{-1/3} \tag{11.63}$$

and

$$\eta_s = 0.38C_n^2\lambda^{-1/3}, \tag{11.64}$$

where σ_s and η_s were defined in Chapter 3. Although these equations were developed in Chapter 3, the function Φ was introduced without proof, whereas hopefully now its origins are clearer.

Note the $\lambda^{-1/3}$ dependence, which arises because of the $k^{-11/3}$ functional form of Φ. A carefully calibrated radar can be used to determine C_n^2 (e.g., *Hocking and Mu*, 1997).

The relation between the cross-section, C_n^2, and the fluctuations in atmospheric quantities like humidity, temperature, electron density, and so forth must also be known for a complete description of the relation between strengths of backscattered radar signal and turbulent energy dissipation rates. This relates to a quantity called the potential refractive index gradient, which has been discussed earlier in this book. These relations are repeated below, for completeness, and are also linked to earlier derivations. We will also expand on some of the terms in the equations, especially γ and F_t, which were introduced earlier but have not been fully developed to date.

First, recall that the energy dissipation rate is related to the potential refractive index structure constant by

$$\bar{\varepsilon} = \left(\gamma \overline{C_n{}^2} \frac{\omega_B^2}{F_t{}^{1/3}} M_n^{-2}\right)^{3/2}, \tag{11.65}$$

where ω_B is the Brunt–Väisälä frequency. The parameter F_t represents the fraction of the radar volume that is filled by turbulence. The quantity γ has been discussed extensively by *Hocking and Mu* (1997), and some of that discussion will be recapped shortly. This equation has appeared as Equation (3.295) in Chapter 3 and as Equation (7.89) in Chapter 7.

The potential refractive index gradient in the troposphere and stratosphere was given in Equations (3.288) and 7.70), but we repeat the expression here for ease of reference, with slight modifications (here we use the potential temperature, and make some other rearrangements):

$$M_n = -77.6 \times 10^{-6}\frac{p}{T}\left(\frac{\partial \ln \Theta}{\partial z}\right) \times \left[1 + \frac{15500 q_{wp}}{T}\left(1 - \frac{1}{2}\frac{\partial ln q_{wp}/\partial z}{\partial ln\Theta/\partial z}\right)\right], \tag{11.66}$$

where again the variables were defined in Equation (3.288). The term in square brackets was denoted as χ by *Van Zandt et al.* (1978); indeed this particular form of the equation was first introduced by these authors (note that the term $\frac{1}{2}$ is actually $\frac{7800}{15500} = 0.503$, which is close to 1/2 anyway). Note that χ tends to 1 as the humidity terms tend to zero.

In the ionosphere, where humidity is no longer important but electron density plays a crucial role, the potential refractive index gradient is

$$M_e = \frac{\partial n}{\partial N}\left[\frac{N}{\Theta}\frac{\partial \Theta}{\partial z} - \frac{dN}{dz} + \frac{N}{\rho}\frac{d\rho}{dz}\right], \tag{11.67}$$

where again the symbol Θ has been used for potential temperature and N is the electron density. This appeared as Equation (7.71) in Chapter 7, along with an alternative but equivalent version in the form of Equation (7.72), derived originally by *Thrane and Grandal* (1981).

The term ρ is the neutral density. The function $\frac{\partial n}{\partial N}$ needs to be determined from electro-ionic theory (see Chapter 3, Equations (3.125) and (3.128) and associated equations).

We now turn to a more detailed discussion about the parameter γ and the fraction F_t. *Wilson et al.* (2005) have developed interesting new techniques for measuring F_t by comparing turbulence backscattered powers using radars with different resolution, and combining simultaneous power and spectral-width measurements of turbulence strengths. The parameter γ has been discussed by a variety of authors and *Hocking and Mu* (1997) have presented a summary of this. Earlier assumptions were that γ is indeed a constant, but it now seems that it is Richardson-number dependent, and also depends possibly on the turbulent Prandtl number. The dependence on the Prandtl number, which describes the relative diffusion rates of momentum and temperature, should not be a surprise, since turbulence intensity depends on momentum transport, and the backscattered power depends on refractive index irregularities, which depend on temperature and humidity transport.

Indeed, γ relates to the relative dissipation of kinetic and potential energies ε_k and ε_p respectively. As was discussed briefly in the introduction to this section on turbulence, these are not the same, and knowledge about their differences is important for a fuller understanding of turbulence. Although it is most common to calculate the kinetic energy dissipation rate ε_K (often denoted simply as ε), it is very important to know that energy is also dissipated as potential energy (and indeed it can be negative). This is often denoted as ε_B, where the B stands for buoyancy, or ε_P, where the P stands for potential. We do not have space here to fully cover the details of potential energy dissipation, but it should at least be noted, and the relevance to γ must be recognized.

Ottersten (1969b) gave $\gamma = \frac{1}{a^2 P_K}\left(\frac{P_K - R_i}{R_i}\right)$, where R_i is the gradient Richardson number, P_K is the turbulent Prandtl number and a^2 is a constant equal to approximately 2.8. *Gossard et al.* (1982, 1984) gave $\gamma = \frac{1}{B_\theta}\left(\frac{P_K - R_i}{R_i}\right)$, where B_θ equals 3.2. *Sengupta et al.* (1987) gave a similar but more complex expression. *Hocking* (1992) gave $\gamma = \frac{3}{22}\left(\frac{|1-R_i|}{|R_i|}\right)$. *Wilson et al.* (2005), following *Lilly et al.* (1974), gave $\gamma = \frac{R_f}{1-R_f}$, where R_f is the flux Richardson number. Since the Prandtl number was involved in many of these formulas, it is important to determine its value. *Hocking and Mu* (1997), in their Figure 2, showed a variety of determinations of P_K and this is reproduced here as Figure 11.26. For $R_i < 0.5$, the inverse turbulent Prandtl number is of the order of 0.5 to 1.4, so that P_K is generally in the range between 0.7 and 2.0.

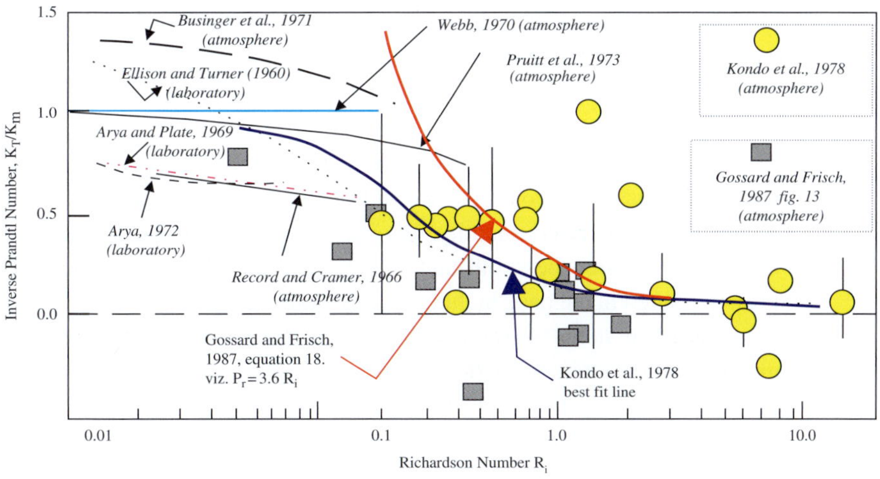

Figure 11.26 Inverse turbulent Prandtl number as a function of Richardson number (from *Hocking and Mu*, 1997).

Despite all of these possibilities, in many cases it turns out to be useful to assume that γ is a constant, numerically equal to about 0.4. *Wilson et al.* (2005) have suggested a value closer to 0.2, and even as low as 0.1. *Hocking and Mu* (1997), in their Figure 3, have discussed the potential problems associated with this, which include potential false identification of layers and underestimation of the strengths of strong layers. Until the various estimates of γ are properly resolved, the assumption of a constant γ may need to be tolerated. For now, we assume a value of 0.4.

11.3.5 Measurement techniques and results for turbulence studies

Two of the most important measurable turbulence parameters are the energy dissipation rate ε and the diffusion coefficient K, since they parameterize heating (energy transport) and diffusion. Determination of ε and K values can be broadly classified into two types:

i) measurements of small scale motions (≤ 5 km) by direct observation, and
ii) large-scale studies of the balance of heat and inert chemical species in the atmosphere.

Radars can contribute in a variety of ways to these studies. Most of these techniques have already been addressed in Chapter 7, and our discussion here will be kept brief in this regard, mainly concentrating on key results.

Essentially, radars can directly measure the intensity of turbulence in the inertial range. If scatter is from scales in the viscous range, the equations need to be modified in some cases, but in those cases the scatter is usually very weak and observations made using viscous scatter are rare. Radars can use either measurements of total scattered power or spectral width methods, as discussed in Chapter 7. From these data, the energy dissipation rate can be determined, and from that, the small-scale diffusion coefficient can be measured through the relation

$$K_M = c_2\varepsilon/\omega_B^2. \tag{11.68}$$

This is a key equation linking heating and diffusion processes. A similitude-type of argument was applied to derive this in Equation (11.46). As a rule, MST radars measure ε, and K_M is a derived quantity. In order to best derive K_M from ε, we need to know a little more about the "constant" c_2. We will also look into other ways that K_M has been measured in the past, including some non-radar procedures, that allow comparisons across different fields. Agreement between estimates by very different techniques allows greater confidence in our data.

The value of c_2 has been derived in many ways, all giving different results. One derivation was given in Equation (11.46), but no real value was presented for c_2. Another approach is to start with the Richardson number. To see this, first recognize that

$$R_i = \frac{\omega_B^2}{(d\overline{u}/dz)^2}. \tag{11.69}$$

But dimensional analysis suggests that we might be able to reasonably write

$$\varepsilon \simeq K_M(d\overline{u}/dz)^2, \tag{11.70}$$

(e.g., *Justus*, 1967), where $d\overline{u}/dz$ is the vertical shear in the mean wind. Of course we need to take care with (seemingly) arbitarily applying dimensional analysis, but let us see where this leads.

Hence we may write

$$R_i = \frac{\omega_B^2}{(\varepsilon/K_M)}. \tag{11.71}$$

Assuming that turbulence only causes diffusion when it exists, it seems reasonable to replace R_i with its critical value for instability, 0.25, so we write

$$K_M = 0.25\varepsilon/\omega_B^2. \tag{11.72}$$

We might therefore propose that $c_2 = 0.25$. But then maybe we should have replaced R_i in (11.71) with $\overline{R_i}$, the mean Richardson number averaged over all turbulent patches? However, since some patches will have negative R_i, that might be questionable too. There are many arguments in the literature which produce expressions of this type – the ones above are only samples. *Weinstock* (1978b) used yet another derivation, and suggested $c_2 = 0.8$. *Fukao et al.* (1994), section 2.3, discusses various possibilities relating to c_2, including one in which $c_2 = \frac{R_f}{1-R_f}$, R_f being the flux Richardson number. Most estimates of c_2 do suggest some relationship with the Richardson number (either flux or gradient Richardson number).

It might be recognized that this discussion has some similarities to the discussion about γ at the end of the previous section, and the possibility exists that c_2 is not in fact a constant at all, but somehow dependent on conditions at the time. In the end, however, we do not have sufficient knowledge to be confident in any chosen non-constant value for c_2, and so we try to find a constant that represents a reasonable compromise – perhaps not an ideal scientific strategy, but one that still has some value at least when dealing with average rates of diffusion.

Equation (11.68) still represents a reasonable relation to use for approximate calculations, albeit with an empirically determined value for c_2. In Chapter 7, Equation (7.64), a value for c_2 of around 0.25 to 0.5 was suggested, and *Fukao et al.* (1994) settled on $c_2 = 0.3$. Based on our interpretation of comparisons of different procedures, we will adopt

$$c_2 = 0.4. \tag{11.73}$$

An example of an extensive set of measurements by this method is that presented by *Fukao et al.* (1994); see Chapter 7 for more details. In due course, we will also present our own summary of values for K_M. Of course it must be remembered that if K_T is needed, the K_M values need to be rescaled by the Prandtl number.

For larger scale diffusion rates, other processes like stochastic diffusion and Stokes' diffusion need to be recognized. Radars can still make useful contributions, but the analysis needs to be adjusted to suit (e.g. *Woodman and Rastogi*, 1984).

Hocking (1987a) gives an extensive discussion of ways to measure the diffusion coefficients K_M (momentum) and K_T (temperature/constituents), and the possibility of Stokes' diffusion (*Walterscheid and Hocking*, 1991) must be added to these. However, since this book is mainly about radar, we will not elaborate on non-radar techniques here. It is sufficient to say that many other techniques exist, like in-situ balloon and rocket measurements, modeling studies (which particularly look at diffusion over much larger scales, such as tens of km vertically and thousands of km horizontally), species concentration studies (e.g. *Johnson and Wilkins*, 1965; *Garcia and Solomon*, 1985), mesopause temperature gradient measurements, computer simulations, and so forth.

One problem with computer-based and theoretical estimates of K_M is that they do not consider the effects of vertical winds. For example, studies of atomic oxygen movement using computer models at 80 to 120 km have been used as one proxy to determine K_M. But such studies assume that all vertical motions are due to diffusion, whereas atomic oxygen could be brought down from 120 km to 90 km by vertical winds at one location, and lifted back up by vertical winds at another. The possibility of such "cells" of circulation was not included in early analyses of this type. Effects like this must be considered in any comparison between techniques, although more recent models are more inclusive of such motions.

There are a few philosophical issues about the nature of diffusion that we will now discuss. Turbulence produces both heating and diffusion, and it is not at all obvious which process dominates. Does turbulence heat (by energy deposition) or cool (by diffusion) its local environment? This was an important argument in the literature for some time in the 1970s. The issue was mainly discussed in the context of mesospheric turbulence, so most of the discussion below will concentrate on that region, although the ideas can easily be extended to the stratosphere and troposphere.

The rate of diffusion of heat depends on both the vertical temperature gradient and the turbulent diffusion coefficient K_M, the former being set up initially by mesospheric solar heating. Both *Johnson* (1975) and *Johnson and Gottlieb* (1970) recognized that similar rates of diffusion and heating should be expected. The question arises as to which is most effective – is heating supplied by the turbulence faster than the rate at which it can diffuse away, or is diffusion more effective, so that turbulence actually diffuses heat across the heat gradients formed by solar effects faster than it causes heating itself (thus cooling the mesosphere)?

The answer to this question depends on the value of the constant c_2 in Equation (11.68), but unfortunately no definitive answer exists at present. We will, however, work through some attempts at an answer, since the process is instructive. One attempt at an answer, presented by *Hunten* (1974), suggests that the rate of transfer of heat through the mesosphere is $F_{xfr} = nH\rho\omega_B^2 K_M$ (where n = 7/5, H = scale height, ρ = density), whilst the rate of loss of heat over one scale height is $P_{\text{heat}} = (R_i)^{-1} H\rho\omega_B^2 K_M$. Thus $P_{\text{heat}}/F_{xfr} = (R_i n)^{-1}$ (the detail of these statements will not be explained here – the reader should consult the original paper for more information). Clearly, heating dominates if $\overline{R_i}$ (also interpreted as c_2 following Equation (11.68)) $\leq$ 0.28, and diffusion dominates if $\overline{R_i} \geq$ 0.28. *Hunten* (1974) claimed that for turbulence to occur, R_i must be less than 0.25, and so heating should dominate, whilst *Johnson* (1975) claimed that whilst R_i must be less than 0.25 to initiate turbulence, turbulence may then persist for values of R_i as high as 1.0. Thus *Johnson* (1975) claimed that $\overline{R_i}$ is nearer 1.0. The estimates suggested earlier for c_2 would imply that diffusion dominates.

Chandra (1980) has presented a more rigorous treatment of estimation of eddy diffusivities to bring into account c_2, and assumed c_2 = 0.6. Meanwhile *Gordiets et al.* (1982) have concluded that K_M has a height dependence of its own, and the answer to the question depends on the height gradient of K_M. They claimed that turbulence heats below about 105 km altitude and cools above.

The CIRA86 model of atmosphere turbulence measurements (*Hocking*, 1990) represented the best data up to 1990. However, a newer and more accurate set of graphs was presented in *Dieminger et al.* (1996), Chapter I.2.2, Figure 9. We now further develop these global models, especially using more recent data from *Hocking and Mu* (1997); *Fukao et al.* (1994); *Nastrom and Eaton* (1997a, b); *Dehghan and Hocking* (2011); *Latteck et al.* (2005); *Lübken* (1997). These improved models are shown below. For conversions between ε and K_M, a value of $c_2 = 0.4$ has been used. It will be seen that there is some degree of consistency between estimates of diffusion coefficents made by these many different methods, suggesting that our treatment has been "about right."

Radar and in-situ comparisons

Finally, before moving on to graphs of typical data, we wish to draw attention to one additional set of methods for measurement of turbulence strengths. This method refers to balloon data, not radar data, but since balloons offer the best options for high-resolution turbulence studies to date in the upper troposphere and stratosphere, it is useful to include these methods here. These are important for understanding the nature of both turbulence and specularly reflecting sheets. *Dole et al.* (2001); *Luce et al.* (1995); *Wilson and Dalaudier* (2003) performed simultaneous radar and balloon measurements in France, while the MUTSI campaign in Japan allowed significant comparisons between balloons and the MU radar (e.g. *Luce et al.*, 2001b, 2002). Various instruments were carried on-board the balloons, especially high resolution temperature probes.

At various points in this book, the issue has been raised as to whether extremely strong layers of turbulence might be invisible to radars. The idea is that intense mixing would drive the mean temperature profile in the turbulent layer towards an adiabatic one, in which case displaced parcels of air would no longer produce a density contrast with the background. Hence no small-scale refractive index gradients would remain from which radiowaves would scatter. Examples of this were shown by *Hocking and Mu* (1997), for example. The matter was also discussed in Chapter 2, Section 2.16, in regard to mixing in the central regions of Figure 7.17, and in Section 11.3.4 in this chapter. It was found that even for strong turbulence, mixing was not quite complete, and furthermore, humidity perturbations generally still remained in the mixed regions, even if the temperature mixing was more complete. The refractive index perturbations from the humidity variations and the remnant temperature perturbations were sufficient to produce detectable radar backscatter. In the case of *Hocking and Mu* (1997), temperature fluctuations were reduced, but were not zero, and in addition the humidity fluctuations were not measured. So one good outcome of the MUTSI experiments was that the possibility of "ghost layers" – that is, layers which are very intense and are so well mixed that they have adiabatic profiles and are invisible to radar – is quite small. It might happen in extremely rare cases, but as a rule some remnant refractive index variability will remain. Likewise in the mesosphere, even if the layer is adiabatically mixed, so that temperature fluctuations are minimal, electron density fluctuations are unlikely to be simultaneously suppressed, so some backscattered signal will be seen.

One of the main foci of the MUTSI campaign was a search for thin sharp steps in refractive index which could produce specular reflections. Such steps were found (see the references above) but their cause is still a mystery.

Finally, it is prudent to briefly describe a newer method for analysis of balloon data. The procedure is based on a sorting algorithm, called "Thorpe sorting." A probe, perhaps on a balloon, moving up through a turbulent layer, measures a quasi-random distribution of temperatures. The profile is divided up into narrow layers, and the radars are adiabatially shuffled around (up and down) in such a way that the stored potential energy is minimized. This means that most layers are located in new positions. The mean layer shift between the initial and final configurations is calculated and referred to as the Thorpe scale or Thorpe length, and the energy adjustment between the initial and final states can be used to give an estimate of the kinetic energy dissipation rate. The method is proving to be quite reliable, and is discussed further in *Wilson et al.* (2010, 2011). It is probably the best hope for more detailed comparisons between radar and balloons.

Some representative data-sets

Here we present typical values of both ε and K_M.

Figure 11.27 shows measurements of turbulence strength deduced largely by non-radar methods. Additional radar data were provided in Figure 7.15 in Chapter 7, and are represented by the box labelled Profiler. Median values from that figure are also shown.

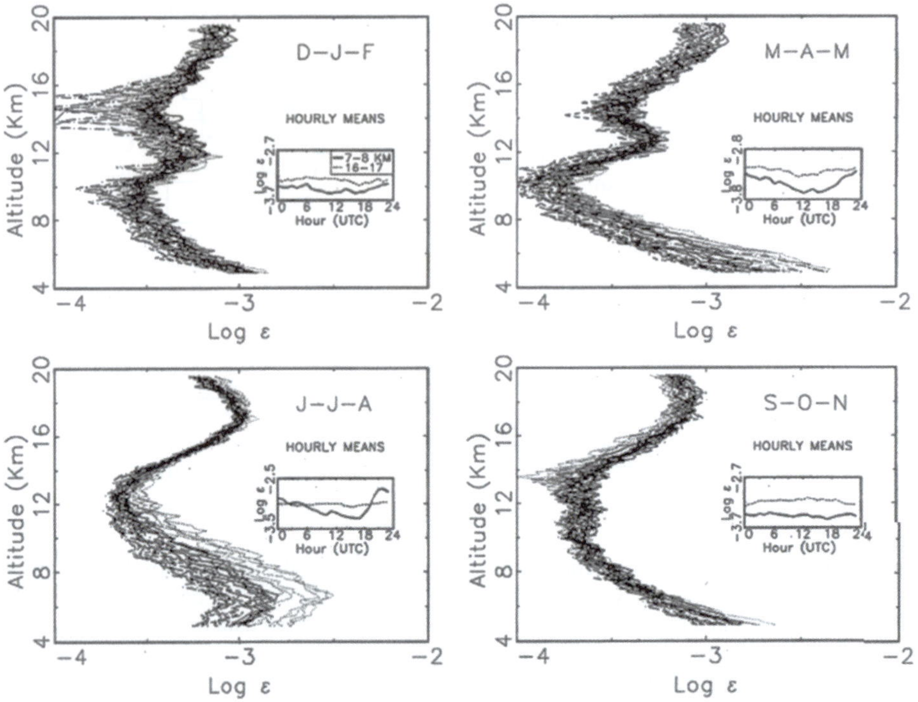

Figure 11.29 Typical turbulent kinetic energy dissipation rates in the lower atmosphere deduced from the White Sands MST radar (from *Nastrom and Eaton*, 1997a). Mean values as high as 10^{-3} W kg^{-1} at 4 km and above 16 km altitude are a little surprising, and are discussed in the text; the values between 8 and 12 km altitude are consistent with other measurements. (Reprinted with permission from John Wiley and Sons.)

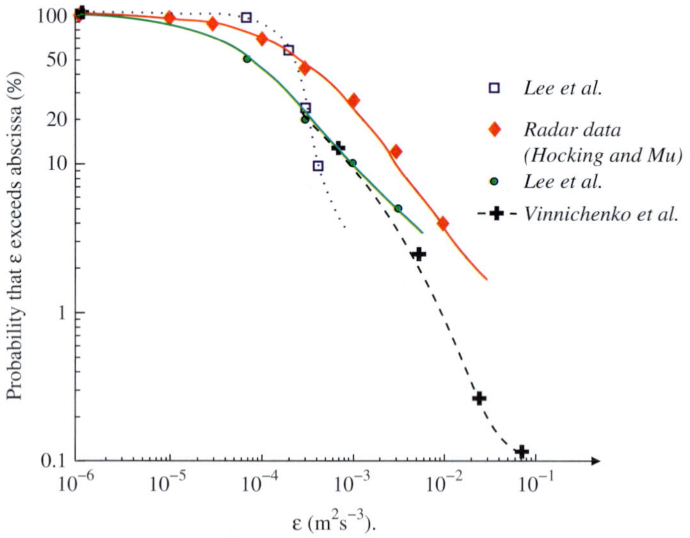

Figure 11.30 Cumulative distribution of turbulent kinetic energy dissipation rates in the lower atmosphere deduced from various sources. See *Hocking and Mu* (1997) for details.

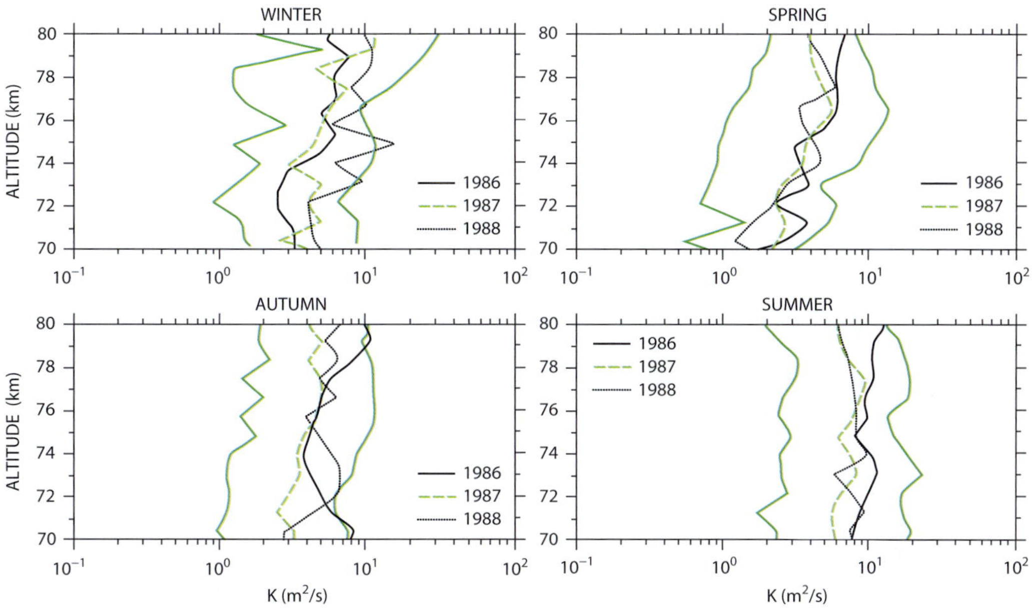

Figure 11.31 Typical small-scale turbulent diffusion coefficients in the mesosphere measured with the Japanese MU radar (from *Fukao et al.*, 1994). (Reprinted with permission from John Wiley and Sons.)

The values of ε may be overestimated at the upper heights because spectral widths were determined by weighted moments, which can result in overestimates when noise is significant. A warning about the perils of this approach was already given in Chapter 7, Section 7.3.2. Those authors also ignored negative values of the energy diissipation rates, which leads to further overestimates of the mean values, as also discussed in detail in Chapter 7, Section 7.3.2. Nevertheless, the average values determined by *Nastrom and Eaton* (1997a) are not too different in typical values to those produced by *Dehghan and Hocking* (2011) and shown in Figure 7.15. Note that the data presented by *Dehghan and Hocking* (2011) and *Dehghan et al.* (2014) have been simultaneously calibrated against dedicated in-situ aircraft measurements made with a specially instrumented Twin Otter aircraft (*Dehghan et al.*, 2014), and so should have good reliability.

Figure 7.15 in Chapter 7 suggests that median values of the order of $1\text{–}2 \times 10^{-4}$ W kg^{-1} are normal in the troposphere. Figures 13 and 14 from *Hocking and Mu* (1997) also suggest that typical values should be of the order of $2\text{–}4 \times 10^{-4}$ W kg^{-1}.

The next suite of figures refer to diffusion coefficients. These have been taken from a variety of references. The different types of diffusion, and their relevance at different scales (as previously discussed), should be borne in mind when comparing the different measurements.

Dieminger et al. (1996), chapter I.2.2, also includes a discussion of the latitudinal and annual variations of K_M, but results are tentative and we will not pursue this matter here. The interested reader is referred to that reference.

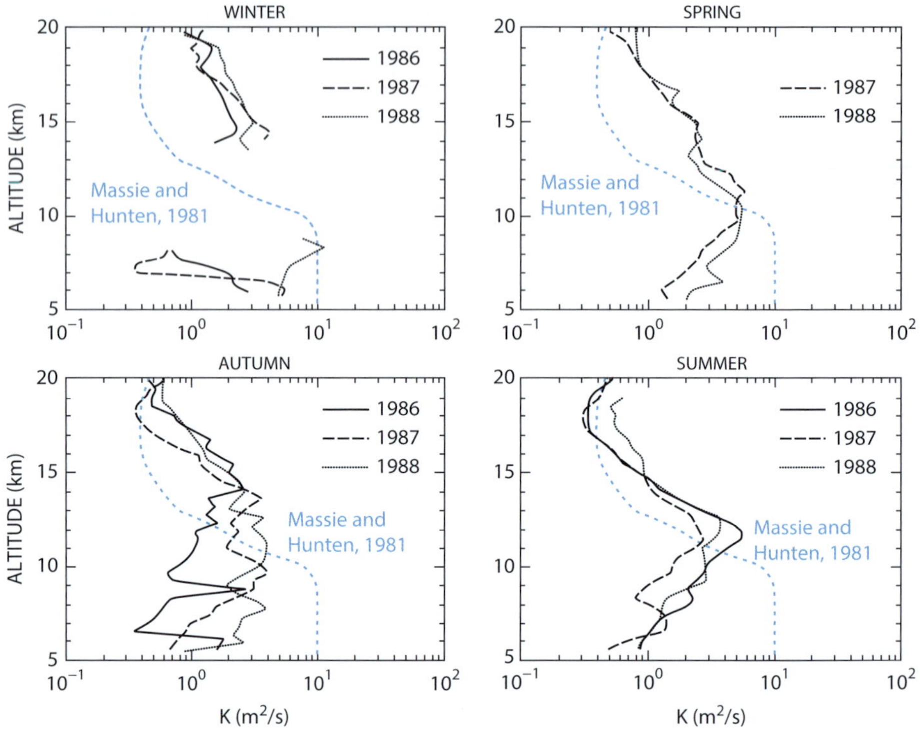

Figure 11.32 Typical turbulent diffusion coefficients in the lower atmosphere measured with the MU radar (from *Fukao et al.*, 1994). (Reprinted with permission from John Wiley and Sons.)

It is fair to say that the energy dissipation rates and diffusion coefficients shown in Figures 11.27 to 11.33 are not at all at variance with the various results from other methods, giving good confidence that radar methods are reliable, and that the various assumptions we have made in the preceding pages are not too unreasonable.

11.3.6 Small-scale structures and anisotropic turbulence

The structure of turbulence at scales less than a few meters in the atmosphere is still an area requiring investigation. The detailed shapes of the scatterers, as well as their statistically averaged shapes, are still unknown. *Hocking and Hamza* (1997) have shown schematically that individual refractivity structures are often elongated, stretched-out, "string-like" structures, but they can take a variety of forms. On average, it is normal to treat the scatterers as ellipsoids, as this form represents a good statistical average, even though almost no scatterers assume this shape. It is important to note that there are differences in shape between the refractive index variations (important for radar scatter) and the structure of the velocity field. Velocity variations do often take quasi-ellipsoidal shapes and are called "vortices." Most studies of turbulence involve correlation functions, structure functions or spectra, which often do have ellipsoidal shapes in space

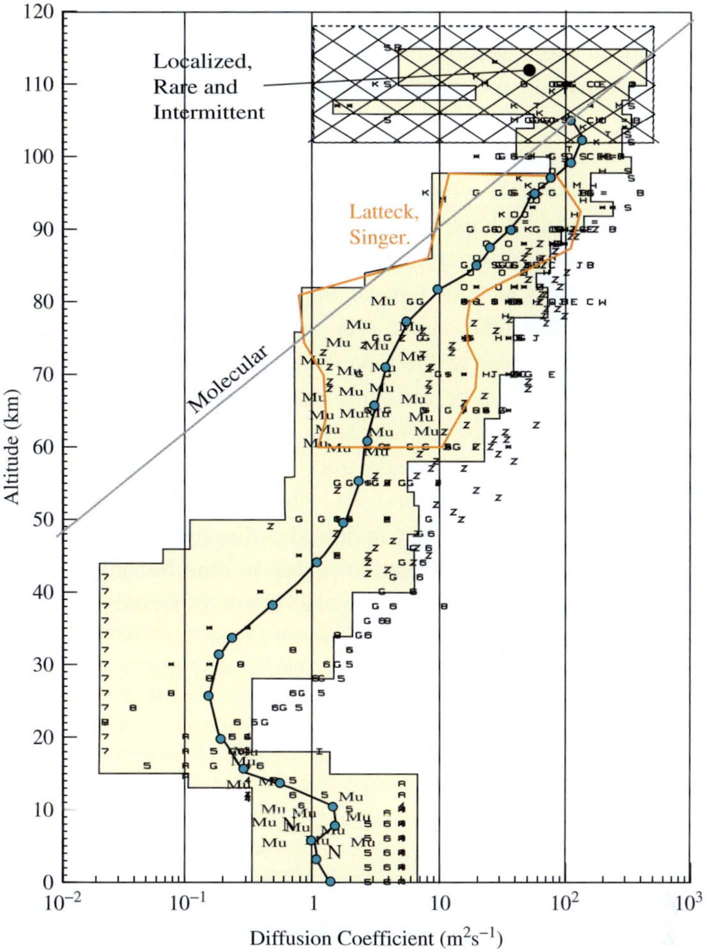

Figure 11.33 Collective of a large number of different estimates of diffusion coefficients for the atmosphere, from the ground to over 100 km altitude. The molecular diffusion coefficient is also shown as the gray line (labelled "Molecular"). When the turbulent and kinematic rates become comparable, turbulence ceases to exist, or at least becomes a rare phenomenon, and energy dissipation can take place by direct viscous dissipation. Adapted and improved from *Dieminger et al.* (1996), Chapter I.2.2, Figure 9. The various symbols refer to measurements by different authors, as described by *Dieminger et al.* (1996), Chapter I.2.2, Table 2, each with different methodologies, from computer models to rockets and balloons, as well as radar. Recent additions include the data labelled "Latteck Singer" (from Figure 11.28) and "Mu" from Figures 11.32 and 11.31 (from *Fukao et al.*, 1994). The heavy dark line shows the weighted average of all of these data calculated by one of the authors of this book (WKH), and the light-gray shaded area shows 16 and 84% percentiles of the distributions.

or wavenumber space, so the assumption of ellipsoidal scatterers, both in refractivity structure and the velocity field, works well when dealing with these functions.

It may seem that a radar with a pulse length of tens and even hundreds of meters cannot contribute a great deal to studies at the scale of a few meters, but this is not true. Radars can contribute significantly in this area, especially through the anisotropy parameter discussed in Chapter 7. *Hocking and Röttger* (2001) have given an extensive discussion of the role radar has played in examining the fine-scale shapes of atmospheric eddies.

For many years, the only theories used for description of inertial range turbulent motions were ones which assumed that the scatterers were statistically isotropic. Yet, as discussed in Chapter 7, radar studies have clearly shown that turbulence is often anisotropic, with scatterers generally stretched out horizontally compared to their vertical extent. Only recently have attempts been made to model atmospheric turbulence and allow for anisotropic eddies. The degree of anisotropy depends on scale. Eddies with vertical scales close to the inner scale are generally isotropic, but scales closer to the buoyancy range can be very anisotropic. *Hocking and Hamza* (1997) have shown how the degree of anisotropy changes as a function of scale, wind-shear, and Brunt–Väisälä frequency, and the detailed spectral form has also been discussed by *Dalaudier and Gurvich* (1997), as well as *Gurvich* (1994), *Gurvich and Kon* (1993) and *Gurvich* (1997). *Gage* (1990) has also briefly reviewed earlier theories of anisotropic turbulence, concentrating especially on correlation scales and the impact on both forward and backward radio scatter. The work of *Staras* (1955) was especially noted in this discussion.

The primary result from *Hocking and Hamza* (1997) is that the ratio of horizontal to vertical length scales ($\ell_x : \ell_z$) for a typical eddy takes the form

$$\frac{\ell_x}{\ell_z} \sim 1 + \gamma\beta^{-1/3}\mathcal{A}, \tag{11.74}$$

where

$$\mathcal{A} = |\frac{d\bar{u}}{dz}|\ell_z^{2/3}\varepsilon^{-1/3} \tag{11.75}$$

is a dimensionless parameter which we will denote as the "eddy anisotropy factor" and γ is a constant of order 1 – perhaps a bit less. The term β is another constant of order unity. Other variations on this formula can be found in *Hocking and Hamza* (1997). The anisotropy clearly depends on the wind-shear, and *Hocking and Hamza* (1997) also discuss the Richardson number dependence.

Although studies of this type are still in their infancy, they have considerable significance. For example, it has already been seen in Chapter 7 how measurements of the anisotropy can give an indication of the degree of convection in the atmosphere and thereby give a predictor for precipitation. Most of the points regarding anisotropic turbulence have been dealt with in Chapter 7, so they will not be repeated here.

11.3.7 Computer modeling of gravity wave breakdown and turbulence production

In Section 11.3.6, the fine-scale structure of turbulence was discussed. This is an important area, and until recently was primarily the domain of experimental studies. *Hocking and Röttger* (2001) is a good summary about some of the small scale details of turbulence.

In the 1990s and the first decade of the twenty-first century, a new tool became available to study these small-scale motions. Computers were developed that were fast enough, and had sufficient memory, to simulate mesospheric turbulence down to scales of the order of a few meters. This technology has given new capabilities to computer simulations of wave breakdown and the resultant turbulence. These important new developments cannot be fully discussed here, but at least a few of the successes can be summarized and some of the more important papers can be noted. These simulations are already beginning to confirm important observations made by radar about the nature of turbulence (e.g., *Hocking and Röttger*, 2001), and are also advancing understanding of the nature of wave breakdown and turbulence. The ability to perform turbulence studies in three dimensions has been found to be especially important, since this allows transverse vortical rolls to develop, which are forbidden with a two-dimensional simulation. *Klaassen and Peltier* (1985a, b) performed important studies of this type, and higher resolution studies were undertaken by *Andreassen et al.* (1994, 1998), as well as *Fairall et al.* (1991); and *Fritts et al.* (1993, 1994, 1996a, b). Later, even higher resolution simulations were performed by *Werne and Fritts* (1999, 2001), and resolutions of a few meters are now possible. The development of counter-rotating vortical pairs in gravity wave breakdown was an important early discovery. One limitation of simulations to date is that they have tended to examine the breakdown of single waves, which break catastrophically; future simulations should include multiple waves interacting together, and ultimately a spectrum of waves, in order to investigate the likelihood of shedding. It also appears that turbulence patches can in turn generate new spectra of gravity waves, an important result in itself, and one previously also proposed by *Weinstock* (1978a). The different modes of breakdown have already been discussed in Section 11.2.12, and *Klaassen* (2003) was especially recommended.

One strong advantage of being able to produce computer simulations at scales of a few meters is that it is possible to better simulate the interaction between electromagnetic waves and the refractive index inhomogeneities, thereby better simulating the radiowave scattering process. *Gibson-Wilde et al.* (2000); *Franke et al.* (2011) were able to do this, although in the earlier publication some extrapolation was needed to get to the scales of 3 meters that were required to simulate VHF radio scatter. The future holds even greater promise for computer simulations, keeping in mind that these should always be monitored and calibrated against experimental studies. More details can be found in *Fritts and Alexander* (2003), section 6, and a relatively recent example of computer modeling of wave breakdown was presented by *Fritts et al.* (2013).

12 Meteorological phenomena in the lower atmosphere

12.1 Introduction

Some of the earliest applications of windprofiler radars were in regard to tropospheric and lower stratospheric studies. The radars developed at the Sunset site near Boulder, Colorado (*Green et al.*, 1979) and in the Harz mountains in Germany (the SOUSY radar (*Czechowsky et al.*, 1976)) were two of the earliest such instruments, and were certainly built with meteorological studies in mind. Some of these radars have already been described in Chapter 2, and the SOUSY radar was extensively discussed in Chapter 6.

The most direct meteorological studies have been in regard to wind motions, but these radars have also been usefully employed in other areas, including studies of turbulence strengths and anisotropy, tropopause height measurements, gravity wave momentum fluxes, precipitation measurements, temperature profile determinations, and various others.

It is impossible to cover all aspects of MST radar applications relating to the troposphere in just one chapter. For this reason, we will concentrate mainly on results, rather than on specific details about techniques. It will be assumed that the techniques have been sufficiently covered in earlier chapters.

The early years of tropospheric studies have been especially well covered in several excellent reviews, including those by *Röttger and Larsen* (1990), *Gage* (1990), *Larsen and Röttger* (1982) and *Balsley and Gage* (1982). Some of the early parts of this chapter will involve a recap of the main results of those publications.

Röttger and Larsen (1990) discussed the origins of VHF MST radar studies in the context of: (i) developments following the use of high-power X, S, and UHF band radars in the United States of America, as well as FMCW (frequency modulated continuous wave) techniques; coupled with (ii) the detection of tropospheric echo fading observed at Jicamarca (Peru) by *Woodman and Guillen* (1974); and (iii) the application of phase-coherent techniques. These events in turn led to the first dedicated VHF-ST radars being built at Sunset, near Boulder, Colorado, and in the Harz mountains of Germany (the SOUSY, sounding system radar). Phase coherent detection was especially important in the development of such systems, for without it, detection of useful tropospheric echoes with VHF systems would be nearly impossible.

While the potential to measure winds, turbulence strengths, and C_n^2 was recognized early in the history of the development of windprofiler radars, it was also recognized that a proper interpretation of the received signals required better knowledge about the entities that scattered the incident radiowaves, and indeed about the scattering mechanisms as well. Considerable discussion arose in the literature about the relative importance of turbulent scatterers versus so-called Fresnel reflectors. These different entities have been discussed already in earlier chapters. The importance of aspect sensitivity was especially studied in Chapter 7 in this book. In the following subsection, we will briefly revisit this aspect of radar theory, since its understanding is so pivotal to interpretation of many subsequent meteorological applications.

12.2 Scattering mechanisms

Two major categories of scatterers are generally considered to exist in the atmosphere – turbulence and so-called specular reflectors. Within each major category various subcategories exist. Turbulence can be either isotropic or anisotropic, and models for specular reflectors vary from individual pancake-like structures to extended undulating sheets and even to horizontally aligned and highly damped atmospheric viscosity waves. These have already been discussed to some extent in earlier chapters, but we will briefly revise our knowledge about them here.

The refractive index in the lower atmosphere was given in Chapter 3, Equation (3.287), and is repeated here for ease of reference in this chapter:

$$n = 1 + 77.6 \times 10^{-6}\frac{p}{T} + 3.73 \times 10^{-1}\frac{e_{wp}}{T^2}, \tag{12.1}$$

where p is the atmospheric pressure in units of millibars (hPa), T the temperature (K), and e_{wp} (hPa) is the water-vapor pressure. Typical deviations of n from unity are of the order of 0–500 $\times$ 10^{-6}, with largest values occurring in the lower tropsphere. Often $N = n - 1$ is referred to as the "radio refractivity."

Radiowave scatter occurs from the atmosphere when the air is mixed (either systematically or quasi-randomly) by various processes, thereby producing three-dimensional structures in the refractive index with Fourier scales of a suitable scale to produce backscatter. The variations in refractive index in the troposphere can be due to either pressure, temperature, or water vapor variability, although usually pressure variations smooth out fairly quickly and are not so important. Water vapor variability is especially important in the lower atmosphere, and temperature variability is important throughout the atmosphere. For a monostatic radar, if the three-dimensional Fourier transform of the mixed refractive index spatial distribution in a specified volume of air includes Fourier components with wave-fronts aligned perpendicular to the radial direction from the radar, and the wavelength of this scale is one half of the radar wavelength, and if this component is strong enough, then the radar will detect a signal backscattered from it. Such Fourier components can arise from random or quasi-random mixing, or from stepped or regular structures. The Fourier scale with this wavelength is called the Bragg scale. The scatter is called Bragg scatter. Turbulence is an example of quasi-random mixing, and sudden sharp changes in refractive index as a function of height (steps) are examples of organized mixing. Both contain Bragg scales.

layers (e.g., *Hocking*, 1985; *Woodman and Chu*, 1989; *Lesicar and Hocking*, 1992). *Worthington et al.* (2001) have produced detailed maps of aspect sensitivity for the MU radar.

There are two main reasons why these reflectors are important in the context of meteorology. The first is that their existence often informs us about the degree of static stability in the local environment, and the second is that they can bias the measurements of atmospheric parameters. This is particularly true in the case of wind measurements by Doppler methods, since the presence of specular reflectors can alter the apparent pointing direction of the radar beam, leading to underestimates of the velocity. This latter point has been discussed already in Chapter 7.

It is the ability of specular reflectors to identify regions of high static stability that is of most interest here.

Regions of stability, frontal passage, and tropopause detection

Although the nature of specular reflectors was not fully understood at the time, the fact that VHF radars were able to detect these reflectors was quickly recognized as an important diagnostic for atmospheric studies. They generally seemed to occur in areas of strong static stability, and it was soon recognized that they could be used at times to identify the tropopause and certain aspects of frontal systems. An example of the vertically backscattered power received by a VHF radar plotted as a function of height and time is shown in Figure 12.2, where a region of stable air is seen to mix into the troposphere below.

Figure 12.3 shows another example over a longer time frame, showing the complex variation of height of the tropopause, and at the same time showing the variation of ozone partial pressure. The "radar tropopause" really does show nicely the transition between the ozone-rich stratosphere and the (generally) ozone-poor region immediately below. These data were supplied from a radar at Montreal in Canada. The ozone measurements were made with a series of closely spaced balloon flights. Some ozone also seems to be leaving the troposphere and flowing to the lower levels of the troposphere (*Hocking et al.*, 2007a), but that aspect will not be discussed here.

Windprofilers are also particularly good for identifying the passage of frontal systems. Figure 7.25 showed an example of a high-level occluded front passing over the McGill VHF radar in Montreal, Canada. The merger of the warm and cold boundaries can be seen clearly just after midnight on the 11th of November at 10.5 km altitude. The boundary with the coldest air continues to rise as time goes by, reaching 13 km at the end of the time shown in the figure. Often these regions of enhanced scatter can be seen to quite low heights in the atmosphere – as low as 5 or 6 km.

A great deal of work has been done in relating radar backscatter to detection of the tropopause (e.g., *Röttger*, 1979; *Gage and Green*, 1982a, b; *Rastogi and Röttger*, 1982; *Larsen and Röttger*, 1983, 1985; *Hocking et al.*, 1986; *Rüster et al.*, 1998; *Browning et al.*, 1998). Many of these earlier works concentrated on the highly aspect-sensitive nature of these echoes. An example was shown in Figure 7.20(a) in Chapter 7.

Gage and Green (1982b) developed a somewhat more objective method for determination of tropopause heights, as described below. However, they have had to revise their

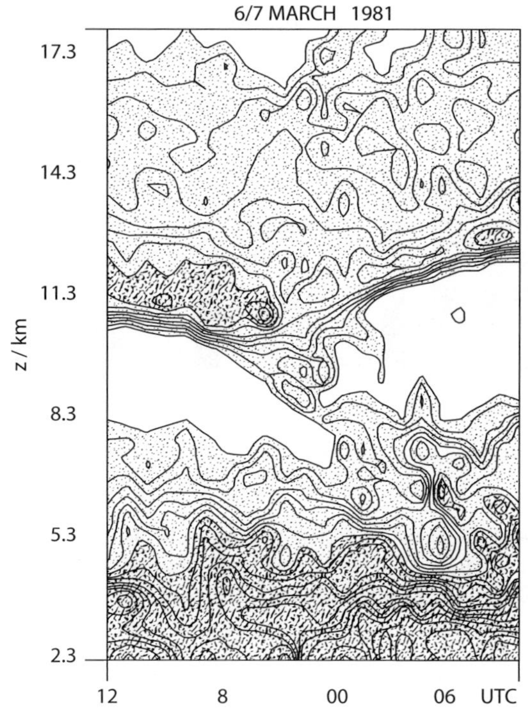

Figure 12.2 Time–height section of backscattered radar power measured with the SOUSY-VHF radar in Germany in 1981. Contours are at steps of 2 dB, and more intense shading corresponds to stronger echoes (taken from *Larsen and Röttger* (1982)). (Reprinted with permission from the American Meteorological Society.)

original proposal because they assumed that the backscattered power should be proportional to the pulse length squared, whereas it was shown by *Hocking and Röttger* (1983) that it should be proportional to the pulse length. Based on this revision, *Gage et al.* (1985) provided a corrected and improved model.

The modified theory goes as follows. It is firstly assumed that scatter from stable regions of the atmosphere (either the troposphere or stratosphere), like the region just above the tropopause, is due to reflection from clusters of horizontally extended specular reflectors. This process is termed Fresnel scatter, which is frequently, but not always, seen. However, we will assume for now that for a significant portion of the time the scatter is due to Fresnel scatter.

Then the effective power reflection coefficient produced by a cluster of Fresnel reflectors within the radar volume is assumed to be given by (*Gage and Green*, 1982a, b; *Gage et al.*, 1985)

$$|\mathcal{R}|^2 = \frac{\left[F_{\mathrm{spec}}(\lambda)\right]^2 (\Delta r)\overline{M_n}^{-2}}{16}, \tag{12.2}$$

where it can be seen that the exponent of Δr is unity, and not 2 (as it was in *Gage and Green*, 1982a, b). $F_{\mathrm{spec}}(\lambda)$ is a calibration constant that needs to be determined from experiment, perhaps using a series of radiosonde launches during radar operation.

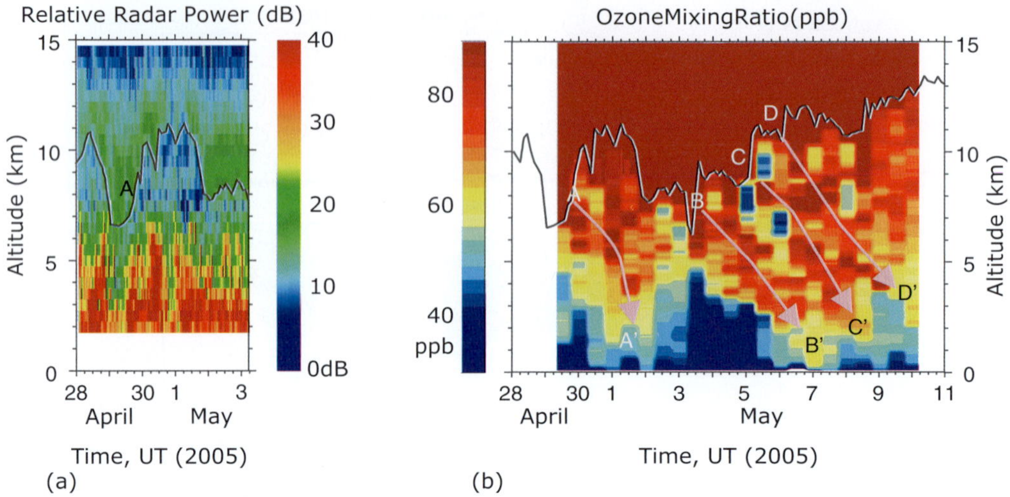

Figure 12.3 Radar backscatter intensity (left) and ozone densities (right) over a period of several days, recorded at Montreal, Canada. The local increase in intensity at heights between 7 and 13 km in the left-hand figure illustrates the radar-derived tropopause, and the ozone densities on the right show an alternative demarcation between the troposphere and stratosphere. The two parameters are well correlated. From *Hocking et al.* (2007a). (Reprinted with permission from Nature Publishing Group.)

The term $\overline{M_n}$ is the potential refractive index gradient, which has already been described in earlier parts of this book., viz.

$$M_n = \frac{-77.6 \times 10^{-6}}{T^2} p \left[1 + \frac{15500}{T} q_{wp} \right] \left[\frac{dT}{dz} + \Gamma_a - \frac{7800}{1 + \frac{15500}{T} q_{wp}} \cdot \frac{\mathrm{d}q_{wp}}{\mathrm{d}z} \right]. \quad (12.3)$$

For regions of the atmosphere close to the tropopause, where we can assume that water vapor content is small, we can write

$$M_n = -77.6 \times 10^{-6} \frac{p}{T^2} \left[\frac{\mathrm{d}T}{\mathrm{d}z} + \Gamma_a \right], \quad (12.4)$$

or if we use the fact that $\frac{\partial \ln \Theta}{\partial z} = \frac{1}{T} \left[\frac{\mathrm{d}T}{\mathrm{d}z} + \Gamma_a \right]$ (*Hocking*, 1985), where Γ_a is the adiabatic lapse rate, then

$$M_n = -77.6 \times 10^{-6} \frac{p}{T} \left(\frac{\partial (\ln \Theta)}{\partial z} \right). \quad (12.5)$$

Any of these forms are suitable for the model discussed below.

Gage et al. (1985) made various adaptations to Equation (12.2). The main modification was that they let $F_{\mathrm{spec}}(\lambda)$ be a function of height, in an attempt to more accurately accommodate the rapid decrease in signal as a function of height. Although $F_{\mathrm{spec}}(\lambda)$ was assumed to be a constant in (12.2), this was only an assumption, but it did not match the rapid decrease in signal with height: there seemed to be extra processes which affected Equation (12.2). For example, if the number density, or average strength, of Fresnel reflectors changes as a function of height for geophysical reasons, perhaps due to systematic variations in the (unknown) generation mechanism as a function of height, then an extra height dependence might exist in addition to the dependence on M_n. These authors concluded empirically that F_{spec} should have an exponential height dependence, and proposed that

$$F_{\mathrm{spec}}^2(z) = F_{\mathrm{spec}}^2(z_0)\mathrm{e}^{\frac{z_0 - z}{H_{\mathrm{spec}}}}, \qquad (12.6)$$

where $z_0 = 10$ km and $H_{\mathrm{spec}} = 6$ km.

With this change they then produced an estimate of M_n as

$$\tilde{M}_n^2 = \frac{16\, P_{Rx}\, z^2\, \lambda^2}{\alpha^2\, P_{Tx}\, A_e^2\, \Delta z\, F_{\mathrm{spec}}^2(z_0)} \mathrm{e}^{\frac{z - z_0}{H_{\mathrm{spec}}}}, \qquad (12.7)$$

where $z_0 = 10^4$ m and $H_{\mathrm{spec}} = 6000$ m, which is an adaptation of Equation (7.69) in Chapter 7. The variables are defined in Equation (7.69), but the height is now z and the pulse-length is Δz. The term $F_{\mathrm{spec}}(z_0)$ needs to be determined by calibration against radiosondes at $z_0 = 10$ km.

Once $\tilde{M}_n$ has been found, either of the expressions (12.4) or (12.5) can be used (in principle) to determine the temperature profile by integrating up (or down) from some pre-measured starting height and temperature. Of course this starting point must be in a region where humidity is negligible if these equations are to be applied, and such a point may not always be accessible.

Of course this all also depends on the assumption that F_{spec} is well-behaved and has a nice exponential decrease with height, and that the scale height H_{spec} really is a constant. The process has been tried, with variable success.

However, even if height profiles of temperature cannot be determined, there is a less challenging thing that can be done. If the profiles of $\tilde{M}_n$ can be found, then climatological values for p and T can be used to specify p/T^2 in (12.4), from which a measurement of the current value of $\frac{dT}{dz}$ can be deduced. This in itself can provide useful information. If the temperature gradient reaches the point where $\frac{\mathrm{d}T}{\mathrm{d}z}$ equals 2 K km^{-1}, then we have reached the tropopause. This is according to definition: according to *Gage and Green* (1982b), the tropopause is "located at the lowest altitude (above the 500 mb level) above which the temperature lapse rate does not exceed 2 K km^{-1} for at least 2 km." So the possibility remained that the technique could be used to determine the tropopause height, under conditions of a stable tropopause.

Gage and Green (1982b) went on to adjust the formula to deal with the case that a radar might measure signal-to-noise ratio, and assumed that galactic skynoise was the main source of noise. We will not discuss this development here because it is a far better procedure to calibrate the radar in an absolute sense and use the procedure described above, principally because this removes the effects of S/N variations which arise due to man-made interference and variations in skynoise.

The model has some interesting ideas, but its usefulness is questionable. The method works largely because of the existence of a secondary minimum in scattered power just above the tropopause (e.g. see Figures 12.3 and 7.25), which means that the tropopause is essentially a region where the scattered power as a function of height shows a local increase. In fact, quite reasonable estimates of the tropopause height can be found simply by identifying the height at which the backscattered power shows this secondary maximum at 7–13 km altitude, with incorporation of some compensation for the pulse-width.

Some additional care is needed with this method. The parameter $F_{spec}(\lambda)$ needs to be confirmed to really be as well behaved as proposed above and not dependent on seasonal factors, or for that matter day-to-day and even hour-to-hour variability. One could easily imagine that ingress of a frontal system, or development of strong convection, or the arrival of a strong gravity-wave, might lead to far more erratic behavior in F_{spec}.

Some caution is also needed in regard to the relationship between strong stability and the tropopause. While it is true that on occasions the tropopause is associated with highly aspect-sensitive scatterers, this is not exclusively so. More recent studies have suggested that the region just above the tropopause can at times be a region of enhanced turbulence, and hence more isotropic backscatter. This can occur especially if upward-propagating waves break just above the tropopause, as for example described by *Van Zandt and Fritts* (1989). Specific examples of such events at the equator were described by *Fujiwara et al.* (2003) and *Yamamoto et al.* (2003), but such events also occur at all latitudes; therefore, some caution is required in regard to interpretation of enhanced radiowave scatter at the tropopause.

Nevertheless, it remains true that enhanced backscatter from near the tropopause is common with windprofiler radars. On some occasions, it may be due to specular echoes and on others it may be due to turbulence, but there is almost always a secondary maximum in backscattered power. Hence windprofiler echoes can be used with good reliability to monitor the height of the tropopause, even just by recording the height of the lower bound to the secondary maximum, and even though the reasons for the enhanced scatter may vary. Whether the mathematical procedures outlined above will help further is still open to debate.

12.3 Wind measurements

12.3.1 The advantages of wind profilers for meteorological studies

Unquestionably, the most notable application of windprofiler radars is the measurement of winds, as can be recognized from their name. Windprofiler radars have become a common instrument in meteorological applications, and it would be impossible to cover all the cases where they have played a significant role. Their use is now quite common in weather forecasting, for example. In this section, we will concentrate on their history and some notable case studies.

In meteorology, as discussed for example in Chapters 1 and 11, atmospheric *mesoscale* motions refer to spatial scales between a few kilometers and one or two hundred kilometers, and temporal scales of the order of minutes to a few hours. Such events include thunderstorms, tornadoes, and various types of local circulations like land, sea and valley breezes. Typical *synoptic* scale events include hurricanes, typhoons, high and low pressure systems, and frontal systems. Windprofiler atmospheric radars have been used to investigate all of these phenomena, (e.g., *Strauch et al.* (1984); *Gage et al.* (1991a); *Webster and Lukas* (1992); *Teshiba et al.* (2001) (and references therein); *Röttger and Larsen* (1990); *Hooper and Pavelin* (2003), among others). Studies of gravity waves have been especially driven by windprofiler research, and these have already been discussed in Chapter 11. For meteorological studies, networks of windprofilers are more useful than individual radars; nevertheless, useful studies have been made with single radars as well, especially in the early days of windprofiler research.

The earliest studies of wind motions using windprofilers were limited by computing power, so that wind velocities were often derived using somewhat primitive algorithms, like weighted spectral moments and auto-correlative techniques. These have already been discussed in earlier Chapters. For example Figure 2.17 in Chapter 2 showed some earlier spectra, and Figure 12.4 shows the results of application of weighted spectral moments. Spectral moment methods and autocorrelative methods are sensitive to noise, and severe noise can lead to very anomalous values of wind speed. These values must be removed by processes like outlier rejection, where data points that are clearly different from their neighbors are identified and removed. This is also called consensus averaging.

With the advent of faster computers, it was possible to apply more complex spectral determination algorithms. These have been discussed to some extent in Chapters 7 and 8, where spectral processing was considered in more detail. Fitting to pre-specified functional forms, such as a Gaussian, leads to less contamination from noise. By limiting the amount of coherent integration, and using filtering processes instead, it is possible to more effectively remove the effects of aircraft, meteors, lightning, and other contaminants (e.g. see Figure 4.22). An example is shown in Figure 12.5, where the upper graph shows the raw wind components recorded in one beam of a Doppler radar. There are far fewer outliers than in the previous figure.

Outlier rejection algorithms have also developed considerably in recent years, and the second graph in Figure 12.5 shows the results of such an algorithm (from *Hocking*, 1997a). In this case, a piece-wise polynomial is fitted between regions of high data density, and large excursions from this fit are removed. Further rejection algorithms have been developed by *Weber and Wuertz* (1991), and most windprofiler radars now employ quite sophisticated algorithms for rejection of anomalous points. Further discussion can be found in *Hocking* (1997a, b), *Wilfong et al.* (1999), *May and Strauch* (1989), *May and Strauch* (1998), and *Anandan et al.* (2005).

Figures 12.6 and 12.7 show examples from a radar in Canada (Walsingham, Ontario) of components of the wind as a function of height and time, and vector plots of winds, for two periods in 2006. These plots are typical of current data quality expected from windprofiler radars.

The previous pictures highlight the good temporal resolution that the radars can achieve. Figure 12.8 demonstrates the capabilities further and shows why the near-continuous capability of these radars is so important. It shows dramatic changes in

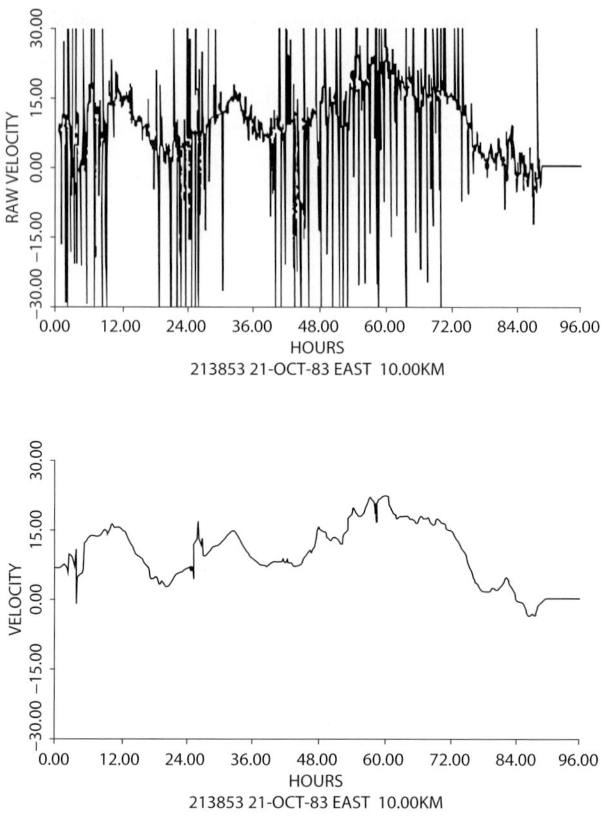

Figure 12.4 Wind velocity determination in the early days of MST radar. The upper graph shows the raw data, which clearly contains many anomalous points. The lower graph shows the time series of wind data after application of a consensus filter (from *Röttger and Larsen* (1990), who adapted the figure from *Chadwick et al.* (1984)). (Reprinted with permission from the American Meteorological Society.)

wind speed (from *Röttger and Larsen*, 1990). In particular, the figure shows a period of rapid wind speed and directional change at 12:00 on 8 February 1982. In this case, the radiosonde data also detected the change, demonstrating that it was indeed real. However, if the radiosonde had been launched 6 hours earlier or 6 hours later, it would never have even detected a change in the wind magnitude. The radar, on the other hand, would have caught it.

Windprofilers have been used for many studies in the last 10 years, and are becoming something of a necessity in many meteorological campaigns. *Teshiba et al.* (2001) and *Sato* (1993), among others, showed the usefulness of such radars during the passage of a typhoon (hurricane) in Japan in 1997 (see Figure 12.9). *Umemoto et al.* (2004) used a coordinated campaign of windprofiler radars, C-band precipitation radars, X-band

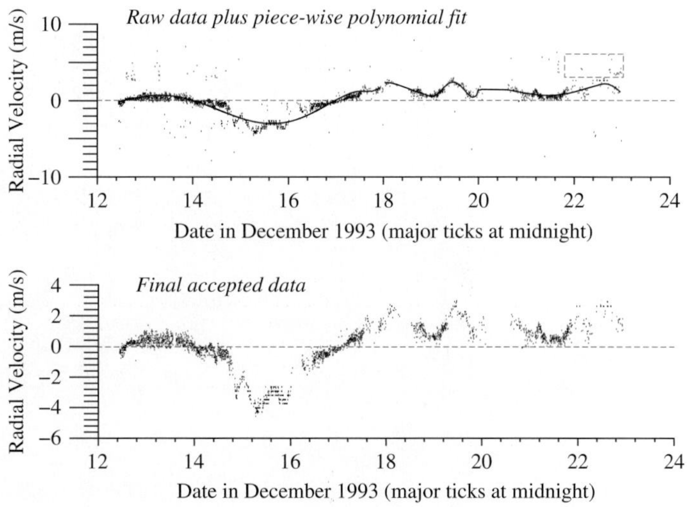

Figure 12.5 Wind velocity determination showing quality control. The upper graph shows the raw data, which clearly contains some anomalous points, but many less than for Figure 12.4. The lower graph shows the time series of wind data after application of a quality-control algorithm. (From *Hocking*, 1997a). (Reprinted with permission from John Wiley and Sons.)

Doppler radars, and rawinsondes to study an orographic rain-band near Kyushu in Japan. Many other examples exist.

12.3.2 Verification of profiler winds

In the previous section, we outlined a variety of special cases that demonstrate the value of windprofilers for wind measurements. Their primary advantages are their ability to record continuously and at high temporal resolution (even in non-cloud clear-air conditions), and their capability to operate without significant user support. However, in order for them to be truly accepted as important instruments for wind measurements, their accuracy needs to be verified by comparison with other reference techniques.

Windprofiler radar winds have been compared with a variety of other reference techniques. Chief among these are radiosondes. Radiosonde-measured winds have been compared with windprofiler winds by many investigators (e.g., *Warnock et al.*, 1978; *Farley et al.*, 1979; *Crane*, 1980a; *Fukao et al.*, 1982; *Larsen*, 1983; *Vincent et al.*, 1987; *Weber and Wuertz*, 1990; *May*, 1993; *Steinhagen et al.*, 1994; *Belu et al.*, 2001; *Reid et al.*, 2005). The precision and relative accuracy of wind measurements with windprofiler radars are discussed in several papers in the literature (e.g., *Strauch et al.*, 1987; *Kudeki et al.*, 1993; *Astin*, 1997; *Reid et al.*, 2005). Generally, agreement is good, and windprofilers are now largely accepted as viable alternative wind measuring instruments (e.g., *Balsley and Gage*, 1982; *Larsen and Röttger*, 1982; *Monna*, 1994). *Reid et al.* (2005) found that spaced antenna winds determined by full correlation analysis tended

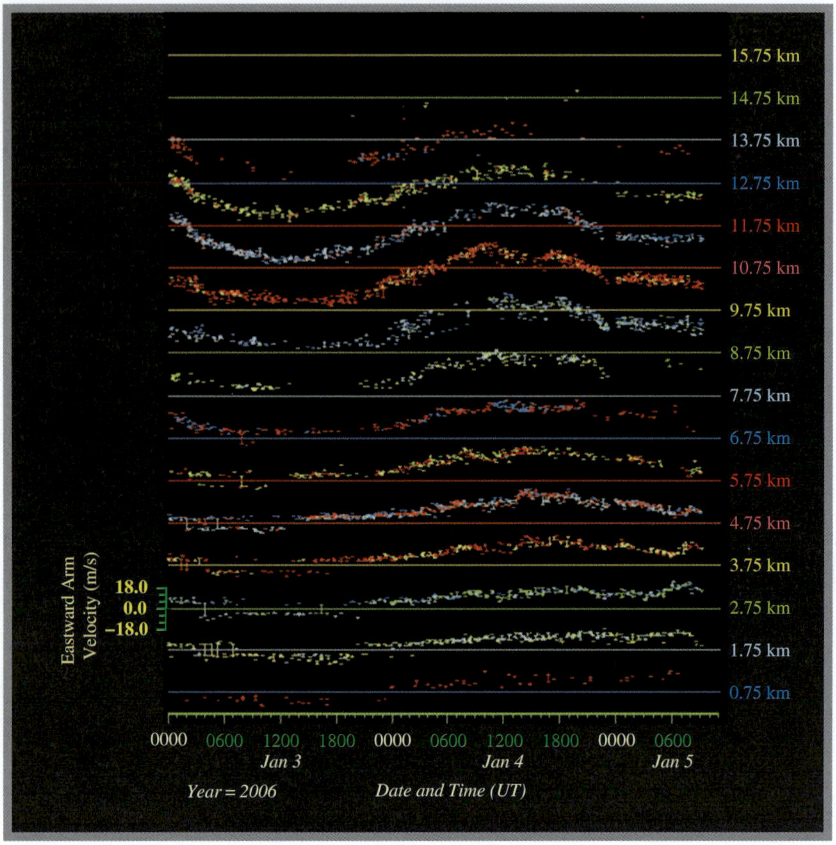

Figure 12.6 Typical wind velocity components (radial velocities) as a function of height and time for a modern radar. Each individual measurement is represented by a single point, and the data quantity is sufficiently high that the lines seem almost continuous. The data were taken with a four-beam Doppler radar. Westward radial velocities have been reversed in sign to make them compatible with the eastward ones. Different colors at each height refer to different beams. Different color schemes have been chosen at each height in order to better distinguish the different data sets.

to underestimate sonde winds by about 10%. Doppler methods can also lead to underestimates, although *Hocking* (2001a) suggested the effect was only a few percent and could be corrected from knowledge of the scatterer aspect-sensitivity. We will not reproduce any of these comparisons here. Comparisons with computer models and analyses such the CMC (Canadian Meteorological Model), the NMC (National Meteorological Center of the United States of America) and the ECMWF (European Centre for Medium-Range Weather Forecasting) have also been carried out (e.g., *Gage et al.*, 1988; *Pauley et al.*, 1994; *Belu et al.*, 2001).

Nevertheless, differences between profiler winds and other techniques do exist, for a variety of reasons. One difference lies in the spatial separation of the instruments. For example, radiosondes can drift tens of kilometers from their launch site by the time they reach upper altitudes. Additionally, often radiosonde launches do not take place at the

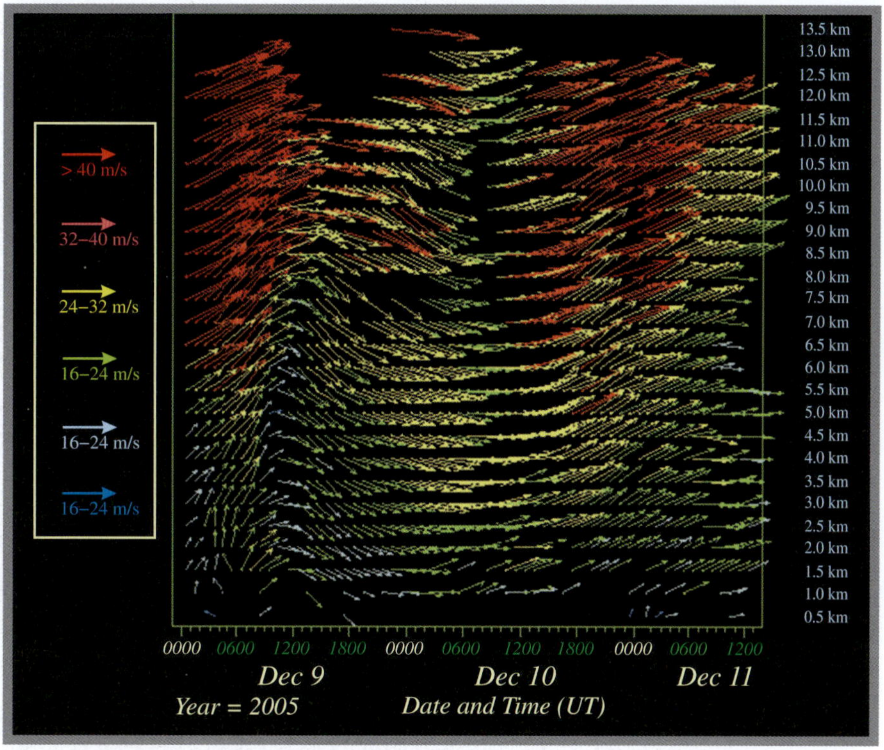

Figure 12.7 Wind velocity vectors over a 2 to 3 day period from Walsingham, Ontario. The radar frequency was 44.5 MHz. The appearance of the jet stream overhead is clearly seen by the strong red arrows.

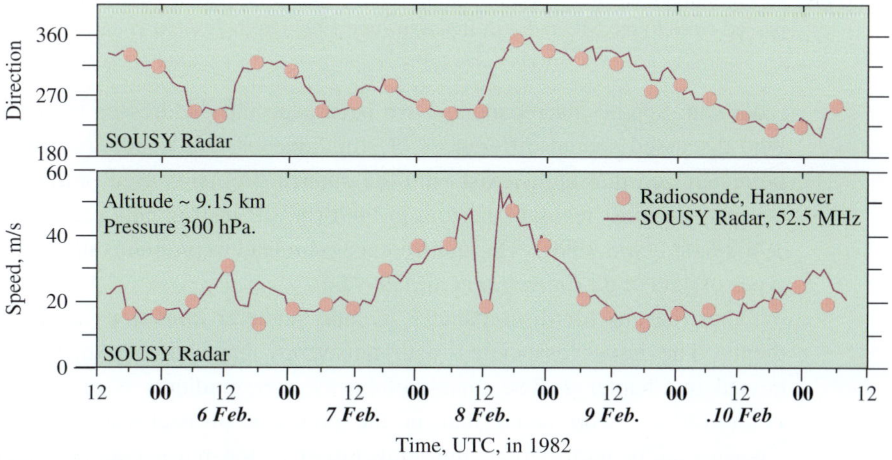

Figure 12.8 Time series of radar measurements of wind speed and direction (solid lines), compared with radiosonde measurements (dots). The radar magnitudes even properly track the sudden drop-out in wind speed at 1200 UT on Feb. 8 (from *Röttger and Larsen* (1990)). (Reprinted with permission from the American Meteorological Society.)

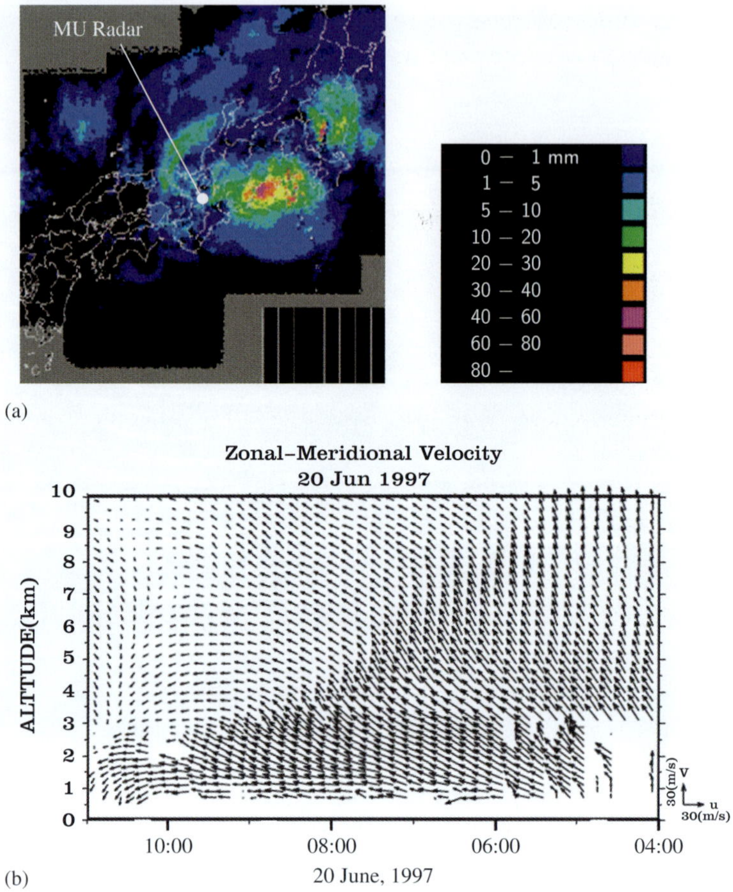

Figure 12.9 Precipitation measurements and windprofiler winds for typhoon 9707 – code-named *Opal*, as it passed close to the MU radar in Japan in June 1997 (from *Teshiba et al.*, 2001).

radar site. Spatial discrepancies have been especially addressed by *Jasperson* (1982), who discussed spatial differences of simultaneous radiosonde ascents. It should also be recognized that faults exist with the other instruments used in the studies, and any comparison must recognize errors in both (or all) instruments used (*Passi and Morel*, 1987; *Rust et al.*, 1990). These differences can lead to problems in integrating different types of data (e.g., *Ciesielski et al.*, 1997).

Other reasons for discrepancies include scatterer anisotropy and backscatter gradients. The issues associated with anisotropy have already been extensively discussed in Chapter 7. The matter of backscatter gradients is discussed by *Johnston et al.* (2002), who noted that in the presence of backscatter cross-sections that change rapidly with height, the convolution of the pulse with the backscatter function can produce peaks that do not appear at the center of the pulse, but are rather shifted to different heights (often lower heights). As a result, winds can be assigned to erroneous heights. This is especially problematic if long pulse lengths are used.

12.4 Winds from windprofiler networks

While windprofilers are powerful tools in stand-alone situations, they have become even more important when operating cohesively in larger numbers. Several countries have developed large networks of these radars and have integrated them into their respective weather offices.

Various frequencies have been allocated for windprofiler research. The World Radio-communication Conference of 1997 (WRC-97) allocated frequencies as described below, according to resolution COM5-5, with footnotes S5.162A and S5.291A (from *Ishihara*, 2005):

- 46–68 MHz : Mesosphere–Stratosphere–Troposphere Radar
- 440–450 MHz : VHF Windprofiler
- 470–494 MHz : VHF Windprofiler
- 904–928 MHz : UHF Windprofiler
- 1270–1295 MHz : UHF Windprofiler
- 1300–1375 MHz : UHF Windprofiler.

Individual countries often have individual additional requirements. Usually, frequencies must be assigned by a suitable frequency allocation agency. The most serious issue associated with profiler research is the wide bandwidth required. A radar with 150 m vertical resolution requires a 3 dB bandwidth of 1 MHz, which is a very wide bandwidth for radars working around 50 MHz. Usually, even an allocation of 250 to 500 kHz is difficult to obtain, since it represents about 1% of the frequency at 50 MHz. Wider bandwidths are easier to obtain at higher frequencies.

The figures that follow show distributions of radars in several networks. Further information can be found in *Dibbern et al.* (2003). Each country (or cluster of countries) uses different philosophies and guidelines for designing the radars in the networks and for supply of operational data to central servers.

The European network, CWINDE (Cost Wind Initiative for a Network Demonstration in Europe, see Figure 12.10) has left the assignment of frequencies to individual groups and organizations, and then simply coordinates output from each of these radars (*Parrett et al.*, 2004). Radars include a mixture of VHF, UHF, and L- and S-band radars. The distribution is shown in Figure 12.10.

In contrast, a network in the USA was organized more tightly by NOAA (National Oceanic and Atmospheric Administration). This network ran till about 2014, after which it was largely closed down, although some portions of it did continue beyond that time. However, we include it for historical reasons. According to *Dibbern et al.* (2003), the majority of radars (32) worked at 404 MHz; three in Alaska worked at 449 MHz, while

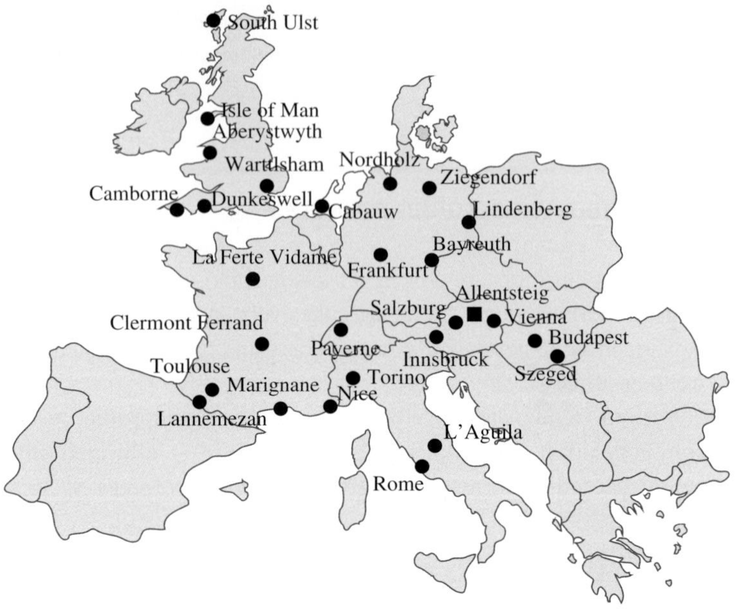

Figure 12.10 Distribution of windprofiler radars in the CWINDE radar network in Europe (from *Hocking*, 2011). (Reprinted with permission from Elsevier.)

a few others worked at VHF – the latter usually being funded by separate organizations. The network has also been discussed by *Law* (1992). A few radars connected with the network operated at frequencies over 900 MHz. The distribution of radars in this network is shown in Figure 12.11. A pacific-wide network, also coordinated by NOAA, also existed for a number of years (e.g., *Gage et al.*, 1991a, b).

The Japanese network has opted to concentrate on lower altitude winds and has built a network dominated by L-band radars (1.3 GHz, see *Dibbern et al.*, 2003), plus a few additional radiosondes and the pre-existing MU radar (e.g. *Fukao et al.*, 1985a, b). This network is shown in Figure 12.12.

The smaller O-QNet in Canada (Figure 12.13) has been designed primarily with radar frequencies in the range 40 to 55 MHz.

The quality of data output and effectiveness of these networks has been examined in a variety of studies (e.g., *Parrett et al.*, 2004; *Benjamin et al.*, 2004; *Ishihara*, 2005). In general, there is agreement that the radars make meaningful positive impact on weather forecasting and analyses. Windprofilers are of special value in the growing area of nowcasting, which refers to forward weather predictions only a few hours, or even a few tens of minutes, ahead in time.

Although never implemented, an interesting proposal was made by *Carlson and Sundararaman* (1982) for a United-States-wide network. They proposed a grid of closely spaced windprofilers distributed throughout the entire continental USA, with the primary objective being to provide wind information to commercial aircraft. As a result of suitable flight planning based on data from these radars, it was proposed that fuel savings could be enormous, since aircraft could avoid flying into

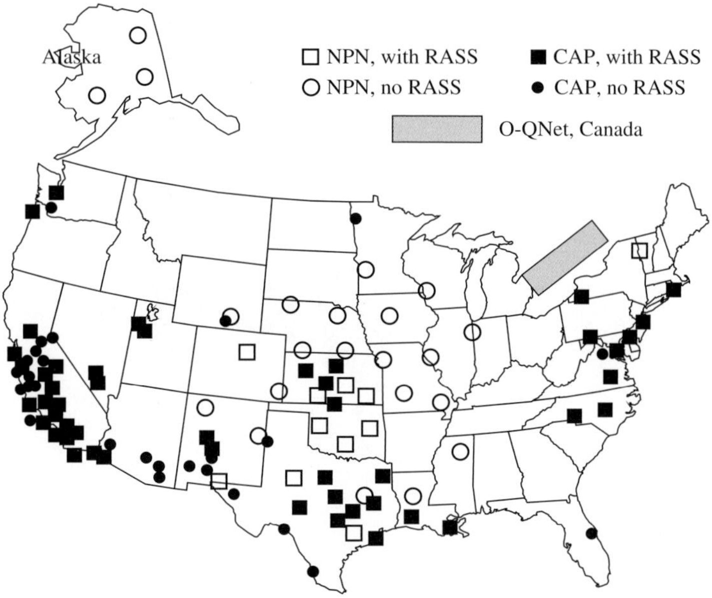

Figure 12.11 Distribution of the windprofilers in the USA in the year 2014. The map is a mixture of CAP profilers (CAP = Cooperative Agency Profilers (mainly 915 MHz, some VHF)) and NPN (NOAA Profiler Network) radars. One additional boundary-layer radar in Novia Scotia, Canada, is not shown. Squares indicate that the radars also have RASS (Radio Acoustic Sounding System) capability. The approximate location of a relatively dense profiler network in Canada – the O-QNet – is also shown along the northern edge of Lakes Ontario and Erie. The NPN profilers originally operated at 404 MHz, but proposals were made to change to 449 MHz. However, before that could eventuate, the USA portion of the network was largely closed down in late 2014, with only a small number of more remote stations kept active. The O-QNet was kept active (from *Hocking*, 2011). (Reprinted with permission from Elsevier.)

heavy upper headwinds and could plan flights that even took advantage of strong tail-winds. It was estimated that the savings in fuel costs could quickly pay for the entire network.

Benjamin et al. (2004) have extensively studied the impact of windprofiler data on weather forecasting in the USA. They compared results of models and analyses in which the windprofiler data were incorporated, with those in which no wind profiler data were used. The latter category was dubbed "profiler data-denial results." Invariably, the windprofiler data helped with predictions, especially in the area of 3 to 6 hour forecasts. In addition, these authors have logged many anecdotal case studies where windprofilers have made significant contributions in areas of significant social and safety impact.

A large list of useful applications of windprofilers to real-life situations was provided by *Benjamin et al.* (2004) in the sidebar on page 1884. Profilers were useful for synoptic analysis, evaluations of forecast models, pre-storm analysis, short-term changes, studies of low level jets (LLJ), supercell prediction, high-wind warnings, cold air surges, and support of fire-fighting and flood warnings, among others.

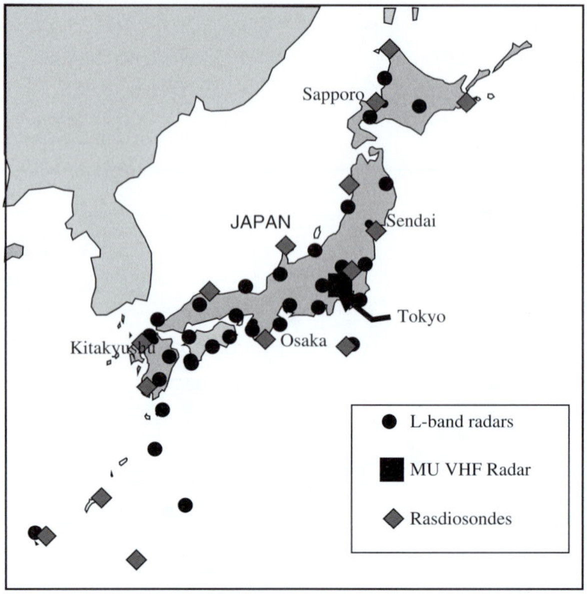

Figure 12.12 Japanese network of windprofiler radars. The radars primarily work at 1.3 GHz (from *Hocking*, 2011). (Reprinted with permission from Elsevier.)

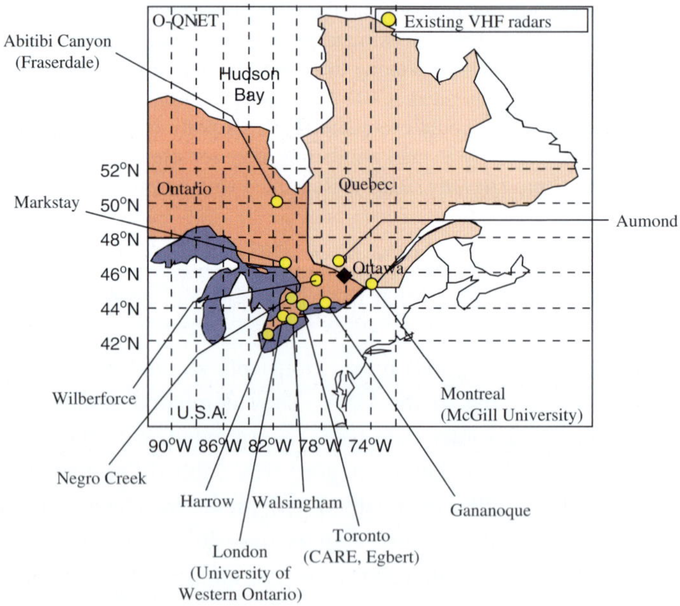

Figure 12.13 A tightly focused radar clustering in Canada, called the O-QNet. Most of these radars are designed to work at frequencies in the 40 to 55 MHz band.

The impact of windprofiler data on improved forecasting cannot be denied, and it is highly likely that many more networks of such radars will be developed in many countries of the world in the next 10 to 20 years.

12.5 Vertical winds

One of the greatest early apparent promises of windprofiler radars lay in the expectation that they would be able to measure vertical motions of the air with high precision, an ability that had escaped most other instruments. Many other radars were designed to look sideways, but because MST and windprofiler radars looked primarily in the vertical direction, it was felt that they should be able to become a principal instrument for studies of vertical motions.

While this has in part turned out to be true, there are limits to the capabilities. Certainly, they have turned out to be effective for studies on scales of a few minutes to a few hours, as demonstrated in Figure 12.14. The radars are very good at showing vertical velocities associated with gravity waves, for example.

However, as longer and longer-term averages are considered, complications arise, and the reliability of the vertical winds can become questionable. Even if one looks at shorter time scales, care is needed, and systematic biases can exist.

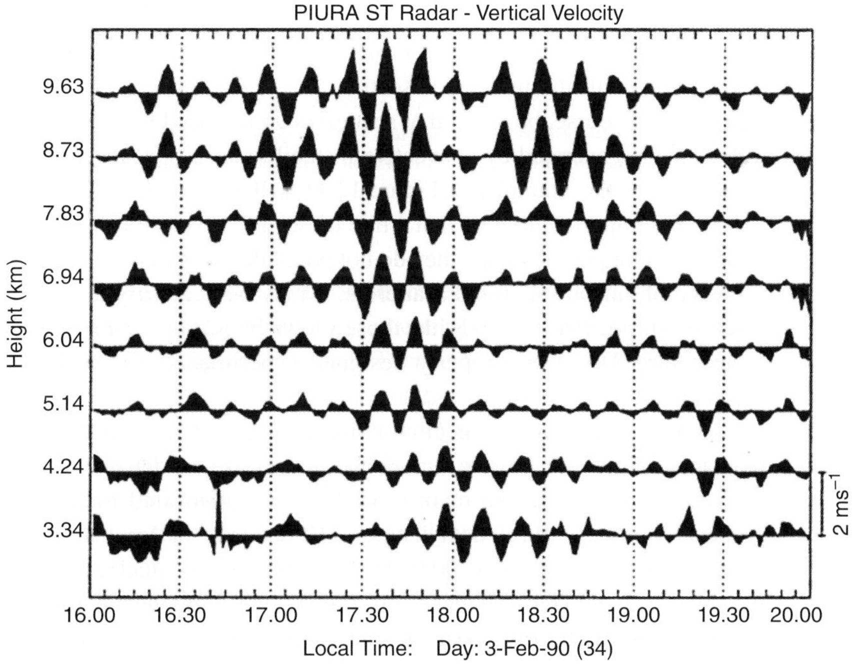

Figure 12.14 Wave-like vertical velocities measured with the Piura radar in Peru (from *Liziola and Balsley* (1997)). (Reprinted with permission from John Wiley and Sons.)

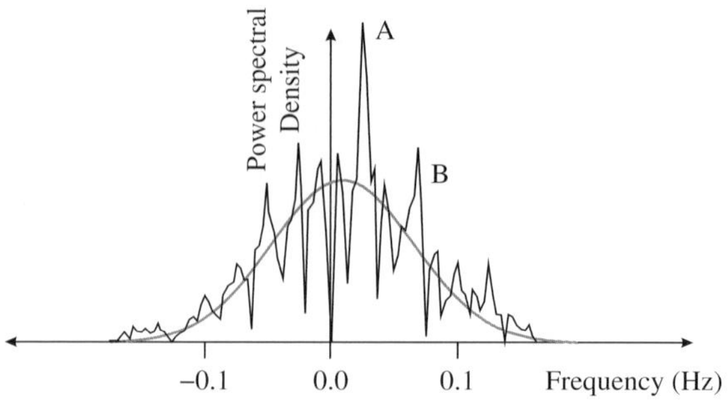

Figure 12.16 A spectrum of the raw in-phase and quadrature components showing the existence of two significant spectral spikes.

In order to obtain reliable vertical velocities, such large spikes need to be removed and then the vertical velocity should be found, either by using weighted moments or spectral fitting of Gaussians or other suitably assumed forms. If the spikes are not removed, erroneous spectral offsets result, but proper removal of them can lead to more reliable vertical velocities, since the impact of specular reflectors has been partially removed.

Even more sophisticated approaches can be applied, such as interferometric techniques, as proposed for example by *Röttger and Ierkic* (1985). However, some such methods come at the cost of longer computing time. Overall, though, the user must decide just what is required from the data, and choose an analysis technique that has a suitable combination of speed and complexity to match the goals of the experiment. With due care, at least some of the in-built biases can be removed.

However, even after all this, some biases may still remain. *Nastrom and Van Zandt* (1994) and *Nastrom and Van Zandt* (1996) have shown that there is a slight tendency for vertically directed radars to measure stronger and more frequent echoes from scatterers embedded in more stable regions of the atmosphere. Gravity waves tend to produce greater stability when their motions are downward, so when long-term averages are calculated, there is a slight tendency for the averages to be downward due to the intrinsic sampling bias associated with more stable regions. This result seems to have been confirmed experimentally (*Nastrom and Van Zandt* 1994; *Nastrom and Van Zandt* 1996; *Hoppe and Fritts* 1995a, b, *Hocking* 1997a). Even horizontal wind measurements can be weakly affected by such biases (*Nastrom and Van Zandt*, 1996). *Muschinski* (1996), *Muschinski et al.* (2001), and *Tatarskii and Muschinski* (2001) have also looked at similar bias effects, and especially the impact of correlations between vertical velocity and the strength of humidity gradients and associated radar scatter, along the same lines as those discussed above.

Nevertheless, provided that the user is aware of these effects and makes efforts to compensate for them that are compatible with the overall objectives of any set of measurements, VHF radars are still useful tools for measurements of vertical atmospheric motions. But caution in interpretation is critical, and fundamental limitations must be recognized.

12.6 Tropospheric temperature measurements

In the earlier section concerning tropopause detection, mention was made of the capability to determine approximate temperature profiles using VHF radar backscatter strengths, and a few methods exist for temperature-profile determination by radar. Some of these were discussed in Chapters 2, 7, and 10. Here, we will briefly recap these methods, but largely refer the reader back to those chapters for more detail. These methods have varying degrees of maturity, and some can be used only in special circumstances. The only method routinely used on an operational basis is RASS (radio acoustic sounding system), and even that is limited by its own noise pollution. It measures the virtual temperature.

One method discussed was determination of vertical velocity fluctuation spectra, and determination of the Brunt–Väisälä cut-off frequency. The method requires that mean winds be very light, which is quite a restrictive assumption. The method has been used in campaigns, but is not used operationally. An example of such data will be shown later in Figure 12.30.

The other method discussed was integration of the potential refractive index gradient, as proposed by *Gage et al.* (1985), but this never reached true operational status. The principle is that Equation (12.4) can be integrated downwards to produce a height profile of temperature in the troposphere, provided that the correct pulse-length dependence of the scattered strengths can be incorporated.

As noted, RASS is used operationally. In this chapter, we may employ results determined by these methods, but will not re-discuss the principles.

12.7 Tropopause determinations

Techniques for determining the height of the tropopause have already been discussed throughout this chapter and elsewhere in the text, but here we will briefly raise some extra points. Apart from the ability to track upper level frontal systems and study tropopause folds and other motions, knowledge of the tropopause height has other applications. One important area of relevance is in regard to satellite retrievals. Often inversion of the data to produce stratospheric temperatures and other measurements requires knowledge of the height of the tropopause, so enhanced scatter from cloud tops does not corrupt the data; windprofilers can provide this information. A second very important application is in regard to studies of ozone transport. *Hocking et al.* (2007a) have shown that ozone intrusions from the stratosphere can be an important contributor to tropospheric pollution (more so than previously realized), and occasions when intrusions are highly likely can be identified by sudden jumps in tropopause height. This gives an important new tool for meteorological studies of pollution in the lower atmosphere. Examples can also be seen in Figure 12.3.

12.8 Mountain waves

Flow of the mean wind over mountains can produce localized effects that can have important effects on local meteorology. Such effects include deviation of the mean flow,

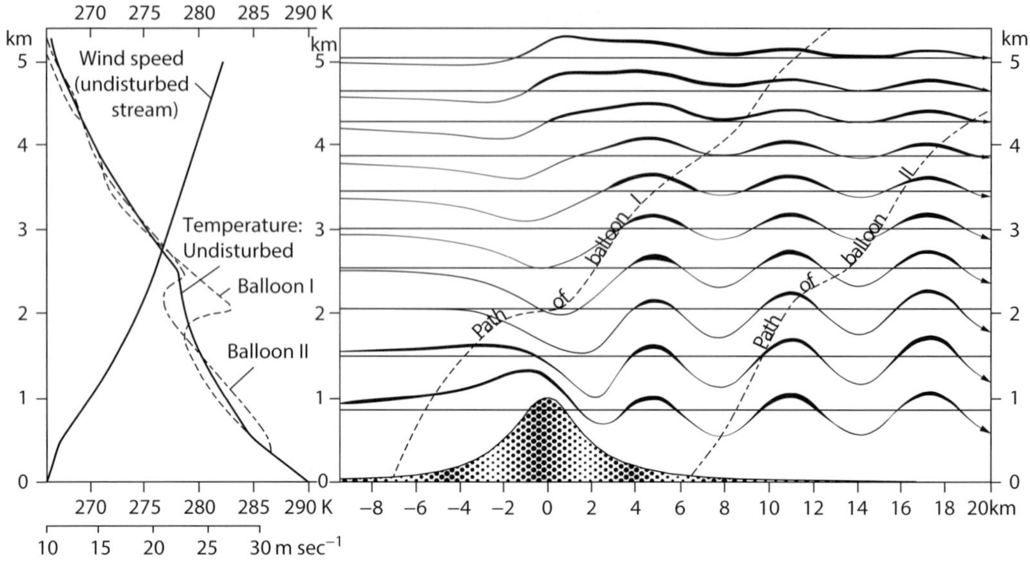

Figure 12.17 Generation of lee-waves by mean wind flow over a mountain, and the accompanying mean wind and temperature structure. Adapted from *Röttger* (2000), who adapted it from *Scorer* (1997).

formation of localized wind gusts, development of localized turbulence, generation of lee-waves and gravity waves, and production of wind pulsations like the Chinook winds. Radars can be well suited to studies of some of these events. However, they can have limitations as well.

Figure 12.17, adapted from *Röttger* (2000) (who adapted it from *Scorer*, 1997), shows generation of a train of lee-waves on the lee side of a mountain.

One point that is immediately obvious from this figure is that the wavefronts in this case show little or no phase change with height. In other words, the ground-based phase speed of the wave is close to zero. (Of course, the intrinsic wave-speed is non-zero, since the wave is propagating against the mean wind.) A radar positioned immediately under the lee-waves, offset from the mountain by say 10 km, looking only vertically, would not see evidence of the wave, but would measure a relatively smooth wind profile as a function of height. However, two or more radars, positioned a few kilometers apart (e.g., *Ecklund et al.*, 1985), would see different mean profiles, and then the wave-like nature of the oscillation might become more apparent. However, it is not necessary (nor even sensible) to use radars exclusively in such studies. Visual observations would clearly show lee-waves in the cloud structure, since clouds occur in the coldest part of the waves (highlighted as dark colors in the figure). Aircraft could also be used, which could measure spatial variability, and even balloons can be useful, as shown in the figure (also see *Caccia et al.*, 1997a). However, because balloons drift with the mean wind, it is harder to untangle the information that they record. Beam-swinging techniques with the radar can also reveal information about the horizontal structure of the lee-waves (e.g., *Worthington et al.*, 1999a, b), as can interferometry (e.g., *Röttger et al.*, 1990a; *Van Baelen and Richmond*, 1991; *Van Baelen et al.*, 1991). In all these studies, the comments made earlier regarding tilting of anisotropic scatterers and specular reflectors must be borne in mind, but, with sufficient care, useful information can be obtained.

An important point about studies of mountain waves is that the background atmospheric environment can be quite complex in structure, and it is no longer possible to assume that the waves are sinusoidal as a function of height. The full solution of the characteristics must be determined through the Taylor–Goldstein equations, as discussed in the chapter on gravity waves, Equation (11.35), viz.

$$\frac{\mathrm{d}^2 W}{\mathrm{d}z^2} + \left[\frac{\omega_B^2}{(\overline{u} - c_\phi)^2} - \frac{1}{(\overline{u} - c_\phi)} \frac{\mathrm{d}^2 \overline{u}}{\mathrm{d}z^2} - k_h^2 \right] W = 0. \tag{12.8}$$

Here, $W(z)$ is the vertical velocity variation as a function of height z, $\overline{u}$ is the mean wind (also a function of height), c_ϕ is the ground speed of the wave (which can be taken to be zero in the case of lee-waves), and ω_B is the Brunt–Väisälä frequency. More complicated versions of this equation can be found in, for example, *Merrill and Grant* (1979).

This equation is clearly of the type $\frac{\mathrm{d}^2 W}{\mathrm{d}z^2} + \kappa_c W = 0$, which, for constant κ_c is just the equation for a mass on a spring (where z becomes time), and has a solution of the type $\exp\{\mathrm{i}\sqrt{\kappa_c} z\}$, i.e. a sinusoidal oscillation in real space with wavelength $k_z = \sqrt{\kappa_c}$. However, the thing that complicates the equation is that κ_c is *not* a constant, and in the general case may be a function of z.

Scorer has developed a somewhat similar set of equations to produce the "Scorer parameter type 1," defined as

$$\ell^2 = \frac{\omega_B^2}{(c_\phi - \overline{u})^2} - \frac{1}{(\overline{u} - c_\phi)} \cdot \frac{\mathrm{d}^2 \overline{u}}{\mathrm{d}z^2}. \tag{12.9}$$

Remember c_ϕ will be zero for stationary lee-waves. Then the instantaneous vertical wavenumber at any height is given by

$$k_z^2 = \ell^2 - |\vec{k}_h|^2, \tag{12.10}$$

where $\vec{k}_h = [k_x, k_y]$ is the horizontal wave vector. However, the term vertical wavenumber is not normally employed, since k_z is height-dependent, so that the vertical variation is not sinusoidal and the effective wavelength changes within its own length.

However, if we persist with the idea of a vertical wavelength for just a little longer, we see that if ℓ^2 is small, then (12.10) is even smaller, or could in fact be negative. Hence the vertical wavelength is either very long, or the vertical profile is evanescent. Evanescence indicates a non-propagating wave, i.e. a trapped wave.

In Figure 12.17, it was noted that the phase showed little or no variation with height. This is common for lee-waves, but is not true for all waves generated by flow over mountains. *Röttger* (2000) and *Röttger et al.* (1981) have analyzed an event that showed clear wave-like structure as a function of height and have used radiosonde measurements to determine the variation of ω_B with height in order to obtain the Scorer parameter of their data. They have concluded that the wave they saw could well have been a quasi-stationary trapped mountain lee-wave, but they could not rule out the possibility that it was an inertial wave. *Hines* (1995a) has shown that wind flow over a mountain range produces a spectrum of waves, with various phase speeds and directions, although there

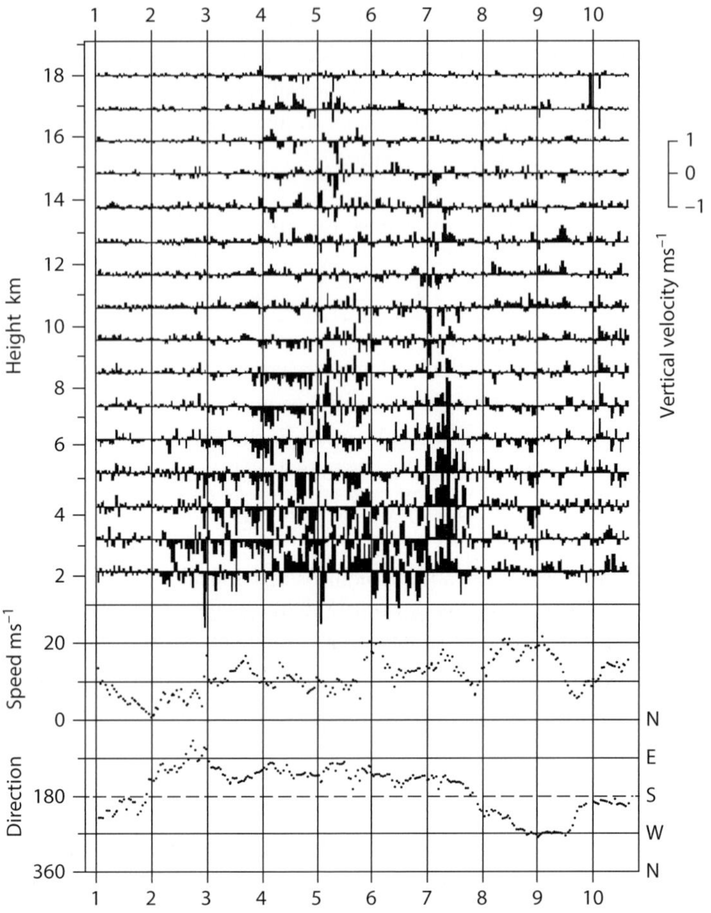

Figure 12.18 Enhanced gravity waves are apparent when the wind is from the east/south-east. Data were recorded with the Aberystwyth radar in Wales (*Prichard et al.*, 1995). (Reprinted with permission from Springer.)

are certain directions in which generation is strongest. Experimental studies suggest that wave generation is strongest when the mean flow is perpendicular to major ridges of a mountain range, and of course is generally stronger when mountains are higher and when mountain ranges have greatest variation in height between mountain tops and valley floors (e.g., *Röttger*, 2000; *Hoffmann et al.*, 1999; *Rechou et al.*, 1999; *Caccia et al.*, 1997b; *Worthington*, 1999). Fig 12.18 shows an example from Wales, in the United Kingdom, where gravity-wave activity is clearly enhanced for certain preferential wind directions. *Nastrom and Fritts* (1992) have also extensively studied gravity waves generated by orography.

Sometimes waves exist which look like lee-waves, but are not. Debate has existed in the literature on occasions about whether some observed lee-waves really are lee-waves at all, or whether a few are in fact inertial gravity waves (e.g., *Cornish and Larsen*, 1989; *Cho*, 1995; *Hines*, 1995b, c). Such arguments have arisen on several occasions,

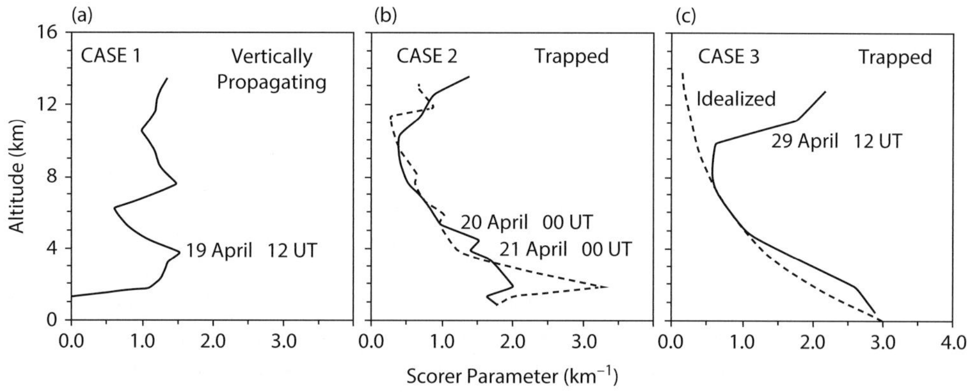

Figure 12.19 Scorer parameter as a function of height for various cases of (a) free, and (b, c) trapped waves. Waves are generally trapped if the parameter becomes small (long effective vertical wavelength). From *Ralph et al.* (1992). (Reprinted with permission from John Wiley and Sons.)

and some cases are still undecided. As a rule, inertia gravity waves have velocity vectors that rotate with increasing height, whereas lee-waves should not, unless they are affected by pre-existing rotational shears in the mean wind. Inertial waves are commonly excited by the jet stream, from where they may propagate both up and down in the atmosphere. We will not elaborate on these discussions here, but refer the reader to *Röttger* (2000), and references therein.

Several well-documented cases of lee-waves do exist, including presentations by *Ralph et al.* (1992), *Worthington and Thomas* (1997), *Prichard et al.* (1995), and *Worthington* (1999). *Worthington and Thomas* (1997) presented four mountain-wave events which maintained approximately constant phase throughout the troposphere, but reflected off the tropopause. Figure 12.19 shows an example from *Ralph et al.* (1992) where the Scorer parameter has been calculated as a function of height. Waves that are generated by mountains may propagate to higher levels, or become ducted (trapped) by suitable mean wind and temperature structures in the upper tropopause. Figure 12.19 shows examples of the structure for trapped and freely propagating waves, with waves being trapped if the Scorer parameter becomes small. As discussed following Equation (12.10), sufficiently small values of ℓ^2 may lead to evanescent waves, which is an indication of wave-trapping.

Mountain waves may also be reflected at the tropopause, returning to the Earth's surface and thereby generating standing waves in the troposphere as the upward and downward wave components interfere. The reflection process can also lead to production of Kelvin–Helmholtz secondary instabilities and turbulence (e.g., *Worthington and Thomas*, 1996; *Yamamoto et al.*, 2003), generated for example at the shear interface between low-level wind jets and upper-level decelerated winds. This is especially so if the wave has large amplitude. Generation of standing waves, whether producing turbulence or not, can lead to the production of downslope wind pulsations, such as the Chinook winds in Western Canada. This process has been discussed in more detail by *Peltier and Scinocca* (1990). Other similar wave-forced flows occur in mountainous or hilly regions elsewhere in the world, though usually the winds are given different, locally dependent, names. Wave breaking may also lead to secondary generation of additional waves (*Satomura and Sato*, 1999), potentially with different properties to the parent waves, and these may in turn propagate to other levels of the atmosphere, depositing momentum and energy when they themselves later break or dissipate.

Mountains do not only generate gravity waves, of course. Large-scale mountain ranges like the Andes and the Rockies may generate planetary waves, which can then propagate into the stratosphere. While these effects are somewhat outside of the realm of traditional meteorology, they still deserve some mention here. Planetary wave propagation is responsible for altering the circulation of the stratospheric polar vortex in mid-winter, especially in the northern hemisphere. The larger percentage of land-mass and mountains in the northern hemisphere relative to the southern hemisphere leads to generation of more planetary waves, which means that the northern polar vortex is more perturbed by planetary waves than the southern vortex. The relative absence of planetary waves in the southern hemisphere allows the stratospheric southern polar vortex to become very strong, with a strong zonal flow and little meridional flow. The limited meridional flow means that mixing between the polar regions and the mid-latitudes is very weak, so that the southern polar stratosphere in mid-winter becomes very cold, allowing development of polar stratospheric clouds and ice clusters upon which heterogeneous ozone reactions can take place as the sun rises in early spring. This results in rapid ozone loss, leading to the infamous southern stratospheric ozone hole.

Both gravity wave and planetary wave production by flow over mountains affect the upper-level circulations in the stratosphere and the mesosphere. Both types carry momentum flux ($\overline{u'w'}$, $\overline{v'w'}$, $\overline{u'v'}$, etc. – in the case of planetary waves $\overline{u'v'}$ relates to the so-called Eliassen–Palm flux – see Equation (1.6)) into the upper atmosphere, where deposition leads to forcing of the mean flow and alteration of the flow from the radiative equilibrium situation (e.g., *Vincent and Reid*, 1983; *Palmer et al.*, 1986; *Fritts et al.*, 1990; *Fritts and Alexander*, 2003; *Thorsen et al.*, 1997).

Further discussion about mountain waves is beyond the scope of this chapter, and the interested reader is referred to *Röttger* (2000) for more extensive discussion and additional references.

12.9 Gravity wave genesis in relation to meteorology

Windprofiler radars also have an important role to play in studies of other types of tropospheric gravity waves. Orographic generation was discussed in the last section. Wave generation by convection and frontal systems and squall lines has been studied via modeling (e.g., *Alexander and Holton*, 2004; *Alexander*, 1996; *Alexander and Rosenlof*, 1996). Often wave ducts appear that are favorable to the resonant growth of gravity waves out of noise, and these can also be a source of wave activity, as discussed for example by *Merrill* (1977) and *Merrill and Grant* (1979). Windprofilers are often the instrument of choice for experimental studies. Radiosondes are also usefully employed for this work.

The ability of windprofilers to detect deep frontal systems, and even occluded fronts, was highlighted in the earlier section on stability. The ability of windprofilers to detect and describe gravity-wave structures was also discussed in considerable detail in that section, as well as in preceding chapters, and these aspects will not be repeated here. However, one area that will be further discussed here is the issue of gravity-wave generation by meteorological systems.

It seems that frontal systems (cold, warm, and occluded) are important sources of gravity waves. Jet streams are another potential source. Both of these types of systems are well suited to windprofiler studies, since windprofilers are not only easily able to identify the events, but can then study the waves generated by them. The capabilities become even stronger when the windprofilers are used in concert with other techniques like radiosondes and microbarographs. Figure 12.20 shows examples of spectra of pressure fluctuations measured by microbarographs for cases where there were no frontal systems near the radar site, for cases when fronts existed within 1000 km of the radar site, and cases when fronts were very close to or over the radar site, for the period of 1996–7. Enhanced variability is seen, and larger maximum values occur, when the frontal systems are nearest the site, especially at the highest frequencies. We assume that these fluctuations are primarily due to gravity waves, as proposed in Chapter 11. Simultaneous radar observations were also obtained in these years. Figure 12.21 shows measurements of the magnitudes of the vertical flux of horizontal momentum measured by windprofiler radar at the same time, throughout 1996–7. The momentum fluxes show similar trends as a function of season to the occurrence of fronts, and although this evidence is not proof alone of the importance of frontal generation (many meteorological features show seasonal variability, of course), this graph coupled with Figure 12.20 certainly is suggestive that frontal generation of gravity waves may be important. The area surrounding the CLOVAR radar is quite flat, so orographic wave generation is less

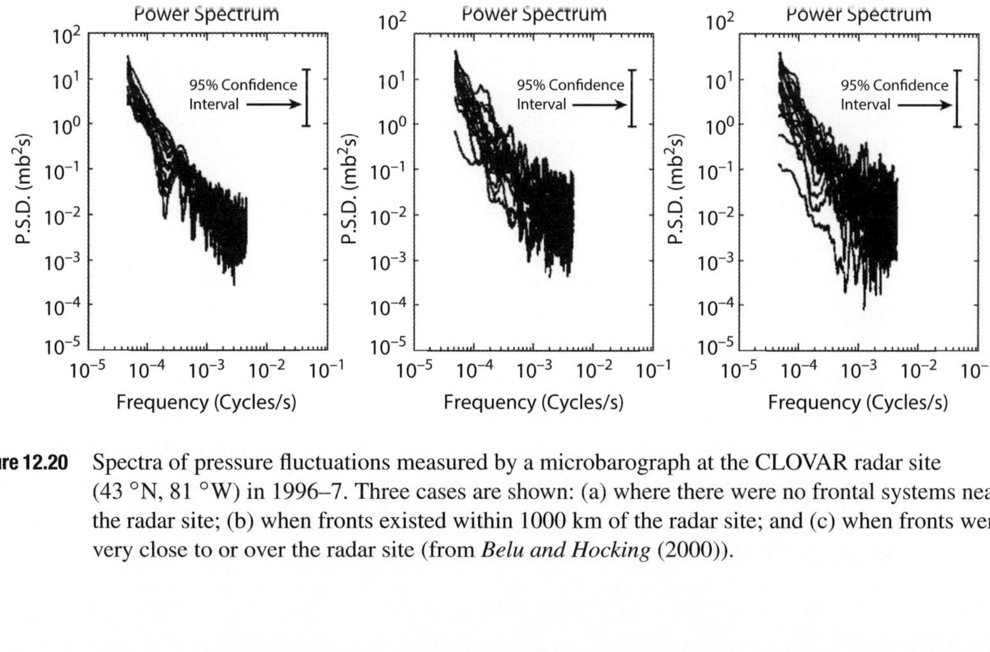

Figure 12.20 Spectra of pressure fluctuations measured by a microbarograph at the CLOVAR radar site (43 °N, 81 °W) in 1996–7. Three cases are shown: (a) where there were no frontal systems near the radar site; (b) when fronts existed within 1000 km of the radar site; and (c) when fronts were very close to or over the radar site (from *Belu and Hocking* (2000)).

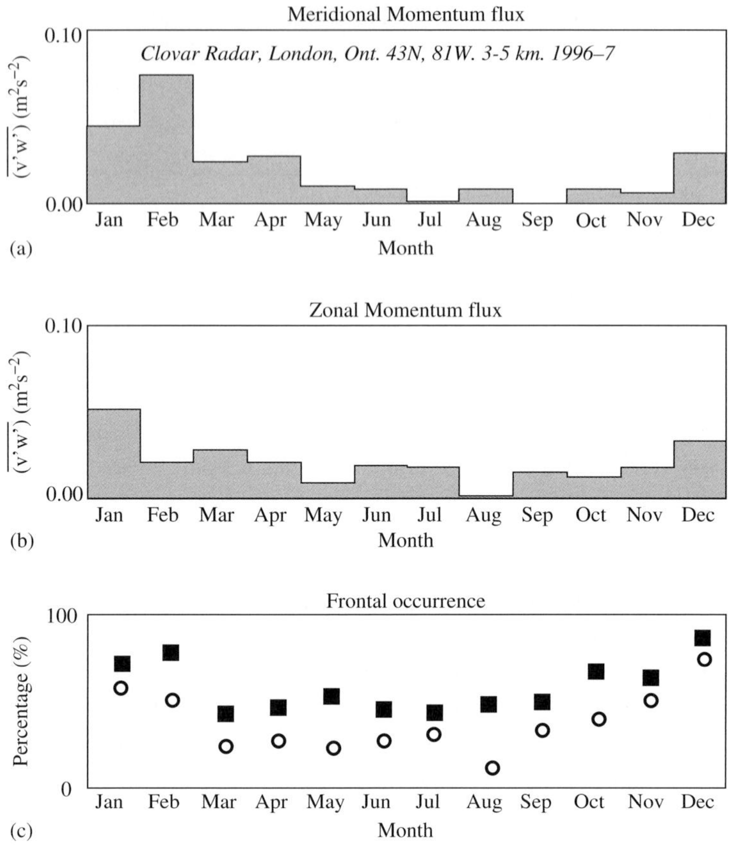

Figure 12.21 Plots of absolute values of the vertical flux of (a) meridional, and (b) zonal momentum flux, divided by atmospheric density, measured as a function of season with the CLOVAR radar in Canada for the period 1996–7. The signs of the momentum fluxes have been removed, since our interest here is only in magnitudes. Graph (c) shows the annual variability of the occurrence of fronts near the radar, with the percentages referring to numbers of days during which fronts were observed near the radar relative to the number of days during which the radar was operating, as deduced from local weather maps. Open circles refer to cold fronts, and filled squares refer to all fronts. From *Belu and Hocking* (2000).

likely, and frontal generation seems to be a strong candidate for wave production in this region.

Ray tracing procedures also often result in upper-level gravity waves being traced back to frontal systems (e.g., *Smith et al.*, 2003) or tornadoes (e.g., *Hung et al.*, 1978) or even solar eclipses (e.g., *Ball*, 1979). More extensive discussions about ray tracing were presented in Chapter 11. Further examples in the literature include *Eckermann* (1992); *Marks and Eckermann* (1995); *Zhong et al.* (1995); *Belu* (1998); *Belu* (1999) and *Belu* (2000). Numerical studies of gravity-wave generation by frontal systems and the relation to frontal collapse, as well as wave generation by squall lines, have also been presented in the literature by, for example, *Ley and Peltier* (1978); *Snyder et al.* (1993); *Alexander et al.* (2004, 1995) and *Alexander and Rosenlof* (1996).

The jet stream is another important meteorological source of gravity waves, not only in the stratosphere, but also in the troposphere. The possibility that some of the gravity waves discussed in the section on mountain waves could in fact have been inertio-gravity waves generated by the jet stream has been discussed in that section (e.g., *Cornish and Larsen*, 1989; *Cho*, 1995; *Hines*, 1995b, c; *Röttger*, 2000). Inertial gravity waves can, for example, be generated by geostrophic adjustment. Important theoretical and experimental study of jet-stream generation include works by *Fritts and Luo* (1992), *Fritts and Nastrom* (1992), *Guest et al.* (2000), *Hertzog et al.* (2001), *Plougonven et al.* (2003); *Plougonven and Snyder* (2005), *Zhang* (2004), *Pavelin and Whiteway* (2002), and *Lane et al.* (2004). *Pavelin et al.* (2001) observed an inertial gravity wave that lasted for 5 days in the lower stratosphere, driven by geostrophic adjustment. Gravity waves may also be generated at the edges of the polar vortices (e.g., *Whiteway et al.*, 1997)). However, it is true to say that a lot more experimental study of jet stream and vortex generation mechanisms remains to be done, and these are excellent future topics for windprofiler research.

12.10 Convection, water, lapse rates, and stability/instability

This next section is a chance to catch up on a variety of issues that relate to basic meteorology in the troposphere, pertaining primarily to dynamical effects associated with heat transport. In this context, water is especially important, as its large latent heat coefficient means that it is a major contributor to the heat budget.

12.10.1 Convection

Another key meteorological area is studies of convection. We have already noted that the degree of anisotropy of turbulence scatterers tells the observer something about the nature of the atmosphere, with quasi-isotropic scatter being especially associated with convection, and anisotropic scatter being more indicative of strong wind-shear. This was already noted earlier (e.g., *Hocking and Hamza*, 1997). Convection was also noted as a key aspect in the report by *Benjamin et al.* (2004), and as a diagnostic for the generation of precipitation in some cases (Figure 12.1).

Profilers are also well suited to the study of convection because of their ability to measure vertical motions. This is especially useful in the case of large convective storms, where vertical motions can reach several meters per second. For example, *Hooper et al.* (2005) have studied such motions and tried to resolve differences between turbulent vertical motions and more organized advective motions.

Convection is also a major source of gravity waves, and many studies have been performed to look for correlations between gravity-wave activity and convective sources. In particular, there have been a large number of studies of convection and gravity-wave generation in the tropics, in which windprofiler studies have played a large role. There have also been a large number of modeling studies of the generation of gravity waves by convection (e.g., *Holton*, 1972; *Alexander et al.*, 2004, 1995, 2000; *Vincent*

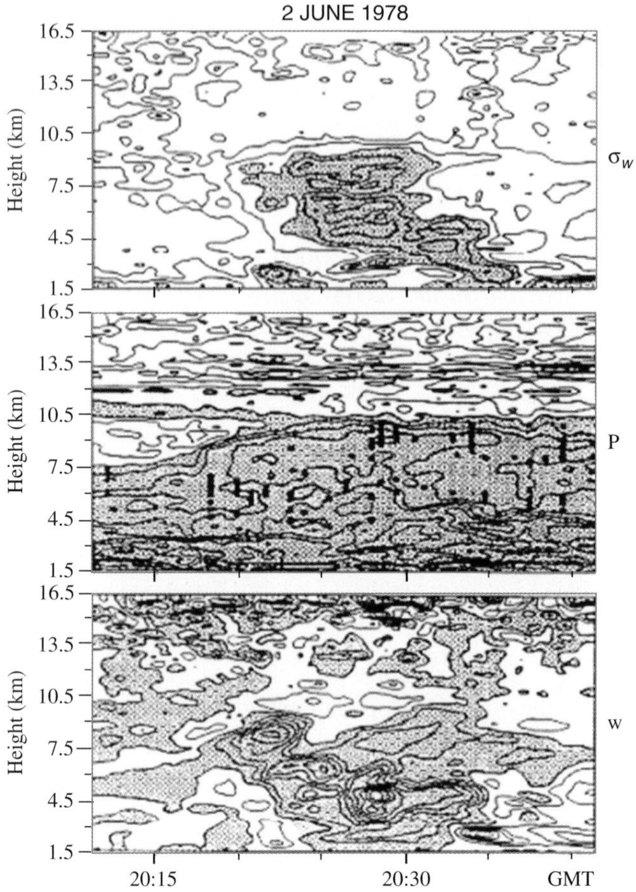

Figure 12.22 Contour plots of spectral width σ_W (upper panel), echo power P (center panel) and radial velocity w in the vertical direction, observed during a thunderstorm passage with the SOUSY radar. The black dots in the center panel indicate occurrence of lightning echoes; lightning detection was discussed in Chapter 10, Section 10.7. The shaded parts identify wide spectra (i.e. strong turbulence), high echo power, and large upward directed velocity respectively. (from *Larsen and Röttger* (1987), adapted from *Röttger* (1980c)). (Reprinted with permission from the American Meteorological Society.)

and Alexander, 2000; *Beres et al.*, 2002; *Holton et al.*, 2002; *Alexander et al.*, 2004; *Alexander and Holton*, 2004; *Beres et al.*, 2004; *Piani et al.*, 2000).

Convective processes can vary from extremely strong, such as in thunderstorms, to gentler convective motions as discussed in relation to Figure 12.1. Refractive index variations due to turbulent humidity and temperature changes in thunderstorms can be fairly strong and enhance the radar echo strength substantially. Figure 12.22 shows an early example of radar observations of a thunderstorm, which passed over the SOUSY VHF radar located in the Harz mountains in northern Germany. Reflectivity layers observed with VHF radars are often stratified, but in this case, they are penetrated by cloud-like expansion in the vertical direction, as seen in the center panel of Figure 12.22. In addition, a strong increase in turbulent and upward directed velocity is apparent.

In the following sections, we will explain how such upwelling and the creation of turbulence develops as seen by ST radar. In meteorological terminology, this is called convection, and we will give a brief background about the fundamentals of atmospheric stability, instability, and convection.

However, before doing that, we need to expand on the concept of the adiabatic lapse rate, which was discussed in Chapter 1. But even before that, we need to more properly define quantities like mixing-ratio, virtual-temperature and moist-adiabatic lapse-rates. This is now done.

12.10.2 Scale height for a multi-species gas

In Chapter 1, we presented a series of Equations (1.23), (1.24), and (1.25), which related to pressure changes as a function of height. These are repeated below:

$$p = p_0 e^{-\int_{z'=0}^{z'=z} \frac{dz'}{H}}, \tag{12.11}$$

where $H = \frac{RT}{Mg}$ is called the scale height. This equation applies for an atmosphere composed of a single species of gas with molar mass M. The constant R is the gas constant in the equation $pV = \eta_m RT$, as taught in beginner courses in physics and chemistry, where V is the volume of gas and η_m is the number of moles in that volume. Of course the atmosphere contains many species of molecules and atoms, so we also define a generalized gas constant, R_*, specified in the following way:

$$R_* = \sum_{i=1}^{i=\nu} \frac{\rho_i}{\rho} \frac{R}{M_i}, \tag{12.12}$$

where the subscript "i" refers to the different species, ρ refers to density and M_i is the molar weight of the species i. At this time (12.12) is just a definition: its relevance will be shown shortly.

Equation (12.11) is a quite general relation, and can be used if the temperature varies with height. For the special case of an isothermal atmosphere, H is constant and so

$$p = p_0 e^{-\frac{z}{H}}. \tag{12.13}$$

Equation (12.12) was presented without proof. In the following pages, we will derive in detail these various equations that deal with multi-constituent atmospheres.

In order to do that, recall that for a single species s of gas in a mixture of gases, the ideal gas equation gives the pressure due to the species s as

$$p_s = \rho_s R_s T, \tag{12.14}$$

where

$$R_s = \frac{K_B}{m_s} = \frac{N_0 K_B}{N_0 m_s} = \frac{R}{M_s}, \tag{12.15}$$

and where M_s is the mass of one mole of species s (= molecular weight). The quantity m_s is the mass of one molecule of species s, and K_B is Boltzmann's constant. R_s is called the *specific gas constant* since it is specific to this particular species.

For a mixture of species, we write

$$p = \rho R_* T, \tag{12.16}$$

where R_* is somewhat similar to R_s. One of our forthcoming tasks will be to show that R_* defined according to (12.16) is conceptually consistent with (12.12). This will be done shortly.

We may now place ρ from (12.16) back into the equation for hydrostatic balance, viz.

$$\frac{\mathrm{d}p}{\mathrm{d}z} = -\rho g, \tag{12.17}$$

to give

$$\frac{\mathrm{d}p}{\mathrm{d}z} = -\frac{p}{R_* T} g, \tag{12.18}$$

or

$$\frac{1}{p}\frac{\mathrm{d}p}{\mathrm{d}z} = -\frac{g}{R_* T}, \tag{12.19}$$

or

$$d(\ln p) = -\frac{g}{R_* T}. \tag{12.20}$$

Taking T and R_* as constants and integrating gives

$$p = p_0 \, \mathrm{e}^{-\frac{gz}{R_* T}} = p_0 \, \mathrm{e}^{-\frac{z}{H}}, \tag{12.21}$$

where $H = \frac{R_* T}{g}$ is called the scale height, as discussed in Chapter 1. Thus the atmospheric pressure diminishes with height according to the above equation.

But we have still not yet properly proven that R_* in (12.16) and in (12.12) are equivalent. We now proceed to do this. We will do this with particular emphasis on water, but the formula we will derive is quite general.

12.10.3 The mixing ratio for water

The water vapor mixing ratio is defined as a measure of water vapor content, through the relation

$$q_{mr} = \frac{\rho_v}{\rho_d}, \tag{12.22}$$

where ρ_v is the density of water vapor and ρ_d is the density of *dry* air (i.e. with water vapor excluded). A very similar quantity is the specific humidity, denoted as q_{wp} and used in the definition of the potential refractive index gradient, M_n. This was defined as

$$q_{wp} = \frac{\rho_v}{\rho_{\text{tot}}}, \tag{12.23}$$

where ρ_{tot} is the total density of air including water vapor. Since water vapor is rarely more than 6% of air by weight, (and very often much less),

$$q_{wp} \approx q_{mr} \tag{12.24}$$

to good accuracy. Of course there are situations where the difference is important, but for many cases we can take them as equal, unless stated otherwise.

We often consider moist air as a mixture of two ideal gases, namely water vapor and the rest. We want to first examine the partial pressure due to the water vapor alone, which we will denote as e_{wp}.

Then for the vapor stage, the water vapor pressure is given by

$$e_{wp} = \rho_v R_v T, \tag{12.25}$$

where R_v is the *specific gas constant* and is the same constant as appeared in (12.15), but in this case M is the mass of one mole of water. i.e.

$$R_v = \frac{R}{M_v}, \tag{12.26}$$

where M_v is the molar weight of pure water vapor. Then $R_v = R/(0.018) = 462$. For the dry portion of the air, we have a similar equation, viz.

$$(p - e_{wp}) = \rho_d R_d T, \tag{12.27}$$

where p is the total pressure and $(p - e_{wp})$ is therefore the partial pressure due only to the dry air.

However, we must now look in more detail at the constant R_d. The same arguments discussed here actually also apply to the constant R_* discussed in Equation (12.16).

To obtain a clearer picture of the value of R_d, we need to look more carefully at the law of partial pressures. This says that if we have a gas comprising ν different *species* of molecules, then the total pressure is given by the sum of the pressures of each species, i.e.

$$p = \left[\sum_{i=1}^{i=\nu}(N_i K_B T)\right]/V. \tag{12.28}$$

Now the mass density of species i is given by the number density, $\frac{N_i}{V}$, multiplied by the mass of a molecule of species i, m_i. Thus $\frac{N_i}{V} = \frac{\rho_i}{m_i}$, and we can substitute this into the previous equation to give

$$p = \sum_{i=1}^{i=\nu}(\frac{\rho_i}{m_i} K_B T). \tag{12.29}$$

Multiplying through top and bottom by Avagadro's number, N_0, we produce

$$p = \sum_{i=1}^{i=\nu}(\frac{\rho_i}{N_0 m_i}(N_0 K_B)T), \tag{12.30}$$

which gives

$$p = \sum_{i=1}^{i=\nu}(\frac{\rho_i}{M_i} RT), \tag{12.31}$$

where M_i is the mass of one mole of species i. We may now multiply the top and bottom lines by the *mean density* of the gas mixture, $\overline{\rho}$, which equals the total mass of the gas mixture contained in a volume V divided by the volume itself. Then we obtain

$$p = \overline{\rho} R_* T, \tag{12.32}$$

where

$$R_* = \sum_{i=1}^{i=\nu} \frac{\rho_i}{\overline{\rho}} \frac{R}{M_i}. \tag{12.33}$$

This is the value of R_* as discussed in Equation (12.12), and so proves the equivalence of the values of R_* in (12.16) and (12.12), as desired.

For the case of dry air, the above sum is over *all* atmospheric constituents *except* water vapor. To a good approximation we can take air to be 80% nitrogen (1 mole weighs 28 grams) and 20% oxygen (one mole weighs 32 grams) by weight, so we can write

$$R_d = R \times \left(\frac{0.8}{0.028} + \frac{0.2}{0.032} \right) = 289. \tag{12.34}$$

Taking the ratio of (12.25) and (12.27), and applying the fact that $\frac{R_d}{R_v} = \frac{289}{462} \simeq 0.62$, we see that

$$q_{mr} = \frac{\rho_v}{\rho_d} = \frac{e_{wp}}{p - e_{wp}} \frac{R_d}{R_v} \simeq 0.62 \frac{e_{wp}}{p - e_{wp}}. \tag{12.35}$$

Since generally e_{wp} is much much less than p (e_{wp} is typically 20 hPa, whilst atmospheric pressure is more like 1000 hPa), we can take $p - e_{wp} \simeq p$ and write

$$q_{rm} \simeq 0.62 \frac{e_{wp}}{p} \simeq q_{wp}. \tag{12.36}$$

This therefore tells us the *mixing ratio* and the *specific humidity*, but meteorologists often like to work with another quantity, namely the so-called *relative humidity*. This is defined as the mixing ratio for the air under study divided by the mixing ratio which *would have* existed if the vapor pressure were *saturated*; i.e., if the vapor pressure reached its maximum possible value. This maximum possible value is a function of temperature; it is determined by finding the vapor pressure at which water starts to condense out of the air as water droplets. (Note we do not consider the possibility of supersaturation in this simple argument.)

Thus the *relative humidity* is defined by

$$r_{wp} = \frac{q_{rm}}{q_s} = \frac{e_{wp}}{e_s}, \tag{12.37}$$

where e_{wp} is the observed water vapor partial pressure and e_s is the water vapor pressure at condensation.

This ratio is often expressed as a *percent*.

How do we find the *saturated water vapor pressure*? We can see this if we consider the phase diagram for water, as shown in Figure 12.23.

Suppose that the partial pressure e_{wp} at the point A is 9 hPa. Then the *saturated partial pressure* can be found by projecting up from the point A along the line of constant

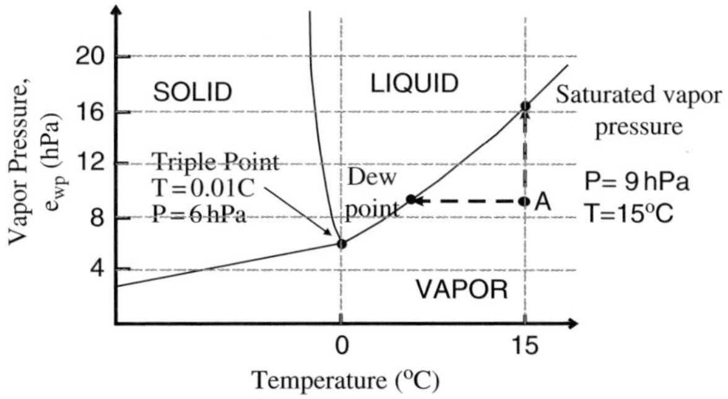

Figure 12.23 Phase diagram of water illustrating calculation of saturated vapor pressure. Note that if the partial pressure is less than 6 hPa, water can only exist as a solid or a vapor – liquid water cannot exist. The dewpoint for point A is also shown for interest.

temperature until the line meets the liquid–vapor interface (the vaporization curve). In our example, this occurs just above 16 hPa. For purposes of illustration we will treat it as exactly 16 hPa.

Thus for a gas at point A, the relativity humidity is $\frac{9}{16}$ multipled by 100, or about 56%.

In air which is at a pressure of 1000 hPa, the mixing ratio at A is

$$q_{mr} = 0.62\frac{9}{1000} \simeq 5.6 \times 10^{-3}, \tag{12.38}$$

i.e. 5.6 *grams* of water per *kilogram* of dry air.

These typical numbers show that water vapor is only a minor atmospheric constituent, but nevertheless a very important one.

12.10.4 Virtual temperature

We have already seen in Chapter 1 and elsewhere that meteorologists often like to deal with *potential temperature*. There is also another form of special temperature which they like to use, and this is called the *virtual temperature*. It is defined in the following way.

Assume that we have equilibrium between the liquid and vapor states, with no *net* condensation or evaporation.

Then we may write

$$p = \rho R_m T, \tag{12.39}$$

where R_m is the specific gas constant for *moist* air, i.e.

$$R_m = \sum_{\text{all species}} \frac{\rho_i}{\rho}\frac{R}{M_i}. \tag{12.40}$$

Note that in this case the sum is over all species in the air *including water vapor*.

We can rewrite (12.39) as

$$p = \rho R_d T^*, \tag{12.41}$$

where R_d is defined by (12.33), but excluding water vapor, i.e.

$$R_d = \sum_{\text{all species except water}} \frac{\rho_i}{\bar{\rho}} \frac{R}{M_i}, \tag{12.42}$$

T^* being *defined* through this relation as the *virtual temperature*. In other words, this is the temperature that dry air *would have* if its pressure and density were equal to the values for our particular sample of moist air.

We see therefore by (12.39) and (12.41) that

$$T^* = \frac{R_m}{R_d} T. \tag{12.43}$$

We may now recognize from (12.40) and (12.42) that

$$R_m = R_d + \frac{\rho_{\text{water}}}{\bar{\rho}} \frac{R}{M_{\text{water}}}. \tag{12.44}$$

But the ratio of the density of water vapor divided by the total density is very nearly q_{mr} (and also nearly q_{wp}), and $\frac{R}{M_{\text{water}}}$ is just R_v from (12.26). Thus we have

$$R_m = R_d \left[1.0 + q_{mr} \frac{R_v}{R_d} \right]. \tag{12.45}$$

Substituting our previous values for R_v and R_d, we have

$$R_m = R_d(1 + 1.61 q_{mr}). \tag{12.46}$$

Substitution in (12.43) then gives

$$T^* = (1 + 1.61 q_{mr})T. \tag{12.47}$$

Maximum possible values of e_s are of the order 60 hPa, so q_{mr} (and q_{wp}) is normally less than about 0.04. Thus generally the difference between the virtual and real temperatures is less than 7 °C. More typically, the value is less than 1 °C. Often T^* is denoted as T_v.

As discussed in Chapter 2 in the section on radio acoustic sounding systems, the speed of sound depends on T_v, and so T_v is the parameter most directly measured by RASS.

12.10.5 The dry and moist adiabatic lapse rates

Although water is a relatively minor species in the atmosphere, it has an impact which far outweighs its relative density. Of course it is important for life on earth, but from the perspective of atmospheric dynamics, it also has a very significant role. This is because of the very high latent heat coefficient of water, so condensation of even relatively small amounts of water can release huge amounts of heat into the atmosphere – and conversely, evaporation of water can absorb large amounts of heat. Hence water vapor carries with it large amounts of "hidden" heat energy in the form of latent heat, which moves almost invisibly but is then released when the vapor condenses or becomes ice. We have already seen that the presence of water vapor alters the speed of sound. It also has a major role to play in its impact on the vertical movement of air parcels, as we will now see. Heat moved in this way is, not surprisingly, called latent heat, or sometimes virtual heat,

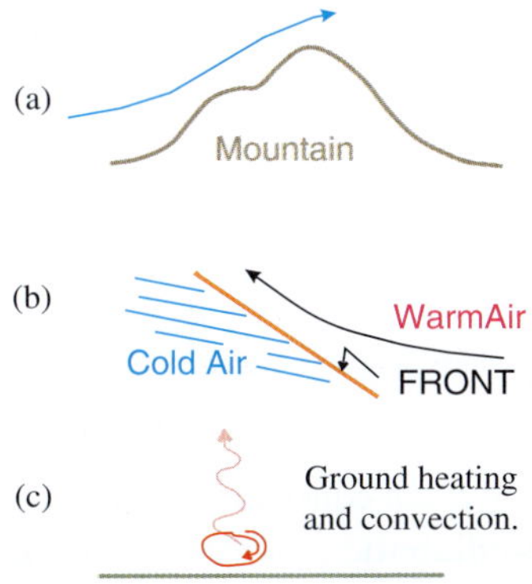

Figure 12.24 Atmospheric processes that produce rising air.

and heat carried by the simple heat capacity of the air is called sensible heat. Note that the virtual heat may be heat of sublimation or heat due to change of phase on freezing/melting, so the term latent heat may be too restrictive.

In Equation (1.42) in Chapter 1, an expression was developed that permitted us to determine the rate of change of temperature for a displaced parcel of air in the atmosphere (the so-called dry adiabatic lapse rate), provided the air was dry. This meant that we could assume that no heat was generated or lost within the parcel. But if a parcel of air contains water, either in vapor, liquid, or ice form, the possibility exists that this water may change phase (e.g. from vapor to liquid), and in the process transfer heat from/to the parcel, due to latent heat release or absorption. So water acts as a hidden source of heat.

For example, if for some reason the parcel of air is forced up, the water vapor may condense and form clouds. Such forcing can occur when air is forced to flow over mountains, when warm air encounters a cold front, or when there is substantial heating at the ground (convection included), amongst other effects (see Figure 12.24).

Unsaturated air will cool, or heat, (depending on whether the motion is upwards or downwards in height), in a very nearly dry adiabatic fashion. But for saturated air, we may have vapor changing to liquid (condensation) or liquid changing to ice, or even vapor changing to ice by sublimation. These processes release latent heat, and this offsets the adiabatic cooling. Let us see if we can determine an expression for the rates of cooling associated with adiabatic ascent of a parcel of air containing condensing water vapor.

[For the record, we will also list some of the latent heat values for water. The typical latent heat for condensation of water is $L_c = 2.5 \times 10^6$ J kg^{-1} at 0 °C. For melting/freezing, the heat of fusion is $L_m = 0.3 \times 10^6$ J kg^{-1} at 0 °C, and for sublimation $L_s = 2.8 \times 10^6$ J kg^{-1} at 0 °C.]

So now consider our parcel of air, with water droplets forming as it rises. The parcel is shown in Figure 12.25.

There are two types of processes we might want to consider:

(a) **The moist-adiabatic process**. In this case, the condensation products remain in the parcel and can re-absorb some of the latent heat released by their own condensation.

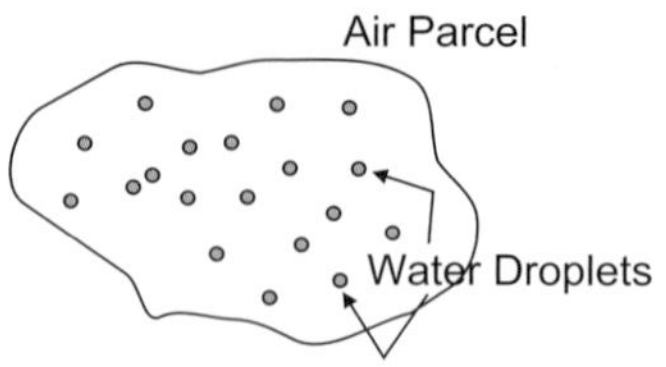

Figure 12.25 An air-parcel with water droplets embedded.

Alternatively we can consider:

(b) **The pseudo-adiabatic process**. In this case, we can assume that as soon as water droplets form, they leave the parcel. Thus all the available latent heat can go into heating the air in the parcel.

The two cases (a) & (b) are extremes, the real case is somewhere in between. However, we find the results are similar. Case (b) is easier to deal with, and is the one we will concentrate on. Note that (b) is an *irreversible process*.

12.10.6 The pseudo-adiabatic process

Take a parcel of *unit mass* of dry air containing a mass q_s of water vapor. Let the initial pressure and temperature be p and T respectively. Now let the parcel undergo a small adiabatic expansion to the state $p + dp$, $T + dT$, and let the new water vapor content be $q_s + dq_s$. (As it turns out, all three of these differential quantities are negative.) The condensation of $-dq_s$ kg of water releases $-Ldq_s$ Joules of latent heat, where L is the latent heat of vaporization. Note that this is a *positive* quantity of heat added to the system, since we *lose* water vapor to water droplets and so dq_s is negative, and the negative value of dq_s multipled by the minus sign in front of the L produces a positive quantity. We assume that *all* this heat released goes into heating the unit mass of dry air.

A digression: The first law of thermodynamics for atmospheric work

Before we continue determination of the moist adiabatic lapse rate, we need to alter some familiar thermodynamical equations to better suit atmospheric studies. Most important of these is the first law of thermodynamics. The normal form is often in terms of C_v, the specific heat per mole per °C at constant volume, but in the atmosphere it is more reasonable to use a form of the first law involving C_p. In the following derivation, we will *not* assume a unit mass, but assume η_m moles of gas.

Recall that the first law says that

$$dQ = \eta_m C_v dT + pdV, \tag{12.48}$$

where dQ is the heat added *to* the system, $dU = \eta_m C_v dT$ is the increase in *internal energy* of the system and pdV is the mechanical work done *by* the system.

Now we can use the ideal gas law ($pV = \eta_m RT$) to write

$$pdV + Vdp = \eta_m RdT \tag{12.49}$$

for a fixed amount of η_m moles of gas. Substituting for pdV in (12.48), we produce

$$dQ = \eta_m C_v dT + \eta_m R dT - V dp \tag{12.50}$$

or

$$dQ = \eta_m (C_v + R) dT - V dp. \tag{12.51}$$

Now C_p, the molar specific heat at constant pressure, is equal to $C_v + R$. (If the reader is not familiar with this, just look at (12.51) and recognize that at constant pressure, dp is zero. Thus $\frac{1}{\eta_m}\frac{dQ}{dT}$ at constant pressure (which is of course the definition of C_p) is equal to $C_v + R$.)

Thus we have

$$dQ = \eta_m C_p dT - V dp. \tag{12.52}$$

This re-states Equation (12.48) in terms of C_p. However, things are still not ideal for atmospheric work. The above equation is really designed for experiments in a closed vessel, but the atmosphere has no boundaries. Hence we prefer to convert to densities, heat per unit mass, etc. Thus we divide (12.52) through by the mass $\mathcal{M}$ of our selected sample of gas, which means we divide through by $\eta_m N_0 m$ where m is the mass of a typical molecule. Then we get

$$dh = \frac{C_p}{N_0 m} dT - \frac{V}{\mathcal{M}} dp, \tag{12.53}$$

where dh is the heat input *per unit mass*. We re-write this as

$$dh = c_p dT - \alpha dp, \tag{12.54}$$

where c_p (note the lower case "c") is the specific heat *per unit mass* at constant pressure, and α is the *inverse* of the mass density, viz. $\alpha = \frac{1}{\rho}$. This is the equation we seek.

Now we return to our original problem involving a parcel with unit mass of dry air.

Back to the pseudo-adiabatic process

The latent heat transferred from the condensation of an incrementally small amount of water vapor $d\mathcal{M}_{vs}$ is $-L.d\mathcal{M}_{vs}$, (in the subscripts, s stands for "saturated", v stands for "vapor") so the amount transferred per unit mass of the air parcel is $-L\frac{d\mathcal{M}_{vs}}{\mathcal{M}_*}$, where $\mathcal{M}_*$ is the total mass of the parcel. This equals $-Ld(\frac{\mathcal{M}_{vs}}{\mathcal{M}_*})$. If we divide the top and bottom lines of the derivative terms by the volume V, and use the approximation $\mathcal{M}_* \simeq \mathcal{M}_d$, where the latter is the mass of dry air in the parcel, then this becomes $-Ld(\frac{\rho_{vs}}{\rho_d})$, which is $-Ldq_{rm}$ from (12.22). Here we are interested in the value of q_{mr} at saturation, which we will denote as q_{mrs}. We will continue using q_s, but remember that q_s and q_{mrs} are now the same quantity and they represent the amount of water vapor embedded in a unit mass of dry air.

Thus substituting for dh with $-Ldq_s$, we obtain for our original parcel of air with *unit* dry mass

$$-Ldq_s = c_p\, dT - \alpha\, dp. \tag{12.55}$$

Because we are assuming water is changing phase, we also assume that the mass of water vapor q_s is equal to the saturated value, so hence forth we consider that $q_s = q_{mrs}$ is the saturated value.

We can now continue our derivation of the equations for the pseudo-adiabatic process. Using the ideal gas law, $p\alpha = R_* T$ (see (12.32) and (12.33)) gives

$$-Ldq_s = c_p dT - R_* \frac{T\,dp}{p}. \tag{12.56}$$

Now recall that $q_s = 0.62\frac{e_s}{p}$ (by (12.36)). Hence we can write

$$\frac{dq_s}{q_s} = \frac{de_s}{e_s} - \frac{dp}{p}. \tag{12.57}$$

But from (12.19) we know that

$$\frac{dp}{p} = -\frac{gdz}{R_* T}, \tag{12.58}$$

so substitution in the previous equation gives

$$\frac{dq_s}{q_s} = \frac{de_s}{e_s} + \frac{gdz}{R_* T}. \tag{12.59}$$

Now substitute for dq_s from (12.59) into (12.56) to give

$$-Lq_s\left[\frac{de_s}{e_s} + \frac{gdz}{R_* T}\right] = c_p dT + gdz. \tag{12.60}$$

Divide through by dz, and replace $\frac{de_s}{\mathrm{d}z}$ by $\frac{de_s}{\mathrm{d}T}\frac{\mathrm{d}T}{\mathrm{d}z}$, to give

$$-Lq_s\left[\frac{1}{e_s}\frac{\mathrm{d}e_s}{\mathrm{d}T}\frac{\mathrm{d}T}{\mathrm{d}z} + \frac{g}{R_* T}\right] = \left[c_p\frac{\mathrm{d}T}{\mathrm{d}z} + g\right], \tag{12.61}$$

where we no longer think of terms like de_s, dT, etc., as incremental units but have converted these terms to differential operators like $\frac{\mathrm{d}}{\mathrm{d}T}$ (which are therefore represented as normal text and not italics). Then extracting $-\frac{\mathrm{d}T}{\mathrm{d}z}$, and replacing q_s with $0.62\frac{e_s}{p}$, we obtain

$$-\frac{\mathrm{d}T}{\mathrm{d}z}\left[L\frac{0.62}{p}\frac{\mathrm{d}e_s}{\mathrm{d}T} + c_p\right] = g\left[1 + 0.62\frac{Le_s}{pR_* T}\right]. \tag{12.62}$$

The term $\Gamma_s = -\frac{\mathrm{d}T}{\mathrm{d}z}$ is called the *moist adiabatic lapse rate*; note the inclusion of the minus sign. Γ_s is the rate at which the temperature of a moist parcel of air *decreases* as it rises. Re-arranging the previous equation, we obtain

$$\Gamma_s = g\left[\frac{1 + 0.62\frac{Le_s}{pR_* T}}{c_p + 0.62\frac{L}{p}\frac{\mathrm{d}e_s}{\mathrm{d}T}}\right], \tag{12.63}$$

where p is the total pressure. This is one form of the equation we seek, but there are better ones. The main unknown is $\frac{\mathrm{d}e_s}{\mathrm{d}T}$, so we need a better expression for this. To obtain that, we revert from e_s to q_s. We can eliminate e_s by using $e_s = \frac{pq_s}{0.62}$. We can also remove the term $\frac{\mathrm{d}e_s}{\mathrm{d}T}$ by using the Clausius–Clapeyron relation. We will not derive this relation here, but it can be found in any reasonable book on thermodynamics, and relates the temperature dependence of the saturated vapor pressure of a substance to the latent heat L by the relation

$$\frac{\mathrm{d}p_v}{\mathrm{d}T} = \frac{L}{TV}. \tag{12.64}$$

This equation is also referred to as the first latent heat equation.

In our application, p_v is replaced with the saturated vapor pressure of water vapor, e_s, and V is replaced using the ideal gas equation. Then

$$\frac{\mathrm{d}e_s}{\mathrm{d}T} = \frac{L}{T}\frac{e_s}{NK_BT}, \tag{12.65}$$

where N is the number of water molecules in a unit mass of water vapor. Note that NK_B can be written as $\frac{N_0K_B}{M_v}$, since the number of molecules in a 1 kg mass is just the number of moles in 1 kg (i.e. $\frac{1.0}{M_v}$ moles) multiplied by N_0. But $\frac{N_0K_B}{M_v}$ is just $\frac{R}{M_v} = R_v$ (see (12.25) and (12.15)). Thus

$$\frac{\mathrm{d}e_s}{\mathrm{d}T} = \frac{L}{T}\frac{e_s}{R_vT}. \tag{12.66}$$

We now use $e_s = \frac{pq_s}{0.62}$, and $R_v = \frac{R_*}{0.62}$, (see the lines of text following (12.33)), and take $R_* \simeq R_d$, to give

$$\frac{\mathrm{d}e_s}{\mathrm{d}T} = \frac{Lq_sp}{R_*T^2}. \tag{12.67}$$

Substitution in (12.63), and recognizing that q_s is the mass of water vapor in a unit mass of dry air at saturation, so that q_s can be replaced with q_{mrs} (the mixing ratio at saturation), as already noted above, then gives

$$\Gamma_s = \frac{g}{c_p}\left[\frac{1+\frac{Lq_{mrs}}{RT}}{1+\frac{0.62L^2q_{mrs}}{c_pR_*T^2}}\right] \tag{12.68}$$

This is slightly better than Equation (12.63) in that it depends only on two variables (T and q_{mrs}), whereas (12.63) is expressed in terms of three variables, namely p, q_{mrs}, and T. Note that in the special case of q_{mr}= 0, we have the simple expression

$$\Gamma_d = \frac{g}{c_p}, \tag{12.69}$$

where Γ_d is the *dry* adiabatic lapse rate (Equation (1.42)). Numerically, this is about 0.01 °C m^{-1}, or about 10 °C per km (more precisely, closer to 9.8 °C per km), as seen earlier.

Note also that in the moist adiabatic lapse rate expression, q_{mrs} is a known function of p and T; the exact relation can be read from the phase diagram, although the relationship is not trivial. Hence (12.68) can be thought of as a function only of p and T. Some typical values of the moist adiabatic lapse rate are tabulated overleaf. The units are °C per kilometer of altitude.

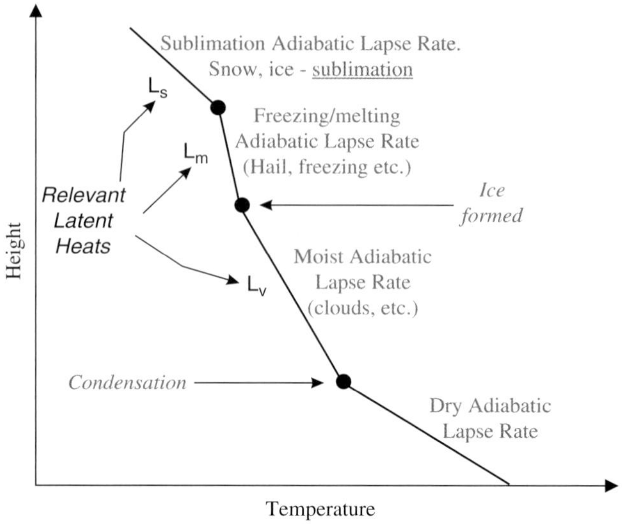

Figure 12.28 Various latent heat transfer processes in the atmosphere.

When an air parcel is displaced vertically by a small increment δz, the buoyancy force on the parcel is due to the difference in the upward differential pressure force on the parcel and its own downward weight, and we may write that the force per unit mass is

$$\mathcal{F}_B = -\omega_B^2 \delta z, \tag{12.70}$$

which can be seen from Equation (1.49). The Brunt–Väisälä frequency ω_B is a measure of the stability of the atmosphere, i.e. of the vertical gradient of the temperature, and was derived in Equation (1.52) in Chapter 1. In dry air it is given explicitly through the relation

$$\omega_B^2 = \frac{g}{T}\left[\frac{\mathrm{d}T}{\mathrm{d}z} + \frac{g}{c_p}\right] = \frac{g}{\Theta}\frac{\mathrm{d}\Theta}{\mathrm{d}z}, \tag{12.71}$$

where g is the acceleration due to gravity, T the temperature, and c_p the specific heat per unit mass at constant pressure p. Θ is the potential temperature, which is the temperature an air parcel would attain if it were moved adiabatically from its present position at pressure p to the level where the pressure is $p_0 = 1000\,\mathrm{mb}$.

In the case of moist air (e.g. fog or cloud), the same expression would be used, but $\frac{g}{c_p} = \Gamma_d$, and the dry adiabatic lapse rate would be replaced by the moist adibatic lapse rate Γ_s.

Θ in dry air is given from Equation (1.66) as

$$\Theta = T\left(\frac{p_0}{p}\right)^{R/C_p}, \tag{12.72}$$

R being the gas constant for dry air and C_p the specific heat per unit mole at constant pressure. The value of $\frac{R}{C_p}$ is numerically equal to 0.283. It is harder to give an exact expression for Θ in the case that moist adiabatic processes are involved in the pathway as the parcel is imagined to move between the height of interest and the ground, so in the following discussions we concentrate on fully dry adiabatic processes. Note that the dry adiabatic theory applies even to cases where the humidity is non-zero – the moist adiabatic cases only apply when changes of phase are taking place (e.g. in cloud or fog).

The atmosphere is stable when $\frac{d\Theta}{dz}$ exceeds zero, and the Brunt–Väisälä frequency ω_B determines the limiting highest frequency of atmospheric buoyancy-driven oscillations, namely gravity waves (see Chapter 11). When $\frac{d\Theta}{dz} > 0$, a parcel of air that has been displaced vertically will return towards its equilibrium position, which is why this condition is referred to as a stable one. In that case the parcel will oscillate, producing wave radiation.

When $\frac{d\Theta}{dz}$ is less than zero, (i.e. $-\frac{dT}{dz} > \frac{g}{c_p}$), the Brunt–Väisälä frequency ω_B is imaginary. A parcel of air placed in this environment will spontaneously rise, and never return to its original position. This means that wave oscillations are not possible, and the atmosphere is statically unstable. This is the condition for convection to arise.

Thus, the negative of the vertical gradient of temperature, $-\frac{dT}{dz}$, which is called the (environmental) lapse rate Γ_e, determines whether the atmosphere is stable or unstable.

This is illustrated in the following way.

The graphs in Figure 12.29 show the temperature T as a function of height z for two extreme cases commonly observed in the troposphere. We choose one height z_0 with temperature T_0 and draw a line through it with a special slope satisfying $-\frac{dT}{dz} = \frac{g}{c_p} = 9.8\,°\text{km}^{-1}$, i.e. Θ = constant. The negative of this special slope, $-\frac{dT}{dz}$, is denoted in the standard atmosphere as Γ_d and is called the dry adiabatic (i.e. without heat exchange in dry air) lapse rate. The negative of $\frac{dT}{dz}$ for the actual atmosphere itself is denoted as Γ_e.

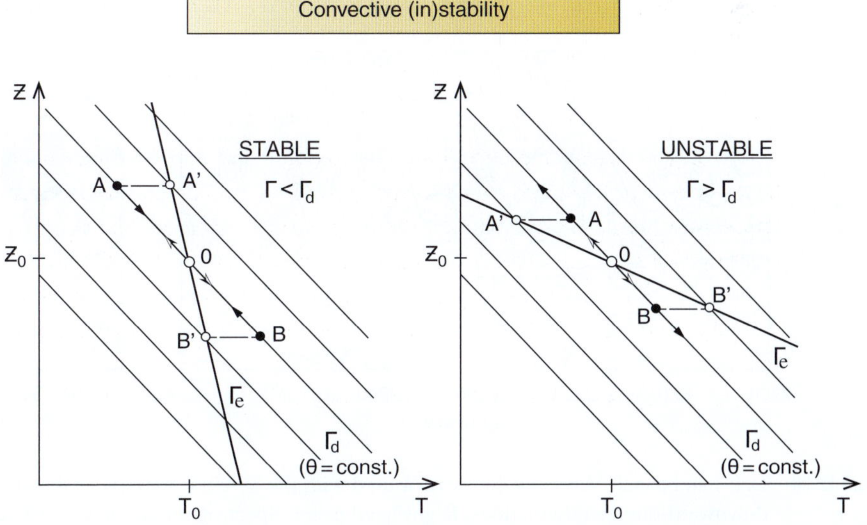

Figure 12.29 Examples of stable and unstable temperature profiles for dry air. This is a slightly more detailed version of Figure 1.27. The parallel lines are lines of constant potential temperature.

A parcel of air displaced vertically would move adiabatically along the line of constant Θ from point O at z_0 and T_0 to point A. The air parcel expands and cools as it rises. The environmental temperature (that is, the temperature of the air immediately surrounding the parcel) follows the line to A′. At A in the left-hand figure, the parcel is cooler and heavier than its environment. The buoyancy force consequently drives it down again along the line of constant Θ to point B where in turn compressional forces heat it to a slightly higher temperature than the environmental temperature. The lighter air parcel consequently moves upward again. This results in a periodic displacement (oscillation) at the Brunt–Väisälä frequency.

Therefore, if the environmental lapse rate Γ_e is smaller than Γ_d, i.e. the negative vertical temperature gradient is smaller than the dry adiabatic lapse rate (the steeper line in the left-hand panel of Figure 12.29), the air parcel will oscillate.

If the parcel were not allowed to oscillate vertically, but was forced to oscillate on another sloping surface, then the frequency of oscillation would be less. Such a scenario is at the basis of the gravity waves discussed in Chapter 11, and in some texts, gravity waves are introduced with this model.

Therefore, these oscillations can only take place in an atmosphere which is stably stratified. Details of these gravity-wave oscillations were described in Chapter 11. Figure 12.30 shows stratospheric–tropospheric radar observations which clearly show the cut-off frequency corresponding to the Brunt–Väisälä frequency deduced from a measured stably stratified temperature. The variation with height arises because the environmental lapse rate changes from the troposphere to the stratosphere. As discussed in earlier chapters (e.g., see Chapter 2, Section 2.16), graphs like this can be used to determine the temperature gradient and hence integrated over height to determine the temperature profile.

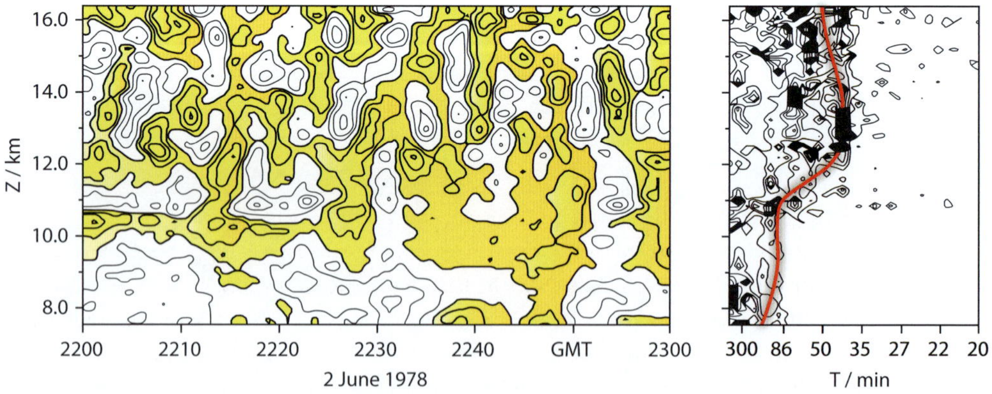

Figure 12.30 Left-hand panel: Vertical velocity measured with the SOUSY VHF radar showing upward- and downward-directed oscillations. Right-hand panel: Spectrogram of the vertical velocity as function of height z. The red line shows the Brunt–Väisälä frequency ω_B calculated from temperature profiles. Wave oscillations are clearly not observed at frequencies larger than ω_B (*Röttger*, 1980b). (Reprinted with permission from Springer.)

The right-hand panel in Figure 12.29 shows the opposite situation, namely the case that the environmental lapse rate Γ_e is larger than Γ_d, which means that the temperature decrease with altitude is larger than the adiabatic case. This is shown by the shallower line passing through the point O at z_0 in Figure 12.29. A parcel of air with temperature T_0 will rise along the line of constant Θ, which means that it soon reaches an altitude where the actual environmental temperature is lower than its own temperature. This gives rise to an even larger upward buoyancy force on the parcel, and the parcel rises continually. Here, the atmosphere is called convectively unstable. Of course, the parcel will expand and cool during its upward convection, but will remain warmer than the temperature of the environment. It will continue to rise for as long as the conditions in the right-hand figure apply. Only if the environmental temperature gradient again starts to exceed the adiabatic rate, and the environmental temperature increases sufficiently that it again reaches that of the parcel, will the case be reached where the parcel may be stopped from its upward motion.

The situation shown in the right-hand graph of Figure 12.29 will not allow the parcel to move downward again, which means wave oscillations are not possible. Generally, the parcel will continue to rise until it finally reaches an inversion layer where the lapse rate is sufficiently smaller than the dry adiabatic lapse rate such that further upward motion ceases. The parcel may even penetrate the inversion layer (increase of temperature over a short altitude range), which is called penetrative convection and is a source for gravity waves propagating horizontally and vertically.

A three-fold scenario therefore exists: (i) an adiabatic atmosphere with $\Gamma_d = \Gamma_e$, which is an unlikely event in nature (at least unlikely over large vertical distances, although it may be valid over vertical distances of a few hundred meters to a kilometer or two, as for example in a well-mixed turbulent layer); (ii) a convectively stable atmosphere with $\Gamma_d > \Gamma_e$; and (iii) a convectively unstable atmosphere with $\Gamma_d < \Gamma_e$.

Besides the convective instability, we are often faced with another instability, namely the shear instability. Both may be described by using the Richardson number, which has also been introduced earlier in this text. In one dimension this is given by

$$R_i = \frac{\omega_B^2}{\left(\frac{\mathrm{d}\overline{u}}{\mathrm{d}z}\right)^2}, \tag{12.73}$$

where $\frac{\mathrm{d}\overline{u}}{\mathrm{d}z}$ is the vertical shear of the horizontal wind velocity $\overline{u}$.

In the following paragraphs, we will give a somewhat more physical view of the reasons for the different categories of R_i than has been given previously. In principle, the larger the Richardson number, the smaller the chance for instability. There are two main cases – R_i is greater than or equal to zero, and R_i less than zero. It should be recognized that the denominator will always be positive, but the numerator may be positive or negative.

If $R_i > 0$, then the numerator must be positive ($\omega_B^2 > 0$). Energy produced by the wind-shear supplies kinetic energy to the system, but since the temperature profile is stable, this kinetic energy essentially goes into storage in the form of potential energy associated with the temperature gradient. However, if the wind-shear is very large, it

The Richardson number R_i is shown by the full line in the upper part of the right-hand panel. The stippled continuation of this line in the lower portion corresponds to an assumption of large $(\Gamma_e - \Gamma_d)$.

Experiment and theory suggest that there is a critical limit at $R_i = 0.25$, which is labeled in the right-hand side panel. When the Richardson number is smaller than 0.25, instability may occur, and at larger values, the atmosphere tends to be more stable (*Miles*, 1961; *Howard*, 1961). However, it should be noted that while $R_i = 0.25$ is a necessary condition for turbulence, it is not sufficient, so that even if R_i falls below 0.25, it is not required that turbulence needs to start. Furthermore, it is possible that as active turbulence decays, the Richardson number may rise above 0.25, but the turbulence may still persist through its own inertia. Hence, the criterion that $R_i = 0.25$ is only a guideline to the existence or otherwise of instability.

Returning to Figure 12.33, we note that dynamic instability occurs in the dark blue region in the right-hand graph. Here, $\Gamma_e - \Gamma_d = 0$, but the wind-shear is large, so the Richardson number approaches zero, i.e. it is smaller than 0.25, but not negative. Indeed, the part of the graph with $0 < R_i < 0.25$, which is labelled with blue stripes, is dynamically unstable. Under these conditions, the so-called shear or Kelvin–Helmholtz instability (KHi) occurs.

When the Richardson number R_i is negative (colored in green), which happens when $\Gamma_e - \Gamma_d < 0$, convective instability arises. This gives rise to the Rayleigh–Taylor instability (RTi).

However, the KHi and RTi instabilities are not the only types of instability; a variety of others exist as well, as discussed in Chapter 11. Other types of instabilities include Holmboe, vortical-pair instability, slant-wise instability (*Hines*, 1988), parametric sub-harmonic instability, Beaumont instability, resonant instability, and oblique instabilities.

The time-lines of development of KHi and RTi instabilities, which occur in the neutral atmosphere, are sketched in Figure 12.34. KHi instabilities start with wave-like

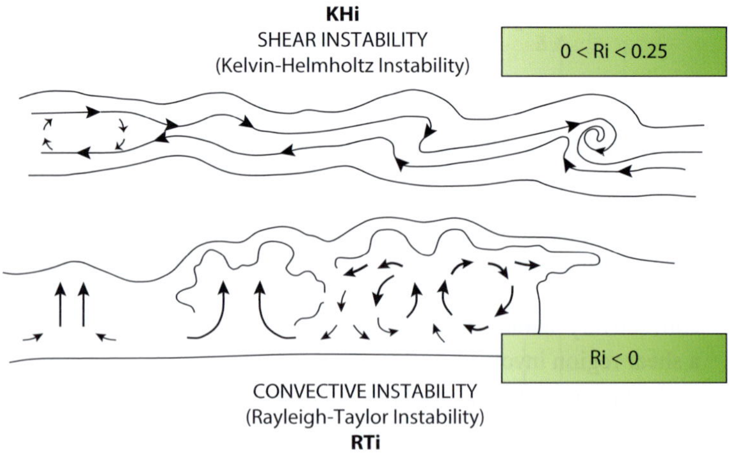

Figure 12.34 Temporal development (time increasing left to right) of the creation of KHi and RTi in the atmosphere, and their final breakdown into turbulence.

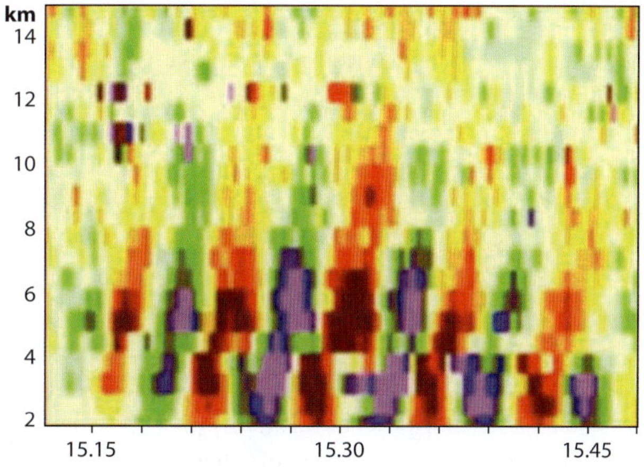

Figure 12.35 Vertical velocities observed with the Chung-Li VHF radar in the troposphere. The phase jump in the vertical direction is indicative of KHi.

structures that in turn develop into rolls and then break down. RTi structures begin with vertical motions which turn into large rolls which then also break down. Further discussion about the development of KHi can be found in the latter sections of Chapter 11, especially in regard to their formation from the breakdown of gravity waves. Further references to follow-up reading can also be found there.

12.10.8 KHi studies by MST radar

Kelvin–Helmholtz structures are often visible with MST radars. They have unique characteristics that can be identified by these radars. For example, Figure 12.35 shows measurements with the Chung-Li VHF radar, showing phase jumps in the vertical direction.

Other observations of KHi with MST radar include studies by *Röttger and Schmidt* (1979), who used deconvolution procedures to study them, and *Chilson et al.* (1997), who were the first to use frequency-domain interferometry to detect Kelvin–Helmholtz billows in the jet stream. *Reid et al.* (1987) detected "cat's-eye" structures indicative of KHi in the mesosphere. Studies of KHi are especially well suited to high-resolution frequency-domain and spatial-domain techniques, and pulse compression methods, as discussed in the earlier chapters of this book. Examples of possible KHi have been presented in Figures 2.2, 9.2 and 10.4. More detailed discussions about these various instabilities can be found in *Fritts and Rastogi* (1985), among other references.

12.10.9 Convection studies with MST radars

In Figure 12.22, we saw some of the parameters that can be measured in association with convection by MST radars. The ability of these radars to measure vertical winds, spectral widths, and turbulence strengths is very important in these studies. Figure 12.36 shows the flow patterns expected in a large cumulo-nimbus cloud, and the vertical flows

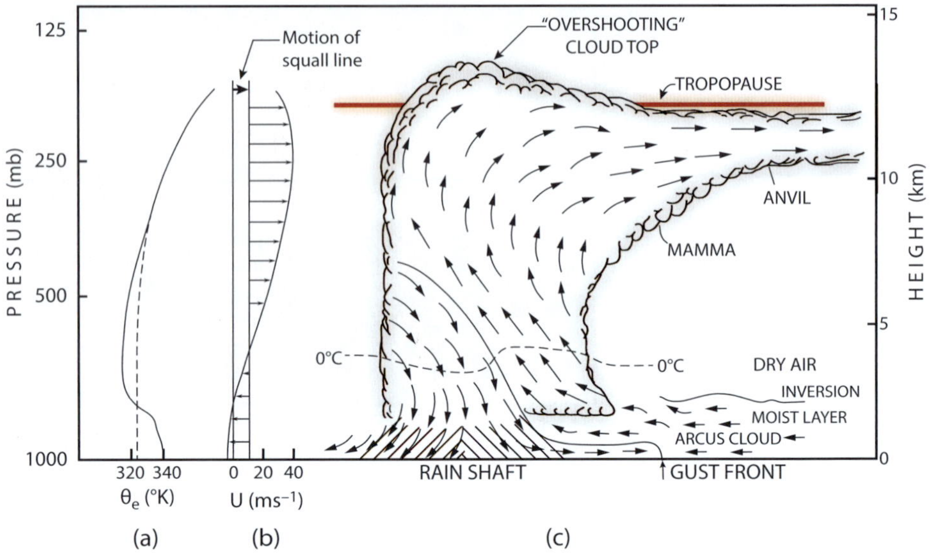

Figure 12.36 Wind flow and shears in a cumulo-nimbus cloud associated with a squall line (from *Wallace and Hobbs*, 1977).

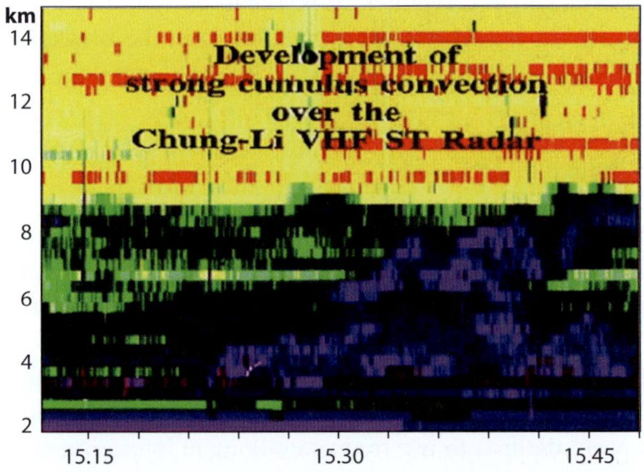

Figure 12.37 Example of radar studies of convection recorded with the Chung-Li MST radar.

are clearly very strong – both updrafts and downdrafts exist. Several papers have used MST radars to study convection in events like this.

Hooper et al. (2003) studied the characteristics of airflows in mid-latitude convection, while *Narayana Rao et al.* (1999), *Jain et al.* (2000), and *Dhaka et al.* (2001) have studied tropical convection. *Hamsen et al.* (2002) have studied the production of gravity waves by convective processes.

Figures 12.37 and 12.38 show examples of vertical motions during strong convection with the Chung-Li MST radar. Other convective studies, during less violent events, have already been discussed earlier in this chapter.

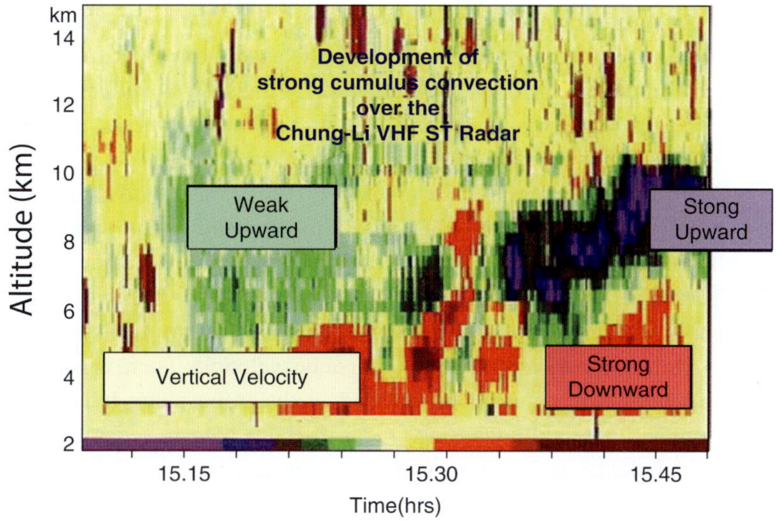

Figure 12.38 Another example of radar studies of convection recorded with the Chung-Li MST radar.

12.11 Turbulence in meteorology

Many structures, including the RTi and KHi discussed in the last section, eventually collapse into turbulence, ultimately heating the atmosphere and causing diffusion. Turbulence studies are therefore an important aspect of MST studies in meteorology.

Turbulence strengths have been dealt with at some length in the Chapter 11. In particular, Figures 11.25 and 11.27 to 11.33 deal with typical values, and some radars now produce turbulence strengths as standard outputs. Figure 7.20(b) showed typical short term variability of ε over a two-day period, and also showed the tendency for values to be large above the tropopause on occasion (remembering that the tropopause echoes seem to be of two different types – extremely stable (specular) or very turbulent). There is little point in repeating these graphs and discussions here.

A more important point is to ask about the reliability of the measurements presented in Chapter 11. In the 1980s, radar measurements of turbulence were considered to be a somewhat new technique and were treated with some caution. In-situ measurements were considered the standard. More recently, through sheer availability of larger data sets, radar measurements have become something of an accepted standard, but verification of their accuracy was, until recently, missing. There had been few substantial campaigns to try to intercompare many different measurements of turbulence strengths, and so some doubts existed about the reliability of the radar methods. Spectral-width methods may be contaminated by non-turbulent vertical motions (e.g., *Hooper et al.*, 2005). Figures 11.27, 11.28, and 11.30 in the last chapter represent cases where in-situ and radar methods were compared, but the results were observed at different times and different locations.

A few authors did try real-time comparisons. *Wilson and Dalaudier* (2003) attempted some preliminary comparisons using the absolute backscatter radar method, and results were encouraging. *Pavelin et al.* (2002) and *Whiteway et al.* (2003) have performed important comparisons with in-situ aircraft measurements, and these comparisons also addressed some interesting physical problems,

However, the paper by *Dehghan et al.* (2014) helped to change this uncertainty. In this paper, a three-way series of measurements using radar, specialist instrumentation on-board a small aircraft, and accelerometers on commercial aircraft, were intercompared. The radar data were verified as accurate, at least to better than a factor of 2. The same paper also revealed that the commercial aircraft instruments were improperly calibrated.

More comparisons would of course be welcome. There is some question as to whether the calibration constants developed in the troposphere by *Dehghan et al.* (2014) are valid at higher altitudes, since radar methods have to separate the relative contributions of turbulence and gravity waves (which can be a height-dependent ratio). Certainly, radar measurements now seem accurate to within a factor of 2 or 3 but better accuracy would be good. In the mesosphere, where useful measurements of any type are difficult to obtain, radar measurements possibly are something of a standard, being as good as rocket data. However, higher precision is still desired at all levels, and further comparisons are strongly encouraged.

12.12 Precipitation and humidity measurements with ST radars

While the main application of windprofiler radars is in measurement of winds, and to a somewhat lesser extent, backscattered power, there are multiple alternative applications. Some of these were discussed in Chapters 2 and 10. Among these topics were the issues of measurement of water-vapor content (humidity) and hydro-meteor (water drops and ice particles) precipitation rates. At present, measurement of these parameters is not a routine application with MST VHF radars and windprofilers, and because these techniques are not generally considered as operational, we left the discussions to Chapter 10. However, here we simply remind the reader of these capabilities, and the possibility that these techniques could be developed further should be borne in mind, as some groups of researchers continue to be actively involved in these areas.

12.13 Boundary layer measurements

Windprofiler research began with VHF radars. These do well above altitudes of typically 1.5 km, but they have always been somewhat limited in their ability to make useful measurements at the lowest altitudes, especially below 1 km. This has, in part, been due to their large physical size, which forces the regions below 1.5 km altitude to be in the Fresnel region, and partly because low altitude measurements require short pulses, requiring bandwidths of the order of 1–2 MHz. Such bandwidths represent a good fraction of the available spectrum, and it is very hard to get frequency allocations of 1–2 MHz at 50 MHz. Therefore, it has been usual practice to build higher frequency radars for low-level studies. This is, in part, why the USA NOAA network was set up with a frequency of 404 MHz.

Recent developments have permitted VHF systems to get much lower in height. For example, *Vincent et al.* (1998) have used small, compact-spaced antenna systems, coupled with rapidly switching passive transmit-receive switches, to achieve altitudes as low as 400–500 m. Lower altitudes have been achieved when using very short pulses (75–150 m), although as noted, this requires very special permission from the radio frequency allocation authorities. *Scipion et al.* (2003) have achieved similar results with similar methods. *Hocking* (2002) has used a different approach, using separate transmitting and receiving antennas (a bistatic system), with specially designed loop antennas for reception, in order to reduce coupling from the ground pulse of the transmit antenna. This approach avoids the need for transmit-receive switches. The design also allows useful low-level wind measurements to be achieved even with a bandwidth of only 500 kHz or less. More recently, Yagi antennas have been successfully used to replace the loop antennas, since they have higher efficiency, and software analysis has been improved so that the receiver can be digitized while the transmitter pulse is still being transmitted (*Hocking and Hocking*, 2010), enabling winds to be measured as low as 400 m altitude even with a pulse that exceeds 400 m in length. This is an important new direction for windprofiler research, since VHF radars are not as susceptible to hydrometeor contamination as higher frequency radars.

12.14 Windprofiler contaminants

All radars are, of course, susceptible to contaminants, but radars used in meteorology can be especially vulnerable. Man-made interference, of course, affects all frequencies, but some forms of contamination are frequency-specific, and stronger in the troposphere. If UHF and L- and S-band radars are used, precipitation can adversely affect clear-air echoes. Of course, with due care, the existence of contamination from hydrometeors can be turned to advantage, as discussed in a previous section on precipitation, but generally with loss of some clear-air information. A more serious problem is birds, which can be especially strong targets at the higher frequencies. Bird migration can lead to artificial "wind jets" being detected by radars, for example. At one time, observations were made of a steady northerly wind jet at about 500 meters altitude, which occurred especially in the northern hemisphere during the Fall months, and especially in the early evening and morning. More intense scrutiny revealed that this was in fact mass bird migration.

Birds are not a serious issue for radars working around 50 MHz, but even these radars are not free of contamination. Such VHF radars often use high pulse repetition frequencies, and so range-aliasing of echoes from more distant targets can be an issue. For example, in polar regions, auroral echoes can at times contaminate the signal and produce artificial winds. If the radar is at a latitude where polar mesosphere summer echoes exist, these can contaminate the signal (this special case was discussed in Section 4.10.1, Chapter 4).

In some cases, the windprofiler radars themselves can contaminate other neighbors. This might be in the form of interference with television reception, or, in the case of RASS, noise pollution. If the effects are too severe, the windprofiler may be closed down.

A multitude of contaminants can affect radars. Any radar system should have personnel on staff who are capable of identifying such contaminants, and who can tell when unusual events are truly unusual, and when they are some form of non-meteorological event.

13 Concluding remarks

13.1 Introduction

It should be clear from the foregoing chapters that the range of applications of MST and windprofiler radar is broad and challenging. Some techniques are mature, some are under development, and some are even no doubt yet to be discovered. Measurements of wind velocities and, by extension, wave motions, wave-mean flow interactions, momentum flux deposition and turbulence, are possible. Capabilities for temperature measurements, and the possibility of humidity measurements, have been discussed. Strange echoes such as polar mesosphere summer echoes have given new insights into the plasma processes of the lower thermosphere. Studies of turbulence anisotropy are possible. We have demonstrated functional radar designs that cost as little as \$100 000 up to many millions of dollars.

We will not dwell on these many achievements, however, which should be self-evident. What is perhaps of greater interest is the future of these instruments, and this will be the main focus here.

13.2 The future

The future harbors both pragmatic and curiosity-driven aspects. From the point of view of the former, networks of radars, providing data for incorporation into computer forecasting and now-casting models, offer the hope of better forecasts. They have been shown to have benefits in forecasting on time-scales from a few hours out to several days, especially with systems deployed in Japan, Europe, and Canada (see Chapter 12). At the time of writing (2015), the European Space Agency is about to launch a specialized satellite instrument (AEOLUS) for measurement of tropospheric winds from space by lidar, and the networks of windprofilers discussed will be crucial tools for validation of these data. However, since the satellite only measures winds at sunrise and sunset, the radars, with their continuous recording capability, will continue to provide valuable input to meteorological models for many years to come.

Accurate records of winds are of course valuable for large-scale forecasts. This can impact aircraft travel, allowing better flight planning. The ability of radars to make reliable measurements of turbulence strengths can also be of value from the perspective of aircraft passenger safety. This capability has not yet been fully employed, but has great potential. Detailed studies of meteorological events can also be of importance in regard to commercial aircraft and passenger safety. Studies of wave breakdown and the subsequent production of clear-air turbulence can help in understanding the details about when and where the velocity fluctuations are most severe. Soliton waves can have substantial pressure gradients across their boundaries, which can lead to sudden altitude loss for aircraft during transit across the boundaries. Since such events can occur in clear air, these radars can be specially useful in their investigation. Radars are also useful for studies of the vertical motions within thunderstorms. Recent studies have also shown the capability of profilers to make useful predictions in regard to tornado formation, and the capability of these radars to recognize jumps in the height of the tropopause indicative of stratospheric ozone intrusion has been discussed.

Some of the above capabilites are not yet operational, but can, and should, be implemented in the future. Such capablities will allow the radars to be used for severe storm prediction, pollution events associated with ozone intrusions, and severe tubulence warnings.

However, it is perhaps in another area that the full suite of radar capabilities can be most effectively employed. This is in the area of space travel. As the future unfolds, it is likely that commercial space travel will become a reality, with passenger flights into space becoming common. In the same way that modern airlines must monitor weather conditions for the safety of their passengers, so future space companies will need to monitor weather coditions between the ground and 100 km altitude. There is no one instrument that can do this. However, some of the various types of MST radars described in this book will play a role. The best MST radars/windprofilers currently can reach altitudes of 25 km or so – particularly the MU radar. Improvements in technology, better digitization, perhaps improvements in signal processing techniques, and more extensive use of distributed transmitters (such as with the MAARSY and PANSY radars discussed in Chapter 10), coupled with greater power, may allow a maximum height of 35 km. These same radars can also probe the regions from 60 to 90 km, albeit intermittently. However, MF radars, using the spaced antenna method, can routinely record winds in the height range 55 to 85 km, as discussed in Chapter 2, and this method is yet to be further developed. Current systems tend to be experimental and of low power, and improvements are feasible. Their biggest drawback is the need for large areas of land. As discussed in Chapter 2, Section 2.5, current systems have been unfairly criticized, and in fact do provide reliable winds up to a maximum height of about 80–85 km. The reasons that heights above this should be avoided were discussed in Section 2.5. However, meteor radars can step in and fill the data between 80 and 95 km altitude. Modern meteor radars can detect 30 000 meteors per day, or more than 1000 per hour, and so can provide good temporal resolution.

Thus an optimal combination of MST/VHF radar, MF mesospheric spaced antenna radars, and meteor radars, can provide real-time winds from the ground to almost 100 km altitude, with hourly updates or better, leaving a gap in the region between 35 and 55 km. This region can be supplemented with lidar information. Such lidars do not currently have the Doppler capability to measure winds, especially in the daytime, but this is likely to be developed in forthcoming years. Hence we expect MST radars to play important roles as forecast instruments for space flight. One of the most important regions is that between 50 and 80 km on re-entry, where astronauts often report "potholes in the sky," these being primarily due to temperature inversions which harbor gravity waves with horizontal wavelengths of a few tens of km, and which appear to a space craft moving at speeds of Mach 18 as a "bumpy road." (Strong turbulence at these heights has little impact – the typical velocities associated even with strong turbulence are a few ms^{-1}, which are trivial in comparison to spacecraft speeds of 6000 ms^{-1}. The main cause of buffetting occurs as the rapidly moving craft bounces off the density perturbations.) The value of meteor radars in studying such events was demonstrated by *Hocking et al.* (2003), who studied atmospheric conditions during the unfortunate destruction of the space shuttle Columbia in February 2003. MST radars have also been used in studies of space debris re-entering the Earth's atmosphere.

Of course one of the mainstays of MST radars is application of the technology to hitherto unknown atmospheric phenomena. One example was the study of PMSEs, as discussed in Chapter 10, but many others have been demonstrated in this book. The potential for radars to make even further discoveries like this is high, and so we rank curiosity-driven applications as important justification for the further use of MST radars.

The MST community has always been quick to capitalize on new technology, and improvements in digitization speeds, and access to multiple-core computers, has allowed new advances in digitization techniques, both in hardware and software. Examples have been discussed (e.g., *Hocking et al.*, 2014; *Yamamoto et al.*, 2014). Distributed transmitters and multi-static modes, such as employed by MAARSY and PANSY, offer potential for further improvements in detectability. Future plans involve extending this concept to transmitter and receiver systems spread over tens and hundreds of km. Such systems require accurately-calibrated atomic clocks, or can use timing from GPS satellites. Passive radars, which do not have their own transmitters, but rather employ transmitters such as those owned by commercial radio stations, are also an interesting future technology. This reduces the operational expense, and removes the need for obtaining frequency licences, but requires large digitization and processing capability, since the received signal must be cross-correlated with the transmitter signal, requiring extensive computing.

In general, the future looks diverse, challenging and promising.

Appendix A Turbulent spectra and structure functions

A.1 Introduction

A full mathematical treatment of turbulence is beyond the scope of this book, even though turbulence is a key cause of atmospheric scatter of radiowaves.

Therefore, in this appendix we will give an overview of some of the main equations used for studies of turbulence. A good additional reference is *Hocking* (1999a), and some of this appendix is adapted from a similar appendix in that article.

The following appendix summarizes the main structure functions and spectra used in turbulence theory, without proof or derivation.

A.2 Velocity structure functions

The first type of function which we will discuss that is commonly used to describe turbulent phenomena is the so-called structure function. There are several of these, but the main ones are $D_\parallel$ and $D_\perp$, which are defined in the following way:

$$D_\parallel(r) = \overline{|u_\parallel(\vec{x}+\vec{r}) - u_\parallel(\vec{x})|^2}, \tag{A1}$$

and

$$D_\perp(r) = \overline{|u_\perp(\vec{x}+\vec{r}) - u_\perp(\vec{x})|^2}, \tag{A2}$$

where we imagine traversing the turbulent medium in a straight line and taking point measurements along the way. Parallel components refer to directions parallel to the direction of traverse, and perpendicular components refer to velocity components perpendicular to this direction. Isotropy has been assumed in this definition, which is why we consider D to depend only on the magnitude r of the vector $\vec{r}$.

Occasionally a 3-D form of the structure function is used, viz.

$$D_{\mathrm{tot}}(\underline{r}) = \overline{|\vec{u}(\vec{x}+\vec{r}) - \vec{u}(\vec{x})|^2}, \tag{A3}$$

where the vector difference between displaced components is used. Because there are two perpendicular components, and one parallel component, we may write

$$D_{\mathrm{tot}} = D_\parallel + 2D_\perp. \tag{A4}$$

For inertial range, homogeneous, Kolmogoroff-style turbulence, we have the following relations:

$$D_{\|} = C_v^2 r^{2/3}, \tag{A5}$$

where $C_v^2 = C\varepsilon^{2/3}$, and C is close to 2.0 (e.g., *Caughey et al.*, 1978; *Kaimal et al.*, 1976; *Paquin and Pond*, 1971). The form of the perpendicular (or transverse) structure function is

$$D_{\perp} = \frac{4}{3} C_v^2 r^{2/3}. \tag{A6}$$

Note that the factor of $4/3$ does *not* indicate anisotropic turbulence. This is the equation for *isotropic* turbulence. The factor of 4/3 arises due to geometrical differences in the way that the different structure functions develop. The interested reader can see the effect by creating parallel and perpendicular structure functions for a random set of suitably structured circular motions. For the case of two-dimensional turbulence, this factor is even bigger, being 5/3.

The total structure function is the sum of one parallel structure function and two perpendicular ones, giving

$$D_{\text{tot}} = \frac{11}{3} C_v^2 r^{2/3}. \tag{A7}$$

There are also a variety of spectral forms, which are used as tools in turbulence studies.

A.3 Spectral forms for velocity measurements

A variety of spectra are used for turbulence studies. These all have different purposes, and are summarized below for Kolmogoroff-type inertial-range turbulence.

The first important expression is

$$\mathcal{F}(\vec{k}) = A\varepsilon^{2/3} k^{-11/3}, \tag{A8}$$

where $k = |\vec{k}|$ and $A = \frac{11\Gamma(\frac{8}{3})\sin(\frac{\pi}{3})}{24\pi^2} C \simeq 0.061C$ (e.g., *Tatarski*, 1971). This is a full three-dimensional function describing the total kinetic energy per unit cell size (due to all three velocity components) in a cell of size $d^3\vec{k}$ at the end of a vector $\vec{k}$ originating from the origin. For homogeneous isotropic turbulence this function is isotropic. Pictorially one can visualize this as a solid sphere in (k_x, k_y, k_z)-space which has highest density at the center, and decreasing density as $|\vec{k}|$ increases, where the density represents $\mathcal{F}$.

Because this function is isotropic, it is often integrated over a shell of radius k to give a new expression, which is

$$E(k) = 4\pi k^2 \mathcal{F} = \alpha\varepsilon^{2/3} k^{-5/3}, \tag{A9}$$

where $\alpha = 4\pi A = \frac{11\Gamma(\frac{8}{3})\sin(\frac{\pi}{3})}{6\pi} C = 0.76655C$ (e.g., see *Tatarski*, 1971; *Batchelor*, 1953). Note that we will largely follow Batchelor's symbol-usage in this document: for example, we use $E(k)dk$ to represent the total energy in a shell in $\vec{k}$-space of thickness

dk, as does Batchelor, whereas *Tatarski* (1961, 1971) used the symbol E to represent the function which we have called $\mathcal{F}$.

If we use $C = 2.0$, then we have

$$E(k) = 1.53\varepsilon^{2/3}k^{-5/3}. \tag{A10}$$

Different authors use different values for the constant 1.53 – anything between 1.35 and 1.53 is common. Note, however, that if one adjusts this constant then the constant C also needs adjustment. We prefer to use $C = 2.0$ because it has at least been measured with good accuracy in the lower atmosphere (e.g. *Caughey et al.*, 1978).

These equations are fairly simple to understand. However, there are more complex variants. An important adjunct (and in fact a more fundamental expression) is the equation

$$\Phi_{ij}(\vec{k}) = \frac{E(k)}{4\pi k^4} \cdot (k^2\delta_{ij} - k_i k_j), \tag{A11}$$

which describes the three-dimensional cross-spectrum between the velocity components in the i direction and the j direction, where i or $j = 1$ mean the x direction, i or $j = 2$ mean the y direction, and i or $j = 3$ mean the z direction. Note that k is the length of the vector from the origin to the point (k_x, k_y, k_z) in k-space, and so $k^2 = k_x^2 + k_y^2 + k_z^2$. For each of these spectra there is a related covariance function; for example,

$$\Phi_{j\ell}(\underline{k}) = \frac{1}{8\pi^3} \int_{-\infty}^{\infty}\int_{-\infty}^{\infty}\int_{-\infty}^{\infty} \mathrm{e}^{-\mathrm{i}\vec{k}\cdot\vec{\xi}} R_{j\ell}(\vec{\xi})d\vec{\xi}, \tag{A12}$$

where $R_{j\ell}$ is the autocovariance function corresponding to $\Phi_{j\ell}$ and where $\mathrm{i} = \sqrt{-1}$ in this expression. We will not discuss these various covariance functions in much detail here; the reader is referred to *Tatarski* (1961, 1971); *Batchelor* (1953) or *Lumley and Panofsky* (1964) for more elaborate discussions.

For cases of isotropic turbulence, we can integrate Φ_{ij} around a shell of radius k, to give (e.g., *Batchelor*, 1953, p. 35)

$$\Psi_{ij}(k) = \oiint \Phi_{ij}(\vec{k})k^2 d\Omega_k. \tag{A13}$$

For homogeneous, isotropic turbulence, we therefore have

$$\Psi_{ij}(k) = 4\pi k^2 \Phi_{ij}(\vec{k}). \tag{A14}$$

$E(k)$ relates to the Ψ_{ij} (and hence to the Φ_{ij}) via the relation

$$E(k) = \frac{1}{2}\left(\Psi_{11}(k) + \Psi_{22}(k) + \Psi_{33}(k)\right). \tag{A15}$$

Note the factor $\frac{1}{2}$; this is introduced so that the integral over all k (i.e. from $k = 0$ to $k = \infty$) gives the kinetic energy per unit mass, $\frac{1}{2}\overline{v_{\mathrm{tot}}^2}$. $E(k)$ is unique in this regard – other spectra have normalizations which do not involve this factor of $\frac{1}{2}$. For example,

$$\int_0^\infty \Psi_{11}(k)dk = \overline{u_1^2}, \tag{A16}$$

where u_1 refers to the velocity component in the x direction.

Sometimes (A15) is also written as

$$E(k) = \frac{1}{2} \oiint_{k_i k_i = k^2} \Phi_{ii}(\vec{k}) k^2 d\Omega_k,$$

where the subscript ii means sum the three terms Φ_{11}, Φ_{22}, and Φ_{33} (e.g., *Lumley and Panofsky*, 1964, p. 28). Likewise $k_i k_i = k^2$ means $k_1^2 + k_2^2 + k_3^2 = k^2$, so the integral is over the surface of a sphere of radius k.

The above spectra are useful from a conceptual viewpoint, but are often hard to determine experimentally, since they require a full three-dimensional description of the turbulent field in all three velocity components. That is, they require knowledge of all three velocity components at all points in space. This is often difficult (if not impossible) to measure.

Therefore, we also look for spectral analogs to the structure functions which were described earlier for a one-dimensional pass through the turbulent field.

To begin, if we have a detector which moves in a straight line through a patch of turbulence, and it records the velocity components parallel to the direction of motion (in analogy to the process described in connection with Equations (A1) to (A3)), and then we Fourier transform the resultant spatial series, we obtain (for Kolmogoroff turbulence) the function

$$\Theta_{11}(k_1, 0, 0) = \alpha'_{11} \varepsilon^{2/3} k^{-5/3}, \tag{A17}$$

where $\alpha'_{11} = \frac{9}{55}\alpha = 0.1244C$. This is in fact a one-dimensional function, which we will denote as ϕ_p, viz.

$$\phi_p(k_1) = \alpha'_{11} \varepsilon^{2/3} k_1^{-5/3}. \tag{A18}$$

It is important to note that this is *not* the same as $\Phi_{11}(k_1, 0, 0)$. Whilst both refer to spectral densities along the x axis, $\Phi_{11}(k_1, 0, 0)$ refers to spectral densities due *only* to "waves" with the phase-fronts aligned perpendicular to the x axis. On the other hand, $\Theta_{11}(k_1, 0, 0)$ refers to the spectral density at wavenumber k_1 due to contributions of "waves" of *all* orientations which cross the x-axis. These concepts are fundamentally different. In fact,

$$\Theta_{ij}(k_1, 0, 0) = \int_{k_2}\int_{k_3} \Phi_{ij}(k_1, k_2, k_3) dk_2 dk_3. \tag{A19}$$

Ottersten (1969a) has also emphasized the differences between these spectral forms, and has highlighted the fact that many experimentalists confuse the different forms, leading to serious misinterpretations of experimental data through the use of inappropriate constants.

Likewise, if we find the spectrum for the velocity components perpendicular to the direction of motion during this traverse, we produce

$$\phi_t(k_1) = \Theta_{22}(k_1, 0, 0) = \alpha'_{22}\varepsilon^{2/3}k_1^{-5/3}, \tag{A20}$$

where $\alpha'_{22} = \frac{4}{3}\alpha'_{11}$.

Additionally, for the choice of $C = 2.0$ described above, we have

$$3\phi_p(k_1) = \Theta_{11}(k_1, 0, 0) = 0.25\varepsilon^{2/3}k_1^{-5/3} \qquad -\infty < k_1 < \infty, \tag{A21}$$

$$\phi_t(k_1) = \Theta_{22}(k_1, 0, 0) = 0.33\varepsilon^{2/3}k_1^{-5/3} \qquad -\infty < k_1 < \infty. \tag{A22}$$

In the case of isotropic turbulence, there is no preferred axis, so that these formulas are not restricted to any particular axis.

Because of the obvious symmetry, many experimentalists often "fold" their negative spectral densities over onto their positive ones. Then we obtain the following functions:

$$\phi'_p(k_\alpha) = 0.50\varepsilon^{2/3}k_\alpha^{-5/3} \qquad 0 < k_\alpha < \infty, \tag{A23}$$

$$\phi'_t(k_\alpha) = 0.67\varepsilon^{2/3}k_\alpha^{-5/3} \qquad 0 < k_\alpha < \infty, \tag{A24}$$

where k_α are wavenumbers along the direction of travel of the probe.

Note that Equations (A21) to (A24) have "$k^{-5/3}$" laws, but so does (A9). However, these equations are conceptually different; (A9) represents an integration over a shell of radius k in three-dimensional k-space, whilst (A21) to (A24) represent spectra determined by a probe moving in a straight line through the turbulence. Nevertheless, it is a common mistake for novice researchers to confuse the two spectra, when they speak of the $k^{-5/3}$ law, which can lead to the propagation of considerable confusion. It is important to conceptually distinguish these spectra.

A.4 Scalar structure functions and spectra

In some studies of turbulence, it is not information about the velocity fluctuations which is sought, but rather density fluctuations associated with certain tracers. One must be careful to choose a good tracer – certainly quantities which react chemically with their surrounds will not obey the following equations (e.g., *Hocking*, 1985).

The structure function is described as

$$D_\zeta(r) = \overline{|\zeta(\vec{x} + \vec{r}) - \zeta(\vec{x})|^2}, \tag{A25}$$

where ζ represents the scalar concentration. For Kolmogoroff inertial range turbulence this is given by

$$D_\zeta(r) = C_\zeta^2 r^{2/3}. \tag{A26}$$

The first important spectral form is $\Phi_\zeta(\vec{k})$, which is the full three-dimensional spectral density function. For Kolmogoroff turbulence, it is given by

$$\Phi_\zeta(\vec{k}) = 0.033C_\zeta|\vec{k}|^{-11/3} \tag{A27}$$

in the inertial range. This function has been chosen to be normalized so that

$$\int_{-\infty}^{\infty}\int_{-\infty}^{\infty}\int_{-\infty}^{\infty} \Phi_\zeta(\vec{k})d\vec{k} = \overline{(\zeta')^2}. \tag{A28}$$

As shown in Chapter 3, Equations (3.292) and (3.293), for the special case of ζ being the refractive index n, the reflectivities can be determined from (A27) to be

$$\sigma_s = 0.00655\pi^{4/3}\lambda^{-1/3} = 0.03014\, C_n^2\, \lambda^{-1/3}$$

and

$$\eta_s = 0.3787\, C_n^2\, \lambda^{-1/3}, \qquad \text{often rounded as} \qquad 0.38\, C_n^2\, \lambda^{-1/3}. \tag{A29}$$

Then for the locally isotropic, homogeneous case we have

$$E_\zeta(k) = 4\pi k^2 \Phi_\zeta(\vec{k}). \tag{A30}$$

Note that E is only a function of the *magnitude* of $|\vec{k}|$. Then

$$E_\zeta(k) = 0.132\pi C_\zeta^2 k^{-5/3} = 0.415 C_\zeta^2 k^{-5/3}, \tag{A31}$$

where $k = |\vec{k}|$.

Finally, we present the spectrum seen if we record along a straight line. This is the spectrum which a probe moving through a patch of turbulence would measure, and is very similar to ϕ_p from Equation (A18) in the section on velocity spectra. This is given by

$$S_\zeta(k_1) = \int_{-\infty}^{\infty}\int_{-\infty}^{\infty} \Phi_\zeta(\vec{k})dk_2dk_3, \tag{A32}$$

which, for the case of Kolmogoroff turbulence, becomes

$$S_\zeta(k) = 0.125 C_\zeta^2 k^{-5/3} \qquad -\infty < k < \infty. \tag{A33}$$

If we fold negative wavenumbers onto positive, we obtain

$$S'_\zeta(k) = 0.25 C_\zeta^2 k^{-5/3} \qquad 0 < k < \infty. \tag{A34}$$

Again (as for the velocity spectra), note that (A31) and (A34) both involve a $k^{-5/3}$ law, but the spectra are conceptually different.

A.5 C_n^2 and ε

The kinetic energy dissipation rate is related to the potential refractive index structure constant by

$$\bar{\varepsilon} = \left(\gamma \overline{C_n{}^2} \frac{\omega_B^2}{F_t{}^{1/3}} M_n^{-2}\right)^{3/2}, \tag{A35}$$

where ω_B is the Brunt–Väisälä frequency. The parameter F_t represents the fraction of the radar volume which is filled by turbulence, while γ is discussed in more detail in the main body of the text.

(Special note: In the paragraphs that follow, we will use Θ_T to represent potential temperature, to avoid conflict with the use of Θ as a spectral function as used above. However, this will only be done in this appendix: throughout the rest of the book, potential temperature is represented by Θ.)

The "potential refractive index gradient" is given in the troposphere and stratosphere by

$$M_n = -77.6 \times 10^{-6}\frac{p}{T}\left(\frac{\partial \ln \Theta_T}{\partial z}\right) \times \left[1 + \frac{15500 q_{wp}}{T}\left(1 - \frac{1}{2}\frac{\partial \ln q_{wp}/\partial z}{\partial \ln \Theta_T/\partial z}\right)\right], \quad \text{(A36)}$$

where z is height, Θ_T is the potential temperature, q_{wp} is the specific humidity, T is the absolute temperature and p is the atmospheric pressure in millibars. The term in square brackets was denoted as χ by *Van Zandt et al.* (1978); indeed this particular form of the equation was first introduced by these authors (note that the term $\frac{1}{2}$ is actually $\frac{7800}{15500} = 0.503$, which is close to 1/2 anyway). Note that χ tends to 1 as the humidity terms tend to zero.

In the ionosphere, where humidity and temperature are no longer important relative to the electron density in producing electromagnetic scatter, we represent the potential refractive index gradient as M_e (to emphasize the dependence of electron density), and

$$M_e = \frac{\partial n}{\partial N}\left[\frac{N}{\Theta_T}\frac{d\Theta_T}{dz} - \frac{dN}{dz} + \frac{N}{\rho}\cdot\frac{d\rho}{dz}\right], \quad \text{(A37)}$$

where again we have used the symbol Θ_T for potential temperature and N is the electron density. The term ρ is the neutral density. The function $\frac{\partial n}{\partial N}$ needs to be determined from electro-ionic theory (e.g., *Sen and Wyller*, 1960; *Budden*, 1965; *Hocking and Vincent*, 1982a). It was also presented in Equations (3.125) or (3.128).

A.6 Understanding M_n

The refractive index of air can be written, based on Equation (3.287) as

$$n = 1 + 77.6 \times 10^{-6}\frac{p}{T} + 3.73 \times 10^{-1}\frac{e_{wp}}{T^2} + f_e(\rho_e, B, \nu_{ec}), \quad \text{(A38)}$$

where we have added a term f_e which depends on the electron density (ρ_e), the electron collision frequency ν_{ec} and the magnetic B-field (B). This is a complicated term, but can be deduced from Equations (3.125) or (3.128) in Chapter 3.

The conversion to Equations (A36) and A(37) above is not as trivial as it may seem. It might seem that we simply need to find the derivative of n as a function of height. This is not so.

In order to determine M_n, it is necessary to look at the *difference* between the refractive index of a parcel of air displaced from an equilibrium at z_0 to a new point $z_0 + \delta z$

and the refractive index of the environment (background) at a height δz above the reference point z_0. It is this difference which defines the changes in refractive index that a radiowave would detect as it encounters the parcel, and which will be responsible for the backscatter.

So the quantity we need is

$$M_n = \left[\left(\frac{\partial n}{\partial p} \frac{\partial p}{\partial z} \right)_{\text{parcel}} - \left(\frac{\partial n}{\partial p} \frac{\partial p}{\partial z} \right)_{\text{env}} \right] + \left[\left(\frac{\partial n}{\partial \Theta_T} \frac{\partial \Theta_T}{\partial z} \right)_{\text{parcel}} - \left(\frac{\partial n}{\partial \Theta_T} \frac{\partial \Theta_T}{\partial z} \right)_{\text{env}} \right]$$
$$+ \left[\left(\frac{\partial n}{\partial e_{wp}} \frac{\partial e_{wp}}{\partial z} \right)_{\text{parcel}} - \left(\frac{\partial n}{\partial e_{wp}} \frac{\partial e_{wp}}{\partial z} \right)_{\text{env}} \right] + \left[\left(\frac{\partial n}{\partial \rho_e} \frac{\partial \rho_e}{\partial z} \right)_{\text{parcel}} - \left(\frac{\partial n}{\partial \rho_e} \frac{\partial \rho_e}{\partial z} \right)_{\text{env}} \right], \tag{A39}$$

where we use the potential temperature Θ_T rather than T. Since the refractive index for electron scatter also depends on collision frequencies and magnetic fields, we should have terms involving these, but they are usually slowly varying with height and generally ignored.

We then recognize that as a parcel moves up or down, its pressure relatively rapidly adjusts to the environmental pressure (this adjustment is the basis of the adiabatic lapse rate, as derived in Chapters 1 and 12), so the terms involving $\frac{\partial p}{\partial z}$ cancel out entirely.

Futhermore, the parcel obeys the adiabatic law, so this means potential temperature is conserved in the parcel. Hence $\left(\frac{\partial \Theta_T}{\partial z} \right)_{\text{parcel}}$ is zero.

We are therefore left with

$$M_n = - \left(\frac{\partial n}{\partial \Theta_T} \frac{\partial \Theta_T}{\partial z} \right)_{\text{env}} + \left[\left(\frac{\partial n}{\partial e_{wp}} \frac{\partial e_{wp}}{\partial z} \right)_{\text{parcel}} - \left(\frac{\partial n}{\partial e_{wp}} \frac{\partial e_{wp}}{\partial z} \right)_{\text{env}} \right]$$
$$+ \left[\left(\frac{\partial n}{\partial \rho_e} \frac{\partial \rho_e}{\partial z} \right)_{\text{parcel}} - \left(\frac{\partial n}{\partial \rho_e} \frac{\partial \rho_e}{\partial z} \right)_{\text{env}} \right]. \tag{A40}$$

In order to deal with e_{wp} and ρ_e, we assume that the *mixing ratio* of the water vapor and the electron density are conserved during parcel displacement. The case of water vapor is discussed in *Tatarski* (1961) p. 57, and the case of the electron density is discussed in *Hocking* (1985), Equations (14) to (21).

We will not repeat these derivations here, but it is through this procedure that Equations (A36) and (A37) are produced.

Appendix B Gain and effective area for a circular aperture

B.1 Introduction

Here, we consider the following question:

Consider a two-dimensional filled circular aperture of the form

$$f(r) = C \quad \forall \mid r \mid < a, \tag{B1}$$

which represents a filled continuous disk of radiators with radiated electric field amplitude C per unit area. We may also consider this as the transmission function for a discrete set of individual radiators, as long as the radiators are sufficiently less than 1 wavelength apart.

Our purpose is to derive the relation between effective area and gain for this system, for the case that the aperture is many wavelengths across (more than say 10).

B.2 Solution

To begin, we remind ourselves of some simple Fourier Theory. *Champeney* (1973), page 46, Equation (3.17) gives the Fourier transform of a general 2-dimensional function $f(\vec{r})$ as

$$\underline{F}_v(\vec{k}) = \int_{-\infty}^{\infty}\int_{-\infty}^{\infty} e^{-i(\vec{k}\cdot\vec{r})} f(\vec{r}) d\vec{r}, \tag{B2}$$

where $\underline{F}_v$ is the most general function in 2-D reciprocal space. For the special case of a circularly symmetric 2-D function f we may write

$$F(k) = \left[\int_0^{\infty} f(r) J_0(kr)\, 2\pi r\, dr\right], \tag{B3}$$

where F is purely real and a function only of the magnitude of $\vec{k}$, i.e. it is radially symmetric.

In the above equations, $\vec{r}$ is the vector in the x-y plane from the origin to the radiator under examination, r is the magnitude of $\vec{r}$, $\vec{k}$ is the wave vector in the direction of interest, and k is its magnitude.

Further expansion for our special case gives (*Champeney*, 1973, page 48, Equation (3.27))

$$F(k) = 2\pi Ca^2 \frac{J_1(ka)}{ka}. \tag{B4}$$

The power radiated is the Poynting vector, which is proportional to $\underline{E}^*\underline{E}$, and the radiated electric field $\underline{E}$ is proportional to F. Some adaptations for the special case of diffraction theory are needed. In diffraction theory, we use $k = \frac{2\pi}{\lambda}$ and $\nu = \frac{\sin\theta}{\lambda}$, θ being the angle from the bore-sight direction. Extra inverse-wavelength effects are needed, as seen in Chapter 3, Figure 3.4. Then following *Champeney* (1973), Equation (11.78), the polar diagram of the radiated power is

$$\mathcal{P}_{Tx}(k) = \alpha k^2 a^4 \left[\frac{J_1(2\pi a\nu)}{2\pi a\nu}\right]^2, \tag{B5}$$

where α is a constant. This equation is a slight re-statement of Equation (11.78), page 154 of *Champeney* (1973), in that we have replaced θ in Champeney's version by $\sin\theta = \lambda\nu$, since the natural unit for diffraction theory is ν. Our expression is the exact one – the one shown by Champeney is an approximation.

We will write that the form of $\mathcal{P}_{Tx}$ as a function of polar coordinates is $P_{Tx}(\theta, \phi)$, which is numerically equal to $\mathcal{P}_{Tx}$ but requires a different representation since it is a function of different variables. Then by definition the gain is

$$G = \frac{P_{Tx}(0,0)}{\frac{1}{4\pi}\int_{\phi=0}^{2\pi}\int_{\theta=0}^{\pi} P_{Tx}(\theta,\phi)\sin\theta\, d\theta\, d\phi}, \tag{B6}$$

where we assume a single wavelength λ.

First, we need to evaluate the term $P_{Tx}(0,0)$, which will be the numerator in the above equation.

To do this, use the identity (available in any reference about Bessel functions, including Wikipedia)

$$J_\nu(x) = \left(\frac{x}{2}\right)^\nu \sum_{k=0}^{\infty} \left(\frac{-x^2}{4}\right)^k \frac{1}{k!\Gamma(\nu+k+1)}, \tag{B7}$$

where $\Gamma(n) = (n-1)!$. Then

$$J_1(x) = \left(\frac{x}{2}\right) \sum_{k=0}^{\infty} \left(\frac{-x^2}{4}\right)^k \frac{1}{k!(k+1)!}. \tag{B8}$$

Taking only the first two terms, since we are only interested in the value at zero for now,

$$J_1(x) = \frac{x}{2} - \frac{x^3}{16} + \cdots \tag{B9}$$

For evaluation at zero, we in fact only need the first term. Substituting into (B5) for $x = 2\pi a\nu$ being small gives

$$P_{Tx}(0,0) = \alpha k^2 a^4 \left[\frac{(2\pi a\nu)/2}{2\pi a\nu}\right]^2 = \alpha \frac{1}{4} k^2 a^4. \tag{B10}$$

The gain is then (from (B6))

$$G = \frac{\alpha \frac{1}{4} k^2 a^4}{\frac{1}{4\pi} \int_{\phi=0}^{2\pi} \int_{\theta=0}^{\pi/2} P_{Tx} \sin\theta d\theta d\phi}, \tag{B11}$$

or for our circularly symmetric case, with the substitution $P_{Tx} = \alpha P_{Tt}$, and recognizing that P_{Tt} is independent of ϕ, so that $\int_{\phi=0}^{2\pi} d\phi = 2\pi$, we have

$$G = \frac{\frac{1}{4} k^2 a^4}{\frac{1}{2} \int_{\theta=0}^{\pi/2} P_{Tt} \sin\theta d\theta}. \tag{B12}$$

Note also that we assume that the polar diagram radiates in the forward direction only, and so have changed the limits of the θ integation to be from 0 to $\frac{\pi}{2}$. The term P_{Tt} is given by (from (B5) and the definition of P_{Tt} in terms of P_{Tx})

$$P_{Tt} = k^2 a^4 \left[\frac{J_1(2\pi a v)}{2\pi a v} \right]^2, \tag{B13}$$

i.e. it is just P_{Tx} divided by α.

We now turn to evaluation of the denominator of (B12), so we need to find

$$I = \int_{\theta=0}^{\pi/2} P_{Tt} \sin\theta d\theta. \tag{B14}$$

To evaluate this, begin by substituting

$$\xi = \sin\theta.$$

Then $d\xi = \cos\theta d\theta = \sqrt{1-\xi^2} d\theta$, or $d\theta = d\xi / (\sqrt{1-\xi^2})$. So now I becomes

$$I = \int_{\xi=0}^{1} k^2 a^4 \frac{J_1^2(\frac{2\pi a}{\lambda}\xi)}{\left(\frac{2\pi a}{\lambda}\xi\right)^2} \frac{\xi}{\sqrt{1-\xi^2}} d\xi. \tag{B15}$$

Writing the term $\frac{2\pi a}{\lambda}$ in the denominator of the first fraction as ka, and partially cancelling with the term $k^2 a^4$ and moving the resultant outside the integral gives

$$I = a^2 \int_{\xi=0}^{1} \frac{J_1^2(\frac{2\pi a}{\lambda}\xi)}{\xi^2} \frac{\xi}{\sqrt{1-\xi^2}} d\xi. \tag{B16}$$

Surprisingly, this rather messy expression has an exact analytical result. The solution can be found in *Prudnikov et al.* (1990). Equation (6) on page 212 of that reference gives

$$\int_0^b \frac{1}{x\sqrt{b^2 - x^2}} J_1^2(cx) dx = \frac{1}{2b} - \frac{1}{2b^2 c} J_1(2bc) \tag{B17}$$

for $b > 0$. In our case, we set $x \equiv \xi$, $b = 1$, and $c = \frac{2\pi a}{\lambda}$. Hence

$$I = a^2 \left[\frac{1}{2} - \frac{1}{2\frac{2\pi a}{\lambda}} J_1\left(2\frac{2\pi a}{\lambda}\right) \right], \tag{B18}$$

or

$$I = a^2 \left[\frac{1}{2} - \frac{\lambda}{4\pi a} J_1 \left(\frac{4\pi a}{\lambda} \right) \right]. \tag{B19}$$

For large arrays, the radius is many wavelengths, so $J_1(\frac{4\pi a}{\lambda})$ tends to zero, leaving

$$I = \frac{1}{2} a^2. \tag{B20}$$

Then from Equations (B12) and (B14),

$$G = \frac{\frac{1}{4} k^2 a^4}{\frac{1}{2} I} = \frac{\frac{1}{4} k^2 a^4}{\frac{1}{2} \frac{a^2}{2}} = \left(\frac{2\pi}{\lambda} \right)^2 a^2. \tag{B21}$$

This can be rewritten as

$$G = \frac{4\pi}{\lambda^2} (\pi a^2). \tag{B22}$$

The term πa^2 is the area of the radar dish, which we will denote as A, so the result is that

$$G = \frac{4\pi A}{\lambda^2}, \tag{B23}$$

as required.

List of symbols used

The following pages summarize many of the variables used in this text. Because the book covers the fields of electromagnetism, radar engineering, signal processing, fluid dynamics, meteorology, and optics, to name a few, it requires well over 400 variables to cover all the mathematics needed. However there are only about 80 distinctively recognizable variables available in the Roman and Greek alphabets, so it is impossible to avoid some degree of duplication in our use of variables. We have partly compensated for this by extensive use of subscripts, but at times the same symbols are used for different applications. If this happens, we avoid doing it in any one problem or solution, and generally avoid duplication in any one chapter.

While certain symbols are well known to represent certain quantities in particular fields, it is necessary in this text to sometimes use unconventional forms. For example, while k often represents Boltzmann's constant in thermodynamics, here k is already used for wavenumber, so we often use K_B instead. Likewise e is commonly used for the charge of an electron in electromagnetism, but e is common in meteorology for water-vapor pressure in air. We choose to leave e for electronic charge, but use e_{wp} for water-vapor pressure.

Note that the reader should look carefully at the font for each variable. Different fonts refer to different representations. For example, F, $\mathcal{F}$, and $\digamma$ are all different. We often use hats and tildes to distinguish characters as well, e.g., $\tilde{\mathrm{A}}$, $\hat{\mathrm{A}}$, and A are all different. Often the tilde ($\tilde{A}$) is used to indicate an *estimator* while the hat ($\hat{a}$) indicates a first order perturbation. Use of a prime (as in $'$) often indicates a perturbation, but in contrast to $\hat{A}$, includes all orders.

Normal (roman) font is used for units (e.g., m, km, s, min, etc.).

Normal font is also used to represent operators (e.g., $\frac{\mathrm{d}}{\mathrm{d}t}$ is a differential operator). Note that in derivations, $d\rho$, dt, etc. are left as italics (i.e. differential elements) until they become part of an operator. For example in a derivation we might write $dp = -\rho\, g\, dz$ where dp and dz are differential elements, but then we would write $\frac{\mathrm{d}p}{\mathrm{d}z} = -\rho\, g$ as the differential equation.

In an integration, $\mathrm{d}z$ is presented in normal font, as it is considered as an operator here (e.g., $\int f(x)\mathrm{d}x$).

Variables like $\mathrm{i} = \sqrt{-1}$ and e (where the latter refers to the natural exponential function) are considered as operators, and hence are presented in normal font. In contrast, e, which has an italic font, is the magnitude of the electronic charge.

Although it adds slightly to the length of the text, we also try to redefine variables within each new application wherever it seems appropriate.

In the following list, the same variable may be mentioned several times. For some cases of duplication, we have tried to indicate the relevant chapter, but sometimes it is obvious which applies. For example, if g appears in a section involving the fluid equations of motion, it is probably the acceleration due to gravity. If g appears in a section on radar receivers, it is likely to be receiver gain. Of course the variables are also defined in the text close to their point of application, so if in doubt, check the local text.

The variables i, j, k, l, m, n, ℓ are often used as dummies of summation. Care has been taken to avoid conflict with other uses of these variables in the same equation.

There are also some other conventions that we adopt:

We choose to distinguish between real and complex numbers by underlining complex numbers. This is not to say that the distinction is absolute. It is quite conceivable that a variable is indicated as a real number, but might be considered as a complex number in some special circumstances. However, we have tried to indicate which variables are *most commonly* considered as complex and which are real most of the time.

We also point out that there are two different types of complex numbers. In one case the representation is something of a "trick," and *in the end we just want real numbers.* This is common with say representation of a propagating wave. This type of complex number representation is designed to allow us to better deal with phase variations. The other type of complex number is quite different, and the main example is representation of the receiver ouptut as real and imaginary components. In this case *both* the real (in-phase) and quadrature (imaginary) components are undeniably real, and must be treated equally. In such cases, we *cannot* simply take real parts at the finish. The same is true if we represent two-dimensional motion as an imaginary number, and also in the case of Fourier transforms. More details about the distinctions are discussed in Chapter 3, Section 3.3.1.

Here are some of the key classes of variable representation:

- underline = complex number, e.g. $\underline{s}$
- arrow over top = vector, e.g., $\vec{r}$
- underline PLUS arrow over top = complex vector, e.g., $\underline{\vec{b}}$
- $y[\,]$ = discrete series (as distinct from $y(\,)$)
- $y(\,)$ = continuous function
- $[\,]$ = matrix (normally) e.g., $[A]$
- use of bold symbols sometimes indicates a matrix (esp. Chapter 8), e.g., $\mathbf{H}$
- $[\underline{x}]$ = complex matrix
- bar over top, or $\langle\,\rangle$, = average (expectation), e.g. $\bar{x}$, $\langle x\rangle$
- hat on top = linearized solution to a nonlinear DE, e.g., $\hat{p}$
- prime = perturbation terms (not necessarily linearized), e.g., u'
- on occasion we distinguish between a cartesian point $(x\ \ y\ \ z)$ and a vector $[x\ \ y\ \ z]$.

e_s = saturated water vapor pressure

e_{wp} = water vapor pressure (In meterological books, it is common to use e, but this represents a conflict with its use for electronic charge here.)

$[e]$ = weighting matrix used in Chapter 9

F

F = solar flux above the Earth's atmosphere

F_d = drag force

Sometimes F is used as a force per unit mass or force per unit volume, e.g.,

F_{vol} = force per unit volume

F_m = force per unit mass at times

At other times it is truly a force.

F_{nett} = nett force

$\mathcal{F}(\ )$: three-dimensional inertial range turbulence function, proportional to $k^{-11/3}$, Appendix A

$F(\ )$: a function – often used as the Fourier transform of f(t)

$\mathcal{F}$ = Eliassen–Palm flux

$\mathcal{F}$ = noise figure of a receiver (Chapter 5)

F_t = fraction of space filled by turbulence

F_{spec} = fraction of specular reflectors in a Fresnel scattering volume

$F_{u,w}$ = wave-spectra for horizontal and vertical velocities used in "universal gravity wave" applications. These may be functions of horizontal or vertical wavenumber, or frequency (Chapter 11).

F: used to represent a Fourier transform in Chapters 8 and 9

FOOR = "frequency of optimum response", used in Capon's method, Chapter 8

F = constant used in describing the 3-D autocovariance function in full correlation analysis in Chapter 9

f

f = frequency; often used with subscripts to indicate central, median, average, special frequencies

$f(\ )$ – function f, e.g., function of time or height

f_c = Coriolis parameter in gravity-wave and atmospheric dynamics theory (often denoted simply as f in many fluid-dynamical texts)

f_{fluct} = spectral half-power-half-width which would be measured due to turbulence alone

$f_{\frac{1}{2}nt}$ = spectral half-power-half-width due to non-turbulent effects (e.g., beam, shear broadening) alone

$f_{\frac{1}{2}\text{expt}}$ = measured (experimental) spectral half-power-half-width

G

G = antenna gain (sometimes with subscripts to differentiate antennas, e.g. G_{Tx}, G_{Rx}, etc.)

$G(\ )$: often Fourier transform of a pulse $g(t)$

G = constant used in describing the 3-D autocovariance function in full correlation analysis in Chapter 9

g

g = acceleration due to gravity of the Earth (sometimes written as a vector $\vec{g}$)
g = receiver gain
g_s = special case of receiver gain (section on calibration)
$g(\ )$: represents shape of a transmitted pulse – either the full pulse or its envelope, depending on application
$g(\ )$ – used as a general function at times

H

H = scale height of atmosphere
H = constant used in describing the 3-D autocovariance function in full correlation analysis in Chapter 9
$\underline{\vec{H}}$ = magnetic field (often complex)
$H(z)$: sometimes used as z-transform receiver filter response, e.g., Chapter 8
$\mathcal{H}(\Omega)$ = z-transform response of the filter discussed in B above, but expressed as a function of normalized frequency
$\underline{H}_\Omega$ = filter function in Capon's method. This is similar in concept to the z-transform $H(z)$ discussed above.
$\underline{\mathcal{H}}_{\Omega_s}(\mathbf{\Omega})$ = generalized filter function in Capon's method, Chapter 8. Similar in concept to the z-transform filter response discussed above.
H_{spec} = a form of scale height associated with the height variation of density and efficiency of specular reflectors in Fresnel scatter (Chapter 12)

h

$h(\ )$ – used as a function at times (e.g. $h(t)$)
$h[\ ]$ = an impulse response used in discussion of z-transform
$\mathbf{h}$ = impulse response matrix in regard to Capon's method in Chapter 8. Related conceptually to $h[\ ]$ discussed above.

I

I = current in electrical circuit
$I(t)$ = in-phase component produced by a radar receiver
$\mathcal{I}(\ )$: used for various specific functions in the text, including Chapter 3 and Appendix B
I_O = outgoing energy intensity radiated from Earth (mostly infrared)
I_A = energy intensity received by Earth after accounting for Albedo
I: one of the Stokes parameters (Chapter 10)

i

$\hat{i}$ = unit vector along x axis
$\mathrm{i} = \sqrt{-1}$: note roman normal font

J

J_ν – Bessel functions (Appendix B)

j

j = current density

K

K = Kelvin (temperature)

K = constant used in describing the 3-D autocovariance function in full correlation analysis in Chapter 9

K = obliquity factor in optical/EM transmission theory (Chapter 3)

K_B = Boltzmann's constant

K_M = turbulent momentum diffusivity (viscosity)

K_T = turbulent heat diffusivity

K_* = constant of proportionality relating the ambipolar diffusion coefficient to pressure and temperature

K_{zz} = vertical diffusion coefficient (same as K_M)

K_{xx}, K_{xy} etc. – various forms of large-scale horizontal diffusion coefficient

k

k = wavenumber of a wave = $\frac{2\pi}{\lambda}$ in electromagnetism

k = wavenumber of a gravity wave, and often the horizontal wavenumber of a gravity wave

k_h = horizontal wavenumber of a gravity wave at times

k_η = Kolmogoroff microscale wavenumber = $\frac{1}{\eta_K}$

k_B = Bragg wavenumber = $\frac{2\pi}{\lambda_B}$

k_{LB} = buoyancy wavenumber in turbulence theory

k_1 = wavenumber along x-axis in turbulence theory (Appendix A)

k_α = wavenumbers along the direction of travel of a probe passing through a turbulent region

L

L = length (either in time or space)

L_{Tx}: stands for "losses" during transmission of a radar. The term is usually expressed as a gain, e.g. if the system loses 15% of its power during transmission to cable, antenna and other losses, $L_{Tx} = 0.85$

L_{Rx}: as for L_{Tx}, but for receiver losses (including losses in antennas and cabling)

L = integer (1 or 2) to discriminate non-complementary and complementary codes (Chapter 5, calibration)

l, *ℓ*

l: used for various lengths

l = length of an antenna element in Chapter 5

ℓ: used for various lengths

ℓ = Scorer parameter

ℓ = typical size of eddy (e.g., viscous eddy, Section 7.3)

ℓ_0 = inner scale of turbulence

M

M = bits in an M-sequence (Chapter 5)
M = number of elements in a pulse code
M_n = potential refractive index gradient (see Appendix A)
$\mathcal{M}$: used to represent the mass of a parcel of air on occasion (see Chapter 12), often with subscripts to discriminate vapor, dry air, etc.
M_v = molar weight of pure water vapor

m

$\underline{m}$ = vertical wavenumber of a gravity wave
m = molecular mass of a molecule or atom, often used with subscripts to indicate species (e.g., m_s)
m_e = electron mass
$\underline{m}_{1,2}$ = roots of a quadratic describing the vertical wavenumber of gravity wave damped by viscosity (Chapter 10)

N

$N = n - 1$ "radio refractivity" (n = refractive index)
N = number density (especially electron number density)
$N_\perp$ = electron number density at point of critical reflection of a radiowave in the ionosphere
N_i = number density of molecules of species i
N = integer – typically number of points in a data-set
N_0 = Avogadro's number
N: in some texts this is the Brunt–Väisälä frequency, but ω_B is used in this text
N = number of coherent integrations

n

n = refractive index(usually of air), sometimes complex
$n_{x,o}$ = X and O characteristic modes of an electromagnetic wave passing through a plasma. Modes are elliptically polarized in the general case, though often considered as circular.
n = frequency (used on rare occasions when there exists a conflict with use of f for frequency, as in $f(n)$)
n_R = real part of refractive index (at times)
n_I = imaginary part of refractive index (at times)
n = integer (number of points in a data-set, or a counter)
n_M = number density of molecules (at times)
$\underline{n}(\,)$ – time series of noise (often complex)
n_e: used on occasion to represent electron density

O

O: used to represents points in a figure, and often the origin

P

P_K = turbulent Prandtl number = $\frac{K_M}{K_T}$
P_{pr} = molecular Prandlt number =$\frac{\kappa}{\nu}$
$P_?$ – various subscripts to describe different types of power
P_r = power at a point $\vec{r}$ (sometimes the receiver)
$\vec{P}$: polarization due to N electrons or a group of charged particles (Chapter 3). Also polarization components in x, y and z directions when used with appropriate subscripts.
$P(\ )$: Fourier transform of a pulse $p(t)$
$P(f)$: filter response of radio-filter
P_{Tx} = transmitted RF power
P_{Rx} = received RF power
P_N = noise level at times
P_{sky} = skynoise power

p

p = pressure (most common use)
p_0 = specific cases of pressure – possibly value at z_0, on occasions the mean pressure
p = momentum
p = polarization due to a single charged particle (often an electron)
$p(\ )$: often used as a radar pulse shape – could be a function of time (t) or range or height

Q

Q = heat content
$Q(t)$ = quadrature component produced by a radar receiver
Q: one of the Stokes parameters (Chapter 10)

q

q_e = electron charge *including* sign
q_{wp} – specific humidity
q_{mr} – mixing ratio of water vapor in air

R

R = resistance
R_r = antenna resistance
R (and various subscripted versions) = ideal gas constant when mass is expressed in moles
R = range, and often a specific range. Often used with various subscripts, e.g. $R_{\min}$, etc.
R: used in some cases as the Fourier transform of the reflection coefficient profile $r(z)$
$\mathcal{R}$ = reflection coefficient of a reflecting layer
$\mathcal{R}$: sometimes reflection coefficient from a load in a circuit, as used to determine the VSWR
R_* = generalized gas constant for the atmosphere when using density
R_v – gas constant for water vapor
$\underline{R}$ = ratio of polarizations in an elliptically polarized wave (Chapter 3)

R_i = Richardson number (gradient Richardson number)
R_f = flux Richardson number
R_e = Reynolds' number
R_E = radius of the Earth
$R(\ )$: is used as a (possibly complex) autocovariance function, especially in Chapters 8 and 9
R_{jk} = various cross-correlation functions used in turbulence studies in Appendix A
$\underline{R}$: used to represent sum of multiple vectors in discussion of Rayleigh distribution
R_1 = ratio of effective tilt angle of beam to bore direction of beam after consideration of impact of scatterer anisotropy (Chapter 2)
R_2 = ratio of effective half-power-half-width of beam to original half-power-half-width of beam after consideration of impact of scatterer anisotropy (Chapter 2)
$\mathbf{R}_y$ = autocovariance matrix (Chapter 8)

r

r = range
r_{wp} = relative humidity of water
$r(\ )$ – reflection coefficient profile as a function of height z
r_{yy}, r_{xy}: used to refer to autocovariance and covariance functions in Chapter 8
r = relative density perturbation $\rho'/<\rho>$ in Chapter 11
r_e = classical electron radius
r_σ = equivalent radius of a single molecule

S

S_{in} = incident wave-energy flux (Poynting vector)
S_r = Poynting vector at the receiver due to electron scatter
S_{Rx} = Poynting vector at the receiver due to N electrons (Chapter 3)
$S(\ )$ = one dimensional turbulent spectrum (Appendix A)
$S(\ )$: often a Fourier transform of a time-series $s(t)$. Also used as output of a convolution in Chapter 4.
$S(\ )$: often a power spectrum, especially in Chapter 8 (S_C, S_P)
S_c = Schmidt number
S = area of a unit antenna-element cell in an array (Chapter 5)
$S_{0,1,2,3}$ = Stokes parameters in polarization studies (Chapter 10)

s

$\underline{s}(\ )$: time series, e.g. $\underline{s}$(t) – often complex, e.g. $I(t) + iQ(t)$
s = Laplace transform variable (Chapter 8)

T

T = period of a wave
T = a length of time
T = temperature
T^* = virtual temperature
T_e = effective temperature of Earth as seen from space

T_B = Brunt–Väisälä period (τ_B is also used at times)
T_s = increment of time, often sampling interval (Chapter 8)
T_s = inter-pulse period (Chapter 5)
$\mathcal{T}_a$: used as atmospheric temperature in Chapter 10 to avoid conflict with temperature
T_{intr} = intrinsic period of a gravity wave
T_c = cutoff period at which gravity-waves can be considered to be heavily damped in the atmosphere (and especially the ionosphere)

t

t is generally the time variable.
t' – often used for retarded time in EM ($= [t - \frac{r}{c}]$)
t_d: used as an eddy decay time

U

U = internal energy of a gas
U: one of the Stokes' parameters (Chapter 10)

u

u = zonal component of wind
u: occasionally used to represent a general velocity
$\vec{u} = u\hat{i} + v\hat{j} + w\hat{k}$ = wind vector
u^* = u-component (zonal) in transformed Eulerian mean (TEM) theory. The * does *not* represent a complex conjugate.
$u(\,)$ = unitary step function (Chapter 8)

V

V = volume
V_s = scattering volume in EM scatter theory (Chapter 3)
V = voltage (often complex)
$\underline{V}_I$ = visibility index (Chapter 9 only)
V_x, V_y = x and y *trace* velocity components in Chapter 9
V: one of the Stokes parameters (Chapter 10)
$VSWR$ = voltage standing wave ratio

v

v = meridional wind component
v_{rad}, v_r: radial velocity (various subscripts used)
v: occasionally used to represent a general velocity
v^* = v-component (meridional) in transformed Eulerian mean (TEM) theory. The * does *not* represent a complex conjugate.
v_{horiz} = horizontal component of wind velocity in azimuthal direction of radar beam
v_{RMS} = root-mean-square radial velocity fluctuations due to turbulence
v_a = radial velocity corresponding to the Nyquist frequency in Chapter 8
v_T = trace velocity in Chapter 9

W

W = amplitude of oscillation, e.g., solution to the Taylor–Goldstein equation
W = power, often with subscripts, e.g. W_t for transmitter power, etc.
W_{Rx} = receiver power
W_N = "fiddle factor" in Chapter 8 (used to explain the FFT)
$W(\Omega)$ = Fourier transform of a window function (Chaper 8)
$\underline{W}, \underline{W}_a, \underline{W}_r$ = forms of polar diagram and radial weighting functions (Chapter 9)

w

w = vertical wind component
w^* = w-component (vertical) in transformed Eulerian mean (TEM) theory. The * does *not* represent a complex conjugate.
w_{mn} = weighting factor for antenna elements in an array (Chapter 5)
$w(\,)$ = window function
$[w]$ = weighting matrix used in Chapter 9

X

X = square of ratio of plasma frequency ω_N to radio frequency ω in Appleton–Hartree and Sen–Wyller equations
$\vec{X}$: used as a point-location at times
X = length scale on occasions
$X(\omega)$: used as a Fourier transform function in Chapter 8
X = imaginary (reactive) part of an impedance Z

x

x: generally x coordinate: often eastward
$\vec{x}$ – often used as a displacement vector in a 1D situation
$x(\,)$ – a function, or series of points, with an argument that is time or space

Y

Y = ratio of gyrofrequency Ω to radio frequency (ω) in Appleton–Hartree and Sen–Wyller equations
Y_L = component of Y for propagation parallel to the magnetic field
Y_T = component of Y for propagation perpendicular to the magnetic field
$Y(\omega)$: used as a Fourier transform function in Chapter 8

y

y: generally y coordinate: often northward
$y(\,)$: often used as a function of time or space

Z

Z = ratio of electron collision frequency to radiowave angular frequency in Appleton–Hartree and Sen–Wyller equations
$\underline{Z}$ = impedance (often with various subscripts e.g. Z_o, Z_L, etc.)

z

z = height (generally)
z = z-variable in the z-transform (Chapter 8)

α

α used as various constants and coefficients, especially in turbulence (e.g., Appendix A)
α = inverse density of air (especially Chapter 12)
α: occasionally used as a tilt angle for the beam (e.g., Figure 7.10)
α = frequency sweep rate in FMCW radar (Chapter 5)

β

β = negative of power law exponent of sky-noise dependence on frequency
β = constant of proportionality used in differential phase experiment (Chapter 10)
$\beta_{O,X}$ = proportionality terms used in the differential phase experiment for O and X modes respectively (Chapter 10)

Γ

Γ = various temperature lapse rates
Γ_a = adiabatic lapse rate
Γ_d = dry adiabatic lapse rate
Γ_s = moist adiabatic lapse rate
Γ_e = environmental lapse rate
Γ: used for gamma function in Appendix B

γ

γ – used as a constant in turbulence theory
γ = ratio of potential and kinetic energy in turbulence theory
γ = ratio of specific heats (C_p/C_v)

Δ

Δ: often a small increment of distance or time, e.g. Δx or Δt
Δ = quantization error in digitization (Chapter 8)
$\underline{\Delta}_A$ = the difference of the polar diagrams of an array with introduced phase errors compared to an ideal array
Δ_ℓ = errors in phases of individual antennas in an array (Chapter 5)
Δ_s = switching time from transmission to reception (Chapter 5)

δ

δ: often used to indicate a small increment of distance or time, e.g. δt or δx
δ: sometimes just a vector difference (especially Chapter 9, where it represents displacements between antennas)
δ = random phase error; phase or phase increment in an array (Chapter 5)
$\delta(\,)$: Dirac delta function
δ_{ij}: Kronecker delta function

ϵ

ϵ = permittivity of a medium
ϵ' = relative permittivity (dielectric constant) – also called κ_e
ϵ_0 = permittivity of free space
ϵ = gas constant divided by mass specific heat = R/c_p
ϵ = random amplitude error (Chapter 5)

ε

ε = turbulent kinetic energy dissipation rate
ε_K = turbulent kinetic energy dissipation rate
ε_P = turbulent potential energy dissipation rate

ζ

ζ = reciprocal-space variable in Fourier theory, viz., 1/ λ
ζ: often used as a length scale in various short-lived contexts, including Figure 7.10
$\zeta()$ – used a function at times

η

η = impedance of free space
η_s = scatter *reflectivity* for dielectric perturbations (especially in Chapter 3). Note that η_s has units of m^{-1}. It is the total backscattered energy radiated into a full sphere (assuming isotropy) per unit volume per unit incident Poynting flux. It is equal to $4\pi\sigma_s$. This is the reflectivity advocated by Ottersten.
η_K = Kolmogoroff microscale
η_m number of moles (Chapters 1 and 12 only)
η_{ij} = y-spacing of antennas in spaced antenna analysis (Chapter 9)

Θ

Θ = potential temperature
Θ_T = potential temperature in Appendix A – used to avoid confusion with the spectral usage of Θ in the appendix
Θ – also used for representation of turbulent spectra in Appendix A, following Batchelor's notation. Often used with subscripts to indicate cross-spectral terms.
Θ_{tt} = transverse turbulence spectral function used in determination of ν_{RMS} for turbulent scatter theory (Chapter 7)

θ

θ = angle from zenith (generally) in radar applications
θ = latitude of a point on the Earth (or rarely, co-latitude)
$\theta_{\frac{1}{2}}$ = half-power half-width 1-way of radar beam
$\theta_{\frac{1}{2}(2)}$ = half-power half-width 2-way of radar beam
θ_h = half-power full width of radar beam
θ_{rms} – standard deviation of one-way beam width
$\theta_{rms2way}$ – standard deviation of two-way beam width

θ_s = anisotropy parameter used in scatter theory to describe degree of anisotropy of the scatterers
$\theta_0 = e^{-1}$ half-width of radar beam
θ_T = zenithal tilt of bore direction of beam from vertical
θ_{eff} = effective tilt of radar beam after compensation for scatterer anisitropy
θ_i = angle of incidence of an EM wave on a refracting surface (especially used in Snell's law)

κ

κ = molecular heat diffusion coefficient
κ_e = relative permittivity in EM (dielectric constant) (Chapter 3) (also denoted ϵ')
κ_m = relative permeability in electromagnetism
κ_{sp} = "spring constant" for bound electrons (Chapter 3)
κ_T – molecular heat diffusivity when there exists a conflict with using κ
κ_ℓ = constant used in relating the inner scale to Kolmogoroff microscale (Chapter 10)
$\kappa_{O,X}$ = absorption terms used in the differential absorption experiment for O and X modes respectively (Chapter 10)

Λ
λ

λ = wavelength (often with subscripts for different types (central, Doppler shifted, etc.). Used for various waves including EM radiation and gravity waves.
λ_v: sometimes used to represent wavelength in a vacuum
$\lambda_{o,x}$ = wavelengths of O and X modes in propagation through plasmas
λ_{grd} = horizontal wavelength of a gravity wave as viewed from the ground (the wavelength is the same independent of the reference frame, but the subscript is added for emphasis)

μ

μ = magnetic permeability
μ_0 = magnetic permeability of free space
μ: sometimes used as a mean (often with suitable subscripts)

ν

ν = molecular viscosity = molecular momentum diffusion coefficient
ν_{ec} = electron collision frequency with neutrals
ν_{eff} = modified electron collision frequency for application with the modified Appleton–Hartree equation
ν = coordinate used for reciprocal space in diffraction theory (generally = $\frac{\sin\theta}{\lambda}$, but on occasion just $\sin\theta$, θ being the angle from the direction perpendicular to the diffracting plane)
$\nu_{x,y}$ = direction cosines in diffraction theory divided by wavelength
ν = number of species of a gas on a few occasions
ν= beam width in one special simulation in Chapter 7 (Figure 7.10)
$\nu_{+\text{ion}}$ = collision rate of a typical positive ion with the neutrals

Ξ

Ξ: used on occasion as a subscript to indicate a special distance, e.g., d_Ξ in calculations of reflection coefficient from a step

ξ

ξ: generally used as either a distance, or a dummy of integration or summation
ξ_{ij} = x-spacing of antennas in spaced-antenna theory (Chapter 9)
ξ: used in Chapter 3 to represent various scalars and vectors for temporary use
$\xi(\,)$: used to represent some general Fourier components in Chapter 3

Π

Π: usually used to represent multiplication of successive terms

π

π = ratio of the circumference of a circle to its diameter in Euclidean geometry.

ρ

ρ = density (often atmospheric density)
ρ_0: used to represent a specific density, e.g., at height z_0, or sometimes a mean value
ρ_d = density of dry air
ρ_v = density of water vapor in air
ρ_{tot} – density of air including water vapor
ρ_{parcel} density of a parcel of air
ρ_{env} = density of environment (i.e. the density of air in the "background" air surrounding a parcel)
$\rho(\,)$ = auto- and cross-correlations (generalized normalized forms). On occasion subscripts are attached to refer to the different data-series being correlated.

Σ

Σ – summation of successive terms
Σ_{total} = electromagnetic power formed by integrating the Poynting flux around a complete sphere (Chapter 3)

σ

σ = standard deviation, used at various places in the text
σ: various subscripted versions used for backscatter cross-sections in scatter theory
σ_e = electron backscatter cross-section
σ_T = Thomson backscatter cross-section
σ_s = scatter *reflectivity* for dielectric perturbations (especially in Chapter 3). Note that σ_e, σ_T above have units of m^2 (area), but σ_s has units of m^{-1}. It is the backscattered energy per unit volume per steradian per unit incident Poynting flux. It is closely related to η_s.
σ_Ω = normalized spectral width in Chapter 8
σ_Q = standard deviation of quantization errors in Chapter 8

τ

τ: often used as a time, or time interval. Especially used as a time-lag in correlation functions and convolutions.
τ = lifetime of turbulent eddy on occasions
τ_B = Brunt–Väisälä period, as an alternative to T_B

Υ

$\Upsilon_{\epsilon''c3}$ = squared magnitude of $F_{\epsilon''c3}$, which represents the energy spectrum. It is not the same as $\Phi_{\epsilon''c3}$, which is the energy spectrum of permittivity fluctuations per unit volume.
Υ: used briefly in Chapter 5 to represent the power polar diagram of a radar antenna

Φ

Φ - represents various types of spectra in scatter theory (Chapter 3)
Φ_N = three-dimensional spectrum of electron density flucuations in a plasma
$\Phi_{\epsilon'}$ = three-dimensional spectrum of *relative* permittivity perturbations in a plasma or gas. Various related forms exist such as $\Phi_{\epsilon''}$ and Φ_{ϵ}, mostly in Chapter 3, where their exact use is explained.
$\Phi_{\epsilon''c3}$ = energy spectrum of relative permittivity per unit volume
Φ_n = three-dimensional spectrum of refractive index fluctuations in a plasma or gas
Φ_ζ = three-dimensional turbulence spectrum of a scalar ζ in isotropic and homogeneous turbulence, proportional to $k^{-11/3}$ for inertial range turbulence (Chapter 3 and Appendix A). Also referred to as $\mathcal{F}$ in Apppendix A.
Φ_{ij} = various types of cross-spectra in Appendix A

ϕ

ϕ_D – representative solution to Maxwell's equations in Chapter 3
ϕ – generally azimuthal direction in polar coordinates
ϕ_p – spectrum of parallel velocity fluctuations produced in a linear path during passage through a patch of turbulence (Appendix A)
ϕ_t – spectrum of transverse velocity fluctuations produced in a linear path passing through a patch of turbulence (Appendix A)

φ

φ: usually used to represent phase delays in a signal. On rare occasions, may be used for azimuthal angle.

χ

χ = radiowave scattering angle (especially in Chapter 3)
χ = relative humidity contribution to M_n (see Appendix A)
χ_a = absorption term differential absorption experiment (see Chapter 10)
$\chi(\,)$: used as a function at times (e.g. ambiguity function in Chapter 4)

Ψ

Ψ_{ij} = various integrated forms of the cross-spectra Φ_ij in Appendix A

ψ

ψ = pressure perturbation normalized to mean density = $p'/ < \rho >$
ψ_p: used on occasion to represent normalized ion collision rate (mainly Chapter 3)
ψ = zenithal angle for polar diagram plots (Chapter 5, e.g., Figure 5.12)

Ω

Ω = gyrofrequency of an electron around Earth's magnetic field
Ω: used for Ohms in electrical circuit theory
Ω = angular rotation rate of Earth
$d\Omega$ used for solid angle in angular integrations
Ω_k: used as normalized angular frequency in Chapter 8
Ω_s: used for frequency of optimum response in treatment of Capon method in Chapter 8
Ω_d = normalized spectral peak in Chapter 8.

ω

ω = angular frequency (main application)
ω_a = cutoff angular frequency of the acoustic branch of the atmospheric internal waves
ω_B = Brunt–Väisälä angular frequency (also N is used to represent the BV frequency in some texts, but we avoid that usage here)
ω_0: various localized applications, e.g., a specific angular frequency, or the angular frequency of an electron "on a spring" – as in electrons bound to a nucleus
ω_p = plasma frequency of a plasma
ω_N = plasma frequency of a plasma of number density N
ω_i = inertial frequency in dynamical and gravity-wave theory, which also equals f_c (radians per second).

Miscellaneous

£: used as a ratio on some occasions
$\otimes$ = convolution
$\angle$ = angle, or a phase term (especially used for phase of a complex number)

References

Ackerman, M., In-situ measurements of middle atmosphere composition, *J. Atmos. Terr. Phys.*, **41**, 723–733, 1979.

Adachi, T., T. Tsuda, Y. Masuda, T. Takami, S. Kato, and S. Fukao, Effects of the acoustic and radar pulse length ratio on the accuracy of radio acoustic sounding system (RASS) temperature measurements with monochromatic acoustic pulses, *Radio Sci.*, **28**, 571–583, 1993.

Adams, G. W., J. W. Brosnahan, and D. P. Edwards, The imaging Doppler interferometer: Data analysis, *Radio Sci.*, **20**, 1481–1492, 1985.

Adams, G. W., J. W. Brosnahan, D. Walden, and S. Nerney, Mesospheric Observations using a 2.66-MHz radar as an imaging Doppler interferometer: Description and first results, *J. Geophys. Res*, **91**(A2), 1671–1683, 1986.

AFC-Laboratories, *Handbook of Geophysics and Space Environments*, U. S. Air Force, Cambridge Research Laboratories, Cambridge, Mass., 1965.

Ahrens, C. D., *Meteorology Today: an Introduction to Weather, Climate and the Environment*, Brooks/Cole, Pacific Grove, CA, USA, 1999.

Aikin, A. C., R. A. Goldberg, W. Jones, and J. A. Kane, Observations of the mid-latitude lower ionosphere in winter, *J. Geophys. Res.*, **82**, 1869–1875, 1977.

Alexander, M. J., A simulated spectrum of convectively generated gravity waves: propagation from the tropopause to the mesopause and effects on the middle atmosphere, *J. Geophys. Res.*, **101**, 1571–1588, 1996.

Alexander, M. J., and J. Holton, Gravity waves generated by a transient localized heat source, *Atmos. Chem. Phys.*, **4**, 923–932, 2004.

Alexander, M. J., and K. H. Rosenlof, Nonstationary gravity wave forcing of the stratospheric zonal mean wind, *J. Geophys. Res.*, **101**, 23 465–23 474, 1996.

Alexander, M. J., J. H. Beres, and L. Pfister, Tropical stratospheric gravity wave activity and relationships to clouds, *J. Geophys. Res.*, **105**(D17), 22 299–22 309, 2000.

Alexander, M. J., P. T. May, and J. H. Beres, Gravity waves generated by convection in the Darwin area during the Darwin Area Wave Experiment, *J. Geophys. Res.*, **109**, D20S04, doi:10.1029/2004JD004 729, 2004.

Alexander, M. J., J. Holton, and D. Durran, The gravity wave response above deep convection in a squall line simulation, *J. Atmos. Sci.*, **52**, 2212–2226, 1995.

Alexander, M. J., *et al.*, Recent developments in gravity-wave effects in climate models and the global distribution of gravity-wave momentum flux from observations and models, *Q. J. R. Meteorol. Soc.*, **136**(650A), 1103–1124,doi:10.1002/qj.637, 2010.

Alexander, S., and T. Tsuda, High-resolution radio acoustic sounding system (RASS) observations and analysis up to 20 km, *J. Atmos. and Oceanic Tech.*, **25**, 1383–1396 doi:10.1175/2007JTECHA983.1, 2008.

Allen, D. C., J. D. Haigh, J. T. Houghton, and C. J. S. M. Simpson, Radiative cooling near the mesopause, *Nature*, **281**, 660–661, 1979.

Allen, K. R., and R. I. Joseph, A canonical statistical theory of oceanic internal waves, *J. Fluid Mechs.*, **204**, 185–228, 1989.

Alvarez, H., J. Aparici, J. May, and F. Olmos, A 45-MHz continuum survey of the southern hemisphere, *Astron. Astrophys. Suppl. Ser.*, **124**, 315–328, 1997.

Amayenc, P., J. Fontanari, and D. Alcayde, Simultaneous neutral wind and temperature oscillations near tidal periods in the F-region over ST Santin, *J. Atmos. Terr. Phys.*, **35**, 1499–1505, 1973.

Amidon, *Iron-Powder and Ferrite Coil Forms*, Amidon Associates Inc., Torrance, California, USA, 1992.

Anandan, V. K., P. Balamuralidhar, P. B. Rao, A. R. Jain, and C. J. Pan, An adaptive moments estimation technique applied to MST radar echoes, *J. Atmos. Oceanic Technol.*, **22**, 396–408, 2005.

Anandarao, B. G., R. Raghavarao, J. N. Desai, and G. Haerendel, Vertical winds and turbulence over Thumba, *J. Atmos. Terr. Phys.*, **40**, 157–163, 1978.

Andreassen, O., C.-E. Wasberg, D. C. Fritts, and J. R. Isler, Gravity wave breaking in two and three dimensions: 1. Model description and comparison of two-dimensional evolutions, *J. Geophys. Res.*, **99**, 8095–8108, 1994.

Andreassen, O., P. O. Hvidsten, D. C. Fritts, and S. Arendt, Vorticity dynamics in a breaking gravity wave. Part 1. Initial instability evolution, *J. Fluid Mech.*, **367**, 27–46, 1998.

Andrews, D. G., and M. E. McIntyre, Planetary waves in horizontal and vertical shear: The generalized Eliassen–Palm relation and the mean zonal acceleration, *J. Atmos. Sci.*, **33**, 2031–2048, 1976.

Andrews, D. G., and M. E. McIntyre, Generalized Eliassen–Palm and Charney–Drazin theorems for waves on axisymmetric mean flows in compressible atmospheres, *J. Atmos. Sci.*, **35**, 175–185, 1978.

Andrews, D. G., J. R. Holton, and C. B. Leovy, *Middle Atmospheric Dynamics*, Academic Press, 1987.

Andrioli, V. F., D. C. Fritts, P. P. Batista, and B. R. Clemesha, Improved analysis of all-sky meteor radar measurements of gravity wave variances and momentum fluxes, *Ann. Geophys.*, **31**, , 889–908, doi:10.5194/angeo–31–889–2013, 2010.

Angevine, W. M., Errors in mean vertical velocities measured by boundary layer wind profilers, *J. Atmos. Oceanic Technol.*, **14**, 565–569, 1997.

Angevine, W. M., S. K. Avery, W. L. Ecklund, and D. A. Carter, Fluxes of heat and momentum measured with a boundary-layer wind profiler radar-radio acoustic sounding system, *J. Appl. Meteorol.*, **32**, 73–80, 1993.

Angevine, W. M., W. L. Ecklund, D. A. Carter, K. S. Gage, and K. P. Moran, Improved radio acoustic sounding techniques, *J. Atmos. Oceanic Technol.*, **11**, 42–49, 1994.

Appleton, E., On some measurements of the equivalent height of the atmospheric ionized layer, *Proc. Roy. Soc.*, **A126**, 542–569, 1930.

Appleton, E., Wireless studies of the ionosphere, *Proc. Inst. Elec. Engnrs (Wireless Section)*, **7**(21), 257–265, 1932.

Appleton, E. V., and M. A. F. Barnett, Local reflections of wireless waves from the upper atmosphere, *Nature*, **115**, 333–334, 1925.

Astin, I., Confidence interval estimation for VHF Doppler radar measurements of wind velocities, *Radio Sci.*, **32**(6), 2221–2231, 1997.

Atlas, D., *Advances in Geophysics*, vol. 10, Academic Press, New York, 1964.

Atlas, D., Tribute to Professor Louis J. Battan, in *Radar in Meteorology*, edited by D. Atlas, pp. xiii–xvii, American Met. Soc., 1990.

Atlas, D., R. C. Srevastava, and P. W. Sloss, Wind shear and reflectivity gradient effects on Doppler radar spectra: II, *J. Appl. Meteorol.*, **8**, 384–388, 1969.

Austin, G. L., and A. H. Manson, On the nature of the irregularities that produce partial reflections of radio waves from the lower ionosphere (70–100 km), *Radio Sci.*, **4**, 35, 1969.

Austin, G. L., R. G. T. Bennett, and M. R. Thorpe, The phase of waves partially reflected from the lower ionosphere, *J. Atmos. Terr. Phys.*, **31**, 1099–1106, 1969.

Avery, S. K., A. C. Riddle, and B. B. Balsley, The Poker Flat, Alaska, MST radar as a meteor radar, *Radio Sci.*, **18**, 1021–1027, 1983.

Baggaley, W. J., R. G. T. Bennett, D. I. Steel, and A. D. Taylor, The advanced meteor orbit radar facility: Amor, *Q. J. R. Astron. Soc.*, **35**, 293–320, 1994.

Bahnsen, A., Recent techniques of observation and results from the magnetopause region, *J. Atmos. Terr. Phys.*, **40**, 235–256, 1978.

Balanis, C. A., *Antenna Theory: Analysis and Design*, 2nd ed., John Wiley and Sons, Chichester, 1997.

Ball, S. M., Atmospheric gravity wave production for the Australian total solar eclipse of 23 October 1976, *Australian J. Phys.*, **32**, 287–288, 1979.

Balsley, B. B., Electric fields in the equatorial ionosphere; a review of techniques and measurements, *J. Atmos. Terr. Phys*, **35**, 1035, 1973.

Balsley, B. B., and K. S. Gage, On the use of radars for operational windprofiling, *Bull. Amer. Meteorol. Soc.*, **63**, 1009–1018, 1982.

Balsley, B. B., and T. J. Judasz, Improved theoretical and experimental models for the coaxial colinear antenna, *IEEE Trans. Antennas Propagat.*, **37**, 289–296, 1989.

Balsley, B. B., W. L. Ecklund, D. A. Carter, and P. E. Johnston, The Poker Flat MST radar: First Results, *Geophys. Res. Lett.*, **6**, 921–924, 1979.

Balsley, B. B., W. L. Ecklund, D. A. Carter, and P. E. Johnston, The MST radar at Poker Flat, Alaska, *Radio Sci.*, **15**, 213–223, 1980.

Balsley, B. B., R. G. Frehlich, M. L. Jensen, Y. Meillier, and A. Muschinski, Extreme gradients in the nocturnal boundary layer: Structure, evolution, and potential causes, *J. Atmos. Sci.*, **60**, 2496–2508, 2003.

Barabash, V., S. Kirkwood, and P. B. Chilson, Are variations in PMSE intensity affected by energetic particle precipitation?, *Ann. Geophys.*, **20**, 539–545, 2002a.

Barabash, V., S. Kirkwood, A. Feofilov, and A. Kutepov, Polar mesosphere summer echoes during the July 2000 solar proton event, *J. Geophys. Res.*, **22**, 759–771, 2002b.

Barat, J., Some characteristics of clear air turbulence in the middle stratosphere, *J. Atmos. Sci.*, **39**, 2553–2564, 1982.

Barratt, P., and I. C. Browne, A new method of measuring vertical currents, *Q. J. R. Meteorol. Soc.*, **79**, 550, 1953.

Bartlett, M. S., Periodogram analysis and continuous spectra, *Biometrika*, **37**, 1–16, 1950.

Barton, B. D., *Modern Radar System Analysis*, Artech House, Norwood, MA, 1988.

Barton, D. K., *Radar System Analysis and Modeling*, Artech House, Norwood, MA, 2005.

Batchelor, G. K., *The Theory of Homogeneous Turbulence*, Cambridge University Press, New York, 1953.

Batchelor, G. K., *An Introduction to Fluid Dynamics*, Cambridge University Press, Cambridge, U. K., 1977.

Becker, E., Dynamical control of the middle atmosphere, *Space Sci. Rev.*, **168**, 283–314, doi10.1007/s11,214–011–9841–5, 2012.

Belova, E., S. Kirkwood, J. Ekeberg, *et al.*, The dynamic background of polar mesosphere winter echoes from simultaneous EISCAT and ESRAD observation, *Ann. Geophys.*, **23**, 1239–1247, 2005.

Belova, E., M. Smirnova, M. T. Rietveld, *et al.*, First observation of the overshoot effect for polar mesosphere winter echoes during radiowave electron temperature modulation, *Geophys. Res. Lett.*, **35**, L03,110 doi:10.1029/2007GL032,457, 2008.

Belrose, J. S., Radio wave probing of the ionosphere by the partial reflection of radiowaves (from heights below 100 km), *J. Atmos. Terr. Phys.*, **32**, 567, 1970.

Belrose, J. S., and M. J. Burke, Study of the lower ionosphere using partial reflection.

1. Experimental technique and method of analysis, *J. Geophys. Res*, **69**, 2799, 1964.

Belu, R., and W. K. Hocking, Gravity wave generation by frontal systems as seen in long-term multi-instrument observations (CLOVAR windprofiler, microbarograph and radiosondes), in *STEP Handbook, Proceedings of the ninth International Workshop on Technical and Scientific Aspects of MST Radar combined with COST76 Final Profiler Workshop*, edited by B. Edwards, pp. 194–197, Toulouse, France, 2000.

Belu, R., W. K. Hocking, N. Donaldson, and T. Thayaparan, Comparisons of CLOVAR windprofiler horizontal winds with radiosondes and CMC Regional Analyses, *Atmosphere-Ocean*, **39**, 107–126, 2001.

Belu, R. G., Ray-tracing of gravity waves through the standard atmosphere: effects of fluctuations and perturbations in the background temperature and wind profiles, in *STEP Handbook, Proceedings of the Eighth Workshop on Technical and Scientific Aspects of MST Radar*, edited by B. Edwards, pp. 167–170, Bangalore, India, 1998.

Belu, R. G., Gravity waves sources and propagation characteristics in the lower and middle atmosphere determined by CLOVAR radar and other ground-based methods, Ph. D. Thesis, University of Western Ontario, Canada, 1999.

Belu, R. G., Sensitivity of ray-tracing models to the fluctuations of the background atmospheric wind and temperature fields, in *STEP Handbook, Proceedings of the ninth International Workshop on Technical and Scientific Aspects of MST Radar combined with COST76 Final Profiler Workshop*, edited by B. Edwards, pp. 206–209, Toulouse, France, 2000.

Benjamin, S. G., B. E. Schwartz, E. J. Szoke, and S. E. Koch, The value of wind profiler data in U. S. weather forecasting, *Bull. Amer. Meteorol. Soc.*, **85**, 1871–1886, 2004.

Benson, R. F., The quasi-longitudinal approximation in the generalized theory of radio wave absorption, *Radio Sci.*, **68D**, 219–223, 1964.

Beres, I., W. K. Hocking, and R. Thomas, Discrimination between lightning-generated RF and radar reflections from lightning, in *Proceedings of the Twelfth International Workshop on Technical and Scientific Aspects of MST Radar*, edited by N. Swarnalingam and W. K. Hocking, pp. 73–76, Publ. by Canadian Assoc. of Physicists, 2010.

Beres, J., M. J. Alexander, and J. R. Holton, Effects of tropospheric wind shear on the spectrum of convectively generated gravity waves, *J. Atmos. Sci.*, **59**, 1805–1824, 2002.

Beres, J., M. J. Alexander, and J. R. Holton, A method of specifying the gravity wave spectrum above convection based on latent heating properties and background wind, *J. Atmos. Sci.*, **61**, 324–337, 2004.

Bianco, L., D. Cimini, F. S. Marzano, and R. Ware, Combining microwave radiometer and wind profiler radar measurements for high-resolution atmospheric humidity profiling, *J. Atmos. Oceanic Technol.*, **22**, 949 965, 2005.

Birner, T., and H. Bönish, Residual circulation trajectories and transit times into the extratropical lowermost stratosphere, *Atmos. Chem. Phys.*, **11**, 817–827, 2011.

Blackman, R. B., and J. W. Tukey, *The Measurement of Power Spectra From the Point of View of Communication Engineering*, Dover, New York, 1959.

Blamont, J. E., and J. Barat, Dynamical structure of the atmosphere between 80 and 120 km, in *Aurora and Airglow*, edited by B. M. McCormac, pp. 156–159, Reinhold Pub. Co., 1967.

Blix, T. A., E. V. Thrane, and O. Andreassen, In-situ measurements of the fine-scale structure and turbulence in the mesosphere and lower thermosphere by means of electrostatic positive ion probes, *J. Geophys. Res.*, **95**, 5533–5548, 1990.

Bohne, A. R., Radar detection of turbulence in thunderstorms, in *Report # AFGL-TR-81-0102 (ADA 108679)*, Air Force Geophys. Lab., Hanscom Air Force Base, Mass., USA., 1981.

Bohne, A. R., Radar detection of turbulence in precipitation environments, *J. Atmos. Sci.*, **39**, 1819–1837, 1982.

Bolgiano, R. J., The general theory of turbulence – turbulence in the atmosphere, in *Winds and Turbulence in the Stratosphere, Mesopshere and Ionosphere*, edited by K. Rawer, pp. 371–400, North Holland, Amsterdam, 1968.

Bonino, G., P. P. Lombardini, and P. Trivero, A metric wave radio-acoustic tropospheric sounder, *IEEE Trans. Geosci. Electron.*, **GE-17**, 179–181, 1979.

Booker, H. G., A theory of scattering by non-isotropic irregularities with application to radar reflections from the aurora, *J. Atmos. Terr. Phys.*, **8**, 204–221, 1956.

Booker, H. G., Radio scattering in the lower ionosphere, *J. Geophys. Res.*, **64**, 2164, 1959.

Booker, H. G., and R. Cohen, A theory of long-duration meteor-echoes based on atmospheric turbulence with experimental confirmation, *J. Geophys. Res.*, **61**, 707–733, 1956.

Borkowski, M. T., Chapter 11 in *Radar Handbook*, in *Solid-state Transmitters*, edited by M. I. Skolnik, pp. 11.1–11.36, McGraw-Hill, New York, 2008.

Born, M., and E. Wolf, *Principles of Optics: Electromagnetic Theory of Propagation, Interference and Diffraction of Light*, 7th edition, Cambridge University Press, 1999.

Bourdillon, A., C. Haldoupis, C. Hanuise, Y. Le Roux, and J. Menard, Long duration meteor echoes characterized by Doppler spectrum bifurcation, *Geophys. Res. Lett.*, **32**, L05,805, doi: 10.1029/2004GL021,685, 2005.

Bowles, K. L., Observations of vertical incidence scatter from the ionosphere at 41 Mc/s, *Phys. Rev. Letters*, **1**, 454, 1958.

Bracewell, R. N., *The Fourier Transform and its Applications*, McGraw-Hill, New York, 1978.

Bradshaw, P., *An Introduction to Turbulence and its Measurement*, Pergamon Press, 1975.

Brasseur, G. P., J. J. Orlando, and G. S. Tyndall, Atmospheric chemistry and global change, in *Topics in Environmental Chemistry Series*, pp. 1–654, Oxford University Press (New York, Oxford), 1990.

Breit, G., and M. A. Tuve, A radio method of estimating the height of the conducting layer, *Nature*, **116**, 357, 1926.

Bremer, J., P. Hoffmann, and T. Hansen, Geomagnetic control of polar mesosphere summer echoes, *Ann. Geophys.*, **18**, 202–208, 2000.

Bremer, J., P. Hoffmann, R. Latteck, and W. Singer, Seasonal and long-term variations of PMSE from VHF radar observations at Andenes, Norway, *J. Geophys. Res.*, **108**, doi:10.1029/2002JD002,369, 2003.

Bremer, J., P. Hoffmann, J. Hoeffner, *et al.*, Long-term changes of mesospheric summer echoes at polar and middle latitudes, *J. Atmos. Solar-Terr. Phys*, **68**, 1940–1951, 2006.

Briggs, B. H., Radar observations of atmospheric winds and turbulence: A comparison of techniques, *J. Atmos. Terr. Phys.*, **42**, 823–833, 1980.

Briggs, B. H., The analysis of spaced sensor records by correlation techniques, in *Handbook for MAP, Ground Based Techniques*, edited by R. A. Vincent, vol. 13, pp. 166–186, SCOSTEP Secretariat, Dept. of Electr. Computer Eng., Univ. of Illinois, Urbana, IL 61801, USA, 1984.

Briggs, B. H., Radar measurements of aspect sensitivity of atmospheric scatterers using spaced-antenna correlation techniques, *J. Atmos. Terr. Phys.*, **54**, 153–165, 1992.

Briggs, B. H., On radar interferometric techniques in the situation of volume scatter, *Radio Sci.*, **30**, 109–114, 1995.

Briggs, B. H., and N. Holmes, Ionospheric observations using ultrasonic image forming technique, *Nature Physical Science*, **243**, 111–112, 1973.

Briggs, B. H., and M. Spencer, The variability of time shifts in measurements of ionospheric movements, in *Report of the Physical Society Conference on Physics of the Ionosphere*, p. 123, Cambridge, 1954.

Briggs, B. H., and R. A. Vincent, Some theoretical considerations on remote probing of weakly scattering irregularities, *Aust. J. Phys.*, **26**, 805–814, 1973.

Briggs, B. H., and R. A. Vincent, Spaced-antenna analysis in the frequency domain, *Radio Sci.*, **27**, 117–129, 1992.

Briggs, B. H., G. J. Phillips, and D. H. Shinn, The analysis of observations on spaced receivers of the fading of radio signals, *Proc. Phys. Soc.*, **63B**, 106–121, 1950.

Briggs, B. H., W. G. Elford, D. G. Felgate, *et al.*, Buckland Park aerial array, *Nature*, **223**, 1321, 1969.

Bringi, V. N., G. J. Huang, V. Chandrasekar, and E. Gorgucci, A methodology for estimating the parameters of a gamma raindrop size distribution model from polarimetric radar data: Application to a squall-line event from the TRMM/Brazil field campaign, *J. Atmos. Terr. Phys.*, **19**, 633–645, 2002.

Brosnahan, J. W., and G. W. Adams, The MAPSTAR imaging Doppler interferometer (IDI) radar: description and first results, *J. Atmos. Terr. Phys.*, **55**, 203–228, 1993.

Brown, P. G., R. J. Weryk, D. K. Wong, and J. Jones, A meteoroid stream survey using the Canadian meteor orbit radar. I: Methodology and radiant catalogue, *Icarus*, **195**, 317–339, doi:10.1016/j.icarus.2007.12.002, 2008.

Browning, K. A., D. Jerrett, J. Nash, T. Oakley, and N. M. Roberts, Cold frontal structure derived from radar wind profiles, *Meteorol. Apps.*, **5**, 67–74, 1998.

Budden, K. G., Effect of electron collisions on the formulas of magnetoionic theory, *Radio Sci.*, **69D**, 191–211, 1965.

Buehler, W. E., and C. D. Lunden, A note on VHF backscatter from turbulence in the upper troposphere, *J. Applied Meteorology*, **4**, 151–152, 1964.

Caccia, J. L., B. Benech, and V. Klaus, Space-time description of non-stationary trapped lee waves using ST radars, aircraft, and constant volume balloons during the PYREX experiment, *J. Atmos. Sci.*, **54**, 1821–1832, 1997a.

Caccia, J. L., M. Crochet, and K. Saada, ST radar evaluation of the standard deviation of the air vertical velocity perturbed by the local orography, *J. Atmos. Solar-Terr. Phys.*, **59**, 1127–1131, 1997b.

Campistron, B., G. Despaux, M. Lothon, *et al.*, A partial 45 MHz sky temperature map obtained from the observations of five ST radars, *Ann. Geophys.*, **19**, 863–871, 2001.

Campos, E. F., W. K. Hocking, and F. Fabry, Precipitation measurement using VHF wind profiler radars: A multifaceted approach to calibrate radar antenna and receiver chain, *Radio Sci.*, **42**, RS4008, doi:10.1029/2006RS003 508, 2007a.

Campos, E. F., F. Fabry, and W. K. Hocking, Precipitation measurements using VHF wind profiler radars: Measuring rainfall and vertical air velocities using only observations with a VHF radar, *Radio Sci.*, **42**, RS3003, doi:10.1029/2006RS003 540, 2007b.

Cane, H. V., A 30 MHz map of the whole sky, *Aust. J. Phys.*, **31**, 561–565, 1978.

Capon, J., High-resolution frequency-wavenumber spectrum analysis, *Proc. IEEE*, **57**, 1408–1419, 1969.

Carey-Smith, T. K., A. J. McDonald, W. J. Baggaley, *et al.*, Antenna beam verification using cosmic noise, in *Handbook for STEP, Proceedings of the tenth International Workshop on Technical and Scientific Aspects of MST Radar*, edited by J. L. Chau, J. Lau, and J. Röttger, pp. 391–394, Piura, Peru, 2003.

Carlson, H. C., and N. Sundararaman, Real-time jet-stream tracking: national benefit from an ST radar network for measuring atmospheric motions, *Bull. Amer. Meteorol. Soc.*, **63**, 1019–1026, 1982.

Caughey, S. J., B. A. Crease, D. N. Asimakopoulos, and R. S. Cole, Quantitative bistatic acoustic sounding of the atmospheric boundary layer, *Q. J. R. Meteorol. Soc.*, **104**, 147–161, 1978.

Cervera, M. A., W. G. Elford, and D. I. Steel, A new method for the measurement of meteor speeds: The pre-t_0 phase technique, *Radio Sci.*, **32**, 805–816, 1997.

Chadwick, R. B., and E. E. Gossard, Radar remote sensing of the clear atmosphere – review and applications, *Proc IEEE*, **71**, 738–753, 1983.

Chadwick, R. B., A. S. Frisch, and R. G. Strauch, A feasibility study on the use of wind profilers to support space shuttle launches, *NASA Contractor Rep.*, **3861**, 1984.

Chakrabarty, D. K., P. Chakrabarty, and G. Witt, An attempt to identify the obscured paths of water cluster ions build-up in the D-region, *J. Atmos. Terr. Phys.*, **40**, 437–442, 1978a.

Chakrabarty, D. K., P. Chakrabarty, and G. Witt, The effect of variations in temperature and nitric oxide density on ion-clustering in the mesopause region during winter anomaly, *J. Atmos. Terr. Phys.*, **40**, 1147–1152, 1978b.

Champeney, D. C., *Fourier Transforms and their Physical Applications*, Academic Press, London and New York, 1973.

Chandra, S., Energetics and thermal structure of the middle atmosphere, *Planet. Space Sci.*, **28**, 585–593, 1980.

Chandra, S., and R. A. Vincent, Remote probing of D-region irregularities, in *Proc. Indian Acad. Sci.*, **A88**, 57, 1979.

Chanin, M. L., and A. Hauchecorne, Lidar observation of gravity and tidal waves in the stratosphere and mesosphere, *J. Geophys. Res.*, **86**, 9715–9721, 1981.

Chao, J. K., F. S. Kuo, I. J. Fu, J. Röttger, and C. H. Liu, The first operation and results of Chung Li VHF radar, in *Handbook for MAP*, edited by S. A. Bowhill and B. Edwards, vol. 20, pp. 359–363, Scostep Secretariat, University of Illinois, USA, 1986.

Chau, J. L., Unexpected spectral characteristics of VHF radar signals from 150 Km region over Jicamarca, *Geophys. Res. Lett.*, **31**, L23,803, doi:10.1029/2004GL021,620, 2004.

Chau, J. L., and B. B. Balsley, Interpretation of angle-of-arrival measurements in the lower atmosphere using spaced antenna radar systems, *Radio Sci.*, **33**, 517–533, 1998a.

Chau, J. L., and E. Kudeki, Statistics of 150 km echoes over Jicamarca based on low-power VHF observations, *Ann. Geophys.*, **24**, 1305–1310, 2006a.

Chau, J. L., and E. Kudeki, First E- and D-region incoherent scatter spectra observed over Jicamarca, *Ann. Geophys.*, **24**, 1295–1303, 2006b.

Chau, J. L., and E. Kudeki, Discovery of two distinct types of equatorial 150 km radar echoes, *Geophys. Res. Lett.*, **40**, 4509–4514, doi:10.1002/grl.50,893, 2013.

Chau, J. L., and R. F. Woodman, Three-dimensional coherent radar imaging at Jicamarca: Comparison of different inversion techniques, *J. Atmos. Terr. Phys.*, **63**, 253–261, 2001.

Chau, J. L., D. L. Hysell, K. M. Kuyeng, and F. R. Galindo, Phase calibration approaches for radar interferometry and imaging configurations: equatorial spread F results, *Ann. Geophys.*, **26**, 2333–2343, 2008.

Chau, J. L., R. F. Woodman, M. A. Milla, and E. Kudeki, Naturally enhanced ion-line spectra around the equatorial 150 km region, *Ann. Geophys.*, **27**, 933–942, 2009.

Chau, T., J. L. Renkwitz, G. Stober, and R. Latteck, MAARSY multiple receiver phase calibration using radio sources, *J. Atmos. Solar-Terr. Phys.*, **118(A)**, 55–63, 2014.

Chen, F. F., *Introduction to Plasma Physics and Controlled Fusion*, Plenum Press, New York, 1984.

Chen, W., Energy dissipation rates of free atmospheric turbulence, *J. Atmos. Sci.*, **31**, 2222–2225, 1974.

Cheong, B. L., M. W. Hoffman, R. D. Palmer, S. J. Frasier, and F. J. López-Dekker, Pulse pair beamforming and the effects of reflectivity field variations on imaging radars, *Radio Sci.*, **39**, RS3014, doi:10.1029/2002RS002,843, 2004.

Cheong, B. L., M. W. Hoffman, R. D. Palmer, S. J. Frasier, and F. J. López-Dekker, Phase-array design for biological clutter rejection: Simulation and experimental validation, *J. Atmos. Oceanic Technol.*, **23**, 585–598, doi:10.1175/JTECH1867.1, 2006.

Cheong, B. L., T.-Y. Yu, R. D. Palmer, *et al.*, Effects of wind field inhomogeneities on Doppler beam swinging revealed by an imaging radar, *J. Atmos. Oceanic Technol.*, **25**, 1414–1422, 2008.

Cherniakov, M. (Ed.), *Bistatic Radar, Principles and Practice*, John Wiley and Sons, Chichester, 2007.

Cherniakov, M. (Ed.), *Bistatic Radar, Emerging Technology*, John Wiley and Sons, Chichester, 2008.

Chilson, P. B., and G. Schmidt, Implementation of frequency domain interferometry at the SOUSY VHF radar: First results, *Radio Sci.*, **31**, 263–272, 1996.

Chilson, P. B., C. W. Ulbrich, M. F. Larsen, P. Perillat, and J. E. Keener, Observations of a tropical thunderstorm using a vertically pointing, dual-frequency, collinear beam Doppler radar, *J. Atmos. Oceanic Technol.*, **10**, 663–673, 1993.

Chilson, P. B., P. Czechowsky, and G. Schmidt, A comparison of ambipolar diffusion coefficients in meteor trains using VHF radar and UV lidar, *Geophys. Res. Lett.*, **23**, 2745–2748, 1996.

Chilson, P. B., A. Muschinski, and G. Schmidt, First observations of Kelvin–Helmholtz billows in an upper level jet using VHF frequency domain interferometry, *Radio Sci.*, **32**(3), 1149–1160, 1997.

Chilson, P. B., E. Belova, M. Rietveld, S. Kirkwood, and U. P. Hoppe, First artificially induced modulation of PMSE using the EISCAT heating facility, *Geophys. Res. Lett.*, **27**, 3801–3804, 2000.

Chilson, P. B., S. Kirkwood, and I. Häggström, Frequency-domain interferometry mode observations of PMSE using the EISCAT VHF radar, *Ann. Geophys.*, **18**, 1599–1612, 2001a.

Chilson, P. B., R. D. Palmer, A. Muschinski, *et al.*, SOMARE-99: A demonstrational field campaign for ultra-high resolution VHF atmospheric profiling using frequency diversity, *Radio Sci.*, **36**, 695–707, 2001b.

Chilson, P. B., T.-Y. Yu, R. G. Strauch, A. Muschinski, and R. D. Palmer, Implementation and validation of range imaging on a UHF radar wind profiler, *J. Atmos. Ocean. Tech.*, **20**, 987–996, 2003.

Cho, J. Y. N., Inertio-gravity wave parameter estimation from cross-spectral analysis, *J. Geophys. Res.*, **100**, 18 727–18 737, 1995.

Cho, J. Y. N., and M. C. Kelley, Polar mesosphere summer radar echoes: Observations and current theories, *Rev. Geophys.*, **31**, 243–265, 1993.

Cho, J. Y. N., and J. Röttger, An updated review of polar mesosphere summer echoes: Observation, theory, and their relationship to noctilucent clouds and subvisible aerosols, *J. Geophys. Res.*, **102**, 2001–2020, 1997.

Cho, J. Y. N., T. M. Hall, and M. C. Kelley, On the role of charged aerosols in the polar mesosphere summer echoes, *J. Geophys. Res.*, **97**, 875–886, 1992.

Cho, J. Y. N., C. M. Alcala, M. C. Kelley, and W. E. Swartz, Further effect of charged aerosols on summer mesospheric radar scatter, *J. Atmos. Terr. Phys.*, **58**, 661–672, 1996.

Choudhary, R. K., J. P. St. Maurice, and K. K. Mahajan, Observations of coherent echoes with narrow spectra near 150 km altitude during daytime a way from the dip equator, *Geophys. Res. Lett.*, **31**, L19,801, doi:10.1029/2004GL020,299, 2004.

Chu, Y.-H., Beam broadening effect on oblique MST radar Doppler spectra, *J. Atmos. Oceanic Technol.*, **19**, 1955–1967, 2002.

Chu, Y. H., and T. Y. Chen, Theoretical study of two-frequency coherence of MST radar returns, *Radio Sci.*, **30**, 1803–1815, 1995.

Chunchuzov, I. P., On the high wavenumber form of the Eulerian internal wave spectrum in the atmosphere, *J. Atmos Sci.*, **59**, 1753–1774, 2002.

Ciesielski, P. E., L. M. Hartten, and R. H. Johnson, Impacts of merging profiler and rawsinsonde winds on TOGA COARE analysis, *J. Atmos. Oceanic Technol.*, **14**, 1264–1279, 1997.

Cohn, M., and A. Lempel, On fast M-sequence transforms, *IEEE Trans. Information Theory*, **IT-23**, 135–137, 1977.

Cohn, S. A., Investigations of the wavelength dependence of radar backscatter from atmospheric turbulence, *J. Atmos. Oceanic Technol.*, **11**, 225–238, 1994.

Cohn, S. A., Radar measurements of turbulent eddy dissipation rate in the troposphere: A comparison of techniques, *J. Atmos. Ocean. Tech.*, **12**, 85–95, 1995.

Cooley, J. W., and J. W. Tukey, An algorithm for the machine calculation of the complex Fourier series, *Math. Comp.*, **19**, 297–301, 1965.

Cornish, C. R., and M. F. Larsen, Observations of low-frequency gravity waves in the lower stratosphere over Arecibo, *J. Atmos. Sci.*, **46**, 2428–2439, 1989.

Coy, L., D. C. Fritts, and J. Weinstock, The Stokes drift due to vertically propagating internal gravity waves in a compressible atmosphere, *J. Atmos. Sci.*, **43**, 2636–2643, 1986.

Craig, R. A., *The Upper Atmosphere: Meteorology and Physics*, International Geophysics Series, Academic Press, NY and London, 1965.

Crane, R. K., Radar measurements of wind at Kwajalein, *Radio Sci.*, **15**, 383–394, 1980a.

Crane, R. K., A review of radar observations of turbulence in the lower stratosphere, *Radio Sci.*, **15**, 177–194, 1980b.

Crochet, M., J. Tabbagh, and N. Makiese, Simultaneous ionospheric drift observations by different techniques at low and mid-latitudes, *J. Atmos. Terr. Phys.*, **39**, 463, 1977.

Croft, T. A., Sky-wave backscatter: a means for observing our environment at great distances, *Revs. Geophys. Space Phys.*, **10**, 73–155, 1972.

Cunnold, D. M., Vertical transport coefficients in the mesosphere obtained from radar observations, *J. Atmos. Sci.*, **32**, 2191, 1978.

Czechowsky, P., and R. Rüster, VHF radar observations of turbulent structures in the polar mesopause region, *Ann. Geophys.*, **15**, 1028–1036, 1997.

Czechowsky, P., J. Klostermeyer, J. Röttger, *et al.*, The SOUSY-VHF-radar for tropo-, strato- and mesospheric sounding, in *17th Conference on Radar Meteorology of the American Meteorological Society (AMS, Oct. 26–29)*, pp. 349–353, Seattle, USA, 1976.

Czechowsky, P., R. Ruster, and G. Schmidt, Variations of mesospheric structures in different seasons, *Geophys. Res, Lett.*, **6**, 459–462, 1979.

Czechowsky, P., I. M. Reid, and R. Rüster, VHF radar measurements of the aspect sensitivity of the summer polar mesopause echoes over Andenes (69 ° N, 16 ° E), Norway, *Geophys. Res. Lett.*, **15**, 1259–1262, 1988.

Czechowsky, P., B. Inhester, J. Klostermeyer, *et al.*, Recent progress with the SOUSY VHF radars, in *Handbook for MAP*, vol. 28, pp. 459–466, Scostep Secretariat, University of Illinois, USA, 1989.

Dalaudier, F., and A. S. Gurvich, A scalar three-dimensional spectral model with variable anisotropy, *J. Geophys. Res.*, **102**, 19 449–19 460, 1997.

Dalaudier, F., C. Sidi, M. Crochet, and J. Vernin, Direct evidence of sheets in the atmospheric temperature field, *J. Atmos. Sci.*, **51**, 237–248, 1994.

Danilov, A. D., Direct and indirect estimates of turbulence around the turbopause, *Adv. Space Res.*, **4**(4), 67–78, 1984.

Davies, K., and G. A. M. King, On the validity of some approximations to the Appleton–Hartree formula, *J. Research of the National Bureau of Standards - D, Radio Propagation*, **65D**(4), 323–332, 1961.

de Wolfe, D. A., A random-motion model of fluctuations in a nearly transparent medium, *Radio Sci.*, **18**, 138–142, 1983.

Defant, F., and H. Taba, The threefold structure of the atmosphere and the characteristics of the tropopause, *Tellus*, **9**, 259–274, 1957.

Dehghan, A., and W. K. Hocking, Instrumental errors in spectral-width turbulence measurements by radars, *J. Atmos. Solar-Terr. Phys.*, **73**(9), 1052–1068, doi:10.1016/j.jastp.2010.11.011, 2011.

Dehghan, A., W. K. Hocking, and R. Srinivasan, Comparisons between multiple in-situ aircraft turbulence measurements and radar in the troposphere, *J. Atmos. Solar-Terr. Phys.*, **118A**, 64–77, doi:10.1016/j.jastp.2013.10.009, 2014.

de Paula, E. R., and D. L. Hysell, The São Luís 30MHz coherent scatter ionospheric radar: System description and initial results, *Radio Sci.*, **39**, RS1014, doi:10.1029/2003RS002,914, 2004.

Desaubies, Y., and W. K. Smith, Statistics of Richardson number and instability in oceanic internal waves, *J. Phys. Oceanography*, **12**, 1245–1259, 1982.

Dewan, E. M., Turbulent vertical transport due to thin intermittent mixing layers in the stratosphere and other stable fluids, *Science*, **211**, 1041–1042, 1981.

Dewan, E. M., and R. E. Good, Saturation and the "Universal" spectrum for vertical profiles of horizontal scalar winds in the atmosphere, *J. Geophys. Res.*, **91**, 2742–2748, 1986.

Dewan, E. M., and R. H. Picard, Mesospheric bores, *J. Geophys. Res.*, **103**, 6295–6305, 1998.

Dewan, E. M., and R. H. Picard, The origin of mesospheric bores, *J. Geophys. Res.*, **106**, 2921–2927, 2001.

Dewan, E. M., N. Grossbard, A. F. Quesada, and R. E. Good, Spectral analysis of 10 m resolution scalar velocity profiles in the stratosphere, [with correction in *Geophys. Res. Lett.*, **11**, 624, 1984], *Geophys. Res. Lett.*, **11**, 80–83, 1984.

Dewan, E. M., R. H. Picard, R. R. O'Neil, *et al.*, MSX satellite observations of thunderstorm-generated gravity waves in mid-wave infrared images of the upper stratsphere, *Geophys. Res. Lett.*, **25**, 939–942, 1998.

Dhaka, S. K., P. K. Devarajan, Y. Shibagaki, R. K. Choudhary, and S. Fukao, Indian MST radar observations of gravity wave activities associated with tropical convection, *J. Atmos. Solar-Terr. Phys.*, **63**, 1631–1642, 2001.

Dibbern, J., D. Engelbart, U. Goersdorf, *et al.*, Operational aspects of wind profiler radars, in *Instruments and Observing Methods Report No. 79, WMO/TD 1196*, edited by WMO, World Meteorological Organization, 2003.

Dieminger, W., On the causes of excessive absorption in the ionosphere on winter days, *J. Atmos. Terr. Phys.*, **2**, 340, 1952.

Dieminger, W., G. K. Hartmann, and R. Leitinger, *The Upper Atmosphere*, Springer-Verlag, Berlin, Heidelberg and New York, 1996.

Dole, J., R. Wilson, F. Dalaudier, and C. Sidi, Energetics of small scale turbulence in the lower stratosphere from high resolution radar measurements, *Ann. Geophys.*, **19**, 945–952, 2001.

Dong, B., and K. C. Yeh, Resonant and nonresonant wave–wave interactions in an isothermal atmosphere, *J. Geophys. Res.*, **93**, 3729–3744, 1988.

Doviak, R. J., and D. S. Zrnić, Reflection and scatter formula for anisotropically turbulent air, *Radio Sci.*, **19**, 325–336, 1984.

Doviak, R. J., and D. S. Zrnić, *Doppler Radar and Weather Observations*, 2nd ed., Dover Publications, New York, 1993.

Doviak, R. J., R. J. Lataitis, and C. L. Holloway, Cross correlation and cross spectra for spaced antenna wind profilers: 1. Theoretical analysis, *Radio Sci.*, **31**, 157–180, 1996.

Doviak, R. J., G. Zhang, S. A. Cohn, and W. O. J. Brown, Comparison of spaced-antenna cross-beam wind estimators: Theoretical and simulated results, *Radio Sci.*, **39**, Art. No.1006, 2004.

Drabowitch, S., A. Papiernik, H. D. Griffiths, J. Encinias, and B. L. Smith, *Modern Antennas*, 2nd ed., Springer, Dordrecht, 2005.

Driscoll, R. J., and L. A. Kennedy, A model for the spectrum of passive scalars in an isotropic turbulence field, *Phys. Fluids*, **28**, 72–80, 1985.

Drob, D. P., J. T. Emmert, G. Crowley, *et al.*, An empirical model of the Earth's horizontal wind fields: HWM07, *J. Geophys. Res.*, **113**, A12 304, doi:10.1029/2008JA013,668, 2008.

Dunkerton, T. J., Wave transience in a compressible atmosphere. Part I: Transient internal wave, mean-flow interaction, *J. Atmos. Sci.*, **38**, 281–297, 1981.

Dutta, G., P. V. Kumar, P. V. Rao, *et al.*, On the optimum radar beam angle to minimize statistical estimation error of momentum flux using conjugate beam technique, *Geophys. Res. Lett.*, **34**, L22,802 doi:10.1029/2007GL030,652, 2007.

Eaton, F. D., S. A. McLaughlin, and J. R. Hines, A new frequency-modulated continuous wave radar for studying planetary boundary layer morphology, *Radio Sci.*, **30**, 75–88, 1995.

Ebel, A., Eddy diffusion models for the mesosphere and lower thermosphere, *J. Atmos. Terr. Phys.*, **42**, 617–628, 1980.

Ebel, A., A. H. Manson, and C. E. Meek, Short period fluctuations of the horizontal wind measured in the upper middle atmosphere and possible relationships to internal gravity waves, *J. Atmos. Terr. Phys.*, **49**, 385–401, 1987.

Eckermann, S. D., Ray-tracing simulation of the global propagation of inertia gravity waves through the zonal averaged middle atmosphere, *J. Geophys. Res.*, **97**, 15 849–15 866, 1992.

Eckermann, S. D., and W. K. Hocking, The effect of superposition on measurements of atmospheric gravity waves : A cautionary note and some re-interpretations, *J. Geophys. Res.*, **94**, 6333–6339, 1989.

Eckermann, S. D., and P. Preusse, Global measurements of stratospheric mountain waves from space, *Science*, **286**, 1534–1537, 1999.

Eckermann, S. D., and R. A. Vincent, Falling sphere observations of anisotropic gravity wave motions in the upper stratosphere over southern Australia, *Pure Appl. Geophys.*, **130**, 509–532, 1989.

Eckermann, S. D., I. Hirota, and W. K. Hocking, Gravity wave and equatorial morphology of the stratosphere derived from long-term rocket soundings, *Q. J. R. Meteorol. Soc.*, **121**, 149–186, 1995.

Ecklund, W. L., and B. B. Balsley, Long-term observations of the arctic mesosphere with the MST radar at Poker Flat, Alaska, *J. Geophys. Res.*, **86**, 7775–7780, 1981.

Ecklund, W. L., B. B. Balsley, D. A. Carter, *et al.*, Observations of vertical motions in the troposphere and lower stratosphere using three closely spaced ST radars, *Radio Sci.*, **20**, 1196–1206, 1985.

Edmon, H. J. J., B. J. Hoskins, and M. E. McIntyre, Eliassen–Palm cross sections for the troposphere, *J. Atmos. Sci.*, **37**, 2600–2616, 1980.

Elford, W. G., A study of winds between 80 and 100 km in medium latitudes, *Planet. Space Sci.*, **1**, 94–101, 1959.

Elford, W. G., Novel applications of MST radars in meteor studies, *J. Atmos. Solar-Terr. Phys.*, **63**, 143–153, 2001.

Elford, W. G., and D. S. Robertson, Measurements of winds in the upper atmosphere by means of drifting meteor trails II, *J. Atmos. Terr. Phys.*, **4**, 271–284, 1953.

Elford, W. G., and A. D. Taylor, Measurements of Faraday rotation of radar meteor echoes for the modelling of electron densities in the lower ionosphere, *J. Atmos. Solar-Terr. Phys.*, **59**, 1021–1024, 1997.

Eliassen, A., and E. Palm, On the transfer of energy in stationary mountain waves, *Geophys. Publ*, **22**, 1–23, 1960.

Ellyett, C. D., and J. M. Watts, Stratification in the lower ionosphere, *J. Res. Nat. Bur. Stand.*, **63DN2**, 117–134, 1959.

Espy, P. J., R. E. Hibbins, G. R. Swenson, *et al.*, Regional variations of mesospheric gravity-wave momentum flux over Antarctica, *Ann. Geophys.*, **24**, 81–88, 2006.

Evans, J., Theory and practice of ionospheric study by Thomson scatter radar, *Proc IEEE*, **57**, 496–500, 1969.

Fairall, C. W., A. B. White, and D. W. Thomson, A stochastic model of gravity-wave-induced clear-air turbulence, *J. Atmos. Sci.*, **48**, 1771–1790, 1991.

Fan, Y., J. Klostermeyer, and R. Rüster, VHF radar observation of gravity wave critical levels in the mid-latitude summer mesopause region, *Geophys. Res. Lett.*, **18**, 697–700, 1991.

Farley, D. T., Faraday rotation measurements using incoherent scatter, *Radio Sci.*, **4**, 143–152, 1969.

Farley, D. T., On-line data processing techniques for MST radars, *Radio Sci.*, **20**, 1177–1184, 1985.

Farley, D. T., B. B. Balsley, W. E. Swartz, and C. La Hoz, Tropical winds measured by the Arecibo radar, *J. Appl. Meteorol.*, **18**, 227–230, 1979.

Farley, D. T., H. Ierkic, and B. Fejer, Radar interferometry: A new technique for studying plasma turbulence in the ionosphere, *J. Geophys. Res.*, **86**, 1467–1472, 1981.

Fejer, J. A., Causality and the Lorentz polarization term, *J. Atmos. Solar-Terr. Phys.*, **47**, 513–516, 1985.

Fenn, A. J., *Adaptive Antenna and Phased Arrays for Radar and Communications*, Artec House, Boston, 2008.

Ferraz-Mello, S., Estimation of periods from unequally spaced observations, *Astronomical J.*, **86.4**, 619–624, 1981.

Flock, W. L., and B. B. Balsley, VHF radar returns from the D region of the equatorial ionosphere, *J. Geophys. Res.*, **72**, 5537, 1967.

Flood, W. A., Revised theory for partial reflection D-region measurements, *J. Geophys. Res.*, **73**, 5585–5598, 1968.

Flood, W. A., Reply (to comments by Holt regarding "Revised theory for partial reflection D-region measurements"), *J. Geophys. Res.*, **74**, 5183–5186, 1969.

Fooks, G. F., Ionospheric drift measurements using correlation analysis; methods of computation and interpretation of results, *J. Atmos. Terr. Phys.*, **27**, 979, 1965.

Forbes, J. M., S. E. Palo, X. Zhang, Y. I. Portnyagin, M. N. A., and E. G. Merzlyakov, Lamb waves in the lower thermosphere: Observational evidence and global consequences, *Geophys. Res. Lett.*, **104**, 17 107–17 115, 1999.

Frank, J., and J. D. Richards, Chapter 13 in *Radar Handbook*, in *Phased Array Radar Antennas*, edited by M. I. Skolnik, pp. 13.1–13.74, McGraw-Hill, New York, 2008.

Franke, P. M., D. Thorsen, M. Champion, S. J. Franke, and E. Kudeki, Comparisons of time and frequency domain techniques for wind velocity estimation using multiple receiver MF radar data, *Geophys. Res. Lett.*, **17**, 2193–2196, 1990.

Franke, P. M., S. Mahmoud, K. Raizada, *et al.*, Computation of clear-air radar backscatter from numerical simulations of turbulence: 1. Numerical methods and evaluation of biases, *J. Geophys. Res. (Atmospheres)*, **116**, 2156–2202, doi:10.1029/2011JD015,895, 2011.

Franke, S. J., Pulse compression and frequency domain interferometry with a frequency-hopped MST radar, *Radio Sci.*, **25**, 565–574, 1990.

Franke, S. J., J. Röttger, and C. LaHoz, Frequency domain interferometry of polar

Fukao, S., T. Sato, T. Tsuda, *et al.*, The MU radar with an active phased array system: 1. Antenna and power amplifiers, *Radio Sci.*, **20**, 1155–1168, 1985a.

Fukao, S., T. Sato, T. Tsuda, *et al.*, The MU radar with an active phased array system: 2. In-house equipment, *Radio Sci.*, **20**, 1169–1176, 1985b.

Fukao, S., K. Wakasugi, T. Sato, *et al.*, Direct measurement of air and precipitation particle motion by very high frequency Doppler radar, *Nature*, **316**, 712–714, 1985c.

Fukao, S., T. Sato, H. Hojo, I. Kimura, and S. Kato, A numerical consideration on edge effect of planar dipole phased arrays, *Radio Sci.*, **21**, 1–12, 1986a.

Fukao, S., T. Sato, H. Hojo, I. Kimura, and S. Kato, Effects of antenna element structure on element properties and array pattern of a planar phased array, *Radio Sci.*, **21**, 56–64, 1986b.

Fukao, S., M. Inaba, I. Kimura, *et al.*, A systemic error in MST/ST radar wind measurement induced by a finite range volume effect: 2. Numerical considerations, *Radio Sci.*, **23**, 74–82, 1988b.

Fukao, S., T. Sato, T. Tsuda, *et al.*, MU radar: New capabilities and system calibrations, *Radio Sci.*, **25**, 477–485, 1990.

Fukao, S., M. D. Yamanaka, N. Ao, *et al.*, Seasonal variability of vertical eddy diffusivity in the middle atmosphere: 1. Three-year observations by the middle and upper atmosphere radar, *J. Geophys. Res*, **99**, 18 973–18 987, 1994.

Fukao, S., H. Hashiguchi, M. K. Yamamoto, *et al.*, Equatorial atmosphere radar (EAR): System description and first results, *Radio Sci.*, **38**, doi:10.1029/2002RS002,767, 2003.

Fukao, S., T. Sato, P. T. May, *et al.*, A systematic error in MST/ST radar wind measurement induced by a finite range volume effect: 1. Observational results, *Radio Sci.*, **23**, 59–73, 1988a.

Furumoto, J., and T. Tsuda, Characteristics of energy dissipation rate and effect of humidity on turbulence echo power revealed by MU radar-RASS measurements, *J. Atmos. Solar-Terr. Phys.*, **63**, 285–294, 2001.

Furumoto, J., K. Kurimoto, and T. Tsuda, Continuous observations of humidity profiles with the MU radar-RASS combined with GPS and radiosonde measurements, *J. Atmos. Oceanic Technol.*, **20**, 23–41, 2003.

Furumoto, J., S. Iwai, H. Fujii, *et al.*, Estimation of humidity profiles with the L-band boundary layer radar-RASS measurements, *J. Meteor. Soc. Japan*, **83**(5), 895–908, 2005.

Furumoto, J., T. Tsuda, S. Iwai, and T. Kozu, Continuous humidity monitoring in a tropical region with the Equatorial Atmosphere Radar, *J. Atmos. Oceanic. Tech.*, **23**, 538–551, 2006.

Furumoto, J., S. Imura, T. Tsuda, *et al.*, The variational assimilation method for the retrieval of humidity profiles with the wind-profiling radar, *J. Atmos. Ocean. Technol.*, **24**, 1525–1545 doi:10.1175/JTECH2074.1, 2007.

Gage, K. S., Radar observations of the free atmosphere: Structure and dynamics, in *Radar in Meteorology*, edited by D. Atlas, pp. 534–565, American Met. Soc., 1990.

Gage, K. S., and B. B. Balsley, Doppler radar probing of the clear atmosphere, *Bull. Am. Meteorol. Soc.*, **59**, 1074–1093, 1978.

Gage, K. S., and J. L. Green, Evidence for specular reflection from monostatic VHF radar observations of the stratosphere, *Radio Sci.*, **13**, 991–1001, 1978.

Gage, K. S., and J. L. Green, Tropopause detection by partial specular reflection using VHF radar, *Science*, **203**, 1238–1240, 1979.

Gage, K. S., and J. L. Green, A technique for determining the temperature profile from VHF radar observations, *J. Appl. Meteorol.*, **21**, 1146–1149, 1982a.

Gage, K. S., and J. L. Green, An objective technique for the determination of tropopause height from VHF radar observations, *J. Appl. Meteorol.*, **21**, 1150–1154, 1982b.

Gage, K. S., J. L. Green, and T. E. Van Zandt, Use of Doppler radar for the measurement of atmospheric turbulence parameters from the intensity of clear air echoes, *Radio Sci.*, **15**, 407–416, 1980.

Gage, K. S., B. B. Balsley, and J. L. Green, Fresnel scattering model for the specular echoes observed by VHF radars, *Radio Sci.*, **16**, 1447–1453, 1981.

Gage, K. S., W. L. Ecklund, and B. B. Balsley, A modified Fresnel scattering model for the parameterization of Fresnel returns, *Radio Sci.*, **20**, 1493–1502, 1985.

Gage, K. S., J. R. McAfee, W. G. Collins, *et al.*, A comparison of winds observed at Christmas Island using wind-profiling Doppler radar with NMC and ECMWF analyses, *Bull. Am. Meteorol. Soc.*, **69**, 1041–1047, 1988.

Gage, K. S., B. B. Balsley, W. L. Ecklund, D. A. Carter, and J. R. McAfee, Wind-profiler related research in the tropical Pacific, *J. Geophys. Res.*, **96**, 3209–3220, 1991a.

Gage, K. S., J. R. McAfee, D. A. Carter, *et al.*, Long-term mean vertical motion over the tropical Pacific: Wind-profiling Doppler radar measurements, *Science*, **254**, 1771–1773, 1991b.

Garbanzo-Salas, M., and W. K. Hocking, Spectral analysis comparisons of Fourier-theory-based methods and minimum variance (Capon) methods, *J. Atmos. Terr. Phys.*, **32**, 92–100, doi:10.1016/j.jastp.2015.07.003, 2015.

Garcia, R. R., and S. Solomon, A numerical model of the zonally averaged dynamical and chemical structure of the middle atmosphere, *J. Geophys. Res.*, **88**, 1379–1400, 1983.

Garcia, R. R., and S. Solomon, The effect of breaking gravity waves on the dynamics and chemical composition of the mesosphere and lower thermosphere, *J. Geophys. Res.*, **90**, 3850–3868, 1985.

Gardner, C. S., Diffusive filtering theory of gravity wave spectra in the atmosphere, *J. Geophys. Res.*, **99**, 20 601–20 622, 1994.

Gardner, C. S., Testing theories of atmospheric gravity wave saturation and dissipation, *J. Atmos. Terr. Phys.*, **58**, 1575–1589, 1996.

Gardner, C. S., Theoretical models for gravity wave horizontal wave number spectra: Effects of wave field anisotropies, *J. Geophys. Res.*, **103**, 6417–6425, 1998.

Gardner, C. S., and M. J. Taylor, Observational limits for lidar, radar and airglow imager measurements of gravity wave parameters, *J. Geophys. Res.*, **103**, 6427–6437, 1998.

Gardner, C. S., C. A. Hostetler, and S. J. Franke, Gravity wave models for the horizontal wave number spectra of atmospheric velocity and density fluctuations, *J. Geophys. Res.*, **98**, 1035–1049, 1993a.

Gardner, C. S., C. A. Hostetler, and S. Lintelman, Influence of the mean wind field on the separability of atmospheric perturbation spectra, *J. Geophys. Res.*, **98**, 8859–8872, 1993b.

Gardner, F. F., and J. L. Pawsey, Study of the ionospheric D-region using partial reflections, *J. Atmos. Terr. Phys.*, **3**, 321, 1953.

Garrett, C., and W. Munk, Space time scales of internal waves, *Geophys. Fluid Dynamics*, **2**, 225–264, 1972.

Garrett, C., and W. Munk, Space time scales of internal waves: A progress report, *J. Geophys. Res.*, **80**, 291–297, 1975.

Garrett, H. B., and J. M. Forbes, Tidal structure of the thermosphere at equinox, *J. Atmos. Terr. Phys.*, **40**, 657–668, 1978.

Gavrilov, N. M., S. Fukao, T. Nakamura, *et al.*, Comparative study of interannual changes of the mean winds and gravity wave activity in the middle atmosphere over Japan, Central Europe and Canada, *J. Atmos. Solar-Terr. Phys.*, **64**, 1003–1010, 2002.

Geller, M. A., Dynamics of the middle atmosphere, *J. Atmos. Terr. Phys.*, **41**, 683–705, 1979.

Gibson-Wilde, D. E., J. A. Werne, D. C. Fritts, and R. J. Hill, Direct numerical simulation of VHF radar measurements of turbulence in the mesosphere, *Radio Sci.*, **35**, 783–798, 2000.

Gnanalingam, S., and K. Weekes, Weak echoes from the ionosphere with radiowaves of frequency 1.42 Mc/s, *Nature*, **170**, 113–114, 1952.

Golay, M. J. E., Complementary series, *IEEE Trans. Inform. Theory*, **IT-7**, 82–87, 1961.

Goldstein, H., D. E. Kerr, and A. E. Bent, Meteorological echoes, in *Propagation of Short Radio Waves*, edited by D. E. Kerr, pp. 588–636, McGraw-Hill, New York, (republished 1990 by Peter Peregrinus Ltd, London), 1951.

Golomb, S., *Shift Register Sequences*, rev. ed., Aegean Park, Walnut Creek, CA, 1981.

Gordiets, B. F., Y. N. Kulikov, M. N. Markov, M. Marov, and J. Ya, Numerical modeling of thermospheric heat budget, *J. Geophys. Res.*, **87**, 4504–4514, 1982.

Gordon, W. E., Incoherent scattering of radio waves by free electrons with applications to space exploration by radar, *Proc. I. R. E.*, **46**, 1824–1829, 1958.

Gossard, E. D., and W. H. Hooke, *Waves in the Atmosphere*, Elsevier Scientific Publ. Co., Amsterdam, 1975.

Gossard, E. E., Measuring drop-size distributions in clouds with a clear-air-sensing Doppler radar, *J. Atmos. Oceanic Technol.*, **5**, 640–649, 1988.

Gossard, E. E., Radar research on the atmospheric boundary layer, in *Radar in Meteorology*, edited by D. Atlas, pp. 477–527, Am. Meteorol. Soc., Boston, Mass., 1990.

Gossard, E. E., and K. C. Yeh, Foreword to a special issue of *Radio Science* on Radar, *Radio Sci.*, **15**, 147–150, 1980.

Gossard, E. E., J. H. Richter, and D. Atlas, Internal waves in the atmosphere from high-resolution radar measurements, *J. Geophys. Res.*, **75**, 3523–3536, 1970.

Gossard, E. E., R. B. Chadwick, W. D. Neff, and K. P. Moran, The use of ground-based Doppler radars to measure gradients, fluxes and structure parameters in elevated layers, *J. Appl. Meteorol.*, **21**, 211–226, 1982.

Gossard, E. E., R. B. Chadwick, T. R. Detman, and J. Gaynor, Capability of surface-based clear-air Doppler radar for monitoring meteorological structure of elevated layers, *J. Climate and Appl. Meteorol.*, **23**, 474–485, 1984.

Green, J. L., and K. S. Gage, A re-examination of the range resolution dependence of backscattered power observed by VHF radars at vertical incidence, *Radio Sci.*, **20**, 1001–1005, 1985.

Green, J. L., K. S. Gage, and T. E. Van Zandt, Atmospheric measurements by VHF pulsed Doppler radar, *IEEE Trans. Geosci. Electron.*, **GE-17**, 262–280, 1979.

Green, J. L., W. L. Clark, J. M. Warnock, and K. J. Ruth, Absolute calibration of MST/ST radars, preprint volume, in *21st Conference on Radar Meteorology*, The American Meteorol. Soc., Edmonton, Alberta, Ca., 1983.

Greenhow, J. S., Systematic wind measurements at altitudes of 80–100 km using radio echoes from meteor trails, *Philosophical Magazine*, **45**, 471–490, 1954.

Greenhow, J. S., and E. I. Neufeld, Diurnal and seasonal wind variations in the upper atmosphere, *Philosophical Magazine*, **46**, 549–562, 1955.

Gregory, J. B., Atmospheric reflections from heights below the E region, *Aust. J. Phys.*, **9**, 324–342, 1956.

Gregory, J. B., Radio wave reflections from the mesosphere: 1. Heights of occurrence, *J. Geophys. Res.*, **55**, 429–445, 1961.

Gregory, J. B., The influence of atmospheric circulation on mesospheric electron densities in winter, *J. Atmos. Sci.*, **22**, 18–23, 1965.

Gregory, J. B., and A. H. Manson, Mesospheric electron number densities at 35 °S latitude, *J. Geophys. Res*, **72**, 1073–1080, 1967.

Gregory, J. B., and R. A. Vincent, Structure of partially reflecting regions in the lower ionosphere, *J. Geophys. Res*, **75**, 6387–6389, 1970.

Groves, G. V., Wind models from 60–130 km altitude for different months and latitudes, *J. Br. Interplan. Soc.*, **22**, 285–307, 1969.

Guest, F., M. Reeder, C. Marks, and D. Karoly, Inertia-gravity waves observed in the lower stratosphere over Macquarie Island, *J. Atmos. Sci.*, **57**, 737–752, 2000.

Gurvich, A. S., A model of three-dimensional spectrum of locally axisymmetric temperature inhomogeneities in a stably stratified atmosphere, *Izv. Atmos. Ocean Phys.*, **30**, 149, 1994.

Gurvich, A. S., A heuristic model of three-dimensional spectra of temperature inhomogeneities in the stably stratified atmosphere, *Ann. Geophys*, **15**, 856–869, 1997.

Gurvich, A. S., and A. I. Kon, Aspect sensitivity of radar returns from anisotropic turbulent irregularities, *J. Electromagn. Waves Appl.*, **7**, 1343–1353, 1993.

Guzmán, A. E., J. May, H. Alvarez, and K. Maeda, All-sky Galactic radiation at 45 MHz and spectral index between 45 and 408 MHz, *Astronomy and Astrophysics*, **525**, A138, 2011.

Hafgors, T., Incoherent scatter radar observations of the ionosphere, in *Handbook for MAP, International School on Atmospheric Radar*, edited by S. Fukao, vol. 30, pp. 333–364, SCOSTEP Secretariat, Dept. of Electr. Computer Eng., Univ. of Illinois, Urbana, IL 61801, USA, 1989a.

Hagfors, T., The scattering of E. M. waves from density fluctuations in a plasma, in *Proc. of the EISCAT Summer School, Tromso, Norway,June 5-13, 1975*, edited by A. Brekke, pp. 15–28, Scandanavian Univ. Books, 1975.

Hall, C. M., T. Aso, M. Tsutsumi, J. Hoeffner, and F. Sigernes, Multi-instrument derivation of 90 km temperatures over Svalbard (78 °N 16 °E), *Radio Sci.*, **39**, RS6001, doi:10.1029/2004RS003,069, 2004.

Hall, T. J., J. Y. N. Cho, M. C. Kelley, and W. K. Hocking, A re-evaluation of the Stokes drift in the polar summer mesosphere, *J. Geophys. Res.*, **97**, 887–897, 1992.

Hamazu, K., H. Hashiguchi, T. Wakayama, *et al.*, A 35-GHz scanning Doppler radar for fog observations, *J. Atmos. Oceanic Tech.*, **20**, 972–986, 2003.

Hamilton, K., High resolution global modeling of the atmospheric circulation, *Adv. Atmos. Sci.*, **23**, 842–856, 2006.

Hamsen, A. R., G. D. Nastrom, and J. A. Otkin, MST radar observations of gravity waves and turbulence near thunderstorms, *J. Appl. Meteorol.*, **41**(3), 298–305, 2002.

Harper, R. M., and R. F. Woodman, Preliminary multiheight radar observations of waves and winds in the mesosphere over Jicamarca, *J. Atmos. Terr. Phys.*, **39**, 959, 1977.

Hauchecorne, A., M.-L. Chanin, and R. Wilson, Mesospheric temperature inversions and gravity wave breaking, *Geophys. Res. Lett.*, **14**, 933–936, 1987.

Havnes, O., F. Melandsø, C. L. Hoz, T. Aslaksen, and T. Hartquist, Charged dust in the Earth's mesopause; effects on radar backscatter, *Phys. Scr.*, **45**, 535–544, 1992.

Havnes, O., J. Trøim, T. Blix, *et al.*, First detection of charged dust particles in the Earth's atmosphere, *J. Geophys. Res.*, **101**, 10 829–10 847, 1996.

Hawkes, R., I. Mann, and P. Brown, *Modern Meteor Science: An Interdisciplinary View*, Springer, Dordrecht, 2005.

Haynes, P. H., C. J. Marks, M. E. McIntyre, T. G. Shepherd, and K. P. Shine, On the "downward control" of extratropical diabatic circulations by eddy-induced mean zonal forces, *J. Atmos. Sci.*, **48**, 651–678, 1991.

Hecht, E., and A. Zajac, *Optics*, Addison-Wesley, Reading, MA, 1974.

H'elal, D., M. Crochet, H. Luce, and E. Spano, Radar imaging and high-resolution array processing applied to a classical VHF-ST profiler, *J. Atmos. Solar-Terr. Phys.*, **63**, 263–274, 2001.

Held, I., and A. Hou, Nonlinear axially symmetric circulations in a nearly inviscid atmosphere, *J. Atmos. Sci.*, **37**, 515–533, 1980.

Held, I. M., The general circulation of the atmosphere, paper presented at 2000 *Woods Hole Oceanographic Institute Geophysical Fluid Dynamics Program* (Available at http://gfd.whoi.edu/proceedings/2000/PDF vol2000.html), Woods Hole Oceanographic Institute, Woods Hole, Mass., USA, 2000.

Hertzog, A., C. Souprayen, and A. Hauchecorne, Observation and backward trajectory of an inertia-gravity wave in the lower stratosphere, *Ann. Geophys.*, **19**, 1141–1155, 2001.

Hertzog, A., G. Boccara, R. A. Vincent, F. Vial, and P. Cocquerez, Estimation of gravity wave momentum flux and phase speeds from quasi-Lagrangian stratospheric balloon flights. Part II: Results from the Vorcore campaign in Antarctica, *J. Atmos. Sci.*, **65**, 3056–3070, 2008.

Hildebrand, P. H., and R. S. Sekhon, Objective determination of the noise level in Doppler spectra, *J. Appl. Meteorol.*, **13**, 808–811, 1974.

Hill, R. J., Nonneutral and quasi-neutral diffusion of weakly ionized multiconstituent plasma, *J. Geophys. Res.*, **83**, 989–998, 1978.

Hill, R. J., and S. F. Clifford, Modified spectrum of atmospheric temperature fluctuations and its application to optical propagation, *J. Opt. Soc. Am.*, **68**, 892–899, 1978.

Hines, C. O., Internal atmospheric gravity waves of ionospheric heights, *Canadian J. Phys.*, **38**, 1441–1481, 1960.

Hines, C. O., Generation of turbulence by atmospheric gravity waves, *J. Atmos. Sci.*, **45**, 1269–1278, 1988.

Hines, C. O., The saturation of gravity waves in the middle atmosphere. Part I: Critique of linear instability theory, *J. Atmos. Sci.*, **48**, 1348–1359, 1991a.

Hines, C. O., The saturation of gravity waves in the middle atmosphere. Part II: Development of Doppler-spread theory, *J. Atmos. Sci.*, **48**, 1360–1379, 1991b.

Hines, C. O., The saturation of gravity waves in the middle atmosphere. Part III: Formation of the turbopause and the turbulence layers beneath it, *J. Atmos. Sci.*, **48**, 1380–1385, 1991c.

Hines, C. O., The saturation of gravity waves in the middle atmosphere. Part IV: Cutoff of the incident wave spectrum, *J. Atmos. Sci.*, **50**, 3045–3060, 1993.

Hines, C. O., Modulated mountain waves, *J. Atmos. Sci.*, **52**, 602–606, 1995a.

Hines, C. O., Comments on "Observations of low-frequency gravity waves in the lower stratosphere over Arecibo," *J. Atmos. Sci.*, **52**, 607–610, 1995b.

Hines, C. O., Reply, *J. Atmos. Sci.*, **52**, 613, 1995c.

Hines, C. O., Nonlinearity of gravity wave saturated spectra in the middle atmosphere, *Geophys. Res. Lett.*, **23**, 3309–3312, 1996.

Hines, C. O., Theory of the Eulerian tail in the spectra of atmospheric and oceanic internal gravity waves, *J. Fluid Mech.*, **448**, 289–313, 2001.

Hines, C. O., G. W. Adams, J. W. Brosnahan, *et al.*, Multi-instrument observations of mesospheric motions over Arecibo: Comparisons and interpretations, *J. Atmos. Terr. Phys.*, **55**, 241, 1993.

Hirota, I., Climatology of gravity waves in the middle atmosphere, *J. Atmos. Terr. Phys.*, **46**, 767–773, 1984.

Hirota, I., Gravity waves, in *Middle Atmosphere Handbook*, vol. 16, pp. 144–148, Scostep Secretariat, University of Illinois, U. S. A., 1985.

Hitschfeld, W., and A. S. Dennis, *Measurement and Calculation of Fluctuations in Radar Echoes from Snow*, Sci. Rep. MW-23, McGill University, Montreal, Canada, 1956.

Hocking, A., and W. K. Hocking, Turbulence anisotropy determined by windprofiler radar and its correlation with rain events in Montreal, Canada, *J. Atmos. Oceanic Technol.*, **24**, 40–51, 2007.

Hocking, W. K., Angular and temporal characteristics of partial reflections from the

D-region of the ionosphere, *J. Geophys. Res.*, **84**, 845–851, 1979.

Hocking, W. K., Investigations of the movement and structure of D-region ionospheric irregularities, PhD thesis, University of Adelaide, Adelaide, Australia, 1981.

Hocking, W. K., On the extraction of atmospheric turbulence parameters from radar backscatter Doppler spectra - I. Theory, *J. Atmos. Terr. Phys.*, **45**, 89–102, 1983a.

Hocking, W. K., Mesospheric turbulence intensities measured with a HF radar at 35 °S - II, *J. Atmos. Terr. Phys.*, **45**, 103–114, 1983b.

Hocking, W. K., The spaced antenna drift method, in *Handbook for MAP*, vol. 9, pp. 171–186, Univ. of Illinois, Urbana, 1983c.

Hocking, W. K., Measurement of turbulent energy dissipation rates in the middle atmosphere by radar techniques: A review, *Radio Sci.*, **20**, 1403–1422, 1985.

Hocking, W. K., Observation and measurement of turbulence in the middle atmosphere with a VHF radar, *J. Atmos. Terr. Phys*, **48**, 655–670, 1986.

Hocking, W. K., Radar studies of small scale structure in the upper middle atmosphere and lower ionosphere, *Adv. Space Res.*, **7**, 327–338, 1987a.

Hocking, W. K., Reduction of the effects of non-stationarity in studies of amplitude statistics of radio wave backscatter, *J. Atmos. Terr. Phys.*, **49**, 1119–1131, 1987b.

Hocking, W. K., Two years of continuous measurements of turbulence parameters in the upper mesosphere and lower thermosphere made with a 2-MHz radar, *J. Geophys. Res.*, **93**, 2475–2491, 1988.

Hocking, W. K., Turbulence in the region 80–120 km, *Adv. Space Res.*, **10**, 153–161, 1990.

Hocking, W. K., The effects of middle atmosphere turbulence on coupling between atmospheric regions, *J. Geomag. Geoelectr.*, **43**, Suppl., 621–636, 1991.

Hocking, W. K., On the relationship between the strength of atmospheric radar backscatter and the intensity of atmospheric turbulence, *Adv. Space Res.*, **12**, 207–213, 1992.

Hocking, W. K., An assessment of the capabilities and limitations of radars in measurements of upper atmosphere turbulence, *Adv. Space Res.*, **17**, 37–47, 1996a.

Hocking, W. K., Some new perspectives on viscosity and thermal conduction waves as a cause of "specular" reflectors in radar studies of the atmosphere, in *STEP Handbook, Proceedings of the Seventh Workshop on Technical and Scientific Aspects of MST Radar*, edited by B. Edwards, pp. 82–85, Hilton Head Island SC USA, 1996b.

Hocking, W. K., Dynamical coupling processes between the middle atmosphere and lower ionosphere, *J. Atmos. Terr. Phys.*, **58**, 735–752, 1996c.

Hocking, W. K., System design, signal processing procedures and preliminary results for the Canadian (London, Ontario) VHF atmospheric radar, *Radio Sci.*, **32**, 687–706, 1997a.

Hocking, W. K., Recent advances in radar instrumentation and techniques for studies of the mesosphere, stratosphere and troposphere, *Radio Sci.*, **32**, 2241–2270, 1997b.

Hocking, W. K., Strengths and limitations for MST radar measurements of middle atmosphere winds, *Ann. Geophys.*, **15**, 1111–1122, 1997c.

Hocking, W. K., The dynamical parameters of turbulence theory as they apply to middle atmosphere studies, *Earth, Planets and Space*, **51**, 525–541, 1999a.

Hocking, W. K., Temperatures using radar-meteor decay times, *Geophys. Res. Lett.*, **26**, 3297–3300, 1999b.

Hocking, W. K., Real-time meteor entrance speed determinations made with interferometric meteor radars, *Radio Sci.*, **35**, 1205–1220, 2000.

Hocking, W. K., VHF tropospheric scatterer anisotropy at Resolute Bay and its implications for tropospheric radar-derived wind accuracies, *Radio Sci.*, **36**, 1777–1793, 2001a.

Hocking, W. K., Middle atmosphere dynamical studies at Resolute Bay over a full representative year: Mean winds, tides and special oscillations, *Radio Sci.*, **36**, 1795–1822, 2001b.

Hocking, W. K., A hybrid Yagi/loop antenna system for VHF boundary layer studies, (invited), session 3Ac3 (Novel radar methods for studying the structure and dynamics of the atmosphere and ionosphere), in *PIERS 2002 (Progress In Electromagnetics Research Symposium)*, Cambridge, Mass., USA, 2002.

Hocking, W. K., Evidence for viscosity, thermal conduction and diffusion waves in the Earth's atmosphere (invited), *Review of Scientific Instruments*, **74**(1), 420–426, 2003a.

Hocking, W. K., A new approach to fast and accurate calculation of spectral beam-broadening for turbulence studies, paper I.3.33, in *10th International Workshop on Technical and Scientific Aspects of MST Radar*, Piura, Peru, 2003b.

Hocking, W. K., Radar meteor decay rate variability and atmospheric consequences, *Ann. Geophys.*, **22**, 3805–3814, 2004a.

Hocking, W. K., Experimental radar studies of anisotropic diffusion of high altitude meteor trails, *Earth, Moon, Planets*, **95**, 671–679, 2004b.

Hocking, W. K., A new approach to momentum flux determinations using SKiYMET meteor radars, *Ann. Geophys.*, **23**, 2433–2439, 2005.

Hocking, W. K., A review of mesosphere-stratosphere-troposphere (MST) radar developments and studies, circa 1997-2008, *J. Atmos. Solar-Terr. Phys.*, **73**, 848–882, 2011.

Hocking, W. K., The atmospheric wave graveyard, (workshop), in *14th International Workshop on Technical and Scientific Aspects of MST Radar plus first joint MST and Ionospheric workshop (MST14/iMST1), May 25–31, 2014*, INPE (Brazilian National Institute for Space Research) Auditorium Fernando de Mendonca, Av. dos Astronautas, 1758 Jd da Granja - Sao Jose dos Campos, Brazil, 2014.

Hocking, W. K., and A. M. Hamza, A quantitative measure of the degree of anisotropy of turbulence in terms of atmospheric parameters, with particular relevance to radar studies, *J. Atmos. Solar-Terr. Phys.*, **59**, 1011–1020, 1997.

Hocking, W. K., and A. Hocking, Temperature tides determined with meteor radar, *Ann. Geophys.*, **20**, 1447–1467, 2002.

Hocking, W. K., and A. Hocking, Diagnostic capabilities of measurements of backscatter anisotropy, (invited), paper II. E. 2, in *10th International Workshop on Technical and Scientific Aspects of MST Radar*, Piura, Peru, 2003.

Hocking, W. K., and A. Hocking, Procedure to extract boundary-layer wind measurements using relatively long pulses, in *Handbook for STEP, Proceedings of the Twelfth International Workshop on Technical and Scientific Aspects of MST Radar, London, Ont., May 17–23, 2009*, edited by N. Swarnalingam and W. K. Hocking, pp. 135–138, Canadian Association of Physicists, Canada, 2010.

Hocking, W. K., and K. L. Mu, Upper and middle tropospheric kinetic energy dissipation rates from measurements of $\overline{C_n^2}$ – Review of theories, in-situ investigations, and experimental studies using the Buckland Park atmospheric radar in Australia, *J. Atmos. Terr. Phys.*, *59*, 1779–1803, 1997.

Hocking, W. K., and J. Röttger, Pulse-length dependence of radar signal strengths for Fresnel backscatter, *Radio Sci.*, **18**, 1312–1324, 1983.

Hocking, W. K., and J. Röttger, Studies of polar mesosphere summer echoes over EISCAT using calibrated signal strengths and statistical parameters, *Radio Sci.*, **32**, 1425–1444, 1997.

Hocking, W. K., and J. Röttger, The structure of turbulence in the middle and lower atmosphere seen by and deduced from MF, HF and VHF radar, with special emphasis on small-scale features and anisotropy, *Ann. Geophys.*, **19**, 933–944, 2001.

Hocking, W. K., and T. Thayaparan, Simultaneous and co-located observation of winds and tides by MF and meteor radars over London, Canada, (43 °N, 81 °W) during 1994–1996, *Radio Sci.*, **32**, 833–865, 1997.

Hocking, W. K., and R. A. Vincent, Comparative observations of D-region HF partial reflections at 2 and 6 MHz, *J. Geophys. Res.*, **87**, 7615–7624, 1982a.

Hocking, W. K., and R. A. Vincent, A comparison between HF partial reflection profiles from the D-region and simultaneous Langmuir probe electron density measurements, *J. Atmos. Terr. Phys.*, **44**, 843–854, 1982b.

Hocking, W. K., and R. L. Walterscheid, The role of Stokes' diffusion in middle atmospheric transport, in *Coupling Processes in the Lower and Middle Atmosphere*, edited by E. V. Thrane, T. A. Blix, and D. C. Fritts, vol. 387 of *C: Mathematical and Physical Sciences*, pp. 305–328, NATO (North Atlantic Treaty Organization), Kluwer Academic Publishers, Dordrecht, Boston and London, 1993.

Hocking, W. K., G. Schmidt, and P. Czechowsky, *Absolute Calibration of the SOUSY VHF Stationary radar*, Max-Planck-Institut für Aeronomie report MPAE-W-00-83-14, Katlenburg-Lindau, FRG, 1983.

Hocking, W. K., R. Rüster, and P. Czechowsky, *Observation and Measurement of Turbulence and Stability in the Middle Atmosphere with a VHF Radar*, University of Adelaide internal report ADP-335, University of Adelaide, Adelaide, SA, Australia, 1984.

Hocking, W. K., R. Rüster, and P. Czechowsky, Absolute reflectivities and aspect sensitivities of VHF radio wave scatterers measured with the SOUSY radar, *J. Atmos. Terr. Phys.*, **48**, 131–144, 1986.

Hocking, W. K., P. T. May, and J. Röttger, Interpretation, reliability and accuracies of parameters deduced by the spaced antenna method in middle atmosphere applications, *Pure and Applied Geophys.*, **130**, 571–604, 1989.

Hocking, W. K., S. Fukao, T. Tsuda, *et al.*, Aspect sensitivity of stratospheric VHF radiowave scatterers, particularly above 15 km altitude, *Radio Sci.*, **25**, 613–627, 1990.

Hocking, W. K., S. Fukao, M. K. Yamamoto, T. Tsuda, and S. Kato, Viscosity waves and thermal-conduction waves as a cause of "specular" reflectors in radar studies of the atmosphere, *Radio Sci.*, **26**, 1281–1303, 1991.

Hocking, W. K., T. Thayaparan, and J. Jones, Meteor decay times and their use in determining a diagnostic mesospheric temperature–pressure parameter: Methodology and one year of data, *Geophys. Res. Lett.*, **24**, 2977–2980, 1997.

Hocking, W. K., B. Fuller, and B. Vandepeer, Real-time determination of meteor-related parameters utilizing modern digital technology, *J. Atmos. Solar-Terr. Phys.*, **63**, 155–169, 2001a.

Hocking, W. K., M. C. Kelley, R. Rogers, *et al.*, Resolute Bay VHF radar: A multipurpose tool for studies of tropospheric motions, middle atmosphere dynamics, meteor physics and ionospheric physics, *Radio Sci.*, **36**, 1839–1857, 2001b.

Hocking, W. K., T. Thayaparan, and S. J. Franke, Method for statistical comparison of geophysical data by multiple instruments which have differing accuracies, *Adv. Space Res.*, **27**, 1089–1098, 2001c.

Hocking, W. K., J. Singer, W. Bremer, *et al.*, Meteor radar temperatures at multiple sites derived with SKiYMET radars and compared to OH, rocket and lidar measurements, *J. Atmos. Solar-Terr. Phys.*, **66**, 585–593, 2004.

Hocking, W. K., P. S. Argall, R. P. Lowe, R. J. Sica, and H. Ellinor, Height dependent meteor temperatures and comparisons with lidar and OH measurements, *Canadian J. Phys.*, **85**, 173–187 doi:10.1139/P07–038, 2007b.

Hocking, W. K., A. Hocking, D. G. Hocking, and M. Garbanzo-Salas, Windprofiler optimization using digital deconvolution procedures, *J. Atmos. Solar-Terr. Phys.*, **118A**, 45–54, doi:10.1016/j.jastp.2013.08.025, 2014.

Hocking, W. K., *et al.*, Applications of a world-wide network of mesospheric radars, with special emphasis on the Columbia Space Shuttle disaster, in *Handbook for STEP, Proceedings of the Tenth International Workshop on Technical and Scientific Aspects of MST Radar*, edited by J. L. Chau, J. Lau, and J. Röttger, pp. 460–460, Piura, Peru, 2003.

Hocking, W. K., *et al.*, The AXONMET – A pole to pole chain of atmospheric meteor radars, in *Proc. of the Twelfth International Workshop on Technical and Scientific Aspects of MST Radar, London, Ont., Canada, May 17–23, 2009*, edited by N. Swarnalingam and W. K. Hocking, pp. 243–246, The Canadian Association of Physics, ISBN 978-0-9867285-0-1, 2010.

Hocking, W. K., T. Carey-Smith, D. Tarasick, *et al.*, Detection of stratospheric ozone intrusions by windprofiler radars, *Nature*, **450**, 281–284, 2007a.

Hodges, R. R., Generation of turbulence in the upper atmosphere by internal gravity waves, *J. Geophys. Res.*, **72**, 3455–3458, 1967.

Hoenders, B. J., The painful derivation of the refractive index from microscopical considerations, in *Proceedings of Light-Activated Tissue Regeneration and Therapy Conference, Volume 12 of Lecture Notes in Electrical Engineering*, pp. 13–26, Springer, Boston MA, USA, 2008.

Hoffmann, P., W. Singer, D. Kueur, and K. Schulz-Schoellhammer, Observations of 3D winds and waves in the tropopause region above Northern Norway with the ALOMAR SOUSY radar during winter 1996/97, in *Proc. European Workshop on Mesoscale Phenomena in Stratosphere*, 1999.

Högbom, J. A., Aperture synthesis with a non-regular distribution of interferometer baselines, *Astron. and Astrophys. Supplement*, **15**, 417, 1974.

Holdsworth, D. A., and I. M. Reid, Spaced antenna analysis of atmospheric radar backscatter model data, *Radio Sci.*, **30**, 1417–1433, 1995.

Holdsworth, D. A., and I. M. Reid, The Buckland Park MF radar: routine observation scheme and velocity comparisons, *Ann. Geophys.*, **22**, 3815–3828, 2004a.

Holdsworth, D. A., and I. M. Reid, Comparisons of full correlation analysis (FCA) and imaging Doppler interferometry (IDI) winds using the Buckland Park MF radar, *Ann. Geophys.*, **22**, 3829–3842, 2004b.

Holdsworth, D. A., R. A. Vincent, and I. M. Reid, Mesospheric turbulent velocity measurements using the Buckland Park MF radar, *Ann. Geophys.*, **19**, 1007–1017, 2001.

Holdsworth, D. A., R. Vuthaluru, I. M. Reid, and R. A. Vincent, Differential absorption measurements of mesospheric and lower thermospheric electron densities using the Buckland Park MF radar, *J. Atmos. Solar-Terr. Phys.*, **64**, 2029–2042, 2002.

Holloway, C. L., R. J. Doviak, S. A. Cohn, R. J. Lataitis, and J. S. V. Baelen, Cross correlations and cross spectra for spaced antenna wind profilers. 2. Algorithms to estimate wind and turbulence, *Radio Sci.*, **32**, 967–982, 1997.

Holt, O., Discussion of paper by W. A. Flood, "Revised theory for partial reflection D-region measurements," *J. Geophys. Res.*, **74**, 5179–5182, 1969.

Holton, J., J. Beres, and X. Zhou, On the vertical scale of gravity waves excited by localized thermal forcing, *J. Atmos. Sci.*, **59**, 2019–2023, 2002.

Holton, J. R., Waves in the equatorial stratosphere generated by tropospheric heat sources, *J. Atmos. Sci.*, **29**, 368–375, 1972.

Holton, J. R., The role of gravity wave induced drag and diffusion in the momentum budget of the mesosphere, *J. Atmos. Sci.*, **39**, 791–799, 1982.

Holton, J. R., The influence of gravity wave breaking on the general circulation of the middle atmosphere, *J. Atmos. Sci.*, **40**, 2497–2507, 1983.

Holton, J. R., P. H. Haynes, M. E. McIntyre, *et al.*, Stratosphere–troposphere exchange, *Rev. Geophys.*, **33**(4), 403–439, 1995.

Hooke, W. H., and R. M. Jones, Dissipative waves excited by gravity-wave encounters with the stably stratified planetary boundary layer, *J. Atmos. Sci.*, **43**, 2048–2060, 1986.

Hooper, D. A., and E. Pavelin, Tropopause erosion by mountain wave breaking, in *Proceedings of the Tenth International Workshop on Technical and Scientific Aspects of MST Radar*, edited by J. Chau, J. Lau, and J. Röttger, Radio Observatorio de Jicamarca/Universidad de Piura, Peru, 2003.

Hooper, D. A., and L. Thomas, Aspect sensitivity of VHF scatterers in troposphere and stratosphere from comparison of powers in off-vertical beams, *J. Atmos. Terr. Phys.*, **57**, 655–663, 1995.

Hooper, D. A., H. J. Reid, and E. Pavelin, The signature of mid-latitude convection observed by MST radar, in *Proceedings of the Tenth International Workshop on Technical and Scientific Aspects of MST Radar*, edited by J. Chau, J. Lau, and J. Röttger, pp. 334–337, Universidad de Piura, Radio Observatorio de Jicamarca, Lima, Peru, 2003.

Hooper, D. A., J. Arvelius, and K. Stebel, Retrieval of atmospheric static stability from MST radar return signal power, *Ann. Geophys.*, **22**, 3781–3788, 2004.

Hooper, D. A., A. J. McDonald, E. Pavelin, T. K. Carey-Smith, and C. L. Pascoe, The signature of mid-latitude convection observed by VHF wind-profiling radar, *Geophys. Res. Lett.*, **32**, L04,808, doi:10.1029/2004GL020 401, 2005.

Hoppe, U.-P., and D. C. Fritts, On the downward bias in vertical velocity measurements by VHF radars, *Geophys. Res. Lett.*, **22**, 619–622, 1995a.

Hoppe, U. P., and D. C. Fritts, High-resolution measurements of vertical velocity with the European incoherent scatter VHF radar: 1. Motion field characteristics and measurement biases, *J. Geophys. Res.*, **100**, 16 813–16 826, 1995b.

Hoppe, U. P., E. V. Thrane, T. A. Blix, *et al.*, Studies of polar mesosphere summer echoes by VHF radar and rocket probes, *Adv. Space Res.*, **14**(9), 138–148, 1994.

Houghton, J. T., *The Physics of Atmospheres*, Cambridge University Press, Cambridge 1977.

Howard, L. N., Note on a paper of John W. Miles, *J. Fluid Mechs.*, **10**, 509–512, 1961.

Huaman, M. M., and B. B. Balsley, Long-term-mean aspect sensitivity of PMSE determined from Poker Flat MST radar data, *Geophys. Res. Lett.*, **25**, 947–950, 1998.

Huaman, M. M., M. C. Kelley, W. K. Hocking, and R. F. Woodman, Polar mesosphere summer echo studies at 51.5 MHz at Resolute Bay, Canada: Comparisons with Poker Flat Results, *Radio Sci.*, **36**, 1823–1837, 2001.

Huang, J., and J. MacDougall, Legendre coding for digital ionosondes, *Radio Sci.*, **40**, doi: 10.1029/2004RS003,123, 2005.

Hung, R. J., T. Phan, and R. E. Smith, Observation of gravity waves during the extreme tornado outbreak of 3 April 1974, *J. Atmos. Terr. Phys.*, **40**, 831, 1978.

Hunten, D. M., Energetics of thermospheric eddy transport, *J. Geophys. Res.*, **79**, 2533–2534, 1974.

Hysell, D. L., Radar imaging of equatorial F region irregularities with maximum entropy interferometry, *Radio Sci.*, **31**, 1567–1578, 1996.

Hysell, D. L., 30 MHz radar observations of artificial E region field-aligned plasma irregularities, *Ann. Geophys.*, **26**, 117–129, 2008.

Hysell, D. L., and J. L. Chau, Inferring E region electron density profiles at Jicamarca from Faraday rotation of coherent scatter, *J. Geophy. Res.*, **106**, 30 371–30 380, 2001.

Hysell, D. L., and J. L. Chau, Optimal aperture synthesis radar imaging, *Radio Sci.*,

Kirkwood, S., I. Wolf, P. Dalin, *et al.*, Polar mesosphere summer echoes at Wasa, Antarctica (73 °S) – First observations and comparison with 68 °N, *Geophys. Res. Lett.*, **34**, L15,803, doi:10.1029/2007GL030,516, 2007.

Klaassen, G. P., A brief overview of gravity wave breaking theory, in *Handbook for STEP, Proceedings of the Tenth International Workshop on Technical and Scientific Aspects of MST Radar*, edited by J. L. Chau, J. Lau, and J. Röttger, pp. 189–193, Piura, Peru, 2003.

Klaassen, G. P., On the viability of Lagrangian theories of internal wave spectra: Implications for Doppler-spread theory and radar measurements, in *Proceedings of the Twelfth International Workshop on Technical and Scientific Aspects of MST Radar*, edited by N. Swarnalingam and W. K. Hocking, pp. 259–265, Publ. by Canadian Assoc. of Physicists, 2010.

Klaassen, G. P., and W. R. Peltier, The onset of turbulence in finite-amplitude Kelvin–Helmholtz billows, *J. Fluid Mech.*, **155**, 1–35, 1985a.

Klaassen, G. P., and W. R. Peltier, Evolution of finite amplitude Kelvin–Helmholtz billows in two spatial dimensions, *J. Atmos. Sci.*, **42**, 1321–1339, 1985b.

Klaus, V., Temperature retrieval with VHF radar using combined techniques, *Ann. Geophys.*, **26**, 3805–3817, 2008.

Klostermeyer, J., On the role of parametric instability in radar observations of mesospheric gravity waves, in *Middle Atmosphere Program Handbook*, vol. 28, pp. 299–308, Scostep Secretariat, University of Illinois, USA, 1989.

Klostermeyer, J., Two- and three-dimensional parametric instabilities in finite-amplitude internal gravity waves, *Geophys. Astrophys. Fluid Dyn.*, **61**, 1–25, 1991.

Klostermeyer, J., and R. Rüster, Radar observation and model computation of a jet stream-generated Kelvin–Helmholtz instability, *J. Geophys. Res.*, **85**, 2841–2846, 1980.

Klostermeyer, J., and R. Rüster, Further study of a jet stream-generated Kelvin–Helmholtz instability, *J. Geophys. Res.*, **86**, 6631–6637, 1981.

Kolmogoroff, A. N., The local structure of turbulence in incompressible viscous fluid for very large Reynolds numbers (Russian), *Proc. the USSR Academy of Sciences*, **30**, 299–303, 1941a.

Kolmogoroff, A. N., Dissipation of energy in locally isotropic turbulence (Russian), *Proc. the USSR Academy of Sciences*, **32**, 16–18, 1941b.

Kolmogoroff, A. N., The local structure of turbulence in incompressible viscous fluid for very large Reynolds numbers (English), *Proceedings of the Royal Society of London, Series A: Mathematical and Physical Sciences*, **A434**, 9–13, 1991a.

Kolmogoroff, A. N., Dissipation of energy in locally isotropic turbulence (English), *Proceedings of the Royal Society of London, Series A: Mathematical and Physical Sciences*, **A434**, 15–17, 1991b.

Kudeki, E., Radar interferometer observations of mesospheric echoing layers at Jicamarca, *Geophys. Res.*, **93**, 5413–5421, 1988.

Kudeki, E., and S. J. Franke, Statistics of momentum flux estimation, *J. Atmos. Solar-Terr. Phys.*, **60**, 1549–1553, 1998.

Kudeki, E., and R. Stitt, Frequency domain interferometry: A high resolution radar technique for studies of atmospheric turbulence, *Geophys. Res. Lett.*, **14**, 198–201, 1987.

Kudeki, E., and R. Stitt, Frequency domain interferometry studies of mesospheric layers at Jicamarca, *Radio Sci.*, **25**, 575–590, 1990.

Kudeki, E., and F. Surucu, Radar interferometric imaging of field-aligned plasma irregularities in the equatorial electrojet, *Geophys. Res. Lett.*, **18**, 41–44, 1991.

Kudeki, E., and R. F. Woodman, A poststatistics steering technique for MST radar applications, *Radio Sci.*, **25**, 591–594, 1990.

Kudeki, E., P. K. Rastogi, and F. Surucu, Systematic errors in radar wind estimation:

Implication for comparative measurements, *Radio Sci.*, **28**, 169–179, 1993.

Kumar, K. K., Temperature profiles in the MLT region using radar-meteor trail decay times: Comparison with TIMED/SABER observations, *Geophys. Res. Lett.*, **34**, L16,811 doi:10.1029/2007GL030,704, 2007.

Kung, E. C., Large scale balance of kinetic energy in the atmosphere, *Mon. Wea. Rev.*, **94**, 627–640, 1966.

Kunkel, K. E., E. W. Eloranta, and J. A. Weinman, Remote determination of winds, turbulence, spectra and energy dissipation rates in the boundary layer from lidar measurements, *J. Atmos. Sci.*, **37**, 978–985, 1980.

Kuo, F. S., C. C. Chen, S. I. Liu, J. Röttger, and C. H. Liu, Systematic behaviour of signal statistics of MST radar echoes from clear air and their interpretation, *Radio Sci.*, **22**, 1043–1052, 1987.

La Hoz, C., Radar scattering from dusty plasmas, *Phys. Scr.*, **45**, 529–534, 1992.

La Hoz, C., J. Röttger, and S. J. Franke, Spatial interferometry measurements with the EISCAT VHF radar, in *Handbook for MAP*, vol. 28, pp. 185–1991, 1989.

La Hoz, C., O. Havnes, L. I. Næsheim, and D. L. Hysell, Observations and theories of polar mesospheric summer echoes at a Bragg wavelength of 16 cm, *J. Geophys. Res.*, **111**, doi:10.1029/2005JD006,044, 2006.

La Londe, L., The design of linearly polarized slotted waveguide feeds for spherical reflectors, *IEEE Trans. Antennas Propagat.*, **27**, 289–293, 1979.

Labitt, M., *Some Basic Relations Concerning the Radar Measurement of Air Turbulence*, Mass. Inst. of Technol., Lincoln Lab., Work. Pap. 46WP-5001, 1979.

Landecker, T. L., and R. Wielebinski, The galactic metre wave radiation: A two-frequency survey between declinations +25 deg and −25 deg and the preparation of a map of the whole sky, *Aust. J. Phys., Suppl.*, **16**, 1–30, 1970.

Lane, T. P., J. D. Doyle, R. Plougonven, M. A. Shapiro, and R. D. Sharman, Observations and numerical simulations of inertia-gravity waves and shearing instabilities in the vicinity of a jet stream, *J. Atmos. Sci.*, **61**, 2692–2706, 2004.

Larsen, M. F., Can a VHF Doppler radar provide synoptic wind data? A comparison of 30 days of radar and radiosonde data, *Mon. Wea. Review*, **111**, 2047–2057, 1983.

Larsen, M. F., and J. Röttger, VHF and UHF Doppler radars as tools for synoptic research, *Bull. Amer. Meteorol. Soc.*, **63**, 996–1008, 1982.

Larsen, M. F., and J. Röttger, Comparison of tropopause height and frontal boundary locations based on radar and radiosonde data, *Geophys. Res. Lett.*, **10**, 325–328, 1983.

Larsen, M. F., and J. Röttger, Observations of frontal zone and tropopause structures with a VHF Doppler radar and radiosondes, *Radio Sci.*, **20**, 1223–1232, 1985.

Larsen, M. F., and J. Röttger, Observations of thunderstorm reflectivities and Doppler velocities measured at VHF and UHF, *J. Atmos. Oceanic Technol.*, **4**, 151–159, 1987.

Latteck, R., W. Singer, and W. K. Hocking, Measurement of turbulent kinetic energy dissipation rates in the mesosphere by a 3 MHz Doppler radar, *Adv. Space Research*, **35**, 1905–1910, 2005.

Latteck, R., W. Singer, S. Kirkwood, *et al.*, Absolute calibration of VHF radars using a calibrated noise source and an ultrasonic delay line, in *Proceedings of the Eleventh International Workshop on Technical and Scientific Aspects of MST Radar*, edited by V. K. Anandan, pp. 301–305, Gadanki/Tirupati, India, 2007.

Latteck, R., W. Singer, R. J. Morris, *et al.*, Similarities and differences in polar mesosphere summer echoes observed in the Arctic and Antarctica, *Ann. Geophys.*, **26**, 2795–2806, 2008.

Latteck, R., W. Singer, M. Rapp, *et al.*, MAARSY: The new MST radar on

Andoya – System description and first results, *Radio Sci.*, **47**, RS1006, doi:10.1029/2011RS004,775, 2012.

Law, D. C., Windprofilers: applications and characteristics, *QST Journal*, **76**, 48–50, 1992.

Lee, Y., A. R. Paradis, and D. Klingle-Watson, Preliminary results of the 1983 coordinated aircraft-Doppler weather radar turbulence experiment, Vol 1., *Tech. Rep.*, Lincoln Lab., MIT, Lexington, Mass., USA., 1988.

Lehmann, V., and G. Teschke, Wavelet based methods for improved wind profiler signal processing, *Ann. Geophys.*, **19**, 825–836, 2001.

Lehtinen, M. S., A. Huuskonen, and J. Pirttila, First experience of full-profile analysis with GUISDAP, *Ann. Geophys.*, **14**, 1487–1495, 1996.

Lesicar, D., and W. K. Hocking, Studies of seasonal behavior of the shape of mesospheric scatterers using a 1.98 MHz radar, *J. Atmos. Terr. Phys.*, **54**, 295–309, 1992.

Lesicar, D., W. K. Hocking, and R. A. Vincent, Comparative studies of scatterers observed by MF radars in the southern hemisphere mesosphere, *J. Atmos. Terr. Phys.*, **56**, 581–591, 1994.

Ley, B. E., and W. R. Peltier, Wave generation and frontal collapse, *J. Atmos. Sci.*, **35**, 3–17, 1978.

Lhermitte, R., Doppler sonar observation of tidal flow, *J. Geophys. Res.*, **88**, 725–742, 1983.

Lhermitte, R. M., and D. Atlas, Precipitation motion by pulse Doppler, in *Proc. Ninth Weather Radar Conf.*, pp. 218–223, Amer. Meteorol. Soc., Boston, 1961.

Li, J., and P. Stoica, An adaptive filtering approach to spectral estimation and SAR imaging, *IEEE. Trans. Signal Proc.*, **44**(6), 1469–1484, doi:10.1109/78.506,612, 1996.

Lilly, D. K., Stratified turbulence and the mesoscale variability of the atmosphere, *J. Atmos. Sci.*, **40**, 749–761, 1983.

Lilly, D. K., Two-dimensional turbulence generated by energy-sources at 2 scales, *J. Atmos. Sci.*, **46**, 2026–2030, 1989.

Lilly, D. K., D. E. Waco, and S. I. Adelfang, Stratospheric mixing estimated from high-altitude turbulence measurements, *J. Appl. Meteorol.*, **13**, 488–493, 1974.

Lindner, B. C., Radio studies of the lower ionosphere, Ph.D. thesis, University of Adelaide, Adelaide, Australia, 1972.

Lindner, B. C., The nature of D-region scattering of vertical incidence radio waves. I. Generalized statistical theory of diversity effects between spaced receiving antennas, *Aust. J. Phys.*, **28**, 163–170, 1975a.

Lindner, B. C., The nature of D-region scattering of vertical incidence radio waves. II. Experimental observation using spaced antenna reception, *Aust. J. Phys.*, **28**, 171–184, 1975b.

Lindzen, R. S., Turbulence and stress owing to gravity wave and tidal breakdown, *J. Geophys. Res.*, **86**, 9707–9714, 1981.

Lingard, D. M., A deconvolution technique for measuring D-region radio wave backscatter, *J. Atmos. Terr. Phys.*, **58**, 1201–1209, 1996.

Liu, C. H., and C. J. Pan, New observational techniques for studying the dynamics of the middle atmosphere using the Chung Li VHF radar, *J. Atmos. Terr. Phys.*, **55**, 1055–1066, 1993.

Liu, C. H., and K. C. Yeh, Scattering of VHF and UHF radar signals from the turbulent air, *Radio Sci.*, **15**, 277–282, 1980.

Liu, C. H., J. Röttger, C. J. Pan, and S. J. Franke, A model for spaced antenna observational mode for MST radars, *Radio Sci.*, **25**, 551–563, 1990.

Liziola, L. E., and B. B. Balsley, Horizontally propagating quasi-sinusoidal tropospheric waves observed in the lee of the Andes, *Geophys. Res. Lett.*, **24**, 1075–1078, 1997.

Lloyd, K. H., C. H. Low, B. J. McAvaney, D. Rees, and R. G. Roper, Thermospheric observations combining chemical seeding and ground based techniques – 1. Winds, turbulence and the parameters of the neutral atmosphere, *Planet. Space Sci.*, **20**, 761, 1972.

Lomb, N. R., Least-squares frequency analysis of unequally spaced data, *Astrophys. Space Sci.*, **39**, 447–462, 1976.

Lopez-Dekker, P., and S. J. Frasier, Radio acoustic sounding with a UHF volume imaging radar, *J. Atmos. Oceanic Technol.*, **21**, 766–776, 2004.

Lothon, M., B. Campistron, S. Jacoby-Koaly, *et al.*, Comparison of radar reflectivity and vertical velocity observed with a scannable C-band radar and two UHF profilers in the lower troposphere, *J. Atmos. Oceanic Technol.*, **19**, 899–910, 2002.

Lothon, M., B. Campistron, S. Jacoby-Koaly, *et al.*, Reply to "Comments on 'Comparison of radar reflectivity and vertical velocity observed with a scannable C-band radar and two UHF profilers in the lower troposphere'," *J. Atmos. Oceanic Technol.*, **20**, 1224–1229, 2003.

Lübken, F. J., Seasonal variation of turbulent energy dissipation rates at high latitudes as determined by in-situ measurements of neutral density fluctuations, *J. Geophys. Res.*, **102**, 13 441–13 456, 1997.

Lübken, F. J., Turbulent scattering for radars:A summary, *J. Atmos. Solar-Terr. Phys.*, **107**, 1–7, doi:10.1016/j.jastp.2013.10.015, 2014.

Lübken, F. J., U. Von Zahn, E. V. Thrane, *et al.*, In-situ measurements of turbulent energy dissipation rates and eddy diffusion coefficients during MAP/WINE, *J. Atmos. Terr. Phys.*, **49**, 763–776, 1987.

Lübken, F. J., J. Giebeler, T. Blix, *et al.*, In-situ measurement of the Schmidt number within a PMSE layer, *Geophys. Res. Lett.*, **21**(15), 1651–1854, 1994.

Lübken, F. J., M. Rapp, T. Blix, and E. Thrane, Microphysical and turbulent measurements of the Schmidt number in the vicinity of polar mesosphere summer echoes, *Geophys. Res. Lett.*, **25**, 893–896, 1998.

Lübken, F. J., M. Zecha, J. Höffner, and J. Röttger, Temperatures, polar mesosphere summer echoes, and noctilucent clouds over Spitzbergen (78 °N), *J. Geophys. Res.*, **109**, 11 203–11 217, 2004.

Lübken, F. J., B. Strelnikov, M. Rapp, *et al.*, The thermal and dynamical state of the atmosphere during polar mesosphere winter echoes, *Atmos. Chem. Phys.*, **6**, 13–24, 2006.

Lübken, F. J., W. Singer, R. Latteck, and I. Strelnikova, Radar measurements of turbulence, electron densities, and absolute reflectivities during polar mesosphere winter echoes (PMWE), *Adv. Space Res.*, **40**, 758–764, doi:10.1016/j.asr.2007.01.015, 2007.

Luce, H., M. Crochet, F. Dalaudier, and C. Sidi, Interpretation of VHF ST radar vertical echoes from in-situ temperature sheet observations, *Radio Sci.*, **30**, 1002–1025, 1995.

Luce, H., F. Dalaudier, M. Crochet, and C. Sidi, Direct comparison between in-situ and VHF oblique radar measurements of refractive index spectra: A new successful attempt, *Radio Sci.*, **31**, 1487–1500, 1996.

Luce, H., M. Crochet, C. Hanuise, M. K. Yamamoto, and S. Fukao, On the interpretation of the layered structures detected by MST radars in dual frequency domain interferometry (FDI) mode, *Radio Sci.*, **34**, 1077–1083, 1999.

Luce, H., J. Röttger, M. K. Yamamoto, and S. Fukao, Scattering layer thickness and position estimated by radar frequency domain interferometry 1. Effects of the limited horizontal extent and advection of the scattering layers, *Radio Sci.*, **35**, 119–131, 2000a.

Luce, H., J. Röttger, M. Crochet, M. K. Yamamoto, and S. Fukao, Scattering layer thickness and position estimated by radar frequency domain interferometry 2. Effects of tilts of the scattering layer or radar beam, *Radio Sci.*, **35**, 1109–1127, 2000b.

Luce, H., M. K. Yamamoto, S. Fukao, D. Helal, and M. Crochet, A frequency domain radar interferometric imaging (FII) technique based on high resolution methods, *J. Atmos. Solar-Terr. Phys.*, **63**, 221–234, 2001a.

Luce, H., M. Crochet, and F. Dalaudier, Temperature sheets and aspect sensitive radar echoes, *Ann. Geophys.*, **19**, 899–920, 2001b.

Luce, H., S. Fukao, F. Dalaudier, and M. Crochet, Strong mixing events observed near the tropopause with the MU radar and high-resolution balloon techniques, *J. Atmos. Sci.*, **59**, 2885–2895, 2002.

Luenberger, D. G., *Linear and Nonlinear Programming*, Addison-Wesley, Reading, Mass., 1984.

Lumley, J. L., and H. A. Panofsky, *The Structure of Atmospheric Turbulence*, John Wiley and Sons, New York, London, and Sydney, 1964.

Maeda, K., H. Alvarez, J. Aparici, J. May, and P. Reich, A 45-MHz continuum survey of the northern hemisphere, *Astron. Astrophys. Suppl. Ser.*, **140**, 145–154, 1999.

Maekawa, Y., S. Fukao, and S. Kato, Vertical propagation characteristics of internal gravity waves around the mesopause observed by the Arecibo UHF radar, *J. Atmos. Terr. Phys.*, **49**, 73–80, 1987.

Maguire, W. B., and S. K. Avery, Retrieval of raindrop size distribution using two Doppler wind profilers: model sensitivity testing, *J. Appl. Meteorol.*, **33**, 1623–1635, 1995.

Mahan, A. I., A mathematical proof of Stokes' reversibility principle, *J. Opt. Soc. Am.*, **33**, 621–626, 1943.

Mailloux, R. J., *Phased Array Antenna Handbook,* 2nd ed., Artec House, Boston, 2005.

Manchester, R. N., Correction to paper by H. K. Sen and A. A. Wyller "On the generalization of the Appleton–Hartree magnetionic formulas", *J. Geophys. Res.*, **70**, 4995, 1965.

Manning, L. A., The theory of the radio detection of meteors, *J. Applied Phys.*, **19**, 689–699, 1948.

Manning, L. A., O. G. Villard, and A. M. Peterson, Meteoric echo study of upper atmosphere winds, *Proc. Inst. Radio Engrs*, **38**, 877–883, 1950.

Manson, A., and C. E. Meek, Partial-reflection D-region electron densities, in *Handbook for MAP, Ground based techniques*, edited by R. A. Vincent, vol. 13, pp. 113–123, SCOSTEP Secretariat, Dept. of Electr. Computer Eng., Univ. of Illinois, Urbana, IL 61801, USA, 1984.

Manson, A. H., and C. E. Meek, Gravity wave propagation characteristics (60–120 km) as determined by the Saskatoon MF radar (Gravnet) system: 1983–85 at 52 °N, 107 °W, *J. Atmos Sci.*, **45**, 932–946, 1988.

Manson, A. H., M. W. J. Merry, and R. A. Vincent, Relationship between the partial reflection of radio waves from the lower ionosphere and irregularities as measured by rocket probes, *Radio Sci.*, **4**, 955–958, 1969.

Manson, A. H., *et al.*, Gravity wave spectra, directions and wave interactions: Global MLT-MFR network, *Earth Planets Space*, **51**, 543–562, 1999.

Manson, A. H., *et al.*, Gravity wave activity and dynamical effects in the middle atmosphere (60–90 km): observations from an MF/MLT radar network, and results from the Canadian middle atmosphere model (CMAM), *J. Atmos. Solar-Terr. Phys.*, **64**, 65–90, 2002.

Marks, C. J., and S. D. Eckermann, A three-dimensional nonhydrostatic ray-tracing model for gravity waves: Formulation and preliminary results for the middle atmosphere, *J. Atmos. Sci.*, **52**, 1959–1984, 1995.

Marple, S. L., *Digital Spectral Analysis with Applications*, Prentice-Hall, Englewood Cliffs, NJ, 1987.

Marshall, J. M., A. M. Peterson, and A. A. Barnes, Combined radar acoustic sounding system, *Applied Optics*, **11**, 102–112, 1972.

Mathews, J. D., The effect of negative ions on collision dominated Thomson scattering, *J. Geophys. Res.*, **81**, 505–512, doi: 10.1029 /JA083iA02p00,505, 1978.

Mathews, J. D., Incoherent scatter radar studies of the mesosphere, in *Handbook for MAP, Ground based techniques*, edited by R. A. Vincent, vol. 13, pp. 135–154, SCOSTEP Secretariat, Dept. of Electr. Computer Eng.,

Univ. of Illinois, Urbana, IL 61801, USA, 1984a.

Mathews, J. D., The incoherent scatter radar as a tool for studying the ionospheric D-region, *J. Atmos. Terr. Phys.*, **46**, 975–986, 1984b.

Mathews, J. D., and B. S. Tanenbaum, A plasma wave and electron-plasma diffusion interpretation of Thomson scattering from a plasma containing negative ions, *Planet. Space Sci.*, **29**, 335–340, doi: 10.1016/0032–0633(81)90,021–0, 1978.

Mathews, J. D., J. H. Shapiro, and B. S. Tanenbaum, Evidence for distributed scattering in D-region partial-reflection processes, *J. Geophys. Res.*, **78**, 8266, 1973.

Mathews, J. D., J. K. Breakall, and M. P. Sulzer, The moon as a calibration target of convenience for VHF-UHF radar systems, *Radio Sci.*, **23**, 1–12, 1988.

Matsuno, T., Lagrangian motion of air parcels in the stratosphere in the presence of planetary waves, *Pure Appl. Geophys.*, **118**, 189–216, 1980.

Matsuno, T., A quasi one-dimensional model of the middle atmosphere circulation interacting with internal gravity waves, *J. Meteorol. Soc. Japan*, **60**, 215–226, 1981.

Matuura, N., Y. Masuda, H. Inuki, *et al.*, Radio acoustic measurement of temperature profile in the troposphere and stratosphere, *Nature*, **323**, 426–428, 1986.

May, P. T., Comparison of wind-profiler and radiosonde measurements in the tropics, *J. Atmos. Oceanic Technol.*, **10**, 122–128, 1993.

May, P. T., and R. G. Strauch, An examination of wind profiler signal processing algorithms, *J. Atmos. Oceanic Technol.*, **6**, 731–735, 1989.

May, P. T., and R. G. Strauch, Reducing the effect of ground clutter on wind profiler velocity measurements, *J. Atmos. Oceanic Technol.*, **15**, 579–586, 1998.

May, P. T., S. Fukao, T. Tsuda, T. Sato, and S. Kato, The effect of thin scattering layers on the determination of wind by Doppler radars, *Radio Sci.*, **23**, 83–94, 1988a.

May, P. T., R. G. Strauch, and K. P. Moran, The altitude coverage of temperature measurements using RASS with wind profiler radars, *Geophys. Res. Lett.*, **15**, 1381–1384, 1988b.

May, P. T., R. G. Strauch, and W. L. Ecklund, Temperature soundings by RASS with wind-profiler radars: A preliminary study, *IEEE Trans. Geosci. Remote Sens.*, **28**, 19–28, 1990.

May, P. T., T. Adachi, T. Tsuda, and R. J. Lataitis, The spatial structure of RASS echoes, *J. Atmos. Oceanic Technol.*, **13**, 1275–1284, 1996.

McAfee, J. R., K. S. Gage, and R. G. Strauch, Vertical velocities at Platteville, Colorado: An intercomparison of simultaneous measurements by the VHF and UHF profilers, *Radio Sci.*, **34**, 1027–1042, 1995.

McLandress, C., On the importance of gravity waves in the middle atmosphere and their parameterization in general circulation models, *J. Atmos. Solar-Terr. Phys.*, **60**, 1357–1383, 1998.

McLandress, C., and W. E. Ward, Tidal/gravity wave interactions and their influence on the large-scale dynamics of the middle atmosphere: Model results, *J. Geophys. Res.*, **99**, 8139–8156, 1994.

McLandress, C., M. J. Alexander, and D. L. Wu, Microwave limb sounder observations of gravity waves in the stratosphere: A climatology and interpretation, *Geophys. Res.*, **105**, 11 947–11 967, 2000.

McNamara, L. F., Statistical model of the D region, *Radio Sci.*, **14**, 1165–1173, 1979.

Mead, J. B., G. Hopcraft, S. J. Frasier, *et al.*, A volume-imaging radar wind profiler for atmospheric boundary layer turbulence studies, *J. Atmos. Oceanic Technol.*, **15**, 849–859, 1998.

Mechtly, E. A., S. A. Bowhill, and L. G. Smith, Changes of lower ionosphere electron concentrations with solar activity, *J. Atmos. Terr. Phys.*, **34**, 1899–1907, 1972.

Medvedev, A. S., and G. P. Klaassen, Parameterization of gravity wave momentum deposition based on nonlinear wave interactions: basic formulation and sensitivity tests, *J. Atmos. Solar-Terr. Phys.*, **62**, 1015–1033, 2000.

Meek, C. E., and A. H. Manson, Use of the full polarization measurement in the partial reflection experiment, *J. Atmos. Terr. Phys.*, **43**, 45–58, 1981.

Meek, C. E., I. M. Reid, and A. H. Manson, Observations of mesospheric wind velocities 2. Cross-sections of power spectral density for 48–8 hours, 8–1 hours and 1 hour to 10 min over 60-110 km for 1981, *Radio Sci.*, **20**, 1383–1402, 1985.

Meikle, H., *Modern Radar Systems*, 2nd ed., Artec House, Boston, 2008.

Melrose, D. B., *Plasma Astrophysics: Non-thermal Processes in Diffuse Magnetized Plasmas, Vol. 1, The Emission, Absorption and Transfer of Waves in Plasmas*, Gordon and Breach Science Publishers, New York, London and Paris, 273pp, 1984.

Merrill, J. T., Observational and theoretical study of shear instability in the airflow near the ground, *J. Atmos. Sci.*, **34**, 911–921, 1977.

Merrill, J. T., and J. R. Grant, A gravity-wave-critical-level encounter observed in the atmosphere, *J. Geophys. Res.*, **84**, 6315–6320, 1979.

Metcalf, J. I., and K. M. Glover, A history of weather radar research in the US air force, in *Radar in Meteorology*, edited by D. Atlas, pp. 32–43, American Met. Soc., 1990.

Miles, J. W., On the stability of heterogeneous shear flows, *J. Fluid Mechs.*, **10**, 496–508, 1961.

Mitra, S. N., A radio method of measuring winds in the ionosphere, *Proc. Instn. Elect. Engrs.*, **96**, 441–446, 1949.

Miyoshi, Y., H. Fujiwara, H. Jin, and H. Shinagawa, A global view of gravity waves in the thermosphere simulated by a general circulation model, *J. Geophys. Res. (Space Phys)*, **119**, 5807–5820, doi:10.1002/2014 JA019,848, 2014.

Mohan, K., D. Narayana Rao, T. Narayana Rao, and S. Raghavan, Estimation of temperature and humidity from MST radar observations, *Ann. Geophys.*, **19**, 855–861, 2001.

Molmud, P., Langevin equations and the AC conductivity of non-Maxwellian plasmas, *Phys. Rev.*, **114**, 29–32, 1959.

Monna, W. A. A., On the use of wind profilers in meteorology, *Ann. Geophys.*, **12**, 482–486, 1994.

Morris, R. J., M. B. Terkildsen, D. A. Holdsworth, and M. R. Hyde, Is there a causal relationship between cosmic noise absorption and PMSE?, *Geophys. Res. Lett.*, **32**, doi: 10.1029/2005GL024,568, 2005.

Morris, R. J., D. J. Murphy, R. A. Vincent, A. R. Klekociuk, and I. M. Reid, Studies of polar mesosphere summer echoes by VHF radar and rocket probes, *J. Atmos. Solar-Terr. Phys.*, **68**, 418–435, 2006.

Morris, R. J., D. J. Murphy, A. R. Klekociuk, and D. A. Holdsworth, First complete season of PMSE observations above Davis, Antarctica, and their relation to winds and temperatures, *Geophys. Res. Lett.*, **34**, doi:10.1029/2006GL028,641, 2007.

Mousley, T. J., D. N. Asimakopoulos, R. S. Cole, B. A. Crease, and S. J. Caughey, Measurement of boundary layer structure parameter profiles by acoustic sounding and comparison with direct measurement, *Q. J. R. Meteorol. Soc.*, **107**, 203–230, 1981.

Muller, G., P. Holloway, F. Henyey, and N. Pomphrey, Nonlinear interactions among internal gravity waves, *Rev. Geophys.*, **24**, 493–536, 1986.

Muller, H. G., Simultaneous observations of meteor winds and ionospheric drifts, *J. Atmos. Terr. Phys.*, **30**, 701, 1968.

Murgatroyd, R. J., The structure and dynamics of the stratosphere, in *The Global Circulation of the Atmosphere*, ed. G. A. Corby, pp. 159–195, Royal Meteorol. Soc., London, U. K., 1969.

Murphy, D. J., and R. A. Vincent, Estimates of momentum flux in the mesosphere and lower thermosphere over Adelaide,

Australia from March 1985 to February 1986, *J. Geophys. Res.*, **98**(D10), 18 617–18 638, 1993.

Muschinski, A., Possible effect of Kelvin–Helmholtz instability on VHF radar observations of the mean vertical wind, *J. Appl. Meteorol.*, **35**, 2210–2217, 1996.

Muschinski, A., Turbulence and gravity waves in the vicinity of a midtropospheric warm front: A case study using VHF echo-intensity measurements and radiosonde data, *Radio Sci.*, **32**, 1161–1178, 1997.

Muschinski, A., and C. Wode, First in-situ evidence for co-existing sub-meter temperature and humidity sheets in the lower free troposphere, *J. Atmos. Sci.*, **55**, 2893–2906, 1998.

Muschinski, A., P. B. Chilson, R. D. Palmer, G. Schmidt, and H. Steinhagen, Boundary-layer convection and diurnal variation of vertical-velocity characteristics in the free troposphere, *Q. J. R. Meteorol. Soc.*, **127**, 423–444, 2001.

Nakamura, T., T. Tsuda, S. Fukao, *et al.*, Comparative observations of short-period gravity waves (10–100 min) in the mesosphere in 1989 by Saskatoon MF radar (52 °N), Canada and the MU radar (35 °N) Japan, *Radio Sci.*, **28**, 729–746, 1993.

Nakamura, T., T. Tsuda, H. Miyagawa, *et al.*, Propagation directions of gravity wave patterns observed in OH CCD images during the SEEK campaign, *Geophys. Res. Lett.*, **25**, 1793–1796, 1998.

Narayana Rao, T., D. Narayana Rao, and S. Raghavan, Tropical precipitating systems observed with Indian MST radar, *Radio Sci.*, **34**, 1125–1139, 1999.

Nastrom, G. D., Doppler radar spectral width broadening due to beamwidth and wind shear, *Ann. Geophys.*, **15**, 786–796, 1997.

Nastrom, G. D., and F. D. Eaton, Turbulence eddy dissipation rates from radar observations at 5–20 km at White Sands Missile Range, New Mexico, *J. Geophys. Res.*, **102**, 19 495–19 506, 1997a.

Nastrom, G. D., and F. D. Eaton, A brief climatology of eddy diffusivities over White Sands Missile Range, New Mexico, *J. Geophys. Res.*, **102**, 29 819–29 826, 1997b.

Nastrom, G. D., and D. C. Fritts, Sources of mesoscale variability of gravity waves. Part I: Topographic excitation, *J. Atmos. Sci.*, **49**, 101–110, 1992.

Nastrom, G. D., and T. E. Van Zandt, Mean vertical motions seen by radar wind profilers, *J. Appl. Meteorol.*, **33**, 984–995, 1994.

Nastrom, G. D., and T. E. Van Zandt, Biases due to gravity waves in wind profiler measurements of winds, *J. Appl. Meteorol.*, **35**, 243–257, 1996.

Nastrom, G. D., D. C. Fritts, and K. S. Gage, An investigation of terrain effects on the mesoscale spectrum of atmospheric motions, *J. Atmos. Sci.*, **44**, 3087–3096, 1987.

Newman, D. B., and A. J. Ferraro, Amplitude distributions of partially reflected signals from the mid-latitudinal D region, *J. Geophys. Res*, **81**, 2442, 1976.

Nicolls, M. J., M. C. Kelley, R. H. Varney, and C. J. Heinselman, Spectral observations of polar mesospheric summer echoes at 33 cm (450 MHz) with the Poker Flat incoherent scatter radar, *J. Atmos. Solar-Terr. Phys.*, **71**, doi:10.1016/j.jastp.2008.04.019, 2009.

Nishimura, K., E. Gotoh, and T. Sato, Fine scale 3D wind field observation with a multistatic equatorial atmosphere radar, *J. Meteor. Soc. Japan*, **84A**, 227–238, 2006.

North, E. M., and A. M. Peterson, RASS, a remote sensing system for measuring low-level temperature profiles, *Bull. Amer. Meteorol. Soc.*, **54**, 912–919, 1973.

Norton, K. A., L. E. Vogler, W. V. Mansfield, and P. J. Short, The probability distribution of the amplitude of a constant vector plus a Rayleigh distributed vector, *Proc IRE.*, **43**, 1354, 1955.

Nussbaumer, V., K. Fricke, M. Langer, W. Singer, and U. von Zahn, First simultaneous and common volume observations of noctilucent clouds and polar mesosphere summer echoes by lidar and radar, *J. Geophys. Res.*, **101**, 19 161–19 168, 1996.

Nyquist, H., Thermal agitation of electric charge in conductors, *Phys. Rev.*, **32**, 110–113, doi:10.1103/PhysRev.32.110, 1928.

Olivero, J., and G. E. Thomas, Evidence for changes in greenhouse gases in the mesosphere, *Adv. Space Res.*, **28**, 937–946, 2001.

Oppenheim, A. V., and R. W. Schafer, *Digital Signal Processing*, Prentice-Hall, Englewood Cliffs, NJ, 1975.

Oppenheim, A. V., A. S. Wilsky, and I. T. Young, *Signals and Systems*, Prentice-Hall, Englewood Cliffs, NJ, 1983.

Ottersten, H., Radar backscattering from the turbulent clear atmosphere, *Radio Sci.*, **4**, 1251–1255, 1969a.

Ottersten, H., Atmospheric structure and radar backscattering in clear air, *Radio Sci.*, **4**, 1179–1193, 1969b.

Owens, J. C., Optical refractive index of air: dependence on pressure, temperature and composition, *Applied Optics*, **6**, 51–59, 1967.

Pace, P. E., *Advanced Techniques for Digital Receivers*, Artech House, Inc., Norwood, MA, 2000.

Pacholczyk, A. G., and T. L. Swihart, Polarization of radio sources. VI - an oscillatory behavior of the intensity in a general solution of the radiation transfer problem in a plasma, *Astrophys. J.*, **196**, pt. 1, 125–127, 1975.

Palmer, R. D., M. F. Larsen, R. F. Woodman, *et al.*, VHF radar interferometry measurements of vertical velocity and the effect of tilted refractivity surfaces on standard Doppler measurements, *Radio Sci.*, **26**, 417–427, 1991.

Palmer, R. D., X. Huang, S. Fukao, M. Yamamoto, and T. Nakamura, High-resolution wind profiling using combined spatial and frequency domain interferometry, *Radio Sci.*, **30**, 1665–1679, 1995a.

Palmer, R. D., K. Y. Lei, S. Fukao, M. Yamamoto, and T. Nakamura, Weighted imaging Doppler interferometry, *Radio Sci.*, **30**, 1787–1801, 1995b.

Palmer, R. D., S. Gopalam, T.-Y. Yu, and S. Fukao, Coherent radar imaging using Capon's method, *Radio Sci.*, **33**, 1585–1589, 1998.

Palmer, R. D., T.-Y. Yu, and P. B. Chilson, Range imaging using frequency diversity, *Radio Sci.*, **34**, 1485–1496, 1999.

Palmer, R. D., P. B. Chilson, A. Muschinski, *et al.*, SOMARE-99: Observations of tropospheric scattering layers using multiple-frequency range imaging, *Radio Sci.*, **36**, 681–693, 2001.

Palmer, R. D., B. L. Cheong, M. W. Hoffman, S. J. Frasier, and F. J. López-Dekker, Observation of the small-scale variability of precipitation using an imaging radar, *J. Atmos. Oceanic Technol.*, **22**, 1122–1137, doi:10.1175/JTECH1775.1, 2005.

Palmer, T. N. G., J. Shutts, and R. Swinbank, Alleviation of a systematic westerly bias in general circulation and and numerical weather prediction models through an orographic gravity wave drag parameterization, *Q. J. R. Meteorol. Soc.*, **112**, 1001–1040, 1986.

Pan, C. J., and J. Röttger, Structures of polar mesosphere summer echoes observed with the EISCAT VHF radar in the interferometer mode, in *Solar-Terrestrial Energy Program: Proceedings of the Seventh Workshop on Technical and Scientific Aspects of MST Radar*, vol. 7, pp. 252–255, SCOSTEP Secretariat, Boulder, Colorado, USA., 1996.

Papoulis, A., *Probability, Random Variables, and Stochastic Processes*, McGraw-Hill, New York, NY, 1965.

Paquin, J. E., and S. Pond, The determination of the Kolmogorov constants for velocity, temperature and moisture from second and third order structure functions, *J. Fluid Mechs.*, **50**, 257–269, 1971.

Parrett, C. A., M. Turp, B. MacPherson, and T. Oakley, Quality monitoring of weather radar wind profiles at the Met Office, in *Proc. of ERAD (Third European Conference on Radar Meteorology and Hydrology)*, pp. 168–173, 2004.

Passi, P. M., and C. Morel, Wind errors using the worldwide LORAN network, *J. Atmos. Oceanic Technol.*, **4**, 690–700, 1987.

Patra, A. K., and N. V. Rao, Further investigations of 150 km echoing riddle using simultaneous observations of 150 km and E region echoes from off-electrojet location Gadanki, *J. Geophys. Res.*, **112**, A09, 301, doi:10.1029/2006JA012,204, 2007.

Patra, A. K., S. Sripathi, V. S. Kumar, and P. B. Rao, Evidence of kilometer-scale waves in the lower E region from high resolution VHF radar observations over Gadanki, *Geophys. Res. Lett.*, **29**(10), doi:10.1029 /2001GL013,340, 2002.

Patra, A. K., T. Yokoyama, Y. Otsuka, and M. K. Yamamoto, Daytime 150 km echoes observed with the Equatorial Atmosphere radar in Indonesia: First results, *Geophys. Res. Lett.*, **35**, L06,101, doi:10.1029/2007 GL033,130, 2008.

Patra, A. K., N. Rao, D. V. Phanikumar, *et al.*, A study on the low-latitude daytime E region plasma irregularities using coordinated VHF radar, rocket-borne, and ionosonde observations, *J. Geophys. Res.*, **114**, A11,301 doi:10.1029/2009JA014,501, 2009.

Pauley, P. M., R. L. Creasey, W. L. Clark, and G. D. Nastrom, Comparison of horizontal winds measured by opposing beams with the Flatland ST radar and between Flatland measurements and NMC analyses, *J. Atmos. Oceanic Technol.*, **11**, 256–275, 1994.

Pavelin, E., and J. Whiteway, Gravity wave interactions around the jet stream, *Geophys. Res. Lett.*, **29**, 2024, doi:10.1029/2002 GL015 783, 2002.

Pavelin, E., J. A. Whiteway, and G. Vaughan, Observation of gravity wave generation and breaking in the lowermost stratosphere, *J. Geophys. Res.*, **106**, 5173 (2000JD900,480), 2001.

Pavelin, E., J. Whiteway, R. Busen, and J. Hacker, Airborne observations of turbulence, mixing, and gravity waves in the tropopause region, *J. Geophys. Res.*, **107**, 4084, doi:10.1029/2001JD000 775, 2002.

Pearman, G. I., and P. J. Fraser, Sources of increased methane, *Nature*, **332**, 489–490, 1988.

Peltier, W. R., and J. F. Scinocca, The origin of severe downslope windstorm pulsations, *J. Atmos. Sci.*, **47**, 2853–2870, 1990.

Peltier, W. R., J. Halle, and T. L. Clark, The evolution of finite-amplitude Kelvin–Helmholtz billows, *Geophys. Astrophys. Fluid Dyn.*, **10**, 53–87, 1978.

Peters, G., H. Hinzpeter, and G. Baumann, Measurements of heat flux in the atmospheric boundary layer by sodar and RASS: A first attempt, *Radio Sci.*, **20**, 1555–1564, 1985.

Peters, G., D. Hasselmann, and S. Pang, Radio acoustic sounding of the atmosphere using a FM-CW radar, *Radio Sci.*, **23**, 640–646, 1988.

Petitdidier, M., and P. Laroche, Lightning observations with the strato-tropospheric UHF and VHF radars at Arecibo, Puerto Rico, *Atmospheric Research*, **76**, 481–492, 2005.

Pfister, W., The wave-like nature of inhomogeneities in the E-region, *J. Atmos. Terr. Phys.*, **33**, 999–1025, 1971.

Phillips, G. J., and M. Spencer, The effects of anisometric amplitude patterns in the measurement of ionospheric drifts, *Proc. Phys. Soc.*, **68B**, 481, 1955.

Piani, C., D. Durran, M. J. Alexander, and J. R. Holton, A numerical study of three dimensional gravity waves triggered by deep tropical convection and their role in the dynamics of the QBO, *J. Atmos. Sci.*, **57**, 3689–3702, 2000.

Piggott, W. R., W. J. G. Beynon, G. M. Brown, and C. G. Little, The measurement of ionospheric absorption, *Annales International Geophysical Year*, **3**, 175, 1957.

Placke, M., P. Hoffmann, R. Latteck, and M. Rapp, Gravity wave momentum fluxes from MF and meteor radar measurements in the polar MLT region, *J. Geophys. Res.*,

120, 736–750, doi:10.1002/2014JA020,460, 2015.
Plane, J. M., Cosmic dust in the Earth's atmosphere, *Chem. Soc. Rev.*, **41**, 6507–6518, doi: 10.1039/C2CS35,132C, 2012.
Plougonven, R., and C. Snyder, Gravity waves excited by jets: Propagation versus generation, *Geophys. Res. Lett.*, **32**, L18,802, doi:10.1029/2005GL023 730, 2005.
Plougonven, R., H. Teitelbaum, and V. Zeitlin, Inertia-gravity wave generation by the tropospheric mid-latitude jet as given by the fastex radiosoundings, *J. Geophys. Res.*, **108**, 4686, doi:10.1029/2003JD003 535, 2003.
Plumb, R. A., Stratospheric transport, *J. Meteor. Soc. Japan*, **80**, 793–809, 2002.
Plumb, R. A., and R. Ferrari, Transformed Eulerian-mean theory. Part I: Nonquasigeostrophic theory for eddies on a zonal-mean flow, *J. Physical Oceanogr.*, **35**, 165–174, 2005.
Pollard, B. D., S. Khanna, S. J. Frasier, *et al.*, Local structure of the convective boundary layer from a volume-imaging radar, *J. Atmos. Sci.*, **57**, 2281–2296, 2000.
Poole, L. M. G., A simplified interferometer design for use with meteor radars, *Radio Sci.*, **39**, doi:10.1029/2002RS002,778, 2004.
Praskovsky, A. A., and E. A. Praskovskaya, Structure-function-based approach to analyzing received signals for spaced antenna radars, *Radio Sci.*, **38**, 1068, doi:10.1029/2001RS002,544, 2003.
Preusse, P., B. Schaeler, J. T. Bacmeister, and D. Offermann, Evidence for gravity waves in CRISTA temperatures, *Adv. Space Res.*, **24**, 1601–1604, 1999.
Preusse, P., S. D. Eckermann, and D. Offermann, Comparison of global distributions of zonal-mean gravity wave variance inferred from different satellite instruments, *Geophys. Res. Lett.*, **27**, 3877–3880, 2000.
Preusse, P., B. Dornback, S. D. Eckermann, *et al.*, Space-based measurements of stratospheric mountain waves by CRISTA, 1. Sensitivity, analysis method and a case study, *J. Geophys. Res.*, **107**, 8178, doi:19.1029/2001JD000,699, 2002.
Prichard, I. T., L. Thomas, and R. M. Worthington, The characteristics of mountain waves observed by radar near the west coast of Wales, *Ann. Geophys.*, **13**, 757–767, 1995.
Probert-Jones, J. R., A history of radar meteorology in the United Kingdom, in *Radar in Meteorology*, edited by D. Atlas, pp. 54–60, American Met. Soc., 1990.
Proctor, R. F., Input impedance of horizontal dipole aerials at low heights above the ground, *Proc. IEEE*, **97**(3), 188–189, 1950.
Prosser, R. T., The Lincoln calibration sphere, *Proc. IEEE*, **53**, 1672–1676, 1965.
Prudnikov, A. P., Y. A. Brychkov, O. I. Marichev, and (translated from the Russian by) N. M. Queen, *Integrals and Series, Vol II, Special Functions*, Gordon and Breach Science Publishers, 1990.
Raab, F. H., Idealized operation of the class E tuned amplifier, *IEEE Trans. Circuits and Systems*, **CAS-24**, 239–247, 1978.
Rajopadhyaya, D. K., P. T. May, and R. A. Vincent, A general approach to the retrieval of raindrop size distribution from windprofiler Doppler spectra: Modeling results, *J. Atmos. Oceanic Technol.*, **10**, 710–717, 1993.
Ralph, F. M., M. Crochet, and V. Vankateswaran, A study of mountain lee waves using clear-air radar, *Q. J. R. Meteorol. Soc.*, **118**, 597–627, 1992.
Rao, P. B., A. R. Jain, P. Kishore, *et al.*, Indian MST radar 1. System description and sample vector wind measurements in ST mode, *Radio Sci.*, **30**, 1125–1–138, 1995.
Rapp, M., and F. J. Lübken, On the nature of PMSE: Electron diffusion in the vicinity of charged particles revisited, *J. Geophys. Res.*, **108**, 8437–8449, 2003.
Rapp, M., and F. J. Lübken, Polar mesosphere summer echoes (PMSE): Review of observations and current understanding, *Atmos. Chem. Phys.*, **4**, 2601–2633, 2004.
Rapp, M., J. Gumbel, F. J. Lübken, and R. Latteck, D region electron number density limits for the existence of polar mesosphere

summer echoes, *J. Geophys. Res.*, **107**, doi:10.1029/2001JD001,323, 2002.

Rapp, M., I. Strelnikova, R. Latteck, *et al.*, Polar mesosphere summer echoes (PMSE) studied at Bragg wavelengths of 2.8 m, 67 cm, and 16 cm, *J. Atmos. Solar-Terr. Phys.*, **70**(7), 947–961, 2008.

Rastogi, P. K., Radar studies of gravity waves and tides in the middle atmosphere: a review, *J. Atmos. Terr. Phys.*, **43**, 511–524, 1981.

Rastogi, P. K., and S. A. Bowhill, Scattering of radio waves from the mesosphere – 1. Theory and observations, *J. Atmos. Terr. Phys.*, **38**, 399, 1976a.

Rastogi, P. K., and S. A. Bowhill, Scattering of radio waves from the mesosphere – 2. Evidence for intermittent mesospheric turbulence, *J. Atmos. Terr. Phys.*, **38**, 449, 1976b.

Rastogi, P. K., and O. Holt, On detecting reflections in presence of scattering from amplitude statistics with application to D-region partial reflections, *Radio Sci.*, **16**, 1431–1443, 1981.

Rastogi, P. K., and J. Röttger, VHF radar observations of coherent reflections in the vicinity of the tropopause, *J. Atmos. Terr. Phys.*, **44**, 461–469, 1982.

Rastogi, P. K., and G. Sobolewski, New quasi-complementary code sets for atmospheric radar applications, *Radio Sci.*, **25**, 1087–1094, 1990.

Rastogi, P. K., and R. F. Woodman, Mesospheric studies using the Jicamarca incoherent-scatter radar, *J. Atmos. Terr. Phys.*, **36**, 1217, 1974.

Ratcliffe, J. A., Some aspects of diffraction theory and their application in the ionosphere, *Rep. Prog. Phys.*, **19**, 188–267, 1956.

Ratcliffe, J. A., *Magnetoionic Theory*, Cambridge University Press, Reading, Mass., 1959.

Ratcliffe, J. A., *An Introduction to the Ionosphere and Magnetosphere*, Cambridge University Press, London, 1972.

Rayleigh, L., *Theory of Sound*, vol. I, 2nd ed., McMillan and Co., New York, 1894.

Rayleigh, L., A mathematical proof of Stokes' reversibility principle, *Phil. Mag. series 5*, **49**, 324–325, 1900.

Readings, C. J., and D. R. Rayment, The high-frequency fluctuation of the wind in the first kilometer of the atmosphere, *Radio Sci.*, **4**, 1127–1131, 1969.

Rechou, A., V. Barabash, P. Chilson, *et al.*, Mountain wave motions determined by the Esrange MST radar, *Ann. Geophys.*, **17**, 957–970, 1999.

Rees, D., R. G. Roper, K. Lloyd, and C. H. Low, Determination of the structure of the atmosphere between 90 and 250 km by means of contaminant releases at Woomera, May, 1968, *Phil. Trans. Roy. Soc. Lond*, **A271**, 631–663, 1972.

Rees, M. H., *Physics and Chemistry of the Upper Atmosphere: Cambridge Atmospheric and Space Science Series*, Cambridge University Press, Cambridge, UK, 1989.

Reid, G. C., The production of water-cluster positive ions in the quiet day-time D region, *Planet. Space Sci.*, **25**, 275–290, 1977.

Reid, I. M., Gravity wave motions in the upper middle atmosphere (60–110 km), *J. Atmos. Terr. Phys.*, **48**, 1057–1072, 1986.

Reid, I. M., Radar observations of stratified layers in the mesosphere and lower thermosphere (50–100 km), *Adv. Space Res.*, **10**(10), 7–19, 1990.

Reid, I. M., and R. A. Vincent, Measurements of mesospheric gravity wave momentum fluxes and mean flow accelerations at Adelaide, Australia, *J. Atmos. Terr. Phys.*, **49**, 443–460, 1987.

Reid, I. M., R. Rüster, and G. Schmidt, VHF radar observations of a cat's-eye-like structure at mesospheric heights, *Nature*, **327**, 43–45, 1987.

Reid, I. M., P. Czechowsky, R. Rüster, and G. Schmidt, First VHF radar measurements of mesopause summer echoes at mid-latitudes, *Geophys. Res. Lett.*, **16**, 135–138, 1989.

Reid, I. M., S. Holdsworth, D. A. Kovalam, R. A. Vincent, and J. Stickland, Mount Gambier (38 °S, 141 °E) prototype wind profiler, *Radio Sci.*, **40**, RS5007, doi:10.1029/2004RS003 055, 2005.

Renkwitz, T., G. Stober, R. Latteck, W. Singer, and M. Rapp, New experiments to validate the radiation pattern of the Middle Atmosphere Alomar Radar System (MAARSY), *Adv. Radio Sci.*, **11**, 283–289, 2013.

Revathy, K., S. R. Prabhakaran Nair, and B. V. Krishna Murthy, Deduction of temperature profile from MST radar observations of vertical wind, *Geophys. Res. Lett.*, **23**, 285–288, 1996.

Rhodes, C. T., X. M. Shao, P. R. Krehbiel, R. J. Thomas, and C. O. Hayenga, Observations of lightning phenomena using radio interferometry, *J. Geophys. Res.*, **99**, 13 059–13 082, 1994.

Rice, S. O., Mathematical analysis of random noise, *Bell Syst, Tech. J.*, **23**, 282–332, 1944.

Rice, S. O., Mathematical analysis of random noise, *Bell Syst. Tech. J.*, **24**, 46–156, 1945.

Rice, S. O., Bell Telephone Monograph B-1589, in *Selected Papers on Noise and Stochastic Processes*, edited by N. Wax, Dover Publ., New York, 1954.

Richards, M. A., *Fundamentals of Radar Signal Processing*, McGraw-Hill, New York, 2005.

Richter, J. H., High resolution tropospheric radar soundings, *Radio Sci.*, **4**, 1261–1268, 1969.

Rind, D., R. Suozzo, N. K. Balachandran, and M. J. Prather, Climate change and the middle atmosphere. Part 1: The doubled CO_2 climate, *J. Atmos. Sci.*, **47**, 475–494, 1990.

Robertson, D. S., D. T. Liddy, and W. G. Elford, Measurements of winds in the upper atmosphere by means of drifting meteor trails I, *J. Atmos. Terr. Phys.*, **4**, 255–270, 1953.

Roble, R. G., and G. Schmidtke, Calculated ionospheric variations due to changes in solar EUV flux measured by the AEROS spacecraft, *J. Atmos. Terr. Phys.*, **41**, 153–160, 1979.

Roger, R. S., C. H. Costain, T. L. Landecker, and C. M. Swerdlyk, The radio emission from the galaxy at 22 MHz, *Astron. Astrophys. Suppl. Ser.*, **137**, 7–19, 1999.

Rogers, R. R., The early years of doppler radar in meteorology, in *Radar in Meteorology*, edited by D. Atlas, pp. 122–129, American Met. Soc., 1990.

Roper, R. G., Atmospheric turbulence in the meteor region, *J. Geophys. Res.*, **71**, 5785–5792, 1966.

Roper, R. G., MWR – meteor radar winds, in *Handbook for MAP, Ground Based Techniques*, edited by R. A. Vincent, vol. 13, pp. 124–134, SCOSTEP Secretariat, Dept. of Electr. Computer Eng., Univ. of Illinois, Urbana, IL 61801, USA, 1984.

Roper, R. G., MWR – meteor radar winds, in *Proceedings of the First GLOBMET Symposium*, edited by R. G. Roper, vol. 25, pp. 1–450, SCOSTEP Secretariat, Dept. of Electr. Computer Eng., Univ. of Illinois, Urbana, IL 61801, USA, 1987.

Roper, R. G., Rocket vapor trail releases revisited: Turbulence and the scale of gravity waves - Implications for the IDI/ISR controversy, *J. Geophys. Res.*, **101**, 7103–7017, 1996.

Roper, R. G., On the radar estimation of turbulence parameters in a stably stratified atmosphere, *Radio Sci.*, **35**, 999–1008, 2000.

Roper, R. G., and J. W. Brosnahan, Imaging Doppler interferometry and the measurement of atmospheric turbulence, *Radio Sci.*, **32**, 1137–1148, 1997.

Roper, R. G., and J. W. Brosnahan, Diurnal variations in the rate of dissipation of turbulent energy in the equatorial upper mesosphere – lower thermosphere, *Radio Sci.*, **40**, 1029–1044, 2005.

Roper, R. G., and W. G. Elford, The seasonal variation of turbulence in the upper atmosphere, *Nature*, **197**, 963–965, 1963.

Roper, R. G., G. W. Adams, and J. W. Brosnahan, Tidal winds at mesopause altitudes over Arecibo (18 °N, 67 °W), 5–11 April 1989 (AIDA'89), *J. Atmos. Terr. Phys.*, **55**, 289–312, 1993.

Röttger, and P. Czechowsky, Clear-air turbulence and tropospheric refractivity variations observed with a new VHF-radar, *Naturwissenschaften*, **64**(11), 580–581, 1977.

Röttger, J., Evidence for partial reflection of VHF radar signals from the troposphere, *J. of Geophysics (Zeitschrift für Geophysik)*, **44**, 393–394, 1978.

Röttger, J., VHF radar observations of a frontal passage, *J. Appl. Meteorol.*, **18**, 85–91, 1979.

Röttger, J., Reflection and scattering of VHF radar signals from atmospheric refractivity structures, *Radio Sci.*, **15**, 259–276, 1980a.

Röttger, J., Structure and dynamics of the stratosphere and mesosphere revealed by VHF radar investigations, *Pure Appl. Geophys.*, **118**, 494–527, 1980b.

Röttger, J., Development of refractivity structures during anticyclonic weather conditions, *Proc. 19th Conference on Radar Meteorology*, 593–598, 1980c.

Röttger, J., Investigations of lower and middle atmosphere dynamics with spaced antenna drift radars, *J. Atmos. Terr. Phys.*, **43**, 277–292, 1981.

Röttger, J., The MST radar technique, in *Handbook for MAP, Ground Based Techniques*, edited by R. A. Vincent, vol. 13, pp. 187–232, SCOSTEP Secretariat, Dept. of Electr. Computer Eng., Univ. of Illinois, Urbana, IL 61801, USA, 1984a.

Röttger, J., Signal statistics of the radar echoes – angle-of-arrival statistics, in *Handbook for MAP, URSI/SCOSTEP Workshop, May, 1984*, edited by S. A. Bowhill and B. Edwards, vol. 14, pp. 84–87, SCOSTEP Secretariat, Dept. of Electr. Computer Eng., Univ. of Illinois, Urbana, IL 61801, USA, 1984b.

Röttger, J., Determination of Brunt–Väisälä frequency from vertical velocity spectra, in *Middle Atmosphere Program Handbook*, edited by B. Edwards, vol. 20, pp. 168–172, SCOSTEP Secr. Univ. of Illinois, Urbana, Ill. USA, Urbana, 1986.

Röttger, J., The instrumental principles of MST radars and incoherent scatter radars and the configuration of radar system hardware, in *Handbook for MAP*, edited by S. Fukao, vol. 30, pp. 54–113, SCOSTEP Secretariat, Dept. of Electr. Computer Eng., Univ. of Illinois, Urbana, IL 61801, USA, 1989.

Röttger, J., Polar mesosphere summer echoes: Dynamics and aeronomy of the mesosphere, *Adv. Space Res.*, **14**, (9)123–137, 1994.

Röttger, J., ST radar observations of atmospheric waves over mountainous areas: a review, *Ann. Geophys.*, **18**, 750–765, 2000.

Röttger, J., VHF radar scatter microstructure measured by combined spatial and frequency domain interferometry – SFDI – A developing approach of three-dimensional interferometry with VHF radar, in *Proc. of the Thirteenth International Workshop on Technical and Scientific Aspects of MST Radar, Kühlungsborn, Germany, May 19-23, 2012*, edited by R. Latteck and W. Singer, pp. 13–26, Distributed by Leibniz-Institute of Atmospheric Physics at the Rostock University, 18225 Kühlungsborn, Germany, ISBN 978-3-00-044654-2, 2013.

Röttger, J., Ionosphere and atmosphere research with radars, in *Geophysics and Geochemistry, 6.16.5.3*, UNESCO *Encyclopedia of Life Support Systems (EOLSS)*, http://www. eolss.net/Sample-Chapters/C01/E6-16-05-03.pdf, Paris, 2014.

Röttger, J., and H. M. Ierkic, Postset beam steering and interferometer applications of VHF radars to study winds, waves, and turbulence in the lower and middle atmosphere, *Radio Sci.*, **20**, 1461–1480, 1985.

Röttger, J., and C. La Hoz, Characteristics of polar mesosphere summer echoes (PMSE) observed with the EISCAT 224 MHz radar and possible explanations of their origin, *J. Atmos. Terr. Phys.*, **52**, 893–906, 1990.

Röttger, J., and M. F. Larsen, UHF/VHF radar techniques for atmospheric research and wind profiler applications, in *Radar in Meteorology*, edited by D. Atlas, pp. 235–281, American Met. Soc., 1990.

Röttger, J., and C. H. Liu, Partial reflection and scattering of VHF radar signals from the clear atmosphere, *Geophys. Res. Lett.*, **5**, 357–360, 1978.

Röttger, J., and G. Schmidt, High-resolution VHF radar soundings of the troposphere and stratosphere, *IEEE Trans. Geosci. Electron.*, **GE-17**, 182–189, 1979.

Röttger, J., and R. A. Vincent, VHF radar studies of tropospheric velocities and irregularities using spaced antenna techniques, *Geophys. Res. Lett.*, **5**, 917–920, 1978.

Röttger, J., J. Klostermeyer, P. Czechowsky, R. Rüster, and G. Schmidt, Remote sensing of the atmosphere by VHF radar experiments, *Naturwissenschaften*, **65**, 285–296, 1978.

Röttger, J., P. K. Rastogi, and R. F. Woodman, High resolution VHF radar observations of turbulence structures in the mesosphere, *Geophys. Res. Lett.*, **6**, 617–620, 1979.

Röttger, J., T. Y. Kang, and M. Y. Zi, Mountain lee waves detected in radar wind profiles, *MPAe-W-00-81-36, Max-Planck-Institut für Aeronomie, Katlenburg-Lindau, Germany*, p. 19, 1981.

Röttger, J., C. La Hoz, M. C. Kelley, U.-P. Hoppe, and C. Hall, The structure and dynamics of polar mesosphere summer echoes observed with the EISCAT 224 MHz radar, *Geophys. Res. Lett.*, **15**, 1353–1356, 1988.

Röttger, J., C. H. Liu, J. K. Chao, *et al.*, Spatial interferometer measurements with the Chung-Li VHF radar, *Radio Sci.*, **25**, 503–515, 1990a.

Röttger, J., M. T. Rietveld, C. La Hoz, *et al.*, Polar mesosphere summer echoes observed with the EISCAT 933 MHz radar and the CUPRI 46.9 MHz radar, their similarity to 224 MHz radar echoes, and their relation to turbulence and electron density profiles, *Radio Sci.*, **25**, 671–687, 1990b.

Röttger, J., C. H. Liu, C. J. Pan, and S. Y. Su, Characteristics of lightning echoes observed with VHF ST radar, *Radio Sci.*, **30**, 1085–1097, 1995.

Rummler, W. D., Introduction of a new spectral estimator for velocity spectral parameters, *Tech. Rep. Tech. Memo. MM-68-4121-5*, Bell Telephone Laboratories, Whippany, New Jersey, 1968.

Rust, D. W., D. W. Burgess, R. L. Madox, *et al.*, Testing a mobile version of a cross-chain LORAN atmospheric (M-CLASS) sounding system, *Bull. Am. Meteorol. Soc*, **71**, 173–181, 1990.

Rüster, R., G. D. Nastrom, and G. Schmidt, High-resolution VHF radar measurements in the troposphere with a vertically pointing beam, *J. Appl. Meteorol.*, **37**, 1522–1529, 1998.

Ryde, J., The attenuation and radar echoes produced at centimetre wavelengths by various meteorological phenomena, *Meteorological Factors in Radio Wave Propagation*, pp. 169–188, Wiley Online Library, 1946.

Ryde, J., and D. Ryde, Attenuation of centimetre and millimetre waves by rain, hail, fogs and clouds, *Tech. Rep. 8670*, General Electric Research Laboratory, Wembley, England, 1945.

Sahr, J. D., Chapter 10, Ionospheric studies, in *Bistatic Radar: Emerging Technology*, edited by M. Cherniakov, pp. 363–387, John Wiley and Sons, Chichester, 2008.

Sahr, J. D., and F. D. Lind, The Manastash Ridge radar: a passive bistatic radar for upper atmospheric radio science, *Radio Sci.*, **32**(6), 2345–2358, 1997.

Sarma, T. V. C., D. Narayana Rao, J. Furumoto, and T. Tsuda, Development of radio acoustic sounding system (RASS) with Gadanki MST radar – First results, *Ann. Geophys.*, **26**, 2531–2542, 2008.

Sasi, M. N., and V. Deepa, Seasonal variation of equatorial wave momentum fluxes at

Gadanki (13.5 °N, 79.2 °E), *Ann. Geophys.*, **19**, 985–990, 2001.

Sato, K., Small-scale wind disturbances observed by the MU radar during the passage of typhoon Kelly, *J. Atmos. Sci.*, **50**, 518–537, 1993.

Sato, K., T. Kumakura, and M. Takahashi, Gravity waves appearing in a high-resolution GCM simulation, *J. Atmos. Sci.*, **56**, 1005–1018, doi:10.1175/1520–0469, 1999.

Sato, K., *et al.*, Program of the Antarctic Syowa MST/IS Radar (PANSY), *J. Atmos. Solar-Terr. Phys.*, **105**, 2–15, 2014.

Sato, T., and R. F. Woodman, Spectral parameter estimation of CAT radar echoes in the presence of fading clutter, *Radio Sci.*, **17**, 817–826, 1982a.

Sato, T., and R. F. Woodman, Fine altitude resolution observations of stratospheric turbulent layers by the Arecibo 430 MHz radar, *J. Atmos. Sci.*, **39**, 2553–2564, 1982b.

Sato, T., T. Tsuda, S. Kato, *et al.*, High resolution observations of turbulence by using the MU radar, *Radio Sci.*, **20**, 1452–1460, 1985.

Sato, T., Y. Inooka, S. Fukao, and S. Kato, Multi-beam pattern measurement of the MU radar antenna by satellite OHZORA, *J. Geomag. Geoelectr.*, **41**, 743–751, 1989.

Sato, T., H. Doji, H. Iwai, *et al.*, Computer processing for deriving drop-size distributions and vertical air velocities from VHF Doppler radar spectra, *Radio Sci.*, **25**, 961–973, 1990.

Sato, T., T. Wakayama, T. Tanaka, K. Ikeda, and I. Kimura, Shape of space debris as estimated from RCS variations, *J. Spacecraft and Rockets*, **31**, 665–670, 1994.

Sato, T., T. Teraoka, and I. Kimura, Simultaneous observation plan of the MU radar with airborne/spaceborne precipitation radars, in *STEP Handbook, Proceedings of the 7th Workshop on Technical and Scientific Aspects of MST Radar*, edited by B. Edwards, pp. 133–136, Natl. Oceanic and Atmos. Admin., Boulder, CO, 1996.

Sato, T., T. Nakamura, and K. Nishimura, Orbit determination of meteors using the MU radar, *IEICE Trans. Comm.*, **E83-B**, (9), 1990–1995, 2000.

Satomura, T., and K. Sato, Secondary generation of gravity waves associated with the breaking of mountain waves, *J. Atmos. Sci.*, **56**, 3874–3858, 1999.

Saxton, J. A., J. A. Lane, R. Meadows, and P. A. Mathews, Layer structure of the troposphere – simultaneous radar and microwave refractometer investigations, *Proc. IEE*, **3**, 275–283, 1964.

Scargle, J. D., Studies in astronomical time series analysis. II. Statistical aspects of spectral analysis of unevenly spaced data, *Astrophys. J.*, **263**, 835–853, 1982.

Schafer, J. P., and W. M. Goodall, Observations of Kennelly–Heaviside layer heights during the Leonid meteor shower of November, 1931, *Proc. Inst. Radio Engrs*, **20**, 1941–1945, 1932.

Scheffler, A. O., and C. H. Liu, On observation of gravity wave spectra in the atmosphere by using MST radars, *Radio Sci.*, **20**, 1309–1322, 1985.

Schlegel, K., A. Brekke, and A. Haug, Some characteristics of the quiet polar D region and mesosphere obtained with the partial reflection method, *J. Atmos. Terr. Phys.*, **40**, 205, 1978.

Schmidt, G., R. Ruster, and P. Czechowsky, Complementary code and digital filtering for detection of weak VHF radar signals from the mesosphere, *IEEE Trans. Geosci. Electron.*, **17**, 154–161, 1979.

Schmidt, H., *et al.*, The HAMMONIA chemistry climate model: sensitivity of the mesopause region to the 11-year solar cycle and CO_2 doubling, *J. Climate*, **19**, 3903–3931, 2006.

Scipion, D., J. L. Chau, and L. Flores, First results of the boundary layer and tropospheric radar systems for ENSO studies in northern Peru, in *Proceedings of the Tenth International Workshop on Technical and Scientific Aspects of MST Radar*, edited by

of SOUSY radar wind measurements with rawinsonde and model data, *Ann. Geophys.*, **12**, 746–764, 1994.

Stockwell, R. G., L. Mansinha, and R. P. Lowe, Localization of the complex spectrum: The S transform, *IEEE Transactions on Signal Processing*, **44**, 998–1001, 1996.

Stoica, P., and R. Moses, *Spectral Analysis of Signals*, Prentice-Hall, Englewood Cliffs, NJ, 2005.

Stolzenburg, M., W. Rust, and T. C. Marshall, Electrical structure in thunderstorm convective regions: Isolated storms, *J. Geophys. Res.*, **103**, 14 079–14 096, doi:10.1029/2003 RS002,907, 1998.

Strauch, R. G., D. A. Merritt, K. P. Moran, K. B. Earnshaw, and D. Van de Kamp, The Colorado wind-profiler network, *J. Atmos. Oceanic Technol.*, **1**, 37–49, 1984.

Strauch, R. G., B. L. Weber, A. S. Frisch, *et al.*, The precision and relative accuracy of profiler wind measurements, *J. Atmos. Oceanic Technol.*, **4**, 563–572, 1987.

Strobel, D., Constraints on gravity wave induced diffusion in the middle atmosphere, *Pure Appl. Geophys.*, **130**, 533–546, 1989.

Strobel, D. F., M. E. Summers, R. M. Bevilacqua, M. T. DeLand, and M. Allen, Vertical constituent transport in the mesosphere, *J. Geophys. Res.*, **92**, 6691–6698, 1987.

Stubbs, T. J., The measurement of winds in the D region of the ionosphere by the use of partially reflected radiowaves, *J. Atmos. Terr. Phys.*, **35**, 909, 1973.

Stubbs, T. J., and R. A. Vincent, Studies of D-region drifts during the winters of 1970–1972, *Aust. J. Phys.*, **15**, 645, 1973.

Swarnalingam, N., and W. K. Hocking, Calibration and calculation of PMSE backscatter cross section using sky noise and calibrated noise source for the Resolute Bay radar, in *Proceedings of the Eleventh International Workshop on Technical and Scientific Aspects of MST Radar*, edited by V. K. Anandan, pp. 328–333, Gadanki/Tirupati, India, 2007.

Swarnalingam, N., W. K. Hocking, and P. S. Argall, Radar efficiency and the calculation of decade-long PMSE backscatter cross-section for the Resolute Bay VHF radar, *Ann. Geophys.*, **27**, 1643–1656, 2009a.

Swarnalingam, N., W. K. Hocking, W. Singer, and R. Latteck, Calibrated measurements of PMSE strengths at three different locations observed with SKiYMET radars and narrow beam VHF radars, *J. Atmos. Solar-Terr. Phys.*, **71**, 1807–1813, 2009b.

Swenson, G. R., and P. J. Espy, Observations of 2-dimensional airglow structure and Na density from the ALOHA, October 9, 1993, "storm flight," *Geophys. Res. Lett.*, **22**, 2845–2848, 1995.

Tabor, D., *Gases, Liquids and Solids*, Penguin Library of Physical Sciences, 1969.

Takao, K., M. Fujita, and T. Nishi, An adaptive antenna array under directional constraint, *IEEE Trans. Antennas Propagat.*, **24**, 662–669, 1976.

Tatarski, V. I., *Wave Propagation in a Turbulent Medium*, McGraw-Hill, New York, 1961.

Tatarski, V. I., *The Effects of the Turbulent Atmosphere on Wave Propagation*, Keter Press, Jerusalem, 1971.

Tatarskii, V. I., and A. Muschinski, The difference between Doppler velocity and real wind velocity in single scattering from refractive index fluctuations, *Radio Sci.*, **36**(6), 1405–1423, 2001.

Taylor, M. J., A. P. V. Eyken, H. Rishbeth, *et al.*, Simultaneous observations of noctilucent clouds and polar mesospheric radar echoes: Evidence of non-correlation, *Planet. Space Sci.*, **37**, 1013–1020, 1989.

Taylor, M. J., Y. Y. Gu, X. Tao, C. S. Gardner, and M. B. Bishop, An investigation of intrinsic gravity wave signatures using coordinated lidar and nightglow image measurements, *Geophys. Res. Lett.*, **22**, 2853–2856, 1995a.

Taylor, M. J., D. N. Turnbull, and R. P. Lowe, Spectrometric and imaging measurements of a spectacular gravity wave event observed during the ALOHA-93 campaign, *Geophys. Res. Lett.*, **22**, 2849–2852, 1995b.

Tchen, C. M., Transport processes as foundation of the Heisenberg and Obukhoff theories of turbulence, *Phys. Rev.*, **93**, 4–14, 1954.

Teitelbaum, H., La mesure de l'échelle interne de la turbulence atmosphérique entre 80 and 100 km d'altitude, *Space. Res.*, **VI**, 438–447, 1966.

Teitelbaum, H., and C. Sidi, Formation of discontinuities in atmospheric gravity waves, *J. Atmos. Terr. Phys.*, **38**, 413–421, 1976.

Teshiba, M., H. Hashiguchi, , S. Fukao, and Y. Shibagaki, Typhoon 9707 observations with the MU radar and L-band boundary layer radar, *Ann. Geophys.*, **19**, 925–931, 2001.

Thayaparan, T., W. K. Hocking, and J. MacDougall, Observational evidence of tidal/gravity wave interactions using the UWO 2 MHz radar, *Geophys. Res. Lett.*, **22**, 373–376, 1995.

Thide, B., H. Derblom, A. Hedberg, H. Kopka, and P. Stubbe, Observations of stimulated electromagnetic emissions in ionospheric heating experiments, *Radio Sci.*, **18**, 851–859, 1983.

Thomas, G. E., Mesospheric cloud and the physics of the mesopause region, *Rev. Geophys.*, **29**, 553–575, 1991.

Thomas, G. E., J. J. Olivero, E. J. Jensen, W. Schroeder, and O. B. Toon, Relation between increasing methane and the presence of ice clouds at the mesopause, *Nature*, **338**, 490–492, 1989.

Thomas, R. J., P. R. Krehbiel, W. Rison, *et al.*, Accuracy of the lightning mapping array, *J. Geophys. Res.*, **109**, doi:10.1029/2004 JD004,549, 2004.

Thorsen, D., S. J. Franke, and E. Kudeki, A new approach to HF Doppler measurements of mesospheric gravity wave momentum fluxes, *Radio Sci.*, **32**, 707–726, 1997.

Thrane, E. V., and B. Grandal, Observations of finescale structure in the mesosphere and lower thermosphere, *J. Atmos. Terr. Phys.*, **43**, 179–189, 1981.

Thrane, E. V., O. Andreassen, T. Blix, *et al.*, Neutral air turbulence in the upper atmosphere observed during the Energy Budget Campaign, *J. Atmos. Terr. Phys.*, **47**, 243–264, 1985.

Thrane, E. V., T. A. Blix, C. M. Hall, *et al.*, Small scale structure and turbulence in the mesosphere and lower thermosphere at high latitudes in winter, *J. Atmos. Terr. Phys.*, **49**, 751–762, 1987.

Titheridge, J. E., The electron density in the lower ionosphere, *J. Atmos. Terr. Phys.*, **24**, 269, 1962.

Tsai, K. C., and P. R. Gray, A 1.9 GHz, 1 W CMOS class-E power amplifier for wireless communications, *IEEE J. Solid-State Circ.*, **34**, 962–970, 2002.

Tsuda, T., T. Sato, K. Hirose, S. Fukao, and S. Kato, MU radar observations of the aspect sensitivity of backscattered VHF echo power in the troposphere and lower stratosphere, *Radio Sci.*, **21**, 971–980, 1986.

Tsuda, T., T. Inoue, D. C. Fritts, *et al.*, MST radar observations of a saturated gravity-wave spectrum, *J. Atmos. Sci.*, **46**, 2440–2447, 1989.

Tsuda, T., T. Adachi, Y. Masuda, S. Fukao, and S. Kato, Observations of tropospheric temperature fluctuations with the MU radar-RASS, *J. Atmos. Oceanic Tecnol.*, **11**, 50–62, 1994.

Tsuda, T., T. E. Van Zandt, and H. Saito, Zenith-angle dependence of VHF specular reflection echoes in the lower atmosphere, *J. Atmos. Terr. Phys.*, **59**, 761–776, 1997a.

Tsuda, T., W. E. Gordon, and H. Saito, Azimuth-angle variations of specular reflection echoes in the lower atmosphere observed with the MU radar, *J. Atmos. Terr. Phys.*, **59**, 777–784, 1997b.

Tsuda, T., M. Nishida, C. Rocken, and R. H. Ware, A global morphology of gravity wave activity in the stratosphere revealed by the GPS occultation data (GPS/MET), *J. Geophys. Res.*, **105**, 7257–7273, 2000.

Tsuda, T., M. Miyamoto, and J. I. Furumota, Estimation of a humidity profile using turbulence echo characteristics, *J. Atmos. and Oceanic Technol.*, **18**, 1214–1222, 2001.

Tsunoda, R. T., and W. L. Ecklund, On the visibility and zenithal confinement of 150 km (F1) radar echoes, *Geophys. Res. Lett.*, **34**, L21,102, doi:10.1029/2007GL031,276, 2007.

Tsunoda, R. T., and W. L. Ecklund, On the sheet like nature of 150 km (F1) radar echoes, *Geophys. Res. Lett.*, **35**, L05,102, doi:10.1029/2007GL032,152, 2008.

Tsutsumi, M., T. Tsuda, T. Nakamura, and S. Fukao, Temperature fluctuations near the mesopause inferred from meteor observations with the middle and upper atmosphere radar, *Radio Sci.*, **29**, 599–610, 1994.

Tsutsumi, M., D. Holdsworth, T. Nakamura, and I. Reid, Meteor observations with an MF radar, *Earth Planets Space*, **51**, 691–699, 1999.

Turek, R. S., R. G. Roper, and J. W. Brosnahan, Further comparisons of simultaneously measured imaging Doppler interferometry and spaced antenna winds, *J. Atmos. Solar-Terr. Phys.*, **60**, 337–347, 1998.

Turtle, A. J., and J. E. Baldwin, A survey of galactic radiation at 178 Mc/s, *Mon. Not. R. Astr. Soc.*, **124**, 459–476, 1962.

Uman, M., *Understanding Lightning*, BEK Technical Publications, Pittsburg, Pennsylvania, 1971.

Umemoto, Y., M. Teshiba, Y. Shibagaki, *et al.*, Combined wind profiler-weather radar observations of orographic rainband around Kyushu, Japan in the Baiu season, *Ann. Geophys.*, **22**, 3971–3982, 2004.

Unitrobe, *Application of Pin Diodes in High Power Duplexers, Application Note N-139*, Unitrobe Corporation, Watertown, MA, USA, 1968.

Valentic, T. A., J. P. Avery, and S. K. Avery, MEDAC/SC: A third generation meteor detection and collection system, *IEEE Transactions on Geoscience and Remote Sensing*, **34**, 15–21, 1996.

Van Baelen, J. S., and A. D. Richmond, Radar interferometry technique: Three-dimensional wind measurement theory, *Radio Sci.*, **26**, 1209–1218, 1991.

Van Baelen, J. S., A. D. Richmond, T. Tsuda, *et al.*, Radar interferometry technique and anisotropy of the echo power distribution: First results, *Radio Sci.*, **26**, 1351–1326, 1991.

Van der Ziel, A., *Noise*, Prentice Hall Inc., NJ, 1954.

Van Zandt, T. E., A universal spectrum of buoyancy waves in the atmosphere, *Geophys. Res. Lett.*, **9**, 575–578, 1982.

Van Zandt, T. E., A model for gravity wave spectra observed by Doppler sounding systems, *Radio Sci.*, **20**, 1323–1330, 1985a.

Van Zandt, T. E., Gravity waves, in *Handbook for MAP*, vol. 16, pp. 149–156, Scostep Secretariat, University of Illinois, USA, 1985b.

Van Zandt, T. E., and D. C. Fritts, Effects of Doppler shifting on the frequency spectra of atmospheric gravity waves, *J. Geophys. Res.*, **92**, 9723–9732, 1987.

Van Zandt, T. E., and D. C. Fritts, A theory of enhanced saturation of the gravity wave spectrum due to increases in atmospheric stability, *Pure Appl. Geophys.*, **130**, 399–420, 1989.

Van Zandt, T. E., and R. A. Vincent, Is VHF Fresnel reflectivity due to low frequency buoyancy waves?, in *Handbook for MAP*, edited by S. A. Bowhill and B. Edwards, vol. 9, pp. 78–80, SCOSTEP Secretariat, Univ. of Illinois, Urbana, IL, 1983.

Van Zandt, T. E., J. L. Green, K. S. Gage, and W. L. Clark, Vertical profiles of refractivity turbulence structure constant: Comparison of observations by the Sunset radar with a new theoretical model, *Radio Sci.*, **13**, 819–829, 1978.

Van Zandt, T. E., K. S. Gage, and J. M. Warnock, An improved model for the calculation of profiles of C_n^2 and ε in the free atmosphere from background profiles of wind, temperature and humidity, in *Preprint volume, 20th Conference on Radar*

Meteorology, pp. 129–135, Am. Met. Soc., Boston, Mass., 1981.

Van Zandt, T. E., W. L. Clark, K. S. Gage, C. R. Williams, and W. L. Ecklund, A dual-wavelength radar technique for measuring the turbulent energy dissipation rate ε, *Geophys. Res. Lett.*, **27**, 2537–2540, 2000.

Van Zandt, T. E., G. D. Nastrom, J. Furunoto, T. Tsuda, and W. L. Clark, A dual-beamwidth radar method for measuring atmospheric turbulent kinetic energy, *Geophys. Res. Lett.*, **29**, 10.1029/2001GL014,283, 2002.

Vandepeer, B. G. W., and I. M. Reid, On the spaced antenna and imaging Doppler interferometer techniques, *Radio Sci.*, **30**(4), 885–901, 1995.

Vanneste, J., Small-scale mixing, large-scale advection, and stratospheric tracer distributions, *J. Atmos. Sci.*, **61**, 2749–2761, 2004.

Villars, F., and V. F. Weisskopf, On the scattering of radio waves by turbulent fluctuations in the atmosphere, *Proc I. R. E.*, **43**, 1232, 1955.

Vincent, R. A., The interpretation of some observations of radio waves scattered from the lower ionosphere, *Aust. J. Phys.*, **26**, 815–827, 1973.

Vincent, R. A., Gravity wave motions in the mesosphere, *J. Atmos. Terr. Phys.*, **46**, 119–128, 1984.

Vincent, R. A., Planetary and gravity waves in the mesosphere and lower thermosphere, *Adv. Space Res.*, **7**(10), 163–169, 1987.

Vincent, R. A., and M. J. Alexander, Gravity waves in the tropical lower stratosphere: An observational study of seasonal and interannual variability, *J. Geophys. Res.*, **105**, 17 971–17 982, 2000.

Vincent, R. A., and S. M. Ball, Tides and gravity waves in the mesosphere at mid and low latitudes, *J. Atmos. Terr. Phys.*, **39**, 965–970, 1977.

Vincent, R. A., and J. S. Belrose, The angular distribution of radio waves partially reflected from the lower ionosphere, *J. Atmos. Terr. Phys.*, **40**, 35–47, 1978.

Vincent, R. A., and D. C. Fritts, A climatology of gravity wave motions in the mesopause region at Adelaide, Australia, *J. Atmos. Sci.*, **44**, 748–760, 1987.

Vincent, R. A., and I. M. Reid, HF Doppler measurements of mesospheric gravity wave momentum fluxes, *J. Atmos. Sci.*, **40**, 1321–1333, 1983.

Vincent, R. A., and J. Röttger, Spaced antenna VHF radar observations of tropospheric velocities and irregularities, *Radio Sci.*, **15**, 319, 1980.

Vincent, R. A., and T. J. Stubbs, A study of motions in the winter mesosphere using the partial reflection drift technique, *Planet. Space Sci.*, **25**, 441–455, 1977.

Vincent, R. A., T. J. Stubbs, P. H. O. Pearson, K. H. Lloyd, and C. H. Low, A comparison of partial reflection drifts with winds determined by rocket techniques, *J. Atmos. Terr. Phys.*, **39**, 813, 1977.

Vincent, R. A., B. Candy, and B. H. Briggs, Measurements of antenna polar diagrams and efficiencies using a phase-switched interferometer, in *Handbook for MAP*, vol. 20, pp. 409–409, Univ. of Illinois, Urbana, IL, 1986.

Vincent, R. A., P. May, W. K. Hocking, *et al.*, First results with the Adelaide VHF radar: spaced antenna studies of tropospheric winds, *J. Atmos. Terr. Phys.*, **49**, 353–366, 1987.

Vincent, R. A., S. Dullaway, A. D. MacKinnon, *et al.*, A VHF boundary layer radar: First Results, *Radio Sci.*, **33**, 845–860, 1998.

Vincent, R. A., S. Kovalam, I. M. Reid, and J. P. Younger, Gravity wave flux retrievals using meteor radars, *Geophys. Res. Lett.*, **37**, L14,802, doi:10.1029/2010GL044,086, 2010.

Vinnichenko, N. K., and J. A. Dutton, Empirical studies of atmospheric structure and spectra in the free atmosphere, *Radio Sci.*, **4**, 1115–1126, 1969.

Von Biel, H. A., Amplitude distributions of D-region partial reflections, *J. Geophys. Res.*, **76**, 8365–8367, 1971.

Von Biel, H. A., A statistical assessment of synoptic D-region partial reflection data, *J. Atmos. Terr. Phys.*, **43**, 225–230, 1981.

Von Biel, A. H., The phase-switched correlation polarimeter – a new approach to the partial reflection experiment, *J. Atmos. Terr. Phys.*, **39**, 769–778, 1977.

Vuthaluru, R., R. A. Vincent, D. A. Holdsworth, and I. M. Reid, Collision frequencies in the D-region, *J. Atmos. Solar-Terr. Phys.*, **64**, 2043–2054, 2002.

Wadell, B. C., *Transmission Line Design Handbook*, Artec House, Boston, 1991.

Wallace, J. M., and P. V. Hobbs, *Atmospheric Science – An Introductory Survey*, Academic Press, London, 1977.

Walterscheid, R. L., Inertio-gravity wave induced accelerations of mean flow having an imposed periodic component: Implications for tidal observations in the meteor region, *J. Geophys. Res.*, **86**, 9698–9706, 1981.

Walterscheid, R. L., and W. K. Hocking, Stokes diffusion by atmospheric internal gravity waves, *J. Atmos. Sci.*, **48**, 2213–2230, 1991.

Walterscheid, R. L., J. H. Hecht, R. A. Vincent, *et al.*, Analysis and interpretation of airglow and radar observations of quasi-monochromatic gravity waves in the upper mesosphere and lower thermosphere over Adelaide, Australia, (35 °S, 138 °E), *J. Atmos. Solar-Terr. Phys.*, **61**, 461–478, 1999.

Wangsness, R. K., *Electromagnetic Fields*, Wiley, New York, 1986.

Warner, C. D., and M. E. McIntyre, An ultra-simple spectral parameterization for non-orographic gravity waves, *J. Atmos. Sci.*, **58**, 1837–1857, 2001.

Warnock, J. M., and T. E. Van Zandt, A statistical model to estimate the refractivity turbulence structure constant C_n^2 in the free atmosphere, *NOAA Tech. Memo., ERL AI-10*, 175, 1985.

Warnock, J. M., T. E. Van Zandt, J. L. Green, and R. H. Winkler, Comparison between wind profiles measured by Doppler radar and by rawinsonde balloons, *Geophys. Res. Lett.*, **5**, 109–112, 1978.

Waterman, A. T., Techniques for measurement of vertical and horizontal velocities; monostatic vs. bistatic measurements, in *Middle Atmosphere Progam Handbook*, edited by B. Edwards, vol. 9, pp. 164–169, SCOSTEP Secretariat, Dept. of Electr. Computer Eng., Univ. of Illinois, Urbana, IL 61801, USA, 1983.

Waterman, A. T., T. Z. Hu, P. Czechowsky, and J. Röttger, Measurement of anisotropic permittivity structure of upper troposphere with clear-air radar, *Radio Sci.*, **20**, 1580–1592, 1985.

Watson-Watt, R. A., L. H. Bainbridge-Ball, A. F. Wilkins, and E. G. Bowen, Return of radio waves from the middle atmosphere, *Nature*, **137**, 1936.

Weast, R. C. (Ed.), *Chemical Rubber Co. Handbook of Chemistry and Physics*, 51st ed., Chemical Rubber Co, Cleveland, 1970.

Weber, B. L., and D. B. Wuertz, Comparison of rawinsonde and windprofiler radar measurements, *J. Atmos. Oceanic Technol.*, **7**, 158–175, 1990.

Weber, B. L., and D. B. Wuertz, Quality control algorithm for profiler measurements of winds and temperatures, *NOAA Tech. Memo., ERL WPL-212*, 32, 1991.

Webster, P. L., and R. Lukas, TOGA COARE: The coupled ocean-atmosphere response experiment, *Bull. Amer. Meteorol. Soc.*, **73**, 1377–1416, 1992.

Weeks, W. L., *Antenna Engineering*, McGraw-Hill, New York, 1968.

Weil, A., and M. I. Skolnik, Chapter 10 in *Radar Handbook*, in *The Radar Transmitter*, edited by M. I. Skolnik, pp. 10.1–10.30, McGraw-Hill, 2008.

Weinstock, J., Nonlinear theory of acoustic-gravity waves, *J. Geophys. Res.*, **81**, 633–652, 1976.

Weinstock, J., On the theory of turbulence in the buoyancy subrange of stably stratified flows, *J. Atmos. Sci.*, **35**, 634–649, 1978a.

Weinstock, J., Vertical turbulent diffusion in a stably stratified fluid, *J. Atmos. Sci.*, **35**, 1022–1027, 1978b.

Weinstock, J., Using radar to estimate dissipation rates in thin layers of turbulence, *Radio Sci.*, **16**, 1401–1406, 1981.

Weinstock, J., Theoretical relation between momentum deposition and diffusion caused by gravity waves, *Geophys. Res. Lett.*, **9**, 863–865, 1982.

Weinstock, J., Simplified derivation of an algorithm for non-linear gravity waves, *J. Geophys. Res.*, **89**, 345–350, 1984a.

Weinstock, J., Saturated and unsaturated spectra of gravity waves and scale-dependent diffusion, *J. Atmos Sci.*, **47**, 2211–2225, 1990.

Werne, J., and D. C. Fritts, Stratified shear turbulence: Evolution and statistics, *Geophys. Res. Lett.*, **26**, 439–442, 1999.

Werne, J., and D. C. Fritts, Turbulence and mixing in a stratified shear layer: 3D K-H simulations at Re = 24 000, *Phys. Chem. Earth*, **26**, 263–268, 2001.

Whalen, A. D., *Detection of Signals in Noise*, Academic Press, New York and London, 1971.

Whitehead, J. D., The quasi-transverse (Q. T.) approximation to Appleton's magneto-ionic equation, *J. Atmos. Terr. Phys.*, **2**(6), 361–362, 1952.

Whiteway, J. A., T. J. Duck, D. P. Donovan, *et al.*, Measurements of gravity wave activity within and around the Arctic stratospheric vortex, *Geophys. Res. Lett.*, **24**, 1387–1390, 1997.

Whiteway, J. A., E. G. Pavelin, R. Busen, J. Hacker, and S. Vosper, Airborne measurements of gravity wave breaking at the tropopause, *Geophys. Res. Lett.*, **30**, 2070, doi:10.1029/2003GL018 207, 2003.

Widdel, H.-U., Vertical movements in the middle atmosphere derived from foil cloud experiments, *J. Atmos. Terr. Phys.*, **49**, 723–741, 1987.

Wilfong, T. L., D. A. Merritt, R. J. Lataitis, *et al.*, Optimal generation of radar wind profiler spectra, *J. Atmos. Oceanic Technol.*, **16**, 723–733, 1999.

Williams, C. R., and K. S. Gage, Raindrop size distribution variability estimated using ensemble statistics, *Ann. Geophys.*, **27**, 555–567, 2009.

Willis, N. J., Chapter 23 in *Radar Handbook*, in *Bistatic Radar*, edited by M. I. Skolnik, pp. 23.1–23.36, McGraw-Hill, New York, 2008.

Wilson, R., and F. Dalaudier, Simultaneous observations of atmospheric turbulence in the lower stratosphere from balloon soundings and ST radar measurements, in *Proceedings of the Tenth International Workshop on Technical and Scientific Aspects of MST Radar*, edited by J. Chau, J. Lau, and J. Röttger, pp. 204–207, Radio Observatorio de Jicamarca, Lima, Peru and Universidad de Piura, Piura, Peru, 2003.

Wilson, R., F. Dalaudier, and F. Bertin, Estimation of the turbulent fraction in the free atmosphere from MST radar measurements, *J. Atmos. Oceanic Technol.*, **22**, 1326–1339, 2005.

Wilson, R., H. Luce, F. Dalaudier, and J. Lefrere, Turbulent patch identification in potential density/temperature profiles, *J. Atmos. Ocean. Tech.*, **26**, 977–993, 2010.

Wilson, R., F. Dalaudier, and H. Luce, Can one detect small-scale turbulence from standard meteorological radiosondes?, *Atmos. Meas. Tech.*, **4**, 795–804, doi:10.5194/amt–4–795 –2011, 2011.

Woodman, R. F., Inclination of the geomagnetic field measured by an incoherent scatter technique, *J. Geophys. Res.*, **76**, 178–184, 1971.

Woodman, R. F., High-altitude-resolution stratospheric measurements with the Arecibo 2380 MHz radar, *Radio Sci.*, **15**, 423–430, 1980.

Woodman, R. F., Spectral moment estimation in MST radars, *Radio Sci.*, **20**, 1185–1195, 1985.

Woodman, R. F., A general statistical instrument theory of atmospheric and ionospheric radars, *J. Geophys. Res.*, **96**, 7911–7928, 1991.

Woodman, R. F., Coherent radar imaging: Signal processing and statistical properties, *Radio Sci.*, **32**, 2372–2391, 1997.

Woodman, R. F., Spread F– an old equatorial aeronomy problem finally resolved?, *Ann. Geophys.*, **27**, 1915–1934, 2009.

Woodman, R. F., and Y. H. Chu, Aspect sensitivity measurements of VHF backscatter made with the Chung-Li radar: Plausible mechanisms, *Radio Sci.*, **24**, 113–125, 1989.

Woodman, R. F., and A. Guillen, Radar observations of winds and turbulence in the stratosphere and mesosphere, *J. Atmos. Sci.*, **31**, 493–505, 1974.

Woodman, R. F., and T. Hagfors, Methods for measurement of vertical ionospheric motions near the magnetic equator by incoherent scattering, *J. Geophys. Res.*, **75**, 1205–1212, 1969.

Woodman, R. F., and P. K. Rastogi, Evaluation of effective eddy diffusive coefficients using radar observations of turbulence in the stratosphere, *Geophys. Res. Lett.*, **11**, 243–246, 1984.

Woodman, R. F., R. P. Kugel, and J. Röttger, A coherent integrator-decoder preprocessor for the SOUSY-VHF radar, *Radio Sci.*, **15**, 233–242, 1980.

Woodman, R. F., B. B. Balsley, F. Aquino, *et al.*, First observation of polar mesosphere summer echoes in Antarctica, *J. Geophys. Res.*, **104**, 22 577–22 590, 1999.

Woodman, R. F., G. Michhue, J. Röttger, and O. Castillo, The MPI-Sousy-VHF radar at Jicamarca: High altitude-resolution capabilities, in *Proceedings of the Eleventh International Workshop on Technical and Scientific Aspects of MST Radar*, edited by V. K. Anandan, pp. 334–337, Gadanki/Tirupati, India, 2007.

Worthington, R. M., Calculating the azimuth of mountain waves, using the effect of tilted fine-scale stable layers on VHF radar echoes, *Ann. Geophys.*, **17**, 257–272, 1999.

Worthington, R. M., Comment on "Comparison of radar reflectivity and vertical velocity observed with a scannable C-Band radar and two UHF profilers in the lower troposphere," *J. Atmos. Oceanic Technol.*, **20**, 1221–1223, 2003.

Worthington, R. M., and L. Thomas, The measurement of gravity wave momentum flux in the lower atmosphere, *Radio Sci.*, **31**, 1501–1527, 1996.

Worthington, R. M., and L. Thomas, Impact of the tropopause on upward propagation of mountain waves, *Geophys. Res. Lett.*, **24**, 1071–1074, 1997.

Worthington, R. M., R. D. Palmer, and S. Fukao, An investigation of tilted aspect-sensitive scatterers in the lower atmosphere using the MU and Aberyswyth VHF radars, *Radio Sci.*, **34**, 413–426, 1999a.

Worthington, R. M., R. D. Palmer, and S. Fukao, Complete maps of the aspect sensitivity of VHF atmospheric radar echoes, *Ann. Geophys.*, **17**, 1116–1119, 1999b.

Worthington, R. M., R. D. Palmer, S. Fukao, M. Yamamoto, and I. Astin, Rapid variations in echo power maps of VHF radar backscatter from the lower atmosphere, *J. Atmos. Solar-Terr. Phys.*, **62**, 573–581, 2000.

Worthington, R. M., A. Muschinski, and B. B. Balsley, Bias in mean vertical wind measured by VHF radars: Significance of radar location relative to mountains, *J. Atmos. Sci.*, **58**, 707–723, 2001.

Wright, J., The interpretation of ionospheric radio drift measurements – I. Some results of experimental comparisons with neutral wind profiles, *J. Atmos. Terr. Phys.*, **30**, 919, 1968.

Yamamoto, M. K., T. Tsuda, S. Kato, T. Sato, and S. Fukao, A saturated inertia gravity wave in the mesosphere observed by the middle and upper atmosphere radar, *J. Geophys. Res.*, **92**, 11 993–11 999, 1987.

Yamamoto, M. K., T. Sato, P. T. May, *et al.*, Estimation error of spectral parameters of mesosphere-stratosphere-troposphere radars obtained by least squares fitting method and its lower bound, *Radio Sci.*, **23**, 1013–1021, 1988a.

Yamamoto, M. K., T. Tsuda, S. Kato, T. Sato, and S. Fukao, Interpretation of the structure of mesospheric turbulence layers in terms of inertia gravity waves, *Physica Scripta*, **37**, 645–650, 1988b.

Yamamoto, M. K., S. Fukao, R. F. Woodman, *et al.*, Mid-latitude E region field-aligned irregularities observed with the MU radar, *J. Geophys. Res.*, **96**, 15 943–15 949, 1991.

Yamamoto, M. K., M. Fujiwara, T. Horinouchi, H. Hashiguchi, and S. Fukao, Kelvin–Helmholtz instability around the tropical tropopause observed with the Equatorial Atmosphere Radar, *Geophys. Res. Lett.*, **30**, 1476, doi:10.1029/2002GL016 685, 2003.

Yamamoto, M. K., M. Abo, T. Kishi, *et al.*, Vertical air motion in midlevel shallow-layer clouds observed by 47 MHz wind profiler and 532 nm mie lidar: Initial results, *Radio Sci.*, **44**, RS004,017R, doi:1029/2008 RS004,017, 2009.

Yamamoto, M. K., T. Toshiyuki Fujita, N. H. B. A. A. Azizi, *et al.*, Development of a digital receiver for range imaging atmospheric radar, *J. Atmos. Solar-Terr. Phys.*, **118A**, 35–44, 2014.

Yamamoto, Y., T. Tsuda, and T. Adachi, Frequency spectra of wind velocity and temperature fluctuations observed with the MU radar-RASS, *Geophys. Res. Lett.*, **23**, 3647–3650, 1996.

Yamanaka, M. D., Formation of multiple tropopause and stratospheric inertio-gravity waves, *Adv. Space Res.*, **12**, 181–190, 1992.

Yau, K. H., G. P. Klaassen, and L. J. Sonmor, Principal instabilities of large amplitude inertio-gravity waves, *Phys. Fluids*, **16**, 936–951, 2004.

Yeh, K. C., and B. Dong, The Non-linear interaction of a gravity-wave with the vortical modes, *J. Atmos. Terr. Phys.*, **51**, 45–50, 1989.

Yeomans, M. E., Chapter 6 in *Radar Handbook*, in *Radar Receivers*, edited by M. I. Skolnik, pp. 6.1–6.51, McGraw-Hill, New York, 2008.

Younger, J. P., I. M. Reid, R. A. Vincent, and D. A. Holdsworth, Modeling and observing the effect of aerosols on meteor radar measurements of the atmosphere, *Geophys. Res. Lett.*, **34**, L15,812, doi:10.1029/2008 GL033,763, 2008.

Younger, J. P., I. M. Reid, R. A. Vincent, D. A. Holdsworth, and D. J. Murphy, A southern hemisphere survey of meteor shower radiants and associated stream orbits using single station radar observations, *Month. Not. R. Astron. Soc.*, **398**, 350–356, doi:10 .1111/j.1365–2966.2009.15,142.x, 2009.

Yu, T.-Y., and W. O. J. Brown, High-resolution atmospheric profiling using combined spaced antenna and range imaging techniques, *Radio Sci.*, **39**, doi:10.1029/2003 RS002,907, 2004.

Yu, T.-Y., and R. D. Palmer, Atmospheric radar imaging using spatial and frequency diversity, *Radio Sci.*, **36**, 1493–1504, 2001.

Yu, T. Y., R. D. Palmer, and D. L. Hysell, A simulation study of coherent radar imaging, *Radio Sci.*, **35**, 1129–1141, 2000.

Zecha, M., J. Bremer, R. Latteck, W. Singer, and P. Hoffmann, Properties of midlatitude mesosphere summer echoes after three seasons of VHF radar observations at 54 °N, *J. Geophys. Res.*, **108**, doi:10.1029/2002JD 002,442, 2003.

Zeller, O., M. Zecha, J. Bremer, R. Latteck, and W. Singer, Mean characteristics of mesosphere winter echoes at mid- and high-latitudes, *J. Atmos. Solar-Terr. Phys.*, **68**, 1087–1104, 2006.

Zhang, F., Generation of mesoscale gravity waves in upper-tropospheric jet-front systems, *J. Atmos. Sci.*, **61**, 440–457, 2004.

Zhang, G., R. J. Doviak, J. Vivekanandan, W. O. J. Brown, and S. A. Cohn, Cross-correlation radio method to estimate cross-beam wind and comparison with full correlation analysis, *Radio Sci.*, **38**, doi: 10 .1029/2002RS002,682, 2003.

Zhong, L., L. J. Somnor, A. H. Manson, and C. E. Meek, The influence of time-dependent wind on gravity-wave propagation in the middle atmosphere, *Ann. Geophys.*, **13**, 375–394, 1995.

Zimmerman, S. P., and E. A. Murphy, Stratospheric and mesospheric turbulence, in *Dynamical and Chemical Coupling between the Neutral and Ionized Atmosphere*, edited by B. Grandal and J. A. Hostet, pp. 35–47, NATO Advanced Study Institute, Spatind, Norway, 1977.

Zimmerman, S. P., C. A. Trowbridge, and I. L. Kofsky, Turbulence spectra observed in passive contaminant gases in the upper atmosphere, *Space Res.*, **XI**, 907–914, 1971.

Zrnić, D. S., Estimation of spectral moments for weather echoes, *IEEE Trans. Geosci. Electron.*, **GE-17**, 113–128, 1979.

Index

If a keyword appears followed by ++ in this index, it indicates that the pages covered are only representative, and many other occurrences may exist. The following keywords particularly will be especially under-represented (or may not exist at all) due to the great frequency of their occurrence, although they will occur selectively with additional descriptors: Backscatter, Correlation, Data, Doppler, Echo, Echoes, Energy, Frequency, Height, Layer, MST, MST radar, Phase, Pulse, Radar, Reflection, Scattering, Spectra, Spectrum, Structure, Temperature, Transmitter, Turbulence, Velocity, VHF, Wavelength, Waves, Spectral.